1 MONTH OF
FREE
READING

at
www.ForgottenBooks.com

By purchasing this book you are eligible for one month membership to ForgottenBooks.com, giving you unlimited access to our entire collection of over 1,000,000 titles via our web site and mobile apps.

To claim your free month visit:
www.forgottenbooks.com/free11179

ISBN 978-0-483-30927-2
PIBN 10011179

THE

JOURNAL OF SCIENCE,

AND ANNALS OF

ASTRONOMY, BIOLOGY, GEOLOGY, INDUSTRIAL ARTS,
MANUFACTURES, AND TECHNOLOGY.

(MONTHLY, FORMERLY "THE QUARTERLY JOURNAL OF SCIENCE.")

———— · ————

VOL. II. (Third Series.)
VOL. XVII. (O.S.)

LONDON:
3, HORSE-SHOE COURT, LUDGATE HILL.

——·◄█►·——

MDCCCLXXX.

INDEX.

AKIN, Dr. K., Baconian philosophy of heat, 686
— report on scientific societies, 481, 543
— spectrum analysis, 735
Almond blossoms, paucity of, 342
Agricultural ant of Texas, Dr. McCook on, 720
American naturalist, 470
Ancestry of man, 340
Anesthésie Chirurgicale, Traité d', by Dr. J. B. Rottenstein, 657
Animal development, rate of, 240
Animals, lower; mind in health and disease, 69
Ant, honey-bearing, habits and anatomy of, by Dr. C. Morris, 430
Anthropology, facts on, 406
Anti-vivisectionism, 730
Antozone, history of, by Dr. A. R. Leeds, 358
Archives de Museu Nacional do Rio de Janeiro II., 1, 2, 3, 4, 594
Argyll, Duke of, his Unity of Nature, by Dr. J. Foulerton, F.G.S., 671
Asceticism and mysticism, 695
Aspirations, flighty, by F. W. Brearey, 381
Astronomy, Elements of, 788
Astronomical observations at Melbourne University, 470
Aurora, the, 376
Australia, Aborigines of, phthisical, 411
Ayrton, Mrs., height and span of the Japanese, 100

BACONIAN philosophy of heat, by Dr. K. Akin, 686
Balbiani, Prof. G., Leçons sur la Generation des Vertébres, 202
Balfour, F. M., F.R.S., Comparative Embryology, 587
Beale, L., F.R.S., How to work with Microscope, 66
Beard, Dr. G. M., English and American physique, 33, 104
— problems of insanity, 351
— sea-sickness, 590

Beauty, meaning of, 342
Bezold, Dr. W. von, Theory of Colour, 392
Bees, Carnivorous, 341
Billing, S., instinct and mind, 228, 438
Bird, C., B.A., F.R.A.S., Lecture Notes on Physics, 209
Birds, food of, 592
Birmingham Philosophical Society, Proceedings of, 267
Bland, John, Disestablishment of Sun, 527
Blowpipe Analysis, Manual of, Col. Ross, 468
Book-binding, art of, 400
Brain, as an Organ of Mind, by Dr. C. Bastian, F.R.S., 648
— dynamics, a point in, by S. Tolver Preston, 532
Brearey, F. W., flighty aspirations, 381
Buckley, Miss Arabella, History of Natural Sciences, 264

CAMERON, C. A., agricultural chemistry and geology, 274
Capron, J. R., F.R.A.S., the Aurora, 376
Cavendish, Hon. Henry, life and character of, 25
Chemical Medicine, Annals of, 68
Chemistry, Agricultural, Johnstone's, by Dr. C. A. Cameron, 274
— Elements of, by W. A. Miller, 661
— Practical, 589
— Treatise on, 136
Clarke, Col. C. B., F.R.S., Geodesy, 405
Coal, formation of, by W. Mattieu Williams, F.C.S., F.G.S., &c., 81
Cobham Journals, 401
Cold, artificial, industrial applications of, 196
Coléoptères de France, by Dr. Sériziat, 654
Colour, theory of in relation to art, 392
Colour sense, 342

VOL. II. (THIRD SERIES).

3 C

Comparative Embryology, by F. M. Balfour, F.R.S., 587

Constitution of the earth, by R. Ward, 395

Corals, tabulate, of the Palæozoic Period, by Prof. W. A. Nicholson, 72

Correspondence, 139, 210, 277, 339, 406, 471, 502, 597, 665, 727, 789

Cotton insects, report on, by J. H. Comstock, 658

Cynolatry, Modern, by Frank Fernseed, 754

DAWSON, Prof. J. W., F.R.S., The Story of the Earth and Man, 271

Degeneration, by Prof. E. Ray Lankester, F.R.S., 465

Development, animal, rate of, 240
— teachings of, 271

Dietsch, Oscar, Food and Beverages, 262

EARTH and Man, Story of, 271

Electricity and Magnetism, Catalogue of Books and Papers on, 463

Embryology, Comparative, Treatise on, 587

Emphasis and symmetry, laws of, 434

Enemies, internal, 305

English and American physique, by Dr. G. M. Beard, 33, 104

Ethnographie, Allgemeine, by F. Müller, 200

Evolution and Involution, by G. Thomson, 334
— Mental, by Dr. J. Foulerton, F.G.S., 554
— of scientific knowledge, by C. Lloyd Morgan, 415

Evolutionism, History of, 1

FALLING motion, origin of, by Dr. C. Morris, 367

Far East, exploration in, 620

Fashion in deformity, by Prof. W. H. Flower, F.R.S. 664

Fermentation, studies on, 138

Fish and Fisheries, United States Report on, 274

Flora of Plymouth, 714

Fluids, abstention from, by S. Billing, 599

Fluorescence, phenomena of, by the Re E R. Hodges, 625

Fog-Lore, 764

Food and beverages, their impurities and sophistications, 262

Forbes, Prof. S. A., The [Food [of Birds, 592

Force, the vehicle of, by Dr. C. Morris, 607

Foulerton, J., M.D., F.G.S., mental evolution, 554
— — on "Unity of Nature," by the Duke of Argyll, 671

Four Forces of Nature, by G. Whewell, F.C.S., F.I.C., 639

Friends, and how we treat them, 215

Frog, curious phenomenon in, 471

Front, change of, 539

Frost as vermin-destroyer, 407

GENERATION des Vertébres, Leçons sur la, par Prof. G. Balbiani, 202

Geodesy, by Colonel A. R. Clarke, F.R.S., 405

Geography, Physical, Lectures on, 398
— and Travel, Stanford's Compendium of, 139

Geological Record for 1877, 338
— Survey of Canada, 531
— — of India, 530

Geology of London, Guide to, 529

Gesammelte Kleinere Schriften, by Prof. Charles Martins, 583

Glacier-Motion, Dr. Croll's Molecular Theory of, by J. J. Harris Teale, 273

Gilbert White reconsidered, 632

HALLEY, Edmund, Life and Works of, by Capt. S. P. Oliver, 91

Haughton, Rev. Dr. S., F.R.S., &c., Lectures on Physical Geography, 398

Heat, a mode of motion, 464
— Baconian philosophy of, by Dr. Akin, 685

Heat and light, by R. Ward, 680

Heath, A., M.D., tuberculosis transmissible through milk and meat, 441

Hincks, Thomas, B.A., F.R.S., On British Marine Polyzoa, 337

Hodges, Rev. E. R., Phenomena of Fluorescence, 623

Hollway's process for reducing sulphides, by W. Mattieu Williams, F.R.A.S., 131

Honeycomb, living—a phase of ant-life, by Dr. C. Morris, 87
Huxley, Prof., F.R.S., &c., Science Primers, 338
Hybrids, fertility of, 341
Hydrogen, peroxide, history of, by Dr. A. R. Leeds, 358

INSANITY and its difficulties, 351

Insect Variety, by A. H. Swinton, 710
Insects of Palæozoic Epoch, 591
Instinct and Mind, by S. Billing, 238, 438
— — —, by R. N. M., 339, 536
Internal enemies, 305
Inventions, new, Reports on, 42, 46, 131, 196
Iron, corrosion of, Prof. Barff's patent for preventing, 46

JAPANESE, height and span of, by Mrs. Ayrton, 100

KANSAS City Review, 272

Kurum Valley, Flora of, 411

LANKESTER, Prof. E. Ray, F.R.S., on Degeneration, 465
Leeds, A. R., Ph.D, history of anto-zone and hydrogen peroxide, 358
— — lines of discovery in the history of ozone, 145
Lepidoptera, Development of, by A. Keferstein, 780
Life, dilemma of, 597
— and Mind, on the Basis of Modern Medicine, by Dr. R. Lewins, 724
—, nature of, 203
Lightning-conductors, by R. Anderson, F.C.S., 265
Lindsay, Dr. W. Lauder, F.R.S.E., &c., Mind in the Lower Animals in Health and Disease, 69
Lippincott, J. S., Critics of Evolution, 470
London, water-supply of, 73

MAN, ancestry of, 340

Mankind, antiquity of, 310
Martins, Charles, Gesammelte Kleinere Schriften, 583
Martyrdom of Science, 150

Mathematics, Synopsis of Elementary Results in, 786
McCook, Dr., Agricultural Ant of Texas, 720
McCoy, Prof. F., Palæontology of Victoria, 596
McMunn, Dr. C. A., Spectroscope in Medicine, 335
Mental Evolution, by J. Foulerton, M.D., F.G.S., 554
Meteorology, modern advances in, 235
— observations in, 272
Microscope, how to work with, by Dr. Lionel Beale, F.R S., 66
Microscopic Science, progress in, by W. T. Suffolk, F.R.M.S , 167
Migrants, Our Summer, by J. E. Harting, F.L.S., F.Z.S., 779
Millennium, sanitary, 701
Minchin, G. M., M.A., Treatise on Statics, 404
Mind in lower animals, by Dr. Lauder Lindsay, 69
Morality and natural science, by S. Tolver Preston, 443
— evils of sectarian, by S. Tolver Preston, 574
Morals of Evolution, 714
Morgan, C. Lloyd, evolution of scientific knowledge, 415
Morris, Dr. C., honey-bearing ant, 430
— living honeycomb, 87
— origin of falling motion, 367
Morse, E. S., Science Department of the University of Tokio, 209
Motion, falling, origin of, 367
Müller, Fr., Allgemeine Anthopologie, 200
Mysticism and Asceticism, by F. Fernseed, 695
Mythologie der Griechen und Römer, by Dr. Seemann, 785

NATURAL Science and Morality, by S. T. Preston, 443
Nature, four fources of, by G. Whewell, F.C.S., F.I.C., 639
Nature's bye-paths, by Dr. J. E. Taylor, 787
— hygiene, by C. T. Kingzett, F.C.S., 644
Nicholson, Dr. H. Alleyne, Manual of Palæontology, 137
— Monograph of Silurian Fossils of Girvan, 208
— Structures and Affinities of Tabulate Corals, 72

Nicols, A., F.G.S., Chapters on Physical History of the Earth, 269
Notes, 142, 210, 280, 344, 409, 473, 537, 602, 667, 729, 794

OLIVER, Capt. S. P., life of E. Halley, 91
— torpedo warfare, 170, 287
Origin of Species, "Egyptian" argument on, 166
Ormerod, Miss E. A., F.M.S., the Cobham Journals, 401
Ornament, discourse on, by H. H. Statham, 651
Ott, Col. C., Studies on the seat of Russo-Turkish War, 273
Ozone, lines of discovery in the history of, 145

PALÆONTOLOGY, Manual of, 137
Palestine, Fauna of, 403
Past year, scientific progress of, by W. Spottiswoode, F.R.S., 8
Pasteur, J., studies on fermentation, 138
Philosophical Society of Washington, 662
Philosophy of War, 205
Phosphorus, elimination of, by W. Mattieu Wiliams, F.C.S., 42
Phrenologists, a query for, 277
Phthisis, influence of climate on, 207
Plymouth, flora of, 714
Polyzoa, British Marine, by T. Hincks, F.R.S., 337
Preston, S. Tolver, evils of sectarian morality, 574
— natural science and morality, 443
— point in brain dynamics, 532
Problems on insanity, reply to criticisms on, by Dr. G. M. Beard, 783

RESEARCH, protection of, 408

Revue Internationale des Sciences Biologiques, 655
Richardson, R., M.A., Nature of Life, 203
Ronalds, Sir F., F.R.S., Books and Papers on Electricity, 463
Roscoe, Prof. H. E., F.R.S., and P. Schorlemmer, F.R.S., Treatise on Chemistry, 136
Ross, Lieut.-Colonel W. A., Manual of Blowpipe Analysis, 468

Royal and Archæological Association of Ireland, Journal of, 663

SALT Industry of England, 154

Schæfer, Prof. E. A., F.R.S., some teachings on development, 271
Science Lectures for the People, 401
— Natural, short history of, 264
— Primers, introductory, 338
Scientific knowledge, evolution of, by C. Lloyd Morgan, 415
— Societies, report on, by Dr. K. Akin, 481, 543
Sea Sickness, Treatise on, by Dr. G. M. Beard, 590
Seasons, exceptional, 750
Sectarian morality, evils of, by S. Tolver Preston, 574
Shipman, James, alluvial and drift deposits of Trent Valley, 596
Smithsonian Institution, Report of for 1878, 713
Sorghum Sugar, Report on, 467
Soul, the, What is it? 283, 298
Sound as a nuisance, 570
Spectrum analysis, Dr. Akin on, 735
Spottiswoode, W., P.R.S., Scientific Progress for Past Year, 8
Statics, Treatise on, 404
Suffolk, W. T., F.R.M.S., progress in microscopic science, 167
Sun, disestablishment of, 527
Synthesis, organic, 222

TASMANIA, Royal Society of, 590

Teall, J. J. Harris, criticism on Dr. Croll's Molecular Theory of Glacier Motion. 273
Telegraphic Engineers, Journal of Society of, 469, 653
Territories, Geological and Geographical Survey of, 403
Thomson, G., Evolution and Involution, 334
— W., F.L.S., Phthisis and Climate, 207
Thudicum, Dr. J. L. W., Annals of Chemical Medicine, 68
Tilden, Dr. W. A., Practical Chemistry, 589
Torpedo warfare, Capt. S. P. Oliver, 170, 287
Tuberculosis, transmission of, 411
Tyndall, Prof., D.C.L., F.R.S., Heat a Mode of Motion, 465
— Lectures on water and air, 49, 112, 179, 246, 317, 510

UNITED States Entomological Commission, 663
Unity of Nature by the Duke of Argyll, Dr. J. Foulerton on, 671
Urania, 276

VACCINATION, truth about, by Ernest Hart, 646
Variety, insect; its propagation, by A. H. Swinton, 710
Vegetation, action of electric light upon, 283
Victoria, mineral statistics of, 663
— Prodromus of Palæontology of, 596
— Mining Surveyor's report, 529
Victorian Review, 276
— Year-book, 1878-79, 275
Vivisection, benefits of, 729
— note on, 602
— Prof. Zöllner on, 501
— Question, present state of, 501

WALLACE, A. R., Stanford's Compendium of Geography and Travel, 135
War, Philosophy of, 205
Ward, R., constitution of the earth, 395, 494

Ward, R., on heat and light, 680
Wasps, carnivorous, 277
Water and Air, lectures on, by Prof. Tyndall, F.R.S., 49, 112, 179, 246, 317, 410
Water-supply of London, 73
Waterton, Charles, by James Simson, 783
What constitutes discovery in Science, by Dr. G. M. Beard, 660
Whittaker, W., F.G.S., Geology of London, 529
Wild Birds' Protection Act, 340, 407, 603
Williams, W. Mattieu, F.R.A.S., &c., Barff's process for protecting iron, 46
— formation of coal, 81
— Hollway's process, 131
— Thomas and Gilchrist process, 42
Wills, Thomas, F.C.S., life of, 652
Working v. fighting, 770
World of the Poets, 745

ZÆHNSDORF, J. W., Art of Bookbinding, 400
Zœllner, Prof. F., Missbrauch der Vivisection, 501

London: Printed by R. J. DAVEY, Boy Court, Ludgate Hill, E.C.

THE

JOURNAL OF SCIENCE.

JANUARY, 1880.

I. THE HISTORY OF EVOLUTIONISM.*

EVOLUTIONISM, in the broad plain sense of the term, as opposed to the doctrine of mechanical-individual creation, may now be hailed as victorious along the whole line. In proof, we need only listen to the voices which are raising the well-known chorus :—" All this we knew long ago! Was there any need of a Darwin or a Wallace to tell us truths which are to be found embodied in classical myths, hinted at by early Christian fathers, and even shadowed forth in Holy Writ ?" We might, indeed, ask the utterers of such voices how it comes that none of them was able to detect the great truth in these old sagas and writings till modern biologists found it in the Book of Nature ? But it is not always good policy to tear away the veil under which neophytes hide the fact of their conversion.

Turning, then, from the initial controversy, we have to organise the territory which we have won. Accepting Evolution as God's way of Creation, we have before us the almost infinite task of tracing out how or by what agencies it is effected? why the organic world is as we find it, and not other? In connection with this undertaking we shall find it highly important to explore all the earlier suggestions and hypotheses which have been put forward as to Descent, its efficient causes, and its laws of operation. Not long ago we had the pleasure of reviewing a thoughtful and suggestive work,† in which the author seeks to prove that

---

* Erasmus Darwin. By ERNST KRAUSE. With a Preliminary Notice by CHARLES DARWIN. London : John Murray.
† Evolution, Old and New. By S. BUTLER.

VOL. II. (THIRD SERIES).   B

Mr. C. Darwin,—and we presume Mr. Wallace,—instead of developing, have rather obscured the doctrines advanced by Erasmus Darwin, Buffon, and Lamarck. The volume now before us is therefore the more opportune. In it Dr. Krause gives a full, an able, and impartial survey of the genius and the researches of the elder Darwin, and assigns him a high place in the history of biological science.

It must not be imagined that prior to the appearance of the "Vestiges" the hypothesis of special creation had remained uncontroverted. Linnæus and Buffon were men of too philosophical intellects to accept as a matter of course the notion of a world mechanically made and peopled with plants and animals by a certain date, as if under contract. But their opinions wavered, perhaps according as evidence on the one or the other side suggested itself to their minds, —perhaps, also, as they were alternately swayed by private conviction or by prudential considerations. These fluctuations are most strikingly shown in the writings of Buffon. Mr. S. Butler solves the difficulty by interpreting all such passages favourable to the Old School as ironical—a somewhat hazardous expedient. But in the works of Erasmus Darwin—who, it must be remembered, is prior in point of time to Lamarck—we find no such wavering. He first established, as Dr. Krause insists, "a complete system of the theory of Evolution." At first sight, indeed, he might seem to have anticipated his illustrious grandson to a very serious extent. He discusses in his works the questions of heredity of adaptation, of the protective arrangements of animals and plants, and of sexual selection. He describes insectivorous plants. He analyses the emotions and social impulses, and seeks to trace out their origin. He suggests that all the limestone rocks in the world "were formed originally by animal and vegetable bodies from the mass of waters." He refers the Fungi to a third kingdom which, like a "narrow isthmus," unites plants and animals. He asks, "Do some genera of animals perish by the increasing power of their enemies? or do they still reside at the bottom of the sea? Or do some animals change their forms gradually, and become new genera?" It is only quite of late that we may venture to reply to the second of these questions in the negative. In tacit but yet unmistakable opposition to the shallow and mawkish teleology of his time, which viewed all nature in relation to man, and affected to know the hidden purposes of God, he asks—"Why has this plant poisonous juices? Why has that one spines? Why have birds and fishes light-coloured breasts and dark backs ?" He treats of the arrangements by which plants defend themselves from unbidden insect guests, thus, in part at least, anticipating a recent and most interesting work by Prof. Kerner. He points out that, in barren moorland districts, horses have learnt how to eat furze—a very nutritious plant—without wounding their mouths. He notices that flower-haunting birds and insects are gaily and vividly coloured, whilst larks, partridges, and hares resemble in their hues the dry vegetation or the earth upon which they rest. He observes that the snake and wild cat and leopard are so coloured as to imitate dark leaves and their lighter interstices. He pronounces the eggs of birds to be so coloured as to resemble adjacent objects. " The eggs of hedge-birds are greenish, with dark spots; those of crows and magpies, which are seen from below through wicker nests, are white, with dark spots ; and those of larks and partridges are russet or brown, like their nests or situations." He suggests that, " like the fable of the chameleon, all animals may possess a tendency to be coloured somewhat like the colours they most frequently inspect, and, finally, that colours may thus be given to the egg-shell by the imagination of the female parent." These suppositions, as Dr. Krause reminds his readers, have lately been proved to be, in many cases, perfectly correct. He recognises the existence and the universality of the struggle for existence —a phenomenon overlooked to this day by a large portion of the intelligent and respectable classes, and sometimes denied even by compilers of books and writers of review-articles. How this contest rages in the apparently peaceful vegetable world Erasmus Darwin has well expressed in the following lines :—

> " Yes ! smiling Flora drives her armed car
> Through the thick ranks of vegetable war ;
> Herb, shrub, and tree with strong emotions rise
> For light and air, and battle in the skies ;
> Whose roots diverging with opposing toil
> Contend below for moisture and for soil."

From such a recognition of the contest waged among all organisms it may seem no very wide step to the hypothesis of Natural Selection as *the* cause—according to the younger Darwin, or, we should rather submit, as *a* cause—of the variation of species. But it was never taken by Erasmus. He further expressed the idea which lies at the foundation of Mr. S. Butler's recent work, " Life and Habit," and which will doubtless effect the solution of all the remaining mysteries of instinct. He regarded the young animal as a continuation, or, as he expresses it, as an " elongation of its parents," and as retaining in consequence the habits of the latter.   He believed that the human race was at one time four-footed, and that hermaphroditism was the general condition even of the higher animals.

We might, indeed, fill much more space than stands at our disposal in showing to what an extent Erasmus Darwin anticipated the most recent biological researches and the most advanced speculations of our own day.

But we have now to meet the two main questions :—Why did he so completely fail to command the assent of the public ? and wherein does his system differ from that of his illustrious grandson ?   That he did not carry conviction to the minds of even a minority of thinkers is undeniable. Even until quite recently the idea of a transformation of species, or of their origin in any other mode than that embodied in Milton's poetical gloss on the Mosaic cosmogony, was, in England at least, branded as philosophically false and theologically impious, and the very name of Darwin had become a bye-word and a reproach.

This failure was due to the combined action of a number of circumstances.   Erasmus Darwin was too far in advance of his own contemporaries to meet with a fair appreciation. As Dr. Krause remarks :—" It is only now, after the lapse of a hundred years, that, by the labours of one of his descendants, we are in a position to estimate at its true value the wonderful perceptivity, amounting almost to divination, that he displayed in the domain of Biology."

Again, very much of the evidence that was needed to convince those capable of judging of the truth of Evolution could not be said to exist.   The disciplines of animal and vegetable geography, which have supplied such a mass of proof in favour of " Transformism," had not been elaborated. Palæontology was also a thing of the future, and no investigator could point to the gradual mutation, *e.g.*, of *Castanea atava* into *Castanea vesca*, or demonstrate the successive stages through which the horse has passed in reaching his present structure.

Embryology, also, was not in a position to speak as she has since spoken.   The collateral evidence in favour of development as the general law of Creation, now furnished by Astronomy in the shape of the nebular theory, was also wanting.   What wonder, then, that the system of the elder Darwin, even had it been much more complete than was actually the case, should be rejected ?

The time, too, was especially unfavourable.   The teleologists were everywhere expounding their favourite dogma of " Man, the measure and the purpose of all things." Foremost among them stood Paley, utterly incapable of rising to any true biological conception, but bold, plausible, and popular, on account of his *à posteriori* demonstration of the existence of God. No small part of his works may be said to have been written at Erasmus Darwin, though the latter is never mentioned by name. Still more unfortunate, in the state of public opinion then prevalent in England, was the fact that Darwin's predecessor, Buffon, and his immediate successor, Lamarck, were both Frenchmen. As such they were at once set down as atheists and "jacobins," and an unmerited and groundless stigma was thus attached to the very idea of Evolution. Dr. Darwin's " Loves of the Plants " was burlesqued by the *Anti-Jacobin* in a humorous effusion entitled the " Loves of the Triangles," and he himself was very openly accused of atheism. The wanton malice or the gross ignorance displayed in this charge must be apparent to all who have taken the trouble to read his works. It would be easy to quote, from the writings of this so-called "atheist," ascriptions of praise and glory to God which almost rise to the fervour and dignity of psalms. But even in our own—as we would fain hope—more candid times we see but too clearly on what slender evidence such accusations are made. Has not the younger Darwin himself been denounced, by some who certainly know better, as the conscious and intentional apostle of infidelity?

A little later on in the century the influence of Cuvier and his school was no less hostile to a candid consideration of the arguments in favour of Evolution. Acute, laborious, and, in some departments at least, a keen and indefatigable collector of facts, the great French professor was wanting in the true philosophic spirit, and tainted to the core with that " aletheophobia " which is the bane of official science. Of him it has been well said that his influence threw back scientific biology for at least one generation.

In England we suffered, in addition, from the predominance of the Quinarian school, as represented by Swainson. Until the atmosphere was cleared of all these mists and clouds no true progress could be effected, and it is therefore no wonder if the clue given by the elder Darwin was neglected and his methods of investigation not followed up.

But we may go further: another cause remains why Erasmus Darwin failed to convince his contemporaries and his more immediate successors, and upon this we cannot touch without showing wherein he differs from his illustrious grandson.

As Dr. Krause very aptly remarks, " It is one thing to establish hypotheses and theories out of the fulness of one's fancy, even when supported by a very considerable knowledge of Nature, and another to demonstrate them by an enormous number of facts, and carry them to such a degree of probability as to satisfy those most capable of judging." Erasmus Darwin, along with a number of most valuable observations and suggestions, lays before us—as was in his day inevitable—not a few puerile and unfounded hypotheses. Thus he puts in the mouth of a philosophic friend the conjecture that the first insects had proceeded from a metamorphosis of the honey-loving stamens and pistils of the flowers by their separation from the parent plant, after the fashion of the male flowers of *Vallisneria*. He believed that, as far as possible, flowers are adapted for self-fertilisation, and even stigmatises cross-fertilisation as " adulterous." Probably, however, his greatest weakness lies in the agency to which he ascribes the gradual transformation of organisms. Like his successor, Lamarck, he depends here on the conscious and intentional attempts of each being to adapt itself to changing circumstances. He declares that all warm-blooded animals have arisen from one living filament which THE GREAT FIRST CAUSE endued with animality, with the power of acquiring new parts, attended with new propensities, directed by irritations, sensations, volitions, and associations ; and thus possessing the faculty of continuing to improve by its own inherent activity, and of delivering down those improvements by generation to its posterity, world without end." This idea is by no means unsuitable as far as animals are concerned ; but with strict logical consistency its author applied it also to the development of plants, and thus became, as Dr. Krause maintains, the most formidable critic of his own system. If we are to suppose plants consciously attempting to adapt themselves to changing conditions, we are ultimately driven to assign them a sensorium, and, as the composite vegetable body is not unlike a coral-stock, this sensorium must be ascribed to every bud. These difficulties, it is scarcely needful to say, the younger Darwin evades by his hypothesis of Natural Selection. The individual which—without any intention or consciousness on its own part, and by a mere accidental variation—is in better accord with external circumstances than are its neighbours, has the better chance of surviving them, and of leaving a progeny. Doubting, as we do, the all-sufficiency of this hypothesis, we must yet admit that it marks out distinctly the interval between Erasmus Darwin and his grandson, and constitutes a most important step in the history of Evolution.

As an introduction to the English version of Dr. Krause's work, Mr. Charles Darwin has contributed a biographical notice of his grandfather, from which we gather interesting facts not a few. He was, for instance, one of the fore-runners both of Sanitary Reform and the Temperance Movement, though free from the savour of quackery and the ultraism by which his successors in both make themselves too often unpleasantly notorious. He was a mechanical inventor, no less than a biologist, and his prophecy as to the future career of steam has been too often quoted to need repetition. He refers in his writings to the value of bones as a manure. He expressed the confidence that microscopic research would lead substantially to the discovery of a new world. Two of his sayings here given are worth quoting :— he delared that "the world was not governed by the clever men, but by the active and energetic," and that "the fool is he who never experiments."

But the most interesting feature of this biographical notice is the light which it throws on the interesting question of heredity. For several generations the Darwin family has been distinguished for an intelligence far above the average, which in two cases at least has risen to the rank of genius. Almost all its members have possessed scientific tastes, and have followed the learned professions, generally with success. We read that Robert Darwin, the father of Erasmus, was a man given to scientific pursuits : he left two sons, Robert Waring, a poet and a botanist, and Erasmus, the subject of this memoir. Of the children of the latter five reached maturity :—Charles, who had already become distinguished as an anatomist, when he died from the effects of a wound received whilst dissecting; Erasmus,. a statistician and genealogist; Robert Waring, a skilful and eminent phy-sician, father of him whom we must designate as *the* Darwin of our own days; Francis, a naturalist of merit; and Violetta, who became the mother of Mr. Galton, the author of the well-known treatise on the "Heredity of Genius." A son of Francis, Captain Darwin, in his "Gamekeeper's Manual" shows "keen observation and knowledge of the habits of various animals." The two sons of Mr. Charles Darwin, George and Francis, have not merely taken part in their father's researches, but have entered into independent scientific investigations.

By way of contrast to the work with which we have been engaged, we cannot help referring to a review article which has recently appeared on the other side of the Atlantic. We have already glanced, in passing, at the attempts made to find the germs of modern discoveries and speculations in very unlikely regions. At one time it was the fashion to declare that the philosophers and the poets of classical antiquity had anticipated all our most valuable ideas. Next came a rage for extracting systems of science from Hebrew roots by dint of high pressure philology, somewhat as soups of doubtful value may be obtained from old bones by the aid of Papin's digester. The latest mania is for seeking out chemical, physical, or biological truths in the writings of *Albertus Magnus*, St. Thomas Aquinas, or St. Augustine. Indeed St. Augustine is declared to have been an Evolutionist.

The article professes to deal with " Malthusianism and Darwinism." In opposition to the former we are told that the lower races of man die out in contact with civilisation, and thus make room for their superiors. But as against the Darwinians we learn, on the contrary, that it is not the higher, but the lower forms that survive. This, we think, is very like self-refutation.

II. SCIENTIFIC PROGRESS OF THE PAST YEAR.

By WILLIAM SPOTTISWOODE, D.C.L., LL.D.

(Being a Condensed Report of the Presidential Address delivered at the Anniversary Meeting of the Royal Society, on Monday, December 1, 1879.)

IN the spring it is our duty to elect into the Royal Society our annual complement of new Fellows, and at our early summer meetings we admit to our ranks the young and vigorous in the career of Science. But in the autumn, or fall, as on the other side of the Atlantic it is so aptly termed, it is our custom to recount the names of those who have dropped from our list. Some of these have fallen in a plenitude of years, and in maturity of mind; others in that early stage of youth where insight and imagination, as yet scarcely differentiated, have both lent their aid to the first flights of thought.

In looking over the list we are reminded in a striking manner of a fundamental difference between the Royal Society and the Academies of the Continent, a difference which may perhaps be the best described by the term "comprehensiveness." For, beside the class of Fellows selected, in accordance with our recent legislation, from the members of the Privy Council, it has always been our custom to gather into our ranks not only men of eminence in Science proper, and in subjects which border on it, but also men of distinction in other paths of life, provided that they have followed those paths on principles which are analogous to our own, and which by no undue strain of the analogy may themselves be called scientific.

In illustration of this remark, I might point in the present list to the man of letters, to the architect, to the politician, to those who have honourably served in various departments of the public service, to the man of wealth who has turned his large means to large-minded purposes for the welfare of the people. And although the act of erasing them from our list marks our loss, yet the fact of having once reckoned them among our number is in itself a gain, and must help to enlist the sympathies of the world outside in our special function, viz., the promotion of natural knowledge, while at the same time it tends to enlarge our own.

To mention briefly a few of these :—In Sir James Matheson we have lost a wealthy and enlightened member, who devoted much of his time, his energy, and his means in promoting the welfare, both moral and intellectual, of the people among whom he made his home.

In the Marquis of Tweeddale we have an instance, happily not singular, of one who, without any professional connexion with the subject, contrived amidst the distractions of active service to lay the foundations of a solid knowledge of one branch of science ; while in later years he became an active collector and the author of valuable contributions to the publications of the Geological Societies over which he presided.

In the category of men of cultivation and leisure, who have turned their attention with good purpose and success to scientific pursuits, we may fairly reckon our late Fellow John Waterhouse, of Halifax.

Sir Thomas Larcom was one of a long series of distin-
· guished men who have been elected Fellows of the Royal

Society from the Corps of the Royal Engineers. Travelling beyond the strict limits of his professional career, he endeavoured to give to his official work a wider range. And it was due to suggestions emanating from him that the Irish Survey was so extended as to make it the opportunity for collecting a great variety of local information, history, language, and antiquities, and that the map became the admiration of scientific travellers.

Mr. Bennet Woodcroft's name will always be associated with the foundation of that important department the Patent Office, with its Library, and Museum, of which he was the first executive officer.

Professor Kelland and Mr. Brooke were among our veterans in the Society, and many of us will long recollect the lively interest in Science which the latter showed during his frequent attendance at our evening meetings.

In Mr. W. Froude, to whom one of the Royal Medals was awarded in 1878, the Society, the public service of the country, and Science in general have sustained a loss which at one time would have been irreparable, and which even now, when his work has become an established science, is difficult to replace.

Of Professor Clifford, and the gap which his death has caused among our friends and the world of science at large, I know not how to speak. His mathematical papers are being collected by a careful and trusty hand, while his philosophical remains have been given to us by one who knew and loved him as he deserved. To the same friend we owe the memoir of his life, written indeed so far as that life can ever be written, but reminding us also at every line that his life was one the full story of which will always seem but half told.

If in Professor Clifford we have lost one of our youngest, in Sir J. G. Shaw-Lefevre we have lost one of the oldest of our mathematical Fellows. Born before the present century, and Senior Wrangler at Cambridge before many of us were in existence, he seems to belong to a past age. But through a long and honourable course of unusually varied and responsible public life he always retained his interest in mathematics, and, himself no mean geometer of the old school, he had in his later years projected and actually begun a new edition of the works of Archimedes. Of his labours in connexion with University College, and afterwards with the University of London, all who are interested in liberal education will retain an appreciative and grateful memory.

Our last loss, that of Professor Clerk Maxwell, has been

in some respects the greatest.   In him the full maturity of
a mind which had suffered no check or shadow of abate-
ment, was devoted to the foundation of a more thorough
knowledge of Molecular Physics and of Electrical Laws
than has hitherto been attained.   The clear insight, the
keen criticism, the just and liberal appreciation of the views
of others, evinced both elsewhere and especially in his
Reports to the Committee of Papers, will long be remembered ; while the more general side of his character, exemplified in the numerous scintillations struck off with his pen,
will be no less cherished by those to whom his scientific
writings were as a sealed book.   To replace such a man
seems at first imposssible : but we should be doing scant
justice to his memory if we did not believe that the good
seed which he so liberally sowed would bear such fruit that
in the time of need there should be no lack.

     *   *   *   *   *   *

An extra volume of the " Philosophical Transactions "
(vol. clxviii.) has been issued, in which the observations
made by the naturalists who accompanied the Transit of
Venus Expeditions to Kerguelen's Land and Rodriguez, and
descriptions of their collections by persons specially acquainted with the several subjects are brought together.
The volume is divided into four sections, viz., the Botany
and Zoology of each of the two islands respectively.

The Botanical collections from Kerguelen's Land, worked
out by Sir J. D. Hooker, Mr. Mitten, the Rev. J. M. Crom-
bie, Dr. Dickie, P. F. Reinsch, and the Rev. M. J. Berkeley,
have added largely to our knowledge of the Cryptogams,
especially the Algæ.   In particular the American affinity of
the Kerguelen Flora, previously established by Sir J. D.
Hooker, in his examination of the flowering plants, is proved
to be also very strongly manifested in the Cryptogams.

Of the Zoological collections from Kerguelen's Land
those of the Mollusca, Crustacea, Arachnida, and Insecta
have yielded the greatest number of novelties ; the two
former offering distinct evidence of the affinity of this Fauna
(as of its Flora) with that of South America.   The Arachnids and Insects of the southern extremity of America are
unfortunately too little known at present to admit of a comparison with the highly interesting new forms discovered by
Mr. Eaton, and described in this section.

In estimating the affinities of the Flora and Fauna of
Rodriguez the authors were under great difficulties, owing
to our imperfect knowledge of the plants and animals of the
other Mascarene Islands.   But almost all their observations

point strongly to the conclusion that the present animals
and plants are the remains of a once more extensive Flora
and Fauna which has been gradually broken up by geological
and climatic changes, and which more recently has been
greatly interfered with by the agency of man. The greater
portion of the botanical section has been undertaken by Dr.
Is. Bayley Balfour, who himself collected the materials.
He estimates the total number of plants found at present in
Rodriguez at 470 species, viz., 297 Phænogams and 173
Cryptogams; and he establishes the remarkable fact that of
the Phænogams not less than 108 are to be regarded as
introduced into the island, and that 189 only are indigenous.
Great interest attaches to remains of the extinct cave-fauna
of Rodriguez; and the collection made by Mr. Slater supplied materials for three papers, by Mr. E. Newton, Mr. J.
W. Clark, and Dr. Günther. With regard to Zoology, the
marine collections were those to which least interest was
attached, as they consisted principally of common forms
spread over the whole of the tropical Indian Ocean. On
the other hand, our knowledge of the Terrestrial Inverte-
brates received considerable additions, especially in the
Myriopods and Arachnids, Coleoptera and Turbellaria,
described by Mr. A. G. Butler, Mr. C. O. Waterhouse, and
Mr. G. Gulliver.

The papers presented to the Society, and read at our
evening meetings, have been more numerous than in any
previous year of our existence, and have during the last
twelve months reached a total of 118. Some of them appear
to have excited unusual interest among the Fellows and
their friends, for on more than one occasion our meeting-
room was filled to an almost unprecedented degree.

But, beside the interest attaching to their reading and
discussion, the papers themselves have offered some very
striking features. It would be as invidious to attempt, as
it would be impossible to establish, any general comparison
of merit among so varied a collection of memoirs; but I
may still be permitted to take this opportunity of expressing
my own impressions of a few which fall, more or less,
within my own range of study.

In purely experimental research—that is, in experiments
guided by a clear conception of what was wanted to be
done, and executed with adequate instrumental appliances
and with the highest manipulative skill—we cannot but be
struck by the assiduity and success with which Mr. Crookes
has continued his labours. He has now brought to a termination his remarkable series of papers on " Repulsion

resulting from Radiation," and has already struck out into a new region, viz., the study of certain Electrical Phenomena which appertain especially to high vacua. In the vacua to which he has now turned his attention the exhaustion has been carried even beyond that in which the phenomena of the Radiometer are produced, and it is by a legitimate and sagacious step that he has now occupied a complete field of inquiry intermediate to the most extreme vacua ever attained, and those in which the stratified discharge is displayed. Of the delicacy and beauty of the experiments it is not necessary here to speak, for they have been already exhibited here, as well as to more than one audience on a large scale. Through this advance on his part a helping hand will doubtless be held out to those who are occupied with the subject of stratification, for in this matter the two advancing powers desire neither a " neutral zone " nor a " scientific frontier " to separate their field of action.

Another communication, full of promise as well of performance, should also not pass unnoticed. I allude to Prof. Hughes's paper on his Induction Currents Balance, the application of which has already branched out and borne fruit in more than one direction. The extreme simplicity of the instruments, and their marvellous adaptation to the purposes for which they were intended, reflect the highest credit on their inventor.

Not unconnected with Mr. Crookes's researches, so far as they are directed towards the ultimate constitution of gases, are those of Prof. Osborne Reynolds. In his remarkable paper " On Certain Dimensional Properties of Matter in a Gaseous State " Prof. Osborne Reynolds has established the fact of what he calls Thermal Transpiration, namely, that when two portions of the same gas are separated by a porous plug, the two surfaces of which are at different temperatures, the condition of equilibrium is no longer that the pressures of the gas on the two sides of the plug should be equal, but that the pressure on the hotter side should exceed that on the colder side by a certain quantity; and that if this is not the case, the gas will transpire through the plug until this condition is satisfied. Prof. Reynolds connects this principle with Mr. Crookes's Radiometer experiments; and the various considerations which he brings to bear on the subject appear to have this idea throughout, that the dynamical similarity of two gaseous systems, bounded by solid surfaces, geometrically similar, depends on the ratio of the homologous lines of the system to a certain quantity, having the dimensions of a line, and which varies inversely as the density of the gas.   Although dimensional considerations have been shown by previous writers to have an important bearing on the Theory of Gases, still his experimental results form an important contribution to our knowledge of the subject ; and in his theoretical investigation he has attacked with vigour, and not without success, some elements of the problem avoided by his predecessors in this field of research.

Photography has of late years become so completely the handmaid of Physical Science, and has so constantly responded to the calls of its master, that its progress, however rapid and however effective, has come to be taken almost as a matter of course.   But the step which Captain Abney appears now to have taken lies entirely out of the ordinary line of advance.   M. Becquerel, as is well known, succeeded some years ago in producing natural colours by the agency of light, one specimen of which—a solar spectrum—was given to me some three years since by M. Cornu, and which, through being carefully preserved in a tin box, retains all its chromatic features.   Captain Abney has now taken up the subject again, and believes it probable that the colours obtained by this process may be preserved unchanged when exposed to ordinary daylight.   His suggestion as to the physical causes of the colour is interesting, and leads us to hope that we may hear more upon the subject.

It was at one time thought that Science and practical life were essentially distinct, and that in their cares and their purposes, in their sorrows and their joys, neither intermeddled with the other.   But in proportion as it has been gradually recognised that Science, and even Philosophy itself, is based upon experience, so has their distinction gradually faded from our view.   And, among many other instances of forecasts in this direction by far-seeing men, I may adduce the dream of Babbage, that mathematical calculation might be reduced to mechanism, and that the data of many problems both of physics and of life might be handed over to an engine which would work out the results. That dream of his was as nearly realised as any dream of his ever could be realised ; for the merit and the charm of the man himself, and the fascination which he threw about all his projects, were derived from the fact that, at whatever rate realisation followed, his ideas flew faster and faster still and outstripped them all.   And if, in the machinery for complicated calculation designed by the Brothers Thomson, much more has been actually achieved than ever before, it is not because this is a generation of more feeble folk, or because there has been any lack of elasticity of mind or of fertility in idea, but because their minds and their ideas have been more under control, and because these men have succeeded in that which their great predecessor always missed, namely, mastery over themselves.

The relations between the Mechanism of the Heavens and the History of the Earth have for some years past been the subject of speculation and of argument amongst many prominent men of science. And calculations have been made, tending more or less to confirm the views of the various writers on the subject. But so vast is the problem, so multifarious in its data, so complicated in its laws, that eager as we are to grasp anything solid or well founded in the inquiry, we are constantly left with a feeling that we may have clutched at a mere floating weed, when we hoped that we had laid hold of something firmly rooted. And on this account we may look with unusual satisfaction at the massive and philosophical research in which Mr. George Darwin is engaged, and of which we have already some substantial instalments. It consists in an investigation on the precession and tides in a viscous spheroid, in which he applies his results to a discussion of the internal condition of the several planets, and of the evolution of the moon and other satellites from the primitive nebulæ.

I now pass on to mention briefly a few other subjects which, outside our immediate sphere of action, form substantial elements of the scientific progress of the past year, and among these I should first mention Meteorology, because the Council, to whom in this country the subject is entrusted, is nominated by the Royal Society.

The science of Meteorology has, during the last few years, attracted an increasing amount of attention, more perhaps from its close relation to the interests of all classes of the community than from the definiteness or novelty of the scientific results to which it has led.

The second International Congress of Meteorology met at Rome in April of the present year, on the invitation of the Italian Government; an interval of nearly five years having elapsed since the date of the first Congress, which was held at Vienna in 1874. The main object of these meetings has been to introduce greater uniformity of method into the meteorological systems of Europe, without attempting, however, to pledge the different Governments to any definite engagements on the subject. With this view a permanent International Committee has been constitued, to furnish a *point d'appui* for common action, and to facilitate, during the intervals between the Congresses, a personal interchange of opinions among the authorities directing the different systems.

A great number of valuable reports, on questions embracing nearly the whole field of Meteorology, were received by the Roman Congress. Among them, special mention may be made of the reports by M. Violle on Solar Radiation; by M. Pernet, on the Determination of the Fixed Points of Thermometers; and by Prof. Everett, on Atmospheric Electricity.

A Conference of a semi-official character, which was originated by the Congress at Rome, was held at Hamburg in the beginning of October, for the purpose of taking into consideration a proposal, emanating from Lieut. Weyprecht and Count Wilczek, to establish for one complete year a circle of meteorological observations round the Arctic regions of the globe. The Conference was attended by representatives of the Meteorological Institutes of Austria, France, Germany, Russia, and the Scandinavian countries; and it was agreed that an effort should be made to establish the proposed circumpolar system of meteorological and physical observations in the year 1881. Some part at all events of this plan will, in all probability, be carried into effect, as the Canadian Government has intimated its readiness to co-operate in it, and as Count Wilczek has offered himself to defray the cost of an Observatory at Nova Zembla.

The unfavourable weather which has characterised the present year has given an additional impulse to the study of English Meteorology: and our own Meteorological Office has ventured to resume the issue of Daily Forecasts of Weather, after they had been discontinued for nearly thirteen years. The attempt has been made under somewhat adverse circumstances, as the uncertainty of the weather of 1879 has even surpassed the proverbial fickleness of the British climate. But the experiment has been liberally supported by the Press, and has been watched with great interest by the public generally. Of its success or failure it would as yet be premature to speak, but it deserves recognition as a serious attempt to apply to one of the most uncertain of the sciences the severe test of prediction.

Transit of Venus (1874) *Reductions.*—The calculations of every kind relating to the five British Expeditions were some time ago concluded. The mass of them refers to the determinations of terrestrial longitudes and the discussion of the measurements of photographs. The preparation of the Astronomical (as distinguished from the Photographic) Observations and Results for Printing is complete for the expeditions to the Hawaiian Islands, Rodriguez Island, and New Zealand. For the other two expeditions much remains to be done—chiefly, however, mere copying. The employ-ment of the staff upon the measurement, re-measurement, and discussion of the photographs, was the most serious cause of the delay that has occurred in presenting to the public the astronomical results. Those who are now engaged upon this subject entertain serious doubts as to whether any result of value, as regards the special object of the expeditions, is to be expected from the photographs, owing to a want of sharpness in the images. Some of the photographs taken by the Russian Astronomers* have been measured, but not very successfully. The telescopic obser-vations of the internal contacts, at many stations in India and Australia, as well as at those occupied by the Govern-ment Observers, in all thirty-one observations of the *Ingress* and forty-eight of the *Egress*, have yielded a value of the mean solar parallax of $8'''85\pm0'''03$.†

The Greenwich Nine-Year Catalogue of 2263 stars, with discussions of the systematic errors of R-A.'s and N.P.D.'s has been published.

The complete reduction of the measures of about 1900 photographs of the sun, taken at Greenwich from 1873 up to the present time, affords data immediately available for determining the distribution of spots and faculæ on the sun's surface, and the position of its axis of rotation. The photo-graphs now regularly taken at Greenwich may be considered a continuation of the Kew series.

The reduction of the Greenwich magnetic observations, 1841—1877, has afforded an opportunity for testing the con-nexion (first suggested by Sir E. Sabine and afterwards investigated as regards the element of declination by Pro-fessor R. Wolf) between magnetic variations and sun-spots. Mr. Ellis has compared the diurnal ranges of magnetic declination and horizontal force with Professor Wolf's curve of sun-spot frequency, with the result that not only is there a general correspondence in the two sets of phenomena, but the minor irregularities of the sun-spot curve are reproduced in the curves of diurnal magnetic range for both elements, and further that the well-marked annual inequality in the latter is itself variable, being greatest at the time of maxi-mum of sun-spots and least at that of minimum.

* At Haven Possuit.
† " Monthly Notices, R.A.S.," xxxviii., p. 455.

The spectroscopic determination of the motions of stars in the line of sight by the displacement of lines in their spectra, has been recently extended at Greenwich to stars of the third and fourth magnitude, raising the number of stars available for the application of this method to 200 or 300. Up to the present time, the motions of 63 stars have been thus determined at Greenwich. Mr. Seabroke, at Rugby, has also applied the method to 28 stars, and his results, though presenting some discordances, *inter se*, generally support those obtained by Mr. Huggins and by the Greenwich observers.*

A new determination of the Figure of the Earth, in which the results of recent measurements of the Indian arc of 24° are included, has been made by Colonel Clarke.†

A connexion between the Sun's Outer Corona and Meteor Streams has been suggested by several observers, Professor Cleveland Abbé, Mr. F. C. Penrose, and others,‡ of the Solar Eclipse of 1878.

From a discussion of the distribution of aphelia and inclinations of cometary orbits, Professor H. A. Newton has come to the conclusion that comets have come to us from the stellar spaces (in accordance with Laplace's hypothesis) and that (with the possible exception of the comets of short period) they have not originated within the solar system as Kant supposed.§

Mr. Gill has deduced from his extensive series of Heliometer Observations of Mars at Ascension, the value $8'''\cdot78\pm0'''\cdot015$ for the Sun's mean Equat. Hor. Parallax.‖ This agrees nearly with the first published result of the British Transit of Venus Expeditions, and also with the result of Cornu's determination of the velocity of light combined with Struve's constant of aberration.

The publication of Lohrmann and Schmidt's Maps of the Moon forms an important contribution to Selenography.

M. Flammarion has published a valuable Catalogue of Binary Stars, giving all available observations (to the number of 14,000) of these objects, with the elements of their orbits, so as to present a complete summary of their history.

Among the more recent Institutions for the promotion of Science, mention should be made of the Observatories

* " Monthly Notices," 1879, June.   1879, October 10.
† " Phil. Mag.," 1878, August.
‡ " Monthly Notices," 1878, November.
§ " Amer. Journ ," 1878, September.
‖ " Monthly Notices," 1879, June.

for the study of Astronomical Physics in France and Germany.

The elder of these in point of date, namely that in France, although founded in 1876, has only quite recently been permanently established in the Chateau at Meudon. The Director has ever since the foundation been principally occupied with solar photography, and he has established the fact that photography is capable of revealing phenomena which must necessarily escape ordinary telescopic observation. In particular, he has succeeded in photographing the granulations previously seen on the solar disk by other astronomers ; and by means of his magnificent pictures, one of which he has recently presented to the Society, he has established the existence of what he terms the *Réseau photospherique ;* in other words, that the surface of the sun is divided into regions of calm and of activity. His present researches will doubtless enrich our knowledge with many new facts.

The German Observatory has been placed at Potsdam. In the selection of the locality, 320 feet above the sea, with a good horizon, and free from the smoke and vibration of the city, and in the furnishing of the Observatory, the German Government has done everything that could be wished. And it may confidently be hoped that in the hands of Professor Vogel and his colleagues much good work will be there done.

The ramifications of Science are now so various in direction, so comprehensive in grasp, and so immediate in their applications, that they have of late years given rise to the establishment of special Institutes and Societies for the study and encouragement of their respective subjects. Among these there may be mentioned the Society of Telegraph Engineers, the Iron and Steel Institute, and others.

At the meetings of these bodies it frequently happens that subjects which may have, in their primary form of a scientific discovery, been laid before the Royal Society, are brought up again and discussed from a different point of view, *e.g.,* in their relation to other subjects, or in their application to the special purposes of life. For instance, in the " Proceedings of the Society of Telegraph Engineers " we find the following :—

1. Wires which run parallel to each other for a long distance are very much troubled by induction between wire and wire. The effect on telegraphs is to diminish speed of working, and on telephones to render speaking impossible. Hughes has studied these effects in his own peculiar way and has shown how, when the number of wires is limited

and the working confined to one direction, they can be entirely eliminated.

2. Mr. Cowper has shown how, by producing the two rectangular components of a plane curve at a distance, it is possible to reproduce handwriting. His instrument works very effectively, but is still perhaps in an early stage of its development.

3. The application of the Duplex system to long cables has been practically and very efficiently carried out by Mr. Stearns on the Atlantic Cable, and by Mr. Muirhead on the long cables of the Eastern Telegraph Company in the Mediterranean and Indian Ocean.

4. The Electric Light is still on its trial for general street illumination in Paris and London; but it has been recently very effectively employed in surveying and sounding the Mediterranean by night, and in the operation of laying and repairing cables by the steam-ship " Dacia," belonging to the Silvertown Company. By its aid ninety-six soundings, made with Sir William Thomson's steel wire apparatus, in water averaging 1500 fathoms depth, were taken in seven days and nights, bottom in every case being brought up.

5. Several improvements in Telephones have been brought out, notably Gower's, in Paris, and Edison's loud-sounding arrangement, and the instruments are gradually finding their way into practical use.

6. Experiments are being made both in England and in France in the transmission of mechanical power by electric currents; and we may hope for valuable results from this field of inquiry.

The Iron and Steel Institute was founded in 1869 under the presidency of the Duke of Devonshire. The Society consists of proprietors and managers of iron and steel works, of metallurgical chemists, geologists, and engineers. The number of its members has grown gradually from 140 at its commencement to its present limit, 1031, of whom 116 are foreigners residing abroad, and comprising the leading metallurgists of America as well as of the continent of Europe. The Association meets twice a year, once (during the spring) in London, and once (in the autumn) by invitation at one of the great industrial centres in England or abroad. In 1873 it met at Liège, in 1878 at Paris, and in 1880 it will meet at Düsseldorf. Each meeting lasts from two to four days, which are spent partly in reading and discussing papers, and partly in visiting works of interest.

The results of the Association have been of considerable utility: (1) by bringing the scientific metallurgist into contact with the man of practical experience, thereby extending the appreciation, and in many instances even the actual cultivation, of Science; (2) by the annual publication of a very valuable record of the progress made in this important branch of applied Science; (3) by acting as a kind of international tribunal on the merits or demerits of new processes, in a manner calculated both to stimulate and to guide further undertakings.

Telegraph   Conference.—When   Telegraphy   began   to assume   an   international   character,   great   inconvenience arose from want of uniformity in the apparatus used, and in the   service   regulations   adopted   by   different   European countries.

Although the construction of the line (whether submarine, underground, or suspended) and the details of construction of the telegraphic apparatus employed may fairly be left in the hands of each country or telegraphic administration, it was soon found to be necessary that a certain standard of efficiency should be insisted upon.   Agreement was moreover requisite in regard to the general character of the instruments to be used in international telegraphy; for instance, whether or not acoustic signals were to be admissible, whether an indelible record should be insisted upon, whether a   type-printer, or an autographic instrument should be used, or a recording instrument representing the letters by dots and dashes, in which case a uniform code would be a matter of necessity.   Beside these points there remained the subject of the tariff to be charged in each country for transit and terminal messages, involving delicate questions, which could be settled only by mutual concessions and general arrangements.

These considerations led the French Government, in 1866, to invite those European Governments who had already taken the administration of the telegraphs in their own hands, to meet at Paris, in order to agree if possible on certain resolutions, subject to ratification by the respective Governments.   England was not represented at this first Conference, because the telegraphs were not yet a Government Department, and it was not thought advisable to invite telegraph companies to take part in the proceedings.

The result of the first deliberations went beyond the mere settlement of pressing questions, and it was decided to hold periodical Conferences at the different capitals of Europe, and to institute a central International "Bureau" (to be located at Berne) to see to the proper application of the resolutions of the Conferences, and to act as an International

Court of Appeal, regarding questions of difference in telegraphic matters arising between the contracting countries.

The second Conference took place at Vienna in 1869, in which several countries, including Turkey, Persia, and British India, took part for the first time. On this occasion it was resolved that the leading telegraph companies should be admitted to the discussions, but should not be allowed to vote ; the power of voting being confined to the Governments, and each country having one vote only, irrespective of its size and importance.

The third Conference was held at Rome in December, 1871, when the British Government (Post Office) was for the first time represented, and also the Indo-European and Cable Companies, which latter are bound by the resolutions of the Conference only in matters of international administration, and of tariff in cases where lines join two countries belonging to the Union.

The fourth Conference met at St. Petersburg in 1876, and the fifth at London in 1879. At the latter meeting difficult questions of tariff arose, which protracted the sittings over nearly two months. At previous Conferences it had already been decided to treat each country as a unit, its particular charge being the same, whether a message had only to cross the frontier or had to pass over long distances into the interior. At the London Conference it was proposed by Germany to adopt a uniform rate per international message for the whole of Europe, analogous to the uniform postal rate for letters, while France went further and urged the adoption of a universal *rate per word*. Neither of these measures, however, was adopted, but a compromise was effected : and after the 1st April, 1880, each word will be charged at a specified rate for each country, five words being added to each message to cover time spent in advising, commencing, and finishing it.

All European countries, as well as Brazil, Persia, India, and Japan, belong to the Conference, and are bound by its decisions.

Any of twelve modern languages using the Roman character may be employed ; Russian, Greek, Turkish, being thus excluded from use.

It would be difficult to over estimate the beneficial results produced by these Conferences in stimulating and perfecting international communication.

Material is not wanting to extend these notices of scientific activity over a wider range ; for our foreign neighbours are always ready to furnish us with information of whatever is new among them either in matter or in method. But a mere enumeration of these things, however interesting in themselves, would form but a dull chapter, unless it should issue in a judicial estimate of results, or a suggestive comparison of national arrangements for the promotion of science. On such an enterprise I am by no means prepared to adventure myself; and even if I should be ever led to make the attempt on some special topic, or some particular aspect of the question, I must, at all events, defer it to a future occasion.

The Award of the Royal Society Medals.—The Copley Medal has been awarded to Rudolph Julius Emmanuel Clausius, Foreign Member of the Royal Society, for his investigations in the Mechanical Theory of Heat.

The Mechanical Theory of Heat as at present understood and taught has been so essentially a matter of growth, that it would be difficult to assign to each investigator the precise part which he has taken in its establishment. It will, however, be admitted by all, that the researches of Clausius rank high among those which have mainly contributed to its development. These researches extend over a period of thirty years, and embrace important applications of the theory not only to the steam engine, but to the sciences of electricity and magnetism.

Even to enumerate those who have contributed to one branch of the subject, viz., the Kinetic Theory of Gases, would be beyond my present purpose and powers; but as Clausius himself states, both Daniel and John Bernoulli[*] wrote on the subject. And, even, to go back to earlier times, Lucretius[†] threw out the idea; while Gassendi, and our own Boyle, appear to have entertained it. Within our own recollection, Joule, Meyer, Kroning, Clerk Maxwell, and others have made invaluable contributions to this branch, as well as to the general subject of the Mechanical Theory of Heat. But however great the value of these contributions, it may safely be stated that the name of Clausius will always be associated with the development of earlier ideas into a real scientific theory.

A Royal Medal has been awarded to W. H. Perkin, F.R.S. Mr. William Perkin has been, during more than twenty years, one of the most industrious and successful investigators of Organic Chemistry.

Mr. Perkin is the originator of one of the most important

---

[*] In the 10th section of his " Hydrodynamics."
[†] " De rerum Naturâ," lib. ii, 111—140.

branches of chemical industry, that of the manufacture of dyes from coal-tar derivatives.

Forty-three years ago the production of a violet-blue colour by the addition of chloride of lime to oil obtained from coal-tar was first noticed, and this having afterwards been ascertained to be due to the existence of the organic base known as aniline, the production of the colouration was for many years used as a very delicate test for that substance. The violet colour in question, which was soon afterwards also produced by other oxidising agents, appeared, however, to be quite fugitive, and the possibility of fixing and obtaining in a state of purity the aniline product which gave rise to it, appears not to have occurred to chemists until Mr. Perkin successfully grappled with the subject in 1856, and produced the beautiful colouring matter known as aniline violet, or mauve, the production of which, on a large scale, by Mr. Perkin, laid the foundation of the coal-tar colour industry.

His more recent researches on anthracene derivatives, especially on artificial alizarine, the colouring matter identical with that obtained from madder, rank among the most important work, and some of them have greatly contributed to the successful manufacture of alizarine in this country, whereby we have been rendered independent of the importation of madder.

Among the very numerous researches of purely scientific interest which Mr. Perkin has published, a series on the hydrides of salicyl and their derivatives may be specially referred to ; but among the most prominent of his admirable investigations are those resulting in the synthesis of coumarin, the odoriferous principle of the tonquin bean and the sweet scented woodruff, and of its homologues.

The artificial production of glycocol and of tartaric acid by Mr. Perkin conjointly with Mr. Duppa, afford other admirable examples of synthetical research, which excited very great interest among chemists at the time of their publication.

It is seldom that an investigator of organic chemistry has extended his researches over so wide a range as is the case with Mr. Perkin, and his work has always commanded the admiration of chemists for its accuracy and completeness, and for the originality of its conception.

A Royal Medal has been awarded to A. C. Ramsay, F.R.S. Professor Ramsay has been for a period of nearly forty years connected with the Geographical Survey of Great Britain, and during by far the greater part of that time either as Director or Director-General of the Survey. During this long period, in addition to his official labours in advancing our knowledge of the geology of this country, he has published works on the " Geology of Arran," " The Geology of North Wales," " The Old Glaciers of North Wales and Switzerland," and " The Physical Geology and Geography of Great Britain," now in its fifth edition. His papers in the " Quarterly Journal of the Geological Society," and elsewhere, are numerous and important, especially those on theoretical questions in physical geology, such, for instance, as " The Glacial Origin of Lake Basins," " The Freshwater Formation of the Older Red Rocks," and " The History of the Valley of the Rhine, and other Valleys of Erosion." There are, indeed, among living geologists few who can claim to have done more to extend our knowledge in the important fields of geology and physical geography.

The Davy Medal has been awarded to P. E. Lecoq de Boisbaudran. The discovery of the metal gallium is remarkable for having filled a gap which had been previously pointed out in the series of known elements. Mendelejeff had already shown that a metal might probably exist, intermediate in its properties between aluminium and indium, before Boisbaudran's laborious spectroscopic and chemical investigation of numerous varieties of blende led him to the discovery and isolation of such a metal.

The separation of the minute traces of gallium compounds from blende is an operation presenting unusual difficulty, owing to the circumstance that compounds of gallium are carried down by various precipitates from solutions which are incapable by themselves of depositing those compounds.

III. THE HON. HENRY CAVENDISH : HIS LIFE AND CHARACTER.

IF there is one scriptural admonition which the scientific workers of the present day fail to obey more rarely than another, it is the one which warns us against the foolishness of hiding our light under a bushel, instead of setting it on a hill so that it may shine before all men.

Every discoverer, now-a-days, whether great or small, as soon as he finds his light,—whether it be a six-thousand-candle electric lamp or only a halfpenny dip,—immediately hastens to place it on the top of the tallest hill he can find, so that it may shine forth literally *urbi et orbi.* Many lights, it is true, give forth only a feeble glimmer; but it is surely better that we should be at times overburdened with crude observations of possibly valueless facts, than that a single particle of truth should be concealed, or its publication delayed even for a day more than is absolutely necessary.

There never was a period in the world's history when scientific observation was so universal as in the present year of grace, and it never before had such a chance of being so thoroughly controlled by publication and criticism. An important discovery in any branch of physical science is now made public with a rapidity that has never before been equalled, and the paper, article, or even telegram containing its history is published and re-published, discussed and criticised, in every civilised language. The observations described are repeated and tested in half a hundred laboratories, and the slightest incorrectness or mis-statement is pounced upon with the utmost eagerness, and published with the same universality as the original researches themselves. The numerous facilities which we possess for spreading and sifting scientific observations are bearing fruit every day, and the scientific press—although its office is to collect and distribute facts rather than to criticise them—has become as great a power in its own particular sphere as its elder sister, the political press, has in the hands of our political fellow-workers.

We have been led into making these remarks upon the great publicity which is now given to scientific researches by a perusal of the excellent sketch of the great Cavendish's life and character, which serves as an introduction to the collection of his electrical researches,* the principal of which are now published for the first time through the liberality of the University of Cambridge. The work has been edited by and brought out under the direction of the late Prof. James Clerk Maxwell, who was lost to science almost before the ink of the last proof was dry. It is indeed a striking example of the fitness of things that these hitherto

* The Electrical Researches of the Honourable Henry Cavendish, F.R.S. Written between 1771 and 1781. Edited, from the Original Manuscripts in the possession of the Duke of Devonshire, K.G., by J. Clerk Maxwell, F.R.S. Cambridge University Press. London: Cambridge Warehouse, 17, Paternoster Row.

hidden documents should have been given to the world as
the last labour of a modern scientist who united in his own
person the same extended mathematical knowledge and
power of acute and accurate observation as was possessed
by the philosopher whose researches he has so ably edited.
It is also equally fitting that the late Prof. Maxwell should,
at the time of his death, have been the first Cavendish professor at the laboratory founded at Cambridge by the present
Duke of Devonshire, in memory and in honour of his illustrious kinsman.

The ¡Hon. Henry Cavendish was the eldest son of Lord
Charles Cavendish, third son of the second Duke of Devonshire, whose family can be certainly traced back to the reign
of Edward III., and with great probability to that of the
Conqueror.   This fact is worthy of notice seeing that, besides Robert Boyle, Henry Cavendish was the only scion of
our English nobility who left his mark on Science.   He was
born at Nice (then in Italy) in 1731, and died at Clapham
in 1810.   Although he had nearly reached eighty years of
age at the date of his death, we know less of his sayings
and doings during that long period than we do of those of
any other modern philosopher of comparable eminence
during a quarter of that time, and this for a simple but
sufficient reason that will presently appear.   From the few
particulars that can be gathered about him from his contemporaries—for his own journals, papers, and correspondence throw little or no light on the matter—Cavendish
appears to have been the very incarnation of shyness, a
quality which seems to have imbued his whole life so thoroughly as to easily account for all his oddities and eccen-
tricities, and to have finally rendered him a perfect type of
a passionless, viceless solitary.   The first we hear of him in
connection with the work-a-day world is in 1755, when, after
three years' study, he left Cambridge without taking his
degree and the honours to which his transcendent abilities
would have entitled him.   This piece of eccentricity is referred by some of his biographers to his conscientious
objection to sign the then necessary religious declarations;
but as he never gave the slightest sign of having either a
conscience or a religion, we think it more consistent to ascribe his flight to his ruling passion, if, indeed, so strong a
term can with any propriety be applied to so passive a defect
as shyness.

During the first twenty years after quitting Cambridge he
appears to have lived at his father's house in Great Marl-
borough Street ; but here again all personal particulars are

wanting beyond the fact mentioned by himself in one of his
papers, that he worked in a back and fore room, one of
which was 14 feet high,—a description which, meagre as it
is, puts out of court those of his biographers who assert
that he lived in a set of stables fitted up for his accommodation. During this period his father, who was a comparatively poor man, allowed him £500 a year. The smallness
of the allowance is supposed by some to have been the
cause of Cavendish's habits of economy and oddities of
character; but this is equally untenable. Riches and poverty are only relative, and the passionless philosopher—
whose apparatus was chiefly made of glass plates, wire, tin-
foil, cardboard, and firewood—must in those days have been
fairly comfortable on such a sum. It is stated that the so-
called meagreness of his annuity was owing to his father's
displeasure at his refusal to enter public or professional life:
whether this be so or not it is impossible to know now,
but the fact nevertheless remains that Lord Charles
was a poor man considering his high connections. He was
a man of Science of no mean ability; and although we have
no means of knowing upon what terms he lived with his
son, he must have appreciated his superior genius.
Somewhere about 1773, about ten years before his father's
death, Cavendish appears to have come into a large fortune
of several hundred thousand pounds, but the source or
sources of it are unknown, except we admit the hackneyed
" rich uncle from India " theory. This fortune he added to,
by his inability to spend, so considerably during the remainder of his life that he died worth a million and a quarter
sterling.

He now seems to have removed to Montague Place, Russell
Square, and to have taken a country house at Clapham,
where he died. His wealth seems to have made no change
in his habits; he still remained the silent investigator,
dragging secret after secret out of his ricketty apparatus,
and carefully hoarding it up. His wonderful habits of precision, however, compelled him to make careful notes of all
that he did, and it is from these notes that we know most
of his greatness as an electrician. The twenty odd packets
of essays and papers—most of them carefully prepared for
reading or printing, but never published—were placed in the
hands of Sir W. Snow Harris by the present Duke of Devonshire when Lord Burlington; but this " valuable mine of
results," as they are called by Sir William Thomson, to
whom they were shown in 1849, has been unworked till
now. As we are not reviewing Prof. Maxwell's work at

present, we must refer to its pages those of our readers who wish to assay these long-concealed nuggets. The collection was partly used by Dr. Grange Wilson in his " Life of Cavendish, written in 1851 for the Cavendish Society, and referred to by Sir W. Snow Harris in his work on " Frictional Electricity" published in 1867, but Sir William did not live to edit them. Under these circumstances the Duke of Devonshire, to whom they belong, placed them in Prof. Maxwell's hands in 1874, with the interesting results that we see. These details are necessary as showing the care with which Cavendish hoarded the written accounts of his labours, which, had they been published in due time, would have added still brighter rays to the already brilliant crown of light that hovers over his name. Investigation was his joy; publication his cross, for it wounded his shyness in its tenderest spot. He undertook the most laborious researches to clear up difficulties and doubts of whose existence and scope he alone was cognisant. The pleasure—his unique one—of solving these problems he must have enjoyed alone, insensible to the desire of communicating their solutions to his fellow men, which appears to have more or less ruled the actions of philosophers of all ages. Of his miraculous precision and accuracy in working we are not called upon to speak, but in his investigation of electrical capacity of a long cylinder his experimental results agree with those worked out mathematically by Prof. Maxwell within a mere fraction, which in one case amounts to only 0·006 −.

So much for Henry Cavendish as regards his own individuality. In his intercourse with the very small world in which he moved he was still more eccentric. Friends and intimates he had none; his acquaintances were few and far between, and even they seem to have consisted only of a small sprinkling of the Fellows of the Royal Society whom he could hardly help associating with. He was elected a Fellow of the Royal Society in 1760, and knowing what we do of his habits, it seems surprising that he should have ever been able to summon up sufficient courage to go through the ordeal of proposal and election. He was now, of necessity, thrown much into contact with Sir Joseph Banks, who was then President of that learned Society, and he appears to have been more inclined to associate with him than with any of his other illustrious companions. He even seems to have forced himself into attending the presidential gatherings at Sir Joseph's house in the then fashionable quarter of Soho Square. Such a concession to gregariousness in this most ungregarious of human beings, as he is justly called by one who knew him, must have been a sore trial to him ; but he, no doubt, consented to undergo it knowing well that he could not carry on his own researches without being acquainted with what some, at least, of the more eminent of his fellow workers were doing.   Hence his visits to Crane Court and Soho Square, and to the Royal Society Club, which he also seems to have frequented occasionally. He also appears to have been a member of the Cat and Bag-pipes Club, which was apparently a kind of Junior Royal Society Club for scientific workers who had not yet earned the privilege of belonging to the more august body.

One of his visits to Soho Square, on the occasion of a larger gathering than usual, is graphically described by one of his contemporaries.   *Ex uno disce omnes.*   He would drive up to the door in his carriage, watch a favourable oppor-tunity, and glide quickly upstairs as soon as he saw the staircase clear.

Reaching the top of the flight, he would hang his three-cornered hat, of the same fashion as the one his grandfather wore, on his own particular peg, and then lose all heart. He would slink nervously about the passage, making several fruitless attempts to enter the drawing-room.   At last he would summon up sufficient courage to turn the handle and creep in.   Those who recognised him knew him too well to welcome him in words, or even by looks.   He would now steal about awkwardly from group to group until he lighted upon one which was conversing on some of the many topics that interested him.   He would listen attentively at a little distance, and occasionally make a remark or ask a question that was always exceedingly to the purpose.   If any one answered him without looking at him, he would turn slightly aside and listen attentively; but if he was addressed per-sonally, he would dart quickly away with a low murmur of displeasure and seek some other and more discreet group of speakers.   One night an event occurred which he must have viewed in the light of a catastrophe.   There was a greater gathering than usual, apparently to do honour to a distin-guished Austrian philosopher who was then visiting this country.   Taking him somewhat unawares, Dr. Ingenhouz, one of his few acquaintances who could presume to take such a liberty, informed him that his learned foreign brother wished for the honour of a personal introduction to him. Poor Cavendish was completely taken aback for the moment. He seemed to have lost all power of speech and motion. He stared helplessly at the Austrian, who by this time had begun a set speech to the effect that the primary object of his visit to England was to be presented to the most learned, the most illustrious, the most— But before he could get any further in his adulatory address the object of it had uttered a sharp cry of pain as if he had been wounded, rushed hatless down the stairs and into his carriage, screaming shrilly to his coachman to drive as fast as possible to Montague Place, where he doubtless solaced himself by making a host of perfectly unnecessary measurements.

His appearance, it need hardly be said, was as singular as his manners. He dressed always in the same bygone fashion. Judging by the only portrait of him in existence —taken stealthily and piecemeal on several occasions by one Alexander, the draughtsman of the Barrow Expedition—he bore but few marks of his rank about him. His costume, which he never altered, winter or summer, consisted of a greyish violet, as much stained and faded as poor Goldy's famous peach blossom velvet ; black breeches, and untidy-gartered white hose and low shoes, the pinched up face and ungainly figure surmounted by a low-crowned three-cornered hat. He spoke in a thin, reedy voice, tremulous from nervousness, but always with extreme appositeness and great gentleness. He was singularly quick of apprehension, and understood everything at a mere word or glance.

How he schooled himself into attending and speaking at the Royal Society meetings is a puzzle to every one acquainted with his singular shyness. Sir Joseph Banks often shielded him from intrusion by warning all strangers of his peculiarity. The only way of entering into conversation with him was to speak with an averted face, as if some one else was being addressed. He would then converse freely ; but as soon as he was in any way taken notice of, he would dart off like a startled fawn.

With all his eccentric ways he was so coldly gentle and kind that his contemporaries tell us that he was not merely looked up to and honoured by his brethren of the Royal Society, but actually loved by many of them. There was nothing of the surly misanthrope about him, and several of his acts seem to indicate that somewhere about his anatomy there was sometimes flowing a tiny warm rivulet of love of his kind. If asked to subscribe to any charity he would look down the list for the largest donation, and immediately sign a cheque for the amount. During his lifetime he gave his splendid library for the use of the public, taking out the books he wanted, and signing a receipt for them. His darling apparatus was also at the service of any deserving student ; and on one occasion, when his curator was driven

to the verge of madness by a student breaking a piece of
Cavendish's most treasured apparatus, the only rebuke that
this gentle solitary could utter was to the effect that young
students could only learn to use delicate apparatus by
breaking it—a story which is surely to the full as pretty as that
of Newton and his little dog Diamond.   On another occasion he was told that a once prosperous acquaintance had
fallen into poverty.   It was suggested to him that a small
annuity might at any rate keep the wolf from the poor
man's door.   Cavendish answered the appeal by tremulously
asking if ten thousand pounds would be sufficient.

It has also been said that he was a misogynist, because
he always communicated with his housekeeper in writing,
and immediately dismissed any female servant who allowed
herself to be met by him in his Clapham house.   These
stories of him, however, are at best nothing but Clapham
gossip, collected some years after his death.   On one occasion this gentle, timid, feeble old man valorously defended
and ultimately rescued a lady from the attack of an infuriated cow,—so at any rate one member of the fair sex could
say he was no misogynist.   At the christening of a young
relation he was told that it was usual to make a present to
the baby's nurse ; he obeyed the admonition by giving her
an uncounted handful of guineas.   That he scrupulously
avoided all communication with the female sex is undoubted,
but he shunned the highest worth and intellect of his time
in the same way.   Had there been a Mrs. Somerville in his
day, and she could have given him any information on any
of the multifarious subjects in which he took an interest, it
is only too probable that he would not have shunned her.

Very few visitors were admitted to Montague Place.   One
of the favoured few reports that the rooms were all apparatus.   Fewer still were admitted to the Clapham house,
where they were always regaled off the same dish,—a leg of
mutton, or, if they exceeded a certain number, two legs of
mutton.   Who can say whether the cold philosopher may
not have sometimes warmed into the genial host at these
symposia, at which one bowl of punch at least must have
been produced in accordance with the custom of the day ?
We are told that at the Royal Society dinners Cavendish
was as cold and reserved as elsewhere ; but is it not possible
that he may have unbent somewhat at those mysterious Cat
and Bagpipes feasts to which he must have gone of his own
free will and choice ?

His death was like his life—solitary.   When he found his
end approaching he called his servant, and strictly com-

manded him to leave him alone until the great event had occurred, and not until then to carry the intelligence to his cousin and heir, Lord George Cavendish.

In presence of the reverence and affection with which he was regarded by his fellow philosophers, he could not be said to have died unwept; he has certainly not died unsung. His deeds will live as long as Science itself.

IV. ENGLISH AND AMERICAN PHYSIQUE.*

By George M. Beard, M.D.

I.

AMERICA is England in a minor key.   All that is good, all that is evil in the United States of the past, or near present, comes directly from Great Britain—the daughter is but a mild type of the mother.   In the angry and inexpert discussions of national characteristics, it is forgotten that the difference between one country and the other is far less than is suggested or commonly credited.   In criticising England for her failures, her weaknesses, her inconsistencies, her preference for the near to the remote, and the practical to the ideal, we are but criticising ourselves, who derive all these traits by direct inevitable transmission from our maternal home.

Even American buncombe is all English : at public banquets, at gatherings of quiet and sober men of science in that empire, I have heard more of self-exaltation, of reciprocal flattery, of glorification, of hosannas over the greatness of themselves, than one should hear in this country at a rustic Fourth-of-July carousal; indeed, we are outgrowing that academic play, which the conservatism of the mother land holds to as with bands of iron.

The delusion that the two nations utterly differ in kind appears in England as well as in America.   This summer the distinguished London physician, Dr. Richardson, gave me an excellent example.   The Doctor, in a lecture before a large audience of working men, quoted a passage from my work on longevity of brain-workers, in which a contrast was

* Communicated by the author, having appeared also in "The North American Review."

made between the environment of the muscle-worker and brain-worker. At the close of the lecture, he was told that if that passage—written by an American—were true, they did not wish to emigrate to America, as many of them had thought of doing.

The laws of nature are not reversed by crossing the Atlantic, or by changing the form of government. Forgetting that the purely practical man can never be a first-class man, Americans are wont to call themselves the practical nation, with an implied criticism on all nations that encourage thought; but all our practicality is English: Yankee ingenuity is simply an importation, and the almighty dollar is sired by the divine guinea. The question, Will it pay? was asked in England before this country was born—the problems of abstract right and truth and justice being turned over to German specialists.

The beliefs, disbeliefs, and unbeliefs of men, the true, the false, and the doubtful, England turns to use; she allows no waste threads, but weaves everything into the pattern of the state; to her, sciences, delusions, religions, are all one; she melts them together and moulds them into cathedrals, railroads, bishops, nobilities, civil liberty, livings, benefices, universities, dignity, solidity, comfort; she travels everywhere and forages in all nations for materials to strengthen and adorn the empire; the blunders of other people become her possession; pioneering, experimenting where great risks to philosophy are involved, she leaves to others, and profits by the results. In this feature we, in a certain way, have followed her example.

With these general resemblances there are phenomenal differences of *physique* and character that are of special interest to students of the nervous system.

American v. English Female Beauty.

While the beauty of the English girl may endure longer than that of her American sister, yet American beauty has this sovereign advantage—that it best bears close observation. The English beauty appears best at a distance, and grows homely as we approach her: the typical American beauty appears more attractive near at hand; in her case, nearness brings enchantment. The American face bears the microscope mainly by reason of its delicacy, fineness, and mobility of expression—qualities that are only appreciated on nearness of inspection. The ruddiness or freshness, the health-suggesting and health-sustaining face of the English girl seem incomparable when partially veiled, or when a few rods away; but as she comes nearer, these excelling characteristics retreat behind the irregularities of the skin, the thickness of the lips, the size of the nose; and the observer is mildly stunned by the disappointment at not finding the nimble and automatic play of emotion in the eyes and features without which female beauty must always fall below the line of supreme authority. The English beauties of national and international fame, at whose feet the empire of Great Britain is now kneeling, are of the American type, and in this country they would be held simply as of average rather than exceptional excellence.

It is no hard task for one travelling in Great Britain or on the Continent to distinguish American ladies from those of any other nationality; the practised observer would make a mistake but rarely. At the great watering-places, as Homburg and Baden-Baden, in the lines where travel is thickest, as on the Rhine and through Switzerland, we may often see a face which, far away, seems to be purely American, but which, as we gain a closer view, is found to be all English; should there be a doubt, the voice—the speaking of a single word—often solves the problem.

Riding, once, from Paris to Calais, there stepped into the coach a lady whom for various reasons I assumed to be English, although her whole appearance—her voice, her conversation—were completely American. I concluded that at last I had found a case where it was impossible to make a differential diagnosis between an American and an Englishwoman; and I very soon found that my reasons for believing her English were not well founded—that she was an American, and a typical American, in her voice, face, expression, gait, and bearing, and even in the functional nervous disease from which she had long suffered.

It were well if these two extremes could be united; an American beauty slowly approaching, an English beauty slowly vanishing, present together a picture of human beauty the fairest that could fall on mortal vision. An American lady who unites the American qualities of intellect, of manners, and of *physique*, and who at one period lived for years in English territory, compresses it all in one sentence: " The English face is moulded, the American is chiselled."

The superior fineness and delicacy of organisation of the American woman, as compared with the women of Great Britain, Germany, and Switzerland, is shown in every organ and function—revealing itself in the play of the eyes, in the voice, in the response of the facial muscles, in gait, and dress, and gesture. The European woman steps with a firmer tread than the American, and with not so much lightness, pliancy, and grace. In a multitude, where both nations are represented, this difference is impressive. In the hourly operation of shaking hands one can tell, in some cases, the American woman of the higher order from a European, Swiss, or German in the same rank. The grasp of the European woman is firmer and harder, as though on account of greater strength and firmness of muscle. In the touch of the hand of the American woman there are a nicety and tenderness that the Englishwoman destroys by the force of the impact. It is probable that the interesting and remarkable feat of muscle-reading, popularly called "mind-reading," would not be so skilfully and successfully performed by English as by American ladies, for the reason that the Americans are physically more delicate and nimble, and their susceptibility to external impressions far greater. This delusion of "mind-reading" was born in this country, and within the past few years. It may be rationally claimed that it could not have originated, or at least have attained so wide popularity in England, Germany, or Switzerland, since not enough could be found there who were capable of performing it to the amusement and astonishment of large audiences.

The voice of the American woman is on a higher key than that of the Englishwoman ; and the partially deaf can hear it more easily.

The attractiveness of American women would appear to be the direct effect of climatic conditions, since beauty of the most precious sort requires fineness of organization, delicacy of features, nimbleness and sprightliness of expression. The same influence that makes the American female more handsome also causes her beauty to decay earlier than in Europe. The Englishwoman is less beautiful, less delicate and attractive between fifteen and twenty-five, yet she retains her beauty longer. Women, like plants, need abundant moisture, else they wither. The rains, the clouds, and the storms that enrobe castles and cathedrals in ivy, and keep the meadows green throughout the year, bring freshness and colour to the face ; so the English matron of forty-five or fifty is, perhaps, sometimes handsomer as well as healthier than at fifteen and twenty.

The Character of Woman as revealed by Dress.

The dress is the woman : all of female character is in the clothes for him who can read their language.

A psychologist of much acuteness once asked me : " Why are bright colours beautiful in the sky, but out of taste in dress ? Why should it be a sign of coarse taste to dress in the most brilliant colours, when all go to see an imposing sunset ? "

The answer is, that high culture and sensitive nerves re-act to slight irritation : while low culture and insensitive nerves require strong irritation. Loudness of dress is, therefore, justly regarded as proof of coarseness of nerve-fibre. The American girl is exquisitely susceptible, is impressed by mild irritation acting upon any of the senses. She dresses in taste, and, where the means are at hand, with elegance, in colours that are quiet and subdued, and noticeable only at a short distance. If we could clothe our-selves in sunsets ; if all this resplendency of crimson and scarlet and gold, and all these varieties of hue and form, could descend upon the delicate maiden, and fall about her in palpitating folds like a rich garment, the eye of that maiden and of those who gazed upon her would soon weary ; the irritation of such splendour would become a pang, and only be worn as a badge and sign of a nature in the lower stages of evolution. Bright colours—scarlet and red, so common in Switzerland and in certain parts of Germany— are never seen in America in any class ; and, among men, the custom of wearing gorgeous and jewelled apparel in public assemblies, as at courts or on occasions of state, is a survival of the barbarian period through which all Europe has passed.

The physiological problem, whether the surface of the eye-ball, independent of the muscles that cover and sur-round it, can express emotion, a near study of the American girl seems to answer quite in the affirmative. The time that nerve-force takes in traversing the fibres from centre to extremity is now mathematically measured, and it is known to vary with the individual, the temperament, and the season ; with race, and climate, and sex, it must also vary. In the brain of the American girl thoughts travel by express, in that of her European sister by accommodation.

America, if archæology is to be trusted, is a modern Etruria, the delicate features and fine forms of prehistoric

Italy emerging from the entombment of ages and reappearing in a higher evolution in the Western Hemisphere.

Certain Nervous Diseases peculiar to America.

A new crop of diseases has sprung up in America, of which Great Britain until lately knew nothing, or but little. A class of functional diseases of the nervous system, now beginning to be known everywhere in civilisation, seem to have first taken root under an American sky, whence their seed is being distributed through the world. A fleet of Great Easterns might be filled with hay-fever sufferers alone; and not Great Britain, nor all Europe, nor all the world, could assemble so large an army of sufferers from this distinguished malady; while our cases of nervous exhaustion would make a standing army as large as that of Russia. Of all the facts of modern sociology, this rise and growth of functional nervous disease in the northern part of America is one of the most stupendous, complex, and suggestive; to solve it in all its interlacings, to unfold its marvellous phenomena and trace them back to their sources and forward to their future developments, is to solve the problem of sociology itself.

A thousand causes have been assigned the task of accounting for this. Among the chief of these accredited causes are fast and excessive eating. Although the Americans are fast eaters, or used to be a quarter or half a century ago, yet in the quantity both of food and drinking they are surpassed both by the English and Germans. Europeans eat oftener than Americans, and eat more, in some cases having four or five meals a day, where the American has but two or three; and consume not only more alcoholic liquors of all kinds, but more fluids of every kind.

The American of the higher class, and these remarks refer only to that class, uses but little fluid of any kind. The enormous quantity of alcoholic liquors, including beer, used in the United States, is used to a large extent by Germans and Irish, and those who live in the distant West and South. There are thousands of Americans who from year to year drink no tea or coffee, and but very little water.

Long since I have surrendered the custom of asking my nervous patients whether they drink coffee, for most of them have been forced to drop the pleasant habit long before they consult me. Through all the Northern States the brain-working classes find coffee in some respects more poisonous than whisky or tobacco, and thousands are made wakeful by even a mild cup of tea. The incapacity for bearing the gentlest wines and beers is for thousands of our youth the only salvation against the demon inebriety. Thus the united forces of climate and civilisation are pressing us back from one stimulant to another, until, like babes, we find no safe retreat save in chocolate and milk and water. In the South, for climatic reasons, these substances are far better endured than in the North; but the very day on which this page is composed I am called to see a Southerner tran siently paralysed, to all appearance, through tobacco alone. Tolerance of stimulants is a measure of nerve; the English are men of more bottle-power than the Americans. To see how an Englishman can drink is alone worthy the ocean-voyage. On the steamer a prominent clergyman of the Established Church sat down beside me, poured out half a tumblerful of whiskey, added some water, and drank it almost at one swallow. He was an old gentleman, sturdy, vigorous, energetic, whose health was an object of comment and envy. I said to him : " How can you stand that ? In America, men of your class can not drink that way." He replied, " I have done it all my life, and I am not aware that I was ever injured by it."

A number of years since I was present in Liverpool at an ecclesiastical gathering composed of leading members of the Established Church, from the bishops and archbishops through all the gradations; at luncheon alcoholic liquors were served in a quantity that no assembly of any profession in this country could have desired or tolerated.

It is with mental work as with drinking: long hours of brain-toil are better endured in Great Britain than in America; there is less exhaustion from the strain of over-work. This fact is observed by men of letters and scholars, and men in public life. Parliamentary leaders, &c., in England, can· do more speaking, more sitting up late at night, as well as more eating and more drinking, than the politicians of America.

It has been said that the strength of a nation is the strength of the thighs rather than of the brain; and an English physician of eminence has observed that the best population of the cities of Great Britain renew their strength from the large-limbed Highlanders of the North, but for whom there would be a constant degeneracy. It would appear, then, that the qualities which are necessary to make a good, strong nation are precisely the qualities which make a good horseman, and that he who can ride well makes a good founder of states. The English as a people have that balance and harmony of temperament that always breeds well. Large families are commanded by unwritten law, and this little island has become the spawning-ground of empires.

The American speaks more rapidly than the Central European; he makes more muscular movements of the larynx in a minute; in his nervousness he clips words, articulating indistinctly, and allowing his voice to fall at the end of a sentence, sometimes so as to be inaudible. The Englishman speaks more slowly, enunciates more clearly, says fewer words to a minute, and, as is well known, keeps the voice up where an American would let it fall. The American says more than the Englishwoman, is easier and more alert for converse, quicker to seize a delicate irony, more facile to respond to a suggestion, than the English lady in the same walk of life. I believe, also, that the English, Germans, and Swiss cannot hear as many words in a minute as Americans, the auditory nerve and the brain behind it being incapable of receiving and co-ordinating as many sounds in a given time. Hence it is necessary to speak to them with more calmness and clearness, whatever language may be employed.

Relation of American Oratory to Climate.

American oratory is in part the product of American climate. For success in the loftier phases of oratory, fineness of organisation, a touch of the nervous diathesis is essential: the masters in the oratorical art are always nervous: the same susceptibility that makes them eloquent, subtile, and persuasive, causes them to be timid, distrustful, delicate, and sometimes cowardly. We blame Cicero for the pusillanimity of his old age, and for his terror in the presence of death, and praise him for his spirit and force and grace in the presence of audiences, not thinking that the two opposite modes of conduct flowed from a single source. A nature wholly coarse and hard, with no thread or vein of nerve-sensitiveness, must always fail in the higher realms of the oratoric art, just as it must fail in all arts. Jefferson, after acting his Rip Van Winkle for years, even now enters upon the stage at each performance with a certain anxiety lest he fail; and of more than one orator has it been affirmed that he always dreaded to speak. "Give me an army of cowards," said Wellington; "it is the man who turns pale in the face of the enemy that will fight to the death." This delicacy of organisation, united with Saxon force, makes America a nation of orators.

Differences in Delusions.

The two nations, England and America, differ in their delusions,—our spiritualism, animal magnetism, and clairvoyance having but slight influence in the mother-country. Delusions, like nervous diseases, are not, however, uniformly distributed in this country, but diminish as we go south. On the warm Gulf Coast the desires to solve the problems of life melt quietly away, the superstitions of the northern type being borne out of sight in the overheated atmosphere; the clairvoyants that grow rich in New York, Brooklyn, and Chicago, would starve in Mobile and New Orleans. A medical patient of mine—intelligent, alert, and observing, well fitted to judge, and of large acquaintance in the southern cities—tells me that one of these seers, who came there, returned without securing any native patronage; the little that came to her was from northern visitors.

The swath that Brown the mind-reader cut across this continent—mowing down professors, scholars, philosophers, editors, colleges, and universities by the score, was in a northern section only; had he gone through the Gulf States his path would have been like that of a ship at sea—a slight ripple, in a moment disappearing. Of the large number of names of universities and schools, of teachers of philosophy and of science, that adorn his placards, like the scalps of the red man as he returns from battle, not one is from the South.

EXPLANATORY REPORTS ON NEW SCIENTIFIC PROCESSES AND INVENTIONS.

[These Reports will not partake of the nature of an Advertisement, but will be impartial descriptions, written and signed by well-known Specialists.— ED. J. S.]

THE THOMAS AND GILCHRIST PROCESS FOR ELIMINATING PHOSPHORUS IN THE MANUFACTURE OF STEEL.

UNTIL recently the puddling of pig-iron was regarded simply as a process of oxidation, and it was not until the practical working of the Bessemer converter revealed the fact that neither sulphur nor phosphorus are effectively removed in it, that the mere oxidation theory was abandoned.

Now we know that puddling does something more, that oxidation alone, though fully effectual in removing silicon and carbon, fails in the elimination of sulphur and phosphorus. It is on this account that special "Bessemer pigs" have to be made from hæmatite or other rich ores nearly free from these impurities. Such are scarce in this country (they amount to only about 12 or 13 per cent of our native ores), while clay ironstones containing much phosphorus are very abundant. Now that mild Bessemer steel is so largely superseding wrought-iron, and the iron-masters of the North have erected huge blast-furnaces in the midst of the impure phosphoric Cleveland ores, very large interests are involved in any effort to render the Cleveland pig available for Bessemer or Siemens-Martin purposes.

The theory of the failure of oxygen in removing phosphorus is not yet fully cleared up, though it has been much discussed. Knowing the extreme readiness with which our old chemical plaything, the stick of phosphorus, unites with oxygen, it was theoretically regarded as the first to be driven off in the course of the vivid combustion of a Bessemer "blow," but such is not the case; the silicon burns freely, the carbon follows, and then the iron itself; while the phosphorus and sulphur remain, as though they were incombustible substances. The probable cause of this anomaly is, that at very high temperatures iron is able to reduce both sulphurous and phosphoric acids, just as carbon at high temperatures reduces the oxides of metals which at lower temperatures have much stronger affinities for oxygen than carbon has. Thus the phosphoric acid may be formed and immediately reduced, and its phosphorus re-combined with the iron as before. But if we can put something else in the way that shall take up this phosphoric acid immediately it is formed, and thereby produce a compound not decomposable by the hot iron, nor capable of combining with it, the original phosphorus may be separated.

This is the aim of the process of Messrs. Thomas and Gilchrist. Some years ago I propounded a theory of the action of puddling in which the removal of the sulphur and phosphorus was attributed to a washing out by the slag or cinder, but at that time I merely regarded the action as a sort of diffusion by adhesion, like the diffusion of the impurities of dirty clothes in the water of the washing-tub. Following up this analogy, I may say that the Thomas and Gilchrist process consists in what is equivalent to the addition of soda by the laundress; a basic slag is produced which is capable of combining with the phosphoric acid and the sulphurous or sulphuric acid as they wash out, and of forming a phosphate, or a sulphur salt, that the iron cannot combine with or decompose.

This idea.of a basic slag is simple enough in theory, but difficult to carry out in practice. The very high temperature of the contents of the Bessemer converter demands a highly refractory lining. The ganisters, &c., commonly used are siliceous,—*i.e.*, acid rather than basic,—and lime, which is cheap, and has the requisite basic character if simply added in the ordinary converter, combines with the silica of the lining, and forms a fusible lime glass, resulting in the destruction of the lining and also of the basic energies of the lime. Soda is still worse, and other bases no better.

Such was the early difficulty encountered by these inventors. What then must be done? The natural answer to this is, that a new lining be found, either basic in itself or allowing the addition of basic material with the charge. As all linings are more or less attacked, and go to supply material for the slag or cinder, the first is the better device, and accordingly the production of a lining that is at the same time basic and refractory has been the object of a long struggle, following the first experiments which satisfied Mr. Thomas that a basic slag is capable of removing the phosphorus.

An account of these experimental struggles would occupy too much space here, and therefore only the final result may be described, which is that Bessemer converters, or the hearths of Siemens-Martin furnaces, are lined with specially made basic bricks, composed of alumino-siliceous magnesian limestone burnt at a very intense white-heat. The best composition of these bricks is found to be 87 to 91 per cent carbonates of lime and magnesia; 3·5 to 6·5 per cent alumina; and 3 to 6·5 per cent of oxide of iron. The presence of not less than 30 per cent of carbonate of magnesia in the raw material is found of great service in preventing the subseqnent disintegration of the burnt bricks by the action of moisture, and also in giving them a highly dense and tough structure. They shrink nearly 50 per cent of their bulk in burning, and thus acquire great density and hardness.

The results have varied with the progress of the experiments conducted for the purpose of battling with the difficulties as they presented themselves. The following are the latest, as stated in a letter I have just received from Mr. Thomas :—

1. " That phosphorus replaces silicon as a heat-giving element with perfect success; thus we have blown at least a thousand tons with under 1 per cent silicon, and much prefer it. We have often much under 0·5 per cent."

In explanation of this I may add that it has hitherto been found desirable to use for Bessemer purposes pig-iron containing exceptionally large proportions of silicon or carbon, or both.

2. " That up to 3 per cent phosphorus can be reduced to 0·05 per cent (if the silicon is low) with regularity."

3. "That when sulphur in pig is over 0·2 per cent, *at least* 50 per cent is eliminated, often 75 per cent. When under 0·1 per cent there seems a difficulty, at present, in reducing to under 0·05 with regularity, though we often remove 80 per cent."

4. " That the phosphorus is removed as phosphate of lime."

5. " That the combustion of phosphorus provides not only sufficient heat to keep the bath liquid, but enough to keep the slag (formed by the 10 to 20 per cent addition of lime) also quite fluid. This slag contains—

Lime and magnesia ...	50 to 65 per cent.
Phosphoric acid	10 to 16 „
Silicic acid	6 to 16 „

The rest oxides of iron and manganese."

6. " We shall be able to run the vessels hard without material repairs for at least a week."

7. " That the bricks can be produced regularly and easily, attention being had to the composition (the burnt brick should contain about 80 to 85 per cent of magnesia and lime, and 15 to 20 of alumina, silica, and oxide of iron), and the firing at a heat much superior to that even for silica bricks."

Mr. Thomas estimates the cost of lining at not more than 1s. 6d. to 2s. per ton of steel " at worst."

I venture to suggest that there still remains open an apparently unexplored field for the application of this process, viz., the production of high quality tool steel from the best hæmatite pigs, as well as producing common rail steel from Cleveland pigs, by the Bessemer process, instead of the costly cementation of Swedish charcoal iron, and its subsequent fusion in crucibles. I am fully aware of the great difficulty of this problem, owing to the stubbornness with which the last traces of phosphorus and sulphur adhere. (2.)

If the basic lining can reduce the phosphorus of average hæmatite pigs as low as that occasionally existing in exceptionally pure samples, fine tool steel may be made in the Bessemer converter. I make this assertion on the practical results obtained about ten years ago, at the Atlas Works, Sheffield, by the late George Brown and myself. I analysed the deliveries of Bessemer pigs, and selected one sample of " Cleator " containing only a trace of phosphorus, and otherwise excellent. With this G. Brown made steel capable of taking 1¼ per cent of carbon, and fairly comparable with " pot steel " made from Swedish charcoal iron.

The importance of this at the present moment is almost incalculable as regards the future of our national steel industry. The temptation to use crop ends of rails and other Bessemer scrap for cutlery and tool-making is now so great that vast quantities are used, and the character of British manufactures is seriously injured. If steel as above described could be freely made as I now suggest, its cost would not exceed that of ordinary rail steel by more than a farthing per pound, which surely would be too little to tempt the meanest of unscrupulous manufacturers.

In conclusion I may add that there is a possibility of development in the opposite direction, viz., that of selecting by preference for ordinary purposes, pig-iron excessively rich in phosphorus, in order to obtain the heat from its combustion in the converter, and also a slag of phosphate of lime rich enough to be used as an artificial manure. (3.)*

W. MATTIEU WILLIAMS.

PROFESSOR BARFF'S PROCESS FOR PREVENTING THE CORROSION OF IRON.

THE most useful of all the metals has an exceptional predisposition to corrosive oxidation,—so exceptional that its durability may be, to a considerable extent, increased by coating it with a thin film of a metal electro-positive to itself. The zinc which covers " galvanised " iron has a stronger affinity for oxygen than the iron which it protects from oxidation.

The source of this anomaly resides in the peculiar chemical and physical constitution of the respective oxides of these two metals. When zinc is exposed to air it combines superficially with an equivalent of oxygen, and then has done with it. This

* Since the above has been in type I learn that this is being done in Germany, and that Mr. Thomas believes that he has perfected a system of turning the slags into a valuable commodity by isolating the phosphate of lime.

film holds the full dose that the metal can obtain by such exposure, and forms a compact continuous varnish, protecting the metal below from further oxidation. Copper and its alloys oxidise in a similar manner, and lead protects itself still more effectually by ultimately obtaining a varnishing film of sulphide. Iron behaves very differently. It is capable of forming a series of oxides, and some of these oxides combine with each other as acid with base, besides which the higher oxides may act as oxygen carriers for the oxidation of the metal or the further oxidation of its lower oxides.

The through rusting of iron depends on these actions :— Exposed to air and moisture the protoxide, or as some affirm the protocarbonate, is first formed ; this rapidly passes on to the hydrated sesquioxide, or well-known red rust. This red compound, besides being a feeble acid, readily gives up some of its oxygen to bodies capable of taking it, and thus a coat of iron rust, instead of protecting the metal below, as in the case of the oxide films of zinc, bronze, &c., has the opposite effect of acting as oxygen carrier. Dr. Miller says that when rusting has begun " moisture is absorbed from the air by the oxide, and thus a species of voltaic action is produced, the oxide performing the part of an electro-negative element, whilst the iron becomes electro-positive and the atmospheric moisture acts as the exciting liquid." Certain it is that, once started, the rusting of iron proceeds inwards, and but limited success has attended the efforts hitherto made to prevent it.

In this, as in so many other instances, the study of natural phenomena might have solved the problem, for there is a natural compound that is not at all addicted to rusting, even under circumstances of the severest exposure. This is the " loadstone " or magnetic oxide of iron of composition intermediate between the protoxide and sesquioxide, and by some regarded as a compound of protoxide and sesquioxide. I have now before me a sample of the ironsand of Taranaki (New Zealand) which, if I am rightly informed, is collected on the sea-shore. It is composed of grains of this black oxide that are separable by the magnet from the associated siliceous grains. The most careful examination reveals no indication of rusting on any of these, in spite of the excessively severe test to which they have been subjected for ages.

Prof. Barff's process consists in forming this oxide upon the surface of manufactured iron articles. Its mere production is easy enough : the essential element of the practical problem is to obtain it as an adherent and perfectly continuous coating, for if it is perforated by the smallest pin-holes, communicating with the metal below, further oxidation will commence at these openings, and the resulting rust will presently undermine the protective coating and spread itself around.

A film of magnetic oxide may be formed by merely heating iron in air ; ordinary iron scale is a rude example of this,

Prof. Barff's method is to subject the manufactured article to the action of superheated steam in a sort of oven chamber or muffle; and he states that one of the conditions of success is that the steam shall be perfectly dry, and another that atmospheric air be excluded. In his earlier experiments it was often found that the black oxide scaled off wrought-iron articles, and this is attributed to an insufficient and irregular supply of steam allowing admittance of air. The presence of moisture is prevented by completely superheating the steam, which is raised to a temperature of about 900° F. for wrought-iron and steel, and from 1000° to 1500° for cast-iron. After describing a long struggle with the difficulties which presented themselves subsequently to obtaining the original patent, Professor Barff told the Society of Arts, on the 26th of March last, that these laborious and costly experiments "have resulted in my being able to state that my process is now commercially perfect, and is waiting the enterprise of gentlemen in the iron trade to take it up and use it." From the prospectus of The Rustless and General Iron Company it appears that this has been done.

The properties of the artificial surface of magnetic oxide which is thus obtained are—

1. The durability of a stable compound not liable to oxidation, nor to the action of ordinary corrosive agents. Prof. Barff has published a number of testimonials from manufacturers and others, who have tested the coating very severely by exposure to weather, to sea water, to alternations of temperature, as in cooking-vessels, &c.

2. The appearance is agreeable, and has the metallic character of iron, but of darker colour. It is totally different from any paint.

3. It gives great hardness. Prof. Barff states that when one-sixteenth of an inch thick or less "an ordinary flat rasp will not remove it without great labour; it resists emery powder.

4. "Substances which adhere to iron, zinc, and enamel will not adhere to it. Saucepans in which arrowroot and other sticky substances are cooked can be cleaned with the greatest ease, after they have been oxidised, a simple wipe removing all dirt." Water was "evaporated in an oxidised pan for six weeks—common tap water; the deposit found was removed with a duster; it did not stick to the iron."

5. "Articles coated can be submitted to a high temperature, even a red-heat, without the coating being injured or disturbed."

6. "The process tightens the rivets, and assists in caulking rivetted iron plates."

It is not applicable to wire or thin plates that have to be bent after treatment, as the coating chips off when bent to a sharp curve.

The above account of the success of the process is based on the statements of the inventor and the testimonials he has received, in reference to which I may add that Prof. Barff, in his published accounts, has candidly admitted the failures of his first attempts and the defects of early specimens, the remedying of which has been achieved by eighteen months' exclusive devotion to experimental investigation of their causes and the means of remedying them.

W. MATTIEU WILLIAMS.

ON WATER AND AIR.*

By John Tyndall, D.C.L., LL.D., F.R.S.,

Professor of Natural Philosophy at the Royal Institution of Great Britain.

LECTURE I.

FIFTY-TWO years ago, commencing on a day in the year 1827, Mr. Faraday gave the first course of juvenile lectures in this place, and the lectures have been continued every Christmas without interruption from that time to the present. For years before his death, when he had given up all other engagements, it was his habit to lecture every year to boys and girls at Christmas, for he thought it a most important thing that they should not grow up in the midst of this wonderful system of Nature in which we dwell without knowing something about it. There cannot, I think, be a doubt that a great deal of the interest that is now taken in the scientific education of the young may be traced back to those juvenile lectures which Faraday delivered in this room.

A knowledge of the phenomena and laws of Nature is sometimes called " natural knowledge." The Royal Society, for instance, which was founded 220 years ago, announced its object to be " the improvement of natural knowledge;" and I trust that we who are here assembled together in the Royal Institution, which was founded 80 years ago, will, for the next fortnight, work cheerfully together in the great field in which natural knowledge is to be gained.

I have chosen for this course of lectures two of our most familiar substances—Water and Air. Water is a very common article of diet. If you take a man weighing 11 stone, and weigh the muscles of that man separately from the bones, they will weigh about 64 pounds; but if you dried those muscles, so as to convert them into a dry mass, they

---

* Being a Course of Six Lectures adapted to a Juvenile Auditory, delivered at the Royal Institution of Great Britain, Christmas, 1879. Specially reported for " The Journal of Science."

would be reduced in weight to 15 pounds, so that, out of the 64 pounds, nearly 50 pounds are pure water. Hence I think you will agree with me that I am not wrong in stating that water is a very important article of food. Every mutton chop and every beefsteak that we consume contains water in this proportion.

But we not only thus feed upon water, but we drink it directly. Whence comes our drinking water? Well, any thoughtful boy or girl here present who had time to think would say that, eventually, it comes from the clouds. How does it get there? That question can be answered thoroughly by-and-bye. At the present time we know, or at all events we assume, that the water comes first of all from the clouds. If you trace the course of the Thames backwards, you find that the very broad river that we have here in London is joined by other rivers right and left, until, finally, you come to the Cotswold Hills, where you find that what is here the Thames comes down to a small rivulet. This rivulet is joined by other rivulets on the right and left as you go down it, and then, at last, the river assumes the breadth at which we have it here when it reaches London. Now, this water which comes from the Cotswold Hills falls on those hills as rain. But not only does the water flow thus over the surface of the collecting valleys, and then flow to the sea in the form of rivers, but it in part sinks into the ground and percolates through the soil, and here and there appears as a pellucid spring. That is the origin of spring-water. It comes originally from the clouds, but it has percolated through the earth and come out somewhere, and there we have our clear spring.

Now, the hardest rocks are more or less soluble in water. You know that sugar and salt are very soluble; but rocks are also more or less soluble in water, so that all our river-water and spring-water has more or less of mineral matter dissolved in it, as sugar is dissolved in a cup of tea. Well, how do we know this? A great portion of the mineral matter may be removed from the water merely by boiling, and it is so removed in all our utensils which are employed in the kitchen. Just before the lecture I went into our own kitchen upstairs, and I looked into this kettle, and I found a thick mineral incrustation. I dare say that I can get this very hard substance out of the interior, and if you look at the interior surface of that kettle after the lecture, you will see that it is covered by this very thick crust, which is so very hard that I can hardly get it to break away. Here it is. There is a crust of this thickness on the interior surface of the kettle, and this is due to the hardness (as we call it) of the water. By boiling, a great amount of the mineral matter is precipitated and is rendered solid, and it settles upon the interior surface of the kettle, and produces that incrustation to which I have referred. Here is a copper tube which belonged long ago to a boiler at the Athenæum Club; and if, after the lecture, you come forward and examine this tube, you will find it coated so that it is almost choked up with a series of beautiful concentric incrustations of the solid matter which was contained in the tap-water supplied to the Athenæum Club and which was deposited upon the interior surface of the tube.

Not only will boiling liberate and re-solidify dissolved matter, but evaporation does the same. If you go to St. Govor's Well, in Kensington Gardens, and look at where the water drips or splashes down from the little outlet, you will see a red mark where it falls. That is oxide of iron, and if you taste the water you will find that it has an inky taste. As the water splashes down, it is in part evaporated, and the iron is liberated there in the form of this oxide of iron. The reason why caverns are usually found in limestone strata is that limestone is more soluble in water than most other rocks; and if you wander, as I have done, in limestone caves, you will usually find in each cave a stream of water which has washed out the limestone and produced the cavern; and sometimes you see from the roofs of those caverns—I was going to say most beautiful icicles—but stalactites hanging down. These are due to the water which has entered into the fissures in the roof above, and percolated through the roof and dissolved some of the limestone, and made its way into the cavern, where the water is in part evaporated; and, in consequence of the evaporation, the solid matter has been deposited, and you find that, as it evaporates, these beautiful stalactites grow longer and longer from the roof towards the floor. Then drops of water fall from the end of the stalactites upon the floor, and there the water is still further evaporated, and a heap is produced called a stalagmite; and. as the water continues to drip, more and more of the solid matter is deposited below, so that the stalagmite grows from below upwards, while the stalactite grows from above downwards, and by-and-bye they meet in the centre, midway, with the point of the stalactite actually in the centre of the stalagmite, and most wonderful and fantastic pillars are thus produced. If you visit any of the great limestone caves you will see examples of this kind. Here are some stalactites from

St. Michael's Cave at Gibraltar. They are so beautiful that it seems a kind of desecration to break them. Here we have beautiful stalactites produced by the evaporation of water containing a mineral in solution, and that mineral is what we call carbonate of lime.

And now we have to examine something about this carbonate of lime. It is a body compounded of carbonic acid and lime. Every boy knows what lime is. I have here some quick-lime in this vessel. In order to show the influence of boiling, I have here water boiling in two different flasks. One of them has, perhaps, been boiling an insufficient length of time to deposit all its solid matter, but I think you will see that there is a very considerable difference between these two flasks. Now, these are two different kinds of water. One of the flasks contains water for which I am indebted to a distinguished engineer, Mr. Homersham, who has made certain water-works at Canterbury. It is the Canterbury water, which has been softened by a process that we shall learn about by-and-bye. You see that the water in this flask is perfectly clear, because all the mineral matter has been removed from it before boiling; but the water in the other flask, which is the tap-water of our house, is thick and turbid. You see that the mineral matter has been let loose, and is forming a kind of mud, in point of fact. If I place the flask in this beam of light those at the right and the left will see that sparkling stuff, which is the mineral matter—the carbonate of lime—which has been liberated by the boiling; and it is this stuff which, when deposited upon the interior surface of our kettle, produces that incrustation to which I have referred. This open vessel merely shows the effect of evaporation. A quantity of water from St. Govor's Well in Kensington Gardens was placed in this basin this morning and evaporated, and there is the substance which gives the water its peculiar taste and its peculiar medical value. This is the substance which, when it is liberated by evaporation, produces that red splash which you see when you look at the well.

Now, we want fully to understand the meaning of this phrase that I have used—carbonate of lime. Carbonate of lime is, as I have said, a mixture of two distinct substances —carbonic acid gas, and lime. I will just remind you of what this carbonic acid is. There is a quantity of it here. My friend, Professor Dewar, is kind enough to help me, and here he has been exhausting a glass globe, and here is another exhausted globe. At the present time the two globes balance each other. Now, if I allow carbonic acid to enter one globe and air to enter the other, you will find that the globe into which the carbonic acid enters will sink down. (Fig. 1.) Why is that? Because the carbonic acid is heavier than

FIG. 1.

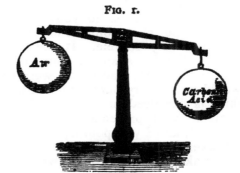

the air; and it is so heavy that if you put it into a vessel it will lie at the bottom.   Here is a vessel which is now full of it, and if Mr. Cottrell gives me a match I will see whether the gas is not there; for, as many of you no doubt are aware, it is a peculiarity of this gas that it will not support a flame.   You see that the flame goes out very soon when I dip it into the gas.

Now, in order to show the weight of the gas, I will blow a soap-bubble, and throw the soap-bubble into this invisible gas.   The gas is there at the present time like a liquid; and, although nobody can see it, I think you will find that when I blow the bubble and throw it into the gas, the gas will be sufficiently heavy to support the weight of the bubble. [The lecturer blew a soap-bubble from a thistle-headed glass tube, and allowed the bubble to fall into a large glass vessel of carbonic acid gas.   The bubble floated about midway down the vessel.]   You see that it does not sink.

Well, now, in order to show you again the heaviness of this gas, I will pour it out before you, although, as it stands there, it is perfectly invisible.   I want you to have a perfectly distinct idea about it.   [The vessel in which the soap-bubble had been floating was then tipped up in front of the electric light, so that its shadow was received on a white screen.   The carbonic acid streamed gently downwards from the lip of the vessel and cast a shadow upon the screen.]   There you see the gas falling, although it is perfectly invisible under ordinary circumstances.   This heavy gas falls down before you like a liquid.

This, then, is one of the constituents of our carbonate of lime.   The other constituent is this ordinary lime.   If you mix this lime with water, the water dissolves a certain amount of lime, and then we have lime-water; and here, thanks to Mr. Homersham, I have a specimen of lime-water which is obtained from the Canterbury works in a way which I will presently describe to you. Lime-water is produced, as I have said, by a solution of ordinary lime in water. Lime is not very soluble. It requires about 70 gallons of water to dissolve a single pound of lime. But here we have our lime-water; and if you were to taste this solution you would find it very pungent to the taste, because of the lime which is dissolved in it. Now, I wish to make carbonate of lime in your presence, and I will take this beaker, as it is called, and I will pour a portion of this lime-water into the beaker. There it is. Mr. Cottrell has here an apparatus for making this substance which we call carbonic acid—that heavy gas which was poured out a moment ago in your presence. I intend to let the carbonic acid bubble through the lime water. You see, I am going to the A B C of the question, and I want to bring you from the A B C as far up as we can climb. But at first I am going to the elements of the question, and those who honour these lectures with their presence will, of course, remember that the lectures are addressed to boys and girls, and that learning or depth would be very much out of place here.

Now, I have the carbonic acid in this vessel, and I will bring it into contact with the lime-water. You will then find that that clear liquid will become milky. Mr. Cottrell will allow the gas to bubble through the water, and you will see in a very short time that the liquid is rendered milky. I have placed this black cloth here [behind the vessel containing the lime-water]. It is one of the devices that we employ in order to make things visible. The change in the appearance of the liquid is due to the formation of this carbonate of lime, which is an insoluble white powder, and which becomes diffused through the water, producing that milkiness which you see before you.

Instead of making the carbonic acid in the way in which it is made there, it may be made by means of marble or chalk, which are themselves carbonates of lime. Carbonate of lime is a compound of lime and carbonic acid; and if you pour upon the marble or chalk an acid which is stronger than the carbonic acid, the carbonic acid will be liberated. Every boy may do this for himself. Here I put a few bits of chalk into this vessel, and if I pour upon them a little acid, you will find that it effervesces. I do this for the

purpose of enabling you to repeat the experiment for yourselves.   Here is a small beaker—a small glass, and we will
put into it some chalk, and pour upon it a little hydrochloric acid, and, as I have said, we shall have effervescence.
There you see the mixture effervescing.   Mr. Cottrell will
pour into a beaker a quantity of lime-water, and allow the
carbonic acid gas to bubble into it.   There you see the gas
is bubbling through the water, and you will find that the
lime-water becomes milky as before.

Here we have a bottle of champagne (B, Fig. 2), and I want
to show you that we can make our carbonate of lime from the
carbonic acid issuing from champagne.   This is a bottle of

FIG. 2.

Mumm's extra dry champagne. I will cut the wire and remove
the cork.   Here is a cork with a bent glass tube passing
through it. We will put it into the bottle. The champagne
is not very well up, so we will stir it a little.   [The bent tube
issuing from the neck of the champagne bottle was held
with its farther end in a glass vessel of lime-water (S), so that
the gas given off by the champagne might bubble into the
solution.   In a short time the lime-water became turbid,
as in previous experiments.]   Here we have got our chalk
produced from the carbonic acid of the champagne.

And now I want to show you that carbonic acid is also a
gas that we exhale from our lungs.   I will take this small
vessel and pour a little lime-water into it, and then simply
blow into the vessel. We inhale the atmospheric oxygen,
and, after this has done its work in the body by burning
part of the body, we exhale carbonic acid.   Here we shall
have the carbonic acid coming from my lungs, which carbonic acid will unite with the lime in that water, and

produce carbonate of lime or chalk in your presence. [The lecturer blew into the lime-water by means of a glass tube.] You see a single inhalation is sufficient to produce this chalk by the carbonic acid from the lungs.

Now I am about to approach a very important point, and that is a point that practically bears upon the question of the solution of minerals in water. This thing which we call carbonate of lime exists in two forms. It exists as a single carbonate, which takes up a certain amount of carbonic acid; and it exists as a double carbonate or bicarbonate, which takes up twice as much carbonic acid. I say that the carbonate of lime exists in two forms—the single carbonate and the bicarbonate; and what you have now to remember is that the single carbonate is almost insoluble in water, and that the bicarbonate is very fairly soluble in water. What is the consequence? The water coming from the clouds has always a certain amount of carbonic acid dissolved in it. When such water falls upon our chalk hills, what occurs? It soaks into the chalk; it percolates through the chalk; it dissolves the chalk; and the chalk so dissolved in the rain-water is present as a bicarbonate of lime. If the carbonate were to remain in the form of single carbonate, the rain-water could hardly dissolve it at all; but, as it is converted by the rain-water into the condition of bicarbonate, the rain-water dissolves a great deal of it. The conversion of the carbonate from the single carbonate to the double carbonate renders it very easily soluble in water. I have here some water from the neighbourhood of Canterbury, where there is a well from which is pumped a million and a half of gallons every day from the body of the chalk. Here is the hard Canterbury water, and I pour a quantity of this hard Canterbury water into the beaker. Now, I say that the lime there is in the soluble form. It is in the form of the double carbonate, or the bicarbonate, which is another expression for the same thing. Now I have here some lime-water. This Canterbury water is beautifully clear and is pleasant to the taste, but you would find it exceedingly difficult to wash in this water. In point of fact, if you operate with soap in this water you find it very difficult to get a lather in it. For a certain time you cannot possibly get a lather. That water contains, I should say, more than 20 grains of carbonate of lime dissolved in every gallon. Where washing operations have to be carried on, on a large scale, with this water, an immense amount of soap is wasted, not to wash, not to cleanse your hands, not to cleanse your linen, but simply, first of all, to

remove the lime from the water before the water can produce
a lather, and before the water can be used as a washing
medium. Well, now, the question is—How is this to be
rendered soft? Or, can this water be rendered soft on a
large scale, or in an effectual way? Yes. That question
has been answered in a most satisfactory manner by Dr.
Clark, of Aberdeen, and he has invented a process for the
softening of water, which, I think, is likely to come into
great request in the future. At first sight it is a very extraordinary process; for what does Dr. Clark do? He takes
the lime from the water by putting more lime into it. Now
that appears to be a strange assertion; but give me your
attention, and there is not a boy here present that will not
understand what I mean by that expression. There [pointing
to a vessel of hard water] is the lime in the form of what we call
bicarbonate. There [referring to another vessel] is the lime,
not united with carbonic acid at all, but as lime-water.
Suppose I pour a quantity of that lime-water upon that
hard water. There is nothing in Science which is more instructive, or more interesting, or more inspiring, than the
power of prevision,—the power of looking in advance,—the
power of allowing the mental eye, so to say, to go before
the eye of the body. Now let us consider what is likely to
occur if I pour this lime water into that hard water. The
lime in that lime-water will take away half the carbonic
acid from the bicarbonate which is present in the hard
Canterbury water. What is the consequence? It will
convert the bicarbonate into simple carbonate. What again
is the consequence? The simple carbonate, being insoluble
in water, will appear as a milky mass in the fluid. I should
not be satisfied if every boy did not understand me in this
most important process. Here we have the lime-water.
By adding this pure lime-water to the hard water we take
away half of that carbonic acid which enables the lime to
be dissolved. Then we convert the soluble bicarbonate into
the insoluble simple carbonate, and down it goes as a milky
mass, and the change is shown in your presence in that way.
There is the lime which has been separated from the hard
water. This is precisely the process used by Mr.
Homersham, at Canterbury, and at Caterham and elsewhere. I have gone specially to all those places to see
the water softened, and a more beautiful process you
could not imagine. Take the case of Canterbury. You
go there into a space which is perfectly clean, covered
from all contact with the dust or *débris* of the air.
You see there three great reservoirs. Each one of these

reservoirs is capable of containing 120,000 gallons of water. You go into a little adjacent, room and you see there a small reservoir, and in it you see a beautifully clear liquid ; but the probability is that in all the reservoirs you will see a beautiful white powder. The small reservoir is used purely for the purpose of making the lime-water. It contains the lime in a very fine state of division, and mixed with water, and therefore called " cream of lime." After being mixed with the water, the lime is allowed to stand, and the solid undissolved lime subsides to the bottom, and you have a beautiful transparent lime-water above. The hard water is pumped from the wells in the chalk, and is then allowed, first of all, to be mixed with a measured quantity of this perfectly transparent lime-water, for the proper quantity is perfectly well known ; and then instantly you have a beautiful fine whitewash produced in the reservoir, some-thing similar, in fact, to this which has been produced in this vessel, and produced precisely in the same way. The mixture of the two transparent liquids produces this white-wash. The mixture is allowed to rest for twelve, or, better still, twenty-four hours. The carbonate of lime which has been liberated sinks slowly to the bottom, leaving a liquid of the most exquisite purity and softness on the top. The carbonate of lime is taken away, and the hard water is in this way converted into soft water. In the bottom you find a powder consisting of carbonate of lime ; and here, thanks to Mr. Homersham, I have some of that beautiful powder. It is as fine as the finest flour. Bakers in general, I believe, are thoroughly honest men, but one would not like this fine powder to fall into the hands of a thoroughly dishonest baker. It is so very like flour that he might materially adulterate his flour by means of this beautiful carbonate of lime. This is the carbonate of lime that has been separated from the chalk water at Canterbury, and deposited at the bottom of the reservoirs.

But now the thought occurs to me, as I am speaking to you, to give you an opportunity of exercising this power of prediction, this power of prevision, of which I have spoken. And here I will take, first of all, a little of the lime-water and mix it with some distilled water. And here is some carbonic acid gas. Now I am going to ask you to turn prophets. When I allow the carbonic acid to enter the mixture of lime-water and distilled water, what you have already observed will occur. We shall have the liquid converted into a milky mass ; but will this be due to the

precipitation of bicarbonate of lime, or to the precipitation of
single carbonate of lime ? Which ? Somebody lisps out
" The single carbonate." I wish he had courage enough to
say it out. It is the single carbonate ; and here is the point
which I want you to predict. Suppose I continue to pour a
sufficient quantity of carbonic acid into the water : suppose
I continue to supply the carbonic acid : what will occur ?
The single carbonate, having plenty of carbonic acid to take
up, will be converted into the bicarbonate, and after having
first of all precipitated the single carbonate, we shall convert
it into the bicarbonate and re-dissolve it. And that is what I
am going to try whether we can do. Here is our liquid, and
here it is being rendered milky—very milky indeed—by
means of the carbonic acid gas. And now, if you will give
me your patience for a time, I am strongly inclined to think
that, by the continued addition of the carbonic acid, we shall
convert the single carbonate into the bicarbonate, and that
that milky liquid will be rendered perfectly transparent by
the solution of the single carbonate. Now, this is one of the
things which men of science, and which boys learning
science, ought to be able to foresee. If our principle be true,
then, by the addition of carbonic acid to the milky liquid,
we shall convert the carbonate into bicarbonate. [The
addition of carbonic acid was continued until the milky-
looking liquid became clear.] This is now bicarbonate. If
I pour into this liquid some lime-water, I shall rob the bicarbonate of half its carbonic acid, and convert it into the single
carbonate. [The experiment indicated was performed.]
Here you see what occurs. You have the single carbonate
in this insoluble form produced thus before you. Now, it is
not for the purpose of amusing you that I bring this before
you. It is for the purpose of bringing before your minds
the manner of the reasoning of scientific men—the strictness
of their logic—the incessant appeal from reason to facts by
which they verify what they conceive.

I will now perform a very homely experiment, but still a
very useful one. I want to give you some examples of the
quality of this Canterbury water. Here are two basins in
which I intend to produce a lather if I can. We will place
in one of those basins some of the soft Canterbury water,
and we will place in the other basin the ordinary tap water
of the Royal Institution, and I will wash my hands with
your permission in both basins, and then we shall see where
the best lather is produced. What you will observe is that,
with a very small amount of soap, I shall be able to produce
a very good lather in the soft Canterbury water ; but I

shall not be able to produce any lather in the hard water of the Royal Institution by the amount of soap that I communicate to it. You will find, perhaps, a little lather—a few bubbles at first, but they will speedily disappear. You will find that upon the soft water a beautiful lather of soap bubbles is formed, and on the other there will be a very greasy mass, which will be due to the combination of a certain acid in the soap with the lime of the water. The acid is called margaric acid, and the curdy mass which is formed is called margarate of lime. Here, from a very small amount of soap, you see a very beautiful lather is produced in a moment in the soft water. A permanent lather covers the bowl of water. I go to the other basin and operate in the same fashion, and I obtain some few soap bubbles at the top, but they will soon disappear ; and I see floating on the top and mixing with the water a quantity of this curdy, unpleasant-looking stuff that I have spoken of as margarate of lime. Now, with a glass tube, I will blow into this lather [that produced from the soft water], and I get a very splendid and permanent heap of bubbles. In the same manner I blow into the basin containing the tapwater, and you see how few bubbles are produced, and their rapid disappearance.

I have now to say a word upon some of the other properties of water, and first we will take its colour. The colour of water is a delicate blue. If you pour water into an ordinary glass you see nothing of the colour of the water, but in large masses you can see the colour, or, at least, the colour can be revealed. For the purpose of showing the colour of water I have here a tube 15 feet long, and that tube is partly filled with water. But before we deal any farther with the colour of water, I should just like to say three or four words with regard to colour in general, and for that purpose I will show you, if I can, on what all colour depends.

I have here in this camera a little artificial sun [referring to the electric lamp]. If we take the light of the sun at noonday on a cloudless day, we call the light which it gives white light. Sometimes, as the sun sinks towards the horizon, his rays become red ; but at noonday, when there are no clouds, the light is white. Our little domestic sun (the electric lamp) is precisely of that kind. We cause the light from our little domestic sun to impinge upon our screen. I take for this purpose simply a little strip of light—a slice of light, if I may use the expression—and there that slice of light is thrown upon the screen and produces that white rectangle which you see. I will interpose in the path of the

light some coloured glasses.   Here I have a green glass.
Well, you see that now the light on the screen is green.   Here
I have a blue glass, and that causes the light to be blue ; and
here I have a red glass, and that causes it to be red.   What
I want you to understand is that the blue and the green and
the red are all in that white light.   And now I want to pull
those distinct colours one from the other.   I want to
separate the colours which produce that white light.   I do
that by sending the light through a body of a certain shape,
which is called a prism.   We will cause the light to pass
through it.   [The experiment was performed, and the usual
spectrum appeared on the screen.]   Is not it lovely ?   I
have seen it a great many times, but never without wonder
at its beauty.   But now I want to make that longer.   I will
send the light through another prism.   Here it is.   You
will observe that by using this other prism I can manage to
make my coloured image a great deal longer.   This coloured
image which I obtain by sending the light through the
prisms is called the spectrum ; and the great Sir Isaac
Newton proved that the slice of white light that you saw a
moment ago upon the screen has contained in it all that
numerous mass of colours which appear in the spectrum.
Now, I wish that I had a ruddy faced boy that I might
cause him to pass through that spectrum.   You would find
his ruddiness varied very much in passing through the
various colours.   I want to bring before you the fact that
the colour of a body is owing to the colour which it throws
back to the eye.   Take a rose.   When the white light of the
sun falls upon the rose, it enters the rose some depth, but,
as it is ejected again from the rose, certain colours are
quenched, and certain colours are sent back to the eye, and
the colour of the rose is due to the colours sent back to the
eye.

Why was the red glass red ?   Because it quenched the
other colours, and allowed only the red to pass.   And what
is true of red glass is true of red cloth, or of any other
colour.   Now, I want to show you that these colours are
thus quenched.   Here I have a beautiful piece of red cloth.
It is now vividly red in the red light of the spectrum.   But
why is this red cloth red ?   It is because it quenches the
other colours, and sends the red colour back to the eye.   If
we pass it on to the green part of the spectrum it will send
back no colour to the eye, and there it is black.   Again, we
will take the opposite experiment.   Here is a piece of green
silk.   The green silk is green because it quenches all the
other colours except green, and sends the green colour back

to the eye. When this green silk is placed in the green part of the spectrum it is, you see, a vivid green. Now I place it on the red part, and there it is simply black. Thus the colour of a body depends partly upon the colours quenched by the body, and partly upon the colours sent back to the eye.

Now, Mr. Cottrell has here a little apparatus—simply a cell—and his object in handing it to me is that I may show you the influence of depth in producing colour. Water is so feebly coloured that, in order to make its colour evident to you, I have been obliged to have that long tube. You see the colour sometimes on the glaciers of Switzerland, in beautiful blue pools of water. Perhaps very many of the boys present will, when they grow older, visit the coast of Naples. Very likely they will go to the end of the Island of Capri, and there they will find coming down the steep precipitous coast, at a certain place, a little arch, just large enough to admit a boat. You pass underneath this arch, and you find yourself under a huge grotto, the walls of which shimmer, in magic beauty, with the most delicate azure light. The reason is that, in consequence of the smallness of the passage, the light is unable to get into the grotto by the entrance, but it first plunges into the blue sea, and then comes up again and illuminates the cavern, so that it has to pass through twice the depth of the water before it reaches the grotto, and, in consequence of the depth of water through which it has to pass, it gets coloured blue. When I say that the light is coloured blue, I mean that its red and yellow and orange and all the other rays except the blue are cut off by the water, and only the blue remains. Now, the experiment I am going to show you is simply to illustrate the influence of depth. Here is a cell composed of a thin chamber; here is one somewhat thicker; and here is one thicker still. The spectrum is on the screen, and if I simply interpose this thin cell, containing a light blue liquid, in the path of the beam you see that it has very little effect indeed upon the spectrum; and the reason is that the stratum is so thin that it cannot rob the light of its rays. Now, I will pour more of this blue liquid into the thicker cell. You observe now that by thickening the layer I have cut away the red ray altogether. You see how little light remains if I make the layer still thicker; and if I went on in this way I should actually cut away the whole spectrum with a substance which, when used in a thin layer, has hardly any effect at all upon the light. In a drinking-glass the water is not in a sufficient mass to give a colour to the light. But in the tanks at Canterbury it is a most delicate sky-blue. If you drop a pin into one of the tanks, which contain 120,000 gallons each, you can see the pin lying at the bottom. I have thrown farthings into the tanks over and over again, and they are scarcely dimmed by the depth of 16 ft. of this beautiful chalk water; and this water has many other things to recommend it. For instance, I have looked into the Report of the Registrar General for information with regard to the influence of temperature upon Thames water, and I have found that, in the early part of the month of June last year, the temperature of the Thames water rose from 60° to 67° or 68°, and that the death-rate from a certain complaint was at first represented by the number 23, but it rose up in three or four weeks to 349, and this increase the Registrar General attributes entirely to the rise in the temperature of the Thames water. Now, one great advantage of this Canterbury water that comes from the body of the chalk is that its temperature is perfectly constant. It is always at an agreeable temperature of 51° to 52°; and if it is properly laid and conducted to the houses it cannot possibly be the propagator of any form of infectious disease.

But now I pass on from the temperature of the Canterbury water to show you the colour of the water. I have filled this long tube to half its depth with the water, so that half the beam of light which passes through the tube will go through the air, and the other half will go through the water. [The 15 feet tube, to which a reference had been made in an earlier part of the lecture, was used in this experiment. The beam of light, after passing through the tube, was received on a white screen. A light green tint was communicated by the water.] There we have a great deal of colour due to the purest water that we can find.

Now I want to point out to you another physical property of water. The liquid state of matter such as water represents is sometimes defined to be that state in which cohesion is absent,—a state in which the particles do not cohere, and in which they are perfectly mobile and are not held together by any force. I would ask you to remember that that is an erroneous definition of the liquid condition. The liquid condition is this:—The little particles of water (and I shall tell you in a future lecture what these particles are called, but for the time being we shall call them " particles ") are gifted with the power of sliding over each other; but, while gifted with that power, they can exercise a very great resistance to being torn asunder. We may measure the cohesion of water by a simple apparatus. Here is a plate of glass, and we can easily balance that plate of glass by a weight (w), and then bring the glass down upon a sheet of water (Fig. 3); then, by adding more weights upon the scale-pan, we can determine the weight necessary to sever one part of the water from the other. Take the case of a soap-bubble. If I had time I could show you the beautiful zones of colour which a soap-bubble produces; and, by means of the zones of colour, we can approximately determine the thickness of the bubble. You sometimes see a heavy drop of water pendant from a soap-bubble; and if you reflect upon this you will see that the film of the bubble must have a tremendous amount of tenacity in order to support the drop of water which hangs from it. Prof. Henry, of Washington, estimated the tenacity of water at several hundred pounds per

FIG. 3.

square inch. But you will say, perhaps, that the soap confers this tenacity upon the water of the bubble. Well, Prof. Henry says that it is not so, and that water, if properly dealt with, is really more tenacious than water with soap in it. But, however true that may be, I want, at all events, to show you that we can get beautiful bubbles and beautiful films of water without any soap whatever.

Here is a tube connected with a cistern at the top of the house, and I am going to make an experiment with it upon a grand scale. It is an experiment which was made by an eminent French investigator, M. Felix Savart. Savart produced films of water by allowing jets of water to fall upon small surfaces. It is wonderful what care we must take in this experiment. The slightest agitation would destroy the beautiful films which it is our object to produce. I will turn the water on and allow it to fall, or to impinge, as we call it, upon this little brass plate.   [A stream of water from the end of a lead pipe was allowed to fall on the centre of a circular flat brass plate, which was supported by a brass pillar standing in the centre of a large concave metallic pan.   The water, upon reaching the brass plate, was spread out in all directions, and fell as a continuous film, the shape of which was, for the most part, dome-like, but occasionally became nearly globular.]   Now you have that beautiful umbrella—that beautiful parasol of water.   You observe the tenacious way in which the water clasps the stem. [The experiment was repeated with two plates of larger sizes.   The falling water was illuminated by a beam of electric light, the beautiful effect of which was varied by the light being caused to pass through coloured glasses.] I see that there are small bubbles of air coming down with the jet of water, and that is the cause of the film breaking so soon ; but still it is very beautiful.   I do not want you to underrate it at all.   Well, this illustrates the tenacity of water where there is no soap mixed with it in order to give it tenacity.

We shall defer the consideration of the other physical . properties of water to our next lecture.

ANALYSES OF BOOKS.

How to Work with the Microscope. By LIONEL S. BEALE,
F.R.S., President of the Royal Microscopical Society.
Fifth Edition.   518 pp.   London: Harrison, Pall Mall.
Philadelphia: Lindsay and Blakeston.

THE reader of this comprehensive treatise on all that relates to
practical research by means of the microscope will hardly recognise, in the present magnificent volume, the little book of
124 pages, bearing the same title, published in 1857. "How to
Work" most fitly explains the character of the book, every page
being eminently practical; that preceding the title-page contains
a note to the reader, in which the author says—"This work may
be read by carefully studying the figures, and then referring to
the text.   A description of every drawing is placed beneath it,
and a reference is given to the page upon which the subject of
the drawing is considered."

The illustrations form a very important part of the book, occupying no less than 98 pages, exclusive of cuts inserted in the
text.   The figures of tissues, especially those very highly magnified, are marvels of delicate wood engraving,—a mode of
reproducing drawings greatly in favour with the author, and exclusively employed throughout the work.   By a very simple and
ingenious process some of them are printed in three colours,
with good effect.   Three electro-copies of the wood block are
taken; from one every portion is cut away except the *bio-plasts*,
coloured red by the alkaline carmine stain; on the second, the
capillaries injected with blue are left, and the remaining block is
arranged to give the outline and shading in black.   The combination of the three blocks presents an accurate view of tissues
finely injected and stained.

The seven parts into which the book is divided comprise :—

1. The Microscope and all necessary apparatus, with the mode
of using the various appliances.   Ten pages are devoted to the
manner of making drawings with the microscope, and the means
of multiplying them by engraving and lithography.

2. On examining, preparing, and preserving objects, including
dissection, cutting thin sections, injecting, and the use of various
staining materials.   This important part occupies 53 pages, and
is so arranged that beginners can readily find the portions useful
to them without being troubled by the very detailed account of
processes more fitted for the advanced student.   Respecting
the use of dyes, the author is especially careful to distin-
gnish between stains used for colouring *bioplasm*, or living

matter, and those employed for demonstrating the structure of formed material; the dyes differ widely both chemically and in the mode of their application.

3. The examination of various substances, both organic and inorganic, is in this part treated still more in detail. The processes used in the examination of rocks and fossils have been contributed by H. C. Sorby, F.R.S., and Frank Rutley, of the Geological Survey. Mr. Sorby's remarks on the use of the Polariscope are worth quoting at length, as the value of this appendage to the microscope is still far from being generally understood. Mr. Sorby writes :—" Polarised light must not be used simply to show structure, or, as is too often the case, merely to show pretty colours, for it is a most searching means of learning the nature and molecular constitutions of the substances under examination. The action of crystals on polarised light, as applied to the microscope, is due to their double refraction, which depolarises the polarised beam, and gives rise to colours by interference, if the crystal be not too thick in proportion to the intensity of the power of double refraction in the line of vision. This varies much, according to the position in which the crystal is cut, and therefore, in a section of a rock, different crystals of the same mineral may give very different results; but still we may often form a good general opinion on the intensity, and may thus distinguish different minerals whose intensity of action varies considerably. But besides this, the intensity, but not the character, of the depolarised light, varies according to the position of the crystal in relation to the plane of polarisation of the light. There are two axes at right angles to each other, and when either of them is parallel to the plane of polarisation the crystal has no depolarising action, and if the polarising and analysing prism are crossed the field looks black. On rotating either the crystal or the plane of polarisation the intensity of depolarising action gradually increases, until the axes are inclined to 45°, and then gradually diminishes till the other axis is in the plane of polarisation. If, therefore, we are examining any transparent body, and find that this takes place uniformly over the whole, we know that the whole has one simple crystalline structure; whereas if it appears as it were to break up into detached parts, each of which changes independently, we know that it is made up of a number of separate crystalline portions, either related as twins or quite independent of each other, as other facts may indicate. By using a plate of selenite of suitable thickness, we may also ascertain in what directions the crystal raises and depresses the tint of colour given by the selenite, and can thus determine the position of the principal axis of the crystal."

4. Treats of Chemical Analysis applied to microscopical investigations. The portion devoted to Spectrum Analysis is from the pen of H. C. Sorby, F.R.S., and scarcely needs comment. A list of works on the subject is appended.

5. Is entirely devoted to Photography. The author has here been assisted by Dr. Maddox and Dr. Clifford Mercer. The instructions given are of the fullest kind, and include some of the most recent improvements.

6. Is particularly Dr. Beale's own, and treats upon original research, especially with the extremely high powers which, combined with the author's manipulative skill, have contributed so much to the determination of the minute details of nerve structure. The special mode of conducting the operations is given at great length, and the interpretation of the appearances presented fully discussed : here the whole resources of the microscope, aided by chemistry, are brought into play, and the author's deductions from his observations concerning structure, formation, growth of tissues, and vitality carefully considered. It is in this portion of the book that the particularly beautiful illustrations before mentioned abound.

7. On the construction of Object Glasses, which appears for the first time in the present edition, explains much that has hitherto been a mystery to microscopists. Little respecting these marvellous optical combinations has till now passed beyond the limits of the workshop; but Mr. F. H. Wenham, F.R.M.S., has kindly placed his published papers and the results of his experience at Dr. Beale's disposal, and given most valuable information as to the manner of working lenses and prisms, with suitable formulæ for their construction. Mr. James Swift has also furnished some manipulative details.

The work concludes with tables for practising the use of the microscope, commencing with very simple operations, and gradually advancing to those of considerable difficulty : the whole are referred to the pages of the book where the description may be found, and also to the plates. This is followed by a list of works and papers on the microscope and matters relating to it, and an Appendix containing an account of apparatus which has appeared during the progress of the work.

Annals of Chemical Medicine, including the Application of Chemistry to Physiology, Pathology, &c. Vol. I. Edited by J. L. W. Thudichum, M.D. London : Longmans. 1879.

This is the first volume of a new scientific periodical which proposes to treat entirely of Chemistry applied to Medicine, and will contain papers on original researches to be carried out in physiological, chemical, and pharmaceutical laboratories in Great Britain and elsewhere. The oftentimes astounding ignorance shown by medical men on chemical subjects is frequently excused on the ground that the latest researches in Chemical

Medicine are scattered abroad in various periodicals and transactions published in different languages and localities.   The
present issue is intended to render this excuse null and void for
the future.   The present volume consists of some twenty-three
articles, most of them being full or summarised papers read by
Dr. Thudichum before the Pathological Institute.   One of the
most interesting of these is the biography of the great German
physicist, R. J. Mayer.   No. XX. is also an interesting paper on
the colouring-matters of bile, which appears to clear up satisfactorily much of the confusion that has hitherto reigned
amongst these interesting bodies.   There are two excellent
name and subject indexes.

We cannot agree, however, with Dr. Thudichum in advocating the introduction of such words as " to fell," "hydrochlor,"
" hydrothion," and " molecle," for to precipitate, hydrochloric
acid, hydrosulphuric acid, and molecule.

Mind in the Lower Animals in Health and Disease.   By W.
    Lauder Lindsay, M.D., F.R.S.E., F.L.S.   London : C.
    Kegan Paul.

We have here two goodly volumes on a subject which, though
hitherto much neglected, must be pronounced to be of the
deepest interest.   It will yet be seen that the attempt to constitute a science of mind from the exclusive study of the human
species is no less a mistake than that of the old naturalists who
began with a description of our bodily structure, and then worked
down to the lower organisms, thus going from the complex to
the simple.

The author tells us that he has studied the subject of mind in
other animals, as compared with that of man, for a series of
years, and that his point of view is that of a " physician-
naturalist."   His professional speciality being the treatment of
mental alienation in man, he passed to the study of comparative
pathology, the result of his inquiries being that the lower animals are subject to the same diseases as we are.   Turning to
psycho-pathology, he extended this conclusion to mental disorders, and was thus finally led to an " investigation of the
normal phenomena of mind throughout the animal kingdom."
Whether this way of entering upon the latter subject is the
happiest is perhaps open to doubt.   He aims at indicating, first,
the spirit and direction in which such an inquiry ought to be
prosecuted ; secondly, its claims on our attention ; thirdly, the
desirability of exactly separating the known from the unknown ;
fourthly, " the new significance of certain facts as interpreted

by the light of modern science;" and lastly, "that facts which controvert popular fallacies are nevertheless facts."

The question of the existence in the lower animals of an immortal soul has been purposely avoided, as not admitting of scientific demonstration. In the first volume the author, after some introductory chapters,—in which we find much to agree with, and also much that is doubtful, and, in our opinion, altogether beside the question,—treats of the dawn of mind in man, the mental condition of children and savages, the evolution of mind in the ascending zoological scale, the alleged intellectual and moral supremacy of man, the inter-relations of instinct and reason, and of unsolved problems in the psychology of the lower animals. He thence passes to morality and religion, to education and its results, to adaptiveness and fallibility, seeking everywhere to demonstrate that the mental phenomena observed in man and in animals are substantially one and the same in kind.

In the second volume he deals with the defects and disorders of mind in man, and in the lower animals.

Finally, he draws certain practical conclusions as to the curability and treatment of animal insanity. An appendix gives the bibliography of the subject, and the scientific names of the animals mentioned in the work.

In so far as Dr. Lindsay is attacking the prejudice of an utter difference of kind between man and the rest of the animal creation, no reader of the "Journal of Science" can doubt that we are with him heart and soul. But we have to ask in how far has he made out his case, and whether he has succeeded in silencing opponents? Here we must admit that we experience no little disappointment. We fear that he has not been sufficiently critical in the selection of authorities. Not to speak of certain literary characters whose training has been anything but scientific, and whom we may well imagine to be occasionally carried away by their feelings, we find occasional references to a certain "Lawson." As he is mentioned in connection with New Guinea, we fear the author must mean Captain Lawson, who invested that island with a fauna totally different from what other travellers have there observed, and who ascended a mountain more than 30,000 feet high in less time than it takes to ascend Etna. We should as soon think of quoting as an authority Lemuel Gulliver.

The anecdote, taken from the "Animal World," that some old rats, finding a young one drowned, "wiped the tears from their eyes with their fore-paws," requires to be taken with a very large grain of salt. The observer must have stood very near to be sure that they were weeping. We have also Watson's story of the blind rat led by a companion by means of a stick, a narrative fraught with internal improbabilities and certain to break down under cross-examination. It must not be forgotten that in several capital cases the author thinks that further observation and experiment is needful, and in his bibliography he calls in question the accuracy of some of his authorities. But would it not have been better to have brought forward none but unimpeachable evidence ?

Just as many readers will conclude that Dr. Lindsay has said more than can be proved in favour of animals, there are others who will contend that he has exaggerated the defects and shortcomings of man. The story of the fight between a man and a bull dog, said to have occurred at Hanley, is scarcely so well-founded as to call for admission in a scientific treatise. As regards the degraded state—moral and intellectual—of certain savages, it is also painted with too gloomy colours.

Sometimes the author seems to have mis-read his authorities. Thus he quotes Belt as speaking of the production of artificial insanity in a white-faced monkey by means of corrosive sublimate. As we remember the passage the animals thus maddened were not monkeys but ants, and so, indeed, Dr. Lindsay quotes this incident in another part of his work.

Perhaps, however, the greatest defect in an otherwise valuable work is the frequent repetition of facts and arguments which have been previously given. Thus pp. 166, 167, 168, and 169 seem mainly a repetition, at somewhat greater length, of matter to be found on pp. 41, 42, and 43. On p. 404, and again on p. 416, we read that a " big dog, after rescuing a little one from drowning, cuffed it first with one paw and then with the other."

A lamentation over the fact that in Glasgow, on account of three deaths from hydrophobia, the authorities—in our opinion most wisely and justly—ordered the destruction of all stray dogs, occurs in vol. i., p. 90, and vol. ii., p. 363, in the latter case being accompanied by some comments which sadly shake our confidence in Dr. Lindsay as a reasoner. Indeed the glorification of dogs and the plea for immunity for their offences is but too prominent. Thus the destruction of dogs for " *mere* biting " (*sic !*) is pronounced " injudicious butchery." How very few dogs in a country like Britain, and especially in towns, are anything but a nuisance, the author does not seem to have asked.

A chapter on animal stupidity—in which the intellectual character of the ass, the goose, and the pig is fully vindicated from the charges commonly brought against them—is introduced among the abnormal manifestations of mind. Why should attributes common to an entire species be considered other than normal ?

The author's repudiation of the phrase " dumb animals," and of the rash assumption that language constitutes a boundary line between man and beast, merits the warmest approval. But it may well be asked why the fact that in the north of Scotland shepherds' dogs accompany their masters to church should be adduced in proof of the religious feelings of animals, or why the rejection of a *Heliconia* by Belt's pet monkey, or of a gaily-coloured frog by ducks, should be cited as instances of individuality of taste and temper. Whilst fully recognising that animals of one species differ among themselves intellectually and morally, even as do men, we cannot see that such facts have the least bearing upon the question.

But we cannot enter further either into the defects or the merits of this work, and we will merely express the hope that it may contribute to weaken the stale prejudice of an antithetical distinction between man and the rest of the animal world.

On the Structure and Affinities of the "Tabulate Corals," of the Palæozoic Period, with Critical Descriptions of Illustrated Species. By H. ALLEYNE NICHOLSON, M.D. Edinburgh and London : W. Blackwood and Sons.

To the generality of readers, even including such naturalists as have never turned their attention to a close and exhaustive study of corals, this volume will be utterly void of interest. To the zoo-phytologist, on the other hand, it will be welcome as conveying the results of careful and accurate research, the more valuable, perhaps, from its special and unobtrusive character. The author is led to agree with Verrill and Lindström on the necessity of abolishing the *Tabulata* as a distinct and separate division of the *Zoantharia*. Under the old name of *Tabulata* he finds included at least twelve distinct groups—some of them *Hydrozoa*, others true *Zoantharia*, others again *Alcyonaria*, and a few forms quite uncertain in their affinities.

The work cannot but increase Prof. Nicholson's high reputation as an able and persevering worker in biological science. The illustrations included in the text are numerous, and, as well as the plates subjoined at the end, are very well executed.

THE

JOURNAL OF SCIENCE.

FEBRUARY, 1880.

I. THE LONDON WATER-SUPPLY.*

BY way of justification for introducing this subject we need merely say that the water-supply of London—a matter affecting the well-being, physical, economical, and in some sense even moral, of more than one-tenth part of the population of the home kingdoms—is no insignificant part of the "condition of England question." As such it may well claim the serious consideration of the public, unless in these days of hysterical sensationalism we have utterly abandoned all pretension to be a practical people. We shall do well to remember, further, that every year aggravates the evils with which the metropolis is now beset, and renders their ultimate abolition more and more difficult and costly.

The water-supply of any community, and certainly of so vast a city as London, should, we submit, conform to the following conditions :—In the first place the available quantity should be far in excess not merely of the actual demand, but of any total likely to be reached in the future, in accordance with the increase of the population. Secondly, the supply should be not intermittent, but continuous. Thirdly, the quality of the water should be unimpeachable. And.

* On the Supply of Water to London from the Sources of the River Severn. By J. F. BATEMAN, C.E., F.R.S., F.G.S., &c. London : Vacher and Sons.

On a Constant Water-Supply for London. By J. F. BATEMAN, C.E., F.R.S., F.G.S., &c. London : Vacher and Sons.

On the Supply of Water to London. By G. A. ROWBOTHAM. London : printed for private circulation.

Water and Water-Supply. By Prof. D. T. ANSTED, F.R.S., F.G.S. London : W. H. Allen and Co.

lastly, the management and property of the supply should be vested in an elective local authority, directly responsible to the rate-payers.

The first of these requirements will not, we presume, give rise to any discussion. All will admit that the consumption of water for purposes domestic, manufacturing, sanitary, and ornamental, is likely to increase even more rapidly than in a direct ratio to the population. One only economy is possible : were Macadamised roads, as they ought to be, banished from all urban districts, the floods of water daily spent in converting their stratum of dust into mire, to the promotion of evil odours, might be a great part saved.

Our second requirement also will scarcely be called in question, at least by the only parties entitled to be heard, namely, the consumers. On this subject Mr. Bateman remarks that, where the constant system is in vogue, " the first cost of introducing the water is reduced to the lowest possible point, and the pollution which more or less commonly attends the storage of water in house cisterns is entirely prevented ; the water is delivered in the purest, freshest, and coolest condition, and very much of the annoyance and inconvenience arising from frozen cisterns and burst pipes is avoided. There is no occasion for exposed pipes in out of the way places—for cisterns, in roofs, or on the top of houses—to be filled with soot and dust in summer, and to be frozen in winter ; and a man may live in tolerable comfort without the dread of the water bursting above his head." Another defect of the intermittent system is that in case of a fire there is often no water in the mains, and precious time is lost while the turncock is being fetched to the scene of action. In a warm summer—an event rare indeed, but still within the range of possibility—we have known the water drawn from a house-top cistern in London to mark 71° F. It is needless to show at length that water exposed at such temperatures to contact with so promiscuous a mixture as town dust cannot be fit for human consumption. As regards those who undertake to supply water, it may justly be said that if they feel unable to conform with indispensable conditions they are bound to withdraw in favour of those who can. But Mr. Bateman shows that in his wide experience a constant water-supply by no means involves waste. We incline to think that the loss involved by the intermittent system is the greater. Ball-cocks are often out of order, or are purposely tied down, so that after the cistern is full the water rushes out at the escape-pipe until it is turned " off " at the main.

On the third condition, the quality of the water, considerable difference of opinion prevails. All authorities appear to agree that the presence of animal and vegetable matter in certain states of decomposition, and that of certain low organisms, such as bacteria, are occasionally injurious, and even destructive, to life. As Science is not yet able to say, with any approach to precision, under what circumstances such putrescent matter and such organisms become dangerous, the only safe rule is to insist, as far as possible, on the total absence of all organic matter whatsoever.

But when we come to consider the mineral or inorganic matters from which no natural water is absolutely free, we find certain chemists, engineers, and physicians who advocate a soft water,—*i.e.*, one in which the mineral matters, and especially the salts of lime and magnesia, are at a minimum. Others, again, no less strongly give the preference to the hard waters containing a large proportion of these same calcareous and magnesian salts.

Mr. Bateman regards softness as a very important requisite for the supply of any town; and after a careful consideration of the various purposes, domestic and manufacturing, to which water is applied, we feel compelled to take the same view. Hard water in the household may be pronounced an almost unmixed evil. It appears probable that in London alone the waste of soap, due to the use of hard waters, reaches the annual value of £500,000. To wash the person with hard water is a true penance. The soap is decomposed, and the lime-soap generated forms a film over the skin and blocks up the pores, so that the effects of the operation are the very opposite of what we aim at in our ablutions. We have the high authority of the late M. Soyer for asserting that in all culinary operations, and especially in the boiling of vegetables, better results are obtained with soft than with hard water.

It is very satisfactory to find that Professors Frankland, Odling, Ramsay, and Way, and Drs. Farr, Lyon Playfair, and Simons coincide in this preference for soft water.

On the other hand, Prof. Ansted considers it "certain that among the healthiest towns in England are some of those supplied with hard water containing both carbonates and sulphates." If we remember, however, that the quality of the water supplied to a town is only one of the many factors on which its sanitary condition depends, we shall scarcely ascribe much weight to what is probably a mere coincidence. It so happens that the towns and cities supplied with soft water—such as Glasgow, Manchester, Leeds, Huddersfield, Halifax—are chiefly manufacturing centres whose inhabitants are liable to suffer from over-crowding, from unhealthy and hazardous employments, and from a polluted atmosphere.

As regards industrial uses, soft water is also generally preferable. For brewing, indeed, the reverse holds good; but it is very easy for brewers to procure their supply of hard water by sinking deep wells. Prof. Ansted, indeed, remarks that " dyeing, bleaching, paper-making, and some kinds of chemical manufactures require the absence of cer-tain ingredients occasionally present in natural water, but none of these manufactures either need or could obtain chemically pure water. Some are much easier and better carried on by water containing much organic impurity." To the ideas expressed in this passage exception must be taken. As a rule chemical manufactures and tinctorial ope-rations succeed better the purer is the water employed. There are certain exceptional cases, such as what was formerly called "madder-work," where the presence of bicarbonate of lime in the water is essential. In dyeing so-called " sad colours " salts of lime and magnesia are also useful as economising dye-wares. But in the majority of cases the dyer, tissue-printer, and bleacher would use chemically pure water were it to be had. Elsewhere Prof. Ansted informs us that " a manufacturer at St. Denis found that for wool-washing muddy and infected water from the Groult was greatly superior to the clear water of an Arte-sian well." At this statement we are by no means surprised; the water from the Artesian well would assuredly be hard, and therefore unfit for washing, whilst the impurities of the river—if of an excrementitious nature—would have a di-rectly detergent effect upon the wool. Finally, we must remember that it is much easier to harden a soft water, if such hardness is required for any special purpose, than to soften a hard one.

Let us now turn to the last consideration—the selection of the hands in which the water-supply of a town is to be vested. Mr. Bateman is an advocate for everything on which the material prosperity of a town depends—such as police, sewerage, water-supply—being in the hands of the inhabitants themselves, and all, we believe, save those who are directly or indirectly interested in the prosperity of water-companies, will cordially support this view. We have the greatest respect for private enterprise whenever and wherever it is kept in due order by the presence, or at least by the possibility, of competition. But we cannot trust it a hair's breadth further. Give an individual, or, what is still worse, a company, a perpetual and indefeasible monopoly, and a career of extortion and of the most high-handed official insolence is at once inaugurated. Of all such monopolies that of a water-company is the worst, because, save by removal from the district, there is no escape from its power. If a gas-company overcharges us or supplies a wretched article we can take refuge in lamps, or in the electric light, or even make our own gas, as is done in many large establishments in the North of England. But for water we can find no substitute. Mr. Shirley Hibberd* proposes, indeed, that the water should be collected on the house-tops, filtered, and kept not in cisterns exposed to air and light, but in underground tanks, lined of course with impervious materials. From a sanitary point of view the plan is feasible, and were it carried out would materially weaken the grasp of those eight old men of the water who are riding about in triumph upon the shoulders of London. But it has one fatal fault ; neither landlords nor tenants will consent to the heavy first outlay. Hence we fear that no appreciable relief to the public can be obtained by this method. As for the metropolitan wells they have long since been condemned as suspicious. Seeing, therefore, that London is and must be solely dependent upon water-supplies drawn from without its ever-widening limits, we shall the better appreciate the wisdom of the legislature in granting to the water-companies a perpetual monopoly. Why such and similar companies should be thus exceptionally favoured we know not. The rights of a patentee lapse at the end of his term of fourteen years. A lessee, at the expiration of his lease, sees the land he has held and the buildings he has erected become the property of the lessor. Water-companies, if sanctioned at all, should have been dealt with in a similar manner. In France they receive a concession for one hundred years, after which their privileges, and even their reservoirs, mains, and other property, revert to the public authorities.

What water-companies will do and what Parliament will sanction may be learnt from the case of Sheffield. When the bursting of the great Bradfield reservoir had drowned more than one hundred people, and destroyed or damaged property exceeding in value the total assets of the local

---

* Water for Nothing : Every House its Own Water-Supply. By SHIRLEY HIBBERD, F.R.H.S. London : Effingham Wilson.

Water Company, they asked for and obtained permission to
borrow a sum sufficient to cover the damages, and to raise
their charges for water, so as to pay the interest of the
loan and gradually to refund the principal.   Thus the
people of Sheffield were taxed for the destruction of their
own property !   This is in principle the same as if the well-
known Glasgow Bank had received power to borrow a sum
sufficient to cover all its liabilities and to recoup itself out
of the pockets of its clients !

It must be remembered that the very fundamental purpose
of water-works differs according as they are vested in a
municipality or in a company. In the former case the object
is to supply the town with the best and most abundant
water at the lowest price, utilising the profits of the undertaking for the reduction of local taxation or for carrying out
needed improvement. The aim of a water-company, on the
contrary, is to make the largest possible profit out of the
job of supplying water, its wholesomeness, quantity, &c.,
being of course very secondary considerations. Anything
more iniquitous than the present method of charging for
water, as pursued in London, can scarcely be imagined.
Your landlord raises your rent, or a parish surveyor, proceeding upon principles fundamentally false,* values your
house at a higher figure, and straightway the water-company—without giving you either more or better water—
exacts a higher figure for the supply. Nay, we know instances where an increase of 20 per cent has been required
without any pretext whatever. Surely such an event is of
itself amply sufficient to ensure the condemnation of the
water-companies. But these monopolist bodies make themselves further notorious by their opposition to every improvement. They are not willing to introduce the constant
system ; they dislike water-meters ; they shrink, in too many
instances, from a thorough filtration of their water or from
softening it by the Clark process. Or if they make any of
these concessions to modern public opinion it is used as a
basis for further extortion.

On the contrary, if we wish to see a satisfactory water-
supply, we must look to those cities and towns where it is
administered by municipalities.

If we next use our four conditions as a standard for
testing the water-arrangements of London, we shall find

---

* That shops, *e.g.*, in the suburbs of London should be valued more highly
in proportion as the neighbourhood grows more populous is intelligible. But
why a dwelling-house should be rated at a heavier sum when dirt, noise, and
traffic increase, and the wealthier inhabitants move away, is an enigma.

them defective in every particular. In the first place, the quantity of water available under present arrangements is not largely in excess of the probable future demand. The daily consumption has grown from 95 million gallons daily in 1869 to 146 million gallons during the summer of 1878. As Mr. Bateman remarks, "there is no adequate source within the means of any of the existing companies to which they can resort " for the needful increase of the available supply. " They will hardly be permitted to increase the draught from the Thames, and any supplies which could still be obtained from the chalk basin on which London lies are altogether insufficient to meet future demands."

That the London water-supply is not continuous, but intermittent, we all know, many of us to our cost. It is not, however, sufficiently known that the city of Manchester, since it adopted a continuous water-supply at high pressure, has not only been able to dispense with fire-engines, but to make a reduction in the property destroyed by fire of 14·3 per cent.*

As to the quality of the London water-supply, the facts furnished by Prof. Frankland and published by the Registrar-General ought to be sufficient. Two companies indeed furnish a water tolerably free from organic impurities, but in return most unpleasantly hard and grossly unfit for washing and cooking.

It may be asked why London has been so long content with the old order of things, and has patiently borne the heavy yoke of the companies? The answer must be sought for in its want of municipal unity. The desultory action of individual " parishes " was utterly unable to cope with the gigantic monopolies which have to be encountered. As far back as 1869 the Duke of Richmond's Commission reported " that the general control of the water-supply should be entrusted to a responsible public body." But no such body has hitherto been constituted.

As to the possible sources of a better and more abundant water-supply there is nothing at all adequate to be found in the south-east of England. The water of the Bagshot Sands is unsurpassed in quality, but its quantity is totally insignificant. The scheme propounded by the Metropolitan Board of Works, and introduced into Parliament in 1878, would have merely furnished 16 million gallons daily,—say

* During the past year there were 60 cases of fire in the metropolis in which, says Captain Shaw, " the water arrangements were unsatisfactory." In the majority of cases the " unsatisfactory arrangements " consisted in the absence or late arrival of the turncock.

about one-twentieth part of what will be needed,—at a first
cost of five and a half millions. It has been said that the
Thames is the "natural" water-supply of London. Granting that this may be true in a certain sense, we have already
drawn upon the Thames too heavily. If we make further
demands we shall seriously interfere with the navigation
above London during average summers. But further, when
we have done our utmost towards the exclusion of sewage,
the waters of a navigable river flowing through a populous
and cultivated country can never be fit for domestic uses.

One outlet remains at present. As was pointed out by Mr.
Bateman as far back as 1865, the head-waters of the
Severn are capable of yielding a supply of 220 million gallons daily, unsurpassed for purity and softness, and not
likely to be polluted by industrial operations or the increase
of local population. This source may, however, soon be
unavailable, Liverpool having turned its eyes in the same
direction. To bring 200 million gallons daily to London
would cost £10,850,000. This is certainly a very serious
sum. But we learn that the Corporation of Manchester do
not hesitate to spend £3,425,000 over bringing an additional
supply of 50 million gallons daily from Thirlmere—a
heavier outlay in proportion to the resources of the northern
city.

The greatest difficulty in the case of the metropolis
springs from the so-called "vested interests" of the eight
water-companies. To buy these bodies out for a sum down
would of course be a most costly undertaking. Mr. Bateman proposes to take over their works, reservoirs, mains,
and privileges, guaranteeing them the same dividends as
they are now receiving, secured as a first charge on the municipal water-rate. Even with this burden it appears that
the burdens of the rate-payers of London would not be
increased.

But there is, or at least may be, a third alternative possible. If a perpetual monopoly of the right of supplying
water to the metropolis has been formally and explicitly
ensured to the companies, then—iniquitous as was the
transaction—it must be strictly carried out. If, however,
no such stipulation has been made, then no considerations
of a romantic or sentimental nature ought to be admitted.
Let the new Board be constituted, carry out Mr. Bateman's
or some other approved scheme, and begin supplying water
in opposition to the companies : they will soon be compelled
to give up the contest, and will be glad to sell their reservoirs and feeding-grounds on easy terms.

II. THE FORMATION OF COAL.

By W. Mattieu Williams, F.R.A.S., F.C.S.

IN the course of a pedestrian excursion made in the summer of 1855 I came ·upon the Aachensee, one of the lakes of North Tyrol rarely visited by tourists. It is situated about 30 miles N.E. of Innspruck, and fills the basin of a deep valley, the upper slopes of which are steep and richly wooded. The water of this lake is remarkably transparent and colourless. With one exception, that of the Fountain of Cyane,—a deep pool forming the source of the little Syracusan river—it is the most transparent body of water I remember to have seen. This transparency revealed a very remarkable sub-aqueous landscape. The bottom of the lake is strewn with branches and trunks of trees, which in some parts are in almost forest-like profusion. Being alone in a rather solitary region, and carrying only a satchel of luggage, my only means of further exploration were those afforded by swimming and diving. Being an expert in these, and the July summer day very calm and hot, I remained a long time in the water, and, by swimming very carefully to avoid ripples, was able to survey a considerable area of the interesting scene below.

The fact which struck me the most forcibly, and at first appeared surprising, was the upright position of many of the large trunks, which are of various lengths—some altogether stripped of branches, others with only a few of the larger branches remaining. The roots of all these are more or less buried, and they present the appearance of having grown where they stand. Other trunks were leaning at various angles and partly buried, some trunks and many branches lying down.

On diving I found the bottom to consist of a loamy powder of grey colour, speckled with black particles of vegetable matter—thin scaly fragments of bark and leaves. I brought up several twigs and small branches, and with considerable difficulty, after a succession of immersions, succeeded in raising a branch about as thick as my arm and about 8 feet long, above three-fourths of which was buried, and only the end above ground in the water. My object was to examine the condition of the buried and immersed wood, and I selected this as the oldest piece I could reach.

I found the wood very dark, the bark entirely gone, and

the annual layers curiously loosened and separable from each other, like successive rings of bark. This continued till I had stripped the stick to about half of its original thickness, when it became too compact to yield to further stripping.

This structure apparently results from the easy decomposition of the remains of the original cambium of each year, and may explain the curious fact that so many specimens of fossilised wood exhibit the original structure of the stem, although all the vegetable matter has been displaced by mineral substances. If this stem had been immersed in water capable of precipitating or depositing mineral matter in very small interstices, the deposit would have filled up the vacant spaces between these rings of wood as the slow decomposition of the vegetable matter proceeded. At a later period, as the more compact wood became decomposed, it would be replaced by a further deposit, and thus concentric strata would be formed, presenting a mimic counterpart of the vegetable structure.

The stick examined appeared to be a branch of oak, and was so fully saturated with water that it sunk rapidly upon being released.

On looking around, the origin of this sub-aqueous forest was obvious enough. Here and there the steep wooded slopes above the lake were broken by long alleys or downward strips of denuded ground, where storm-torrents or some such agency had cleared away the trees and swept most of them into the lake. A few uprooted trees lying at the sides of these bare alleys told the story plainly enough. Most of these had a considerable quantity of earth and stones adhering to their roots: this explains the upright position of the trees in the lake.

Such trees falling into water of sufficient depth to enable them to turn over would sink root downwards, or float in an upright position, according to the quantity of adhering soil. The difference of depth would tend to a more rapid penetration of water in the lower parts, where the pressure would be greatest, and thus the upright or oblique position of many of the floating trunks would be maintained till they absorbed sufficient water to sink altogether.

It is generally assumed that fossil trees which are found in an upright position have grown on the spot where they are found. The facts I have stated show that this inference is by no means necessary, not even when the roots are attached and some soil is found among them. In order to account for the other surroundings of these fossil trees a very violent hypothesis is commonly made, viz., that the soil on which they grew sunk down some hundreds of feet without disturbing them. This demands a great strain upon the scientific imagination, even in reference to the few cases where the trees stand perpendicular. As the majority slope considerably the difficulty is still greater. I shall presently show how trees like those immersed in Aachensee may have become, and are now becoming, imbedded in rocks similar to those of the Coal Measures.

In the course of subsequent excursions on the fjords of Norway I was reminded of the sub-aqueous forest of the Aachensee, not by again seeing such a deposit under water, for none of the fjords approach the singular transparency of this lake, but by a repetition on a far larger scale of the downward strips of denuded forest ground. Here, in Norway, their magnitude justifies me in describing them as vegetable avalanches. They may be seen on the Sogne fjord, and especially on those terminal branches of this great estuary, of which the steep slopes are well wooded. But the most remarkable display that I have seen was in the course of the magnificent, and now easily made, journey up the Storfjord and its extension and branches, the Slyngsfjord, Sunelvsfjord, Nordalsfjord, and Geirangerfjord. Here these avalanches of trees, with their accompaniment of fragments of rock, are of such frequent occurrence that sites of the farmhouses are commonly selected with reference to possible shelter from their ravages. In spite of this they do not always escape. In the October previous to my last visit a boathouse and boat were swept away; and one of the recent tracks that I saw reached within twenty yards of some farm buildings.

What has become of the millions of trees that are thus falling, and have fallen, into the Norwegian fjords during the whole of the present geological era? In considering this question we must remember that the mountain slopes forming the banks of these fjords continue downwards under the waters of the fjords which reach to depths that in some parts are to be counted in thousands of feet.

It is evident that the loose stony and earthy matter that accompanies the trees will speedily sink to the bottom and rest at the foot of the slope somewhat like an ordinary sub-aërial talus, but not so the trees. The impetus of their fall must launch them afloat and impel them towards the middle of the estuary, where they will be spread about and continue floating, until by saturation they become dense enough to sink. They will thus be pretty evenly distributed over the bottom. At the middle part of the estuary they will

form an almost purely vegetable deposit, mingled only with
the very small portion of mineral matter that is held in suspension in the apparently clear water. This mineral matter
must be distributed among the vegetable matter in the form
of impalpable particles having a chemical composition similar to that of the rocks around. Near the shores, a compound deposit must be formed consisting of trees and
fragments of leaves, twigs, and other vegetable matter mixed
with larger proportions of the mineral *débris*.

If we look a little further at what is taking place in the
fjords of Norway we shall see how this vegetable deposit
will ultimately become succeeded by an overlying mineral
deposit which must ultimately constitute a stratified rock.

All these fjords branch up into inland valleys down which
pours a brawling torrent or a river of some magnitude.
These are more or less turbid with glacier mud or other
detritus, and great deposits of this material have already
accumulated in such quantity as to constitute characteristic
modern geological formations bearing the specific Norsk
name of *ören*, as *Laerdalsören, Sundalsören,* &c., describing
the small delta plains at the mouth of a river where it enters
the termination of the fjord, and which from their exceptional fertility constitute small agricultural settlements
bearing these names, which signify the river sands of *Laerdal,
Sundal,* &c.   These deposits stretch out into the fjord,
forming extensive shallows that are steadily growing and
advancing further into the fjord.   One of the most remarkable examples of such deposits is that brought by the
Storelv (or Justedals Elv), which flows down the Justedal,
receiving the outpour from its glaciers, and terminating at
Marifjören. When bathing here I found an extensive sub-
aqueous plain stretching fairly across that branch of the
Lyster fjord into which the Storelv flows. The waters of
the fjord are whitened to a distance of two or three miles
beyond the mouth of the river. These deposits must, if the
present conditions last long enough, finally extend to the
body, and even to the mouth of the fjords, and thus cover
the whole of the bottom vegetable bed with a stratified rock
in which will be entombed, and there be well preserved, specimens of the trees and other vegetable forms corresponding
to those below, which have been lying so long in the clear
waters that they have become soddened into homogeneous
vegetable pulp or mud, only requiring the pressure of solid
superstratum to convert them into coal.

The specimens in the upper rock, I need scarcely add,
would be derived from the same drifting as that which pro-

duced the lower pulp; but these coming into the water at the period of its turbidity and of the rapid deposition of mineral matter, would be sealed up one by one as the mineral particles surrounding it subdivided.   Fossils of estuarine animals would of course accompany these, or of fresh-water animals where, instead of a fjord, the scene of these proceedings is an inland lake.   In reference to this I may state that at the inner extremities of the larger Norwegian fjords the salinity of the water is so slight that it is imperceptible to taste.   I have freely quenched my thirst with the water of the Sör-fjord, the great inner branch of the Hardanger, where pallid specimens of bladder wrack were growing on its banks.

In the foregoing matter-of-fact picture of what is proceeding on a small scale in the Aachensee, and on a larger in Norway, we have, I think, a natural history of the formation not only of coal seams, but also of the coal measures around and above them.   The theory which attributed our coal seams to such vegetable accumulations as the rafts of the Mississippi is now generally abandoned, mainly because it fails to account for the state of preservation and the position of many of the vegetable remains associated with coal, which show no signs of river transport.

There is another serious objection to this theory that I have not seen expressed.   It is this:—Rivers bringing down to their mouths such vegetable deltas as are supposed, would also bring considerable quantities of earthy matter in suspension, and this would be deposited with the trees.   Instead of the 2 or 3 per cent of incombustible ash commonly found in coal we should thus have a quantity more nearly like that found in bituminous shales, viz., from 20 to 80 per cent.

The alternative hypothesis now more commonly accepted —that the vegetation of our coal-fields actually grew where we find it—is also refuted by the composition of coal-ash.   If the coal consisted simply of the vegetable matter of buried forests its composition should correspond to that of the ashes of plants; and if so the refuse from our furnaces and fireplaces would be a most valuable manure.   This we know is not the case.   Ordinary coal-ash, as Bischof has shown, nearly corresponds to that of the rocks with which it is associated; and he says that " the conversion of vegetable substances into coal has been effected by the agency of water;" and also that coal has been formed, not from dwarfish mosses, sedges, and other plants which now contribute to the growth of our peat bogs, but from the stems and trunks of the forest trees of the Carboniferous Period, such as *Sigillariæ*, *Lepidodendra*, and *Coniferæ*.* All we know of these plants teaches us that they could not grow in a merely vegetable soil containing but 2 or 3 per cent of mineral matter. Such must have been their soil for hundreds of generations in order to give a depth sufficient for the formation of the South Staffordshire 10 yard seam.

All these and other difficulties that have stood so long in the way of a satisfactory explanation of the origin of coal appear to me to be removed if we suppose that during the Carboniferous Period Britain and other coal-bearing countries had a configuration similar to that which now exists in Norway, viz., inland valleys terminating in marine estuaries, together with inland lake basins. If to this we superadd the warm and humid climate usually attributed to the Carboniferous Period, on the testimony of its vegetable fossils, all the conditions requisite for producing the characteristic deposits of the Coal Measures are fulfilled.

We have first the under-clay due to the beginning of this state of things, during which the hill slopes were slowly acquiring the first germs of subsequent forest life, and were nursing them in their scanty youth. The deposit, then, would be mineral mud with a few fossils and that fragmentary or fine deposit of vegetable matter that darkens the carboniferous shales and stripes the sandstones. These characteristic striped rocks—the " linstey " or " linsey " of the Welsh colliers—is just such as I found in the course of formation in the Aachensee near the shore, as described above.

The prevalence of estuarine and lacustrine fossils in the Coal Measures is also in accordance with this : the constitution of coal-ash is perfectly so. Its extreme softness and fineness of structure ; its chemical resemblance to the rocks around, and above, and below ; the oblong basin form common to our coal seams ; the apparent contradiction of such total destruction of vegetable structure common to the true coal seams, while immediately above and below them are delicate structures well preserved, is explained by the more rapid deposition of the latter, and the slow soddening of the former as above described.

I do not, however, offer this as an explanation of the formation of *every kind of coal*. On the contrary, I am satisfied that cannel coal, and the black shales usually associated with it, have a different origin from that of the ordinary varieties of bituminous coal. The fact that the products of distilla-

* HULL, On the Coal-fields of Great Britain.

tion of cannel and these shales form different series of hydro-carbons from those of common coal, and that they are nearly identical with those obtained by the distillation of peat, is suggestive of origin in peat-bogs, or something analogous to them.

III. LIVING HONEYCOMB:
A NOVEL PHASE OF ANT LIFE.*

MUCH as has been written about the marvels of instinct, there are still discoveries of great interest to be made in this prolific field. Particularly in the domain of those insect Yankees, the ants, with their wonderful ingenuity and human-like manners and customs, there is room for extended observations.

Some lately-discovered facts in relation to them are so curious and interesting that it may be advisable to give them greater publicity than they have yet obtained. Some of these facts have long been known to the world of Science, but not to the public. Others are new discoveries. As a whole they form one of the most surprising chapters in the history of animal life and contrivance.

Varied as are the social habits of the ants, it is generally considered that social bees surpass them in one particular, namely, their mode of storing supplies of winter food, the storehouses of ant-food having no contrivance similar in ingenuity to the honeycomb, with its rich supply of the sweets of life.

But the truth is that certain tribes of ants are well aware of the value of nature's sweetmeats as articles of food, and have developed a mode of storing up their winter honey still more curious than that practised by the bees. They possess, in fact, what may be called living honeycombs; perambulatory cells filled with distilled sweetness. We refer to the honey-bearing ants of New Mexico, concerning which some very interesting facts have been brought to light during the past summer.

The Rev. Dr. M'Cook, of Philadelphia, a noted observer of ants and ant-life, has been interviewing these honey-bearers, and his results differ so widely from the ordinary facts of insect instinct that they cannot but prove of general interest. These ants had been previously known only in

* This account of Dr. M'Cook's results has been kindly communicated to the Editor by Mr. C. Morris, Philadelphia.

New Mexico, but he discovered them in Colorado, inhabiting the locality known as the " Garden of the Gods," their nests being excavated in the stony crests of low ridges which run through this mountain-girt Paradise.

The ridges are composed of a friable sandstone, into which our minute masons mine deeply, digging galleries which sometimes run for several feet into the rock. The nest, outwardly, is some 10 inches in diameter by from 2 to 3½ inches in height, composed of sand and bits of stone carried from within, some of which seem large enough to defy a regiment of ants to move them.

Inside the nests successive chambers are excavated, connected by galleries, the floors of the chambers being comparatively smooth, while the ceilings are left in a rough state. But this roughness is no evidence of carelessness in the builder. It has, on the contrary, an important object : this is to furnish foothold for the clinging feet of certain extraordinary-looking creatures, which form the living honeycombs of which we have spoken.

Fancy an animal with the head and thorax of a small ant, but with all the posterior portion of the body converted into a round sac, of the size of a large pea, and of a rich translucent amber hue,—it being, in fact, distended into a reservoir of honey. This honey-bag is immense when compared with the size of the ant, the unchanged parts of which might pass for a black pin's head attached to the side of a marrowfat pea. These odd-looking creatures cling to the roof of the chamber with their feet, the distended honey-bag hanging downwards like an amber globe. On seeing them we instinctively imagine that their leg-muscles must be developed in a fashion to put to shame those of human athletes, since it is no light weight which they are thus forced to continuously support.

In each chamber of the nest about thirty honey-bearers are found, making some three hundred to the complete nest. Besides these there are hundreds or thousands of others, workers and soldiers, lords and queens, to whom the honey-bearers serve as storehouses of winter food.

Dr. M'Cook succeeded in bringing some of these home with him alive, providing them with nest-building materials, and with sugar for nutriment. He has one very interesting nest in a glass bottle, with its interior chamber well displayed. The roof of this is covered with depending globules of honey, so large as almost to conceal the minute clinging insect of which they really form part.

But the marvellous feature of the case yet remains to be

described. Not only is the abdomen of the ant converted into a receptacle for honey, but the whole internal economy of the body is transformed for this purpose. All the organs of the abdomen have quite disappeared: viscera, nerves, veins, arteries, have alike vanished; and there remains only a thin transparent skin, which is capable of great distension. It is thus in reality a honey-cell, and much stranger than that of the bee, the waxen walls of the latter being replaced in this case by the tissues of a living animal. The creature can afford to dispense with the abdominal organs, since its life-duties are so metamorphosed that it has henceforth to act only as an animated sweetmeat.

Dr. M'Cook's observations enabled him to discover that the working ants, returning from their out-door foraging, with their bodies distended with the honey they have somewhere harvested, enter the chambers of the nest and eject this sweet fluid from their own mouths into the mouths of the honey-bearers, whose bodies become greatly distended with the delicious food. In other cases he perceived hungry ants seeking for a meal from the food thus generously stored up. The honey-bearer seemed to slightly contract the muscles of the abdominal skin, forcing from its mouth minute globules of honey; these clung to the hairs of the under lip, and were eagerly lapped up by the hungry ants waiting to be fed. It is probable, however, that these supplies are principally intended as winter-stores for the workers, for the feeding of the larvæ, and for the dinner-table of the queen, who is, as usual, too proud or too dignified to do her own foraging.

The working ants take great care of their helpless honey-bearers. When one, through some convulsion of Nature,—occasioned perhaps by the tap of a gigantic human finger,—looses its hold, and drops to the floor of its chamber, it is at once picked up by a worker, and carried back to its old foothold on the roof of the apartment. How this minute creature can drag up a perpendicular wall a mass twenty times its own size and weight is only less surprising than it would be to see an adroit climber of the human race ascending the face of a precipice and pulling after him a ton weight.

With regard to the source of the honey, these ants are not known to feast on flowers, like bees and some of our home ants, nor could any evidence be found of the presence of the *Aphis*, or ant-cow, which many of our ants milk for its honey.

The honey-gatherer is difficult to observe. It is a nocturnal ant, keeping out of sight of the sun during the day, and only venturing forth at nightfall in search of food.   Dr. M'Cook observed them, in the summer twilight, marching outward from the nest in long columns, and pursuing night after night the same paths.   He watched them for a considerable time before he succeeded in finding the goal of these nightly expeditions.   At length, discovering some ants on the twigs of a species of scrub oak, which grew abundantly at the foot of the ridge, he observed that they showed a marked preference for certain small oak-galls which were ranged along the sides of the twigs.

The next thing to be done was to examine these galls. We are accustomed to associate galls with the idea of bitterness only, yet they proved to be the true honey-yielders.   On the round green masses minute drops of a sweet juice were found : this the ants eagerly licked up, passing from gall to gall until fully laden, or returning to the original gall at a later hour when fresh sweetness had exuded from it.

The gall-nut, it is well known, is an excrescence upon the leaves of a species of oak ; it is produced by the puncture of a small hymenopterous insect for the purpose of depositing its eggs.   A minute grub lies in the centre of the soft mass which composes the gall.   Whether the sweet juice came from this grub, or from the sap of the tree, was not readily to be discovered, though it was most likely an exudation of the sap.

All night the busy gatherers of sweets were occupied in collecting honey from the galls.   Towards morning they were seen in great numbers returning to the nest, their bodies swollen with the night's harvest of honey, which, as we have said, is given to the living honeycombs within, being forced from the bodies of the workers and into the mouths of the honey-bearers, until, by the time the season is over, they present a remarkable distension.

This is about all that is known at present concerning the habits of these strange ants.   They very likely have other sources of honey at other seasons; but the most interesting fact is the surprising mode of storage of this sweet food.

In New Mexico the inhabitants put these ants to a very peculiar use, supplementing their dinners with a plateful of honey ants for dessert.   The overladen insects wait in enforced patience while the preceding courses of the dinner are being eaten. The mode of partaking of this strange dessert is to pick up an ant, nip the honey-bag with the teeth, forcing its sweet contents into the mouth, while the remainder is thrown away. We are told that this is not so disagreeable a habit as it might at first sight seem, the skin surrounding the honey being reduced to a thin transparent membrane, with nothing necessarily unpleasant in its character. Nevertheless, most of us will prefer to continue indebted to the bee for our supply of honey, leaving the ants to enjoy the fruits of their own labours.

IV. EDMUND HALLEY: HIS LIFE AND WORK.

ON June 12th, 1874, Prof. Adams, President of the Royal Astronomical Society called the attention of the members at the meeting to the original labours of Jeremiah Horrocks in connection with the then forthcoming transit of Venus; the result being the erection of a memorial to Horrocks in Westminster Abbey. Several of those who took then such a keen interest in the noblest of astronomical problems (one of whose keys was about to be unlocked by the keen observers of various nationalities) have passed away; but the veterans who survive, together with an ever-increasing crowd of junior recruits, are now looking forward to the next recurrence of the infrequent transit, and on this occasion we may well devote a page to an outline of the life of the famous Dr. Halley, whose name is so well known, but whose history has yet to be written. For beyond more or less unsatisfactory and inauthentic sketches of his career, no actual scientific biography of our grand countryman exists. The want of such a work will, we believe, be ere long supplied by the Savilian Professor of Astronomy at Oxford, in whose hands the mass of material collected by the late Prof. Rigaud rests; so that the result of the editorial labours of the Reverend Professor Pritchard may be looked for at an early date. Meantime we may briefly recapitulate what is generally known of the indefatigable and long-lived cosmographer to whom it is now proposed to erect a fitting monument on the site of his deserted station at St. Helena.

Edmund Halley was born at Haggerston, near London, on the 29th of October, 1656, and was the only son of a soap-boiler. He was educated at St. Paul's School, and at an early age evinced a peculiar aptitude for mathematics. He first applied himself to the study of languages and sciences, and at sixteen years of age made original observations on the movements of the magnetic needle before he left St. Paul's School. In 1673 his university career commenced at Queen's College, Oxford, where, at the age of twenty, he made careful observations of the spots on the sun, the first results of his study of solar physics, made during July and August, 1676, being published in conjunction with those of Flamsteed.

Young Halley's first ambitious scheme was to make an accurate general catalogue of the stars, and as Hevelius and Flamsteed were employed in cataloguing the stars of the Northern Hemisphere, interest was made with the then reigning monarch, Charles II. (who appears to have been much interested with the ardent enthusiasm of the young astronomer), who permitted Halley to proceed, *at his own expense*, to the nearest available spot under the southern tropic, viz., St. Helena, which small island had been retaken by Sir Richard Munden from the Dutch in 1673, and re-granted to the East India Company by a new charter, and a European garrison was raised for its defence under Captain Gregory Field. The educational state of the island in those days may be judged from the fact that three of the few members of Council were unable to write, and could only affix their marks to the proceedings of the Board. The latitude and longitude of the island even were imperfectly known, and the means of ascertaining the latter were so inaccurate on board ship that at least one ship returned to England with the intelligence that the island had disappeared altogether. Halley's journal or log-book of the voyage is still extant, and what is now a matter of a fortnight's steaming in a Union Company or Donald Currie steamer was then a formidable voyage of two or more months (the *Northumberland*, under Rear-Admiral Sir George Cockburn, leaving the Start on the 9th of August, reached St. Helena on the 15th of October, 1815, with Napoleon on board), and we can imagine the enjoyment he must have felt during the tropical nights on the Atlantic, and the first indelible impressions with which he beheld the gradual rising of the Centaur, Argo, Crux, and the Nubeculæ of Magellan, whose pale phosphorescent radiance contrasts with the dark spot through which the terrestrial spectator catches a glimpse of the ultra-planetary depths of space.

Arrived in St. Helena, then overgrown with forests and rare shrubs, now extinct for ever, he appears to have landed at James's Fort, in Chapel Valley, late in the year, and must have found no small difficulty in the selection of a site for his temporary observatory; but doubtless tempted by the unusual clearness of the season, he selected a magnificent situation on the northern side of the central mountain, to which he must have had considerable difficulty in transporting his instruments, as even with the present modern roads wheeled transport is most limited and precarious. Under Governor Field, however, with slave labour and royal auspices, Halley industriously set to work towards the completion of the catalogue of the southern stars, and delineated a planisphere of their projection. It was here, says Prof. Forbes, that " Halley was the first to see clearly what a powerful means of determining the sun's parallax an observation of contact really is. So far as I can discover, he first mentions the method in a letter to Sir Jonas Moore, written at St. Helena in 1677,* just after having seen a transit of Mercury. The exactness with which he believed the time of contact to be determinable led him frequently afterwards to urge his countrymen to make every effort to utilise the method on the occasion of the transits of 1761 and 1769, when he should be dead.† And thus, in addition to his celebrated prediction of a comet, he left a second legacy to his successors, who, as Englishmen, might be entitled to be proud of his foresight, though he could not live to reap the glory of it."—(" The Coming Transit of Venus," by Prof. George Forbes, " Nature," April 23, 1874.

" Halley saw (what many people fail to see even now) that the great accuracy of the method consists in this, that in one second of time Venus moves over about 0·02" ; and if we can determine the time of contact, with an error of no more than a second, we are measuring the sun's parallax with an error of no more than 0·02 of a second of arc. Halley even pointed out the best stations for observations."—(*Ibid.*)

Ninety-four years subsequently to this letter Dr. Maskelyne

* Hooke's Lectures and Collections. 1678.
† Catalogus Stellarum Australium; also Phil. Trans., 1694 and 1715.

and Mr. Waddington were able to take advantage of Halley's legacy, and to partake in their share of its fruition.

The result of Halley's observations in the Southern Hemisphere, made in spite of the clouds which hung over the heights of the rocky island (caused by the counter-current of air above the trade-winds), was the publication of the positions of 350 stars which he had fixed in his " Catalogus Stellarum Australium." (See Humboldt's " Cosmos," vol. iii., 207.)

Halley returned to England in 1678, and at the early age of 22 years was elected a member of the Royal Society. On his return to England he was not idle, for we soon find him at Dantzic engaged in settling a friendly and international controversy between the Royal Astronomer of England and Dr. Hevelius. He was travelling on the Continent with his companion Mr. Nelson when—between Calais and Paris—he made his second (December) observations of the remarkable direct elliptic comet of 1680, the length of whose tail was calculated at 100 millions of miles, with a period of 8813 years, having observed it before its perihelion passage the previous month.

During a portion of 1681 Halley spent a considerable time in Paris with Cassini at the Observatory, and subsequently went to Italy, where we find a record of him in a first edition of a copy of the " Principia " of his friend Newton, bearing an autograph inscription from the donor to the Abbé Nazari, who edited a scientific journal at Rome at this period. The following is the entry in Halley's handwriting* :— ·

> ILLUSTRISSIMO DNO
> DRO ABBATI NAZARIO
> ROMÆ HUMILLIME OFFERT
> EDM. HALLEY.

During the year 1682 Halley observed and determined the parabolic elements of the retrograde orbit of the remarkable comet of long period which bears his name. The results of his calculations made it evident to him that this comet must have been identical with the comet observed in 1607 by Kepler and Longomontanus, and also with one whose elements were calculated from observations by Apian, at Ingoldstadt, in 1531, and which previously appeared in 1456 and 1378, as calculated by Pingré and Langrier. Halley predicted the return of this comet at the close of 1758 or

* See *Nature,* March 6, 1879, p. 422.

beginning of 1759—the first prediction of the return of a comet.

This comet, as is well known, actually passed its perihelion on March 12th, 1759, just one month before the time calculated by Clairaut, Lalande, and Madame Lepaute ; and again on November 15.95, 1835, within five days of Rosenberger's elaborate calculation. We now await the next advent of Halley's comet in 1909-10, its perihelion passage being fixed by the late Count G. de Pontécoulant (as communicated by him to the Paris Academy of Sciences in 1864) to May 23.87 (Paris time ?), 1910 ; it having been last glimpsed by Dr. Lamont with the Munich refractor on May 17th, 1836. (See "Nature," February 11, 1875, p. 286.)

An account of his next research is given by K., in a paper on the "Early History of Magnetism," in "Nature," April 27, 1876, p. 523, as follows :—" In 1683 Halley presented a paper of great importance to the Royal Society of London, entitled 'A Theory of the Variation of the Magnetical Compass.' In this communication he states that the 'deflection of the magnetical needle from the true meridian is of that great concernment in the art of navigation that the neglect thereof does little less than render useless one of the noblest inventions mankind ever yet attained to,' and gives as the result of 'many close thoughts' the following explanation of the variation of the compass :—

" ' The whole globe of the earth is one great magnet, having four magnetical poles or points of attraction, near each pole of the equator two ; and in those parts of the world which lie near adjacent to any one of those magnetical poles, the needle is governed thereby, the nearest pole being always predominant over the more remote.' He remarks that the positions of these poles cannot as yet be exactly determined, from want of sufficient data, but conjectures that the magnetic pole which principally governs the variations in Europe, Tartary, and the North Sea is about 7° from the North Pole of the earth, and in the meridian of the Land's End ; whilst the magnetic pole which influences the needle in North America, and in the Atlantic and Pacific Oceans, from the Azores westward to Japan, is 15° from the North Pole, and in a meridian passing through the middle of California. The variation in the South of Africa, in Arabia, Persia, India, and from the Cape of Good Hope over the Indian Ocean to the middle of the South Pacific, is ruled by the most powerful of all these magnetic poles, which is situated 20° from the South Pole of the earth, and in a meridian passing through the island of

Celebes; in the remainder of the South Pacific Ocean, in South America, and the greater part of the South Atlantic Ocean, it is governed by a magnetic pole 16° from the South Pole in a meridian 20° west of the Straits of Magellan."

" On this hypothesis Halley explains the variation observed in different places, and among others cites the two following instances :—On the coast of America, about Virginia, New England, and Newfoundland, the variation was found to be west, being above 20° in Newfoundland, 30° in Hudson Strait, and 57° in Baffin's Bay. On the coast of Brazil, on the contrary, it was found to be east, being 12° at Cape Frio, and increasing to 20¼° at the Rio de la Plata, thence decreasing towards the Straits of Magellan. Thus, almost in the same geographical meridian, we find the needle at one place pointing nearly 30° west, at another ⋅ 20½° east : this is explained by the north end of the needle in Hudson Strait being chiefly attracted by the North American magnetic pole, whilst at the mouth of the Rio de la Plata the south end is attracted by the south magnetic pole situated west of the Straits of Magellan. Sailing north-west from St. Helena to the equator, the variation is always in the same direction, and slightly east. Here the South American is the chief governing pole, but its power is opposed by the attraction of the North American and Asian South Poles, the balance as you recede from the latter being maintained by approach to the former."

" During the next nine years Halley continued his investigation of the causes of magnetic variation, and in 1692 he made another communication to the Royal Society, in which he endeavoured to meet two difficulties he had always felt in his former explanation ; one, that no magnet he had ever seen or heard of had more than two opposite poles; the other, that these poles were not—at least all of them—fixed in the earth, but slowly changed their positions. The following observations are cited by Halley in proof of the motion of the magnetic system :—At London, in 1580, the variation was 11° 15' E., in 1622 it was 6° E., in 1634 it was 4° 5' E., and in 1657 there was no variation ; whilst in 1672 it was 2° 30' W., and in 1692, 6° W. At Paris the variation was 8° or 9° E. in 1550, 3° E. in 1640, 0° in 1666, and 2° 30' W. in 1681. At Cape Comorin it was 14° 20' W. in 1620, 8° 48' W. in 1680, and 7° 30' W. in 1688. Halley considered the external parts of our earth as a shell, separated by a fluid medium from a nucleus or inner globe, which had its centre of gravity fixed and immovable in the common centre of the earth, but which rotated round its axis a little slower than the superficial portions of the earth. The nucleus and the exterior shell he regarded as two distinct magnets, having magnetic poles not coincident with the geographical poles of the earth. The change observed in Hudson's Bay being much less than that observed in Europe, Halley concluded that the North American pole was fixed, while the European one was movable; and, from a similar observation on the coast of Java, he considered the Asian South Pole as fixed, and the pole west of the Straits of Magellan to be in motion. The fixed poles he regarded as those of the external shell, and the movable as those of the inner nucleus. Of these latter the one placed by him in the meridian of the Land's End was ascertained, in the present century, to have moved to Siberia in 120° E. long., and that placed by him 20° from the Straits of Magellan to have moved between 30° and 40° west of this position; while those poles regarded by Halley as fixed were found but slightly altered in position since his time. It is extremely interesting to find that not only modern observations of declination, but also those of dip and magnetic intensity, have received their best explanation on the assumption of four magnetic poles. Much, however, that is mysterious remains unsolved, and Halley's remarkable words may even now with truth be quoted:—" Whether these magnetical poles move altogether with one motion or with several, whether equally or unequally, whether circular or libratory, if circular about what centre, if libratory after what manner, are secrets as yet utterly unknown to mankind, and are reserved for the industry of future ages."—(" K." *ibid.*)

Intent on still further carrying out his magnetic researches, Halley was successful in an application to King William to fit out a scientific expedition, and in 1698, as Capt. Halley, R.N., received his appointment to the command of H.M.S. *Paramour*, in which vessel, a Pink (?), his sailing orders directed him to seek by observations the discovery of the rule of variations of the compass, and as well to survey and lay down the latitude and longitude of His Majesty's American settlements. The *Paramour* left England in November, 1698, but was absent only for seven months, on account of sickness breaking out among the crew during their stay in the equatorial calms, and his Lieutenant mutinying. On their return home the Lieutenant was court-martialled and cashiered; and Capt. Halley, in no way disheartened, and with his expedition augmented by the presence of a smaller consort placed under him, set off in September, 1699, and voyaged throughout the Atlantic Ocean from the Arctic to the Antarctic regions, until the ice stopped his progress. He revisited the well-known scenes of his early scientific success on the isolated rock of St. Helena, which doubtless inspired him with the thoughts of old Robert Herrick, whose lines aptly apply to this now barren spot :—

> " Rockie thou art, and rockie we discover
> Thy menne, and rockie are thy wayes all over.
> O menne! O manners! now, and ever knowne,
> To be a rockie generatione!
> A people currish, churlish as the seas,
> And rude almost, as rudest salvages ;
> With whom I did, and may resojourne whenne
> Rocks turn to rivers, Rivers turn to menne."

The voyage of the *Paramour* included the coasts of the Brazils and the West Indies on the West, together with the West Coast of Africa, Madeira and the Canaries, &c., and lasted a whole year : the result was speedily made public by the production of the much-cited " Halley's Magnetic Chart" (on which the isogonal magnetic curves or " Halleian lines " were laid down), the original of which is in the British Museum. By the assistance of Mr. Winter Jones, the principal librarian, copies in a reduced size have been produced from this chart, and they were inserted as an Appendix in the Magnetical and Meteorological volume of the Astronomer Royal's Report at Greenwich for the year 1869.

In April, 1701, Capt. Halley was employed as Hydrographer to His Majesty, and surveyed the British Channel, an accurate chart of which was published under his superintendence, besides a systematic series of observations on the tides was undertaken with satisfactory conclusions. By this time the fame of Halley as astronomer, discoverer, and engineer, had spread throughout the civilised world, and the Emperor of Germany (Leopold), desirous of forming a harbour for shipping in the Adriatic, applied for and obtained from Queen Anne the services of Captain Halley, who embarked for Holland on November 22nd, 1702, and proceeded to Istria. In consequence of the political opposition of the Dutch, Halley's designs were not carried out as to the construction of two ports in Dalmatia ; but the approval of the Emperor Leopold was signified by the present of a diamond ring and an especial letter of recommendation to the Queen of England, who again sent him to Istria, where, in conjunction with the Emperor's engineer, he added to the fortifications and harbour-works of Trieste. On this occasion, passing through Hanover, he supped with the Electoral

Prince (subsequently our King George I.) and his sister the Queen of Prussia, and at Vienna he was personally presented to the Emperor Leopold.

In 1703 Capt. Halley returned to his native country, and was made Professor of Geometry in the University of Oxford, in the place of Dr. Wallis, and the degree of LL.D. was also conferred upon him. The professorship of Astronomy at the University of Oxford was withheld from him, it is said, at the instigation of the orthodox Stillingfleet—another instance of the numerous attempts to stifle true scientific investigation by narrow-minded and bigoted divines.

The late Captain, now Doctor Halley, as indefatigable as ever, did not seek in his professorial chair the *otium cum dignitate :* only forty-seven years of age, he began to translate into Latin, from the Arabic, " Apollonius de Sectione Rationis," and to restore the two books " De Sectione Spatii " of the same author (which are lost), from the account given of them by Pappus ; and he published the whole work in 1706. Afterwards he took a leading part in preparing for the press and editing " Apollonius' Conics," and he successfully supplied the whole of the eighth book, the original of which is lost. He likewise added Serenus on the Section of the Cylinder and Cone, printed from the original Greek, with a Latin translation, and published the whole in folio.

In 1713 he was made Secretary of the Royal Society, and on May 3rd, 1715, with the astronomer De Louville, he made his observations of the celebrated total solar eclipse from the house of the Royal Society, in Crane Court, Fleet Street, which he has so well recorded in the " Philosophical Transactions " of the Society.

In 1720 he was appointed Astronomer Royal at Greenwich, in room of his old friend and colleague Flamsteed ; and nine years afterwards had the distinction of being chosen as a foreign member of the Academy of Sciences at Paris.

Full of years and honours this great man died at Greenwich, in his eighty-seventh year, after a long career of such scientific benefits to his countrymen as few of our greatest philosophic worthies can boast. Besides his magnetic works and astronomical discoveries—among which was the discovery of the acceleration of the moon's mean motion, and the method of finding longitude at sea—his principal works were the " Tabulæ Astronomicæ,' and an " Abridgment of the Astronomy of Comets," in addition to his " Catalogus stellarum Australium " before mentioned. We are also indebted to him for the publication of several of the works of Sir Isaac Newton, who had a particular friendship for him, and through whom some of his most important discoveries were made. Thus, for instance, the first evaluation of the mass of Jupiter is that of Newton in the Cambridge edition of the " Principia " (1713), inferred from Halley's observation of an emersion of Jupiter and his satellites from the moon's limb, giving for the denominator of the fraction 1033. (See " Nature," September 23, 1875.)

Such is a brief account of the talented Edmund Halley, the details of whose busy and varied life must be awaited with the utmost interest. It will, we are assured, be a source of gratification and pride to astronomers, and indeed to all scientific men, throughout the Anglo-Saxon dominions to unite in raising such a memorial as has been suggested by the organiser* and observer of the Mars Expedition in 1877, and at last we may expect to see Halley's Mount crowned with an appropriate memorial of the astronomer.

V. THE HEIGHT AND SPAN OF THE JAPANESE.

WE well remember that quaint little group of two-sworded, strangely-dressed, men, who in the 1862 Exhibition were pointed out as Ambassadors from that then almost *terra incognita*—Japan. The curious would saunter past these eastern islanders in order to form a comparative idea of the height of men whose ample skirts made their height appear greater than their, in truth, diminutive stature warranted. China has since sent us for exhibition the gigantic " Chang ;" but Japan, though puzzling us with its clever legerdemain and fascinating us with beautiful *objets d'art*, has as yet not shown us that it can produce men of fine growth. Possibly the Japanese agree with Shakspeare, that " Small herbs have grace," whilst " ill weeds grow apace," for it would appear that many of the probable causes of their smallness are directly due to their own agency.

The data that we have for estimating the height of the Japanese are more exact now than the rough measurement above alluded to. Mrs. Chaplin Ayrton, M.D., has recently

* DAVID GILL, Esq., F.R.A.S., Astronomer Royal at the Cape.

published* the results of nearly three hundred observations of the height and span of the Japanese. She found the average height to be 5 feet 3 inches, and their span 4 feet 11 inches. In the case of twenty-four women, taken at random, the tallest was a trifle over 5 feet 2 inches, and the average was 4 feet 8 inches, with an average span of 4 feet 6 inches. The shortness of the span as compared with the height is a general characteristic that is especially marked in the case of the women. This gives rise to the theory whether the habit of raising the shoulder-pole for carrying burdens, and the universal practice of tying the infant to the back, may not—by making the arms unused to great muscular exertion—arrest, in a measure, their development.

From the tables of height given the Japanese appear to be not only much smaller than the western nations, but also less variable in height. This may possibly be due to a less great mixture of races than now exists in Europe. During four years Mrs. Ayrton appears to have only seen one Japanese who struck her as exceptionally tall, a circumstance that in any European capital could not have failed to happen more frequently. Taking the percentage of the observations, 60 per cent of the Japanese observed had the span shorter than their height, 33 per cent had the span greater than their height, whilst in only 6·8 per cent was the ideal of proportion realised, namely, that the height and span shall be equal. It is to be regretted that the more complex measurements of the head, &c., were not taken, but possibly the docility of the subjects experimented upon would not have admitted of very detailed examination, as it seems the Japanese have a strong superstitious dislike to be measured, and even object frequently, on the same grounds, to be photographed.

In seeking the explanation of the smallness of the Japanese it is proverbially known that climate is often called to account. Doubtless climate has much to do with the stunted growth of Laplanders and Esquimaux, but Japan occupies a temperate zone. Granted, however, that the climate of Japan is magnificent, still the seasons are more marked and the variations of temperature more sudden than with us, and the question arises whether the sudden demands made upon the organism to accommodate itself to such variable atmospheric conditions may not be unfavorable to harmoniously continuous growth. The clothing of a nation illustrates very

* "Recherches sur les Dimensions Générales et sur le Developpment du corps chez les Japonais."—Thesis for the degree of Doctor of Medicine presented and sustained by Mrs. Chaplin Ayrton before the Faculty of Medicine of Paris.

fairly the conditions under which they live, and in Japan we find the system of dress to consist essentially of many layers of similarly made garments, the characteristics of which are that, being loose and having one tie each, they can be readily taken on and off; and the abdomen is protected by a long and thick sash.   In all variable climates, as in Spain, great advantage has been found to result from keeping the middle of the body well protected.   The Japanese of the upper classes usually peel themselves of layers of their loose clothing towards the middle of the day, resuming the garments as evening closes in; whilst the coolie, whose wardrobe is less extensive, casts towards noon his cotton dress in favour of his tattooed skin, and is to be seen in the evening only comfortable in a wadded coat.

The use of charcoal braziers is probably an unhealthy custom; for although Japanese rooms are habitually open to the air in the daytime, still the custom of sitting for many hours with the mouth almost directly over the brazier cannot but lead to a considerable amount of the poisonous oxides of carbon being inhaled.

The physique of the Japanese does not tend to inspire confidence in a vegetarian *régime*, upon which, with fish, this people may be said to live.   However, they also eat their vegetables highly salted; and we know that sailors, who, under otherwise healthy conditions, yet eat much salted food and live in an atmosphere impregnated with salt, are usually of a short square build.   Thus it may possibly be the salted nature of the food, rather than the absence of meat, that is deleterious to growth.   Again, with regard to diet, the growth of the children appears to bear a favourable comparison with European children in infancy, when they are nursed even up to three years old, but to become unsatisfactory during the later period of childhood, seeming to indicate that the food supply—although equal to the demand for simple existence—is insufficient during the period of the body's most active growth.   It seems probable that the direction in which the food is insufficient is in fatty matters, for, as regards the nitrogenous element, beans, which contain much vegetable albumen, are largely consumed by the Japanese.   Of course, however, many of the additional causes of the smallness of the Japanese may be so remote as to cease to affect the nation except by hereditary influence.   It appears that authors are agreed that the Japanese are a mixed race. Hilgendorf states that at least one-third of the Japanese skulls are provided originally with a double jugal bone, which remains more or less at a later period, and which might almost be called the " os Japonicum." We know of no observations that connect this cranial peculiarity with the type of face to which it belonged during life ; but if, as Mrs. Ayrton thinks, the Malay, Chinese, and Aino types can in the modern Japanese be detected by practice and attentive observation, it might be possible, and would be most interesting, to ascertain to which national element might be attributed a peculiarity which so careful an observer as Hilgendorf considers so indubitable a characteristic of such a large portion of the people of Japan.

This thesis of Mrs. Chaplin Ayrton brings us back to the history of the attempt made by women during the last ten years to obtain a higher education. In 1869 the University of Edinburgh opened its matriculation examination to women as well as to men, and five ladies passed this examination, and became each a " Civis Edinensis." But to pass the various examinations at the University of Edinburgh, not only is a certain standard of knowledge necessary, but the candidate must produce certificates of having attended certain courses of lectures within the walls of the College. Joint classes of men and women were, however, specially forbidden by the Faculty, and separated courses for so few women were unattainable. Hence after a four years' residence—during which the five and some other girls who had now joined them had, in open competition with the men students, obtained prizes and scholarships, not only at the University, but also at the Royal College of Surgeons, Edinburgh—the small band of women dispersed, many with their well-earned laurels, but all without the much-coveted but unattainable degree. One of them, the writer of the thesis we are reviewing, went as a pioneer to Paris, to see whether there she could obtain the medical education denied her in her own country. In recognition of her four years' academical course at the University of Edinburgh, the Paris Faculty awarded her the honorary degrees of " Bachelier ès Sciences " and " Bachelier ès Lettres," and allowed her to commence the complete course, which, after her subsequently passing the necessary seven examinations, has now been concluded by the award of the degree of Doctor of Medicine on her presenting and sustaining her thesis.

VI.   ENGLISH AND AMERICAN PHYSIQUE.*

By George M. Beard, M.D.

II.

Civilisation, Climate, and Teeth.

THE American dentists are the best in the world; not necessarily because the Americans have greater mechanical skill than Europeans, but because the early and rapid decay of teeth in Americans of the better class have compelled them to give special attention to dentistry.

This quick decay of teeth in America, like various forms of nervous diseases that go with this decay, is the result not of climate alone, but of climate combined with civilisation : the confluence of these two streams is necessary.   Irregularities of teeth, like their decay, are the product primarily of civilisation, secondarily of climate ; they are rarely found among the Indians or the Chinese, and, according to Dr. Kingsley, are rare even in idiots ; the cretins of Switzerland, the same authority states, have " broad jaws and well-developed teeth."

Another fact of much instructiveness is, that decayed teeth in Indians and negroes are less likely to annoy and irritate than the same amount of decay in sensitive, nervous, and finely organised whites of any race.   Coarse races and peoples, and coarse individuals, can go with teeth badly decayed without being aware of such decay from pain, whereas in a finely organised constitution the very slightest decay in the teeth excites pain which renders filling or extracting imperative.   The coarse races and coarse individuals are less disturbed by the bites of mosquitoes, by the presence of flies or of dirt on the body, than those in whom the nervous diathesis prevails.

Nervous force travels more slowly, the reflex irritation is less perceptible by far, in the dark races and those who live out-doors than in those who live in-doors and are of a nervous frame.   In the strong and coarsely built, local irritation remains local, and does not reverberate through the body ; while, on the other hand, in the feeble, the sensitive, and the highly and finely organised, any local excitement is

* The first part of this article appeared in the January number of the " Journal of Science."

speedily transmitted and puts the whole system into disturbance.

The simple operation of sneezing illustrates this law in a most interesting and significant manner. It is said, for example, of the negroes of the South, that they rarely if ever sneeze. It is certain that the nervous, feeble, sensitive, and impressible of any race are far more likely to be provoked into sneezing from slight irritation of the nasal passages than those of an opposite temperament. In hay-fever sneezing is one of the leading symptoms, and is provoked by irritations in themselves of the most trifling character, which those not victims of the disease can only be forced to believe by a personal battle with this enemy of the race.

Differences of Climate.

These psychological differences come mainly from differences of climate, and secondarily from institutions. In Great Britain and Central Europe there is no summer and no winter, as we in America are accustomed to understand those terms. Warm days they have, but not, as with us, a succession of days that are hot and oppressive during all the twenty-four hours. In the valleys of Switzerland, and in Great Britain, there are days that are called very warm, but which we in America would regard as simply comfortable, followed by nights of agreeable and delightful coolness; and this coolness comes on as early as four or five o'clock in the afternoon: people do not suffer from the continuity of heat and deprivation of sleep. A well-known physician of London told me that he made no change in his clothing all the year round, dressing in August very much as in February. One who should attempt this in New York would desire to perish.

The European climate allows more out-door life than American—not only in Paris, where many pass the larger portion of their time in the open *cafés* and on the boulevards, but in Ireland, England, and throughout Germany, men, women, and children pass far more time in the out-door air all the year than in the United States. The climate allows them to do this, and encourages it, while in America the winters are so cold and the summers so hot, and the twilights so short, that we are forced to stay under a roof. We do have a certain number of days in June and October when it is pleasant and inviting out of doors, when it is possible to sit in the open air, after the European mode;

but these days are so small a minority of the whole year that they do not foster or inspire a habit of out-door life; thus, we stay in-doors even more than is necessary in our own climate. One of the great advantages, possibly the chief advantage, in many cases, change of air for consumptives, is that they live out-doors; and, provided they can get an abundance of out-door air, it matters less than many suppose where they go—whether to the mountains or to cold or warm climates.

Not only are there many more days of rain in Great Britain than in America, but there are more clouds in the sky, even when it does not rain. Clouds, by well-known physical laws, interfere with evaporation; and thus the dampness remains longer in the earth than in a land where sunshine is more free. Thus, the number of days of rain and the amount of rain being the same in Great Britain and America, Great Britain would be more moist. This persistent moisture, as is well known, is the cause of the greenness and long-continued beauty of the foliage of Great Britain, of Ireland, and of the Scottish lakes. Certain threads and cloths, I am told, can only be manufactured at the highest advantage in this moist atmosphere.

My friend Professor Ball, of Paris, told me there is a great difference between Great Britain and France. In Paris, at least, where the sky is far clearer, more like that of America, the streets dry up much more quickly after a rain. The French, as also is well known, are more nervous in some respects than the English, with a finer type of organisation, more nearly resembling Americans.

Either climate, that of Great Britain or of America, is hard at first to bear: when we become worn to either we prefer it to the other. I am told by one who well knows that quite a number of Englishmen and Scotchmen who have lived in this country and acquired property, returning to their homes, after a time came back to America; they missed the noise, the hurry, the struggle to which in the life of business they had become accustomed here, and the mother-land seemed dull and cold.

Americanisation of Europe.

Observations in both continents bring into view two processes that are of supreme import in their relation to the future of mankind. One is, the Americanisation of Europe, the other the Germanisation of America. That Americans were more rapid in their movements, more intense in their whole life, and concentrated more activity in a certain period of time than any other people, has been the faith of all travellers, and this belief has a foundation of reality; but in Europe at least, and to a less degree in Continental Europe, we now observe the same eagerness, intensity, feverishness, and nervousness that have hitherto been supposed to be peculiarly American.

Particularly was I amazed by this when I was in Cork during the present year, attending a meeting of the British Medical Association. The labour of a month was compressed into a week. Every one was in haste—officers and members having only bits of time to breathe or speak; a procession of suppers, breakfasts, balls, banquets, scientific orations, garden-parties, and excursions at every point of the compass, crowded so closely as to tread upon each other's heels; after such a vacation one needed a vacation. At no gathering outside of political assemblages in America have I seen such excitement, such hurryings, such impatience, such evidences of imminent responsibility as among the leaders and officers of that meeting.

This Americanisation of Europe would seem to be the complex resultant of a variety of influences—the increase of travel and trade, and concentration, and intensifying of activity required by the telegraph, railway, and printing-press—the endosmosis and exosmosis of international life—a reciprocity of character. It is clear that even in Europe each generation becomes on the whole rather more sensitive than its predecessor, and in this pathological process even Germany shares; Switzerland, perhaps, being less affected up to the present time than almost any other part of Central Europe.

The nervousness of the third generation of Germans is a fact that comes to my professional notice more and more. Men whose parents on both sides were born in Germany, here develop the American type in all its details—chiselled features, great fineness and silkiness of the hair, delicacy of skin, and tapering extremities. Such persons have consulted me for all phases and stages of functional nervous trouble. Indeed, I have seen in my professional experience no more severe examples of nervous suffering than in this class. Englishmen, even those who were born in England, develop either in their own country, or in this, the land of their adoption, many of the prominent symptoms of functional nervous diseases that are supposed to be especially and pre-eminently American. I am told by one of the leaders of German science, Professor Erb, of Heidelberg, whose opportunities for getting facts on this theme are exceptionally good, and whose capacity for observing and for reasoning justly from his observations is very great, that in nearly all parts of Germany there can be found at the present day, and that too without very much seeking, cases of functional nervous disease in all respects the types of what we see in America; and there has been of late years an increase in these disorders. Even Irishmen born in this land or brought here early are not entirely safe from the chances of nervous contagion.

Prose style is dying or sleeping in Great Britain; the countrymen of Milton and De Quincey must cross the Channel if they would seek for living models of the literary as for the dramatic art. Literature takes its inspiration from the multiplication table, the newspaper supplying at wholesale the words, phrases, and witticisms with which the authors clothe their borrowed thoughts. Suggestions, intimations, and adumbrations of the literary art are seen, but they are crushed under mountains of everydayisms. What everybody will read within twenty-four hours, what nobody will read after twenty-four hours, is the motto that rules the best periodicals in Great Britain: each issue washes out the preceding; the monthlies follow each other with haste, like waves beating upon the shore, and, like them, are quickly lost in the sea of forgetfulness everlasting.

Science, which in its highest phases is but poetry and philosophy in harness, is, in Great Britain, better than its literature; but, in nearly all the great realms of science, England would starve were she not kept constantly nourished at the breast of Germany. Outside of the circle of men of pure genius, like Crookes, the scientific men of England feel that they have reached the highest possibilities when they have given popular lectures on what Germany discovered from five to twenty-five years ago. The profession of medicine—a part of science—lies near the bottom of the middle class, buried under successive strata from royalties and nobilities through the church, the army, the navy, the bar, and successful trade. The descendants of Young and Newton and Harvey are organising to drive a part of experimental physiology from the empire.

As literary art declines in England, the oratorical art seems to rise. Even speakers of but little fame are, many of them, easy and flowing, at times rapid as well as clear in their utterances; so much like Americans that only peculiarities of speech suggest the land of their birth.

Fear of new sciences and philosophies is a most interesting

evidence of the Americanisation of certain classes in Great Britain.   This stout and virile people, so bold in adventure and in battle, tremble in the presence of new ideas—as the savages whom they conquer and subdue, on the approach of a storm.   The scientific discoveries of Germany are a terror; they fear an irruption from the Continent which may over- whelm and bear away their philosophic heirlooms; and their great effort is to erect and guard a line of defence, to keep back the surges of new truths, as the Dutch make dykes against the inroads of the sea.

Thus it is that men in the highest stations, either in church or politics, are always under arms, expected to do duty against the invasions of Continental philosophy.   Half the literature of the last quarter century of Great Britain, outside of fiction, is devoted to proving the truth of the untrue and the undemonstrable.   Their very best men will probably soon begin to see that this chronic alarm is needless.   Supersti- tion is always safe—ignorance everywhere is its own protec- tion; in all the conflicts between intellect and emotion, intellect—with here and their an interlude—is almost sure to be worsted and trampled upon.   The demonstrably false can always be trusted to take care of the things of itself; it is truth alone that has cause to be afraid.   Science, along all its lines, is open to attack, in peril even of its existence —none of its facts being so walled about as the unproved and the untrue.   Truth is a plant as sensitive and tender as it is precious and rare; like all noble and highly developed organisms, it is liable to fatal injury and quick decay: bitten by frost, choked by weeds, broken by storms, the object of attack for all the non-expertness of mankind; error alone has the elements of enduring life.   On this toss- ing sea of humanity, families, tribes, peoples, nations, and empires rise and fall in endless pulsations—arts, literatures, sciences, discoveries, philosophies appearing for a moment on the crest of the waters, then sinking into the fathomless depths; but over all, unchanged save in form, rests a cloud that through the ages never lifts, and a darkness that is scarcely illumined but by the momentary lightnings that flame on its borders, and fugitive glimpses of the distant stars.   It is the undemonstrable, and the demonstrably false, that have ever ruled mankind, and are destined to rule it.   The superstitions and mythologies of Egypt are read in the hieroglyphics on her temples and pyramids and monuments, but the arts that reared those structures have passed for ever from the possession of men; knowledge dies, while delusions live.

Germanisation of America.

The Germanisation of America—by which I mean the introduction, through very extensive immigration, of German habits and modes of life and pleasure—is also a phenomenon which can now be observed, even by the dullest and nearest-sighted, in the large cities of the northern portion of our country. As the Germans in their temperament are the opposite of the Americans, this change promises to be in most respects beneficial, encouraging in every way out-door life and amusements, tending to displace pernicious whisky by less pernicious beer and wine, setting the example of coolness and sobriety, which the nervously exhausted American very much needs. Quite true it is that the second and third generations of Germans do themselves become Americanised, through the effects of climate and the contagion of our institutions; but the pressure of immigration provides, every year, a supply of phlegmatic temperament.

The American race, it is said, is dying out; but there is no American race. Americans are the union of European races and peoples, as lakes are fed by many streams, and can only disappear with the exhaustion of its sources. Europe must die before America. In sections of America as in New England, and in large cities, the number of children to a family in certain classes is too small for increase of population; but these classes are a minority in society, and immigration is as certain as the future. Malthus forgot that the tendency of all evil is to cure itself; the poison and the antidote are rooted in the same soil. The improvement in the *physique* of the Americans of the most favoured classes during the last quarter of a century is a fact more and more compelling the inspection both of the physician and the sociologist. Of old it was said that the choicest samples of manly form were to be found in the busy hours of the Exchange at Liverpool; their equals, at least, now walk Broadway and Fifth Avenue. The one need for the perfection of the beauty of the American women —increase of fat—is now supplied.

The true philosophy on this as on all themes is neither optimism nor pessimism, but *omnism*, which sees both the good and the evil in nature, and aims to make the best of both. America is now on the borders of its golden decade, in which the forces that renovate and save will be far mightier than the forces that emasculate and destroy.

The typical American of the highest type will in the near

future be a union of the coarse and the fine organisations ;
the solidity of the German, the fire of the Saxon, the delicacy of the American flowing together as one—sensitive,
impressible, readily affected through all the avenues of
influence, but trained and held by a will of steel; original,
idiosyncratic, learned in this—that he knows what not to
do, laborious in knowing what not to do ; with more of wiri-
ness than of excess of strength, and achieving his purposes
not so much through the absolute quantity of his force as in
its adjustment and concentration.

ON WATER AND AIR.*

By JOHN TYNDALL, D.C.L., LL.D., F.R.S.,

Professor of Natural Philosophy at the Royal Institution
of Great Britain.

LECTURE II.

YOU must be disposed to imagine that water, because it
is so easily movable or so easily tossed aside by the
hand in passing through it, would be, like air, very
compressible, and that you could squeeze it into a smaller
space.   Now this would be a gross error.   Notwithstanding .
this wonderful mobility that it possesses, it is almost incompressible.   The first attempt to compress water was made
in the year 1620, by the illustrious philosopher Bacon.   He
wanted to ascertain whether water was, like air, capable of
compression, and he operated thus :—He got a globe of lead
similar to that which I have here.   This was made for me
by my exceedingly obliging friend, Prof. Abel, of Woolwich.
This is exactly the first form of Bacon's experiment when
he attempted to compress water.   He filled the globe of
lead with water, and he placed it upon an anvil.   He then
took a great sledge-hammer, as I do now, and endeavoured
to squeeze the globe of lead into a smaller shape : he beat
the lead again and again, and tried to compress the water.
Well, he did compress it to a certain extent ; but he found,
as he himself in his own quaint language expresses it, that
the water, impatient of further pressure, finally exuded
through the lead and covered the surface of the globe like a
dew.   Here, then, we have an instance of the incompressi-
bility of water as evidenced in Bacon's first experiment.
Nearly fifty years afterwards a similar experiment was made
by a member of that great Academy, the Accademia del
Cimento, of Florence.   They used a silver globe instead of a
lead globe, and, on account of the Academy having published a description of the experiment, it usually goes by
the name of the Florentine experiment.   But it has been

* Being a Course of Six Lectures adapted to a Juvenile Auditory, delivered
at the Royal Institution of Great Britain, Christmas, 1879.   Specially reported for " The Journal of Science."

shown by my excellent friend, Mr. Spedding, and his distinguished and eminent colleague, Mr. Ellis, that nearly fifty
years prior to the experiment in Florence it had been made
by Bacon; and therefore, instead of being called the Florentine experiment, it ought really to be called the Baconian
experiment.

In our last lecture we had some experiments illustrating
the tenacity of water. Well, this tenacity is enormously
increased if you repel the air from the water. Take a flask
of water, heat it over a spirit-lamp, and you will always
find, after a certain time, bubbles coming from the water,
and condensing like bubbles of steam. They are small
bubbles of air held in solution by the water. If you rid the
water of this air, if you get the air from between the particles of water, you enormously increase its tenacity, and it
has very many peculiar properties when it is thus rid of the
air. For instance, here is a quantity of water contained in
this tube, and if I bring the water into one end of the tube
and turn it down you hear that the water rings like a solid.
[In this experiment the water was contained in a glass tube
called the water-hammer, from which the air had been expelled.] This is water that has been deprived of its air by
boiling. The water was placed in this tube, and one end of
the tube was drawn out nearly to a point, for the steam and
air to escape through the orifice, and after a certain amount
of boiling the orifice was hermetically sealed. This water,
as I have said, has far greater tenacity than ordinary water,
and in order to demonstrate this I have here a glass V-shaped
tube (A B C, Fig. 4) which illustrates a series of experiments
made by a distinguished man in Belgium, M. Donny. He
showed that when water is deprived of its air the tenacity
of the water is enormously increased. You see that, as I
turn the tube, the water rests at the same level in the two
legs of the tube. If I hold the tube in one way the water
goes into one leg, and if I hold it in another way the water
goes into the other leg. Now this tube has been deprived
of its air in a somewhat similar manner to the water-
hammer, as it is called, that I have just shown you, and
you hear that the water rings when I bring it into contact
with the interior surface of the tube.

Now what I want to show you is that, if I really bring
that water into close contact with the interior surface of the
tube, it will not flow back: there it will cling, and the particles of water will so hold on to the glass that the water
will not flow from one end to the other. When I first turn
down one end of this tube you observe that there is a little

ring. The sound is not quite dead. [The water having
been brought into the leg of the tube (A B), the tube was
gently tapped on the lecture table, so as to cause the water
to adhere more closely to the interior.] Now the sound is
very dead indeed ; and I have got the water in contact with
the tube that contains it. Now I will lift it up carefully,

FIG. 4.

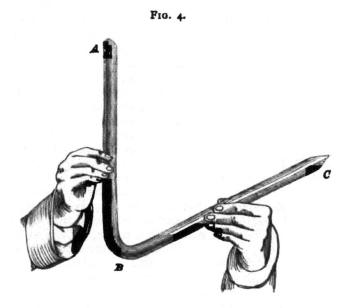

and you observe that the water no longer stands at the same
level in both tubes, but fills the leg of the tube A B altogether
and leaves the other quite empty, and that is because of the
tenacity of the particles of the water which are clinging to
the sides of the tube, the water having been deprived of
its air.

Now, let me be perfectly exact. I have said that the
water gets rid of this air by boiling ; but some years ago a
series of experiments was made by Sir William Grove,
which seemed to show, or which at least justified him in
saying, that he believed that we never saw boiling water at
all in the proper sense of the term. He has sometimes
boiled water not for hours only, but for several days. When
you rid water of its air, as M. Donny did, you can heat it
to a temperature far beyond the ordinary temperature of
boiling water, and from time to time it gives as it were a

sudden leap. When you attempt to boil it, it leaps almost out of the vessel in which it is contained, if it be freed from the air. Well, Sir William Grove found that he got, not this regular, tranquil, continuous boiling, but a series of sudden leaps in his boiling, due to a sudden generation of a quantity of steam ; and he found that he never could get a bubble of steam that condensed entirely in cold water or in the cold oil that he had above the water. There was always a little air ; and when he examined it he found it to be a particular portion of air,—that is, the nitrogen,—so that he never could get a bubble of steam without this bubble of gas. He came to the conclusion that if you could entirely free water of its air you never could get boiling at all. It would be heated up until it would be chemical)y decomposed instead of undergoing ordinary boiling. Well, so much for the tenacity of water and its non-compressibility.

I have now to say one or two words more with regard to the colour of water hinted at in our last lecture. You know that on that occasion I formed a brilliant spectrum upon the screen, and that I took this vessel—which consists of three cells,—and filled the thin cell with a slightly coloured liquid, which had hardly any effect upon the spectrum. I then filled another cell, and that cut off a third of the spectrum. I then filled the thickest cell, and it cut off practically the whole of the spectrum. You had there an example of the effect of the deepening of the layer of water in quenching the light. Now I am going to introduce to you a problem which is not usually introduced into lectures, but I think that it is one which you will easily understand, and which is worth understanding. Here is a black liquid. It is, in fact, a quantity of ink. The light shines upon it from the top of the house, and it strikes the surface of the liquid, and as I look into it I see the image of the gaslights overhead : that is due to what is called surface reflection. The ink reflects light from its surface, but this is simply some of the light that falls upon it. The light that gives a body its colour must go into the body to a certain depth, and be ejected from the body minus some of the constituents of white light ; but the light which comes from the surface of the body is simply white light. Why is this ink black ? Simply because there is no ejection of light from the interior of the liquid. No light comes from within. You have only surface reflection, and therefore the liquid is black.

And now let us reason together. Suppose you stand upon a ship's side and look down into water two or three thousand fathoms in depth—the deep, deep, perfectly pure sea. That

water is perfectly clear. I mean that it is water without
any mud or suspended matter in it, but still capable of
colouring light, as was the water which was shown here in
our last lecture. Well, I say take water pure enough and
deep enough to allow the whole of the light to plunge
into it until the whole spectrum is extinguished. You
see plainly that that water would not differ from our
ink. You would have surface reflection, but you would have
no light from the interior of the water. The colours of the
spectrum would be quenched, and the consequence is that
you would have practically a black liquid, no matter how
transparent the water might be. Now, if any of you ever
cross the Atlantic, or even cross the Bay of Biscay, you will
find that in certain deep portions of the sea the water approaches a fine shade of indigo, and it is sometimes almost
as black as ink. I remember the late most excellent Mr.
Charles Kingsley writing to me about this. He said that
nothing impressed him more, in crossing the Atlantic, than
the wonderful blackness of the sea. But you see that the
cause of this is that the water is exceedingly pure and exceedingly deep,—so deep that the sunlight, when plunged
into it, is actually quenched within the body of the water,
and none is sent back to the eye. The same thing occurs
practically in some of the moraines in the Swiss glaciers,
where the ice is exceedingly pure and exceedingly deep.
Looked at superficially that ice is almost as black as pitch.
Break a piece off it, and look through it, and it is almost as
transparent as crystal. And so with regard to the Atlantic
water, which looks so inky black. If you take it in a glass
you will find that it is as transparent as possible.

Now, whence come the varying colours of the ocean
water? You sometimes see it a vivid green, and sometimes
a yellow-green. In 1870, when I went to Algiers to try to
observe a total eclipse of the sun, I tried to solve this
question, and in order to do so I took this white plate that
you see here before you, which you see is weighted with a
leaden weight ; and here is a portion of a long rope which
was used in the experiment. My object was to account for
the extraordinary changes that I observed in the colour of
the sea. I have observed them since in crossing the Atlantic
from the banks of Newfoundland. For instance, you pass
from water which is of this deep indigo, or almost as black
as ink, into water of a vivid green ; and in going from
Gibraltar to Spithead the variations in the colour of the
water were extraordinary. I operated in this way in order
to instruct myself upon the subject, and in order to ascer-

tain why one water should be green and another almost
black, or, at the least, of a shade of indigo. A very good
fellow, named Thorogood, was lent to me by Capt. Sanderson, who was the commander of the vessel in which I sailed.
The man, who was accustomed to heave the lead, used to
get into a boat near the front of the vessel, and I went
towards the stern, and this man then used to heave the plate
out. Being weighted it sank rapidly, and by the time it
came to the stern, where I was standing, it was at a good
depth. Immediately after it entered the water it became
green: that green deepened, and approached a black; but
I never saw it properly blue. It was a blue-green at the
most, and that was at the utmost depth. That, then, was
the colour of the water at the particular depth to which the
plate reached.

Now I think that you will be able to follow me. The
water at a certain depth coloured the plate of a vivid green.
Supposing that that plate had been broken into fragments,
and the fragments thrown into the water, each fragment
would have sent a modicum of green light to the eye; there
would be seen a number of green spots in the water. Suppose that, instead of being broken into small fragments, the
plate had been ground into a powder so fine that it would
remain suspended in the water; every particle of that powder would send its modicum of light to the eye, and the
consequence would be that the water with such particles
diffused through it would appear to be green water. You
see, by the experiment which I made in the blue-black water
in the Bay of Biscay, that I produced a green colour by
means of the white plate. I am now going on, by the process of reasoning, to deduce the fact that if I ground the
plate into powder, and diffused that powder through the
water, the water would appear green by reason of the light
reflected by the little particles of powder to the eye.

Now this completely accounts for the colours of the ocean
observed between Gibraltar and Spithead. It was sometimes a dull yellow-green, sometimes a bright emerald-green,
sometimes a cobalt-blue, sometimes an indigo colour, and
sometimes almost as black as ink. I took samples of all
these waters in clean stoppered bottles, and by sending a
powerfully concentrated beam through them, here, in London, examined their condition. The blackest water, which
was taken from the deepest portions of the Bay of Biscay,
was the purest. After it, in order of purity, came the indigo,
cobalt, bright green, and yellow-green water. The water
last named was thick with suspended particles, while the

Bay of Biscay water was almost wholly free from them. In the green water the particles were very fine, but so numerous as to send an apparently unbroken green light to the eye. In deep water the particles rest tranquilly at the bottom, in shallow or shoal water they are stirred up by the agitation of the sea. Shoal water, therefore, is always green to the eye. Thus we connect the colours of the sea with its mechanically suspended matter.

The following table clearly shows the connection between the colour of the sea and its suspended matter :—

No.	Locality.	Colour of Sea.	Appearance in Luminous Beam.
1.	Gibraltar Harbour	Green	Thick, with fine particles.
2.	Two miles from Gibraltar	Clearer green	Thick, with very fine particles.
3.	Off Cabreta Point	Bright green	Still thick, but less so.
4.	Off Cabreta Point	Black-indigo	Much less thick, very pure.
5.	Off Tarifa	Undecided ..	Thicker than No. 4.
6.	Beyond Tarifa	Cobalt-blue	Much purer than No. 5.
7.	Twelve miles from Cadiz	Yellow-green	Very thick.
8.	Cadiz Harbour	Yellow-green	Exceedingly thick.
9.	Fourteen miles from Cadiz	Yellow-green	Thick, but less so.
10.	Fourteen miles from Cadiz	Bright green	Much less thick.
11.	Between Capes St. Mary and Vincent	Deep indigo	Very little matter, very pure.
12.	Off the Burlings	Strong green	Thick, with fine matter.
13.	Beyond the Burlings ..	Indigo.. ..	Very little matter, pure.
14.	Off Cape Finisterre ..	Undecided..	Less pure.
15.	Bay of Biscay..	Black-indigo	Very little matter, very pure.
16.	Bay of Biscay..	Indigo.. ..	Very fine matter, iridescent.
17.	Off Ushant	Dark green..	A good deal of matter.
18.	Off St. Catherine's ..	Yellow-green	Exceedingly thick.
19.	Spithead	Green	Exceedingly thick.

Now I want to show you the effect of this suspended matter, and I cannot do it better than by showing the effect of our beam upon the dust of this room. The atmosphere always contains some suspended matter, and even rain-water and the clearest ice are not absolutely pure.

[In illustration of the purity of water taken from melted ice, the apparatus (Fig. 5) was shown and explained. A glass funnel, passing air-tight through a glass plate, has attached to it a glass bulb B, fitted with a stopcock. A block of ice, of great purity, had some days previously been placed in the funnel, and covered air-tight by a glass receiver. This receiver was connected to an air-pump, and the air removed from the interior of the apparatus. Air was then allowed to enter the apparatus by passing slowly through a tube containing cotton-wool. By this means the entering air was completely filtered of its floating matter. The ice melted, and the water trickled down the funnel into the bulb. When the latter was full the stopcock was opened, and it was emptied, the water carrying with it in suspension any contamination that it had obtained from the sides of the funnel and bulb. This washing process was repeated several times, when, from the very heart of the ice, the

FIG. 5.

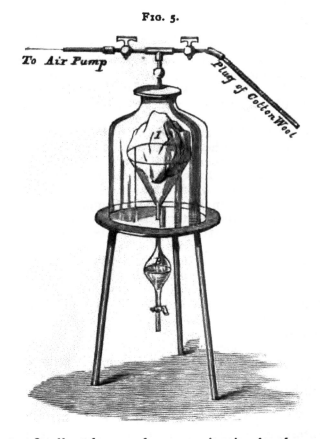

water was finally taken, and on examination by the concentrated beam it was found to be of a most delicate blue colour, proving it to be of the greatest purity.]

If we now send a beam from our electric lamp across the room, we shall illuminate the suspended dust. [An electric beam was made to traverse the theatre from the lamp at L, and brought to a sharp focus at *b*, Fig. 6.] Here we have our beam; that dust which you see in its path is nearly all combustible matter. The beam is fairly vividly shown upon the dust of the room. Now I will take something that will enable me to do away with the dust of the air. [A colourless gas flame was held underneath the beam at the focus *b*, immediately destroying its luminosity and producing a space of intense blackness.]   Now you see these extraordinary black wreaths that go up into that beam.   What are they ? You would imagine that they were the smoke of the flame, would you not ? But they are not at all. If I were to throw smoke into the beam, the smoke would appear quite white. I will try.   We will strike a match and allow it to burn beneath the beam.   [The smoke of a burning match was allowed to ascend into the electric beam, and appeared as a white cloud.]   The blackness which you saw was not smoke at all.   The black appearance was due to the air having been deprived of its floating matter.   That floating matter is combustible, and when it is burnt away, and the air from which it has been burnt ascends through the beam, this blackness is produced as if there were a dense black smoke, although, as I have said, it is not smoke at all, for smoke actually produced whiteness instead of blackness.

Now, I want to enable you to compare the water of Canterbury with the water from our own cistern.   I explained to you the nature of the Canterbury water in our last lecture.   We have some of it in this flask, and in this other flask we have some of the water from our tap which is supplied, I believe, by the Grand Junction Water Company.   For a long time past we have had but very little rain until to-day, and the consequence is that the tap water at the present time is exceedingly good.   It is comparatively clear, for we have it now at its best.   If you read the reports in " The Times " and in other newspapers, you will find that the water supplied to London is now very good indeed, and very free from suspended particles.   A beam of light will be made to pass through these flasks, so that you may judge of the comparative purity of the water.   I think you will find a great predominance of purity on the part of the Canterbury water.   I am now going to light a very famous candle—the Jablochkoff candle as it is called. Right and left of the candle are placed two lenses so as to concentrate the light within the flasks.   [The two flasks were illuminated as described, the course of the rays of light being perceptible in both flasks.]   If the water were absolutely pure you would see no track of the beam in it. The green which you see in the flasks is due to the suspended matter of the water.   In the Canterbury water you have a considerable amount of suspended matter, but there is less than in the tap water, notwithstanding the fact that the tap water is now in its very best condition.

I have now to refer to water in motion—to jets or veins of water, as they are sometimes called; and in order to begin

Fig. 6.

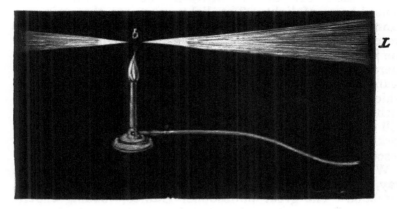

Fig 7.

this, I will try to show you a jet of water coming out of this reservoir which is placed behind me.   Mr. Cottrell is going

up to where that vessel stands.   He has placed behind it an
electric lamp.   It gives an intense beam of light.   I want to
show you, first of all, simply the vein of water flying out of
this vessel.   Mr. Cottrell will send into that vein a condensed beam of light.   The light goes into that vein and
cannot get out of it, and will be carried down with the
water, and I hope that you will see the vein illuminated by
the internal reflection of the light.   At the back of the vessel
there is a little glass window, and through that window the
beam is sent, so that it will fall exactly upon the orifice at
which the vein of water will run out.   Mr. Cottrell will
withdraw the cork from the vessel and the illuminated water
will rush down.   [The theatre was darkened, and upon the
cork of the vessel being withdrawn the illuminated water
descended like a stream of liquid fire.   Fig. 7.]

Well, here you have a continuous vein flowing out in this
way, and forming the curve which mathematicians call a
parabola.   The great philosopher Galileo, making experiments at Pisa on the celebrated leaning tower of which all
of you have heard, established what we call the laws of
falling bodies.   He determined the laws obeyed by bodies
when they fall, and his celebrated pupil, Torricelli, connected
those laws of Galileo with the outpour of a vein of water of
this kind.   You will understand immediately the law which
rules this vein.   If I allow a leaden ball to fall from a certain
height to the ground it reaches the ground with a certain
velocity.   If I allow it to fall from a greater height it reaches
the ground with a greater velocity.   If I allow it to fall from
the top of the house it would reach the ground with a still
greater velocity; and it is on this account that bodies falling
from a great height are more destructive than bodies which
fall from a lesser height.   Now, Torricelli proved that the
velocity of the water issuing from that orifice is exactly that
which would be acquired by a body falling from the top of
the vessel to the orifice.   From that velocity it is able to
describe this beautiful parabolic curve.

Now we have to consider other veins which, unlike that
we have just seen, are not continuous.   If you allow water
to flow from an aperture in the bottom of a tin vessel, you
will find that the vein of water divides itself into two distinct
portions.   You have first of all a portion that appears perfectly tranquil—almost as tranquil as a glass rod.   A little
below that part you will find the rest of the vein unsteady;
but the vein of water appears always continuous throughout,
although one part is steady and the other unsteady.   Now,

that continuousness of the vein is an illusion, and it has
been the subject of examination by very distinguished men.
What is the cause of this change in the character of the
vein from steadiness to unsteadiness? Mr. Cottrell has
here arranged an experiment which shows a very remarkable
property of fluid bodies, and which illustrates the shape
which they assume if left free to choose their own shape.
He is going to place some oil in a liquid in this vessel. It is
very difficult to arrange this experiment. The liquid ought
to be of exactly the same weight or specific gravity as the
oil, and if it be so arranged you will find that when the olive
oil is placed in this liquid it will exactly float in the liquid,
no matter where you place it, without sinking and without
rising, and you will find that the olive oil, free to choose its
own shape, becomes a perfect sphere. The rain which you
saw descending to-day when you were coming here appeared
to be descending in lines or streaks. If such a rain shower
could be illuminated with a single flash of lightning on a
dark night you would see the rain drops, as perfect spheres,
standing still in the air. The reason that they appear as
streaks in the air is that the impression made upon the
retina of the eye by every drop of rain lasts for a little time.

I have to thank my excellent friend Professor Dewar for
giving you an illustration of what I now want to show you.
There is an egg. That egg would sink in water. It would
swim upon salt and water or brine, that is if the brine were
made very strong. He has here made some brine, and that
mixture is of exactly the same weight as the egg. The egg
at present floats at the junction of the two liquids (Fig. 8),
but it would float in any part of the column of brine where
it might be placed. [From the bottom of the glass vessel
to B was filled with the mixture of salt and water; from B
upwards was filled with ordinary water; the egg is shown
floating at the junction of these two liquids.]

Here is the olive oil which is arranged in the way that
has been described. It has been dropped into a mixture of
alcohol and water, and there you see they assume the
globular form. If put into alcohol alone the oil would sink
to the bottom, as it is heavier than alcohol. If poured into
water the oil would rise to the top. Hence the necessity of
getting a liquid with a specific gravity something between
that of alcohol and that of water, in order to cause the
drops of oil to float in any part of this liquid.

Now here I have to notice the investigation of that old
blind philosopher, as Faraday used to call him, Plateau of
Ghent. He has been blind, I believe, for nearly half a

K 2

century, and he has worked at these subjects and produced
the most beautiful results from his investigations simply by
having a good assistant, and directing his assistant to make
his experiments for him, and to describe the results of those
experiments.   We might occupy a course of lectures in
describing the experiments of Plateau.   This experiment
was devised by him for the purpose of withdrawing the oil
from the "pull of gravity," as he calls it.   A most extraordinary exhibition of the tendency of drops of water to
assume this globular form is, I think, to be seen in the

Fig. 8.

whirlpool rapids some two or three miles above the Falls of
Niagara.   There you have a stream that drains half a
continent.   Those rapids are about a hundred yards across,
and at part of the rapids you have great masses of rock
on each side.   The water impinges upon these masses of
rock and waves are formed, and sometimes two waves
coalesce in the centre of the river.   One wave gets upon
the other, and sometimes the contact is so violent that the
crest of the waves is thrown into the air, and these beautiful water globules are produced.   [A photographic view of
the rapids was thrown upon the screen.]   It is, of course,
difficult to catch a picture at the very moment, but you can

imagine how the waves meet. This spray is due to the coalescence of two waves, both adding their force together, and tossing their crest into the atmosphere. It is simply a bundle of most beautiful spheres of water.

Now, as we must make our footing sure as we proceed, we will pass on from this to the consideration of a vein of water issuing from an orifice in the bottom of a vessel. As I said, this vein is composed of two distinct parts. One part is steady and the other unsteady, but the vein appears continous throughout. We can by a device which was made by that able man to whom I referred in our last lecture—Felix Savart—so illuminate a jet of water as to reveal the drops into which it resolves itself. The unsteady portion of the vein, though it appears continuous, is the place where the vein resolves itself into distinct spheres or globules of water. Savart made very particular observations upon this subject, and he found that this unsteady portion of the vein was composed of two different kinds of drops—large drops and small drops. Between every two large drops he found one or more exceedingly small drops. Now, I want to connect this appearance which was noticed by Savart with the explanation of this appearance which was given by the old blind philosopher, Plateau. Here is a little wire stand, and with the greatest ease we can put a quantity of oil upon that stand and cause it to assume a spherical form in this mixture of alcohol and water. Here is a wire formed into a circle. We will bring the wire down upon the sphere of oil and draw out the sphere into a cylinder. You can draw the cylinder up to a certain point, but infallibly at a certain point it cuts itself into two and forms two globes. By properly operating you can go on until you make the length of the cylinder a little more than three times its diameter. Beyond that point you can not go. The cylinder then infallibly nips asunder, as I have said, and forms globules. This is the limit of stability of a liquid cylinder. Now when a jet of water falls from an orifice in the bottom of a vessel, you have there a liquid cylinder; but you have that cylinder at a certain point resolving itself into drops. It will nip itself, and the nipping begins very close to the orifice which is left by the liquid. But it takes a certain amount of time to form the drops, and in the interval the vein is transported over a certain distance from the orifice from which the water issues. Now imagine this liquid cylinder nipping itself. The space between the two drops becomes a thinner and thinner cylinder. One drop separates, as it were, from the other, leaving a little thin or narrow cylinder between them. That little narrow cylinder finally breaks when the drops are perfectly formed, and it forms a little drop between the two larger drops. We have an apparatus which will enable me to show you this. Into the tall glass cell, c (Fig. 9), Mr. Cottrell has placed a mixture of alcohol and water. At the top of the cell he has placed a funnel, F, with its end dipping into the mixture. The funnel, fitted with a tap, contains a quantity of oil, and by turning on the tap

Fig 9.

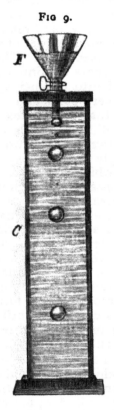

we will allow the oil to issue from the funnel and slowly fall through the alcohol and water. There is a drop of oil coming from the funnel, and now you will see that it will become narrower and narrower, and that little central cylinder becoming very fine will break, and you will have a little sphere between two larger drops. The appearance of these spheres was explained not by Savart but by old blind Plateau. This is what Savart noticed in his first experiments upon liquid veins. I want to show you these drops, if I can ; but before I do that I will tell you the meaning of what I have said about the persistence of impressions upon the retina. Every boy has burnt a stick and twirled it round so as to form a circle of fire in the air. The appearance of that circle is due to the impression on the retina. If you complete the circle in the fifth or the eighth of a second—for the time differs in different eyes—you see not simply the burnt end of the stick, but a circle of fire. The impression produced upon the eye when the end of the stick begins the circle is not extinguished until the end comes round again to its starting point; and thus you have a continuous circle of light. We can in this way mix colours in the eye. You saw the spectrum on the last occasion, and I told you then that if we were to mix all these colours together we should reproduce the white light from which that spectrum came. Newton's way of doing it was to colour a disc with what he called the primitive colours, and

FIG. 10.

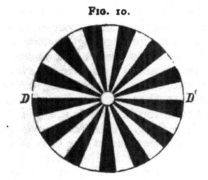

then to impart the motion of rotation to the disc. If the motion be quick enough to enable the eye to retain the impression of all the colours during a revolution of the disc, then all the colours will be shown, as it were, simultaneously to the eye, and the colour will disappear, and the disc will be rendered white. This disc will now be turned, and you will see immediately that the colours will disappear. [A disc of card containing the prismatic colours in due proportions was caused to rotate with great rapidity so that the individual colours could no longer be distinguished, and the disc appeared white.] It is an experiment which I dare say most of you understand perfectly. Now, in place of the colour disc we will use a pasteboard circle with black and white segments. We might do this in a more perfect form with a larger apparatus, but I want to do it in a simple manner.

We shall illuminate this rotating circle with a series of flashes, and I will show you what you would see if you were to observe such a rotating disc by means of a single flash of lightning. The effect produced by flashes of lightning will be imitated here. When the disc turns round, the black and white will entirely disappear. [The black and white disc was caused to rotate, and at first illuminated by an uninterrupted beam from the electric lamp. Under these conditions the surface appeared of a grey tint.] You see that the black and white blend together in virtue of the persistence of the impression on the retina. I dare say some of the boys looking at it may turn their eyes or shake their heads in such a way that the disc may be resolved into its black and white sectors. We will now illuminate it with a distinct succession of flashes, and you will then see that it is no longer of a uniform grey. You see that every flash brings out the sectors so that you can see them separately as if the black and white disc were standing still.

Now if we take our liquid vein and illuminate it in that way, you will see it resolved into a series of beautiful liquid globules, and we are now going to try to illuminate a descending jet of water by means of a succession of flashes. These flashes are obtained by a disc of zinc, which is cut into radial slits, and is rapidly rotating before the electric lamp. You will see that as the jet of water descends a certain portion at the top is dark (*a n*, Fig 11), and the lower portion appears less dark. The less dark part is the unsteady portion. We see that at the less dark part the unsteadiness begins. That is the point at which the jet resolves itself into drops. [The jet of water was allowed to descend, and its shadow was thrown upon the screen.] Now we will try to illuminate the jet by means of a succession of flashes. If the repetition of these flashes of light be at the proper rate we shall be able to resolve the jet into drops. [The beam from the electric lamp was then caused to flash as indicated by means of the zinc screen.] You now see that the jet is resolved into drops, as I predicted (Fig. 12). I will now show you the effect of sound upon the jet. I blow a certain pitch with an organ-pipe, and that enables the vein of water to considerably shorten its continuous part. You see that the top of the vein shrinks when I blow a note. In this way we get an apparently continuous vein resolved into its constituent drops.

I have here in front of the table an arrangement which will enable me to show you the influence of sound upon the continuous portion, and also on the unsteady or discontinuous portion of the vein. If I turn the water on you see how

silently it flows.   [A steady stream of water was caused to
descend from a tap into a cup filled with
water, about 2 feet below.]   You see that
beautiful stalk of water.   The least agitation
will break that vein, but at the present
moment you have a continuous jet which
enters the water below perfectly silent.   I
will now try the effect of sound.   This vein
of water is very sensitive indeed.   It is very
near the point at which it will resolve itself into drops.   Now I simply draw my
fiddle-bow across a tuning-fork, and instantly
you see that the vein shortens,—several
swellings and contractions form out of it,
and we cause the drops to be formed high up
in the vein, and you hear them breaking and
bubbling in the basin below.   If I stop the
sound the vein becomes steady once more.
If I draw the bow again over the fork you
see how the vibrations of the sound cause the
vein to resolve itself into beautiful liquid
spheres.   We have a resolution of the vein
into drops by means of these musical vibrations.   Savart, as I have said, investigated
these subjects, but some of these things were
not noticed by him.

The particular effect which I now wish to
show you is a very beautiful one indeed, and
which, I believe, we noticed here ourselves
for the first time.   We will allow a jet of
water to issue from an iron nipple.   Here
we have our lamp, which will enable us to
illuminate the jet as before.   It requires very
great care in order to cause the water to
issue from the nipple with the proper velocity.   [A jet of water was projected from the
iron nipple (Fig. 13) upwards at an angle of
about 60°, and fell over in the form of an arch,
the further end of which became sprayed
out as it approached the floor.   The shadow
of the jet was thrown upon the screen.]   I
will blow this organ-pipe near the jet, and
you see that the broken portion of the jet is
gathered up into a few small beautiful drops
(Fig. 14).   If I blow properly you see how the drops are
gathered up altogether into one continuous line.   [By varying the conditions the jet may be made to divide into two or

Fig. 12.   Fig. 11.

more bands, as shown in Fig. 15.]   We will now make the
shadow of the drops as large as possible by bringing the jet
as closely as we can to our lamp.

I want now to illuminate that jet by means of our
rotating slits.   [The zinc screen mentioned in the former
experiments was again put into action for the purpose of
breaking the beam of light into flashes.   The sprayed or
unsteady portion of the falling jet was seen to consist of
globules.]   Now you see the drops very distinctly.   Those
who look at the drops themselves will see a greater amount

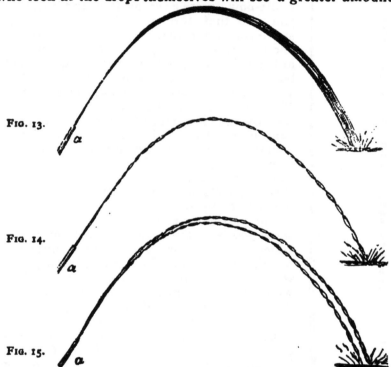

FIG. 13.

α

FIG. 14.

α

FIG. 15.

α

of beauty than those who look at the shadow on the screen.
You observe that when I sound a note the drops become
more distinct.   [The note produced from the organ pipe not
only rendered the drops more distinct, but caused the whole
of the jet of water to change itself into a single line of
globules.]

I have introduced these subjects to your attention in order
to show you how beautiful are the commonest objects that can
engage your attention when they are subjected to scientific
examination.   We shall pursue our work in the next
lecture.

EXPLANATORY REPORTS ON NEW SCIENTIFIC PROCESSES AND INVENTIONS.

[These Reports will not partake of the nature of an Advertisement, but will be impartial descriptions, written and signed by well-known Specialists.—
Ed. J. S.]

Hollway's Process for Reducing Metallic Sulphides without External Fuel.

Excepting in gunpowder, matches, and the Bessemer converter, the only elements hitherto used on a practical scale as artificial fuel have been carbon and hydrogen. This affirmation must, however, be limited in its application to the world in which we at present reside, as we are told of another place where the chief fuel is sulphur. Prometheus scaled Olympus, and from its celestial heights brought down upon lower earth the fire of the gods. Mr. Hollway has travelled in the opposite direction, and now shows us how we may here make use of sulphur as a fuel for the important work of smelting metals.

The combustibility of sulphur having been so long and so popularly understood, and the fact that it is so large a constituent of so many of the commonest ores of common metals being so well known, it is strange that this revelation should have been delayed so long.

The principle of the process is extremely simple. It consists merely in applying a small quantity of ordinary fuel to a natural compound of metal and sulphur, and having thus started the combustion, just as we use paper and firewood to start a coal fire, allowing it to proceed with the aid of nothing more than an efficient blast of air. Thus the sulphur which stands in the ore as the impurity to be removed effects its own removal by a fiery immolation, and the heat it gives out in this suicidal proceeding fuses, and in some cases completes the reguline separation of the metal.

This extreme simplicity lies, however, all in the principle; the practical and profitable applications demand some ingenuity in devising the necessary apparatus and arrangements for conducting the process with the greatest possible economy, and utilising the by-products, the value of which constitutes an essential commercial feature of the invention. Unlike coal and other hydrocarbons, sulphur in burning produces a compound more valuable than the combustible itself. Besides the sulphurous acid produced by the combustion of the sulphur of the metallic sulphides, a quantity of sublimed and unburnt sulphur is also thrown off, and this, being commercially valuable, has also to be collected and commercially shaped. Hence the prac tical problem has demanded considerable expenditure of inge- nuity, labour, and capital. I have watched Mr. Hollway's struggles with these difficulties for some time past, with that hopeful sympathy which is so justly due to every genuine inventor, and therefore have received with much pleasure his latest reports of very promising commercial beginnings. These reports are accompanied with drawings of smelting-works erected in a country that has lately suffered so largely from the destructive combustion of sulphur that there will be poetical justice if a new and beneficent combustion of sulphur should develop its important mineral resources. The Bakarnitza Smelting-works, to which I allude, are at Madianpek, Belgrad, Servia. The general plan of the whole of these works need not here be reproduced, but the section of the Hollway oxidiser (see Figure) recently added to them elucidates at a glance the details of the process.

The sulphides to be reduced are spread upon the charging- floor on the right, and shovelled in sufficient quantity into the funnel-shaped depression of the floor. From this they are screwed down by the " automatic feed " arrangement, which can be obviously made to deliver at varying speeds, as may be required, and shuts off the escape of gases like the cup and cone of a blast-furnace. They thus trickle down into a sort of hearth or crucible, resembling the Bessemer converter in re- ceiving a blast below the surface of melted metal. (It is lettered " Bessemer " in the figure.) Instead of turning over and dis- charging all its contents at the end of the spasm of violent combustion, or " blow," as in the Bessemer process, the Hollway Bessemer is fixed, and its action is continuous. The " slag " (I use this word under protest—it should be " cinder ") of course floats on the top of the metal, and overflows into the settling basin, and from thence into the slag waggons, as shown in the figure ; and when the metal reaches the mouth of the crucible that also flows into the basin, from which it is tapped as this basin fills.

The arrangements of the two blasts—one below for effecting the combustion of the sulphur in the midst of the metal, as the silicon and carbon are burnt in the Bessemer converter ; and the cold blast above for starting the condensation and increasing the flue draught—are shown and explain themselves in the figure ; also the flue which carries the sulphur vapour into the sulphur chambers on the left, where the condensation is com- pleted. The partitions, $p\,p$, increase the length of the journey of the outgoing vapours, and afford additional surface for the condensation of the sulphur, which deposits itself on the cham- ber walls in the usual crystalline form of crude sulphur.

It is obvious that "lead houses" and other ordinary sulphuric
acid plant may be added to this, in order to utilise the 14 or
15 per cent of sulphurous acid that otherwise escapes by the
chimney, this sulphurous acid being identical with that produced
in the kilns of the vitriol-makers by roasting pyrites.

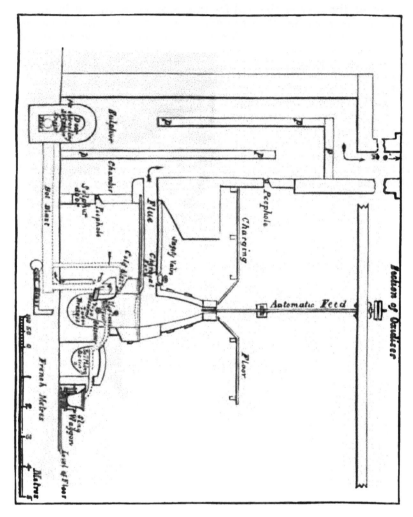

We have here an English invention which, at first, is likely to
be better appreciated abroad than at home, seeing that it offers
such special advantages in countries that have rich deposits of
sulphides, but little or no coal.

The smelting of copper appears to afford the widest field for

Mr. Hollway's operations, the most important ores of that metal being sulphides, and they commonly occur far away from coal-fields.

One of the merits claimed for this process is, that it is applicable to the working of very poor as well as very rich ores, or, as a preliminary operation, for obtaining the regulus of mixed ores, such as those containing copper, nickel, cobalt, silver, &c. This is of considerable importance, inasmuch as there are vast quantities of such poor ores in mountain and other regions of difficult access, which ores, on account of their large bulk in proportion to the metal they contain, will not bear the cost of carriage to this country for smelting. If, however, the metal can be brought into the state of a crude regulus, or semi-metallic mass, by this operation of self-supporting combustion, the regulus will be sufficiently valuable to bear the cost of transport to Swansea, or any other metallurgical centre where the fuel and other requisites required for its final purification are abundant.

W. MATTIEU WILLIAMS.

ANALYSES OF BOOKS.

Stanford's Compendium of Geography and Travel. Based on
Hellwald's " Die Erde und Ihre Vœlker." "Australasia."
Edited and extended by ALFRED R. WALLACE, F.R.G.S.
With Ethnological Appendix by A. H. KEANE, M.A.I.
London : E. Stanford.

THIS volume, unlike the rest of the series to which it belongs,
consists almost entirely of new matter, the account of Malaysia,
Australia, and the Pacific Islands, as found in Hellwald's original
work, being far too meagre.   It can scarcely be necessary to say
that to bring up the treatise to a level with the expectations and
the wants of the British public no more competent editor than
Mr. A. R. Wallace could have been found, specially qualified as
he is by prolonged travels in a large portion of the islands
described.

The work embraces an account of the Malay Archipelago,—
taking the term in its widest sense,—of Australia proper, of
New Zealand, and of the island-groups extending to the eastward more than half-way across the Pacific.   These regions,
though in actual land-area little exceeding Europe, extend in
longitude over more than one-third the circumference of the
globe, and present diversities both of physical features and of
organic life hardly to be found elsewhere.

After a general survey of the peculiarities—geographical, geological, botanical, and ethnological—of this extensive region, the
author proceeds to a systematic account of the several islands,
beginning with Australia proper as the largest and most important.   As may be expected, the natural history of the country
is described briefly, yet with great care and evident accuracy.
It may interest the reader to know that one species of *Eucalyptus*,
growing about 40 miles to the east of Melbourne, is the largest
tree in the world, surpasssing even the far-famed Wellingtonias
(Sequioas) of California.   A fallen specimen of the length of
480 feet has been actually measured.   Such a tree, if it could
grow in St. Paul's Churchyard, would wave its crest high over
the cross at the summit of the cathedral.   It is remarkable that
the alpine flora of Australia contains thirty-eight species not
merely representative of, but actually identical with, forms now
occurring in Europe.   The plants of Australia are by no means
so exceptional as its animals.   No widely distributed order is
absent from Australia, and, on the other hand, no leading Australian order is wanting in the rest of the world.   Sir Joseph D.
Hooker, in his interesting treatise on the Flora of Australia,
suggests that its antecedents may have inhabited an area to the

westward of the present Australian continent, and that the three southern floras—the Antarctic, Australian, and South African—may all have been members of one great vegetation which may once have covered an area as large as that of Europe.

It is interesting to notice that, as far as Australia is concerned, the epoch of subsidence in the southern hemisphere has passed its maximum, and that elevation has been traced at various places along the coast.

It must not, however, be supposed that the entire work is devoted to strictly scientific considerations. Both as regards Australia and the smaller islands we have full accounts of the progress of discovery and colonisation, of commerce, agriculture, and mining.

In the account of Borneo Mr. Wallace once more pays a well-merited tribute to the memory of the late Sir James Brooke, and triumphantly refutes the vile calumnies that have been again vented against him, both in and out of Parliament. Never before was a ruler, alien in race, language, and religion, so beloved by his subjects. He has left behind him, as Mr. Wallace truthfully observes, " over the whole of northern Borneo, a reputation for wisdom, for goodness, and for honour, which will dignify the name of Englishman for generations to come."

Our author, it must be noted, is—not without reason—somewhat sceptical as to the propriety and prudence of suddenly forcing our civilisation upon simple and ignorant populations. The general results of such attempts is that a few merchants and planters accumulate large fortunes, whilst harmless races are " improved off the face of the earth."

Perhaps the only point upon which we must differ from Mr. Wallace is where he, arguing from Java to England, recommends irrigation. He forgets, we fear, that in our climate—where evaporation as compared with rainfall is at a minimum—water, instead of being, as under the brilliant suns of tropical or semi-tropical regions, the one thing needful to ensure fertility, is with us, in average seasons and for our most important crops, the great cause of the farmer's troubles.

We need proceed no further with our survey of this most valuable book. Were we to notice even a tithe of the important facts which it contains we should fill up an entire issue of the " Journal of Science." It may truly be said of Mr. Wallace that " *Nihil tetigit quod non ornavit.*" The present work will only extend and enhance the reputation he has already won.

A Treatise on Chemistry. By H. E. ROSCOE, F.R.S., and C. SCHORLEMMER, F.R.S. Vol. II., Metals (Part 2). London : Macmillan and Co. 1879.

WE have already reviewed two parts of the work which concludes with the volume before us, and we can only repeat what we then said, that all students will welcome the work of Professors Roscoe and Schorlemmer as a masterly treatise on the newest discoveries and the most recent methods of modern chemistry. This concluding volume contains lucid descriptions of the metals of the iron, chromium, tin, antimony, and gold groups. Three supplementary chapters are devoted to spectrum analysis, —a subject in which Prof. Roscoe is one of the foremost and most successful workers,—the natural arrangement of the elements-and the periodic law of Mendeleef, and an account of the processes employed by MM. Pictet and Cailletet for the liquefaction of hydrogen, oxygen, and other gases which have hitherto been considered permanent. No pains have been spared to perfect the descriptions of the various technical processes by the aid of beautifully executed engravings. There are, in our opinion, several inaccuracies in the nomenclature, and of course a few clerical errors; but the work has, notwithstanding these, been most carefully prepared, and will take a high position as a standard chemical treatise.

A Manual of Palæontology for the Use of Students. With a General Introduction on the Principles of Palæontology. By H. ALLEYNE NICHOLSON, M.D., D.Sc., &c. Second Edition, revised and greatly enlarged. Edinburgh and London: W. Blackwood and Sons.

WE have great pleasure in welcoming a second edition, revised and enlarged, of Dr. Nicholson's excellent Manual. All the more important discoveries made from 1872 down to near the end of 1878 have been incorporated in the work; the section on Stratigraphical Palæontology has been entirely omitted, as requiring to be dealt with separately; and the nnmber of illustrations has been greatly increased. The features remaining unchanged are the spirit of accuracy and thoroughness, and the sound judgment which pervade all of Dr. Nicholson's writings. We can hence recommend him to the student as a safe guide, who, whilst always willing and ready to theorise upon a well-ascertained basis, never gives too free rein to the imagination, or seeks to be wise beyond what is demonstrated. In an age given to rash speculations such a teacher is invaluable.

The remarks on the conclusions to be drawn from fossils should be read and re-read by the young geologist. It is exceedingly important that so highly-disciplined and judicious a thinker as Dr. Nicholson considers that the existence of man may be traced back to the Post-pliocene, possibly even to the Miocene, Age.

Studies on Fermentation, the Diseases of Beer, their Causes, and the Means of Preventing them. By L. PASTEUR. A Translation made with the author's sanction, by F. FAULKNER and D. CONSTABLE ROBB. London: Macmillan and Co.

THIS work is an English version of Pasteur's well-known " Etudes sur la Bièrre," a work whose scope is far wider than its title would convey. In addition to a scientific examination of the process of fermentation, in the ordinary sense of the word, and to the application of the principles arrived at to the practice of brewing, the author surveys all those chemical changes—whether in dead or living matter—which are due to the operation of certain minute organisms. He shows further that this fermentative character may, under a certain well-defined condition, belong to every animal or vegetable cell.

The question of abiogenesis, or so-called spontaneous generation, is also touched upon. But in these days there is really very little need for Pasteur and Tyndall to demonstrate, by special experiments, that where all germs or spores are rigorously excluded life fails to manifest itself, even in the most favourable medium. Commerce is every day making the experiment on a far wider scale, and always with one uniform result. Millions of tins of preserved meat, milk, fruits, vegetables, soups, &c., are stored up for a length of time, and are conveyed half round the globe. Yet when opened they are invariably found free from low organisms, save when air has by accident or negligence been allowed to penetrate. If it is objected that the temperatures to which the contents of such vessels have been subjected has rendered them unfit for the support of bacteria, vibriones, &c., we need merely open a tin and expose it to the air, when we shall soon find it swarming.

Hence we hold that refutations of Dr. Bastian are scarcely necessary.

The work will be of very great value to English brewers, though we do not believe that they will be prevailed upon to adopt Pasteur's favourite system of bottom fermentation, nor that the British public would thank them if they did.

CORRESPONDENCE.

DO BIRDS FEED ON BLACKBERRIES AND RASPBERRIES?

To the Editor of The Journal of Science.

SIR,—In your notice of the Proceedings and Report of the Royal Society of Tasmania (vol. i., third series, p. 755) I quote as follows :—"The question was raised whether birds ever feed upon the blackberry or raspberry. With Mr. Spicer, we think that they do not in Europe, but we are by no means certain." In my garden (Sunbury, Middlesex) the blackbirds and thrushes carry them off in thousands. So voracious are they that unless disturbed they would carry off the whole crop. I have seen them jump at the trees, and carry off the fruit. Such good judges are they that the ripest and best are sure to be selected. I can therefore say, whatever may be their habit in other parts of Europe, if it be as you and Mr. Spicer seem to think, my raspberries are exceptions.

S. B.

ACADEMIES AND SOCIETIES.

ROYAL SOCIETY, *January* 15.—The papers read included one entitled "Results of and Inquiry into the Periodicity of Rainfall," by G. M. Whipple, B.Sc., F.R.A.S., Superintendent of the Kew Observatory. The exceptionally heavy rainfall of the past spring and summer directed a large amount of attention to the records of rainfall in this country, and more than one investigator stated that he had found a certain periodicity existing in the quantity of rain annually collected. Dr. Meldrum, Prof. Balfour Stewart, Mr. Hennessey, Prof. Stanley Jevons, Dr. Hunter, and others, have also widely published theories based upon the assumption that the variation in the yearly amounts of fall depends in some manner upon solar phenomena as exhibited by the changes in the appearance of the sun's surface, thereby indicating a cycle of approximately ten or eleven years' duration ; but even amongst the supporters of this so-termed " sun-spot " theory of rainfall there are differences of opinion as to the exact nature of the influence an increase of sun-spots would exert upon the rainfall of any locality. Mr. G. J. Symons,* has partially investigated these theories, and shown the ten-year period does not obtain universally. With a view of dealing with the largest mass of material possible, the author took the long series of rainfall observations made at Paris from 1689 to 1875, published by M. Marie Davy, in the " Annuaire de l'Observatoire de Montsouris." Starting with an assumption of a period, which he first made five years in length, and subsequently extended, he grouped all the observations together, first in five-year groups, then in six-year, then in seven-year, and so on, year by year, until he reached thirteen years. The means of these furnished a set of curves, showing the variation from the mean in the amount of annual rainfall for each of the years composing the series under consideration. The author's results lead him to conclude that all predictions as to rainy or dry years, based upon existing materials, must in future be considered as utterly valueless.

PHYSICAL SOCIETY, *January* 24.—Mr. Grant read a paper and exhibited experiments on " Induction in Telephonic Circuits."

* Nature, vol. vii., p. 143.

Herr Faber exhibited his new speaking machine, which is designed to imitate mechanically the utterances of the human voice by means of artificial organs of articulation made on the human model, and actuated by an operator who depresses certain keys, as in playing a musical instrument. The organs are a bellows made of wood and india-rubber, which answers to the lungs ; a small windmill placed in front of the latter, to give the " r " or trilling sounds ; a larynx, made of a single membrane of hippopotamus hide and india-rubber, to give the " drone " or basic tone of the voice ; a mouth with two lips ; a tongue ; and a nose or proboscis, made of india-rubber tubing placed below the mouth, but curving up towards it. Fourteen distinct vocal sounds can be uttered by the instrument, but in combining these any word in any language can be played by the keys. Thus Herr Faber caused his machine to say such words as " Mariana," " Eliza," " Philadelphia," " Constantinople," and various sentences in French, English, and German, more or less distinctly. Laughing and whispering were also produced, and the voice of the instrument, which was ordinarily loud and clear, and resembling that of a girl, was lowered in pitch and loudness to a more masculine tone.

NOTES.

Mr. D. Winstanley has described to the Manchester Literary and Philosophical Society a new instrument for recording sunshine. A differential thermometer with a long horizontal stem—in which latter is contained throughout the greater portion of its length some fluid intended to operate by its weight—is attached to a scale beam or some equivalent device which also carries a lead pencil by means of which the record is made. The whole is so arranged that in its normal state it rests gently—upon that side to which the pencil is *not* attached—on an embankment provided for that end. Close beneath the pencil point a disc of metal rotated at the proper speed carries a paper dial whereon marks and figures are engraved corresponding with the hours at which the sun may shine. When using this instrument it is enclosed within a box which permits one bulb only of the thermometer—that most distant from the clock—to be affected by the radiance of the sun, which when it shines expands the air contained therein, forces the fluid along the tube, and by altering the equilibrium of the beam brings some portion of its weight to bear upon the pencil point, and so the record is commenced. When the sun becomes obscured, the air expanded by his rays contracts, the fluid in the tube returns, the normal equilibrium is restored, and the pencil ceases to produce its mark. The stem of the thermometer is 18 inches long and about the eighth of an inch in bore. Mercury, in consideration of its weight, is the fluid Mr. Winstanley employs, and in conjunction with it some sulphuric acid is enclosed, because of the mobility which is thereby gained. The bulbs of the thermometer are 2 inches in diameter or thereabouts, and that they may be more rapidly affected the glass thereof is thin. Both are blacked, and the one intended to receive the radiance of the sun projects above the box in which the apparatus is contained into a dome of glass.

At a subsequent meeting Capt. Abney, R.E., F.R.S., exhibited his photographs of the ultra-red portions of the solar spectrum, and first of all showed that the light transmitted by ordinary bromide of silver was of an orange tint, showing absorption in the lowest end of the spectrum. He then explained how he had tried to load the molecules comprising this bromide of silver by means of gum resins, and that he had thus been enabled to photograph slightly beyond the lowest limit of the visible spectrum. Further researches proved that bromide of silver could be prepared in two molecular states—one that already shown, and the other in which absorption takes place in the red as well as in the blue. This was found sensitive to every radiation. He pointed out that the blue form of the silver bromide could be converted into the red form by simple friction, and that after friction it was insensitive to the ultra-red radiation. Prof. Roscoe here exhibited the different preparations of gold in minute division made by Faraday himself, some of which transmitted blue light and others red, showing that at all events two cases of molecular condition exist in the case of metallic gold. Capt. Abney then threw upon the screen photographs of the prismatic spectrum, in one of which the lowest limit of the prismatic spectrum was reached. Various photographs of the ultra-red portion of the diffraction spectrum, extending from 7600 to about 11,000, were exhibited. Those from which Capt. Abney has made his final map were taken on double the scale, with twice the amount of dispersion. Capt. Abney showed various prismatic spectra, exhibiting different states of atmospheric absorption, in one of which Piazzi Smyth's rain-band was markedly visible. Some photographs of the spectrum in natural colours were also exhibited.

In many public aquaria, as at the Crystal Palace and Westminster, and also at that recently opened at the Aston Lower Grounds, near Birmingham, unchanged sea-water is employed; and such water has been found by experience to be fit for an almost indefinite period for maintaining marine animals in health. The principle adopted is to keep the bulk of the water in underground reservoirs, and to constantly pump from them a fresh supply into the show tanks, the delivery of the water being made through a series of fine jets. The great object of a constant supply, delivered in such a manner, is to force air along with fresh cool water into the tanks containing the animals. By that means the water is maintained of an almost constant temperature, and even if temporarily rendered cloudy by the presence of decaying fragments of food, &c., the rapid oxidation resulting from the air forced in along with the fresh water from the jets, soon renders the water quite clear. At Brighton and some other places the sea-water is intended to be used for a limited time only, a fresh supply being pumped in from the sea as often as required. Since it would be impracticable for inland aquaria to be supplied even at distant intervals with fresh sea-water, on account of the great cost incurred in collecting and for carriage, the system of unchanged water has of necessity to be adopted. Mr. H. Williams Jones examined a sample of water which, after being used for eight years in a large public aquarium, was quite bright, and appeared to answer perfectly the requirements of the animals. The results obtained we take from the " Chemical News," vol. xl., p. 282. It was inferred that a very large amount of nitric nitrogen would be present as a result of the oxidation of the organic matter thrown off by the animals, or left in the water in the form of uneaten and unremoved dead food. As anticipated, a very large excess over what exists in actual sea-water fresh from the ocean was detected. The following figures show the difference between the aquarium-water and sea-water taken about the same date from Brighton, and which fairly represents the amount present naturally in sea-water.

	Nitric Nitrogen. Parts per 100,000.
Recent sea-water	0 0325
Aquarium-water	12·0498

In all well-stocked aquaria the number of living animals of a large size in relation to the bulk of water is greatly in excess of what exists in Nature; and the food of such animals kept in confinement instead of being produced in the water in which they live is received from an outside source in the shape of food formed in the ocean. The nitric nitrogen is formed in aquaria faster than it is removed by vegetation, which is usually limited to *confervæ*, and so it accumulates.

At the November Scientific Meeting of the Royal Microscopical Society, an improved form of Microtome (section cutter), contrived by F. H. Ward, M.R.C.S., was exhibited. The most noteworthy addition is the means of releasing the screw, when it is required to move the plug rapidly, thereby saving wear of the delicate screw, and also the time required in carrying the imbedded tissue through a long distance, either for removal or bringing to the top of the well. The bottom of the well is removable, and is retained in place by two bayonet catches. When removed from its fitting the bottom is separable into two halves, so as to release the screw: the continuity of thread in the internal screw is maintained in the two halves, when in position, by means of two metal pegs on the face of one half accurately fitting into holes on the face of the other. The screw has thirty-six threads to the inch, and is attached to a head having thirty-five notches on its circumference, into which a spring catch falls. The cutting surface is of plate glass, and can be removed at pleasure for the purpose of cleaning. The whole apparatus is extremely compact, and is provided with a screw clamp to fix it firmly to the table.

Mr. Crookes exhibited his experiments on Radiant Matter at the Ecole de Médecine, Paris, on the 8th and 11th ult.; at the Observatory on the 15th, and at the Société de Physique on the 16th of January.

THE

JOURNAL OF SCIENCE.

MARCH, 1880.

I. THE LINES OF DISCOVERY IN THE HISTORY OF OZONE.

By Albert R. Leeds, Ph.D.

I. *Its Original Discovery, Sources, and Properties.*

THE history of ozone begins with the clear apprehension, in the year 1840, by Schönbein, that in the odour given off in the electrolysis of water, and accompanying discharges of frictional electricity in air, he had to deal with a distinct and important phenomenon. Schönbein's discovery did not consist in noting the odour; that had been done by Van Marum, more than half a century before, but in first appreciating the importance and true meaning of the phenomenon. For while Van Marum, Cavallo, and others who followed them, connected the odour with the electricity, calling it the "electrical odour" or "aura electrica," and thus made it the property of an imponderable agent, Schönbein ascribed it to the peculiar form of matter operated upon. The hypothesis of Van Marum necessarily remained ·barren of fruits; that of Schönbein speedily enriched chemical science with a host of acquisitions. Clinging tenaciously to the doctrine that there could not be a variety of origin for one and the same odour, and that the kind of matter producing it in every case must be identical, Schönbein fixed his discovery by giving to that one and certain kind of matter the name of Ozone. By adhering to this guiding clue, he added, as a third source of ozone, the action of moist phosphorus upon air (1840 to 1843), and since that time, besides electrolysis, electrical influence, and the action of air upon moist phosphorus, no other sources of ozone of practical utility have been discovered.

The fact that Schönbein so stoutly insisted on, and eventually so triumphantly established, the *identity* of the ozone from whatever source derived, must not be lost sight of in any estimate of his merits as a discoverer. The earliest attack came from De la Rive, who attributed the odour to metallic oxides set free from the metals used as electrodes, or as terminals in electric discharges. But Schönbein pointed out that besides the improbability of an odour arising from solid bodies, this hypothesis required that solid bodies should have the property of indefinite suspension in the atmosphere instead of being deposited or washed down by water (1840 to 1843).

The next attacks came from Fischer, who regarded Schönbein's ozone as probably peroxide of hydrogen, and from Williamson, who thought there were two kinds of ozone,—one the ozone given off in electrolysis, and which he regarded as a higher oxide of hydrogen, differing from the previously well-known peroxide, and the other formed by the action of phosphorus on moist air. But Schönbein disposed of both objections : the first, by showing that the chemical and physical properties of ozone are not the properties of peroxide of hydrogen ; the second, by demonstrating that whatever might be the true nature of ozone, the gaseous matter obtained in the electrolysis of water was in all respects identical with that formed by the action upon air of moist phosphorus (1844 to 1845).

During these first five years Schönbein was busily engaged in ascertaining the properties of ozone. Since no peculiar methods were employed in the furtherance of these discoveries, they need not detain us here further than briefly to summarise them, and to point out what corrections have been rendered necessary by the labours of subsequent investigators. They are—1st. Its eminent oxidising powers, as shown by its ability to transform most metals into their higher oxides, and to raise the lower oxides into the condition of peroxides. Certain of the non-metals—phosphorus, chlorine, bromine, and iodine—are similarly oxidised. Schönbein's statement that it does not unite with nitrogen under ordinary circumstances, but enters into combination when alkali is present, has been abundantly disproved— among others, by Berthelot (1878), who has shown that no combination occurs even when alkali is present. It oxidises sulphites and nitrites into sulphates and nitrates, and many sulphides into their corresponding sulphates. It destroys (as has since been more elaborately demonstrated by Houzeau, 1872) many gaseous compounds of hydrogen, like those with sulphur, selenium, phosphorus, iodine, arsenic, and antimony. It discharges vegetable colours, and powerfully attacks many organic bodies. The nature of its action in the latter case has been more extensively studied by Gorup-Besanez (1863), and he has described the products of the reactions which occur when ozone is allowed to act upon organic substances, alone or in presence of alkali. 2nd. According to Schönbein ozone is insoluble in water. The observations of subsequent experimenters conflict on this point, but there appears to be much evidence to show that it is soluble in water, though only in small degree. 3rd. Schönbein pointed out that atmospheric air strongly charged with ozone acts powerfully on the mucous membranes and produces symptoms of catarrh. This and his analogous discovery that ozone is present in the atmosphere, and plays there a very important part, attracted to the subject not only great popular attention, but enlisted as observers a multitude of students of medicine the world over, who hailed the newly-discovered body as an invaluable therapeutic agent, and rushed forward to establish by sufficiently numerous observations the relations between its presence or absence in the atmosphere, and the kind and prevalence of disease. Thirty years have passed away, and neither anticipation has been realised. Indeed, at the present hour, the possible value of ozone as a therapeutic agent is obscured by its having fallen into the hands of empirics; and the multiplication of inexact observations, and the crude and hasty generalisations therefrom, have covered with a sort of scientific opprobrium the whole subject of Atmospheric Ozone.

What causes have led to these lamentable results in the past; what prospects are there that both subjects can be reinstated in good scientific standing in the future?

And first with regard to ozone as a therapeutic agent. Without considering at present the unsettled questions of a medical character, as to the proper mode or amount or propriety of application, we apprehend that there have been hitherto three grave instrumental difficulties: — 1st. To obtain ozonised air or oxygen of known strength and of adequate purity. 2nd. It is doubtful whether in one form in which the attempt has been made to employ ozone in medicine, that of "ozonised water," any ozone whatever has been present. Such was the case with the "ozone-water" of Krebs, Kroll, and Co., in which Rammelsberg found *chlorine.* Since ozone is so slightly soluble in water at common temperatures that it is extremely difficult to demonstrate the fact of solution, the proposition to employ " ozonised water " as a remedial agent opens a wide door to quackery.   3rd. It is certain that from the mixture of potassium permanganate and sulphuric acid, which has been and is recommended as a convenient source of ozone for medical use, no ozone, but merely chlorine and oxides of chlorine (due to impurities in the permanganate) are derived.

These errors have been exposed and the difficulties overcome.   There is no obstacle to having in the office of the physician, the sick-room of the patient, or the wards of the hospital, ozonisers suitable to each place, and adequate to supply ozonised air or oxygen of known strength and purity. This being the case it remains for the therapeutist to do his part of the work, and to discover when and how ozone is to be employed in legitimate practice.

Second, to detect the amount of ozone present at any time or place in the atmosphere, and the *rôle* this atmospheric ozone plays as a disease excitant or prophylactic. The objections which vitiate the observations hitherto made are two in number :—1st. The ozonoscopes hitherto employed, Houzeau's and the thallium-test included, are all affected by some one of the gaseous bodies possibly present in the atmosphere, as well as by ozone.   2nd. The method of conducting the observations is in its nature inexact, and variations in wind, temperature, humidity, &c., are allowed to increase the resultant errors.

Advance in this direction is to be looked for only when the methods at present in use are abandoned in favour of others more in harmony with those pursued in other branches of gas analysis, and when reagents are employed which will assign true values to the amounts of ozone determined.

II. *The Nature of the Constituent Matter of Ozone.*

In his speculations upon the nature of ozone, Schönbein was far less fortunate than in his multiplied inquiries into its sources, properties, and applications.   The difficulty at that time of procuring air or oxygen containing more than a minute percentage of ozone, and of manipulating it when obtained, was very great, so that precise quantitative investigations were attended with formidable obstacles, and probably for that reason were rarely instituted by Schönbein.

He brought forth a variety of hypotheses, thus introducing great uncertainty into a confessedly difficult subject, and necessitating the labours of chemists during nearly a quarter of a century for their complete overthrow.

His earliest hypothesis was that ozone is a compound consisting of hydrogen and oxygen. This, in 1844, he abandoned in favour of the theory that ozone itself is elementary, and along with hydrogen enters into the composition of nitrogen, which is a compound substance. The following year he reverted to his original hypothesis, and, while maintaining strenuously that ozone is not peroxide of hydrogen, he nevertheless upheld the view that it is composed in certain unknown proportions of hydrogen and oxygen.

The second hypothesis was overthrown by the experiments of Marignac and De la Rive, who showed that ozone could not be derived from the decomposition of nitrogen, inasmuch as they obtained it by passing electric sparks through perfectly pure and dry oxygen. They proved the resultant body to be ozone, by causing it to react on moist silver and potassium iodide with the formation of argentic peroxide and iodate of potassium. They explained these reactions by supposing that under the influence of the electric discharge the oxygen had acquired an electrified condition, with exalted chemical properties : in other words, that ozone is oxygen, and oxygen only, but oxygen in an electrified state. Plausible as was this explanation, there was nothing in the experiments—water having been present in the reaction upon silver and potassium iodide—to confute the different interpretation brought forward by Schönbein, that ozone was oxygen to which in some way was added the elements of water. Nor was this point settled by a more elaborate experiment of the same nature, instituted by Fremy and Becquerel in 1853, who demonstrated that when a certain volume of oxygen is confined over an aqueous solution of potassium iodide, moist silver, or mercury, *all* of the oxygen undergoes absorption by the reagent under the influence of a sufficiently prolonged series of electric sparks.

The first to abandon the theory that hydrogen is a constituent of ozone was Schönbein himself (1849). He employed air, ozonised as strongly as possible by moist phosphorus, and afterwards dried by passage through a sulphuric acid drying tube. That water was employed in the generation of the ozone was not from Schönbein's point of view an essential element in the problem ; it was whether this ozone after drying still contained the elements of water or hydrogen.

Three hundred litres of the desiccated air were passed through a narrow glass tube heated to redness, in order to decompose the ozone, and then through a second sulphuric acid drying tube. Since the latter in repeated experiments showed no increase of weight, Schönbein regarded the absence of hydrogen in ozone as conclusively proven. At the same time he did not accept the views of Marignac and De la Rive, declaring that to him the existence of an allotropic modification of a gaseous body was inconceivable. For a long time, however, the theory that ozone was a compound of hydrogen and oxygen prevailed. It derived great weight from the experiments which had been made by Williamson in 1845. He prepared ozone by electrolysis, and to avoid obtaining any hydrogen along with the electrolytic oxygen, used oxide of copper dissolved in sulphuric acid as the electrolyte. The gas was dried over calcium chloride, and then passed over ignited copper turnings into a second drying tube : this uniformly showed an increase of weight. The copper previous to ignition had been reduced by carbonic oxide, and not by hydrogen, in order to prevent the possibility of any occluded hydrogen being given up, on ignition, to the stream of ozonised oxygen.

These views were apparently confirmed by Baumert's experiments (1853). He passed the electrolytic oxygen evolved in such a manner as to exclude the presence of hydrogen, through a very long sulphuric acid drying tube, and thence into an absorption apparatus containing potassium iodide and provided with a sulphuric acid bulb-apparatus, to condense evaporated water. In case the matter of ozone and oxygen were identical, the weight of oxygen equivalent to the weight of iodine set free by the ozone should have been equivalent to the total gain in weight by the absorption apparatus. But, according to the experiments, this weight was less, and the numbers found apparently assigned to electrolytic ozone the formula H_2O_3. And since Baumert found that ozone prepared by the electric discharge could not be made to yield up the elements of water on strong heating, while that prepared by electrolysis could, he regarded the two as different bodies—the former as allotropic oxygen, the latter teroxide of hydrogen.

Thus the old hypothesis, against which Schönbein had so long striven, that there were two (and possibly more) bodies of the nature of ozone, was rehabilitated. It was finally overthrown by Andrews (1856), who showed that the preceding experiments on electrolytic ozone had been vitiated by the presence of a small but appreciable quantity of carbonic acid, which, unless very great precautions be taken, is always present in the evolved gas. In very numerous experiments he showed that the weight of active oxygen equivalent to the weight of the iodine set free in the absorption apparatus, was equal to the entire gain in weight of the apparatus, and therefore no hydrogen as well could have been present; also that the properties of electrolytic ozone, and that obtained by the action of the electrical spark on pure and dry oxygen, were identical. More especially, it was shown that both were converted into ordinary oxygen at a temperature of about 237° C.; and from the whole investigation the author drew the conclusion, which was confirmed by the still more elaborate experiments of Soret in 1863, and is now universally adopted, "that ozone, from whatever source derived, is one and the same substance, and is not a compound body, but oxygen in an altered or allotropic condition."

III. *The Exact Nature of the Relations existing between Ozone and Ordinary Oxygen.*

We have seen that Marignac and De la Rive, as the result of their experiments performed in 1845, had enunciated the view that ozone was oxygen in a peculiar electric state. They proposed to abandon the name "ozone," which assumed an independent chemical existence for this body, and to call it merely "electricised oxygen." This view of the constitution of ozone was one not readily susceptible of investigation by usual chemical methods. But the case was different with the hypothesis which was shortly afterwards advanced by Dr. T. Sterry Hunt, in 1848. Since his intuition of a truth, not fully demonstrated until twenty years later, is of a very striking character, it will be interesting to quote it as originally announced. In a paper on the anomalies presented in the atomic volume of sulphur and nitrogen, Dr. Hunt says—"In considering such combinations as SO_2 and SeO_2, which contain three equivalents of the elements of the oxygen group, it was necessary to admit a normal species which should be a polymer of oxygen, and be represented by $O_3 = (OOO)$. The replacement of one equivalent of oxygen by one of sulphur would yield sulphurous acid gas (OOS), and a complete metalepsis would give rise to (SSS). The first compound is probably the *ozone* of Schönbein, which the late researches of Marignac and De la Rive have shown to be in reality only oxygen in a peculiarly modified form, &c."

The hypothesis herein stated, that ozone is triatomic oxygen, necessarily involved the assumption of such a corresponding difference in density, and other physical properties,—differences admitting of exact quantitative proof or disproof. Such were the experimental difficulties in the way, however, that it was not until 1860 that an investigation was made into the volumetric relations of ozone to oxygen. The experiments of Professors Andrews and Tait then resulted in establishing that when perfectly pure and dry oxygen is converted into ozone, under the influence of the silent electric discharge, it becomes more dense, the amount of contraction being proportional to the quantity of ozone produced; also that when ozone thus condensed is exposed for a short time to a temperature of 270° to 300° it expands to its original volume. That the increase in density was exactly proportional to the amount of ozone formed was proven by an analysis of the contracted gas by means of potassium iodide. The amount of iodine in every case set free was precisely equivalent to the weight of a volume of oxygen equivalent to the volume of the contraction which the oxygen had experienced in the process of ozonation. The same laws were demonstrated to hold good with regard to electrolytic ozone, not only by these authors (1860), but also by Von Babo and Claus and by Soret (1863).

Andrews and Tait found great difficulty in reconciling the theory of the allotropism of ozone with their experiments, inasmuch as the oxidation of a body like mercury, potassium iodide, &c., was effected without any diminution in the volume of the contracted gas. In other words, the density of the allotropic oxygen concerned in this oxidation was apparently infinite. They sought therefore to explain the origin of ozone by the assumption of a decomposition of the oxygen.

But in 1861 Odling put forth the interpretation that ozone was a compound of oxygen with oxygen, the combination being attended by a contraction. Hence if one portion of the combined or contracted oxygen were absorbed by an oxidisable body, the other portion would be set free, and by its liberation might expand to the initial volume. He likewise suggested that this contraction might consist in the condensation of three volumes of oxygen into two volumes, not because this ratio was the only one which would explain the volume and density relations, so far as then known, but because, on the hypothesis of the dual nature of oxygen, this was their simplest possible explanation. Four years later Soret discovered that a very remarkable reaction occurs when electrolytic ozone is allowed to act upon oil of turpentine. Its volume is diminished by a volume equivalent to twice that of the oxygen, corresponding to the iodine set free on passing the ozonised oxygen into a solution of iodide of potassium. The latter, it will be remembered, is the same as the diminution in volume which the oxygen undergoes in ozonation, and may be called the contraction-volume. Hence the two volumes of ozonised oxygen absorbed in Soret's experiments contained not only their own volume of oxygen, but that contained in the contraction-volume, or in all three volumes of ordinary oxygen. The density of ozone, therefore, was to the density of oxygen as three to two, or 1·6584, the density of ordinary oxygen being 1·1056.

Soret inferred rather than demonstrated these relations, inasmuch as in his first set of five experiments the ratio of the total volume of ozonised oxygen absorbed by the turpentine to the contraction-volume was 2·4, and in his second set of seven experiments 1·81, both of these results being far from 2, the theoretical number.

However, in 1872, Sir Benjamin Brodie, by the introduction of methods of exact volumetric character, supplied a rigorous experimental demonstration. He obtained in a set of eight concordant experiments made with oil of turpentine, for the ratio between the whole diminution in the volume of the original oxygen, to the diminution in the volume of the ozonised oxygen, as a mean result, 3·02 to 2·02. Operating in the same manner with a neutral or slightly alkaline solution of sodium hyposulphite, he obtained, as a mean result of twenty-seven concordant experiments, the same ratio of 3·02 to 2·02. In these experiments the actual weight of the oxygen absorbed could not be determined otherwise than by calculation from the alteration in volume. But by the oxidation of stannous chloride, under proper conditions, he effected a direct determination, and found that the weight of the oxygen absorbed from the ozonised oxygen by the stannous chloride was almost exactly three times the weight absorbed from the same gas by potassium iodide. At the same time the volume in the first case was almost exactly twice the contraction volume, as determined by the latter reagent.

II. SALT INDUSTRY OF ENGLAND.

I. BRINE PITS.

WO methods generally are adopted for obtaining salt supplies in this country, viz., from brine springs and from rock salt. The former is by far the more ancient of the two, as will be presently shown. At the earliest time of which we possess records on the subject wherever there was a brine spring or a manufacture of salt, the place appears to have been called *Wich*. Thus the name Droit-wich was probably originally *Wich*, and it is supposed that the prefix *Droit* was given to designate a certain legal or allowed brine pit. We have also Nantwich, Middlewich, Shirleywich, and so on.

Some of the earliest records of the brine springs relate to those of Droitwich. From these it appears that in the year 816 Kenulph, King of the Mercians, gave Hamilton and ten houses in Wich, with salt furnaces, to the Church of Worcester; and about the year 906, Edwy, King of England, endowed the same church with Fepstone and five salt furnaces, or scales. In Domesday Book, when, between the years 1084 and 1086, William the Conqueror caused an inquiry to be made into the names of the several places, by whom they had been held in the time of Edward the Confessor, the last hereditary Saxon king, and who held them when the inquiry was made, the Wiches and salt houses then in operation are respectively recorded; and it is clear that in these times, as regards Cheshire and the detached parts of Flintshire called Maylor, the rights of property were fully exercised over the brine springs and salt works, and and that there then existed certain well defined customs with regard to them. In King Edward's time there was a Wich in Warmundestron hundred in which there was a well for making salt, and between the king and Earl Edwin there were eight salt houses, so divided that of all their issues and rents the king had two parts and the Earl the third. From our Lord's Ascension to Martinmas anyone having a salt-house might carry home salt for his own house; but if he sold any of it either there or elsewhere in the county of Cheshire, he paid toll to the king and the earl. Whoever after Martinmas carried away salt from any salt-house except one devoted to the earl's private use paid toll, whether the salt was his own or purchased. In the year 1245 it is

recorded that Henry the Third caused the brine springs to
be destroyed to prevent the Welsh, with whom he was at
war, from getting supplies of salt.   The brine springs of
Cheshire were described in 1808 as being found in the valleys
of the Weaver and Wheelock, and nowhere else except at
Dirtwich, on the border of the detached part of Flintshire,
and a weak spring at Dunham, near the river Bollin.   In
the Dane valley springs are also supposed to have once
existed in the neighbourhood of Congleton, as some of the
inclosures were named Brinefield, Brinehill, &c.   In addition
to these places, Mr. W. H. Ormerod, in 1848, on the salt
field of Cheshire, adds discoveries of brine at Acton, Broad-
lane, and Hatherton, near Nantwich, the site of the viaduct
of the Manchester and Crewe Railway over the Wheelock,
Red Lane, Elton, the Flint Mill near Middlewich, the west
side of Hartford Bridge, Eaton, opposite Vale Royal, and
the finding of brackish water at Minshall Vernon.   At
present the brine springs worked in Cheshire are in the
same valleys as in the olden times, but the principal works
are confined to the lines of the river, canal, and railway
communication.   Brine still flows to the surface at Brine
Pits Farm and at Shewbridge, between Audlem and Nantwich; but in 1808 there were no manufactures above
Nantwich; and at Nantwich itself, one of the most ancient
of the Wiches, salt ceased to be manufactured about the
year 1847.   At Dirtwich, the higher Wich on the Cheshire
side ceased to be worked about the year 1830, and the lower
Wich, in Maylor hundred, Flintshire, ceased about the year
1856.   In Staffordshire, at Shirleywich and Weston-upon-
Trent, brine has been used from early times.   There appear
to have been numerous ancient pits along the banks of the
Trent, but the top water having got in, they became waste.
In modern times, the late Mr. Stevenson, C.E., was called
in to advise, and he put down a shaft and tubed it with iron
castings; but this also proved a failure, and at present there
is only one shaft at Weston and one at Shirleywich.   In
Shropshire, at Adderley Wood, between Adderley and
Audlem, a trial shaft was sunk in the blue lias and perforated
into red ground.   The water found in this shaft was salt, as
also in another shaft close by which had been sunk many
years before.   Brine springs have also been found in other
parts of England: as in Somersetshire, Westmoreland,
Durham, Lancashire, and Yorkshire; but they were either
weak, like those occasionally met with in coal mines, or
were situated where fuel was scarce, so that they have not
been much noticed.

Until the discovery of rock salt, it was only supposed that
brine springs were formed by the solution of rock salt.   It
is now known that water permeates from the surface, and
that, until recently, all the brine springs came from the top
of the rock salt, or as it is locally termed, the "rock-head."
In addition to this source of supply, brine is now obtained
in large quantities in the neighbourhood of Northwich, from
the old rock-salt mines, into which water from the surface,
together with brine from the rock-head, finds an entrance.
The weak solution coming in contact with the old pillars
and the remainder of the rock salt that was left unworked,
is formed into pure brine ; and the chambers being pierced
where the brine is fully saturated, the pumping goes on
without the separating barrier being dissolved.   There are,
therefore, two sets of modern brine pits in existence, those
of the rock-head and those of the inundated old workings.

The precautions necessary for securing the brine shafts
are, in many respects, identical with what is required in the
rock salt pits, and having been earlier in point of time, the
necessities seem to have been met as they arose.   It seems
that in sinking to many of the springs the supply of brine,
when cut into, was so copious that the sinkers had to escape
for their lives, sometimes rising up the shaft amongst the
brine without any opportunity being afforded of seeing what
was underneath, which accounts for the lateness of the discovery of the rock salt.   In these sinkings, when it is still
unknown at what depth the brine is likely to be met with,
there seems to be no entire remedy against these sudden
entries of brine.   But in the proved districts it is now
observed that before reaching the top of the rock salt, where
the rock-head brine flows, there is often a bed of hard marl-
stone, called "the flag," and that for a few feet above it the
marl is of a granular structure, called "horsebeans."
Therefore, when these indications are observed, and the
brine is expected to be found at a high pressure, the practice
is to case the shaft sides carefully down to the flag, to keep
them secure, and prevent surface water from entering.   The
flag is then either blown through with powder, or bored
through with boring rods.   One of the best methods of tapping the brine, when under pressure, is to sink the shaft
nearly as deep as to where the brine is expected, and then
to case it with iron cylinders, having an iron bottom to the
lowermost cylinder, with two holes in it to which pipes may
be attached.   From each of these two pipe holes a column
of pipes, of about four inches in diameter, is erected inside
the cylinders, either to the top, or as high as the brine is

expected to rise, the bottom pipe having holes in it to let out the brine when it is tapped. Being thus equipped, a set of boring rods is let down each pipe, and the remaining strata at the bottom of the cylinders are bored through into the brine. On being thus tapped, the brine rises up the bore hole, and, entering the pipe, passes through the holes near the bottom into the cylindered shaft, where it rises to its level. Provision is also sometimes made by having a plug, or a tap, at the bottom, below the holes in each pipe, so that if needed the entry of the brine into the shaft may be stopped, and the shaft emptied. In the brine shafts, where brine is pumped out of the old rock salt mines and is met with at a much higher pressure than in the rock-head brine shafts, the tapping has been attended with extraordinary difficulties. The brine in these old workings rises to as high a level as the brine that is found at the rock head, and as it has to be tapped through a pillar near the bottom of the old workings, the pressure is proportionately higher.

It seems to have been long noticed that in Cheshire the Northwich brine contained a trace of iron, and that the earthy salts were the same which were held in solution by sea water, being principally chloride of magnesium and sulphate of lime; the proportion of the earthy salts to pure chloride of sodium in sea water being greater than that which prevailed in the brine. Analyses made in 1808 of various brine springs showed that the percentage of chloride of sodium and of earthy salts in one pint of brine varied from 26·566 to 21·250 of the former, and from 2·500 to 0·625 of the latter.

The brine, on being pumped from the pits, is run into large cisterns, or into reservoirs, made sufficiently high for the brine to flow by gravitation through pipes as it is required in the evaporating pans. It is there evaporated upon one general principle. The heat for the evaporation is usually supplied from coal fires beneath, but sometimes the spare heat from a steam boiler is used after it has passed the boiler, or the discharged steam from an engine is sometimes utilised in the same way; and, occasionally, there are pipes with steam in them amongst the brine in the pans. In this way, according to the different degrees of heat applied, the product is large or small grained salt. For what is called lumped, or fine grained salt, the brine in the pan is brought to a temperature of 226° Fahrenheit, which is the boiling point for brine. Crystals soon form on the surface, and after skimming about a little they subside to the bottom. Each crystal appears granular or a little flakey, and is in the form of a small quadrangular though irregular pyramid. For common salt the temperature is 160° to 170°. The salt thus formed is close in texture and clustered together in larger or smaller pyramids according to the heat applied. For large-grained flakey salt the temperature is 130° to 140°. For large grained fishing salt the temperature is 100° to 110°, the slowness of the evaporation allowing the salt to form in large cubical crystals, although it appears they are not perfect cubes. To produce these kinds, foreign matter supposed to be of a harmless kind, such as the white of eggs, calves and cows feet, ale, flour, resin, butter, alum, &c., have long been added to the brine for clarifying and to promote crystallisation.

The earthy matter contained in the brine is got rid of in the manufacture by its adhering to the pans in the form of scale, called pan-scale, or pan-scratch. There is also the chloride of magnesium, called bittern, which remains in solution after the chloride of sodium, or common salt, is formed. This is often purposely allowed to flow away by having the floor, called the hurdles, on to which the salt is lifted from the pans, lower than the top of the pan. The pans are of various sizes, the only limitation being that they must not be too wide for a man to draw out the salt with a ladle. They are commonly made of wrought iron, three eighths of an inch in thickness, and about 50 or 60 feet in length by 24 or 25 feet in breadth, and 2 feet in depth; but some of the new pans are 140 feet by 30 feet by 2 feet.

In the early manufacture of salt, it appears that evaporation was by the heat of the sun, and the operation of the air. The brine or sea water was run into shallow pits or reservoirs, where it evaporated to a certain degree, and was afterwards completed by pouring it upon twigs, and sometimes, it is said, by pouring it upon burning wood, and collecting the salt deposited upon the ashes. Until long after historic times wood was the only fuel used, and the large consumption for this purpose seems to have been early complained of. It was not until the year 1656 that the substitution of coal at Nantwich is mentioned as a novelty. The brine springs, which flowed naturally into the valleys, would probably be used first, and when they began to fail, or become insufficient for the increasing wants, it appears they were followed down with buckets and with pumps worked by hand, water-wheel, and windmill, until now, when the only method used for pumping is the steam engine.

III. THE MARTYRDOM OF SCIENCE.*

By J. W. SLATER.

THE history of human progress presents no feature more interesting yet more commonly overlooked and mis-represented than the treatment of discoverers and inventors. That these men have, as a rule, fared ill at the hands of their species is carelessly or grudgingly admitted. But the questions by whom have they been persecuted, and what may have been the motive of their enemies, are avoided even in works where we might expect them to be carefully discussed and fully answered. Such omission may be especially charged against Sir D. Brewster. His treatise is merely a biography of certain astronomers who have been, for anything the reader learns to the contrary, incidentally and casually afflicted by their contemporaries, and it omits the most striking instances of persecution. Nay, the very term " martyrs of science " is applied quite vaguely, and is made, *e.g.*, in the work of M. Tissandier, to include three distinct classes of men. We have on the one hand personages whose love for research has cost them health and even life itself. We find physicists like Richmann, chemists like Gehlen, Mansfield, Chapman, who have been struck dead whilst engaged in some hazardous experiment. We read of naturalists like Marcgrave and the elder Wallace ; geographers, navigators, and travellers, such as Cook, La Perouse, Franklin, Livingstone ; meteorologists like Crocé-Spinelli, who in their ardour for discovery have succumbed to ungenial climates, to the attacks of savages, to hunger, tempest, or to an irrespirable atmosphere. All honour to these men, and to the noble army of which they may be taken as representatives. They have fallen in the cause of science, but they have undergone no persecution, and may hence be regarded as victims rather than martyrs.

We turn to another class : illustrious inventors and dis-coverers not a few have been clearly and decidedly perse-cuted ; hunted down by mob-violence, imprisoned, or even judicially murdered ; but these inflictions are to be traced

---

* Les Martyrs de la Science. By GASTON TISSANDIER. Paris: Maurice Dreyfous.
  Heroes of Invention and Discovery. Edinburgh: W. P. Nimmo and Co.
  The Martyrs of Science. By SIR D. BREWSTER.
  History of the Inductive Sciences. By PROF. W. WHEWELL.
  Hypatia. By the REV. C. KINGSLEY.

not to their scientific discoveries, speculations, and writings, but to their religious or political opinions. When the house of Priestley was sacked and burnt by the rabble of Birmingham, and when his very life was endangered it was not the chemist and physicist but the so-called "Jacobin" and Socinian whom Midland roughdom sought to crush. It is not, we believe, generally known that the attack on Priestley's house was headed by the town-crier, a man of the name of Sugar, who rang his bell and exclaimed :—

> " Pile up the wood higher,
> I am Sugar, the crier ;
> By my desire
> This place was set on fire !"

This man and his doggerel are only worth our notice as proof of the official countenance lent to the outrage. It is utterly incredible that a town crier would thus avowedly act as the ringleader of a mob unless sure of the connivance of his superiors.

If Campanella was put seven times to the torture, on one occasion for forty hours in succession ; if he passed twenty-seven of the best years of his life in loathsome dungeons ; if, after his release, he narrowly escaped the rage of a brutal populace, it was not as the champion of the Copernican system of astronomy, the refuter of mediæval Aristotelianism, but as a patriot who longed to deliver southern Italy from the tyranny of Spain, that he suffered. Still we may concede that like all the reformers of science he must have aroused the hatred and jealousy of many of the learned, who would doubtless use against him whatever influence they possessed.

Servetus was certainly a learned physician, and is by some ranked as one of the forerunners of Harvey. But his judicial murder by Calvin was due solely to his theological opinions. The merits of Bernard Palissy, not merely as the creator of modern fictile art, but as an able physicist, chemist, and geologist, cannot be contested. He shocked the philosophasters and sophists of his day by maintaining that fossil shells were not, as was then supposed, mere freaks of nature, but the remains of extinct animals. He dared to deny that stones were capable of growth. He pointed out the possibility of artesian wells. With an almost prophetic insight he foretold the evil consequences of the destruction of forests, and in our day not merely meteorologists and farmers, but governments find that he was in the right. But in spite of all his innovations in science and in industrial art—or rather in consequence of those very innovations—he was honoured and protected by Catherine of Medicis and Henry III. That he was at last arrested, condemned to death, and allowed to die in the Bastille, was the consequence of his firm adherence to the doctrines of the Huguenots. Had it not been for his scientific greatness he would have perished earlier.

If Lavoisier perished on the scaffold amidst the storms of the first Revolution, he merely shared the fate of his colleagues the *fermiers généraux*, none of whom were men of science. It is true that "the brutish idiot into whose hands the destinies of France had then fallen"—as Prof. Whewell justly remarks—declared that "the republic had no need of chemists." But these foolish words give us no right to assert, as a modern writer has done, that Lavoisier suffered death for his chemical ideas.

If Bailly likewise perished upon the scaffold, and if Condorcet poisoned himself to escape a similar fate, they died not as philosophers and mathematicians, but as victims of indiscriminate popular frenzy.

There are many other men whose names we are thus compelled to erase from the list of the martyrs of science— men whose inventions and discoveries have been of the highest order, but whose sufferings and death cannot be justly looked on as a consequence of their achievements.

But there still remains a third and a too numerous class: thinkers and discoverers who have been persecuted in many cases to the death, not incidentally but because of the very services they have rendered to science. Their persecutions have differed very much in nature and degree according to the age and the country in which they lived. In the dark ages it was practicable to arrest a troublesome thinker, and to put an end to his researches, or at least to their promulgation, by the straightforward means of imprisonment, torture, banishment, and even death at the stake. Hypatia of Alexandria was seized by a mob of infuriated monks, who literally tore the flesh from her bones with fragments of pots, dragged her mangled remains outside the city, and there burnt them. The Bishop Cyril, who had instigated the outrage, endeavoured to screen the malefactors from justice. Virgilius, Bishop of Salzburg, was burnt by Boniface, the Papal legate, for asserting the existence of antipodes. Cornelius Agrippa, after much persecution, died at last of actual famine. Roger Bacon, perhaps the mightiest philosopher of the middle ages, of whom it has even been said that could he revisit the earth he would shake his head at the slowness of our progress since his death, suffered bitterly. He was first prohibited from lecturing at the University of Oxford and from communicating his researches to any one. The accession of Clemens IV. to the Papal chair gave the illustrious sage a short respite, of which he availed himself to draw up three works, and to publish one of them, the *opus majus.* Scarcely was this effected when the enlightened pontiff died, and his successor was indifferent, if not formally hostile. Roger Bacon was summoned to appear at Paris before the legate Jerome of Ascoli, was convicted of heresy and witchcraft, and sentenced to imprisonment for life. His works were also condemned as impious, and all persons were forbidden to read them under pain of excommunication. It is certain that he remained ten years in a loathsome dungeon, and that his treatment, even in that rude age, was considered exceptionally harsh. Some say that he died in prison ; others, that he was at length set free at the intercession of certain powerful nobles, and ended his days in England. He is said to lie buried at Oxford. We can wish that ancient university no greater boon than that his spirit may ever rest upon its professors.

Three centuries later Rome witnessed one of the foulest murders ever committed. Giordano Bruno, for upholding the teachings of modern astronomy, and especially for maintaining the immensity of the universe and the plurality of worlds, was burnt to death in the Campo di Fiore on February 16, 1600. The words of the sentence passed upon him are significant :—" Ut quam clementissime et citra sanguinis effusionem puniretur." Not less memorable was the reply of the hero-philosopher : " You feel more fear in pronouncing this sentence than I do in receiving it !"

One of the greatest merits of Bruno is his enunciation of the doctrine that on all scientific questions the Scriptures neither possess nor claim any authority, but embody merely the opinions current at the times when they were written. This proposition, from which follows as a corollary that the Church can have no claim to pronounce on the truth or falsehood of scientific theories, was afterwards enforced at length by Galileo in his celebrated letter to the Dowager Grand Duchess Cristina of Tuscany. We cannot help regretting that he, when brought before the Inquisitors in the Convent of Minerva, did not act up to his profession by denying *in toto* the authority of the court. Had he done so his life would doubtless have been in great peril, but the enemies of science would have been deprived of much scope for sophistry. " E pur si muove " was well, but " non coram judice " would have been infinitely better. It is worthy of note that, unless we are misinformed, St. Augustine had warned the clergy against the attempt to exercise a jurisdiction over science.

As we approach modern times a change becomes manifest. Ecclesiastical bodies in the more civilised parts of Europe were deprived of civil power, and could no longer imprison, torture, or burn inventors and discoverers. But the old spirit faded away very slowly, and even in our days it still occasionally comes to light. Men of science, scientific works, and learned societies were, and still are, traduced, denounced, and held up to public hatred. Scarcely a capital step has been taken in any branch of research but it has been branded as atheistic. Dean Wren, the father of the celebrated architect, upheld the geocentric theory of the universe and the immovability of the earth in a strain worthy of Caccini or Scioppius. It was objected against the Royal Society that its members neglected the wiser and more discerning ancients and sought the guidance of their own unassisted judgments, and that by admitting among them men of all countries and religions they endangered the stability of the English Church." It was urged that experimental philosophy was likely to lead to the overthrow of Christianity and even to atheism. Among these writers a prominent place belongs to Henry Stubbs, of Warwick, and the Rev. Richard Cross, of Somerset, the latter of whom charged the Fellows of the Royal Society with "undermining the universities, destroying Protestantism, and introducing Popery!"

It would have been fortunate for Bruno, Galileo, and not a few of their colleagues, if the Inquisition and the Order of Jesus had taken the same view of the tendency of their researches. The discoveries of Sir Isaac Newton excited an outburst of hostility very similar to that which has in our own times greeted the theory of organic evolution. Then geology became the great bugbear, then followed the nebular hypothesis, till, as we have just hinted, anti-scientific jealousy concentrated itself upon the views of Darwin, Wallace, and their followers. If we read the controversial literature which has issued from the English press within the last half century, and note the motives therein imputed to men of science, we can scarcely doubt what would have been the fate of Buckland, Lyell, Sedgwick, Oken, Carus, Richard Owen, Darwin, had their enemies possessed as much power as malice. It must also be remembered that the practical applications of science, and all attempts at its extension among the public, have been met with an hostility no less pronounced. Franklin's lightning conductor, and Jenner's discovery of vaccination, have been condemned from the pulpit as impious and blasphemous attempts to set aside the decrees of heaven. A similar condemnation has since been pronounced against the use of anæsthetics, especially in midwifery.

The late Society for the Diffusion of Useful Knowledge, the London University, and the British Association, have each in turn passed through a tempest of abuse. The last mentioned body, indeed, is still regularly " preached at " in every town which it visits.

In France the Chancellor, D'Aguesseau, refused a licence to print Voltaire's " Letters on England," because the author therein expounded the discoveries of Newton, and disproved the vortex theory of Descartes. For adopting Locke's denial of innate ideas, a " lettre de cachet " was issued against Voltaire, and he was compelled to seek safety in flight. More recently the freedom of science seems to be recognised in France, Germany, and even in Italy. We must not, however, forget that the Roman Church has never formally retracted her claim to adjudicate upon scientific truth. An " Index," of proscribed books is still issued, and within the present century Pope Gregory, in an encyclical letter, characterised the freedom of the press as " deterrima illa ac nunquam satis execranda et detestabilis libertas artis literariæ."

In Britain the anti-scientific spirit still lingers more decidedly than elsewhere. Its chief lurking-places are sometimes said to be among the clergy and country gentlemen. We are not sure that this view is correct. Passing through a street in one of our northern manufacturing towns, the present writer once heard a demagogue addressing a crowd on something which he contended must be put down. That something was science! We are bound to say that his listeners gave every mark of sympathy and approval. The manner in which inventors have often been treated in different parts of England seems to show that such feelings are widely spread. The country which first wins over her working-classes to favour invention, and to become themselves inventors, will command the industrial supremacy of the world. America is fast attaining this object by her patent system, which enables even a poor man to secure his property in an invention. Our statesmen, Whig and Tory alike, can scarcely be restrained from laying additional difficulties in the way of patent right.

If we now, summing up, seek to know who have been the

chief persecutors of science, we shall find the conventional answers too narrow. Many persons have laid the chief blame upon Roman Catholicism. It is very questionable, however, whether other churches, if they had been as widely spread, and had possessed as great civil power and authority, might not have equalled or even exceeded Rome. The religious bodies of Britain, established or dissenting, have certainly been unsurpassed in the virulence of their attacks upon geology and upon the New Natural History. We strongly suspect that the Church of Rome will be the first religious body to admit that the doctrines of evolution and of the high antiquity of the human race are not necessarily opposed to the teachings of the Scriptures. So-called infidels of various grades of opinion have contended that Christianity in any and every form is the persecutor of science. We would submit, on the contrary, that discovery was persecuted in heathen and democratic Athens, where all the influence of Pericles barely sufficed to save his friend the philosopher Anaxagoras from a worse fate than banishment. Nay, we may even venture to predict that modern "freethought" will before long appear as the adversary of science, and if sufficiently powerful, as her persecutor.

The jealousy of the industrial classes we have already glanced at.

Lest we should feel tempted to ridicule the suicidal folly of the working-classes in thus seeking to repress improvement, let us remember that science is sometimes her own persecutor. Men who have gained a high official position in universities and academies are often actuated by a jealousy very similar to that which we have traced among ecclesiastics. They establish a certain scientific orthodoxy, based often to a great extent on mere conjecture and assertion, and seek to frown down and to silence the unknown outsider who calls in question one of their dogmas, or who discovers a truth which they have overlooked. That any region of research should be officially tabooed is a humiliating circumstance. The dread of truth, the jealousy of discovery, is not confined to the Holy Inquisition, and no disestablishment of Churches, no secularisation of schools and colleges, not even the suppression of every religion—were such a step possible—would put an end to its action.

IV. THE "EGYPTIAN" ARGUMENT ON THE ORIGIN OF SPECIES.

By J. W. SLATER.

AMONG the scientific results of the French Expedition to Egypt, under the first Napoleon, was a series of observations of the birds and mammals sculptured on the most ancient monuments or actually preserved in the state of mummies. On a comparison of these images and remains with their living representatives as existing in our own days, a very close resemblance, or perhaps even an absolute identity, was perceived. The sacred ibis of this nineteenth century might have served for the original of the representations of his ancestors chiselled out at least three thousand years ago. Hence the inference was drawn by the Cuvierian or official school of French naturalists, that, if the characteristic features of animals undergo no change in so long a time, species must be regarded as incapable of transmutation, and their respective differences must be absolute and primordial.

As this contention has lately been resuscitated by Dr. F. Bateman, of Norwich, it may need a brief reconsideration.

It cannot escape our notice that this argument rests upon certain assumptions which, to say the least, have never been proven. The Cuvierians take for granted that because a certain phenomenon cannot be shown to have happened within a certain interval of time, and in one particular locality, it can never have happened at all. They tacitly hold that the modification of species, if it occur at all, must proceed constantly at a uniform rate, at all times and in all places. They assume that the space of three thousand years forms a sufficient portion of the life of the organic world to decide the question. They forget that in Egypt the climate and outward circumstances generally have undergone no perceptible change within historical ages.

All these considerations have been urged by various writers with a view of showing that the evidence of the old Egyptian monuments cannot be taken as decisive.

But there is a further objection which, so far as I am aware, has not yet been raised, and which to me appears utterly fatal. If we take a further glance at the oldest

Egyptian monuments we find there pourtrayed men of different races, or, if the term is preferred, varieties, possessing the very same distinctive characteristics which we observe in their descendants in our own times. The negro of to-day is the exact counterpart of the negro figured in the most ancient of these sculptured records.

If, then, permanence of type for the term of let us say three thousand years proves that certain animal species are immutable and primordially distinct, surely the same conclusion must be extended to the varieties of the human race, and, *e.g.*, the Negro, the Arab, and the Jew must be traced each to a distinct and independent origin. But if M. Flourens and Dr. Bateman consider that these well-marked —and, as we have just seen, permanent—forms have been produced from one common stock, it is hard to see how they can consistently regard a similar permanence in other animal forms a proof of radical diversity and of unalterable fixity.

V. PROGRESS IN MICROSCOPICAL SCIENCE.*

By W. T. Suffolk, F.R.M.S.

THE Fellows of the Royal Microscopical Society are certainly possessed of a "Journal of which any learned body might be proud : the first volume, issued at the end of 1878, merely professed to be a Journal of the Society," with other microscopical information. In the present volume a most important addition has been made. No less than two hundred and forty-eight English and foreign scientific periodicals are regularly examined, and their contents —so far as they come within the scope of the Royal Microscopical Society's work—indexed under the author's names. The value of this arrangement to students scarcely needs comment : the difficulty of knowing what has been written and where to find it is one of the rocks ahead to all those

* Journal of the Royal Microscopical Society, containing its Transactions and Proceedings, and a Record of Current Researches relating to Invertebrata, Cryptogamia, Microscopy, &c. Edited, under the direction of the Publication Committee, by Frank Crisp, LL.B., B.A., F.L.S., one of the Secretaries of the Society; assisted by T. J. Parker, B.Sc., A. W. Bennett, M.A., B.Sc., and F. J. Bell, B.A., Fellows of the Society. Vol. ii., 1010 pp. London : Williams and Norgate. 1879.

engaged in any kind of scientific research. The " record of current researches " has also been greatly extended, and the contents of the whole volume rendered most acceptable by copious indexing, which takes the following forms :—A carefully arranged table of contents occupying eighteen pages, lists of plates and woodcuts, and a very full alphabetical index, in which the repetition of subjects under different letters has not been spared. The energetic Secretary of the Society, finding the work of conducting such a publication beyond the power of any single individual, has called to his aid the three gentlemen mentioned in the title, whose united efforts have made the work an almost perfect record of microscopical and biological research.

The Royal Microscopical Society, desirous of making others partakers of its good things, has elected the presidents of eighty-six kindred societies, at home and abroad, *ex officio* Fellows : these, with the numerous Honorary Fellows, place the Society in communication with observers in most places throughout the world.

A careful series of experiments, to determine the thermal death-point of known Monad germs when the heat is endured in a fluid, have been carried out by the Rev. W. H. Dallinger, F.R.M.S., so well known for his researches in the life-history of Monads, in conjunction with Dr. John James Drydale, of Liverpool. The processes employed are necessarily elaborate, and need special appliances for the purpose of microscopical observation, the whole of which are described in detail in the paper communicated to the Royal Microscopical Society.* As might be expected the adult organisms were less capable of enduring a rise of temperature than the germs, a heat of from 140° to 142° F. being always fatal. With respect to the germs, development took place—after an exposure to 265°—about three and a half hours after the heating process: a temperature above 267°, however, proved fatal to the germ. The heat endured dry was much greater, a spore germinating after exposure to a temperature of 250°, a temperature of 212° in fluid being fatal.

A paper by Mr. J. W. Groves, on the " Preparation of Stained Sections of Animal Tissues," appears in the " Transactions of the Quekett Microscopical Club " (vol. v., p. 231). The author remarks that although the staining of tissues by the more simple dyes is very easy, yet of the large number prepared scarcely any are ever so stained as to show

---

* Journal R. M. S., vol. iii., p. 1. January, 1880.

the structures to the best advantage; indeed many of them are utterly useless. The subject is treated of under the heads of " Material—Methods of Preserving and Hardening it," " Cutting of the Sections," " Tingeing Agents," " Method of Staining." The paper is valuable on account of the large number of formulæ given, and the numerous practical hints upon points where the experimenter is likely to meet with difficulties or take the wrong way of operating. The whole concludes with the following summary :—

" 1. Let the material be quite fresh.

2 *a*. Take care that the hardening or softening fluid is not too strong.

b. Use a large bulk of fluid in proportion to the material.

c. Change the fluid frequently.

d. If freezing be employed take care that the specimen is thoroughly frozen.

3 *a*. Always use a sharp razor.

b. Take it with one diagonal sweep through the material.

c. Make the sections as thin as possible, and—

d. Remove each one as soon as cut, for if the sections accumulate on the knife or razor they are sure to get torn.

4 *a*. Do not be in a hurry to stain, but—

b. Remember that a weak colouring solution permeates the section better, and produces the best results ; and—

c. That the thinner the section is the better it will take the stains.

5 *a*. Always use glass slips and covers free from scratches and bubbles, and chemically clean.

b. Never use any but extra thin circular covers, so that the specimens may be used with high powers.

c. Always use cold preservatives, except in the case of glycerine jelly, and never use warmth to hasten the drying of balsam or dammars, but run a ring of cement round the cover.

6. Label the specimens correctly, and keep them on the flat and in the dark."

Herr Zeiss, of Jena, has produced a new low-power objective, which, by altering the distance of its combinations by means of a rotating collar, gives a range of magnifying power varying from that of a glass of 4 inches focus to one of 2 inches. This means of varying the power at pleasure is valuable in dissection, especially when used with an erecting binocular microscope.

VI. OFFENSIVE AND DEFENSIVE SUB-MARINE WAR.*

By Capt. S. P. Oliver, (late) R.A.

Part I.

IN quaint old " Pepys's Diary," under date of March 14, 1662, we find the following entry :—" Home to dinner. In the afternoon come the German Dr. Knuffler to discourse with us about his engine to blow up ships. We doubted not the matter of fact, it being tried in Cromwell's time, but the safety of carrying them† (*sic*) in ships; but he do tell us, that when he comes to tell the King his secret, (for none but the Kings, successively, and their heirs must know it,) it will appear to be of no danger at all. We concluded nothing : but shall discourse with the Duke of York to-morrow about it." And again, in 1663, as follows :—" Novbr. 11th. At noon to the Coffee-house where with Dr. Allen some good discourse about physick and chemistry. And among other things I telling him what Dribble the German Doctor do offer of an instrument to sink ships; he tells me that which is more strange, that something made of gold, which they call in chymistry ' *Aurum Fulminans*,' a grain, I think he said, of it put into a silver spoon and fired, will give a blow like a musquett, and strike a hole through the silver spoon downward, without the least force upward ; and this he can make a cheaper experiment of, he says, with iron prepared."—(" Memoirs of Samuel Pepys, Esq., F.R.S." Edited by Lord Braybrooke.)

The doctors " *Knuffler* " and " *Dribble*," probably partners if not identical, do not seem to have been more successful

---

* Torpedoes and Torpedo Warfare, by C. W. Sleeman, Esq., late Lieut. R.N., and late Commander Imperial Ottoman Navy. Portsmouth : Griffin and Co. 1880.

† " *Them*," *i.e.*, the said engines, which were evidently buoyant torpedoes.

than Lord Dundonald was, nearly two centuries later, in introducing their infernal machine to the notice of the Admiralty under the Duke of York, whose secretary Pepys was, for we hear no more mention of this important secret which could only be communicated to royalty and heirs apparent. It is fortunate indeed that Mr. Whitehead has not made this condition a portion of his compact, preferring a royalty in hard cash for his non-exclusive secret; for otherwise we should be obliged to send our young princes through a course of torpedo instruction at Fiume, preparatory to their having command of the *Vernon** and *Hecla*,† *et hoc genus omne*, at Portsmouth.

The above notice of the contemplated use of torpedoes in wars is not alluded to in Commander Sleeman's lately-published manual, which, however, mentions the attempt of Lambelli to destroy a bridge across the Scheldt, in the neighbourhood of Antwerp, during the siege of that city in 1585. But all of the foregoing were most probably drifting and floating explodable magazines, not differing in principle much from fire-ships, whereas the essence of torpedo war is subaqueous as opposed to subaërial explosion: it is therefore not until 1775 that we find the present system of submarine warfare dates its inception.

To Capt. Bushnell, of Connecticut, belongs the credit of first inventing the method of igniting charges of gunpowder under water, and he also first devised a submarine boat, with which, however, he failed to destroy the hostile British cruisers.

Next in order of succession come the submarine bombs and carcasses of Robert Fulton, which were tried and rejected by Napoleon; and although encouraged by Mr. Pitt, "England at that time being mistress of the seas, it was clearly her interest to make the world believe that Fulton's schemes were impracticable and absurd." Earl St. Vincent, in a conversation with Fulton, told him in very strong language "that Pitt was a fool for encouraging a mode of warfare which, if successful, would wrest the trident from those who then claimed to bear it, as the sceptre of supremacy on the ocean." Last of all, Fulton submitted to the American Congress his elaborate schemes for rendering American harbours impervious to British attacks. These schemes included drifting, harpoon, and spar torpedoes, together with block-ships, stationary mines, and cable-

* H.M.S. *Vernon*. The Torpedo School-Ship in Portsmouth Harbour.
† H.M.S. *Hecla*. Torpedo ship, now experimenting at Spithead.

cutters; but owing to the unfavourable report of Commodore Rodgers, who Lieut. Sleeman characterises as unfair and prejudiced, Fulton's submarine inventions were not adopted, and the skilled inventor turned his ingenuity to the more humanitarian improvement of the steam-engine.

In fact as long as clockwork, originated by springs, weights, levers, &c., was employed, torpedo science was practically a failure; and until after the introduction of voltaic electricity submarine batteries were regarded merely as futile experimental toys. In 1839 Colonel Pasley, R.E., used galvanic electricity to blow up the wreck of the *Royal George*, and Colonel Colt, whose name will ever be connected with the revolver,* first publicly essayed to explode a case of gunpowder by electricity at a distance, in 1842, and he accomplished the destruction of a vessel under weigh by an insulated electric cable in 1844. Two years subsequently the two great explosive compounds of modern times, viz., *gun-cotton* and *nitro-glycerin*, were discovered; and in the next war that took place stationary buoyant mines under water, fired on contact or observation by means of electricity, were used for defensive purposes by the Russians on a large scale, both in the Baltic and Black Sea, thereby rendering inactive all naval operations on the part of the British squadrons, as much by their moral as by their physical effects. During the Baltic campaign the Russian ports were rendered impregnable against naval attack, if we except distant bombardment, such as Sveaborg underwent, and the capture of Bomarsund, which was practically undefended.

The explosions from which H.M. ships *Merlin* and *Firefly* narrowly escaped deterred the near approach of Admiral Napier's fleet, and his famous signal of "*Lads sharpen your cutlasses*" was thereby rendered a bye-word of bombast.

The breaking out of the Civil War in America brought the science of torpedo warfare more prominently forward, and, as might have been expected, was first used for defence by the Confederates on the Savannah River, causing thereby great delay in the advance of the Federal gun-boats; and in December, 1862, the iron-clad ship *Cairo* was shattered and sunk in twelve minutes by two stationary torpedoes in the Yazoo River—the first instance in history of a vessel of war being thus destroyed in actual war. Subsequently several instances are cited in Commander Sleeman's work

---

* Colt, when a mere lad, commenced experiments with submarine mines as early as 1829.

of destruction and damage caused by the Confederate torpedoes, but almost all by defensive mines, whereas nearly
all the offensive attacks by means of torpedo-boats failed.

Thus there was a total of thirteen iron-clads, monitors,
and gun-boats,* besides the aforenamed *Cairo*, all belonging
to the Federals, which were sunk by stationary *defensive*
submarine mines, without counting others† severely damaged. The only successful *offensive* torpedo attack was
fatal to both sides, the Confederate submarine boat being
sunk, together with the Federal antagonist, by running into
the hole caused by the explosion of her own torpedo.

> " 'T is sport to have the engineer
> Hoist with his own petard."
>
> (*Hamlet*, Act iii., Scene 4.)

As our immortal Shakspeare has it.

On the other hand, the failures of offensive attack were
conspicuous,‡ and three Confederate vessels were destroyed
by their own mines, owing to the shifting of the position of
the barrel torpedoes.

The only Federal torpedo success during the war was that
of a boat armed with the Wood and Lay disconnecting-spar
torpedo, by which the Confederate iron-clad *Albemarle* was
sunk.

Later the *Paraguayans*, in 1866, completely destroyed the
Brazilian war-steamer *Rio Janeiro*, by a stationary torpedo,
one of the Brazilian fleet which was bombarding Curru-
paity. This was the end of the second epoch of active
warfare in which torpedoes had been engaged, and an interval of over ten years elapsed before submarine warfare was
again employed by belligerents; but this time was not idly
employed by the submarine engineer, for in 1864 Captain
Lupuis, an Austrian, in association with Mr. Robert Whitehead, the Superintendent of Iron-works at Fiume, constructed
the "fish torpedo," with which such highly successful experiments were made, that various nations were justified in
purchasing the patent on certain conditions advantageous
to the patentee; the English Government paying £17,500
for the secret, on the recommendation of a committee of

---

* Sunk :—Iron-clad, *Cairo*; gunboat, *Baron de Kalb*; transport, *Maple
Leaf*; iron ship, *Commodore Jones*; monitor, *Tecumseh*; steamer, *Otsego*;
steamer, *Bazeby*; monitor, *Patapsco*; steamer, *Harvest Moon*; and two monitors and three gunboats unnamed.

† Severely damaged :—Monitor, *Montauk*; gunboat, *Commodore Barney*.

‡ Boat-torpedo attacks failed :—Ship, *Ironsides*; ship, *Memphis*; ship, *Min-
nesota*; frigate, *Wabash*.

officers, who rightly advised the Government not to purchase the *exclusive* secret, for which Mr. Whitehead demanded £54,000, an incident unnoticed by Commander Sleeman, who also passes over in silence the following facts which certainly pertain to the history of this subject. For torpedoes were, alas! not to be confined to civilised belligerents; and as there are in all times and ages uncrupulous individuals who do not hesitate to make war on mankind, so the commercial world was startled by a communication, published in June, 1873, by Mr. H. F. Hemming, Consul for Venezuela, stating that small torpedoes, made to look like lumps of coal, had been actually despatched from France to a Venezuelan port, not named, where a steamer was to be laden with goods of little value, but heavily insured, and this vessel would be sent to sea, having these coal torpedoes on board, in the expectancy that she would be lost and the speculators gain a large sum. A few days later the "Times" published a letter from "Warhawk" on the same subject, and the French Minister of Marine issued a warning circular on the same subject.* Fortunately this timely information put the insurance agents on the *qui vive*, and no explosions or total loss of insured vessels could be traced to this cause. But it does not follow that none have taken place since the scare has been forgotten, and thus it is not out of place to remind the public of such hidden dangers; and the underwriters at Lloyd's should remember that "*Magna pericla latent*" is the apposite motto adopted by the torpedo school-ship, H.M.S. *Vernon*. These *coal torpedoes* or *coal-shells*, models of which were actually exhibited in London (?), were made of gun-metal, and probably contained dynamite with fulminate of mercury so arranged that they would explode in a certain fixed time after being thrown into a steamer's furnace, which would give time for the operator to escape, or, if placed in the coal bunkers, would find their way in due course of time to the fires during the voyage: thus the expenditure incurred in fitting out and loading the insured vessel would be not only recouped, but a handsome sum in insurance gained besides.

In 1875 another species of torpedo was invented, and, like the coal-torpedo above, it was rather sub-aërial than submarine. This was the infernal machine employed by Thomas, and the premature explosion of which, at Bremerhaven, in 1875, caused such destruction,† although its results were

---

* See "Saturday Review," December 25, 1875, p. 809, from which we extract most of these particulars.

† One hundred and twenty-eight persons killed, besides the wounded, about sixty in number.

inferior to what was expected of it by its suicidal inventor, who, fortunately for humanity, survived his self-inflicted wound long enough to confess his futile machination. This machine of Thomas was not unique, for similar engines were in existence, indeed a *model* of one was obtained in Paris a month before by the gentleman whose knowledge of coal-shells was so intimate, and who by this time has probably perfected with greater accuracy such inartistic inventions. The dynamite in this apparatus was to have been exploded by the concussion of a hammer 30 lbs. weight, set, like an alarum, to fall after a given period--a devilish alarum indeed.

We now come to the latest epoch of torpedo warfare, viz., that inaugurated by that nearly fatal *fiasco*, when H.M.S. *Shah* aimed the first Whitehead torpedo which had ever been fired in earnest against the Peruvian iron-clad *Huascar*, since captured by the Chilians. The attempt was a failure, owing to the latter vessel (fortunately for both parties) altering her course at the instant of discharge. This occurred in May, 1877, and on December 20th, of the same year, " the Russians made an attack with Whitehead torpedoes on an Ottoman squadron lying in the harbour of Batoum ; but owing to a want of practical knowledge of the manipulation of such weapons, no vessels were sunk or damaged, but two fish torpedoes—one in perfect condition— were found the next morning high and dry, on the beach at that place." Thus the Turkish Government managed to get the Whitehead secret and torpedo without payment.*

The third and last, and only occasion on which a fish tor-pedo has been fired successfully against an enemy's ship in actual warfare was on January 26th, 1878, when the Russian steamer *Constantine* fired a Whitehead against a Turkish guard-ship, and completely destroyed her.

From the foregoing we may safely conclude that at pre-sent offensive torpedo warfare is most uncertain, the odds being in favour of the defence. But at the same time offensive torpedo operations have hardly yet had their fair trial ; the art of torpedo attack is still in its infancy, but its age of puberty is not far distant, and we see the tendency in the numerous newly-invented torpedo-boats, which are ever on the increase, in size, in power, and in speed.

Owing to Commander Sleeman's absence from England, in China and Japan, where he is attempting to induce the Orientals to purchase Capt. McEvoy's lately improved duplex spar-torpedoes, and other patented auxiliaries of

* A lost Whitehead torpedo is advertised for by Capt. Singer, of H.M.S. *Hecla*, since February 5th ; an offer of £5 being made for its recovery.

destruction, he has only brought down the history of torpedo-boats to the American inventions of Messrs. Herreshoff, of Rhode Island, U.S.A., whose novelty consisted chiefly in its coil boiler, by which a good working pressure can be obtained within a few minutes of lighting the fire, whilst the fire and the boiler can be blown off in a few seconds. But this advantage has been comparatively nullified by the fact that the Thorneycroft and Yarrow boats can be supplied with steam from the main boilers of the ships to which they are attached. Thus in a late experiment at Stokes Bay a Thorneycroft (No. 56) was supplied with steam from the *Lightning*, beginning to move in 9 minutes 35 seconds, with a pressure of 60 lbs., whilst the *Herreshoff* got under weigh in 8 minutes 2 seconds, with a pressure of 90 lbs.; but after trial on the measured mile the former attained a speed of 14·168 knots, and the latter only 12·972 knots. It was at this trial that the *Herreshoff* boat ran into the *Manly* tug, her sharp nozzle going completely through her side, happily above the water-line; and although the *Herreshoff*'s bow was stove in, owing to her *five* water-tight bulkheads, she floated easily, the breaking of her glass steam-guage being the only damage done to the machinery by the concussion.

Messrs. Yarrow are now building, for the French Government, boats to go backwards and forwards at the rate of 36 knots per two hours,—*i.e.*, an average of 18 knots either way, what is lost in going ahead being gained in going astern. It will be interesting when this guaranteed rate is attained.

The same builders are also building a larger class of torpedo vessels which will supersede the *Lightning* class (first class), measuring 84 feet in length, with a beam of 7 feet 6 inches. The new vessels are to be constructed of a more powerful type, viz., 100 feet in length, with beam of 12 feet 6 inches, and they are to carry enough coal to enable them to steam 1000 miles.

Some boats have also been tried, made by Messrs. Maudslay, Sons, and Field, of plates of manganese bronze, whose thickness does not exceed from one-eighth to one-sixteenth of an inch : they are not quite so stiff as steel plates of the same thickness, and vibrate when the pulsations caused by the propeller are isochronous with those produced by the spring of the boat going at from 10 to 12 knots. This vibratory quivering, however, disappears when a rate of 16 knots is attained.

It is greatly to be hoped that these inventors, in attempting to obtain various improvements in speed and strength of material, combined with capability of rapid manœuvring, for war purposes, will also serve the ends of peace by promoting novel contrivances adapted to more humanitarian progress.

As Robert Fulton had ultimately to turn his ingenious mechanism from the infernal machine towards the improvement of steam-ships, so we shall soon see all the new improvements of torpedo-boats carried out in more efficiently built vessels for the trade and passenger service of maritime nations. So if the ends justify the means, even extreme "*jingoism*" will have had its share in hastening the progress of modern engineering science.

Since writing the above the news has reached England of the terrible attempt by revolutionary conspirators to destroy the Emperor of Russia in his Winter Palace, on the evening of the 17th of February. General Gourko, who has been charged with the special investigation of this incident, officially reports that, from an examination of the area of destruction and the appearance of the demolition and *débris*, the explosion must have been caused by several pounds of dynamite, ignited in all probability by electricity. The charge appears to have been placed in the cellars of the basement where fuel is stored for the heating apparatus, and this would seem as if the dynamite had been concealed in coal-torpedoes, such as those mentioned in the preceding pages. Although it is thought by some that the mine was sprung prematurely, in accordance with a designed purpose of terrifying the Czar, it seems more probable that an improperly adjusted time-fuze was in reality the cause of the explosion anticipating the instant calculated by the conspirators. It appears that ten lives were sacrificed on the spot, whilst forty-five were wounded, of whom several have since died from the effects, making a total of fifty-five victims on this occasion. It is also suggested that a detonator may have been set in motion by machinery similar to that of the Thomas infernal machine ; anyhow the significance of this occurrence is great, and warns us of the presence of secret means of destruction whose origin and design we can only guess at, and in respect to which our knowledge and consequent legislation are alike at fault. It may also serve to remind us of what is yet a very moot question, viz., whether in sanctioning the manufacture of *secret* destructive weapons —nay, in actually encouraging their invention and improvement for purposes of legitimate warfare—we are not thereby unconsciously incurring a certain responsibility as to the use of these same machines by illegitimate belligerents engaged in internecine war. Lord Vincent's warning at the commencement of the present century is not without point at the present day.

No one can deny that in our time there is an absurd anomaly, which would be ludicrous were it not so terrible, in that civilised nations have, by an international convention, stigmatised the use of an explosive bullet, however openly it may be used (say for the purpose of exploding an ammunition or limber box), because it may mangle or injure a single individual without killing him ; and yet these same nations do not hesitate to introduce into their service as an accredited weapon, whose machinery is secret (or at all events presupposed so), a torpedo whose immediate purpose is to utterly annihilate a man-of-war and her crew of nigh a thousand souls. Submarine mines are defensible on all grounds ; your antagonist runs his head against them at his own risk ; but the offensive torpedo is in the same category as the explosive bullet over which it takes higher rank.

ON WATER AND AIR.*

By John Tyndall, D.C.L., LL.D., F.R.S.,

Professor of Natural Philosophy at the Royal Institution
of Great Britain.

Lecture III.

WANT you to be able to associate with every term
that I employ in these lectures a perfectly definite
meaning. If I speak of things in science, I wish you
to be able to present before your minds an image of
what I am speaking of. Now, there are various terms that
I am afraid I must employ, and therefore it is my duty to
try and make clear to you exactly what I mean by those
terms.

After a vast amount of experiment and of thought, philosophers have come to the conclusion that all bodies are composed of what we call "smallest parts." An ancient philosopher and poet used to call them "first beginnings." I
say that all bodies are composed of these exceedingly small
parts—the smallest into which a body can be divided. They
are beyond the range of all microscopes, but still we see
them with the mind's eye, though they cannot be seen with
the eye of the body. These parts are called "atoms." I
shall have immediately in this jar a collection of atoms of
oxygen gas, which I will pour into the jar from this bottle.
The atoms will fill the jar. Mr. Cottrell will open the bottle
so as to allow a quantity of the atoms to issue. The jar is,
as you see, placed upright, and the gas being slightly
heavier than air falls into it, and it is rapidly filled. I will
now make it evident to you that I have in the jar the thing
which we call oxygen gas. [The jar was filled with oxygen
as indicated. A piece of wood was then ignited and plunged
into the gas, with the usual result of a strongly intensified
combustion.]

If I dip this glowing splinter of wood into the jar, you see
that it immediately bursts into flame. The oxygen is
capable of producing this combustion.

* Being a Course of Six Lectures adapted to a Juvenile Auditory, delivered
at the Royal Institution of Great Britain, Christmas, 1879. Specially reported for " The Journal of Science."

I will now take another jar containing a bundle of atoms
of another gas (if I may use the term). This gas is
hydrogen. In the case of hydrogen, I take the jar which is
to receive the gas and hold it upside down ; because the
gas, being very much lighter than air, will rise into the jar,
like a light body in water, and rapidly fill it.

The gas is now entering the jar, and I trust that we shall
displace the air in this manner. [The jar having been filled,
the hydrogen was ignited.] If I apply a light you will see
that I have gathered a combustible gas which burns with
the flame which you see before you. If you were close to
this jar you would see that it is covered with moisture.
What was really done when this flame was produced ? In
point of fact, the hydrogen burnt in the oxygen of the air.
And what was the result of the union of the oxygen with the
hydrogen ? The result of their union was the production of
our familiar substance, water ; so that water is composed of
these two substances—oxygen and hydrogen, combined
together in definite proportions, as chemists say. The
atoms of hydrogen and the atoms of oxygen coalesce, and
they form little groups of atoms. And here I shall have to
use another term of which you ought to know the meaning.
The manner in which these things combine is this: two
hydrogen atoms single out one oxygen atom, and unite with
it, so that the smallest particle of water is composed of a
group of three atoms, two of which are hydrogen and one of
which is oxygen. This group of atoms has a different name
from the atoms themselves. It is called a "molecule."
Hence we have two "atoms" of hydrogen and one "atom"
of oxygen, and we have one "molecule" of water produced
by their union. We can take water and produce from it
these gases very easily indeed. The ordinary way of producing
this hydrogen gas is to pour a little dilute sulphuric acid
upon zinc. Here are fragments of zinc, and here is sulphuric
acid diluted with water. The zinc has, as chemists say, a
power of attraction which it exerts upon oxygen, and when
this liquid is poured upon zinc, the zinc seizes upon the
oxggen and allows the hydrogen to go free. I will now pour
into this globular vessel, containing fragments of zinc
(Fig. 16), some dilute sulphuric acid, and you see we get a
stream of hydrogen gas issuing from the liquid.

By applying a light to the end of the jet attached to the
apparatus the issuing gas can be ignited, and it burns with
a pale, lambent flame.

In this case we use granulated zinc; that is, zinc broken
up into small pieces, for the purpose of enabling the acid

more easily to act upon the metal. Other metals will act in the same way. For instance, here I have a curious metal which was discovered in this place by the famous Sir Humphry Davy. It is a metal called potassium, which is

Fig. 16.

extracted from ordinary potash, and the attraction of this metal for oxygen is so powerful that, no doubt, we shall be able to burn the metal even upon ice : for ice, as you know, is solid water. If we place a little piece of this metal upon the ice, you see that it actually burns upon the ice. How is it that it burns? It burns because it seizes upon the oxygen of the water and allows the hydrogen to go free ; and that purple flame which is produced is due to the liberated hydrogen which has been ignited by the heat produced by the chemical union between the potassium and the oxygen.

Thereis another thing with which I should like, at all events, to begin to familiarise your minds, and that is the nature of what we call heat. You saw it developed here by this combination. It is a notion universally accepted at the present day that this thing which we call heat is a brisk motion or vibration—a quivering, so to say—a tremor of the

atoms and molecules of bodies. I have explained to you
already what I mean by atoms and molecules. As I said,
this opinion is universally entertained at the present day;
but it is curious to observe, in the course of history, how
this opinion has developed itself, and how early the minds
of sagacious, penetrating men saw and seized upon this idea
that heat was what we call atomic or molecular motion.
At the time of Sir Isaac Newton there were men who seized
this idea that heat was a motion; but there was, in particular, one man of the greatest genius, named Robert
Hooke, who endeavoured to give his mind a definite image
to rest upon with regard to what we call fluidity—that condition which we get when this substance, ice, is converted
into liquid water. Hooke's image is so quaint and so pene-
trative that I think I will read his account of his conception
in his own words. He says:—" First, what is the cause of
fluidness ? This I conceive to be nothing else but a certain
pulse or shake of heat: for heat being nothing else but a
very brisk and vehement agitation of the parts of a body (as
I have elsewhere made probable), the parts of a body are
thereby made so loose from one another, thát they easily
move any way and become fluid." So with the particles of
our ice when they are converted into water. He continues:
" That I may explain this a little by a gross similitude,
let us suppose a dish of sand set upon some body that is
very much agitated and shaken with some quick and strong
vibrating motion, as on a millstone turned round upon the
under-stone very violently whilst it is empty, or on a very
stiff drum-head which is vehemently or very nimbly beaten
with the drumsticks. By this means the sand in the dish,
which before lay like a dull and inactive body, becomes a
perfect fluid, and ye can no sooner make a hole in it with
your finger but it is immediately filled up again, and the
upper surface of it levell'd. Nor can you bury a light body,
as a piece of cork, under it, but it presently emerges or
swims as 'twere on the top; nor can you lay a heavier on
the top of it, as a piece of lead, but it is immediately buried
in sand, and (as 'twere) sinks to the bottom. Nor can you
make a hole in the side of the dish but the sand shall run
out of it to a level. Not an obvious property of a fluid body,
as such, but this does imitate; and all this meerly caused
by the vehement agitation of the conteining vessel, for, by
this means, each sand becomes to have a vibrating or
dancing motion, so as no other heavier body can rest on it
unless sustain'd by some other on either side: nor will it
suffer any body to be beneath it, unless it be a heavier than

itself." You see the image before his mind; and, in teaching the young, I always like to keep before them certain definite, clear images that will answer to the terms which we employ. Hooke's image of the vibrating sand particles may, perhaps, help you to conceive of the vibration of the particles of water which takes place when ice is converted into a liquid. Well, Hooke's image of fluidity answers very well for our conception of the mere liquid state. But now suppose that you have a liquid such as I have here, with a free surface—with those molecules and atoms vibrating in this way. I have now to appeal to the eye of the mind. I can not show these things to the eyes of your body, but I am sure that every boy here present who will give me his attention will be able to see mentally what I mean. Just as in the case of the Whirlpool Rapids which I showed you in the last lecture, where two waves coalesced and tossed the crest of the wave high into the air, so you can imagine the vibrations sometimes so coalescing at the free surface of the liquid that they shall toss the molecules of the water into the air. If you have seized this image, you have now what we call evaporation. The particles of liquid are loosened by the vibrations produced by heat, and they are incessantly jerked into the atmosphere above the liquid; and therein consists all evaporation which will reduce the whole of the liquid to vapour if you only allow it sufficient time. This, then, is our conception of evaporation.

Now I approach the subject with which we started—the conversion of the waters of the ocean into vapour, that vapour from which all our rains and rivers are derived. You may say to me, "The water of the ocean is salt, and the water coming from the clouds is perfectly sweet and without salt. How can you get the sweet water from the salt water?" Quite easily, I answer. There, for instance, is a flask (F, Fig. 17) containing a very strong brine—so strong that after boiling it for a time, by means of a gas-flame, a large quantity of the salt is rendered solid in the bottom of the vessel. We kept that boiling yesterday for a considerable time. We led the steam from that flask through the worm-pipe, placed in the glass vessel C, that you see there. We have surrounded the worm with cold water, kept cold by ice. The steam passes over from that flask of salt water, and, after passing through the cool worm-pipe, it is received in the form of water in this flask, F'. There is the fresh water extracted from the brine. This water is perfectly sweet—sweet as any spring-water. This shows you that in the process of distillation, whether it is artificial or is pro-

Fig. 17.

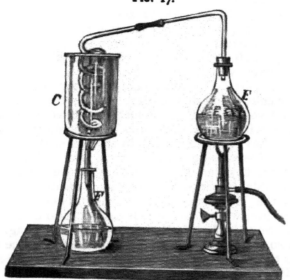

Fig. 18.

duced by the sun, you liberate not the molecules of the salt, but the molecules of the water only, and thus from the salt ocean you get sweet water.

You must, at the same time, have perfectly definite ideas of the meaning of vapour. Here is a boiler, B (Fig. 18), and beneath it we will now start a flame so as to make the boiling of the water go on vigorously; and there you have, issuing from the jet J,—what? Will you call it steam, or will you call it vapour? If you called that cloud either steam or vapour you would be in error. That cloud that you see issuing from the boiler is not vapour. It is composed of fine particles of water. It is, as I have elsewhere expressed it, a kind of dust of water,—particles of water finely divided and precipitated, so as to produce this appearance. But if I take a flame and bring it underneath the cloud, at a short distance from the jet, the cloud entirely disappears. I can cut it off in this way, at the very nozzle of the tap, and no cloud whatever is seen. Now that invisible thing into which I have converted the visible cloud is true vapour. The vapour of water is as invisible as the air that you breathe. This cloud that you see is the vapour of water precipitated so as actually to form water; and in that way our clouds are produced.

Now I want your patience to accompany me while I make a thorough investigation of the manner in which this vapour is raised from the ocean—an investigation of the agent which is instrumental in producing this vapour.

Before I do that, however, here is an experiment which my friend Prof. Dewar has arranged for me, and which enables me to follow up the idea which I threw out before you—that heat consists of this intense vibratory motion to which I have referred. I mentioned in our last lecture the name of Sir William Grove,—Mr. Justice Grove,—and I told you how he had operated upon boiling water. He made a series of extremely interesting experiments many years ago upon the action of heat upon water. He placed pieces of intensely incandescent platinum, or platinum wire, in water, and he found that the atomic vibrations of which the intense heat consisted were so violent that they actually shook asunder the molecules of water, and reduced them to their constituent atoms of oxygen and nitrogen. Here we have an apparatus which will show you this action (Fig. 19). At the present moment Prof. Dewar is going to ignite, by means of electricity, the spiral of platinum wire contained in the glass tube s, by attaching a battery to its terminal wires. Observe what is going on. The water is boiling vigorously

in the flask F, and steam is passing over this incandescent
spiral.   The mind of every boy here present must figure the
atoms of the spiral as vibrating in the most intense manner
and coming into collision with the molecules of steam which
are passing over the spiral.   You must imagine that the
vibrations of the incandescent spiral are shaking those molecules of water asunder, and are converting the water into its

FIG. 19.

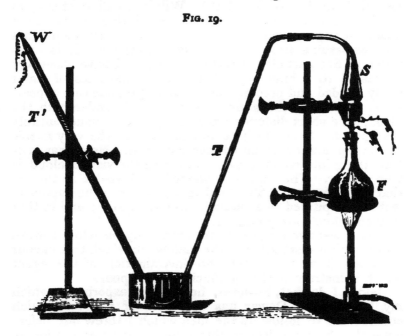

constituent atoms of oxygen and hydrogen.   You see those
bubbles that are being carried down the tube T, and are
rising in the tube T' and collecting in the upper part of the
tube.   These bubbles that you see rising are not bubbles of
steam, and they are not bubbles of air, but they are bubbles
of the mixed gases, oxygen and hydrogen, which are the
constituents of water.   Here the gases are collecting, and
if we allow the operation to go on for a minute or so we
shall obtain a certain amount of gas at the upper part of the
tube.   We can cause the two mixed gases to combine immediately, simply by sending an electric spark between the two
points of platinum which are sealed into the top of the tube,
and you see the bright flash of light produced at the moment
of their combination.   Thus, by this simple experiment, we
obtain, mechanically speaking, a shaking asunder of the

molecules of water and a reduction of the water to its constituent parts. I could decompose water and reduce it to
its constituent elements in another way,—by electricity,—
but in this case it is simply done by heat.

Now I want to show you the decomposition of water in
another way. I have here an apparatus (Fig. 20) by means
of which I can send an electric current through the water,
and cause it to be resolved into its constituents oxygen and
hydrogen. In the glass vessel D are two platinum plates,
which are completely surrounded by water. The upper part
of the glass vessel is contracted into a narrow tube, s, which

FIG. 20.

is bent over, and the end immersed in a solution of soap-
suds contained in a basin. Here is one end, or, as we call
it, one pole of a voltaic battery; and here is the other pole.
If we now connect this battery to the platinum plates contained in the vessel D, we immediately obtain a stream of
gas rising from the plates, and passing over through the tube
s, collect in bubbles on the surface of the solution of soap-
suds.

We will take some of these bubbles on the hand, and see
whether they do not contain a mixture of oxygen and hydrogen. After a short time all the air that was in the apparatus in the first instance is entirely expelled, and we have
nothing but the mixed gases issuing from the apparatus. [A
few bubbles of the mixed gases having been collected on the
solution of soap, a light was applied to them, and they exploded with a loud detonation.] You may take these bubbles
on your hand and explode them without any evil effect
whatever. But you have just heard the noise which accompanies the combination of the oxygen and the hydrogen.

We will now pass on to the consideration of other things. I trust that you have a clear conception of what is meant by evaporation and by vapour. I want to proceed to the investigation of the particular agent which produces the aqueous vapour of our air, and which produces all our rains, snow, and hail. Of course I need not tell you that the agent that raises the vapour from the surface of the sea is the sun. The sun pours down his heat upon the sea and warms the water, and this water rises into the air as vapour, and when it is condensed in the air it falls as rain.

The investigation which I propose to place before you at the present moment is this :—How does the sun bring this about ? Are the whole of the sun's rays influential in producing this evaporation ? Are those rays which enable us to see bodies influential in producing it ? Are those rays that produce vision influential in evaporating the water of the ocean ? All these questions we have now to examine ; and I trust that you will give me your patience if I try to examine them to the very bottom.

In the first instance I will try to produce before you what I will call an artificial sun. [A powerful electric current from a magneto-machine was set in action, producing an intense light and heat.] This, then, is a little artificial sun with which we are going to operate, and that it may not annoy you too much we have enclosed the lamp which regulates this current in this kind of camera. This light is produced by a gas-engine downstairs, which works a beautiful machine which has been kindly presented to the Institution by Mr. William Siemens. The rays from this little sun shall impinge upon a reflector such as I hold in my hand. They will be gathered up by the reflector and thrown forward. Here is our camera (Fig. 21), and I will place within it this reflector M, and through the aperture in front of the camera we will reflect the rays emanating from the intensely-heated carbon-poles P and P', and there I obtain an intense focus of heat and light at F. Now I want to show you that that beam is capable of producing ignition. For instance, I hold this piece of paper at the focus F. [The paper immediately took fire.] Look at that ! That is done by the pure radiation—not by any flame. The paper is ignited by the condensed radiation from our little domestic sun. Here is a metal, and I dare say that this metal will burn just like the sheet of paper, if I place it in the focus of the light. [A piece of thin sheet zinc, upon being held at the focus, immediately burnt with its characteristic greenish blue flame.] We have the zinc burning like so much commön paper. The heat is so intense that I could have readily ignited a diamond. A very famous experiment of the celebrated Florentine academicians was to ignite a diamond by means of a tremendous heat. Another way of converging the beam is by means of what is called a conjugate mirror. I will send a wide beam out through the aperture in our camera, and will then receive that beam upon another mirror. The second mirror will gather up the beam to a point or focus, and at that point we shall obtain all these tremendous effects to which I have been referring. I obtain in front of the lamp a fairly parallel beam, which, in the first instance, is not collected to a point or focus. But it is collected to a focus by the second mirror. [A focus having been produced, pieces of paper and sheet zinc were ignited therein, as in the former experiment.]

I now want you to observe a further experiment. I have here a substance which, practically, does not interfere at all with the passage of the light which emanates from the

Fig. 21.

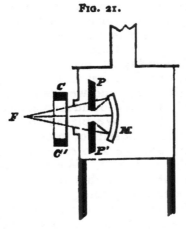

carbon-points. The light goes through it with the greatest ease. [A flat glass cell (c, c', Fig. 21), containing a solution of alum in water, was interposed in the track of the converging rays, in such a position that the focus fell beyond the cell.] I now place in the focus a piece of clean white gun-cotton. You see that the gun-cotton remains perfectly unignited, although it is a very inflammable substance. Now we will remove the cell containing the solution of alum, and you will at once notice what occurs. [The solution was removed, and the gun-cotton instantly exploded.] The solution of alum, being a perfectly transparent thing, does not

interfere with the luminous rays coming from our little sun.
It must have been something else than the luminous rays
which caused that gun-cotton to ignite. But now let us see
whether these luminous rays do not possess heat. You will
find that they do, for if I darken the gun-cotton with a black
powder you will find that the luminous rays are quite competent to ignite it. [The cell of alum water was replaced in
the track of the beam.] We will place our blackened gun-
cotton in the focus, just as in the former case ; and you will
see that, although this beam of light had no effect whatever
upon the white gun-cotton, yet when the gun-cotton is
blackened you have an explosion, although the light has
passed through the solution of alum.

Now this leads me to say a few words for the purpose of
inquiring into the cause of these differences ; and I think
that if you follow me you will understand it perfectly well.
I will tell you an experiment that I made yesterday. That
is an exquisite plate of glass of the purest materials prepared
for me by Mr. Chance, of Birmingham. It is perfectly
white glass, altogether uncoloured, and you would imagine
that it would be perfectly transparent to the radiation from
our electric light. Here is another plate [a plate of rock
salt] not more transparent than the plate of glass, if looked
at by the eye. Yesterday we placed it in the path of the
beam. What occurred ? First of all, the glass plate was
placed in front of the lamp, and we examined the beam that
had passed through it. We found it enfeebled by its passage
through the glass plate. We placed in the same position
this other plate, which consists of rock salt, which is perfectly transparent, and we also examined the beam that
passed through the rock salt. We found the beam much
less enfeebled than it was by this transparent glass plate.
I then felt the rock salt by placing my hand upon it, and it
was almost perfectly cold. I touched the glass plate, and I
almost burnt my hand ; so that the glass had the power of
intercepting the heat and lodging that heat in the body of
the glass, which power was not possessed by the rock salt.
In scientific language, we say that this glass is a better
absorber of heat than the rock salt. The rock salt transmits
the heat and does not absorb it : the glass absorbs the heat
and does not transmit it. The effect of this difference is
that a body that does not absorb heat can not be heated.
A perfectly transparent body, when it is operated upon by
the rays of pure light, can not be heated by those rays of
light. For instance, if I place my alum cell in front of the
lamp and allow the heat to pass through it, I might place

behind that alum cell a sheet of paper, a sheet of glass, or a sheet of ice, sensitive as this substance is, and the rays that pass through the alum cell would be perfectly incompetent to warm the water or to melt the ice; so that these luminous rays are perfectly incompetent to heat transparent bodies. But if a body can

selves manifest
manifest. And hence it is that when I operated upon clean

did not at all a

from our lamp
when I allowed
the total radiation from the lamp to fall upon it, then I exploded it. And also when I blackened the gun-cotton, those luminous rays were competent to explode it. You might suppose that rays that had passed through water or through ice had no power to heat anything afterwards.

FIG. 22.

B

C

You might imagine that in passing through cold water or ice the beam would be entirely robbed of its heat; but I want to show you that that is a delusion. Here, for instance, blackened gun-cotton as before, and I
hope to   tiful lens of ice.
There it   ice.
I intend to make use of that ice as a burn   The
luminou   the
ice, and   pect
able to explode the
gun-cotton, c. Inasmuch as these are luminous rays, the gun-cotton is blackened in order to enable it to absorb them. We will, first of all, introduce a cell of water between the camera and the ice lens. You see that here we have a beautiful parallel beam, and Mr. Cottrell will interpose the ice

lens in the path of that parallel beam. Those persons in front will observe the convergence of the rays by the ice lens. The lens brings the rays to a focus, and I will now remove our cell of water and there we have instantly the explosion of our gun-cotton.

Before coming into the room, I thought that I would try and hasten the next experiment [that of making water boil by means of the beam from the electric lamp] by heating in a test glass the water that I am going to use before placing it in the focus; but I think that it will be better and fairer for the experiment if I begin with cold water, operating with that cold water as the sun operates with the cold water of the ocean. Mr. Cottrell will now give me a little cold water from the tap, and I will converge the beam so as to produce an intense focus in front of the lamp. I will now bring this cold water in a glass tube in front of that focus, and we will place across the beam the cell containing the solution of alum. The cold water may remain in the focus till dooms-day without the light producing any sensible heat in the water. The beam of light having passed through the cell of alum water, all the rays that are capable of heating water are completely taken away from the beam; so that I might hold the water in a focus for any length of time without its even becoming warm. We will now take away the cell of alum and water, and we shall soon see that the water in the tube will be raised to ebullition. In order to make the experiment as conclusive as possible, it is worth while to begin with cold water. In a very short time, if you give me your patience, I trust to be able to boil the water in your presence. At the present time I see bubbles of steam being generated, and after a little while you will find it visibly boiling before you. I hear it beginning to boil already. I now feel the dancing of the water. [After a short interval.] Now I think that those in front can observe that the water is actually boiled by radiant heat. Mr. Cottrell will now introduce a cell containing alum water. The boiling instantly ceases. The light is no longer capable of boiling the water. We take the cell away, and the boiling recommences imme-diately. Now I will ask him to interpose a substance which will entirely cut away the light, and you will find that the ebullition of the water will not cease. In fact, to speak particularly, the emanation from this artificial sun of ours, just like that from the natural sun, consists of two classes of rays, one capable of exciting vision, and the other, far more powerful, capable of exciting heat and not vision; and here, in this case, the alum cell or the water cell cuts off a great portion of these rays that are not capable of exciting vision. But you might infer from the experiment that we have made that there is something besides the light-giving rays that produced the effects that you have seen—the burning of the paper and the exploding of the gun-cotton.

Now we want to detach from the electric light this particular portion of the emanation that is effective in producing those combustions and in boiling the water; and for this purpose we must discover a substance which shall c ut off all the luminous rays, and allow to pass only those rays that are called invisible rays. Here I have such a substance. Mr. Cottrell has here a cell containing it. He will throw upon the screen the image of the carbon points from which our light is produced, and here is the substance to which I have referred. It is perfectly dark. If you look through it at the sun at noon-day you will find that this substance will cut off the rays of light. You will see that the opaque liquid contained in this cell will completely cut off every

Fig. 23.

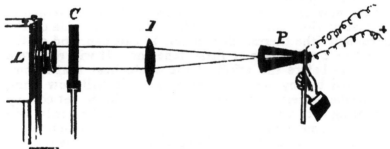

trace of the light from our little domestic sun. No trace of that light can pass through the liquid. Now what I want you to understand is that through this opaque solution a certain portion of the radiation from our domestic sun will pass absolutely with perfect freedom, and we shall be able to cause water to boil in the focus. When that has been done, I will interpose the alum cell in the path of the beam instead of the opaque solution, and you will see that when the alum cell is used the boiling immediately ceases. You will see that although all the light has been cut off by the solution, the water will still continue to boil. I said that we might have inferred from our former experiments that there was in the beam, besides the light, something which caused the water to boil ; but inferences ought never to be accepted in science if we have an opportunity of converting those inferences into facts. Perhaps it is worth while to make an experiment. By means of this larger lens—I (Fig. 23)—I will converge upon the beam emanating from the camera L the dust of the room. You see the beam passing through the air of the room. If we bring the lens forward a little from the light, you have a conical beam, and here you have the point of convergence of the beam at P. I will just mark that point of convergence. I have here an instrument (held in the hand at P) which will enable me to test whether or not there is any heat in that beam. Mr. Cottrell has associated this instrument with the magnetic needle which you see suspended there before you ; and if I breathe for a moment on this instrument, you will see the effect which the warmth of my breath produces upon the needle. It is an instrument used for the purpose of converting heat into electricity, and causing the magnetic needle to move. The warmth of my breath will be sufficient to produce a large deflection of the needle. I breathe, and there you see the needle moves away, and, as I have said, a large deflection is produced.

Well, now, here we have our beam of light converged in the dust of the room. I will place upon this side of the instrument [the thermo-electric pile P] this cap, and on the other face I will place a little reflector. It will require a little time before the effect of the waste heat that I communicated by my breath disappears. We will now bring the needle to rest. We will cause the pile to get cool by presenting it to a cool part of the room, and the needle will gradually come back to zero. If I present the pile towards the piece of ice, perhaps it will be still better. It will cause the needle to come back still more rapidly towards zero. If I place this instrument in the path of this beam of light, you will get a deflection of the red end of the needle towards me. Now I want you to see that if we place this opaque solution (contained in a cell, c,) in the path of the beam and entirely cut off the light, still the heat will pass through, and fall upon the pile, and you see that effect produced which is indicated by the movement of the needle.

Now I want to collect these heat rays, and show you that they are capable of boiling water. We have here our focus (P, Fig. 21), and I will boil the water as before. Mr. Cottrell will go to the other lamp and cause its light to fall upon the boiling water, for I want you to see the experiment. [The tube of water which had been used in the previous experiments was again caused to boil in the electric beam. The cell (c, c′, Fig. 21) containing the opaque solution was then interposed.   The opaque liquid consisted of a solution of iodine in bisulphide of carbon.]   You saw a moment ago that the interposition of the alum cell completely stopped the boiling; but now, when we put in this cell of opaque liquid, the boiling continues as before.   So here we are boiling the water by the invisible rays and not by the visible rays, and this is the point that I am labouring to bring you to.   I say it advisedly, for it has been some labour.   I have been labouring to bring you up to this point—that it is not the luminous rays of the sun that cause the ocean to evaporate and that cause the vapour to go into the atmosphere, but that it is the invisible heat-rays of the sun that produce this effect.   If I wished to show you various experiments in connection with these non-luminous rays, I could show you that they have power to produce all the effects of ignition and combustion which we have already produced at the focus of the full beam.   Just allow me to make one such experiment.   Mr. Cottrell will give me a beam.   Here is a focus produced in the air by the reflection of the rays.   If now I take a piece of black paper, and hold it in the focus of the rays, you see that it instantly bursts into a blaze.   We will now take the opaque solution which will destroy all the light, and place it in the beam.   I now interpose a piece of black platinum, and hold it in the perfectly dark focus of the rays, where there is no luminosity whatever, and you find that the platinum is raised to vivid incandescence.   So if I place a piece of blackened zinc in the same position, it immediately bursts into flame.   Again we place our blackened paper in this invisible focus, and it is immediately set on fire.   You see the paper bursts into flame at this focus, which is entirely without light.

And here we come to the conclusion, with regard to the origin of our rains, and hails, and snows, that all these are due to the sun, and that they are mainly and almost exclusively due to that portion of the sun's rays which does not at all administer to vision—the invisible rays of the sun of which you would not have known except by means of the experiments which we have made, and which were first made by the celebrated Sir William Herschel.

EXPLANATORY REPORTS ON NEW
SCIENTIFIC PROCESSES AND INVENTIONS.

[These Reports will not partake of the nature of an Advertisement, but will
be impartial descriptions.—ED. J. S.]

APPLICATIONS OF ARTIFICIAL COLD IN INDUSTRIAL
CHEMISTRY.

No fact is more familiar to scientific men than that the behaviour
of substances brought in contact is powerfully modified by the
temperature to which they are submitted, the results produced
being affected not merely in degree, but sometimes even in kind.
But whilst in manufacturing operations temperatures ranging
from the average heat of the atmosphere up to the melting-point
of platinum have stood at the disposal of the technologist, and
have been applied as requisite, it is only of late that what is
familiarly known as *cold—i.e.*, temperatures below the freezing-
point of water—has been made available on a practical scale.

The following examples will at least show that the freezing-
machine, as the antipodes of the furnace, has a future before it
of which inventors will do well to take heed :—

M. Georges Fournier, of Paris, has devised a method for obtaining sodium sulphate, commonly known as Glauber's salt,
depending on this novel industrial agency :—In various parts of
France, especially in the neighbourhood of Rheims and in
Picardy, there occur extensive deposits of pyritic shales far too
poor in sulphur for use in the kilns of sulphuric acid works, and
only occasionally utilised in the manufacture of copperas and of
alum. But now the latter product is generally prepared from
china clay, or from bauxite, and now the works at La Tolfa
have been revived by means of French capital, and are sending
to France and elsewhere large quantities of the much esteemed
" Roman alum," the makers of alum from pyritic shales—despite
the cheapness of their raw material—find themselves driven out
of the market.

The problem, then, which M. Fournier undertook to solve is
this :—How are such shales to be utilised, and especially how
can the sulphur which they contain be made to effect the
decomposition of common salt and the production of sodium
sulphate ?

He takes the shales in the state in which they were formerly used by the alum and copperas makers. On prolonged exposure to air, namely, the sulphur of the pyrites becomes oxidised to sulphuric acid, and, attacking both the iron and the clay with which it is intimately associated, it forms a mixture of iron and aluminium sulphates. The shales are then lixiviated with water, and as the solution generally contains an excess of acid it is neutralised, as far as possible, by being allowed to stand in contact with scrap-iron. Being then run off into crystallising tanks a very considerable quantity of copperas is deposited. So far, it will be clearly understood, M. Fournier is merely following in the footsteps of the manufacturers of copperas and alum. But now comes the new feature of the process. After the copperas has been deposited, the mother-liquor, containing all the aluminium sulphate and a portion of iron sulphate, is run off, and mixed with common salt in such proportions that there may be sodium enough to combine with all the sulphuric acid, and chlorine enough to take up all the aluminium and the iron. The mixed solution, which should not exceed 23° Baumé in specific gravity, is then exposed to a temperature of $-2°$ to $3°$ C. In what manner this temperature is produced is a matter of indifference as far as the principle of the invention is concerned. When I saw the process worked on the large scale at the establishment of M. Raoul Pictet, in La Chapelle, the sulphurous acid freezing-machine was used with very good results. In a short time double decomposition takes place, and we have in solution—instead of sodium chloride and aluminium sulphate— sodium sulphate and aluminium chloride. As, however, at temperatures bordering on and a little below 0° C., sodium sulphate is almost insoluble, it is deposited in the state of a fine crystalline sediment of the deca-hydrated salt (the ordinary form of Glauber's salt), whilst the aluminium and iron remain in solution as chlorides. The mother-liquor is rapidly run off, and the deposit is freed from the chlorides mechanically lodged among its crystals by washing with brine cooled down to 0° C. by an ingenious application of what might be called the waste cold. This brine afterwards serves towards the decomposition of the next lot of sulphates of aluminium and iron, so that nothing is lost. The crystals after washing are freed from moisture in a centrifugal machine, and are then fit for any purpose to which Glauber's salt is applicable. I have lately learnt from the inventor that he has succeeded, at a very trifling cost, in using this salt as a raw material for the production of caustic soda.

We return now to the mother-liquor—a solution of chlorides of aluminium and iron. One of the guiding principles of industrial chemistry is that every result of a decomposition should be brought out in a merchantable form. How far can this be carried out in the present instance? To obtain chloride of aluminium in a hydrated state a very costly and circuitous process has

hitherto been employed. The commercial sulphate of alumina
—so called cake-alum—is dissolved, precipitated with an alkaline
carbonate, and the precipitate re-dissolved in hydrochloric acid.
Hence its price has naturally been greater than that of the sulphate, and the latter salt has therefore been employed in preference. There is, however, one class of purposes for which the
chloride of aluminium would at equal prices be preferred to the
sulphate. Several years back it was very justly remarked by
Mr. Crookes that the settlement of the sewage question must to
a great extent turn upon cheap soluble alumina. The power of
alumina in such states to seize upon organic matter not merely
when suspended, but when dissolved in water, and to precipitate
it in combinations of which " lake-colours " are the most familiar
type, is well known to practical men. How many of the every-
day operations of the dyer and the colour-maker entirely or
mainly depend upon this very power I need not stay to point out.
That the action of aluminous salts upon fœcal matter is closely
analogous to their behaviour with tinctorial principles must be
admitted when we see that, according to the recent official investigations of Dr. R. Angus Smith, the " albuminoid ammonia,"
or organic nitrogen, in the effluent waters from the Aylesbury
Sewage-works, was found as low as 0·024 grain per gallon,—a
little more than one-fifth of the limit proposed as admissible by
the late Royal Rivers Pollution Commission.

Yet so long as the soluble salts of alumina were high in price
results such as the above, however interesting from a theoretical
point of view, were of little value to the practical sanitarian.
Soluble alumina, in the form of sulphate, has indeed experienced
a remarkable reduction in price during the last eight years ; but,
thanks to M. Fournier, its future cost in the state of chloride
will be even lower than it is at present. Nor should it be overlooked that, other things being equal, chlorides are decidedly
preferable to sulphates for sanitary purposes.

The aluminium chloride prepared as described above is not
applicable to dyeing and to the destruction of vegetable matter
mingled with wool, on account of its contamination with iron,
an impurity which in the treatment of sewage and night-soil is
by no means objectionable. But for the preparation of a pure
aluminium chloride M. Fournier proposes to use the ordinary
cake-alum as a starting-point. This salt is dissolved, mixed
with a proper proportion of brine, and refrigerated as described
above. The results are Glauber's salts and aluminium chloride
fit for the so-called "carbonisation " of wool, and of course cheaper
than that obtained in the ordinary manner.

Another French inventor, M. Pechiney, of Salindre, in order
to utilise the " sel mixte " of the salt-works on the Mediterranean
coast, a mixture of common salt and of magnesium sulphate,
exposes the lyes by means of Carré's ice-machine, to a temperature of − 6° C. The results are analogous to those obtained in

the Fournier process ; Glauber's salt is deposited in a fine crystalline mass, whilst chloride of magnesium—an article now largely consumed for weighting textile fabrics—remains in solution.

M. Pechiney proposes a very ingenious process for converting this hydrated sodium sulphate into the anhydrous salt, known in the trade as salt-cake. About 1500 kilos. of the crystalline paste, previously drained, are placed in a wooden cistern fitted with an agitator, and dissolved in a little of the mother-liquor with the aid of steam introduced by a copper worm. When the whole is dissolved, 250 kilos. of common salt are added. An exchange now takes place ; the salt gradually dissolves, and the sulphate is deposited in its stead, but in an anhydrous condition. This sediment is then transferred to a filtering-vat with a double perforated bottom, and allowed to drain. It is thus obtained free from all foreign matter, save about 5 per cent of water and 0·5 of common salt.

It is highly probable that the processes here described may point the way to further applications of low temperatures in the chemical arts.

J. W. SLATER.

ANALYSES OF BOOKS.

*Allgemeine Ethnographie.** Von FRIEDRICH MÜLLER. Wien:
Alfred Hölder.

WE have here a luminous, compact, and yet comprehensive
survey of an important science, furnished by a masterly hand.
In the outset the author points out the distinction between
ethnography (ethnology) and anthropology. The latter science
regards man, physically and psychically, as a specimen of the
zoological species *Homo.* The former views him as an individual
belonging to a certain society, depending on custom and descent
and united by a common language. After sketching the history
of the science and explaining the reason of its absence in
classical antiquity, he explains the meaning of the terms "race"
and "people," which may be summed up in the words that race
is an anthropological, but people an ethnographical, conception.
He remarks that though we now find no men not belonging to
some people, we must still admit that there were at one time no
peoples, but merely races.

Proceeding to the systems of anthropology and ethnography,
he gives a summary of the classifications of Blumenbach,
Cuvier, Retzius, and Haeckel, the latter of which, mainly
founded upon the properties of the hair, is expounded at some
length. According to this arrangement man is primarily divided
into Ulotriches and Lissotriches. In the former division each
hair is a flattened band, having an ovoid section. In the Lissotriches the hairs are cylindrical, and their section is a circle. To
language the author—in our opinion very judiciously—assigns
merely a subordinate value in the diagnosis of peoples.

We next meet with the question, What is the position—we
might say the valence—of human races in zoological system-
atics ? Herr Müller, in acccord with Darwin, regards them as
sub-species. We must here, however, point out that the mutual
fertility of different human races can scarcely be taken as a
ground for denying them the rank of distinct species, when we
find the American bison and the domestic cow—though belonging
to separate genera—still capable not merely of interbreeding,
but of producing a permanently fruitful progeny. The various
races he considers have arisen by natural selection from one
primitive type, which has long since disappeared, and which
branched out first into the Ulotriches and Lissotriches.

* Universal Ethnography.

As regards the origin of man the author declares himself an Evolutionist, whilst fully admitting that neither of the existing theories of the origin of animal life can be regarded as free from difficulties. The existence of mankind upon the earth he traces back to dates far earlier than those of the vulgar chronology. The dawn of Egyptian culture may be placed about B.C. 5500, whilst a " Mediterranean race " probably existed as far back as 9000 to 10,000 years B.C. The original home of man he seeks in the warmer regions of the eastern hemisphere. We may particularly notice here the following passage :—" Another circumstance which points to a warm climate is the general nakedness of man, which cannot be explained by the subsequent use of clothing, since savages who go about naked, far from being more densely haired, are, with few exceptions, characterised by the baldness of those parts which in civilised man are covered with hair." The author here seems to overlook the fact that in one of the lowest existing types of mankind, the Australian aborigines, the back is clothed with dense hair.

The question whether or no mankind is descended from an original pair, Herr Müller justly sets aside as totally unscientific.

The author next points out the peculiarities, physical and psychical, in which man differs from the rest of the animal world. He next examines the distinction between the various human races, and insists strongly upon their permanence. On the oldest Egyptian monuments the negro is depicted as we find him to-day, an interval of from 4000 to 5000 years having failed to modify his structural characteristics. We may here remark that in spite of this persistency the advocates of mechanical creation consider him as of one species with the European, and as descended from the same first parents. Yet with scant consistency they take a similar permanence, as seen in the domestic animals of ancient Egypt as a proof of their original and permanent distinctness and of the invariability of species ! The mixture of races of approximately equal rank yields, he considers, an improved offspring. The contact of very unequal races is, however, fatal to the weaker.

In succeeding chapters we find an examination of the influence of soil and climate upon human culture, an account of the ancient centres of civilisation, of its conditions, and of the part played by the different human races in its development, where the lowest rank is assigned to the Australian. The wanderings of nations are not investigated. It is remarked that very few tribes can be rationally considered as aborigines of the countries which they now occupy. The Caffres and Zulus, for instance, have entered South Africa as intruders from distant regions in the north-east, and their rights are simply those of conquest—a fact which our professional philanthropists would do well to consider. The original seat of the Indo-Germanic (Aryan) race the author seems to seek in South-eastern Europe. As an argument in favour of this view, he contends that the linguistic remnants common to all the subdivisions of the Aryan race betrays no acquaintance with the fauna and flora of Asia. It may be replied that the fauna of Asia, excluding the " Indian region " of Wallace, differs little from that of Europe, and that anciently the difference was probably less than at present.

The remainder of the work is devoted to a description in detail of the various races and peoples.

We can decidedly recommend this treatise to all students who wish to acquire a comprehensive knowledge of ethnology as now understood, without being compelled to wade through a mass of learned journals and transactions.

Leçons sur la Génération des Vertébrés. * Par G. Balbiani, Professeur au College de France. Paris: Octave Doin.

These lectures, collected by Dr. Henneguy, form part of the course of comparative embryogeny delivered at the College of France during the winter term of 1877-78, and embody both a general survey of the present state of science in this direction and the results of the author's own researches. To give within the available space a full analysis of a work so abounding in details is of course impossible, but we shall endeavour to select the more salient points.

Proceeding from the theory of Knox, subsequently demonstrated by the researches of Waldeier, that every embryo is at first bisexual, he succeeds in throwing a welcome light upon the obscure phenomenon of parthenogenesis. He shows that, *e.g.*, in the Aphides, the ovum or female element is fecundated by the action of an embryogenous vesicule, which is the homologue of a spermatoblast of the male sexual gland, and from this fecundation there results the development of the egg up to the formation of a perfect animal. Hence, a female *Aphis* is histologically, though not morphologically, a hermaphrodite. Judging from the frequence of parthenogenesis among other invertebrate animals, a similar hermaphroditism is far from exceptional. In the silkworm a certain number of non-fecundated eggs undergo their normal development. Barthelemy records that on one occasion all the eggs laid by a virgin female proved prolific. Among other Lepidoptera it is certain that there is only a small number of males.† Among the Psychidæ parthenogenesis is an

* Lessons on the Generation of the Vertebrata.
† We are here on debateable ground; many authorities declare that among butterflies the males predominate in number.

ordinary fact. Among the Hymenoptera numerous species of *Cynips* have, according to the author, no known males. Dzierzon has detected parthenogenesis in the hive-bee. Even among vertebrate animals unfecundated *ova* undergo a partial development. The segmentation of such *ova* was observed by Bischoff in the frog, the bitch, and the sow, and these observations have been verified by more recent authors. Agassiz and Burnett recognised evident traces of segmentation in certain American species of *Gadus.* In the common hen, Oellacher discovered that the non-fecundated eggs experience a beginning of segmentation in the oviduct. But there is no known instance among vertebrate animals of the development having proceeded to the formation of a perfect animal.

It must be admitted, in the meantime, that this explanation of parthenogenesis raises a number of further questions. Thus the last generation of *Aphides* produced every season is composed of individuals of both sexes. Yet, in a heated conservatory, *Aphis rosæ* has continued to reproduce itself by parthenogenesis for four years in succession. Why is this? Why, among these insects, are the individuals produced by parthenogenesis brought into the world living, whilst those normally generated appear as eggs?

Concerning the spermatozooids, Professor Balbiani does not unhesitatingly admit the modern theory, which regards them as mere histological elements, analogous to the blood-corpuscles. He considers that there is something to be said for the old doctrine of their vitality. He holds that their movements display a certain degree of purpose, being directed to a definite end. Thus in *Bombyx mori* at the moment of intercourse the seminal fluid of the male is received in a special vesicle, which the next day is found completely empty, all the spermatozooids having emigrated into another capsule, which opens into the oviduct opposite to the former capsule, and where they await the *ova*, to fecundate them in passing. But the walls of the former receptacle have no contractile elements, so that the passage of the spermatozooids from one capsule to another can only be due to a spontaneous inward movement. In the former receptacle there remain only a few misformed and motionless spermatozooids. There exists consequently a kind of "natural selection," even among the sexual elements.

On the Nature of Life : An Introductory Chapter to Pathology. By RALPH RICHARDSON, M.A. Second Edition. London : H. K. Lewis.

WE have here an exposition and a criticism of the three prevalent theories of life, the so-called physical, the vital, and the natural. The first, with which the name of Prof. Huxley is especially connected, views life as "a mere form of energy," analogous in its operation and relations to heat, light, magnetism, or electricity. At the same time it is regarded as "a property of the protoplasm, similarly as we ascribe ordinary physical properties to inorganic and inert substances."

On the other hand, the "vital theory," as advocated by Dr. Lionel Beale, regards "life as a power, force, or property of a special and peculiar kind," temporarily influencing matter and its ordinary forces, but entirely different from and in no way correlated with any other.

Finally, the "natural theory," advocated by John Brown in 1770, by Fletcher in 1826, and by the author in the present work, views life simply as "the sum of the actions of organised beings."

It is a characteristic and valuable feature of Dr. Richardson's work that he protests against that vagueness of language, which is especially dangerous in biological speculation. How often is the distinction between a property, a power, and an action forgotten! The author contends that "a property signifies only a susceptibility of motion; a power, only a means by which this susceptibility may be called into action, whilst an action designates the phenomena resulting from the two (power and susceptibility) in co-operation." Again, in his Harveian Oration for 1870, Sir W. Gull remarked—"The vegetable kingdom is no more than *an expression* in a higher form of the terrestrial conditions which even common experience proves to be in a general way necessary to vegetable life," and, further, the organisation of our bodies is "the expression of the highest correlation of these external conditions." Dr. Richardson asks for the precise meaning of this word "expression," and shows that in its mathematical sense it is quite inapplicable to individual objects. He then quotes Dr. Beale's arguments in opposition to Huxley's theory, none of which have been controverted. The phenomena of assimilation, secretion, and reproduction "differ absolutely from any actions known to occur in any kind of non-living matter whatever."

Nor is the vitalist theory, in turn, found more satisfactory. It is unsusceptible of demonstration. That life is required for the first formation of organised tissues cannot be denied; but it is the life or living action of the thing organising, *i.e.*, of the progenitor of such organism, and not of some Vital Force, Power, or Influence.

We think this work may be very profitably studied as an antidote to the modern diseases of vagueness in language, and of rashness and dogmatism in theory.

The Philosophy of War. By JAMES RAM.   London: E. J. Davey.

THIS is a difficult book for a critic who, like ourselves, is at once by inclination and by duty barred from any excursion into the thorny fields of party politics.   The author holds that the hand of God has been shown in times past in the chastisement of nations and their rulers, in the transfer of empire from one people to another, in the very substitution of race for race—all by war. He considers, therefore, that war is an ordained method of procedure on the part of Nature ; that its sufferings and miseries are but a small item in the incessant sacrifice of life, whether human or brute, which is constantly going on; that individual happiness is not so much the aim of Providence as is the elevation of the type. He believes that man, like every other organism, is raised and ennobled only or mainly by the struggle for existence of which war is merely one of the more open manifestations.   It is difficult to see how anyone who believes in the benefits of competition can come to a different conclusion. For in truth all competition is war.   If A, by superior strength, bravery, activity, cunning, or practice in the use of arms, kills B outright, or reduces him to serfdom, the "friends of peace" express their horror; but they fail to see how exactly similar is the process when A, by dint of a larger capital, of greater skill and tact, perhaps even of a more unscrupulous disposition, undersells B, takes his business from him, and reduces him to poverty, perhaps to pauperism or to starvation.   In the latter case the suffering is less acute, but more prolonged.

One argument in favour of the idea that war is specially sanctioned and ordained by Providence has escaped the author. The very arrangement of land and water on the surface of the globe is eminently favourable to aggression and invasion. Suppose, for instance, there had been a " silver streak of sea " between France and Germany, how the history of Europe, from the days of Louis XIV. downwards, would have been modified. Suppose that another channel of salt water joined the Adriatic to the Black Sea, cutting off the Balkan peninsula from North-eastern Europe, would there ever have been an " Eastern question " at all ?

But we may go still further : unless geologists are utterly mistaken, the successive revolutions of the earth's surface have tended in the direction of opening roads for armies and closing the paths of commerce.   Thus there is every reason to believe that in pre-historical ages the Red Sea joined the Mediterranean. Had this feature remained there would have been no Suez Canal requiring a watchful guardianship, and no possibility of Egypt being one day overrun from the north. We are told that a vast inland channel extended from the Black Sea to the Pacific somewhere about the mouth of the Amur. Had this region not undergone elevation, so as to be converted into dry land, there would have been no need of any Afghan war.

But does the struggle for existence really lead to the elevation of individuals, of races, or of species? It is a remarkable fact that when man undertakes the improvement of any species his very first act is utterly to stamp out that competition from which certain theorists expect so much. He supplies the wants of each individual so fully that it does not require to expend any part of its energies in any struggle. We have seen this question happily illustrated by reference to a turnip-field. The farmer tolerates no competition either between individual turnips or between the latter and any interlopers in the shape of weeds. The consequence is, not merely a higher general average size among the roots, but the production of some individuals far surpassing any that can be developed in a competitive—*i.e.*, an over-crowded and unweeded—field.

Nor when a struggle takes place are we at all sure that the survivor, though possibly the one most in harmony with the surrounding circumstances, is by any means the best. What, for instance, are the plants which multiply and spread the most rapidly, and which, so far from needing our fostering care, bid defiance to all our efforts for their destruction? Is it the wheat, or the vine, or the rose, which thus obtrudes itself everywhere? One of the most obvious characteristics of rampant weeds is their generally low position in the scale alike of utility and of beauty.

In the animal kingdom we find something altogether similar. The commonest insects are, as a rule, the most mischievous and the ugliest. In proof we refer to the locusts, the phylloxera, the plant-lice and scale-insects, the various mosquitoes and sandflies, the common house-fly, the wireworm beetles, the cockchafer, and many more. Now it is generally admitted by modern naturalists that a numerous species is one which is gaining ground in the struggle for existence, whilst a scarce species, on the contrary, is one which is verging upon extinction.

The same rule prevails among birds. The commonest species in Europe, the sparrow, is precisely the one for which least can be said.

After extending, as we might, this survey through other departments of animated nature, we are surely entitled to ask if the struggles between different human races and nations are quite as ennobling and elevating as our author believes, and whether war is quite so sure to end in the triumph of the best?

Whilst unable to refrain from the expression of these doubts, we hold that Mr. Ram's treatise contains many valuable lessons for a nation like ourselves, in whom the " tribal instinct " and the very impulse of self-preservation, if not decaying, are being systematically attacked and denounced as criminal and selfish.

On Phthisis and the Supposed Influence of Climate. Being an Analysis of Statistics of Consumption in this part of Australia. By WILLIAM THOMSON, F.R.C.S., F.L.S. Melbourne: Stillwell and Co.

CLIMATIC influence upon phthisis and tuberculosis is a question which is daily growing in importance in the scientific world. So much of the mysterious is still involved in the pathology of the disease that any straw is eagerly clutched at, and too many are willing to attribut to fog and mist attributes to which they have no just claim. Phthisis is admittedly hereditary; is popularly supposed to be influenced by climate; and generally understood to be non-contagious, and amenable only to prophylactic and palliative treatment. Are these surmises correct? That it is hereditary may be taken for granted. Is it non-contagious? If so, why is it that the apex of the lung is most liable to be attacked? Is it possible that true phthisis arises in the first instance from inhalation of living organisms that propagate in the tissues? If so, does not the root of the etiology of phthisis lie rather in sanitation than in climatic influence? And, further, if this be granted, is it not rational to suppose that direct contagion may take place? That direct contagion is possible is tacitly admitted, if not openly confessed, by the frequent counsel given by the profession to patients of a scrofulous tendency to remove to open spaces in hilly districts where the climate is equable, dry, and bracing. Here, avoidance of over-crowding is the clearest benefit obtained, climate having but little apparent influence. In the case of Switzerland, for instance, Prof. Parkes says—"Although on the Alps phthisis is arrested in strangers, in many places the Swiss women on the lower heights suffer greatly from it: the cause is a social one; the women employed in making embroidery congregate all day in small, ill-ventilated, low rooms, where they are often obliged to be in a constrained position; their food is poor in quality. Scrofula is very common. The men who live an open-air life are exempt."—(*Practical Hygiene*, 1878, p. 440.)

The remarks thus applied to Switzerland are apparently equally applicable to Victoria, and the colony of New South Wales generally. Hitherto it has been *de rigueur* to prescribe emigration to this colony as a prophylactic of phthisis; but a recently published work of Mr. William Thomson, F.R.C.S., of South Tana, goes far to dispel this illusion. In this exhaustive statistical treatise Mr. Thomson shows that during the years 1876, 1877, and 1878, 3003 persons died of phthisis in the colony of Victoria, of whom 762, or 25·37 per cent, were born in Australia; 2003, or 66·70 per cent, had resided there upwards of five years; 112, or 3·73 per cent, had lived there upwards of two years; and only 126, or about 4 per cent, had lived there less than two years.

The first assumption of theorists upon reading, in the "Victorian Year-Book for 1878," that phthisis had "resumed its place at the head of the list of causes of death," was that the influx of persons more or less affected with tubercular disease would account for the increase; but the figures of Dr. Thomson prove that this surmise is incorrect. What deduction are we to draw from this? The climate of Victoria is, theoretically, specially adapted for such as suffer from pulmonary complaints: neither the open spaces nor Eucalyptus trees are wanting, and the death rate in town is not strikingly greater than in the country. Thus during the year 1878, of all deaths from phthisis in Victoria, only thirty-six more occurred in town than in the country. From this it would seem that the influence of climate has been considerably over-rated when dealing with the etiology of phthisical disorders, and from hereditary transmission and non-contagion attention must be turned to generation and direct contamination. In this research the worker may be aided by the recognised value of sea voyages, or the possible influence of bromine and iodine upon persons of scrofulous or consumptive tendency. By analogy it might be argued from this that the minute organisms giving rise to phthisis are parasitic, since bromine and iodine are most destructive to parasitic life of every form.

The argument thus resolves itself into the question, Is phthisis a parasitical disorder? If so, spontaneous generation, within the control of sanitation, and direct contagion follow as inevitable corollaries. Whilst the pathology of the disorder remains as at present, so obscure, any indication of direction in research will be welcomed by scientific investigators.

A Monograph of the Silurian Fossils of the Girvan District in Ayrshire. With Special Reference to those contained in the "Gray Collection." Fasc. II. By H. ALLEYNE NICHOLSON, M.D., D.Sc., &c., and R. ETHERIDGE, Jun., F.R.G.S. Edinburgh and London: W. Blackwood and Sons.

WE have here a continuation of the valuable monograph which we have previously had the pleasure of noticing. The present part is entirely occupied with the Crustaceans from the Silurian rocks of Girvan, and treats of the Trilobita, the Phyllopoda, the Cirripedia, and the Ostracoda. It is illustrated with six plates, and contains a table of the geographical distribution of the crustaceans in the Girvan district.

Memoirs of the Science Department, University of Tokio, Japan.
Vol. I., Part 1. By E. S. MORSE. Tokio: published by
the University.

THIS volume is devoted to an examination of the shell-mounds
of Omori. These mounds, in addition to shells, contain bones
of monkeys, deer, boars, wolves, and dogs, and give plain evidence of the former prevalence of cannibalism.

The author has also detected the jaw of a large baboon-like
ape, possibly a Cynopithicus, which now occurs in the Philip-
pines. Japanese tradition tells of monstrous apes formerly found
in the islands.

We may mention that the paper, the letter-press, and the illustrations have all been produced by native labour.

Lecture Notes on Physics. By CHARLES BIRD, B.A., F.R.A.S.,
Second Master in the Bradford Grammar-School. London:
Simpkin, Marshall, and Co. Bradford: T. Brear.

THIS little book is essentially examinational. More than one-
third of its bulk consists of a list of the "questions set in the
Science and Art Examinations" from the years 1867 to 1879
inclusive. Thus both those who are performing and those who
are undergoing the operation known in polite circles as "preparing
for an examination"—but described by horrid, plain-spoken people
in a word of one syllable—may form a tolerably accurate notion
of what they may expect. It must not for a moment be supposed that we attach any blame to Mr. Bird. He is merely
accommodating himself to existing circumstances. Unless he
acts thus his pupils might possibly "know," but they would fail
to "pass," and he himself would suffer in reputation accordingly.
He therefore does what we must all do more or less until we
become enlightened enough to make a clean sweep of the whole
system. Will this be done before it has quite eaten out the
intellectual life of the nation?

The notes are clear, intelligible, and accurate, and, if properly
filled up with details and illustrated with the necessary experiments, will form a good course on physics.

CORRESPONDENCE.

THE FERTILITY OF HYBRIDS.

To the Editor of The Journal of Science.

SIR,—In a contemporary of yours a correspondent ventures upon the bold statement that "of the 20,000 species of animals" no two species have been found to produce fertile hybrids. Passing over the facts that the known species of animals approach 200,000 much more nearly than 20,000, and that not in one case out of one hundred has the fertility or non-fertility of hybrids been fairly studied, I would refer this writer to Mr. J. A. Allen's "History of the American Bison," published in the "Ninth Annual Report of the United States Geological and Geographical Survey." On page 585 he will read that the "buffalo" (bison) interbreeds freely with the domestic cow, and that the half-breeds are fertile.—I am, &c.,

TRUTH-SEEKER.

BAMBOOS.

To the Editor of the Journal of Science.

SIR,—In Wallace's "Tropical Nature" (p. 53) we read as follows :—"The gigantic grasses called bamboos can hardly be classed as typical plants of the tropical zone, *because they appear to be absent from the entire African continent.*" This assertion, when the name of Mr. Wallace is connected with it, appears most extraordinary.

On the West coast of Africa, and far in the interior, the huts of the natives are composed of split bamboos. The dwellings of those in the British Settlements certainly are. In a journey I made into the interior I found both cane and clay dwellings ; the fences of the private enclosure were almost invariably, I may say, of interlaced split cane. In the forests I met not only with hollow-jointed bamboos, with the peculiar glazed flint surface, but long trailing canes interlacing the forests, of many—perhaps more than a hundred—feet long. How long I do not know; they seemed interminable. In one place I came to a cane brake said to be 16 miles long, through which passed the road, and which, for the mile or two I rode in it, was interlaced overhead by the canes—a species of natural arbour. If I am not mistaken De Chaillu speaks of the canes interlacing the forests he visited. Either Mr. Wallace must be in error or the flinty grasses I saw were not bamboos, yet they exactly agree with the description contained in Baird's "Student's Natural History."

I have no doubt Mr. Wallace is most accurate in his descriptions of Malayan and South American tropical products, but he seems to have fallen into an error in respect to Tropical Africa. And this is the more curious if there be any reality in his half-uttered hypothesis (Malay Archipelago), that Celebes was once connected with the African continent.—I am, &c.,

S. B.

EDMUND HALLEY.

"In the 'Journal of Science' (formerly called the 'Quarterly Journal of Science,' but now, having become monthly, styled as above) for this month we were surprised to read, in an article on 'Edmund Halley, his Life and Work,' that 'in 1720 he was appointed Astronomer-Royal at Greenwich, in room of his old friend and colleague Flamsteed.' Irony would indeed be out of place on such a subject, but we trust that the biography of the second Astronomer-Royal, which Prof. Pritchard is understood to be preparing, will throw some additional light upon the unfortunate and ever-to-be-regretted disputes between him and Flamsteed, which must always form a sad episode in the history of astronomy, leading as they did to feelings, on one side at least, the reverse of friendly."—(*Athenæum*, No. 2729, Feb. 14, 1880.)

To the Editor of the Journal of Science.

Sir,—The above paragraph in the "Athenæum" has been brought to my notice, and as the writer of the article in question I would take some notice of it.

As you are aware, I did not pretend to give anything like an exhaustive monograph of Edmund Halley. The sketchy article was only intended to help to draw public attention to the proposed erection of a monument to Dr. Halley in St. Helena, which proposition was first mooted in the pages of "Nature," on January 29th, 1880. Being absent from London, and unable to consult the "Philosophical Transactions," &c., I expressly asked assistance from Prof. Pritchard, the Savilian Professor in whose hands the Halleian papers have been since 1875, *i.e.*, over four years since. I was emboldened to do this from the great courtesy I had experienced from other astronomers, who heartily have approved of my friend Mr. Gill's scheme; but the Professor of Astronomy at Oxford did not answer my letter.

I judged that a former and early friendship had existed between Flamsteed and Halley, from the latter having been employed to settle an *amicable* discussion between the second Astronomer-Royal, or rather King's Astronomer, and Cassini, according to the account given in the "Encyclopædia Britannica" (ed. 1828).

I shall be only too happy to have more mistakes and errors pointed out in my account of Halley, as the more notice attracted to the memory of the great man the better, and my intended object will have been accomplished. The Astronomer-Royal has signified his approval of the scheme, which will shortly be carried out.—I am, &c.,

THE WRITER OF THE ARTICLE.

PS. Anyhow attention has now been drawn to the inexcusably long time during which Prof. Rigaud's materials have been delayed from publication by the Rev. Chas. Pritchard. The public have been very patient, but the scientific world has a right to demand the fulfilment of the expectations held out in 1875.

NOTES.

Mr. E. A. Thompson writes to the "American Naturalist" that certain moths, *Plusia precationis*, having been caught by their tongues in the pollen-pockets of *Physianthus albens*, an Asclepiad plant, were stung to death and devoured by what were supposed to be ordinary honey-bees. Dr. Hermann Müller considers the fact of the moths being thus entrapped new and interesting; but mentions that his brother, Fritz Müller, in South Brazil, has observed bees eagerly licking the juice dropping from pieces of flesh which had been suspended to dry in the air. Mr. Darwin suggests that the bees may possibly tear open the bodies of the moths in order to get at the nectar contained in their stomachs. Both these distinguished naturalists recommend further observation. It is stated by Prof. A. J. Cook, of the Michigan Agricultural College, that bees kill the drones not by stinging, but by tearing with the mandibles.

According to Prof. Church, withered leaves of the usual autumnal colours—yellow, red, or brown—can be rendered green again by steeping in water along with a little zinc-powder.

Dr. Auerbach, writing to the "Chemiker-Zeitung," mentions as a curious fact that during an entire summer he observed water-beetles—probably *Gyrinus natator*—living in tanks of a saturated solution of Glauber's salt. When alarmed the beetles took shelter under the crystals, just as they do in ordinary circumstances under water-plants. A little of the liquid so harmless to insects, having found its way by leakage into an adjoining river, proved fatal to multitudes of fish.

Dr. Yandell, in a letter to the "Louisville Medical News," speaks of a fertile female mule, now to be seen at the Jardin d'Acclimatation, Paris. She has brought forth no fewer than six foals—some by zebras, some by an ass, and some by a stallion.

Dr. Erlenmayer, in the "Medical and Surgical Reporter," gives it as his opinion that the Semitic nations, including the ancient Hebrews, were left-handed, and that this peculiarity was the reason why they wrote from right to left. He adduces evidence in favour of his theory both from the Talmud and the Old Testament. We must, however, remember that in one of the earlier books of the latter (Judges, iii., 15) left-handedness is mentioned as a personal peculiarity.

The chloride of methyl is now used in extracting the odoriferous principles of plants. It enables the manufacturer to dispense with the tedious process of *enfleurage*.

Dr. S. Schneck and Dr. Elliott Coues communicate to the " American Naturalist " decisive instances of the mischief done by the European sparrow, which has been rashly acclimatised in America, and of its utter failure as a remedy for the cabbage caterpillar (*Pieris rapæ*).

W. H. Ballon communicates to the " American Naturalist," on what appears to be good authority, a case of a considerable number of swallows being found torpid in the hollow of a tree, in April.

According to Dr. Staub's observations, published in the " Botanische Zeitung," the flowering period of trees is hastened only when the mean temperature for the month is at least $3.5°$ F. higher than the average : a smaller increase does not affect vegetation. On the other hand, the smallest fall in the temperature of the month occasions a retardation.

Papilio alexanor, a very local species of butterfly from the South of France, is considered by many entomologists to be a hybrid between *P. machaon* and *P. podalirius*.

Prof. W. J. Beal considers that, in countries where they occur, humming birds must play a very important part in the fertilisation of flowers.

According to Dr. Jousset de Bellesme, cocoons do not serve insects as a protection against the cold. The pupa resists congelation by reason of a continuous liberation of heat due to the destruction of the muscular system of the larva, which is much more considerable than that of the mature insect. The large quantity of uric acid discharged at the time of metamorphosis is a proof of the extent of the organic transformations which have taken place.

, From observations on the cattle of Brittany, M. Bellamy is led to question the common doctrine that in-breeding is injurious.

THE

JOURNAL OF SCIENCE.

APRIL, 1880.

I. OUR FRIENDS AND HOW WE TREAT THEM.

ONE of the most unpleasant phenomena characteristic of the present age is the variety and number of minute but formidable enemies which assail us, either directly through our persons or indirectly through our means of support. Insects and Fungi seem to rival each other in the amount of depredation they can cause. We have, for instance, the *Oidium*, the *Phylloxera*, and the *Phona uvicola*, successively ravaging the vineyards of France, and bringing in their train distress and destitution. We have the redoubted Colorado beetle (*Doryphora*) taking a heavy percentage of the potato-crops of Canada and the United States, and the locust of the Rocky Mountains (*Caloptenus spretus*) widely destroying the harvests. We see the coffee-shrub withering away under the inroads of *Hemileia vastatrix*. Even useful insects, such as the silkworm and the honey-bee, succumb to parasitic diseases. Lastly, to complete this very brief sketch of minute plagues, we have flies which transfer from victim to victim the poison of disease, and make themselves the bearers certainly of carbuncle and of ophthalmia, and too probably of fever, cholera, and small-pox. There exist, we know, a certain class of writers who from time to time labour to prove, with an amount of sophistry which the most unscrupulous criminal advocate might envy, that all these scourges are blessings somewhat thickly disguised. Leaving these champions of nuisances to be dealt with by the broad common-sense of mankind, we may ask why should such tiny pests, animal or vegetable, have become so much more prominent in our times than in the days of our forefathers ?

Our first reply must amount to a partial denial of the alleged facts. Throughout history we can trace the existence of blights and mildews, and visitations of locusts, followed up in due course of famine. The point of distinction is that in our days any new depredator or *vastatrix* that makes its appearance is at once examined, described, and classified under a name which is safe not to err on the side of brevity. And whilst the *savants* are thus busy with their microscopes the ravages of the pest are duly chronicled in the press, discussed on 'Change in a hundred cities, and made the basis for speculative operations.

But there are veritable causes at work rendering it very probable that the evils in question may have seriously increased. The day of small things is over, whether for states, for manufacturing establishments, or for fields. We have vast areas of land, formerly occupied with a promiscuous vegetation, now covered with multitudes of one and the same species,—vines, coffee-trees, wheat, or potatoes, as the case may be. This is not without its meaning. Just as we can never have a widespread conflagration in a rural district where the houses stand far apart from each other, so epidemics do not arise where the species which they might attack, animal or vegetable, are few and far between, but rage more often and more severely just in proportion as such species become numerous and densely concentrated. This rule holds good with all classes of destroyers, from the jaguar and the wolf to the *Doryphora* and the *Phylloxera*, and even to those minute organisms which appear to be the causes of infectious diseases. They increase with their pabulum or their prey. Hence our modern agriculture has fostered its own enemies.

There is still a further cause for the spread of all such blights as are occasioned by insects. We have interfered with the balance of Nature, frequently destroying the very species whose especial task seems to be the repression of locusts and caterpillars, of gnats and carbuncle-flies (*mouches charbonneuses*). This task, as naturalists have again and again pointed out, though so far with very little effect, is mainly performed by the small soft-billed birds,—species which subsist almost exclusively upon an animal diet. It may be taken for granted that just in proportion as a bird feeds upon insects, rather than upon seeds or fruits, it must of necessity be harmless, and even beneficial to man. Yet in spite of such considerations scarcely any class of animated beings have been so ruthlessly and wantonly persecuted.

Setting on one side, therefore, all other supposed causes of the increase of noxious vermin, and all proposed agencies for their destruction, it will not be useless to inquire into the growing scarcity of small birds in this country, and into its origin.

Foremost among the enemies of our feathered auxiliaries stands that unlovely being the bird-catcher or bird-fancier. His mission appears to be the sale of all such birds as possess anything attractive either in their plumage or in their song. He has a den—perhaps in the Seven Dials or in Whitechapel—where a number of unhappy goldfinches, linnets, larks, blackbirds, nightingales, and the like, are committed to close confinement until purchased by some man or boy of pseudo-ornithological propensities. As the captives are rarely supplied with proper food, or otherwise treated at all in accordance with their wants, their cage-life, if not merry, is at least short, and there is a brisk and constant demand for victims new. To ensnare these is one of the main points of the bird-fancier's craft. For this purpose he undertakes expeditions in all suburban districts, and is sometimes found as far as 50 miles out of town in search of his prey. He has his decoys and his lime-twigs, and his cunningly-devised "calls" for imitating the note of the species he hopes to entrap, and his success is unfortunately much greater than he deserves. His calling, be it remarked, is perfectly legal, except where he commits acts of trespass, carries off the eggs of pheasants and partridges, &c.—sins for which he has a very strong inclination.

Some years ago an Act was indeed duly passed by the Legislature for imposing a little wholesome and much-needed restraint upon the doings of this worthy. The disciples of the late Charles Waterton pleaded that, if these serviceable little creatures are to exist at all, a "close time" during the breeding season must be allowed them, just as in the perfectly similar case of game. During this time, if we rightly understand the provisions of the statute, neither shooting, snaring, nor nest-taking was to be permitted. Never has a parliamentary enactment proved such a dead letter. No one seeks or cares to enforce it. It is simply ignored. It so happens that at this very time, during which the snaring of birds is legally prohibited, and when it is especially fatal to the perpetuation of the species, their capture is particularly easy. Some of the victims sought for are birds of passage, and, if they enjoy immunity whilst nesting, the time during which their capture in England is possible is greatly curtailed. Hence the interests of the bird-catcher are completely opposed to those of the birds and of the general public. Now the bird-catcher looks carefully after number one, whilst the birds go unprotected, on the good old principle that what is everybody's business is nobody's. Throughout the theoretical "close season" disreputable-looking lads and men, carrying upon their shoulders large, but evidently light, packages, discreetly covered with calico or baize, may be seen wending their way into the country in the early morning and returning at various times from 10 o'clock till noon. These raids are made by preference on Sunday mornings. If the observer goes a little farther afield he will see the covered cases unpacked and placed ready for action, whilst the operator crouches behind a bush or hedge to await the result. If successful the bird-catcher packs up his cages and returns home, walking calmly and fearlessly past the policeman on duty, who never wastes a glance upon him, as we have repeatedly witnessed. Perhaps the whole soul of the constable is concentrated upon that mystery the "*bona fide* traveller." Perhaps he knows nothing at all about the Act in question. Indeed, from "information which we have received," we suspect that this is too·often the case. There being no associations, local or national, who make it their business to see the Act put in force, the attention of the police authorities is never called to the infractions which are daily going on. We cannot lay our hands upon a single instance of an offender being prosecuted. Our natural-history societies from time to time express their regret at the havoc which is taking place, and a few letters on the subject occasionally appear in the provincial papers and in some of the scientific journals; but there the matter ends. There is no regular concerted action, such as may be observed in many matters of much less importance. We demand such action. We are not aware whether the Royal Society for the Prevention of Cruelty to Animals can, in accordance with its constitution, here interfere, or whether its functions are limited to the protection of domestic animals. We have indeed stated the claims of our clients, the birds, merely from a utilitarian point of view, regarding them as the friends of the farmer and the gardener: but on the score of humanity the complete suppression of bird-catching during the nesting season is no less urgently called for. How many broods of young birds perish miserably of cold and hunger, whilst their parents are struggling and wailing in the cages of the bird-fancier, can scarcely be conceived. It is plain that the majority of the birds trapped during the spring months must leave a nest in a state at least of partial orphanhood, and we must remember that to supply the young with food the constant labour of both parents is generally requisite.

The question then is, shall the Act for the preservation of birds not included in the game-list remain, as it now is, a mere impotent protest which the nation is too careless and indolent to put in force, or is it to be made a reality? The detection of the offenders would be ridiculously easy. They carry with them all the evidence needed for their own conviction. Cannot the police be got to examine these men when returning with their packages, and to take proceedings whenever they find traps, decoys, and captive birds during the close season? Could not country gentlemen see the propriety of putting their gamekeepers on the alert, especially as the bird-fancier, if not himself a poacher, is sure to be a friend and accomplice of poachers? Could not our natural-history societies take simultaneous action, at least in the shape of memorials to the local authorities, pointing out how completely the law is left in abeyance, and urgently calling for its enforcement? Might not the societies form in every district a nucleus round which the friends of humanity and the lovers of all that is beautiful in Nature might rally?

Then as regards the kindred evil of bird-nesting, how much more might be done for its suppression than now takes place! In France, where the extirpation of small birds has proceeded to a greater length than with us, and where a much larger amount of injury has been experienced in consequence both to property and persons, energetic attempts are now being made by the Government and the local authorities to lead the public, and especially the young, into a different course. "*Ne denichez pas*"—don't rob nests—is a maxim strenuously enforced upon the schoolboy. Lists of useful and harmless birds, reptiles, and insects have been drawn up by the Minister of Agriculture, under the advice of eminent naturalists, and posted up in every village,—and all such creatures the countryman is officially exhorted to spare and protect.

In America there are even State-naturalists whose duties are to ascertain what birds, insects, &c., are harmless and what are injurious, and to recommend the former for preservation. We do not see that any corresponding steps are being taken in England. We fear that in most of our schools in country districts, whether " board " or " denominational," the children are allowed to remain in ignorance of the claims of the lower animals upon man. In many places—we may mention Aylesbury—the harmless newt is looked upon as a loathsome and venomous monster, which may be justifiably put to a death of torture. If one is found it is placed in any convenient little hollow of the ground, covered with petroleum, and then set on fire. Surely such ignorance is a national disgrace. We do not, indeed, recommend the appointment of "county naturalists." Such posts, if created, would be filled by poor helpless examinees who would be better acquainted with every subject than with zoology, and who in the field would be unable to distinguish a nightingale from a wagtail.

There is another cause which leads to the decrease, and must ultimately bring on the extirpation of many rare, curious, and beautiful species. This cause is peculiar to our country, and in the eyes of all foreigners it appears almost beyond the ridiculous. We refer to the " British " mania among the collectors of birds and insects. All scientific naturalists in these days, worthy of the name, consider the distribution of any living species on the surface of the globe as one of the most important points in its history. Consequently it is essential to know what species occur in every country, and even every district. But the mania in question is something quite different. Take, for instance, one of those butterfly-hunters who fancy themselves entomologists. He buys a cabinet, procures a set of printed labels of " British " species, fixes them duly in the drawers, and then endeavours to add to each label one or more specimens of each kind; this he effects by capture, exchange, or purchase. As regards the rarer forms, he is willing to pay the most preposterous price for a poor shabby specimen which has been—or is said to have been—captured within the United Kingdom, whilst he will scarcely accept as a gift one of the same species of avowedly Continental origin. Nothing of the kind, to our belief and knowledge, prevails in any other country. We have met with German collectors of insects exceedingly scrupulous as to the exact locality of every specimen admitted into their cabinets; but we never knew a higher price offered for a German specimen of any insect, *as* German, than if it had been taken in France, or Switzerland, or Hungary.

As regards birds the case is precisely analogous. The consequence is that as soon as a rarity is seen every marksman in the district is on the alert to bring it down and offer it for sale or exhibition as an authentic " British specimen."

Hence rare kinds are speedily blotted out, and others which probably have in former ages been inhabitants of our country, and which might re-establish themselves if left unmolested, have no chance of so doing.

The mania for British specimens tends, therefore, to impoverish the British fauna. We may here remark that of the birds shot and the insects captured, not one in a hundred serves for any truly scientific purpose. Many of them are simply wasted; many employed for barbaric decorations or worked up into uncouth devices. It is sad to see, in some cottage or country inn, a roller, or a bee-eater, or a hoopoe stuffed out of all approach to its natural shape, and to be told that it was killed in the neighbourhood. It is equally sad to read in the "Loamshire Herald" that a "rose-coloured pastor was seen several times last week, and was duly shot by Mr. Addlepate, of this town."

We loathe the bird-catcher and the bird's-nester, but we entertain quite as little affection for those who seek to destroy every beautiful species under the false pretext of "collecting." The true naturalist, whatever be his favourite department, is chary of animal life, and kills no specimens save the few required for determination, for dissection, or for experiment.

There is yet another agency, quiet but effectual, by which the numbers of our insect-eating birds have been thinned. We refer to the agricultural reform, so-called, which reduces our fences to a mere nominal line of stumps, or replaces them with wire; which seeks to abolish hedge-row trees, and lays small plots of land together into one huge field. We know that a certain amount of soil is thus economised, but there is the other side of the account to be considered. These very improvements deprive the birds of cover for their nests, and compel them to go elsewhere. The natural consequence is that the caterpillars, the aphides, the weevils, *Halticæ*, and the like, in these model fields are not sought for and destroyed. If we require evidence for this assertion it may be found in the fact that where hedge-rows and scattered trees are wanting, as in most parts of France and Spain, or where wire fences and railings are used in their stead, as in the Western States of America, there the damage from vermin is greatest. Then man—with Paris green, and fire, and traps, and cunningly-devised sweeping-machines—strives, at great expense and trouble, to undo the mischief which has sprung from his own short-sighted greediness.

We have referred to the waste of birds and insects for

personal decoration, as fashion now sanctions. This folly, for we can give it no milder name, has little effect upon our native species, though we have heard, on apparently good authority, that a lady once wore at a ball a dress trimmed with the skins of a multitude of redbreasts. But we greatly fear that not a few rare and magnificent natives of tropical regions are doomed to extirpation in virtue of this very whim. We should recommend a law prohibiting, under very stringent penalties, the use of any humming-bird, sun-bird, trogon, or bird of Paradise, or any part or appendage of such bird, as ornaments or accessories of human attire, or their sale by any plumassier or milliner. Such an enact-ment would not be, as some persons imagine, a "sumptuary" law, since its object would be to restrict not luxury, but the wanton slaughter of beautiful, harmless, and in many cases useful creatures. We do not think that there is any present hope of such a measure being even introduced into Parlia-ment, but it is our duty to point out its imperative necessity.

II. ORGANIC SYNTHESIS AND ITS SOCIAL BEARINGS.

NATURE, if we may use the expression, places before man not a few dilemmas which are scarcely perceived in his earlier career, but which are painfully felt in proportion as population thickens and civilisation advances. Of these, one of the foremost relates to the prime necessa-ries of life and their production. If bread, meat, butter, vegetables are to be had at a low price, the farmer and the grazier complain that their business is unremunerative, and can only be carried on at a loss. If the same articles are dear, the trading and manufacturing classes suffer. Hence we find that a season of general plenty is not regarded by the cultivator of the soil as an unmixed blessing. If his crops are greatly increased their money-value will fall in propor-tion, and the main result of the bountiful harvest will be, as far as the farmer is concerned, more work for the same profit. Hence we find, in various departments of the food-trade, that prices are systematically kept up by the intentional waste of a portion of the available supplies. This is notoriously the case, as far as England is concerned, with fish. Those who rule the trade prefer to sell 50 tons at 6d. per lb. in place of 100 tons at 3d., and with an article so perishable they find little difficulty in carrying out their schemes. A great misfortune is that whilst in the manufacture of hardware, or of textile and fictile goods, the progress of invention has increased the profits of the producer, nothing similar holds good in the growth of food. Strange as it may sound, the rudest agriculture is relatively the most lucrative. If a man roughly breaks up a fruitful piece of ground, scatters seed over it, and, without troubling himself much with its after-nurture, gathers in the crop, his gross produce is almost exclusively net profit. But if he is obliged to expend a large amount of labour, and to employ costly appliances, mechanical and chemical, though his gross returns are very much larger than those of his rude forerunner, his percentage of profit is necessarily much lower, and he may be driven—as are many English farmers of the present day—to reflect whether his capital might not be more usefully invested. To this comes the further fact that the agriculturist is exceptionally exposed to unforeseen disasters from the seasons. Hence we find, in brief, that the interests of the food-producer and of the food-consumer are in antagonism. The question arises which of these two bodies is to be sacrificed to the other? This question has now been debated in England for many years, to the great glorification of the speech-makers, statesmen, agitators, and economists of high or low degree. Extensions of franchise, protection, free trade, trades-unionism have been prescribed and tried, but none of them has healed the disease. Is it, then, too much to hope that John Bull may begin to look for help in a quite different quarter, like the piper's cow?

> " The coo considered wi' hersel'
> That wind wad never fill her."

May we not look for relief to inventors and discoverers rather than to orators?

We use for our support plants and animals. We eat such beings either entirely or in portions, or we extract from them by mechanical and chemical processes certain substances needful for our maintenance. Why do we make use of their intervention? Simply because we cannot derive nourishment in a direct manner from the soil, from the air, and the water.   Our food must consist of matter in certain organic combinations; and though we can find the single ingredients of such combinations in minerals, in the air, and in water, we cannot assimilate them and make them part and parcel of our tissues.   But this task, which is impossible for our bodily organs, plants effect with ease.   They create nothing, but they put together the elements which they obtain from the dead matter around them, and form fruits, roots, leaves, &c., which we can eat, digest, assimilate, and be thereby nourished.   The animals which serve us for food support themselves mainly upon plants, and do little more than concentrate the nutritious matter found therein, and render it perhaps more easily assimilated.   Their part is therefore of secondary importance, and need not further engage our attention.

Let us now take as one of the simplest cases a useful plant— the sugar-cane—and enquire what is the service which it renders to mankind?   Simply this : it takes the three well-known and plentiful elements, carbon, hydrogen, and oxygen, and combines them together in such proportions and with such an arrangement of particles as to constitute sugar. Every cane, therefore, is a workshop in which sugar is pro-duced, and all that remains for man to effect is to separate out the sugar, which in Nature exists dissolved in the juices of the plant, and preserve it in a state of purity ready for use.   But let us now suppose that we could take these same three elements—carbon, oxygen, and hydrogen—as they exist in air and water, and combine them together in the right proportions, thus forming sugar by the synthesis— the putting together—of its ultimate materials.   We should thus dispense with the soil required for the growth of the crop; with the manure for its fertilisation; with the labour needed for planting, tending, and reaping the canes ; with the plant and machinery used for the extraction of the sugar, and with the attending waste; and lastly, but not least, with the time which now elapses from first planting the cane to the day when the sugar is ready for the market, and with the interest on the capital sunk.   Add that the supply of sugar would be henceforth quite independent of hurri-canes, floods, plagues of rats and termites, and all the various contingencies, living or lifeless, which now from time to time blast the hopes of the planter.   The conse-quence would surely be a benefit to all parties concerned, greater profit to the producer, and easier prices to the consumer.

Let us now suppose other articles of food generated in a

similar manner from lifeless matter.   Surely in such a case the anxiety of life would be lessened, and the margin between every man and actual want would become indefinitely wider than at present.   Should we not here have a " reform " wider, deeper, and more satisfactory than statesmen of any school ever dream of ?

But when suggestions of this kind are laid before the public it must not be assumed that they meet with universal approbation.   A friend of ours once broached, in a mixed company the notion of the synthetic production of food.   A reverend gentlemen present gazed on him with a look of horror, and said he little thought he should ever have " lived to hear such an awfully wicked and blasphemous proposal." On this outcry no comment is needed.   It is merely a repetition of the opposition with which the lightning-conductor, the discovery of vaccination, and the use of anæsthetics were greeted.   It is strange, however, that " ministers of all denominations " never reflect how by such opposition they play into the hands of the so-called " free-thought " party.

Another note of opposition is raised by some who affect to speak as the representatives of the working man, probably by the same right that the ichneumon might represent the caterpillar.   They denounce us indignantly for seeking to feed the poor on " drugs and chemicals."   Their eloquence, being solely the outcome of their ignorance, needs no refutation.   Still we may show that this kind of opposition is in part to be traced to the circumstance that not a few of the outside public confound the proposal which we are making with certain fraudulent and semi-fraudulent practices which now play but too prominent a part in the production and sale of food.   Time was--at least so it is charitably supposed—when our articles of food and drink were all what they professed to be ; when bread was made from the powder of wheat, with no other additions than water, salt, and yeast ; when wine was the fermented juice of the grape ; when milk was a secretion of the cow, and butter a constituent separated from such milk.   In virtue of that very food dilemma of which we spoke at the outset, this is no longer the case.   Butter, cr at least what is commonly sold as such, may consist of lard, tallow, rape oil, and the like. But synthetical butter, such as we suggest, will be made not from the fat of any animal or vegetable, but from the very same fats which exist in natural butter, prepared from inorganic materials.   The possibility of the presence of any morbid or infectious matter would thus be excluded.   Between artificial food and adulterated food there can be no analogy.

From a third quarter come sceptical voices of a more intelligent and matter-of-fact nature. Say they:—All this is very well to propose, and would doubtless prove exceedingly useful if carried out, but what is the prospect of its ever becoming a practical reality? Can sugar, or any other article of food, be produced artificially from its inorganic elements? So far, we must certainly admit that this triumph has not been achieved. But we must keep it clearly in mind that chemists are, it might almost be said day by day, learning how to prepare artificially substances which a little time ago were obtainable only through the intervention of animal or vegetable life. There is in particular one instance so significant and instructive that, though the facts of the case are perfectly well known to all students of chemistry, we cannot refuse it a brief notice. From time almost immemorial a certain substance, now known under the name of alizarin, has been employed for producing exceedingly fast and beautiful colours upon cotton and other vegetable materials. Till within a few years this dye was only obtainable from certain plants, and above all others from madder, which on this account was very extensively cultivated in the neighbourhood of Avignon, in Alsace, and Zealand. There is, we may remark in passing, reason to believe that it was formerly grown to a considerable extent in Norfolk. It is certain that a street in Norwich retains to this day the name of the " Madder Market." Latterly, however, chemical research pointed out the way of preparing it artificially from some of the constituents of coal-tar. At first this artificial alizarin was a mere curiosity, but gradually the process for its preparation has been simplified and rendered cheaper, until at last cotton goods can be dyed a Turkey-red with this manufactured product at less expense than with the root of the madder-plant, and the cultivation of the latter is therefore now substantially a mere matter of history. Now we must especially beg the non-chemical reader to bear in mind that this artificial alizarin is not a substitute for, or an imitation of, the natural article, as is, for instance, " butterine " of butter. It is in every one of its properties, physical and chemical, identical with the product of the madder-plant. Place a sample of pure alizarin in the hands of the most able chemical analyst, and he will be unable to tell whether it has been obtained from coal-tar or from the plant. Hand the same sample over to the practical dyer or calico-printer, and he also is at a loss.

There is, however, an objection which may here be urged. This same alizarin is also composed of the three elements, carbon, oxygen, and hydrogen, though in very different proportions and in a very different arrangement from what we find in sugar.   But we do not form it artificially from these elements as they occur in a free state, or in mineral compounds, but from coal-tar.   Now coal is well known to be the residuum of vegetable matter, so that our artificial alizarin is, after all, a plant-product, like that obtained from the madder-root.

To this objection it may be replied that alizarin, like all the coal-tar colours, can be obtained from the mineral oils of the Caucasus, and these—according to no less an authority than Prof. Mendelejef—are derived not from the decomposition of any animal or vegetable remains, but from the reactions of purely mineral matter, in which graphite—a form of inorganic carbon—is supposed to have played a leading part.   Hence, then, alizarin obtained from the refuse of the petroleum of the Caspian would be in its origin quite independent of organic life.

There is another quite recent instance of the artificial preparation of a so-called organic body which is free from all objections, though its utility as yet is but limited. Formic acid was at one time regarded as a secretion of certain ants, beetles, and other insects, and as obtainable only through their mediation.   Then a method was discovered of preparing it artificially by distilling vegetable substances, such as tartaric acid, with oxidising agents.   Here, however, we are still employing organic matter, and a final step was still wanting.   This has now been taken : it is found that if the gas formerly known as carbonic oxide, but now termed carbon monoxide, is passed over a mixture of caustic soda and lime with certain precautions, formic acid is generated in such abundance that if required it would be available for industrial purposes.   Now, as carbon monoxide is not merely an inorganic compound, but can be prepared without employing any ingredient of organic origin, the problem is solved, and the animal compound, formic acid, is obtained synthetically from inorganic matter.

In face of such facts there is, we submit, no reason to set aside the notion of the synthesis of food-products as chimerical.   It will very probably be found a long and difficult task, but in consideration of its paramount importance men of science will not be justified in despairing.

There is, however, one limitation to which especial attention must be drawn.   The useful bodies which we obtain from the animal and vegetable kingdoms may be very naturally divided into two great classes.   On the one hand, there are so-called chemical individuals—such as sugar, starch, tartaric acid, glycerin, gelatin, albumen, and many more. On the other, there are organised bodies, possessing a definite structure, and composed of a mixture of two or more of the chemical individuals aforesaid. Now there is no *à priori* reason why we should not ultimately be able to produce artificially any and every chemical individual which Nature affords. But there is no reason to expect that we can ever succeed in forming even the simplest part of any plant or animal. Thus we may hope to form artificially albumen, with all its natural properties, and to use it as a valuable article of diet, but it will never be in our power to manufacture an egg. We may make the sugar, the tartaric acid, and every other compound found in a grape, and by combining these in the right proportions we may obtain a sum-total which shall have the taste, the smell, and the physiological action of a grape, but we can never give it the structure of a grape.

Here, then, is a rubicon which we can never cross so long as Life approves itself a something distinct from heat, light, electricity, and all the other phases of force or energy.

III. INSTINCT AND MIND.

By S. BILLING.

THE question whether the character of mind found in animals and that displayed by man is the same, both in its bases and characteristics, appears to be generally interesting; and it must be admitted that with the scientific and the *quasi* scientific the greater number incline to the idea that between them there is no specific difference. I will attempt to show not only that there is a specific difference, but that mind *in its true significance* has no place with the animal: to go further, if man were but an animal it would have no place with him. Elsewhere I have contended* that man is an animal and something more, for he has a character of intelligence which as an abstract metaphysic may be inferred to be the *soul*, and herein is the

* Scientific Materialism and Ultimate Conception.

essential difference between man and the lower animals. It appears to me that the true distinction between animal instinctives—great as they are—and mental power is lost sight of, or not understood or ignored, by the many narrators of pretty instances of animal sagacity, and by those sages who insist as a scientific dogma that instinct and mind are common to animals and to man.

In my work I have treated the subject in a discursive and fragmentary manner : the distinctions appeared so marked and decisive that I refrained from giving the subject a special chapter; the incidental observations seemed to me conclusive of the subject, and were really interfused for the purpose of showing the fallacy of the materialism so broadly interpolated in many scientific memoirs. I am impressed with astonishment to find some of our greatest biologists and so-called metaphysicians insisting on a mentality (if it be one) wholly due to sensory impressions as synonymous with the reasoning faculty of cultured man. How this arises it is difficult to understand, for the reasoning on the subject appears to be wanting in those characters of induction and deduction so necessary in the discrimination of scientific theses, and also in that analytical acumen which alone can be determinative of the problem.

I begin my special observations by an admission so large in character that some may consider it as conclusive of the question. I concede as animal instinctives, consciousness, association, appreciation, comparison, contrivance, constructiveness, will, and memory, and all the inferences appertaining to these faculties which have their bases in the senses. All of them are powers which are necessary for the enjoyment of animal existence, and are displayed in the many amusing and instructive anecdotes of animal life. When I shall speak of insect life the instinctive capacity amounts to a *prevision*, this prevision greatly exceeding the instinctives appertaining to the more highly constituted animals, not excluding man. I mean by a prevision the power which some insects possess of foreseeing conditions necessary to the preservation and to the perpetuation of the particular species. It cannot be argued that this faculty could have arisen from any process of reasoning, it having relation to acts of which the parents could not be cognizant) as in the majority of the instances the parents have no experience to direct them, they having ceased to exist before their progeny has being. It is therefore an innate and hereditary instinct, and has application to the particular varieties of insect life.

The characteristic of mental power is that of an individualism : there is no individualism in the animal instincts, however much there may be of character. The argument would be equally imperative even if cultured man was eliminated. Savage man has abstract ideas of the equities, —*i.e.*, the particular relations of man to man, a sense of justice which controls his innate sensuous propensities ; and has besides a conscience, of which, in its true abstract consideration, there is no instance in animals. It would seem, the fact that no animal has been discovered which is a fireraiser or a fire-preserver should settle the distinction between instinct and mind. Monkeys will assemble around a fire in the woods left by a hunter or wayfarer, but there is no record of their ever having added a stick to the fire, imitators as they are, in order to continue the warmth they so much affect ; even the anthropomorphous apes, which our *savants* mark out as the immediate ancestors of man, must be included. The preservation of the fire, the connection of the sticks with the warmth, is a reasoning process to which they have not attained. Surely, if the animal mentality and that of man were the same, the isolation in respect of fire would not have continued for the hundreds of thousands of years through which fire has been the familiar implement of man, and which he has possessed, if geological evidences are to be admitted, before the pre-glacial period. Geikie and his companion, in the pre-glacial boulder clay at Brandon, in Suffolk, found charred sticks and flint implements.

In examining the question of Instinct and Mind it is necessary to define the meaning intended to be conveyed when the words are used. Instincts are inherited faculties appertaining in particular degrees to all animal organisms, and have their origin in the senses, and their enjoyment in sensuous gratification and in the perpetuation and preservation of their species. All mentalities which are said to pertain to animals have their expression in sensuous aptitudes, and are directed to the gratification of the selfish instincts. If the various anecdotes denoting animal mentality are thoroughly examined they will all be found directed to some sense gratification, and that when the animals subject them to control it arises from the memory of some former discipline. The study of biological history proves that the animal instincts are graduated in accordance with the necessities of existence : this can be traced throughout the range of animation commencing with the protamœba, whose only instinctive faculty appears to be directed to

nutrition and to perpetuation. To go beyond the protamœba to the blood corpuscules, unless they are considered to
be amœbas, we have a collective community possessing no
independent life. Dumas, when experimenting on the blood
corpuscules after the method of Berzelius and Müller, found
they were subject to decomposition, but which he obviated
in filtration by passing a current of air through the liquid,
thus renewing the condition in which they exist in arterial
blood. This would seem to show that they have respiration,
and it has been shown they are propagated by fission. If
they change their forms, and breathe, they have animation,
but to them cannot be conceded those conditions named
instincts. From the blood corpuscules, supposing an independent animation up to and including man, we have
graduated instinctive aptitudes exactly apportioned to the
place in Nature which each animated variation holds. If
we are to argue that the manifestation of particular instincts
in particular animals shows mind, it may be claimed for
plants,—*e.g.*, the *Valisneria spiralis*. The female flowers
float on the surface of the water ; the male plant is chained
to the bottom : when the period for fecundation arises, the
peduncles of the male flowers break, mount to the surface,
and float around the female flowers, whereby fecundation is
effected. The *Utricularia* has even a more interesting process. These plant facts are quite as extraordinary as any
instances of mental aptitudes recorded of animals. The
plants manifest desire, purpose, and will, and the instances
go to prove that creation or animation—which you will—is
an extended chain with innumerable links. Regarding man
in his animal nature, there appears to be no distinction between him and the higher order of animals in the gratification of the instincts of his nature, excepting so far as the
sensory selfism is directed by a higher order of instinctive
adaptability.

When we define mind we define a higher power than instinctive aptitudes ; call it intelligence, we will not pause
for a word : it is that power which enables man to conceive
ideas, and to reason on subjects which in no wise are included
in the category of the senses, or have their bases in them—
abstract ideas, the true use of which is to subject the animal
selfishness, the outbirth of the sensory instincts, to the
power of these abstractions, whereby an equitable relation
is instituted between man and man and between him and
the lower creations, his being is elevated, and his brute
instincts subjected to reason. It does not follow because
man rarely reduces these equities into practice that therefore

they do not exist. I do not care to enter upon abstract
questions of religious belief, although in some form they
have their phase in the minds of all men, and therefore
speak of social equities. It is obvious the higher the mental
abstractions are carried, the farther mind is removed from
instinct.

As I define instinct or animal mentality, it is directed to
sensuous gratification, and has its bases in the senses, and
thence result emotions and the simulation of reasoning.
Intelligence is the concentration of all mental aptitudes in
which arise abstract principles leading to the contemplation
of ideal sentiments and sympathies, teaching the exact relation of thing with thing, not only discerning phenomena,
but ascertaining their origin and bases, and, by a parallelism
of principles, adducing mathematical precision and metaphysical exaltations. The ice fades from the view; it is a
sense perception of the phenomenon, but the cause of its
fading is ascertained by a mental abstraction; thus we
should say we perceive effects, but by mind we find their
causes, and conclude the distinctive powers of the mind have
higher bases than sense gratifications and sense emotions.

There is another distinction between Instinct and Mind
which in the consideration is most important; even if it
does not conclude the argument, it marks the line of demarcation so broadly that it becomes difficult to imagine
how the boundary is to be passed. This I call the *tribal.*
On consideration it will be found that any peculiar characteristic observed in creatures,—whether it be of a class,
genus, species, or variety,—that each member of the particular denomination, be it genus, species, or variety, is
possessed by the communal tribe as an innate inherited
power. If we take dogs, the species is divided into varieties
or the genus into species; whether the proper phrase be
species or variety, each has its peculiar distinction—thus,
the pointer, the gazehound, the shepherd's dog, the retriever, &c. The aptitude of each is tribal; the gazehound
has not the instinct of the pointer, or the shepherd's dog
that of the retriever, &c. Each variety has the aptitude of
the instinct without instruction. Whenever tricks are
taught to animals the means are either the incitation of
appetite or the infliction of pain, the memory of which
keeps the creature steadfast to its teaching, until it becomes
as it were ingrafted into its nature. The tribal characteristic is found in the habits of wild animals; particular
species of birds have the same constructive instinct, and
varieties of the same species differ in the constructive

faculty; and whenever a variation occurs in a particular species or variety, each member of its class practises the same peculiarities. In some rare cases are found adaptabilities to particular circumstances : not to multiply instances may be noted the jackdaw which built its nest in the window-ledge of a tower, and piled up sticks to make the basis of its nest more stable. The constructive instinct of birds is most diversive in character, but the tribal character is always maintained. It is the same when the *prevision* of insects is noticed : the water-spider (Naiadeæ) weaves a water-tight cone beneath the water, attaching it to the bottom with stays of silk, in order to provide a shelter for its young. Its labour would be useless unless the creature could empty the cone of the water with which it is filled. Walcknaer, who critically observed their habits, says they fix a globule of air to a hair of their bodies, and descend with it to the nest and eject it beneath the cone, by which means they force back the water, and fill it with "respirable gas." "The spider comes to the surface of the stream, takes a bubble of air under its abdomen," and carries it beneath the water : the voyages are repeated till the bell is completely filled with air. Thus we may say Nature invented the diving-bell; but man, the copyist, has not equalled the inventor : the insect begins and completes its dwelling below the water, and afterwards expels the water, replacing it with air. In all animal mechanics the principle is the same ; it is the whole tribe or species which possess and practice the aptitude to which culture could not add; the instinct is the inheritance of the particular tribe or variety. In all this there is no individualism ; we find only a class distinction. The sassafras bombyx deposits its eggs on the leaf of the tree, and then interweaves the leaf-stalk with the stem of the tree so strongly that the leaf, laden with its living freight, remains on the tree, braving the blasts of the winter. This process could be conceived to be a reasoned abstraction (although related to the affectional instinct), but the same fatality as an adverse argument exists—the distinction is tribal. The larva of the ant-lion in dry fine sand hollows a funnel-like inverted cone, and makes the slopes so regular, and at such an inclination, that they cannot be climbed : the larva buries itself at the bottom of the trap ; an insect stumbling on the edge of the pit falls into it ; the yielding sand carries it to the foe in waiting, and, more, the creature ejects from its lair the *débris* of its feasts,—using its head by way of a catapult, it launches the refuse far from the snare : still the instinct is tribal. Some ichneumons, or *vibrating flies*, pierce the outer skin of a caterpillar and insert their eggs beneath it ; when hatched the larvæ distinguish between the fat and the vital integuments, and it is only at the close of their existence they assail the vital organs, and then issue through the openings in the skin, and spin their cocoons on the surface of the corpse which had been their home and their aliment. In these instances—thousands more could be adduced— there appears to be a working and applied intelligence, and if they were the result of culture could not be distinguished from reason. But in these cases of prevision, as with other inherited charaĉteristics, they are innate and tribal.

The illustrations above adduced are so broad and dis- tinĉtive as should suffice to exclude the idea that animal mentality and mind, as presented by man, are the same. The first has its base in the senses, and is common to the species or varieties by which it is exhibited, *i.e.*, tribal. In man's mentality there is an individualisation. The inseĉt and the animal gets from its tribal instinĉt that which the races of man acquire by culture, and in the beyond of the sense mentality man has the abstraĉt and the ideal.

There are doubtless many extraordinary tales told of the doings of animals, but on a careful analysis they may all be traced to affeĉtional emotions and to sensory impulses ; many probably are unintentional exaggerations arising from imperfeĉt observations, and are sometimes amplified by the astonishment excited by what appears an unusual incident. I could tell extraordinary tales of my dog Grip, as could other possessors of dogs ; but in all the peculiarities there is always a something wanting—that missing link which shows the distinĉtion between mind and instinĉtive adapt- ability.

I have endeavoured comprehensively to treat this most interesting subjeĉt, avoiding the introduĉtion of anything extraneous into the argument. To me the problem appeared to be a simple one ; but when I find a great diversity of opinion enveloping it, and ponderous tomes presented as- suming to be analyses of mind and instinĉt, I pause in wonder. The works, for the most part, are confined to anecdotes of animals, amusing, but very vague when used as the bases for argument.

In reviewing the subjeĉt I cannot but feel that irrelevant and ill-digested comments have built up an argument which appears to be wanting in a fundamental principle. In the diorama of the universe the most modest of inorganic products, the merest speck of animation, through its cycle by development, reaches up to man. Each plays its part in the grand Kosmic picture : each is an adjunct in one grand scheme ; the inorganic is succeeded by the organic, so instinctive capacity is succeeded by mental power, for some comprehensive end, although its purpose may be concealed from our purview. Had there not been a purpose to be accomplished, other than merely the successions or mutations of the inorganic and the organic, Mind becomes a needless feature in the grand panorama ; but as there are no wastes in Nature, we must conceive Mind in the vast projection of creation plays its part as a distinctive capacity, or we must conceive its institution to be but a wasted energy. This is inconceivable, for each part in the Kosmic idea has its place ; and impossible, for if intelligence designed, intelligence completed its project. The Kosmos discloses a tremendous mechanism, diversified yet homogeneous ; and as in the march of events we find that there are no mechanisms but which owe their institution to intelligence, we then can view the universe as the condensation of a thought—the material presentment of an immaterial energy.

IV. MODERN ADVANCES IN METEOROLOGY.

NONE of the Sciences, save perhaps Electricity, has taken such rapid strides during the last fifty years as the much-despised science of Meteorology. Strange to say the advance of the one has been closely bound up with the progress of the other in a much more intimate manner than the casual observer would suspect. For the greatest advances in electrical science have been the result of the improved methods of measuring electrical quantities which were necessitated by the growth of the telegraph ; and the modern advances which have been made in meteorological science consist of deductions drawn from observations simultaneously made over large areas, and transmitted by telegraph to central stations. The entire modern system of storm-warmings—whether we refer to those issued by our own Meteorological Office, or to the predictions of " Old Bob " out in the Far West—dates from the era when the preparation of daily weather-charts was rendered possible by the use of telegraphic signals. To a large section of the public, however, the process by which these deductions and predictions are arrived at is a complete mystery. They hear of cyclones and anti-cyclones, and areas of depression travelling over land or sea, and think that the invisible and intangible entities thus described are something quite beyond their comprehension. While it seems wonderful enough that Captain Saxby can predict a high tide in the Thames, or the American meteorologists " cable " us a storm-warning two or three days in advance, the occasional failure of a prediction or the unexpected and unannounced advent of a storm continues to puzzle the good folk, who once more take refuge in the notion that the governance of the weather has not yet been entirely handed over to the calculators, who must therefore be regarded as only one degree more entitled to respect in their predictions than are Zadkiel and Old Moore.

Very wisely, therefore, it was determined some months ago, by the Council of the Meteorological Society, to arrange for the delivery of a course of popular lectures by some of the most accomplished of modern meteorologists, who should, as it were, take the public into their confidence as to the method in which the observations of the weather are made all the world over, and how these records can be *used* for the benefit of mankind. These lectures have recently been published in a handy little volume entitled " Modern Meteorology."* Anyone desiring to possess, in simple language and accessible form, an accurate and reliable account of the modern methods of weather-lore, cannot fail to be pleased with this little work.

The first two lectures of the series are of an introductory character, and deal with the Physical Properties of the Atmosphere, and with the Range and Distribution of Temperatures in the Air: they are by Dr. R. J. Mann and Mr. J. K. Laughton respectively. Admirably do they set forth the necessary bases of observation of climate and weather. The nature and composition of air, its compressibility, its expansion by heat, its weight and pressure, are all dealt with in turn, and the relation of the invisible water-vapour it contains to the production of mist and rain explained. The question of the temperature of the air at different

* Modern Meteorology: a Series of Six Lectures delivered under the auspices of the Meteorological Society. (London: Edward Stanford. 1879.)

points of the globe is treated of with equal clearness, and the causes that tend to modify it are set forth.   The influence of exposure in promoting rapid and extreme changes of temperature, and, on the other hand, of shelter from radiation in hindering such extreme changes of climate, are commented on ; and it is shown how we may explain one of the marvels of Arctic voyagers, who have frequently observed the intense heat of the direct sunshine blistering black paint, and making the pitch boil and bubble up in the seams of the ship's deck, whilst all around the snow lies thick over the frozen earth, and the temperature in the shade is many degrees below freezing-point.   Our old friend the Gulf Stream, and the service it performs in carrying heat across the ocean, thereby rendering habitable these islands, which would otherwise be as inhospitable as Labrador, claims attention as being but one of a very large number of similar ocean-currents, all performing the same great function of equalising extremes of temperature and softening down the otherwise unendurable rigours of torrid and arctic climates.   The account given of the warm and dry winds often experienced on the lee-side of a mountain, or range of mountains, is extremely instructive.

The third lecture, on the Barometer and its Uses, is by Mr. R. Strachan.   Some of Mr. Strachan's remarks anent the popular notions concerning the predicting power of the weatherglass are so excellent that we quote them piecemeal.

" The barometer has always been vaunted as a weather oracle, but it really has no pretences to such a dignity.   It simply shows the statical pressure of the atmosphere above it.   This pressure must change before there can be any variation of the barometer, supposing of course that it is kept in the same temperature.   The air puts the mercury in motion; hence inertia and friction must cause the column to change *after, not with, certainly not before*, the air pressure varies.   If it be admitted that every wind has its weather, then to observe the direction and the force of the wind is the first step to observe the weather.   Now the wind is a dynamical condition of the air, one which we are accustomed to roughly estimate from our sensation, and it is very advantageous to have in addition an exact measure of the statical condition of the air at the same time, which the barometer gives us.   We gauge then not only the horizontal movement of the air, but its vertical mass as well, and with the two estimates we are better enabled to judge of proximate changes than with either alone.   However, the practice has been, for the most part, to relate changes of wind and weather to the state of the barometer; and innumerable rules have been propounded to enable everyone to become weather-wise by the aid of a barometer. . . . A sort of elixir of them has been distilled and bottled up on the barometer scale, to satisfy the craving for condensation, for knowledge formulated—in short for Science, which humanity manifests. I refer to the lettering on the scale, with which we are all familiar, which, however little objectionable in cities like London and Paris, when it got abroad among the hills and mountains in the interior of continents, and away on oceans, appeared sometimes grotesque, not to say absurd."

Now the great advance from these untrustworthy guesses from solitary observations arose from the systematic observation over large areas. Such observations revealed that over a storm area the atmospheric pressure was distributed in a singular and symmetrical way; that while all great storms and tempests whirl round a calm centre or nucleus, this nucleus, or " eye " of the cyclone, has the lowest barometer, while the pressure increases all around in widening circles to the outmost regions of the storm. Moreover, it is known that in the northern hemisphere all such storms rotate round the centre of depression in the opposite direction to the hands of a watch. The relation to these cyclones of certain centres of greater barometric pressure, or anticyclones, which usually bring fine weather, and around which the winds sweep in an opposite direction, is carefully explained.

One other quotation we must make from this excellent lecture, respecting the well-known hypothesis of a connexion between the cycles of sun spots and the cycles of weather and storms :—

"As regards annual values of barometrical observations, no periodicity has yet been traced for them. They appear to me to afford the most precise data for investigating the sun-spot cycle as connected with weather. There are no *à priori* reasons known to me for supporting that theory, and I consider the great amount of labour expended in bolstering it up as very much misapplied, and the whole thing as a wild-goose chase, on such paths as the rainfall, the black-bulb thermometers, and auroræ. Only I would remark that if it be proved that the sun-spot maximum coincides with a barometer minimum, and the sun-spot minimum with a barometer maximum in India, there must be some region or regions where the law is reversed, otherwise we should have to acccount for an abstraction from our atmosphere in some years, and an accession to it in others."

The fourth lecture, on Clouds and Weather-signs, by the Rev. W. Clement Ley, is a most readable and lively chapter, and its value is enhanced by a beautifully coloured plate. We do not remember any published work which gives with such admirable clearness an account of the relation subsisting between different kinds of cloud and the lines of barometric pressure and of cyclonic movement. The description in particular of the "anti-cyclonic stratus" frequent during a spell of fine weather in winter, and often accompanied by dense ground fogs, is extremely interesting and valuable.

The lecture on Rain, Snow, Hail, and Atmospheric Electricity, by Mr. G. J. Symons, is a clear summary of the most recent information on these heads, while Mr. R. H. Scott, of the Meteorological Office, sums up the whole matter with a discourse on the Methods and Objects of Meteorology, showing why a scientific knowledge of the weather is sought, and what incalculable value it will be to mankind. To every profession and calling of mankind, to the farmer, the cotton-grower, the sugar-maker, the soldier, and the physician, the power of foreseeing and foreknowing the conditions of time and climate for a few months or weeks ahead would be of priceless moment. The skein to be unravelled is truly a tangled one, but thread by thread it is being tracked out and unfolded. But three centuries ago the advent of comets and the occurrence of unusually high tides were regarded as phenomena outside the realm of exact knowledge, and as defying all calculation. Science, which steadily lays hold of the uncorrelated facts of Nature and marshals them into ordered symmetry, has long ago conquered the comets and the tides, and has even taken in hand the star-showers. Who shall doubt her power also to seize upon the laws of wind and weather, tracking them to their hidden sources, and making them her own. The modern meteorologist is of this faith, and works patiently to this end. With Newton he believes that "the whole difficulty of philosophy seems to me to lie in investigating the forces of Nature from the phenomena of motion, and in demonstrating that from these forces other phenomena will ensue." And with Newton he can add—" I would that *all* other natural phenomena might similarly be deduced from mechanical principles."

IV. THE RATE OF ANIMAL DEVELOPMENT.

By J. W. SLATER.

" Consider young ducks."

ONE of the attempts which have been made to establish the existence of a " great gulf " between man and beast may be pronounced exceptionally curious as an instance alike of careless and defective observation and of rash conclusions. That by such arguments men of eminence could really mislead themselves, and succeed for a length of time in misleading the outside public, is deeply humiliating. Prof. St. George Mivart suggests* that a book should be written on the " stupidity of animals." We are far from denying that such a work would be useful; but should the needful companion volume on the " stupidity of man " make its appearance in due course, it might not unfittingly open with the reasoning we are about to quote.

To begin then : the slow bodily development of the human infant and its prolonged helplessness are matters far too familiar to require proof, or even illustration. No less familiar and universally admitted is the rapidity with which foals, calves, lambs, kids, chickens, and ducklings acquire the use of their limbs and other organs. These facts could not fail to come under the notice even of the most careless observers. But who could have imagined that the said facts would be, without further enquiry, at once seized hold of as a theme for stilted declamation, and be elevated to the rank of a fundamental distinction between man and the lower animals. Yet this strange error has actually been committed, not merely by men of words, like Addison, Paley, and Whewell,—which is surely sad enough,—but even by a man of things, like Sir Humphry Davy. The great chemist attempts to show that man does not use his limbs instinctively like other animals. Says he :—

" Man is so constituted that his muscles acquire their power by habit,† but in the colt and the chicken the limbs are formed with the power of motion, and these animals walk as soon as they have quitted the womb or the egg.

---

* **Lessons from Nature.**
† To speak of acquiring a power by habit is scarcely rational. The power must exist before the habit can be formed.

" *Physicus.*—I think I have observed that birds learn to fly and acquire the use of their wings by continued efforts in the same manner as a child does that of his limbs.

" *Ornither.*—I cannot agree with you. Young birds cannot fly as soon as. they are hatched, because they have no wing-feathers ; but as soon as these are developed, and even before they are perfectly strong, they use their wings, fly, and quit the nest without any education from their parents."*

Very similar assertions are found in a laborious attempt made by the late Prof. Whewell† to set aside the palpable fact that man, like every other animal, has an instinctive— or we might perhaps better say a hereditary—knowledge of the functions of his voluntary organs.

Said the Professor :—" The child learns to distinguish forms and positions by a repeated and incessant use of his hands and eyes : he learns to walk, to run, and to leap by slow and laborious degrees ; he distinguishes one man from another and one animal from another only after repeated mistakes. Nor can we conceive this to be otherwise. How should the child know at once what muscles he is to exert that he may stand and not fall, till he has often tried ? How should he learn to direct his attention to the differences of different faces and persons till he is roused by some memory or hope which implies memory ? It seems to me as if the sensations could not, without considerable practice, be rightly referred to ideas of space, force, resemblance, and the like. Yet that which thus appears impossible is, in fact, done by animals. The lamb, almost immediately after its birth, follows its mother, accommodating the action of its muscles to the form of the ground. The chick just emerged from the shell picks up a minute insect, directing its beak with the greatest accuracy. Even the human infant seeks the breast and exerts its muscfes in sucking almost as soon as it is born."

So, after all, "that which thus appears impossible " is, in fact, done not by "animals " only,. but by man also ! The concession contained in the last sentence is simply fatal to what has gone before. To be consistent the learned Professor ought by all means to have asserted that infants learn to suck only " by slow and laborious degrees," and after its sensations have been rightly referred to appropriate " ideas." It would scarcely be a more unwarrantable assumption than

* Collected Works. Vol. ix. Salmonia, p. 105.
† Philosophy of the Inductive Sciences, ii., p. 616.

those he has indulged in abundantly in the course of his argument.

In the same vein as Davy and Whewell, teleologists and natural theologians, when enlarging upon the marvels of instinct, have seldom failed to " trot out" the colt, the calf, or the lamb, to invite our consideration to the chickens and the "young ducks," and to erect upon the precocity of these creatures—as compared with the tedious development of our own species—a fancied wall of demarcation between man and beast. Had they been really actuated by a scientific spirit they would have felt it their bounden duty to ascertain whether *all* the lower animals were, in contrast to man, able to use their limbs soon after their birth. Had they done so they might have met with evidence similar to what is thus given by an actual observer* in describing an infant orang-utan which had come into his possession :—" The Mias like a very young baby, lying on its back quite helpless, rolling lazily from side to side, stretching out its hands into the air, wishing to grasp something, but hardly able to guide its fingers to any definite object, and when dissatisfied opening wide its almost toothless mouth, and expressing its wants by an almost infantile scream. . . . When I had had it for about a month it began to exhibit some signs of learning to run alone. When laid upon the floor it would push itself along by its legs, or roll over, and thus make an unwieldy progression. When lying in the box it would lift itself up to the edge into almost an erect position, and once or twice succeeded in tumbling over."

Thus we see that the nearer brutes approach to man in their structure the more gradual is their development. The process which in the colt and the lamb is contracted so as to escape observation is here shown at very considerable length. That the child, especially in the higher races of mankind, makes a still more gradual progress, is plainly a mere question of degree.

The young ape which Mr. Wallace observed was, beyond all reasonable dispute, acquiring the use of its limbs precisely in the same manner as a human child. If the latter learns, by slow and laborious degrees, what muscles he must exert in order to effect any desired movement, so does the young ape. If the child cannot judge of the position and distance of objects till it has by considerable practice learnt to refer its sensations to appropriate " ideas," the same must be said of the young Mias. But if the young

* A. R. WALLACE, Malay Archipelago, p. 45.

apes, and indeed all other young animals, inherit from their forefathers a latent knowledge of the use of their organs, which is called into activity as soon as their muscular and nervous systems are sufficiently developed, the same holds good of the human infant.

Of course it would be unfair to demand of such men as Prof. Whewell that before theorising and dogmatising they should go forth to the forests of Borneo in search of facts. As for Davy, his splendid achievements in chemistry may cover his failure in biology. But surely every man in Europe, though he may never have met with infant apes, must have seen how kittens, when beginning to walk, totter, stagger, and roll over just like young children; how they pat at, and endeavour to touch, objects beyond their reach; and how, even after the fore legs have gained a considerable degree of firmness and obey volition, the hinder extremities remain feeble, and are often for a time trailed helplessly along. Thus, then, we see that in the Mammalia, instead of man standing alone, sharply contrasted to the rest of the class, he merely occupies one extremity of a series towards the other end of which stand our much-talked-of friends the lamb and the foal, whilst the carnivorous animals and the apes occupy intermediate positions. Some very plain reasons why this should be the case will follow in due course.

But what are the facts concerning birds? Are they all able, as soon as hatched, to direct the beak with perfect accuracy, to select suitable nourishment, and to flutter about awaiting merely the growth of their wing-feathers before they can take flight? Davy's "Ornither" must have been either a wilful sophist or a most egregious goose. Had he been an accurate and conscientious observer he must have been aware that what he predicates of birds in general is true, in any sense, merely of the Gallinæ, Grallæ, Anseres, and Struthiones, and assuredly not of the Passeres, Picariæ, Columbæ, Psittaci, and Raptores. Did any of the authors to whom we have been referring, before indulging in platitudes on young ducks, ever take the trouble to "consider" young hawks, young thrushes, or young canaries? Had they done so they would have seen that such nestlings, instead of being able to "direct the beak with the greatest accuracy," can merely sit in the nest with open mouth waiting to be fed! A young canary, so far from being able to stand or walk, seldom fails to break its legs if startled and induced by fright to attempt leaving the nest. Such facts as these are known to every bird-fancier,—nay, we might say to every rustic youth who has ever robbed a nest and has attempted to bring up the callow young by hand. They are not known, it appears, to men of erudition.   It was, we think, the Prime Minister of Gustavus Adolphus, of Sweden, who said to his son—" Thou knowest not with how little wisdom the world is governed."   In like manner, and with even more truth, it might be said that we know not with how little accurate thorough knowledge books are compiled, the world is misinstructed, and imposing reputations are built up.

We do not demand original observation from Professor Whewell.   Everyone knows that the possessors of inherited wealth are apt to despise the man who has acquired a fortune by his own exertions.   But there is a class of men—more numerous we fear in England than in any other civilised country—who, with a still more unjustifiable prejudice, contemn all knowledge that has not been derived from book, and scorn original research and discovery.   Still it is strange that none of these writers should have met with the following observation from Gilbert White :*—" On the fifth of July, 1775, I again untiled part of the roof over the nest of a swift.   The squab young we brought down and placed upon the grass-plot, where they tumbled about and were as helpless as a new-born child.   When we contemplated their naked bodies, their unwieldy, disproportionate abdomina, and their heads too heavy for their necks to support, we could not but marvel."

Davy and Whewell might, further, have found in Erasmus Darwin's " Zoonomia"† some remarks on the different stages of maturity which animals of different species have reached when they are first brought into the world.   The author uses these very words :—" The chicks of the pheasant and the partridge have more perfect plumage, more perfect eyes, and greater aptitude for walking, than the callow nestlings of the dove or the wren.   It is only necessary to show the first their food and teach them how to pick, whilst the latter for days obtrude a gaping mouth."   Would it have been too much trouble for a man of such extensive reading as Prof. Whewell to have run his eyes over the passage above quoted ?   Being, moreover, a German scholar,—at least to the extent of an occasional mistranslation from the language,—the Professor might have read that Lorenz Oken divided the class Birds into two main subdivisions, nest-

---

* Natural History of Selborne.   Letter XXI.
† Vol. i., pp. 187—194.

sitters and nest-quitters (*nest-hocker und nest-flüchter*), according as when hatched they remain helpless in the nest or are at once able to run about and seek food for themselves.

Davy, by the mouth of "Ornither," gives a very lame explanation of the fact that the majority of birds cannot fly as soon as hatched. Before they can take flight they have to await not alone the growth of their wing-feathers, but the simultaneous development of the muscles. The Raptores, Passeres, &c., are, as we have already seen, unable to walk as well as to fly. Does this inability depend upon the want of feathers? The fact that parent birds educate their young is clearly established by the interesting observations of Dr. C. Abbott.*

In the case of birds of prey the process of education is somewhat prolonged, even after leaving the nest. It is thought by many that Deuteronomy xxxii., v. 11, is a description of the manner in which eagles train their young to fly; "stirring up" the nest, *i.e.*, shaking and disturbing it so as to compel the nestlings to leave their cradle; "fluttering" over them and "bearing them on her wings," —that is to say, following and intercepting their downward movement, and aiding them to re-ascend.

Thus we see that the condition of the young of the lower animals is after all analogous to that of the human infant. The child, indeed, is still slower in learning to walk than the kitten or the young ape, not because he has to learn in a different manner, but because the development of his muscles and joints is much more gradual; because his head is relatively heavier; because he has to support himself on one pair of limbs only, thus rendering his base much narrower and his centre of gravity higher from the ground; and because, as we have already pointed out in the case of the kitten, the hinder extremities gain strength more slowly than the anterior.

Surely, therefore, the helplessness of the human infant can no longer be regarded as an exceptional phenomenon, and all conclusions based upon it by rhetoricians may be safely dismissed to dream-land, whence they came.

---

* Quarterly Journal of Science, vol. vi., p. 361.

ON WATER AND AIR.*

By JOHN TYNDALL, D.C.L., LL.D., F.R.S.,

Professor of Natural Philosophy at the Royal Institution of Great Britain.

LECTURE IV.

IN our last Lecture I tried to explain to you that the rays of the sun that are really influential in producing eva- poration are rays which are entirely incompetent to excite vision. They are what are called invisible rays, and, lest I should have been misunderstood, I think I will dwell for a moment upon this point of visible and invisible rays. I will take here, as I have already done, a slice of white light, and you shall see it, in the first instance, upon the screen. There it is; and now we turn this beam aside and send it through a body which we call a prism. That will separate the different colours of that white light the one from the other, and you will have upon the screen that beautiful exhibition that we call the spectrum. There it is. Now I will add another prism to this single one, and in that way I hope to get, as we have formerly done, a larger spectrum. [A second prism was introduced.] Here, then, is this beautiful display of colour produced by a slice of ordinary white light. Now, I want you to understand that all these lovely colours emanate from this thing that I have called our little " domestic sun," the so-called electric light. Now, suppose that I were not gifted with the sense of vision, but suppose me to be gifted with an extreme sensitiveness as regards heat; then I might walk through that spectrum, and report to you what I should experience supposing my- self to be a blind man. I will simply walk through, begin- ning at this dark portion of the spectrum [adjacent to the red],—or, at least, beyond the spectrum altogether. Sup- posing my sense of warmth to be exceedingly delicate, I should say to you that I am now receiving a very powerful amount of heat from that electric lamp. You do not see

* Being a Course of Six Lectures adapted to a Juvenile Auditory, delivered at the Royal Institution of Great Britain, Christmas, 1879. Specially re- ported for " The Journal of Science."

the rays, but the heat falls there far beyond the red. A greater amount of heat falls there than falls in any part of the visible spectrum which you see before you. Let me suppose that I commence here at this distance. I should feel a certain amount of warmth here. I come along; the warmth augments. I come farther along, and the warmth is greater still. In this position I should have the maximum amount of warmth falling upon my face. Then, as I pass on, I come to the visible portions of the spectrum,

FIG. 24.

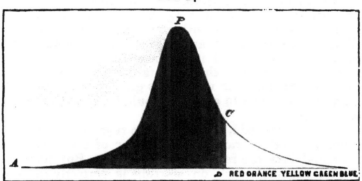

and I find that I am less warm here. I pass into this brilliant red colour, and I still find a less amount of heat falling upon me. I go on through the entire spectrum in this way until I reach the extreme blue end, and a gradually decreasing amount of heat falls upon my face. I go even beyond this extreme blue end of the spectrum into the dark space beyond it, and allow rays which have no power at all to excite vision to fall upon my face, and yet I should be able to report to you that far beyond the visible eye a gradually decreasing warmth was falling upon my face. Thus to a blind man this whole range of radiant power would represent itself as heat. Of this vast range along which I have gone, the eye, that wondrous organ, selects a little space as available for vision: but beyond this space we have, in both directions, a vast amount of radiation, and so strong are the radiations in the invisible portion that those radiations beyond the red end have about seven or eight times the radiant power of the whole of the visible spectrum. In this diagram (Fig. 24) the area of that dark mountain (c p) represents the power of the invisible radiation to which I have referred. The other portion of the diagram represents the power of the visible spectrum; so that you have invisible rays equal to between seven and eight times the visible in point of energy.

In our last lecture I tried in your presence to separate one class of rays from the other. Here is a very beautiful clear liquid, called bisulphide of carbon. Here is an element called iodine. You see it here in the form of crystals. If I pour some bisulphide of carbon into this glass cell, and drop into it some of our crystals of iodine, you will see that the liquid instantly becomes dark. This is the solution which I used in our last lecture for the purpose of cutting off the visible rays and allowing the invisible rays to pass through. The sharpness with which that solution cuts off the light of the spectrum, and allows all that mountain of heat, as I have expressed it, to pass through, is amazing. As I have said, this is the preparation which was used to sift or filter the radiation which was given forth by our little sun, and to enable us to see what could be done by the invisible rays alone.

Well, as I have said, these invisible rays are those which are instrumental in raising the vapour of the ocean into the air and into the region of the clouds. But how is it that this vapour, which on starting is perfectly invisible, becomes cloudy? That you will understand immediately. You saw in our last lecture that, by the application of heat, we were able to convert a visible cloud into invisible vapour; and now I want to show you the effect of the simple expansion of air. The air here, at the lower portions of our atmosphere, has to bear the weight of all the atmosphere above it. The consequence is, that it is compressed or closely compacted together; and when that air ascends, carrying with it a load of aqueous vapour into the upper regions of the atmosphere, the pressure upon it becomes less and less, and the consequence of this is that it expands more and more, and by its expansion it chills itself. The chilling thus produced is competent to cause the invisible vapour which started from the ocean to curdle up as cloud. Let us take, for instance, the plains of Lombardy, in Italy. I have sometimes been there when the heavens overhead were of the deepest blue, and not a cloud was to be seen. I have been there when the wind has blown towards Monte Rosa, in the Alps. You can see the effect splendidly from the top of Milan Cathedral. I have seen the air which passed over the plains of Lombardy without a single trace of cloud; and when it came in contact with the mountain it was tilted up to a higher level and chilled, and the Alps became covered with cloud, though while the air was passing over the plains of Lombardy it was perfectly clear.

Mr. Cottrell has, by means of a pressure syringe, compressed a quantity of air into this iron box (v, Fig. 25), and

FIG. 25.

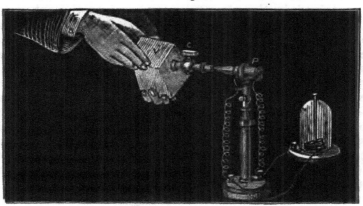

I want to show you now that, simply by being allowed to expand, the air in this box will chill itself. You see that when I chill the face of this instrument which is connected with the galvanometer, the red end of the needle goes away from me. The smallest contact with the chilled blade of a knife causes the red end of the needle to move in that way. I will warm the instrument a little, so as to bring the needle on the side which indicates heat. Now, what I want you to understand is this:—Here I have compressed air. At the present moment that air is warmer than the air in this room. I will now cause the air to expand by simply turning the stopcock and allowing it to issue. It will force itself out, and a chilling will be produced. The work done by the air in forcing itself out of this box will cause a chill in the air which forces the first portions of the air out. Thus the last portions of the air will be so chilled that you will find that the red end of the needle will go across zero and up on the other side. [The compressed air was allowed to escape from the iron box, v, and impinge upon the face of the thermo-electric pile, P, connected to a galvanometer.] Although the air when compressed was warm, yet it has become chilled by its own expansion, and has chilled the pile. You see that the needle comes down from the direction of heat, and I have no doubt that it will cross zero and go up considerably upon the other side, proving the chilling

T 2

of the air by its own expansion.   [The needle followed the course indicated by the lecturer.]

In this glass tube I have a certain quantity of moist air. Mr. Cottrell will now, by means of the air-pump, exhaust this long cylinder of the air which it contains, and then I will ask him to turn on a stopcock so as to allow the moistened air contained in the glass tube to pass into the exhausted cylinder.   If I mistake not, the simple expansion of that moistened air by this means will cause a chilling to occur in the glass tube, and that chilling will be sufficient to bring down a visible cloud of moisture.   When I tell him he will turn on the cock.   Some of the moistened air will rush from the tube into the exhausted cylinder in connection with the air-pump, and will fill that cylinder, and, by virtue of the expansion, the air which is left behind will be chilled exactly like that which was contained in the iron box, and in consequence of that chilling we shall have a cloud precipitated before our eyes.   [The experiment was then performed, the interior of the tube being illuminated by the electric light.] There you see a beautiful cloud precipitated the moment he turns the stopcock.

Well, this shows you how clouds can be generated simply by the expansion of the air.   When the air ascending up to a higher region is thus expanded, in virtue of the lessening of the pressure upon it, the aqueous vapour contained in the air comes down in this way as cloud.   Well, these clouds, of course, descend upon us as rain ; but sometimes they descend upon us as snow.   The aqueous vapour, when thus condensed in the upper regions of the atmosphere, is not only converted into particles of water, but if the chill be sufficient it is converted into a solid ; and nothing can be more beautiful than those crystals of snow which are produced in the air in calm weather, when the aqueous vapour of the atmosphere is congealed.   A great many years ago Dr. Scoresby observed the beautiful forms exhibited by the descending snow in the Arctic regions.   Mr. Glaisher has also made beautiful drawings of those snow crystals as found in England.

In the winter of 1859 I wished to ascertain the motion of a glacier in mid-winter.   I think I was beside the Mer de Glace on the 28th December, 1859, and during a portion of the day heavy snow fell, and I think that in about half an hour my hat was covered for an inch thick with snow.   On examining this snow I found every particle of it was found to consist of wonderfully beautiful crystals.   I am now going to show you some of these drawings (Fig. 26), but these cannot be compared in point of beauty with the snow crystals that fell upon my hat at that time. I say that this is a rude and crude form in comparison with the beauty of the crystal itself. I wish you to remember that in all cases the crystalline form is preserved. You always have six of those leaves, and no deviation from that number. You may have deviations in the form of the leaves, but there is no deviation whatever from the six-leaved form. Mr. Cottrell

Fig. 26.

will show you another, and you will find in every case this six-leaved stellar form of the snow. There is no deviation from this rule.

Now I want to try to make clear to you how it is that scientific men explain to themselves this wondrous architectural power, if I may so speak, displayed in the building up of these crystals. We have a number of these crystalline bodies here. For instance, this block is a piece of rock crystal, and here we have a specimen of artificial crystal. Here is a crystal of sulphate of copper, and in the front here you have a mass of alum crystals; and here, thanks to Messrs. Hopkin and Williams, I have a number of artificial crystals. Now I want you to be able to realise the mode in which scientific men represent to their minds how these wonderful crystals are built up. Here I have what every boy or girl is more or less acquainted with, a bar magnet; and here is a magnetic needle. If I hold this magnet thus

in my hand, and draw the magnetic needle towards it, you
see that the red end is attracted. There is a certain vivacity
of vibration which becomes more and more languid as the
centre of the magnet approaches the needle, and directly I
come above the centre of the magnet you have the thing
reversed. The blue end of the needle is now attracted instead of the red. If I bring the magnet down the needle
turns round, and the red end is again attracted. Thus you
see that we have one-half of this magnet attracting one end
of the needle and the other half attracting the other end,
and we have always attraction and repulsion exerted together.

Now I wish to define a term that I have to use with
reference to this double force. This simultaneous exertion
of attraction and repulsion is what we call "polarity."
Magnetism is called a polar force because of this duplex
action. In the case of the attraction of gravitation we have
nothing of that kind. The sun attracts the planets, and
every particle of matter attracts every other particle without
the exercise of this double force. The attraction of gravitation is not a polar force. The attraction of magnetism is
a polar force.

Now Mr. Cottrell will throw upon the screen the image
of a small magnet. There you see it (Fig. 27); and almost
every boy has made experiments with iron filings upon magnets. Here we have our magnet, and I will shake some
iron filings over it, and we shall see that those iron filings

FIG. 27.

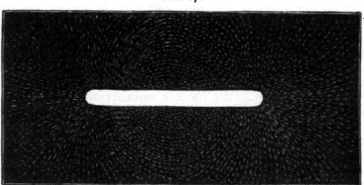

arrange themselves in a certain definite fashion over this
magnet. You see how beautiful it is; that is lovely. You
see how beautifully they arrange themselves. Every little

particle of iron filings here has two poles, and in virtue of this double polar action they so act upon each other as to arrange themselves in this particular way. We have here an example of the action of what we call a polar force. We see that this polar force is capable of causing these small particles of iron to arrange themselves in definite curves. Now what I want you to understand is this :— Starting from this conception (the notion of a polar force given to those iron filings) the scientific man, who uses the eye of his mind far more than the eye of his body, sees in the power that builds those snow crystals together a kind of polar force. Were it not that in a room so warm as this it is difficult to operate upon ice, I could produce before you those beautiful crystals of water which, as I have said, are so lovely when you see them upon your window-panes on a cold morning. We have been operating upon these crystals, and have produced them in a variety of ways. But instead of water I will operate upon something different. Here we have various kinds of beautiful crystals. This is a crystal of our rock salt that we had in our last lecture ; and if you look at this you will see, as in a diamond, planes of shining surfaces. Those planes are the planes of cleavage, and we can cut the crystal parallel to those planes of cleavage and produce these beautiful shining surfaces. This rock salt is cleavable into beautiful cubes ; so with the diamond also. It is in the same way cleavable into masses corresponding to its own crystalline form.

I will take a film of liquid, which I trust we shall be able to crystallise. Mr. Cottrell has here some very clean plates of glass, and he will pour upon one of the plates of glass a film of liquid that will be capable of crystallising. We may use water as the liquid in this experiment, but it will be more convenient to experiment with sal-ammoniac, or, as it is called, chloride of ammonium. Mr. Cottrell will pour a film of this liquid upon a glass plate, which he has now in his hand ; and he warms the plate for a little while in order to hasten the evaporation. Then he will place the plate in front of a microscope, and he will throw the image upon the screen. Here you see the particles suspended in the liquid, and I have no doubt that when the evaporation has been going on for a time you will find crystals starting through this mass. You will find an illustration of that wonderful architecture, by which the snow crystals are built up. There, you now can see the crystals forming. See how they dart across the field—those wonderful crystalline spears, which are illustrations of this building-power to which I have re-

ferred, and which depends, as I have said, upon this polar
force exerted by the molecules of matter. We will let the
crystallisation go on until it overspreads the entire screen.
The crystallisation of water upon the window pane on a
frosty morning is quite as beautiful—perhaps more beautiful
—than this. This effect is exceedingly good. Mr. Cotɪrell
has given himself great trouble to become master of this. I
have rarely seen it better than the exhibition that you have
now before you. It is wonderful to see how the crystallisation flashes forth, as it were, by sudden bursts of power, and
falls into these beautiful shapes. I will content myself with
one more sample of this architectural power exerted by the
ultimate particles of matter. Here is a small cell which
contains a solution. In our last lecture you saw that by
means of our voltaic battery we decomposed water into its
two constituent parts, and converted the water into oxygen
and hydrogen. I might have collected the separate gases
before you in your presence, and shown you that these gases
were oxygen and hydrogen. In the same way we shall try
to decompose this solution of acetate of lead ; and here,
instead of the plates of platinum which I employed last time,
we have these little rods of platinum wire. They are very
clearly shown on the screen. We will now send a voltaic
current across this solution in which these two wires are
immersed, and you will find that on one of them the lead
will be liberated ; but, instead of forming a kind of sand
heap—a confused mass of the atoms of lead—those atoms
of lead, as they are liberated, will build themselves up into
most beautiful crystalline ferns.

Mr. Cottrell has now sent the current through the liquid,
and he has done so with caution, in order that you may see
the growth of those little ferns of lead sprouting from that
wire. You have a beautiful growth of lead crystals, thus
showing, as I have said, the wonderful architectural power
possessed by the atoms of lead when they are liberated.
Nothing can be more beautiful than that slow growth.
These beautiful fern-like masses resemble a form of vegetation growing in your presence. I will now ask Mr. Cottrell
to reverse the direction of the electric current, and you will
find these masses disappear, and, on the other side, you
will find that the lead will be thrown out first of all in films
like a spider's legs. The crystals first formed will melt
away, and the ferns will be produced on the opposite pole of
the battery. [The current was reversed, and the crystals of lead
disappeared from one pole and were formed on the other.]

This force of crystallisation which we have here mani-

fested and illustrated comes into play when water passes
from the vaporous condition into the solid condition in our
atmosphere; and, not only so, but it comes into play in
every mass of ice upon our lakes. There are before you
some splendid masses of ice. This is not natural ice, but
it is ice which has been artificially produced by the General
Ice Factory Company, a company that has instituted itself
for the production of this article, ice, in London. Now, I
want to show you how it is that these masses of artificial
ice are produced. I have never seen such a large block of
artificial ice as this before. This ice is produced very
cheaply—so cheaply as to compete successfully with the ice
which is brought from the lakes of Norway. The cheapness,
however, you and I care very little about as regards that
about which I now wish to speak to you. Here is a clear
liquid called sulphuric ether, and upon the face of this
thermo-electric pile I will place the copper basin, B (Fig. 28).

Fig. 28.

I will warm the basin a little so that the red end of the
needle, connected to the pile by the wires G, shall come well
towards me; and now I pour a little of this sulphuric ether
into the copper basin, and I will warm the liquid by simply
rubbing it round with my warm fingers. I want you to
notice the effect of the evaporation of that ether. It is a
very volatile substance, and, when it evaporates, the evaporation is accompanied by the chilling of the ether, so that,
although, in the first instance, the needle will rest in its
present position upon the side of heat, still, after a little
time, in virtue of the evaporation of the ether, which evaporation will infallibly be associated with a chill, the needle
will come down, and will pass to the other side. You will
see that in this way, by the simple evaporation of the ether,
we have produced a very great amount of cold. There, you
see that we have got an amount of cold which has carried
the needle from where it was down across zero, and far away
in the direction of cold. Now, this artificial ice is produced
exactly by such means. A mass of ether is caused to sur-
round vessels in which is contained brine, or salt and water.

This mixture is chilled by the evaporation of the ether, and
the chilled brine is carried round vessels containing water,
and then the water is converted into the solid form by the
chill produced by the cold brine. And now, not only are
those beautiful crystalline forms that you see upon the
London pavement, and upon your own windows on a frosty
morning, evidences of this wonderful building power
possessed by the ultimate particles of matter, but every
particle of ice that spreads over the Serpentine, or covers
the lakes of Switzerland and Norway at the present moment,
is built upon this wondrous plan; and I want, if I can, to
dissect in your presence a plate of ice. I see that Mr.
Cottrell has got a plate of ice there before you, and I will
tell you how it was that this dissection, or analysis, if you
like to call it so, occurred. I remember walking through
Kensington Gardens, I think, in the year 1858. Many of
you were not born then. And I thought to myself, "Ice
absorbs or stops a very considerable amount of heat. What
becomes of that heat?" I knew that the ice, when once
raised to the point when it begins to melt, could not be
raised higher in temperature.

You may remember that in our last lecture I placed a plate
of rock salt and a plate of glass in the path of our strong
beam. The plate of rock salt was not warmed. The plate
of glass was heated. And then we inferred from that fact
that a certain amount of heat was stopped by the glass, but
was not stopped by the rock salt. I asked myself, in walking
through Kensington Gardens, "What occurs in the melting
of a block of ice when a sunbeam has gone through it?" A
good deal of the sunbeam is intercepted: What becomes of
it? It cannot warm the ice: then it must liquefy it in some
peculiar way within the mass of ice. Well, I got a clear
block of ice; and I went into the room through which you
passed in coming into this theatre, and the sun was shining
brightly: and I placed a lens in the path of the sunbeams,
and there I could see, as you have seen over and over again
in this room, the track of the sunlight depicted upon the
floating dust of the air. I fixed the point where the light
was converged, and I put my block of ice so that the focus
of the lens should be in the very heart of the ice. Immediately I saw a number of little specks of silvery brightness in
the very body of the ice. That did not at all solve the problem I had in hand. I asked myself, "What has become of
the heat?" I took up the ice, and looked at it by means of
a pocket-lens, and I found that round about each one of
these spots I had a most delicate and beautiful little liquid

flower. The heat had expended itself in reducing the ice
to the liquid condition—simply in undoing the very work of
crystallisation that you have seen exhibited before you on
the screen. The particles of ice had been built together
upon a certain plan. The sunbeam took those particles or
molecules asunder, just reversing the process of crystallisation. And round about those little spots I found a beautiful
six-petalled flower of liquid.

And now we will try to show you, in a more or less perfect
form, some of those liquid flowers produced by the liquefaction of the ice. We will pass a beam from our electric light
through the ice and focus these flowers, as they are formed,
upon the screen. Observe those beautiful shapes that are now
forming (Fig. 29). Do you see that every one of them has six

FIG. 29.

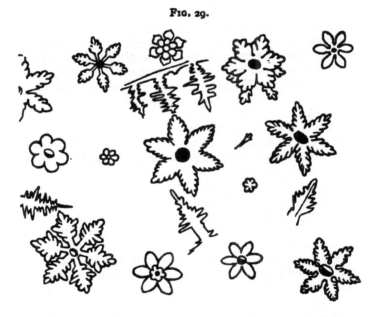

petals. That is the re-solution of the ice. The ice has
stopped a portion of this beam, and that portion of the beam
has expended itself in producing these beautiful forms.
Some of them are beautifully crimped with fern-like
crimpings.

Well, crystals of snow form in this way in calm weather
upon Alpine summits, and there they are gathered and collected year after year; and there is a certain elevation
where the quantity of snow that falls every year is in excess

of the quantity melted. I had a letter from Switzerland yesterday morning, and at the present time the quantity of snow upon the mountains at Brieg, in the valley of the Rhone, is comparatively small, although the cold is very great. That is an important point to notice. In point of fact, there is not sufficient aqueous vapour in the air to produce snow, however cold the weather may be.

And here let me draw attention, in passing, to a point which has misled many not only eminent, but very illustrious, men. If you go to Switzerland, or even to some parts of England, you find traces of ancient glaciers. If you go to Cumberland, or to Wales, or to the south-west coast of Ireland, you find traces of ancient glacier action which are just as pronounced and clear as the traces of glacier action that are going on to-day in Switzerland. In trying to account for the period in which glaciers extended over nearly the whole of Europe some eminent men have put forth their suppositions. One celebrated notion is that the radiation of the sun, by some means or other, was diminished—that the solar power became diminished, and that, in consequence of the chill, we had the Glacier Epoch. There is also another supposition : it is more than a supposition. It has been thought that the whole of our solar system is moving through space ; and it was supposed that, in passing through space, the solar system has passed through spaces which have a very low temperature, and that during its passage through those spaces of low temperature the Glacier Epoch, as it is called, was produced. Now I think that every boy in this assembly will see the defect of this reasoning. At the present moment, as recorded to me yesterday by a letter from Switzerland, the cold is intense, but the quantity of snow which produces glaciers is very little. Why? You are prepared with your answer. The very material of which glaciers are formed is absent. The air is dry, notwithstanding the great cold, and the consequence is that you have very little precipitation of snow, and very little of that nutriment, or of that material, which produces glaciers. I want you to understand that these vast masses of ice which come down the Alpine valleys are entirely due to the congelation of aqueous vapour and its precipitation as snow ; so that it is the aqueous vapour which produces the snow. I hope you will see my reasoning. The very agency which gives birth to these large masses of ice which form the glaciers is heat. You could not have those vast masses of ice without having vapour raised from the tropical ocean. This vapour is carried northward ; it is congealed in the atmosphere, and it comes down as snow. The snow becomes compacted together as glaciers, and the whole process is due to the action of heat. Let me try to impress this thing upon your minds still more fully. Take the case of a great glacier near which, I am happy to say, I live for six or eight weeks every year—the great Aletch glacier. I look down upon that mass, all frozen as it is, forming a great river of ice, and I ask myself, What amount of solar heat has been necessary to produce that glacier? Understand me—What amount of solar heat has been necessary to evaporate the water of the ocean so as to produce this glacier? This amount of heat. Suppose a mass of iron five times the weight of that Aletch glacier. Suppose the mass of the glacier quintupled—rendered five times heavier than it is : suppose that mass of iron raised to the fusing-point of cast-iron— white hot. The amount of solar heat which was necessary to produce the present Aletch glacier would be exactly equal to that which would be competent to raise to its fusing-point a mass of iron five times the weight of the glacier—competent to raise the iron to a white heat. Hence you see here what a power the sun exerts in the production of those Alpine glaciers. In point of fact, those who have speculated upon this subject forgot altogether that the case was one of distillation. We had a case of distillation here in our last lecture. I had some salt water boiled, and we condensed the steam from the salt water in another vessel, thus producing sweet water : and that is precisely what is done by the distillation of the ocean. The ocean sends up its aqueous vapour into the air—sweet vapour without any salt. This vapour is again condensed and congealed in the higher regions of the atmosphere, and the process, as I have said, is one of natural distillation. In distillation heat is just as important as cold. You must have the heat to produce the vapour : you must have the cold to act as a condenser. And this is the action of the mountains of Switzerland upon the aqueous vapour of the air.

Observe here this beautiful flask covered with a white coating, which is due to the coagulation of the aqueous vapour of the breath that has come from your lungs. You have been breathing out into this atmosphere and rendering the atmosphere humid. I put a very cold mixture into that flask, and the vapour which came from your lungs has become congealed into hoar-frost on the surface of the flask. And I want to show you that the luminous rays of our domestic sun have no power whatever to melt the hoar-frost. And so it is with regard to the glaciers of the Alps.

The power that melts them is those invisible rays that we referred to in our last lecture.

Now, I also want to show you a peculiar action on the part of water, which is of the utmost consequence in nature. Almost all bodies are expanded upon being heated. Mr. Cottrell has prepared a flask of water, and he intended me to heat it with a spirit lamp, so that you would see the expansion of the water in the tube which goes through the cork of the flask. The water would have expanded, and would have risen in the tube, and you would have seen on the screen the image of the water trickling over the top of the tube. I would have then removed the spirit-lamp which caused the water to expand, and would have surrounded the flask with a freezing mixture. Very soon you would see the water contracting in consequence of the cold. It would gradually contract and go down to a certain point, and then it would cease to contract. Then, if I continued the application of the cold, the water would begin again to expand. Up to a certain point the water would be contracted by cold. Beyond that point, the cold would act exactly as heat would act, and cause expansion; and I intended to lift this column of water by expanding it by means of cold, so that you would see it trickling over the top of the column through the action of cold instead of through the action of heat, as previously. Water contracts until the temperature gets down to about 4° C., or 39° F. That is about 7° above the point where it freezes. That point corresponds to what they call the maximum density of water. Then if you continue the cold, it expands. What does this expansion mean? It is simply a preparation for the act of crystallisation which converts water into ice. At this particular point, 39° F., the molecules, although still in the liquid condition, are beginning to act upon each other, so as to try to build themselves, as it were, into the crystalline form. The crystal of water requires more room than the water itself. When you lower the temperature so as to bring it down to 32° F., at which point water freezes, then the expansion is sudden and so powerful as to rend everything opposed to it. Here is a thick bombshell of cast-iron, for which I am indebted to my friend, Professor Abel. Similar bombshells were filled with water and screwed tightly down, and placed in a freezing mixture in front of the table before the lecture commenced; and those explosions which you have heard during the lecture are, I anticipate, due to the bursting of those bombshells in consequence of the force exerted by the water in passing from the liquid condition into the solid condition.

There is also an iron bottle similar to what used to be operated upon by that glorious old fellow, Faraday. He had these cast-iron bottles made. They are half an inch thick. The bottles were filled with water, and by placing them in a freezing mixture he used to burst those bottles. I think that we shall find that the bombshells and the bottles have been burst. [Upon the vessel containing the freezing mixture being uncovered and searched, it was found that the bombshells and iron bottles had been burst by the freezing of the water contained in them. One of the shells measured $3\frac{3}{4}$ inches in internal diameter, and varied in thickness from three-quarters to seven-eighths of an inch.] This will give you an idea of the irresistible power exerted by the molecules of water in passing from the liquid condition into the solid condition.

ANALYSES OF BOOKS.

The Most Important Articles of Food and Beverages, with their Impurities and Sophistications. A Practical Guide to their Detection.* By OSCAR DIETZSCH, Chemist to the Industrial Museum of Zurich. Zurich: Oscar Füssli and Co.

OF all departments of chemistry that which bears upon public health, and upon the detection of adulterations and impurities in food and in other articles of general consumption, has received in England the largest relative share of attention. Hence we doubt not that medical practitioners, chemists, and even the outside public, will take an interest in a foreign work which in a short time has reached its third edition, and which, as we learn from the Preface, is about to be translated into the English, French, and even the Russian languages. The author, who evidently possesses very extensive experience in this branch of analysis, addresses himself solely to professional men, and does not waste space on the task of teaching outsiders how to perform operations which unless done accurately had much better never be attempted at all.

The first article examined is milk. Here we notice with regret the great weight which the author lays upon the specific gravity of the sample, and consequently upon the use of the lactometer. We had hoped that the arguments against this instrument were so widely known as for ever to deprive it of any higher rank than that of a scientific toy. Dr. Dietzsch points out, however, very judiciously, that statements made concerning the milk of individual cows may have a certain value from a physiological or a pathological point of view, but are utterly useless as regards market-milk, which in these days, in towns at least, is a mixture of the milk of many cows, where individual and exceptional variations are of course lost in the average. If, as is alleged, MM. Vernois and Becquerel, when examining different sorts of market-milk in Paris, found a sample of sp. gr. 1·016, and, without ascertaining anything concerning its origin, analysed it as a genuine milk and inserted the results in their tables, they were guilty of a piece of negligence which cannot be too gravely reprehended.

The author does not give any opinion as to the constancy of

---

* Die Wichtigsten Nahrungsmittel und Getränke, deren Verunreinigungen und Verfalschungen. Praktischer Wegweiser zu deren Entdeckung.

the " solids minus fat," upon which Prof. Wanklyn and his followers lay much weight. The sophistications of milk customary in Switzerland appear to be substantially the same as those prevalent in England, *i.e.*, removal of the cream and addition of water.

Dr. Dietzsch considers milk-sugar the most constant component of milk, varying merely from 4½ to 5½ per cent. He does not, however, accept the determination of this constituent as a rational process of milk analysis, since it may easily be added to an impoverished milk. As a test for soda and for borax he mixes 100 c.c. of the sample with 0·1 grm. tartaric acid, and applies heat. If the sample does not coagulate, one or the other of these salts is present. An exceedingly valuable feature of this chapter is the account of certain pathological modifications of milk, which are much more to be dreaded than any adulteration. Unfortunately the detection of morbid products in mixed market-milk is exceedingly difficult. Pus is recognised under the microscope as pale, round, faintly-granulated vesicles of different magnitudes, which when touched with acetic acid swell up and lose their granulated surface. We feel tempted to ask whether a spectroscopic examination might not lead to the detection of disease-products in milk ?

From milk we pass, by a natural sequence, to water. Here we are somewhat surprised to find English authorities completely ignored. Neither of the rival processes of Professors Frankland and Wanklyn, for the determination of organic and especially nitrogenous matter, is even mentioned. Animal substances are to be detected by the smell of burnt horn given off on the ignition of the dried residue, and approximately determined by the use of potassium permanganate in its most primitive form.

For the detection of ammonia the author Nesslerises the water in its original state. For a quantitative determination he acidulates with hydrochloric acid, concentrates, adds caustic soda, and distils, condensing the fumes in dilute hydrochloric acid. He then evaporates the distillate to dryness in a tared porcelain capsule, and weighs the ammonium chloride thus formed.

On the sophistications of wine the author's views are exceedingly judicious. He considers that nothing should be sold as wine save the liquid produced by the alcoholic fermentation of grape juice without any addition. " Plastering " he condemns, but he is scarcely correct in asserting that this malpractice is chiefly confined to the South of France. It prevails also in Spain, Sicily, even Greece, and has, we regret to say, extended even to Australia. Thousands of bottles of Malaga, he states, are artificially concocted at Cette which never touched the soil of Spain. The artificial colouration of wines, it appears, is defended by merchants on the ground that it is convenient for their customers, the retailers, who can then add water with less fear of detection. The use of glucose obtained from potatoes he considers most injurious both in compounding wines and in brewing, as a considerable quantity of amylic alcohol is thus introduced into the wine or the beer. Impure glycerin, containing formic and butyric acids, is also injurious. The physiological action even of pure glycerin is far from being accurately known.

As regards vinegar we cannot entirely agree with the author. He is certainly in the right in giving a low rank to malt vinegar, which is absurdly preferred in England; but fruit vinegars—especially the American apple vinegars—retain the delightful aroma of the fresh fruit, and are scarcely inferior to wine vinegars.

The addition of alum to bread is not lawful in England, as the author seems to believe, nor is it quite so common as was the case half a century ago. For its detection he recommends the logwood process.

Upon the " surrogates " of coffee he is deservedly severe. None of them, he remarks, contains either the nitrogenous constituent, caffeine, nor the ethereal oil, nor the tannin. Chicory in particular occasions determination of blood to the head and injures the sight. Our legislature never committed a greater mistake than in sanctioning the sale of coffee mixed with chicory.

It is interesting to learn that the tea imported by way of Russia—the so-called " caravan tea "—is largely mixed with the leaves of *Epilobium angustifolium.*

We should be glad, did space permit, to extend further our survey of this valuable work, with which we hope our public analysts and medical officers of health will before long become acquainted. Its weakest portion, we think, consists of the chapters on milk and on water, and its strongest, perhaps, of those on wine and on beer. The chapter entitled " The Chemical Expert in Court " must be especially commended to the notice of the profession.

A Short History of Natural Science, and of the Progress of Discovery from the time of the Greeks to the Present Day. By ARABELLA B. BUCKLEY. Second Edition. London: Edward Stanford.

WHEN the first edition of this work made its appearance we could do no other than admit that, notwithstanding certain errors of detail which an industrious critic might here and there disco. ver, the authoress had given a broad, clear, and mainly correct outline of the progress of scientific research and discovery. In the present edition various omissions have been supplied, as, for instance, the chapter on Sound. Sundry errors have been rectified, and we may now, therefore, more emphatically repeat the recommendation which we gave after examining the former edition.

The work, though mainly intended for children and young persons, may be most advantageously read by many persons of riper age, and may serve to implant in their minds a fuller and clearer conception of " the promises, the achievements, and the claims of Science." The manner in which the theory of Evolution is presented to the reader is admirably judicious. Scrupulous persons who have shrunk from the new Natural History as having an atheistic tendency may here see how utterly groundless were their fears.

A very useful feature of Miss Buckley's work is the bibliographical note placed at the end of each chapter, indicating to the student what books to consult in order to fill up the outline here placed before him.

Lightning Conductors: their History, Nature, and Mode of Application. By RICHARD ANDERSON, F.C.S., F.G.S. London: E. and F. N. Spon.

WE have here a most able and complete monograph, historical, theoretical, and practical, on a subject of great importance, and yet generally neglected. Of course in these days of general enlightenment everyone knows that the lightning-rod was invented by Benjamin Franklin nearly 130 years ago. We can all quote the " eripuit cœlo fulmen," and can point a moral by referring to the opposition encountered by the novel and daring project. But side by side with all this, our paper knowledge, stands the grim fact that in England, at least, from one-half to two-thirds of our public buildings, including churches and chapels, are without any protection against lightning. Of private houses the author considers that not five out of every thousand are fitted with conductors. So rare are they, in fact, that when seen they excite curiosity as a something strange and fantastic. Yet we cannot, in defence of our national apathy, represent the danger from lightning to persons and property as a something infinitesimal and unworthy our notice. Mr. Anderson shows that the deaths from this cause in Great Britain and Ireland average not less than one hundred yearly. Of the damage to live stock, farm premises, and other property, no accurate statistics are attainable, but it is by no means trifling. Still, with a fatalism almost Turkish in its character, we refuse to apply the safeguard which Science has placed in our hands. We submit to lightning as we do to water-companies, macadamised streets, competitive examinations, chimney-pot hats, and a hundred other constituted nuisances.

Mr. Anderson's work opens with a history of the lightning-rod, or, we might rather say, of the progress of electrical science with especial reference to conductors. That Dr. Franklin's invention met with the opposition usually meted out to all discoveries and inventions, or rather to their originators, is matter of history. None of the various forces usually active on such occasions was here wanting. We trace scientific and academic jealousy, as embodied in the Abbé Nollet ; popular ignorance and suspicion displayed by the mob in Geneva, in St. Omer, in Siena, in Padua, and elsewhere. There was ecclesiastical hatred not less shown in Puritan New England than in Catholic Italy ; and, lastly, there was royal ill-will in the person of George III. It will scarcely be credited that this monarch was a very acrimonious partisan in a controversy which raged on the question as to whether a lightning rod should terminate in a ball or in a point ; that he actually demanded from the Royal Society a formal condemnation of points, and when very justly informed by their president, Sir J. Pringle, that the laws of Nature were not alterable at royal pleasure, he demanded the resignation of the speaker. With this intimation Sir J. Pringle, finding but very lukewarm support on the part of the Fellows, found it necessary to comply. We know no second instance, in any civilised country, of Science being thus formally insulted in the person of her official representative, till we come to the speech delivered in the House of Commons by Acton Smee Ayrton, in justification of his misconduct towards Sir William Hooker.

To return, after this historical survey : Mr. Anderson treats of metals as conductors of electricity, the character of lightning and of thunderstorms, and then enters into the *rationale* of lightning protection. The theory, he shows, is simplicity itself. All that is required is to establish such conduction as to lead away any flash of lightning that may strike the house or other building to be protected, without giving it the opportunity for damage. In other words, the conductor must form the " line of least resistance " between the summit of the building and the ground. But to carry out this principle into practice requires no small amount of skill and judgment. Where these essentials are wanting, as is too often the case, the conductor will be faulty, or may even be worse than none at all. Among the besetting sins of conductors we find it pointed out that the metal may be deficient in thickness; that its continuity may be interrupted ; that in obedience to a gross error it may be isolated from the building it is to protect by means of glass supports ; that it may not be placed in connection with metallic masses in or upon the building, such as leaden or zinc roofing and gutters, water and gas pipes, iron balustrades, bells, iron cramps and leadwork used in securing the masonry of spires and pinnacles, safes, machinery, &c. All such objects, the author shows, should be united in one metallic system, so that wherever the lightning may fall it may find a direct path down into the ground. In this last respect, also, too many conductors are faulty. The rods, wire-ropes, &c., often dip merely a few inches into earth which may become dry, and consequently non-conductive. "The earth-connection," Mr. Anderson declares, "is the alpha and omega of lightning protection." Further, a conductor originally perfect may be rendered useless by corrosion, or by damage inflicted by careless workmen. Hence these appliances should be subjected to an occasional inspection, and their conductivity tested by means with which every electrician is well acquainted.

These lessons are enforced by practical examples, in which the damage or destruction of buildings supposed to be protected is explained by some deviation from the principles above laid down.

An interesting feature of the work is a catalogue of public buildings struck by lightning from 1589 to September, 1879, and of twenty-four cases of the explosions of powder magazines from the same cause. One of these, at Brescia, in August, 1769, proved fatal to three thousand persons. There is also a very elaborate bibliography of the subject from Hieronymus Cardanus downwards.

The book is plentifully illustrated, and will prove a most valuable guide to engineers, architects, builders, and to all persons entrusted with the care of public buildings, magazines, shipping, &c.

Proceedings of the Birmingham Philosophical Society. 1878-79. Vol. i., No. 3. Birmingham : Corns, Sheriff, and Rattey.

THIS volume is very much more valuable and satisfactory than the average proceedings of our provincial learned Societies, and Birmingham has fair reason to feel proud of the result. The simple fact is that we find here a very fair proportion of actual sterling scientific work instead of mere words and dreams.

Dr. Norris's paper on the " Development of Mammalian Blood " is exceedingly interesting. He finds that there exist in mammalian blood numerous corpuscles incapable of being seen by the microscope—not from their minuteness, but because they have the same refractive index and colour as the liquor sanguinis in which they float. They are colourless biconcave disks, between which and the red biconcave disks other corpuscles can be detected possessing every gradation of tint. The morphological elements of the lymphatic glands and spleen, when examined with the needful precautions, instead of being found identical with the ordinary white blood corpuscles, are found to be disks of the same size as the red corpuscles which are gradually becoming biconcave. As the body which undergoes conversion into the red corpuscle must be regarded as the analogue of the white corpuscle of the ovipara, it follows that in mammals this is the corpuscle of the glands and spleen and their equivalent, the invisible corpuscle of the blood, and not the ordinary white corpuscle as supposed. The latter is probably a mere accidental aggregate of adhesive lymph corpuscles.

Mr. H. W. Crosskey communicates some notes on glacial phenomena in the Vosges Mountains, which are ubiquitous and well characterised.

The Rev. H. W. Watson gives a very fair survey of the researches of Mr. Crookes on molecular motion in rarefied gases.

Dr. Blake's paper on the connection between general and technical education contains many valuable observations, and gives a brief but valuable account of industrial training in Belgium, France, Prussia, and Austria. He rightly sees that the apprenticeship system has had its day, and survives now as a mere nuisance. But the author is strangely mistaken in saying that " a premium is seldom paid, and consequently no one feels it to be his business to look after or to inform the beginner." The very contrary is the fact : premiums are more common now than in the days when apprenticeship was a useful reality, but the money thus paid is in the majority of cases literally obtained under false pretences. The chief danger to be apprehended in all attempts made to give the British workman a higher and more scientific training is that, instead of teaching him to do anything thoroughly, we shall cram him to pass an examination in it.

Under the title " Our Philosophy and our Life," Mr. J. Kenward gives an estimate of the influence of scientific research on prevailing thought and character in England. Whilst admitting the learning, and in some respects the depth of thought, shown in this Essay, we can by no means give it our unqualified approval. The author takes decidedly too rose-coloured a view of the relative position of England in the culture of Science. Our discoverers, indeed, are surpassed by none in the height to which they have risen, but in comparison with those of Germany they are few indeed. Few they must remain so long as we formally discourage original investigation by refusing to accept it as a passport to honours and positions in our universities and museums. Mr. Kenward's survey of the state of Science in England,—which, by the way, contrasts strangely with that given by Dr. G. M. Beard in our February number,—if perused by anyone not conversant with the history of discovery, would lead him to believe that our country can claim the merit of the recent

liquefaction of the permanent gases, as well as of other discoveries to which we have only a very partial right.   On the other
hand, some of the men who have done us most honour are
damned with faint praise on the authority of Gwyn Jefferson !
"A reaction of opinion in favour of simple Devolution," we
read, "may not be far distant." The precise sense in which the
term "Devolution" is to be taken does not appear; but if Mr.
Kenward seeks an antithesis to Evolution, surely Revolution is
the word to be used—a consideration which we hope will console
Prof. Virchow.   Much of the Essay touches, if it does not go
beyond, the scope of our journalistic competence, and must be
handed over to the organs of party politics.   To his anti-vivisectionist utterance we must reply that the very definition of
cruelty is the infliction of pain and death without a sufficient
motive.   As regards what Mr. Kenward and some others call
the "subjection of women," we must remind him that it is
merely a case of differentiation, or, in other words, the division
of labour.   If any woman, by reason of an exceptional mental
character, feels disposed to devote herself to scientific research,
we are ready and willing not merely to permit, but to applaud.
To foster, however, what seems to us an anomaly, to aim at the
formation of whole bodies of *savantes*, seems to us a grave error.

A much more valuable paper is Mr. Lawson Tait's researches
on the digestive principles of plants.   The author has obtained
full proof of the digestive process in *Cephalotus, Nepenthes*, the
Droseraceæ, and the *Dionæa muscipula*, and confirmed the conclusion drawn by Mr. Darwin from his researches in this direction.
He has discovered and examined two substances, to which he
gives the names of droserine and azerine.   The former of these
compounds plays a part resembling that of pepsine in the human
stomach.   Azerine has also a digestive action, an apparent
power of attracting insects, and a singular tendency to absorb
moisture from the air.   If such papers as this and that of Prof.
Norris continue to appear in the "Proceedings of the Birmingham Philosophical Society, they will soon be regarded in the
learned world as an authority of no mean rank.

Chapters from the Physical History of the Earth.   An Introduction to Geology and Palæontology.   By ARTHUR NICOLS,
F.G.S.   London : C. Kegan Paul and Co.

WHEN we meet with an elementary work on any department of
Science which contains no reference to the object of passing
examinations, and is not fitted up with strings of questions, we
utter an ejaculation of thankfulness, and our heart warms towards the author.   Mr. Nicols seeks to prove useful to the

student by presenting him with a comprehensive outline of the earth's history, and also to hold out a helping hand to those who with very inadequate means and opprtunities are striving after self-education. In this endeavour he is by no means unsuccessful. The physical history of the earth, and man's gradual comprehension of its successive phases and of their lessons, is given here clearly, fairly, and judiciously. Minute detail in a volume which does not extend to three hundred pages is of course out of the question.

In a few cases we can scarcely admit the author's views without some limitation. Thus the student may probably rise from the perusal of the introductory chapter with the impression that Aristotle's philosophy was the cause of the so-called " stationary epoch " in the career of Science. We submit it was rather the bad use to which the writings of the Greek sage were put by the learned world of the Middle Ages. Instead of using his observations and conclusions as a starting-point for further research, it took them merely as the basis for debates, discussions, and commentation, and had it possessed Newton's " Principia " or Darwin's " Origin of Species " it would have misapplied them in the same manner. In this respect the learned of the Dark Ages differed, after all, little from us modern Englishmen. We, too, love words rather than things, and we would rather " be examined " in books than test their statements or follow out their indications. Nor are we quite certain that " Science owes much to the Reformation." Protestant theologians have denounced the advances of Science with an energy never surpassed.

In the author's survey of the leading truths of geology and palæontology we notice with pleasure that, whilst he states facts and their interpretations, he is free from rash and presumptuous dogmatism. Where there is room for a difference of opinion he states the evidence on either side, and suspends judgment. Such being the author's disposition, it is the more satisfactory to find that he pronounces for the continuity of life in opposition to the Cuvierian doctrine of successive devastations and re-creations. He shows that a great physical break in the strata is accompanied by a corresponding gap in the succession of organic life, and *vice versa*. He contends that the testimony of the rocks, though often defective, yet in no known instance contradicts the law of continuity. He does not find that palæontology in " an unhealthy region for Darwinism," as Prof. Swallow has recently declared. He admits, further, the vast antiquity of the human race, and he recognises the difficulty of drawing " a scientific boundary " between man and the lower animals.

In short, we must pronounce Mr. Nicols's work, accurate, judicious, and moderate, and can recommend it to both the classes for whom it is designed,—to the student as affording a good oversight of regions which he is about to survey in detail, and to the general reader who merely wishes for a correct summary of the results of geological and palæontological research.

The Story of the Earth and Man. By J. W. DAWSON, LL.D., F.R.S., F.G.S. Sixth Edition. London: Hodder and Stoughton.

WHEN a work has, like the one before us, reached its sixth edition, the task of the reviewer is greatly simplified. To give an analysis of the book, to expound the author's views, and to point out what in our judgment is deserving of approval or of censure, is no longer needful. We can merely say that to us it is painful in the extreme to find a man whose abilities are indisputable, and whose services to science we are happy to recognise, so led away as to offer a protracted and disingenuous opposition to the greatest reform in Science which has been effected since the overthrow of the Ptolemean astronomy. How anyone with the remotest claims to candour can still look upon Evolution as opposed to creation is a problem we are unable to solve. No less painful are the personalities directed against opponents. Thus Prof. St. George Mivart, being a Catholic, is encountered in "No Popery" style, almost worthy of Lord George Gordon and his followers. If Darwinism is wholly or in part a mistake, if Evolutionism is altogether a delusion, let this be shown in a scientific manner and by arguments analogous in their spirit to those used when chemical and physical theories are under discussion. Principal Dawson, by his constant appeal to the *odium theologicum*, puts himself completely out of court.

Some Teachings of Development. Being the Substance of the last two of a Series of Twelve Lectures on Animal Development, delivered at the Royal Institution during the months of January, February, and March, 1879. By E. A. SCHAFER, F.R.S., Fullerian Professor of Physiology. London: H. K. Lewis.

WE have here, in the brief compass of three-and-thirty pages, a sound and valuable antidote to the reckless and passionate assertions of such men as Prof. Swallow and Principal J. W. Dawson. The author shows clearly and convincingly that the successive phases in the development of the individual represent similar phases in the process of formation or development of the race to which the individual belongs. Development, therefore, represents descent. The ancestors of every animal have successively exhibited structural conditions which in an abridged form are represented by the successive stages of development of the individual. This, the author justly says, is "the only logical conclusion to which the study of animal development leads."

The misfortune is that we have to deal with men who obstinately keep their eyes closed. They are the rightful heirs of the sages who refused to look through Galileo's telescope, and who, when he performed his celebrated experiment at the leaning tower of Pisa, refused to accept the evidence of their own senses.

The Kansas City Review. Vol. iii., No. 10, February, 1880.

THE principal paper in this issue, "Evolution in Creation," by Prof. G. C. Swallow, is one upon which we cannot congratulate the author. It is fundamentally vitiated by the assumption that Evolution and Creation are antagonistic ideas. Now we hold, as do multitudes of other naturalists, that Evolution is simply the manner in which Creation was effected. That there are Evolutionists who seek to dispense with God is granted. But in like manner there have been, and doubtless still are, believers in the permanence of species who hold that each form of organic life came into existence spontaneously, or else has existed from all eternity.

Results of Observations in Meteorology, Terrestrial Magnetism, &c. Taken at the Melbourne Observatory during the Year 1876, under the Superintence of R. L. J. ELLERY, F.R.S., Government Astronomer. Vol. v. Melbourne: J. Ferres.

THIS volume is a valuable collection of observations, but contains little calling for especial remark. The mean temperature of the air during fourteen years is 57·0° F. for spring, 65·3 for summer, 58·7 for autumn, and 49·2 for winter, the latter figure nearly agreeing with the average for the entire year in England. The yearly mean temperature for Melbourne is 57·6. The highest temperature recorded in the shade is 111·2, and the lowest 27·2. The mean hourly velocity of the wind is 10·5 miles, certainly too much to be pleasant. Fog is not entirely absent. Thus at Wilson's Promontory we find mention of three successive days of "dense, wet, driving fog," and this in the middle of October, which answers to our April.

Studies on the Seat of the Russo-Turkish War of 1877-78.* A
Report of Swiss Engineer Officers on their Mission to the
Seat of War in 1878, as presented to the Swiss War De-
partment. By C. OTT, Colonel of Engineers. Zürich:
Orell Füssli and Co.

WE have here a minute criticism of the operations of the con-
tending powers from a purely professional point of view. The
author infers that the defence of a country requires fortified
central positions, in order that all may not be lost with the fall
of external lines of defence, and that a prolonged resistance and
a resumption of the offensive may be possible. These chief
places of arms serve both for the protection of important towns
and for supporting the movements of the army. The chief defence of such positions must lie in detached forts and outworks,
the strength of the fortifications be reduced internally. The
fundamental idea is the establishment of fortified radii. It is
economical, as far as possible, to substitute inanimate masses
for living defences,—a view which has its especial signification
for Britain, whose armies are relatively small in comparison with
its financial resources. Such external detached forts, Col. Ott
considers, should be prepared in time of peace. He holds that
in case of war the Swiss *Landwehr* would find its proper function
in the defence of such entrenched positions.

This work evidently merits the careful attention of military
men in England.

A Criticism of Dr. Croll's Molecular Theory of Glacier Motion.
By J. J. HARRIS TEALL, M.A., F.G.S. London: Simpkin
and Marshall.

THE author contends that in his work " Climate and Time " Dr.
Croll uses terms—such as conduction, radiation, and molecule—
in various senses in which they are not employed by physicists,
and " applies the results of physical research, expressed in language in which the terms are used in one definite sense, in support of a theory in the statement of which the same terms are
used in a different sense." He contends that the entire theory
is based upon an ambiguous use of the word " molecule." Dr.
Croll attributes, *e.g.*, to the molecules the property of capillarity.

---

* Studien auf dem Kriegs schauplatze des Russisch-Türkischen Krieges.

Elements of Agricultural Chemistry and Geology. By the late
    Prof. J. F. W. JOHNSTON and Dr. C. A. CAMERON. Tenth
    Edition. Edinburgh and London: W. Blackwood and
    Sons.

JOHNSTON's "Agricultural Chemistry" is so well known, and so
deservedly valued by that part of the public who take an especial
interest in the first and most important part of the useful arts,
that neither an exposition of its nature and contents nor general
laudatory phrases can be at all admissible. The present edition
has been elaborated by Dr. Cameron in accordance with the discoveries made since the death of Prof. Johnston. The chemical
nomenclature has also been altered—a step, in our opinion, of
doubtful expediency. Practical men never will speak of sulphuric
acid as "dihydric sulphate." We fear that the terminology of
"new chemistry" is responsible for no small amount of confusion, and has seriously tended to frustrate the diffusion of a
knowledge of chemical facts among the industrial classes.

  We are sorry also to find that in our cold, rainy, over-watered
climate Dr. Cameron considers irrigation as the best method for
disposing of town-sewage, and that too without any previous
defecation.

  Apart from these drawbacks the work appears entitled to our
almost unqualified commendation.

United States Commission on Fish and Fisheries. Part V.
    Report of the Commissioner for 1877. Washington:
    Government Printing-Office.

WE have here a goodly volume devoted to an inquiry into the
decrease of food-fishes, and to considerations on the propagation
of such fish in the waters of the United States. It is well stored
with facts interesting both to the ichthyologist and to all practically engaged in the culture and catching of fish. In connection
with the carriage of fish from the place of capture to distant
markets, we find a very elaborate account of the various contrivances for producing artificial cold. There is also a very
complete series of annual reports of the Norwegian fisheries
near the Loffoden Islands.

  We notice here an interesting essay by Karl Dambeck, on the
"Geographical Distribution of the Gadidæ, or Cod Family, in
its relation to Fisheries and Commerce."

  The manufacture of manure from the refuse of the fisheries
appears to be rapidly assuming more extensive proportions on

both sides of the Atlantic. As far back as 1870 the refuse of from 4 to 5 millions codfish was annually worked up into " fish-guano " on the Norwegian coasts, the product containing 8 to 10 per cent of nitrogen and 10 to 15 per cent of phosphoric acid. The manure finds a ready sale in the Scandinavian peninsula, in Germany, and in France.

On looking over the facts and figures contained in this volume —the vast quantity of human food obtained from the sea, together with the supplies of oil, of glue, and of manure—we may well question the economic advantages which certain enthusiastic advocates of " dietetic reform " allege as likely to result from the disuse of animal food.

Victorian Year-Book for 1878-9. By H. HEYLYN HAYTER, Government Statist. Melbourne : Ferris. London : Robertson.

WE have here an elaborate assortment of statistics, vital, financial, agricultural, criminal, moral, religious, and intellectual, the whole enabling us to form a very clear idea of the colony of Victoria as it now is.

Turning to those portions which appeal most to our notice, we are glad to perceive that the Mining Schools of Ballarat and Sandhurst are making good progress, though something remains to be done before they can at all rival, *e.g.*, the Polytechnic School of Zürich. Surely Victoria ought not to be beaten by Switzerland. The Melbourne University will not, we fear, under its present constitution, prove a "researching" body ; it is modelled too closely on the English system, even to the extent of granting doctorships of music. It is a pity that Berlin, Bonn, or Jena was not taken as a model.

The agricultural statistics show some unsatisfactory facts. The yield of wine has fallen off by 410,333 gallons, as against 755,000 in 1875. The loss is partly ascribed to the *Phylloxera*, and partly to the introduction of the English sparrow, which here, as well as in the United States, has proved, as any judicious naturalist would expect, a capital blunder.

The cultivation of the olive and the mulberry—the former for oil and the latter for silk—progress but slowly. Australia ought to be one of the greatest producers of wine, raisins, oil, and silk in the whole world. Tomatoes only occupy 3 acres, whilst that nuisance chicory has extended to 155 !

The Victorian Review.   No 2, December 1, 1879.

WE must plead guilty to no small amount of disappointment on examining the goodly number before us.   Save a brief account of the results of a journey of exploration in Western Australia, there is not here a single paper which, even if we were desirous, we could discuss in the "Journal of Science."   We cannot help expressing our deep' regret at finding that the deluge of political agitation which has for some years overspread the home kingdoms, diverting public attention from discovery and from invention, has extended even to Australia.

Urania.   A Monthly Journal of Astrology, Meteorology, and Physical Science.   Vol. i., January and February.

IT may perhaps stagger the British public no little to learn that in this much belauded but somewhat sceptical nineteenth century there is actually a journal published which earnestly, soberly, and with evident sincerity, claims for Astrology the rank of a true Science.   Some of us will perhaps feel indignant at the fact, and will begin to reflect whether—according to the new system which regards an appeal to the police-courts as the *ultima ratio* in treating a scientific difficulty—the publication and sale of such a work may not be brought within the provisions of a well-known statute anent " palmistry " and subtle devices.   But how many of us in these days could render a reason for rejecting astrology and for pronouncing it in the orthodox manner a mere tissue of imposture and delusion ?   Let us take the simplest case :—It is asserted that the weather is influenced by the conjunctions of the planets, or by their position in their orbits, a proposition surely which, *à priori*, commands neither our assent nor our dissent.   How many of us could meet it by an appeal to facts, and prove from actual and accurate observation that no such influence exists ?   The fact is that we are too rash alike in our beliefs and disbeliefs.   If, however, the writers of " Urania " are not in error, it would surely be easy for them to produce evidence which cannot be gainsaid.   A forecast of the weather for the next twelve months, which should be found to agree with facts, will serve their purpose better than any quantity of discussion.

CORRESPONDENCE.

A QUERY FOR PHRENOLOGISTS.

To the Editor of The Journal of Science.

SIR,—I believe it has been said, by an authority no less competent than Professor Owen, that to distinguish between the brain of man and that of the anthropoid apes is the "anatomist's difficulty." On the other hand, it is maintained by phrenologists that the organs of the "moral sentiments," and those sentiments themselves, are wanting in the lower animals. If this assertion is true, there ought to exist a very broad and well-marked distinction between the human and the simian brain. Again, few naturalists of the New School will accept the dogma that the so-called moral sentiments are confined to man. I certainly maintain that the manifestations of benevolence, veneration, and consciousness can be distinctly traced, *e.g.*, in dogs.

As there appear some symptoms of a recrudescence of phrenology, this difficulty, I think, calls for an explanation from its advocates.—I am, &c.,

R. N. M.

ARE WASPS CARNIVOROUS?

To the Editor of the Journal of Science.

SIR,—A correspondent in a contemporary, after stating that in Sutherland he had seen a wasp "in the act of devouring a caterpillar which was still alive, but considerably mangled by the mandibles of a wasp," enquires "Is it not unusual for the common wasp to eat living creatures of any sort?" Had he directed his attention to wasps when in a room he would have seen them chase flies, catch (I cannot say devour), kill, and suck their juices. In Greece there is a wasp which makes a prey of spiders, tracking them as a spaniel will track a hare or a partridge. Dr. Erasmus Darwin ("Zoonomia," vol. i., p. 263) says, "Wasps are said to catch large spiders, and to cut off their legs, and carry their mutilated bodies to their young.—' Dict. Raison,' tome i., p. 152." Further he says, "One circumstance I shall relate which fell under my own eye, and showed the power of reason in a wasp as it is exercised among men :—A wasp on a gravel-walk had caught a fly nearly as large as himself. Kneeling on the ground, I observed him separate the tail and the head from the body part, to which the wings were attached. He then took the body part in his paws, and rose about 2 feet from the ground with it; but a gentle breeze wafting the wings of the fly turned him round in the air, and he settled again with his prey on the gravel. I then distinctly observed him cut off, with his mouth, first one of the wings and then the other, after which he flew away with it unmolested by the wind." —I am, &c.,

S. B.

[Wasps are not merely carnivorous, but are even guilty of cannibalism. In the summer of 1878, having killed one which persisted in settling upon the stage of our microscope, we were surprised shortly after to see another wasp alight upon the body, make incisions into the thorax and abdomen, and feast upon the juices.—Ed. J. S.]

A CASE OF HEREDITY.

To the Editor of the *Journal of Science*.

SIR,—As you have from time to time dealt with the question of heredity you may feel an interest in the following facts, which I quote from an American journal of standing :—"Elias Phillips, of Freetown, Mass., who recently appeared as a witness in a burglary trial, having turned State's evidence, is a great-grandson of the notorious criminal Maltone Briggs, who was in prison with seven of his sons at one time. Brigg's ancestry is traced back to a noted pirate in the time of Earl Bellamont, and his family has for over a century furnished notorious criminals in every generation."

Surely such facts can require no comment.—I am, &c.,

A JURIST,

DEVELOPMENT AND DIFFERENTIATION.

———

To the Editor of the Journal of Science.

SIR,—We generally find that the processes of development and differentiation go hand in hand, the former being scarcely conceivable without the latter. Thus animal species which are scarcely distinguishable when in the embryo state become more and more unlike each other as they come to maturity. But there are some cases which I, at least, feel unable to bring under this rule. Thus there are certain Lepidoptera whose caterpillars are not only as distinct, respectively speaking, as the mature insects, but where the diagnosis of the latter is scarcely possible without reference to the caterpillar. Hence the structure of the larvæ is more and more taken into account in determining and classifying insects.—I am, &c.,

AN ENTOMOLOGIST.

——— ———

A DIFFICULTY TO SOLVE.

———

To the Editor of the Journal of Science.

SIR,—Turning over some old notes I came upon the following record of facts, which may perhaps interest some of your readers :—A wealthy landowner was seized with the whim of converting a village upon his estate into a town. Among other steps to this end he erected at one entrance a gateway modelled after those of mediæval fortified cities, with an arch over the road surmounted by several rooms, which I suppose were to represent the dwelling of the warder. It soon became noised abroad that whoever slept in one of these rooms, no matter what might be his constitution or his habits, was, in nine cases out of ten, attacked with nightmare of a very violent kind. I heard all the particulars from a scientific gentleman who had taken much pains to trace the cause of this singular phenomenon, and had passed many nights in the room, but without being able to find the slightest clue.—I am, &c.,

SENEX.

NOTES.

Dr. W. B. Carpenter, F.R.S., in a Lecture delivered at the Royal Institution, January 23rd, maintained, in accordance with Prof. Geikie, that the present Continental ridges have probably always existed in some form, and that the present deep Ocean-basins likewise date from the remotest geological antiquity. He considers it, however, possible that at some period a band of connection may have existed between Europe and America, and that New Zealand, Tasmania, and South America may have been linked together by ridges of dry land, whilst Madagascar may have been joined in a similar manner to the African continent, and even with Asia. He does not consider that there is satisfactory evidence for the supposed continent Atlantic being between Africa and America.

On March 4th Mr. G. J. Romanes, F.R.S., delivered a Lecture in the Glasgow City Hall on "Mental Evolution." He urged that the burden of proof lay with those who maintain that the human mind has not been evolved, but has been produced in an exceptional manner. He considered that the theory of Evolution had nothing to fear and everything to hope from the advance of psychological science.

Mr. E. D. Cope delivered last autumn a most interesting Lecture before the California Academy of Sciences, on the Modern Doctrine of Evolution. Whilst accepting the theory in its widest sense, and admitting Natural Selection as a *vera causa*, he contends that "Selection" and "Survival" do not go to the root of the matter, *i.e.*, the origin of the fittest.

A Mr. T. Warren O'Neill, of the Philadelphia Bar, has published a "Refutation of Darwinism," and meets himself with a neat refutation in the "American Naturalist." The author believes that the present condition of animals is one of degradation from a state of original perfection, which has been brought about by the severity of the struggle for existence. The reviewer shows that this view, be it true or false, leaves the origin of species untouched, whilst palæontology shows no evidence of that primitive physiological integrity which Mr. O'Neill assumes.

Count d'Ursel, in a recently published work on South America, asserts that in Bolivia, Peru, Chili, La Plata, and Brazil, he has met with an insect which after its death is transformed into a plant. He describes and figures this creature as a thick, hard grub, with distinct articulations. When about to die it buries itself some centimetres deep in the earth, and there gradually increases in circumference till it presents an appearance very like that of a potato. A stem is then put forth, which in the spring months bears a crop of blue flowers ! Without at all impeaching the veracity of the Count, we fear there is here scope for error of observation.

M. le Bon, in a memoir recently crowned by the French Academy of Sciences, shows that the differences in the cranial development of individuals of one and the same race become greater the higher the race rises in the scale of civilisation. Hence, far from tending towards equality, men tend, on the contrary, towards increasing differentiation.

M. Torrès Coicedo, the Ambassador of Salvador, has presented to the Society of Acclimatisation, of Paris, two curious plants, the guaco and the cedron, said to be successfully used in South America as an antidote to the poison of serpents and to that of rabies. In both respects their reputed efficacy requires confirmation.

M. Jousset de Bellesme contends that luminous insects, such as the Lampyridæ, have not a luminous substance stored up, but that their special secretion is consumed as fast as it is produced.

According to MM. Brongniart and Cornu multitudes of insects of the species *Syrphus mellinus* have been destroyed by a microscopic fungus, of the genus *Entomophthora*. M. Pasteur suggests the propriety of experimentation with a view to discover some fungus capable of destroying the *Phylloxera*.

According to Dr. Parrot the right hemisphere of the brain is the earliest developed. Contrary to what holds good in the white race, in the negro the posterior part of the cranium is the earliest to become ossified.

A subsidy of 900 francs allowed the first year of the French Government to the Zoological Society of France has been withdrawn.

Bœttger, in his Monograph of the Reptiles and Batrachians of the South of Portugal, describes twenty-one species, five of which are new to the country and one is entirely new to Science.

Prof. Von Tieghem has shown that *Bacillus amylobacter*, which is now the great agent in the decomposition of cellulose, appears to have exercised the same function in the Carboniferous Age.

According to M. G. Carlet, insects when walking move their feet in the following order :—

1—4
5—2
3—6

Thus always resting on a triangle, formed by the front and hind foot of one side and the middle foot of the opposite side. In the Arachnides, which have four pairs of legs, the order of movement is—

$$
\begin{aligned}
&1\text{—}5\\
&6\text{—}2\\
&3\text{—}7\\
&8\text{—}4
\end{aligned}
$$

A species of *Curculio* (*Phylonomus punctatus*) is committing great ravages in the clover-fields of Lombardy.

New enemies of the vine have been recently recognised :— *Phona uvicola* near Narbonne, and *Cladosporum viticolum* and *Peranospora viticola* in North America.

A Biological Curiosity.—I open *the book* at page 184,* and I read :—" On placing a *decapitated* frog at the bottom of a vessel filled with water, the animal rises to the surface and keeps itself there, *with its head in the air*. If the frog is placed in the same vessel, under an inverted jar filled with water, it *behaves in the same manner.*"—*Revue Internationales des Sciences Biologiques*, 3e Année, p. 187.

According to Dr. Cohn, Prof. Virchow, and Dr. Almgvist, the colour-sense in uncivilised nations—such as the Nubians, Lapps, and Tschutksches—is very well developed, though they may be deficient in words to express the different shades. This is further proof of the incorrectness of the view put forward by Magnus, Gladstone, and Geiger, that a lack of accurate colour-names indicates the want of accurate colour-perception.

Mr. L. F. Ward ("American Naturalist") contends that hermaphroditism is a thraldom, necessary at the outset, from which all living things are seeking to escape. The animal kingdom has for the most part thrown off this yoke, and the vegetable world looks to insects as its liberators.

E. L. Ellicott, in the same journal, has observed hundreds of snakes (*Eutænia sirtalis*) heaped together in a mass. On another occasion he noticed " a ball of black snakes (*Bascanion constrictor*) rolling slowly down a hill-side."

[We have heard rumours of mass-meetings of snakes taking place in various parts of Eastern Europe ; but we had never the good fortune to meet with such a case, and our informants were not the most trustworthy.—Ed. J. S.]

A Strange Psychological Theory.—At the last Congress of German Naturalists and Physicians a certain Prof. Jäger, amidst the laughter of the assembly, propounded the theory that the

---

* H. MILNE-EDWARDS, Leçons sur la Physiologie et l'Anatomie, comparée de l'Homme et des Animaux, vol. xiii., p. 184, line 8.

soul of every man and animal is to be found in the specific odour which each exhales. The Boston "Journal of Chemistry" quotes from the Berlin "Gegenwart" a report of some alleged experiments made by one Dunstmaier to test the accuracy of this hypothesis, and which gave an affirmative result, even to the isolation of such principles as timidity and courage dissolved in glycerin.

Hereditary Effects of Extirpation of the Spleen.—According to "Medicinische Neuigkeiten," Dr. Masoire is engaged with a series of experiments upon rabbits for the purpose of testing this question. Already the weight of the spleen in the offspring of the animals operated upon has been reduced to one-half. The experimentalist hopes to ascertain not merely the functions of the spleen, but to trace in how far the removal of an organ during a series of generations may affect its development.

Prof. Huxley, on behalf of Baron Ettinghausen, communicated to the Royal Society a valuable report on the Fossil Flora of Alum Bay. He has recognised at least 116 genera and 274 species, distributed into 63 families. Many of the forms indicate that the climate was at least subtropical. The number of palms, however, is small as compared with those found in the Sheppey deposits. Many of the species form connecting-links uniting characters which have in later epochs become separated. There are indications that certain Miocene genera had not become differentiated in the Eocene period. Thus *Castanea*, perfectly developed in the Miocene, seems to be represented in the Eocene by a *Castanea*-like oak, *Quercus Bournensis*, which combines in itself the characters of both genera, now no longer found united. More than fifty of the Alum Bay species are common to Sotzka and Haering, whilst a lesser number are common to Sézanna, the Lignitic of America, and to other Floras.

Dr. W. H. Gaskell communicates to the Royal Society an account of researches on the "Tonicity of the Heart and Arteries." He finds that the heart, like the arteries, possesses what may be called tonicity, the variations in which play an important part in determining the features of the cardiac beat. The tonic condition is probably due to the alkalinity of the blood, as it is lowered by dilute acid solutions.

Dr. C. William Siemens, F.R.S., has communicated to the Royal Society an account of some most interesting investigations on the Action of the Electric Light upon Vegetation. His experiments lead to the following conclusions: that electric light is efficacious in producing chlorophyll in the leaves of plants, and in promoting growth; that an electric centre of light equal to 1400 candles, placed at a distance of 6½ feet from growing plants, appeared to be equal in effect to average daylight at this season of the year, but that more economical effects can be attained by more powerful light centres; that the carbonic acid and nitrogenous compounds generated in the electric arc produce no sensibly deleterious effects upon plants enclosed in the same space; that plants do not appear to require a period of rest during the twenty-four hours of the day, but make increased and vigorous progress if subjected during daytime to sunlight and during the night to electric light; that the radiation of heat from powerful electric arcs can be made available to counteract the effect of night frost, and is likely to promote the setting and ripening of fruit in the open air; that while under the inflence of electric light, plants can sustain increased stove heat without collapsing, a circumstance favourable to forcing by electric light; that the expense of electro-horticulture depends mainly upon the cost of mechanical energy, and is very moderate where natural sources of such energy, such as waterfalls, can be made available. Dr. Siemens has since learned from the observations of Dr. Scübeler, of Christiana, on the effects of uninterrupted sunlight during the long summer days of the Arctic, that plants under the influence of continuous light not merely grow rapidly, but develop more brilliant flowers and larger and more aromatic fruit than under the alternate influence of light and darkness. The formation of sugar depends chiefly upon temperature. Dr. Siemens mentions the interesting fact that the human skin is blistered by the action of the electric light, though without any sensation of excessive heat—an effect analogous to that produced in a clear atmosphere of the sun's rays.

We beg to call the attention of our readers to the arrangements of the Royal Institution for the second quarter of the year. When we state that Prof. Huxley is announced to lecture on the "Coming of Age of 'the Origin of Species,'" Prof. Tyndall on "Light," and Mr. G. H. Romanes on "Mental Evolution," no one can doubt that the Council have provided for the members, subscribers, and their friends an intellectual treat of no mean order.

Two of our weekly contemporaries propose a conference of curators and other officials of museums. Whilst warmly approving of the suggestion, we hope that the curators, should they meet, will not fail to protest against the present regulations concerning admission into their honourable body.

The "Medical Press" describes and figures a suction instrument for poisoned wounds, which may prove very useful in case of the bites of serpents, rabid dogs, &c. As it is now well known that the poison, *e.g.*, of the cobra is dangerous if it comes in contact with a mucous membrane, suction with the lips in such cases cannot be recommended.

We learn that a mischievous propensity is springing up among local sanitary authorities to remunerate their medical officer of health merely by fees for being called in " when the authority sees fit "—which may be never. Those who have not forgotten the lamentable circumstances which led to the resignation of Dr. C. Fox will see reason for vigilance lest the medical officers of health should be reduced to mere nonentities in favour of persons interested in the conservation of nuisance.

Prof. Roscoe has recently exhibited some specimens of finely divided gold, prepared by Faraday, proving that metallic gold may exist in two distinct molecular conditions.

Prof. Metschnikoff proposes a new method of dealing with noxious insects. The application of yeast to shrubs and plants thus affected has often proved successful, and is recommended by Prof. Hagen, of Cambridge, Mass. But according to Prof. Metschnikoff (see " Zoolog. Anzeiger," No. 47) the action of yeast, when successful, is due to the accidental presence in it of the fungus known as " green muscardine " (*Isaria destructor*), which are exceedingly deadly to insects. He has succeeded in cultivating the muscardine, and hopes that by its means the *Phylloxera*, the Colorado beetle, &c., may be eradicated. It is to be hoped that the new power we are thus invoking may prove amenable to control, or perhaps the remedy may prove worse than the disease. What if this fungus prove deadly to the silk-worm, the honey-bee, and, worse still, to those species—chiefly Hymenopterous—which effect the fertilisation of flowers ?

According to the " Kansas City Review " remains of gigantic race of men, considerably advanced in civilisation, have been discovered " in a cave on the old Smith farm in Tiffin Township, Adams County, Ohio." The cavern had at one time been a burial-place, and contained many tombs adorned with bas-reliefs of a very high artistic character. One of the tombs, when broken open, was found to contain the mummy of a man 9 feet 1 inch in length. Several tools and weapons of copper were found in the tomb, hardened to such a degree that a file will barely scratch the lance. It is to be hoped that further examinations will be undertaken by fully qualified persons. Perhaps at last a light will be thrown on the early history of the western hemisphere.

The same Review mentions, on the authority of Dr. F. A. Ballard, that in Jackson County, Missouri, has been found the tusk of a mastodon over 14 inches in diameter and 12 feet in length.

The " Leadville Herald " announces the discovery of a true glacier in the Mosquito Range of the Rocky Mountains.

In a memoir on magnetic circuits in dynamo- and magneto-electric machines, communicated to the Royal Society by Lord Elphinstone and C. W. Vincent, it is shown that any interference with the lines of force about a magnetic circuit means an interference with the magnetic circuit itself, and points to the possibility of building up magnetic force of magnets by the mere movement of wires in these lines of force, though the coils moved need not of necessity be connected with the helices surrounding the magnets.

Mr. F. Geddes has laid before the Royal Society a paper on the coalescence of amœboid cells into plasmodia, and on the so-called coagulation of invertebrate fluids. The author considers it may be safely asserted that all the evidence we possess points to the conclusion that the power of coalescing with its fellows, under favourable circumstances, to form a plasmodium, is at any rate a very widely spread—if not a general—property of the amœboid cell.

Dr. MacMunn, according to a memoir communicated to the Royal Society, has succeeded in isolating urobilin from urine, and determining some of its principal optical and chemical properties. It is an amorphous, nitrogenous, brownish red pigment, derived from one of the colouring-matters of bile.

Dr. Thudichum has laid before the |Royal Society a paper on the modifications of the spectrum of potassium effected by the presence of phosphoric acid, and on inorganic salts and bases found in combination with educts of the brain. He combats the evidence recently advanced by Prof. E. Roscoe in favour of the existence of " protagon " as a definite chemical compound. He also points out errors in the determination of inorganic matter in the brain, and in organic bodies generally.

M. E. A. Schäfer, F.R.S., communicates to the Royal Society the results of an elaborate investigation of the immature ovarian ovum in the common fowl and in the rabbit, with observations on the mode of formation of the discus proligerus in the rabbit and of the ovarial glands in the dog.

According to the researches of C. DeCandolle and Raoul Pictet the seeds of plants lose little of their germinating power even by prolonged exposure to intense cold. The refrigeration was carried down to $-80°$.

ERRATA.

Page 168, line 3, *for* " acceptable " *read* " accessible."
    ,,   line 25, *for* " Drydale " *read* " Drysdale."

THE

JOURNAL OF SCIENCE.

MAY, 1880.

—————————

I. OFFENSIVE AND DEFENSIVE TORPEDO WAR.

By Capt. S. P. OLIVER, (late) R.A.

PART II.

IN the last paragraph of my former paper I had only just time, before the MS. went to press, to allude to the tragical event which occurred at the Winter Palace on the 17th of February. In connection with this affair it may be as well to remind ourselves that it is not always Imperial and dynastic representatives who are the destined victims of assassination by means of infernal machinery, as we are too apt to think, as such means have been used by the Royalist party when in a minority against a Republican chief; witness the attempt against Buonaparte by the Chouanists, which in many respects is a parallel case to the recent incident at St. Petersburg. We may mention briefly the circumstances as stated by Napoleon himself subsequently at St. Helena, to the Count de Las Cases, on the 28th October, 1815, and which is thus recorded in the journal of the latter :*—" In the course of the conversation this day the Emperor adverted to the numerous conspiracies which had been formed against him. The infernal machine was mentioned in its turn. This diabolical invention, which gave rise to so many conjectures, and caused the death of so many victims, was the work of the Royalists, who obtained the first idea of it from the Jacobins. The Emperor

---

* Memoirs of the Life, Exile, and Conversations of the Emperor Napoleon By the Count DE LAS CAS. Vol. i., p. 238. (Colburn, 1836.)

stated that a hundred furious Jacobins—the real authors of the scene of September, the 10th August, &c.—had resolved to get rid of the First Consul, for which purpose they invented a 15- or 16-pound howitzer (*sic*) [howitzer-shell ?], which on being thrown into the carriage would explode by its own concussion, and hurl destruction on every side. To make sure of their object they proposed to lay *chevaux de frise* along a part of the road, which, by suddenly impeding the horses, would of course render it impossible for the carriage to move on. The man who was employed to lay down the *chevaux de frise*, entertaining some suspicions of the job which he had been set upon, as well as of the morality of his employers, communicated the business to the police. The conspirators were soon traced, and were apprehended near the Jardin des Plantes, in the act of trying the effect of the machine, which made a terrible explosion. The First Consul, whose policy it was not to divulge the numerous conspiracies of which he was the object, did not give publicity to this, but contented himself with imprisoning the criminals. He soon relaxed his orders for keeping them in close confinement, and they were allowed a certain degree of liberty. In the same prison in which these Jacobins were confined some Royalists were also imprisoned for an attempt to assassinate the First Consul, by means of air-guns. These two parties formed a league together; and the Royalists transmitted to their friends out of prison the idea of the infernal machine, as being preferable to any other plan of destruction.

It is very remarkable that, on the evening of the catastrophe, the Emperor expressed an extreme repugnance to go out. Madame Buonaparte and some intimate friends absolutely forced him to go to an oratorio. They roused him from a sofa where he was fast asleep ; one fetched his sword, and another his hat. As he drove along in his carriage he fell asleep again, and awoke suddenly, saying that he had dreamed he was drowning in the Tagliamento. (To explain what he alluded to it is necessary to mention that some years previously, when he was General of the Army of Italy, he passed the Tagliamento in his carriage during the night, contrary to the advice of everyone about him. In the ardour of youth, and heedless of every obstacle, he crossed the river, surrounded by a hundred men armed with poles and torches. His carriage was, however, soon set afloat ; Napoleon incurred the most imminent danger, and for some minutes gave himself up for lost.) At the moment when he now awoke, on his way to the oratorio, he was in the midst of a conflagration, the carriage was lifted up, and the passage of the Tagliamento came fresh upon his mind. The illusion, however, was but momentary; a dreadful explosion immediately ensued. "We are blown up!" exclaimed the First Consul to Lannes and Bessieres, who were in the carriage with him. They were for stopping the carriage, but the First Consul enjoined them not to do it on any account. He arrived safe, and appeared at the Opera as though nothing had happened.' He was preserved by the desperate driving of his coachman. The machine injured only one or two individuals who closed the escort.

.The most trivial circumstances often lead to the most important results. The coachman was intoxicated, and there is no doubt that this proved the means of saving the life of the First Consul. The man's intoxication was so great that it was not until next morning he could be made to comprehend what had happened. He had taken the explosion for the firing of a salute. Immediately after this event measures were adopted against the Jacobins who had been convicted of meditating the crime, and a considerable number were banished. They, however, were not the real criminals, whose discovery was brought about by another very singular chance. Three or four hundred drivers of *fiacres* subscribed a louis or twelve francs each, to give a dinner to the First Consul's coachman, who had become the hero of the day and the boast of his profession. During the feast one of the guests, drinking to the health of the First Consul's coachman, observed that he knew who had played him the trick, —alluding to the explosion of the machine. He was immediately arrested; and it appeared that on the very night, or the night preceding the explosion, he had drawn up his *fiacre* beside a gate, whence had issued a little cart that had done all the mischief. The police proceeded to the place, and it was found to be a coach-yard, where all kinds of vehicles were lent on hire. The keepers of the yard did not deny the fact; they pointed out the stall in which the cart stood: it still presented traces of gunpowder. The proprietors declared that they were given to understand the cart had been hired by some Bretons who were concerned in smuggling. The man who had sold the horse, together with every individual who had participated in the affair, were easily traced out, and it was proved that the plot had been formed by some Chouan Royalists. Some active and intelligent men were despatched to their head-quarters in Morbihan. They took no pains to conceal their share in the transaction, and only regretted that it had not succeeded. Some of them were apprehended and brought to punishment.* It is said that the chief conspirator afterwards turned Trappist, and sought to expiate his crime by religious austerities."

Such were two of the many plots against Napoleon, who, in adverting to the many attempts that had been made to assassinate him (such as those disclosed during the enquiry into the conspiracy of Charles d'Hosier, Moreau, Georges, and Pichegru), observed that he was bound in justice to say that he had never detected Louis XVIII. in any *direct* conspiracy against his life, though such plots had been incessantly renewed in other high quarters. The following is significant :—" If," said Napoleon, " I had continued in France in 1815, I intended to have given publicity to some of the later attempts that were made against me. The *Maubreuil* affair, in particular, should have been investigated by the First Consul of the Empire, and Europe would have shuddered to see to what an extent the crime of secret assassination could be carried."—(*Id.*, vol. ii., p. 226.)

In a Lecture given before the United Service Institution, February 15th, 1878, by Admiral the Right Hon. Lord Dunsany, " On the Laws and Customs of War, as limiting the use of Fire-ships, Explosion Vessels, Torpedoes, and Submarine Mines," we were told " Some of the old jurists of the highest authority justify assassination and poisoning." Wheaton, a standard American authority, tells us—" Even such institutionary writers as Bynkershoek or Wolf, who lived in the most learned and not the least civilised countries of Europe, at the commencement of the 18th century, assert the broad principle that everything done against an enemy is lawful ; that he may be destroyed, though unarmed and helpless ; that fraud, and even poison, may be employed against him ; and that an unlimited right is acquired by the victor to his person and property. Such, however, was not the sentiment and practice of enlightened Europe at the

---

* The two principal conspirators were Arena and Cerachi, although they were doubtless inspired on this occasion, in 1800, by far higher and aristocratic Royalist plotters. Since this attempt forty-one murderous attacks have been made on Royal personages and Rulers of States, including three attacks on our own Queen, four on the person of the Emperor of Germany ; five of these attempts were directed against the present Czar—in 1866, 1867, and three times within the last twelve months have they been renewed. The assassination of Napoleon III. was intended on seven occasions, and the life of his predecessor, Louis Philippe, was attempted six times within eleven years. Very few assassinations were successful, viz., those five murders which reckon as their victims Paul II., Emperor of Russia, strangled in 1801 ; Prince Obrenovitch, of Servia, killed in 1868 ; President Lincoln, shot in 1865 ; President Morenos, assassinated in 1875 ; and President Gill, of Paraguay, murdered in 1877. The execution of the Emperor Maximilian, of Brazil, can hardly be included in this category.

time they wrote, since Grotius had long before inculcated a more humane principle." In the same lecture Lord Dunsany alludes to the use of the coal-torpedoes, mentioned in my previous notice (see page 174), as a variety of warfare practised by the Confederates "which would certainly be more honoured in the breach than in the observance. That we find detailed in the ' Edinburgh Review ' (October, 1877). I call your attention to it because it certainly seems to be on the verge of the lawful,* if it is not beyond it. The particular means of explosion consisted of a hollow lump of iron filled with a charge of dynamite. It was rubbed over with coal-tar and dust, and exactly resembled a large lump of coal. I am not sure whether it was used with success, but it certainly was used by the Confederates." In the discussion which followed the reading of the above-named paper Major E. H. Cameron, Instructor Royal Laboratory Dept., Woolwich, said as follows:—"He" (Lord Dunsany) "had also instanced explosive coal lumps, and I think gives the credit of that, or the discredit, to the Confederates. I can only say I have had one of those lumps in my own hands which came from the Northern side, and was intended by Northern cruisers to have been put on board blockade-runners, who were to have been allowed to proceed on their mission, to discharge their cargo into the bunkers of Confederate cruisers, and thus the explosive coal-lump was to have produced its unhappy results when least expected." Since I wrote my paper (in the No. for March) I have myself obtained one of these coal-shells from the gentleman who, under the pseudonym of "War-hawk," warned the public seven years since of their manufacture. I shall place it in the Museum of the Royal United Service Institution for public inspection. Has anyone seen the so-called *bullets-asphyxiants ?*

There is no doubt that publicity to all secret contrivances is the best method of combating their use for mischievous purposes ; thus, for instance, a brass-founder when required to make certain casts would have his suspicions aroused did the models ordered resemble blocks of coal, had he heard of these diabolical bombs. It is only on this plea of public utility that I have ventured to bring to notice in the pages of the " Journal of Science " such apparently digressive

---

* The Right Honble. Montague Bernard, the distinguished jurist, referring to the above, said, " For my part I confess that the sending an infernal machine that looks like a mere piece of coal into an enemy's country is worse to my mind than even poisoning an enemy's officer, because you cannot foresee what destruction it may cause or who may be the sufferers. I can conceive nothing that falls more thoroughly within the description of an unfair mode of warfare than that."

matter, and, after apologising for the interpolation, now resume my remarks upon the more legitimate methods of torpedo warfare.

In the former article, called forth by the publication of Commander Sleeman's work, I alluded chiefly to the offensive attack by torpedo-boats armed with the spar and Whitehead or locomotive torpedoes, instancing the few occasions on which such attacks had been successful in real warfare ; and I now proceed to point out that, whilst the means of active attack and defence—especially whilst *on the move*—have lately been greatly developed, still the improvements have added yet more to the odds on the side of defence than in proportion to the new modes of offence.

We may take the latter first.

In the " Révue Maritime et Coloniale " there have been some interesting papers by Le Lieutenant Chaband Arnault, translations from which have been published by Lieut. E. Meryon, R.N., in the " Journal of the Royal United Service Institution," in the last number of which we find, among the results arrived at, the following conclusions, which the author considers as established from "the conscientious examination of the different sea-fights in which torpedo-boats have taken part."　I propose to take them *seriatim.*

(A.) "An attack made against a ship by means of a tor-pedo-boat does not require more exceptional circumstances nor more absolute devotion than any other operation of naval war."　Granted, with a reservation best expressed in the following remark of Sleeman, p. 8 :—" The cause of the want of success in war-time with offensive torpedoes lies in the fact that during peace-time the experiments and practice carried out with them are done so under the most favourable circumstances, that is to say, in daylight, and the nerves of the operators not in that high state of tension which would be the case were they attacking a man-of-war on a pitch-dark night, whose exact position cannot be known, and from whose guns at any moment a sheet of fire may be belched forth, and a storm of shot and bullets be poured on them whilst on actual service ; this would in nine out of ten in-stances be the case.　Some uncertainty must and will always exist in offensive torpedo operations when carried out in actual war, where, as in this case, the success of the enter-prise depends almost wholly on the state of a man's nerves; yet this defect, a want of certainty, may to a considerable extent be eradicated were means to be found of carrying out in time of peace a systematic practice of this branch of torpedo warfare under circumstances similar to those experienced in war-time; and this is not only possible, but practicable." Here it should be observed that it is not only *pluck* and *devotion*, but a steadiness of nerve, that is most important. There is many a brave man capable of leading a forlorn hope in storming a breach, and hand-to-hand combat whilst boarding an enemy's ship or blowing in a stockade, and such other feats in which excitement and passion of combat are incentives to daring deeds; but it requires a different class of cold-blooded, phlegmatic temperament to manipulate delicate adjustments with a quiet pulse, free from all excitement or apprehension of the immediate neighbourhood of sudden death. It is like the comparison of the spirit of a prize-fighter with that of a surgeon during a mortal operation.

(B.) "An attack of this description, where the assailant only risks forces relatively small, offers such chances of success that it should always be tried when the occasion presents itself." This is a self-evident proposition, as the *matériel* and *personnel* exposed by the assailants to the fire of the enemy is hardly worth mentioning in comparison with the amount of damage to be perpetrated; for instance, a small Yarrow boat *versus* the *Inflexible!*

(C.) "One single torpedo-boat of moderately good quality is sufficient to surprise at night a vessel at anchor; a day attack on a vessel under weigh requires the co-operation of several torpedo-boats, especially built for the purpose." Lord Dunsany instances the exploit of the Russian steamer yacht *Constantine*, who sent off her torpedo-cutter against the Turkish ironclad guardship at Batoum, and destroyed the latter. He says, "A point I would call your attention to is the very small risk to the torpedo crew at which the wholesale destruction is effected. The huge guns of the present day make excellent practice; but they fire very slowly, and, so far as I could judge, the chances of hitting a torpedo-boat in motion would be *nil*."[*] The *Duilio* would be nigh powerless against a small flotilla of swift Yarrow or Thorneycroft boats with bow-rudders, able to dart their Woolwich torpedoes from their outriggers on all sides. Behemoth amid a shoal of threshers and swordfish must soon succumb.

(D.) "Under all circumstances where such boats, collected into a flotilla, have to attack one or more ships, their attacks should be simultaneous." The majority of actions fought by American and Russian torpedo-boats have been fought

---

[*] In the Navy estimates for 1880-81, just submitted to Parliament, we see that fifty-six torpedo-boats are to be constructed by private firms for the British Navy. We should like to see double the number provided for.

without anything like concerted action ; but it is manifest
that several boats which unite their endeavours distract the
attention of their enemy, and augment, if they do not ensure,
their chances of success. Hence, says Lieut. Arnault, " A
day attack requires special boats, fast, turning rapidly, presenting as small a target as possible to the enemy's gunners,
and the crew of which should be absolutely protected against
musketry. Grape and case can pierce the boat ; but the
experience gained at Rustchuk and Nicopolis shows us that
if the engines remain unharmed the boat escapes by keeping
up a high speed. Hence great lightness of hull, rapidity of
movements and turning, small dimensions, musketry protection limited to the machinery and the fighting posts of the
crew—such seem to us to be the principal requirements
which boats specially built for torpedo fighting should fulfil.

(E.) " The use of ' diverging torpedoes ' in boats should be
discontinued." Witness the failure of the diverging torpedoes despatched from the *Tchesmé* against the *Assar-i-
Chefket*, at the Sulina mouth of the Danube. The fact is the
slightest floating protection disarms all towing divergent
torpedoes ; even a mere rope stretched between two booms
will suffice to throw off the tow-line of a Harvey or other
similar weapon.

(F.) " Spar torpedoes are usefully employed by boats attacking on a very dark night, or with a slightly agitated sea,
a ship moored amongst strong currents, or which may be
expected not to be protected by external defences." Commander Sleeman says, " The spar, the Whitehead fish, and
the Harvey towing torpedo have each been subjected to the
test of actual service, the former weapon being the only one
that has under those conditions been successfully used."
Taking this fact into consideration, also the high pitch of
excellence that has been attained in the construction of steam
torpedo-boats, and also the results of the numerous exhaustive experiments that have been from time to time
carried on in England, America, and Europe, with various
modifications of the locomotive, towing, and spar torpedoes,
there can be no two opinions as to which of the numerous
species of offensive submarine weapons is the most practicable
and effective,—and that is the spar or outrigger torpedo.
To manipulate successfully locomotive and towing torpedoes
in an attack against hostile vessels, the operators must be
not only unusually fearless and self-possessed, but must also
possess a thorough practical knowledge of the complicated
method of working and manœuvring those weapons—in fact
they must be specialists ; whilst in the case of the spar
torpedo, which may be fired by contact, it is only necessary

to employ men capable of handling a boat well, and possessed of dash and pluck, to ensure an attack by such means being generally successful.

(G.) " Whitehead torpedoes will be employed by preference when the night is comparatively clear, the sea very calm, the current slight, and when one may suppose the enemy's ship to be provided with obstacles." The fish torpedo fired from a boat in close proximity to the attacked vessel, in *smooth* water and unmolested, would sink a vessel, which under the same circumstances, owing to her being protected by booms, might prove impregnable to a spar torpedo attack, but such favourable conditions will not often occur in war-time.

(H.) " Against a vessel under weigh the two species of torpedoes will both respectively find their value according to circumstances. It would appear, then, to be best to combine in a flotilla intended for the defence of harbours and road-steads both spar and Whitehead torpedo-boats. Finally, in certain circumstances, a ship under weigh may defend her-self advantageously against torpedo-boats by booms and nets." Such are a few of the principles deduced by Lieut. Arnault from the published accounts of the employment of torpedo-boats against ships, which summed up would seem to give the odds in favour of the attack, but if we examine more closely we shall see that there is something to be said on the other side; for instance, in all the examples, the attack of boats on large ships has been met in all cases by merely passive resistance : in future this will never be the case. It will be a case of " Greek meeting Greek," for every large ship of war will now have its own attendant flotilla of guard-boats, themselves torpedo-vessels,* and an element which we have never yet seen discussed really now enters upon the scene, and must be taken into account— *i.e.*, the manœuvring of torpedo-boat *versus* torpedo-boat. One exciting tournament of this description has already been rehearsed, although the incident did not form part of the programme, the other day, or rather night, at Spithead. It is thus described :—" An accident, which very nearly termi-nated in the destruction of a couple of torpedo-boats, oc-curred between 7 and 8 o'clock on the night of the 5th inst., at Spithead. Four second-class Thorneycroft torpedo-boats, attached to the *Hecla*, torpedo store ship, Capt. Morgan Singer, now lying at Spithead, were despatched from the

* On the 27th February, 1880, the Peruvian monitor *Mancocapa* came out o Arica and engaged the famous Chilian *Huascar* at close quarters : the *Huascar* feared to ram her antagonist, "*on account of her having a torpedo-boat in tow* " ! The further details will be interesting.

ship soon after dark for practice, each boat taking a different course and being under independent control. The boats of this class, when fully accoutred, carry each three Whitehead locomotive torpedoes, which are discharged in a line with the keel from slings at the side, and the object of the practice, which was of the ordinary routine character, was to endeavour to approach within striking distance of the *Hecla* without being discovered by those on board. It need scarcely be remarked that as the attack was a sham one, and was made during the night, there were no explosive projectiles used on the occasion, the whole practice being confined to evolutions under steam. The hostile craft approached from various points of the compass at about half-speed, or from 6 to 8 knots an hour. Two of the boats, a black one and a grey one, began the attack from the eastward, and when the former drew within a quarter of a mile of the *Hecla* she was noticed and ruled out of the attack. At this moment, while she was lying with her machinery at rest, the grey craft was heard coming at a rapid rate in her direction, those on board apparently not being able to discern her black hull in the water. She was hailed, and though her engines were promptly reversed she struck the stationary boat a violent blow on the port side, breaking one of the plates and doing considerable damage to the angle frames or stiffeners. The water rushed in, but was happily confined to the second or smoke-box water-tight compartment, whereby the buoyancy of the craft was preserved. Had she been struck a couple of feet or so farther aft the engine-room would have been flooded, and probably some of the men severely scalded. Considering that these boats are mere steel shells, not more than one-sixteenth of an inch in thickness, it seems surprising that the stricken boat was not cut in two ; and this would undoubtedly have been the case had the momentum of the second boat not been arrested by the wire-rope stay which surrounds the craft at the water-line, and of which only a strand or two were cut. The attacking boat was even worse damaged, her sharp stem being broken and distorted, and the bow plates bent and bulged from their fastenings. The forward collision bulkhead, however, performed its work admirably, and did not permit even so much as a weeping of water to penetrate beyond the foremost compartment. Though both vessels were disabled, no one was injured. The boats were about to be forwarded in the Wivern for use in China, and others will now have to be substituted."

We may expect to see these torpedo-boats used as auxiliary to defence as well as for offensive purposes, and they will probably develop an armament,* being fitted with the Hotch-kiss gun, which fires a shot weighing a pound, penetrating an inch and a half plate at 300 yards; and as the torpedo-boats are seldom more than three-sixteenths of an inch thick, we shall hear no more of the shot-holes being water-proofed by rapid steaming through the water, and even the Fosberry patent india-rubber protective will be of little avail. "*Kieselghur*" is used with success for *extempore* plugging of shot-holes.

There is one observation which should be made before taking leave of this position of the subject, and that is the reply to the question put by Admiral Sir Frederick W. E. Nicolson, Bart., C.B., when presiding at the discussion of Lord Dunsany's paper before mentioned. "The Chairman: I should like to ask Lord Dunsany a question. As I gather from your Lordship's argument, you would treat with stern justice the persons employed in a torpedo-boat. I want to ask you this: that torpedo-boat I presume you send away from a vessel. Supposing subsequently you captured that vessel, in what manner would you treat the Captain or the Admiral flying his flag in the vessel from which the torpedo-boat had been sent?" The answer was skilfully evaded; but this is certain, that on board of all vessels carrying tor-pedoes and torpedo-boats the crews who man them should be perfectly aware of the risks to which they are exposed at the enemy's hands if captured, in the same way that the crews of fire-ships of olden times perfectly appreciated that they were entitled to no quarter from the enemy. And, moreover, distinct orders should be given by the Admiralty as to the usage to be practised towards all torpedo oppo-nents. Have the Admiralty given any such orders? or is the "*custom of war in like cases*" relied upon as covering all discretion on the subject.†

In the articles of war a paragraph must soon be inserted as to what torpedist operators are to expect at our hands, and if it happens to be inserted "*shall suffer death*" all am-biguity will be avoided without circumlocution. Lord Dun-sany said, "One sees many occasions on which a sentimental answer can be given in a moment—'*Hang him.*'" *Verbum sap.*

* In view of these hand-to-hand combats, small hand torpedoes, composed of india-rubber waterproof bags containing gun-cotton (oz.), are designed, with detonators and wires, to be thrown and detonated as hand-grenades.

† It is perfectly possible for a torpedo-launch to leave Portsmouth after dark, and to destroy ironclads inside the breakwater at Cherbourg, and return to Portsmouth before daybreak, without discovery on a dark winter's night, with a *minimum* risk of detection, and without exhibiting its nationality.

II. THE SOUL: WHAT IS IT?

CONCERNING the constitution of man there are three distinct theories. The first regards him as composed simply of a body, actuated for a time either by the ordinary forms of energy or by some modification thereof not yet recognised, and as losing at death his personal individuality. The second and more popular view acknowledges in him a double nature, comprising, in addition to the palpable, ponderable, and visible part or body, an invisible and immaterial principle, known promiscuously as " soul " or " spirit." But there is yet a third theory, which considers man as a threefold being, made up of body, soul, and spirit. It is no part of our present purpose to define the exact sense in which these last two terms are used. It may suffice to say that by the ordinary advocates of the triplicity of human nature the " soul " is supposed to be the purely immaterial element, whilst the " spirit " forms a connecting-link between the two, and, if not purely incorporeal, possesses none of the ordinarily recognised properties of matter.

An author* whose speculations we are about to examine exactly reverses these two terms, and looks upon spirit as a something absolutely immaterial and transcendent, whilst the soul, the seat of the will, the passions, and emotions, is perceptible by one, at least, of our senses, and is even capable of being experimentally isolated and obtained in solution.

We well know that the orthodox method of treating such an announcement is either by contemptuous laughter or by the " conspiracy of silence." We, however, hold that Science has nothing to lose, and may have much to gain, by a dispassionate examination even of the wildest theories. By so doing she will probably fare as well as did the fabled brothers who, in searching for a supposed buried treasure in their estate, marvellously enhanced its fertility.

We find ourselves confronted by a number of facts, hitherto without explanation and without connection. Among these must rank the phenomena of sympathy and antipathy as between different individuals, human or brute. On first meeting with some person of whom we have no previous

* Professor Jäger.

knowledge, we often experience a strong liking or a violent dislike, for neither of which we can render any definite reason. As a rule women and children are more frequently impressed in this manner than are adult men. It very often happens, too, that if we suppress and overcome these sudden prepossessions, we find in the end that they were justifiable, and that second thoughts were not best. What is the key to that strange personal ascendancy which some men seem to possess over their associates? This mastery is not necessarily connected either with physical or intellectual superiority, nor certainly with rank or position. There are characters who are obeyed, even by their official superiors or their employers. There are others who cannot uphold authority over inferiors without incessant punishments.

Further, the emotions and passions of men assembled together are infectious, passing from one to another more rapidly than bodily diseases. From one or from a few energetic individuals enthusiasm may be diffused through a senate, a regiment, or a ship's crew. On the other hand, a few terrified or bewildered persons may spread a panic among thousands. It is commonly said that emotions propagate themselves, but we wish to know in what manner and by what means this is effected.

Again, domestic animals very often display a sympathy for some persons, and a hatred for or a fear of others, which are very hard to explain. We have known men who could approach the most ferocious dog without any fear of injury, whilst others can scarcely walk along a public thoroughfare without being barked at by every cur they meet. Why will a horse obey implicitly one stranger and become unruly and refractory if touched by another? Why will cows or oxen, when driven along a road, make a rush at some one passenger, after letting scores of others go by unharmed? The reasons commonly given for these differences of behaviour will not bear close inspection. Dog-worshippers, forgetting such cases as that of the robber and assassin Peace, insinuate that their pets have an instinctive repugnance for a man of bad character. Others assert that dogs attack only the timid. We knew a gentleman, exceptionally courageous, who was a particular object of the ill-will of these animals. He was often first made aware of the presence of a dog by a volley of yelps and snarls close behind him, so that the attack cannot have been provoked by any demonstration on his part. It is said that animals fawn on such as like them, but flee from, or, if strong enough, attack such as view them with dislike. We can contradict this assertion from personal experience. We are by no means fond of cats, yet these creatures approach us without hesitation, spring upon our knee, rub their heads against our face, and can scarcely be made to understand that such attentions are far from welcome. A very remarkable fact is the influence which some men possess over horses, and which seems rather a personal peculiarity than any secret that could be communicated to others. It is said that such men have been known completely to subdue a vicious horse by blowing into his nostrils.

We find, again, sympathies, and especially antipathies, which may be traced between entire species of animals, and which some of us seek to explain by the indefinite and long-suffering word " instinctive." If a dog has been stroked with a gloved hand, and if the glove is then held to the nose of a young kitten, still blind, the little creature begins to spit in anger. How is this fact to be explained ? The kitten has never yet seen a dog, but by the mere odour it recognises a hostile element. Heredity ? True, but how is the antipathy handed down from generation to generation ? By what sign does the blind animal detect the presence of an enemy ? Very similar is the dread or disgust felt by mice in the neighbourhood of a cat. If one of these animals is kept in a house, no matter how lazy and sleepy she may be, the mice generally withdraw to safer quarters. Shall we suppose that they have all seen or been chased by this enemy, or that those who have fared thus spread the news to their companions ?

There is still a further phenomenon which may be looked on as a heightened antipathy—fascination. We all know that very intense fear, instead of prompting to flight, may paralyse. It is said that certain rapacious creatures, especially serpents, have the power of producing in their intended victim a kind of torpor, so that it helplessly and passively awaits certain destruction. We never had the good fortune to witness an incident of this kind, but Knapp, in his " Journal of a Naturalist," gives a case as from his own observation.

We come next to a class of phenomena on which accurate observation and careful experiment are still more needful. It is asserted by popular tradition, and is half admitted by Dr. O. Wendell Holmes, that certain animal secretions if introduced into the body of some other animal, of the same or of a different species, may have a strangely modifying action upon the individual thus inoculated. This is said to have occurred in cases where the bite of a rattlesnake has not proved mortal.

Taking a general view of all these phenomena, in so far as they are actually established, it would seem that animals, including man, must throw off from their surfaces some emanation capable of acting upon other animals and men with whom they come in contact or in near proximity. This supposed emanation may vary in its character in one and the same individual, according to its psychical condition. If the vapours or gases thus emitted by two animated beings are in harmony, the result is sympathy or attraction. If they disagree, the consequence is antipathy, showing itself as hatred in the strong and as fear in the weak. This, it will be doubtless admitted, is a possible explanation of some of the phenomena above noticed ; but is it the true or the only one ? Do such emanations really exist ? It is, we think, certain that many animals become aware of the presence either of their prey, of an enemy, or of a friend, by the sense of smell, even at very considerable distances. Our lamented friend Thomas Belt was led to the conclusion that ants are able to communicate with each other by means of this sense, and have in fact a smell-language. Unfortunately the sense of smell is so weak in man that it becomes very difficult for us to decide.

Prof. Jäger holds that certain decompositions take place in the animal system in strict accord with psychic changes. All observers, he tells us, agree that muscular exertion effects but a very trifling increase of the nitrogenous compounds present in the urine. On the other hand, Dr. Boecker and Dr. Benecke* have proved that intense pleasurable excitement effects a very notable increase of the nitrogenous products in the urine, derived, as a matter of course, from the decomposition of the albuminioid matter in the system. Prout and Haughton have made a similar observation concerning the effects of alarm and anxiety. Hence, therefore, it would appear that strong emotion involves an extensive decomposition of nitrogenous matter, and in particular of its least stable portion, the albuminous compounds. But does the whole of the matter thus split up reappear in the urine ? Prof. Jäger thinks that a portion escapes in a volatile state, forming the odorous emanations above mentioned. This portion he considers is the soul, which exists in a state of combination in the molecule of the albumen, and is liberated under the influence of psychic

---

* **Pathologie des Stoffwechsels.**

activity. Hence his soul, like the body, is not a unitary entity, called once for all into existence, but is a something perpetually secreted, and as perpetually given off. It pervades the entire system. Each organ has its distinct psychogen, all of which, however, are merely differentiations of the one primary ovum-psychogen. Further modifications take place from time to time, in accordance with the mental condition of the man or other animal. It will here be remembered that, according to Haeckel ("Die Heutige Entwickelungslehre in Verhältnis zur Gesammt-wissenschaft "), all organic matter, if not matter altogether, is be-souled. Even the "plastidules"—the molecules of protoplasm—possess souls.

In support of the assumption that a volatile something is given off from albumen, Prof. Jäger gives the following delicate experiment :—If we prepare, from the blood or the flesh of any animal, albumen as pure as possible, and free from smell and taste, and treat it with an acid, there appears a volatile matter which is perfectly specific, differing in the case of each animal species. But this odour varies according to the intensity of the chemical action. If this is slight we perceive the specific " bouillon odour " which the flesh of the animal in question gives off on boiling. On the contrary, if the reaction is violent, the odour given off is that of the excrement of the species. Here, then, we have the two main modifications of psychogen, the sympathetic and the antipathetic form.

Dr. O. Schmidt, Professor of Chemistry and Physics at the Veterinary College of Stuttgart, has repeated these experiments upon the brains of animals. The odoriferous principle is here evolved much more easily than from egg albumen. Immediately on the addition of an acid an offensive odour appears, which vanishes as rapidly, and cannot be caused to reappear. Nor has it been found possible to elicit from brain the more agreeable odour.

It will doubtless be granted that certain yet unexamined specific odours are given off by living animals ; that these odours may be repulsive or attractive to other species ; that they may be liberated more abundantly under mental excitement. But where is the proof that these odours are the soul in any condition ? May they not be regarded merely as an effect which psychic emotion, along with other agencies, produces in and upon the body ?

We will therefore, though not without misgivings, quote an experiment to which Duntsmaier ataches much importance. He placed in a large wire-work cage a number of hares, and allowed a dog to run round this prison, snuffing at the inmates, and attempting to get at them for about two hours. It need scarcely be said that the hares were in a state of great terror. At the end of that time the dog was killed ; his olfactory nerves and the interior membranes of the nose were taken out with the least possible loss of time, and ground up in glycerin. The clear liquid thus obtained contained the souls of the hares, or at least portions of them, in an intense state of painful excitement. Every animal to whom it was administered, either by the mouth or by injection under the skin, seemed to lose all courage. A cat after taking a dose did not venture to spring upon some mice. A mastiff similarly treated slunk away from the cat. Now we are here confronted by a serious difficulty : if a second dog was rendered timid by merely a small portion of this extract of fear, how is it that the first dog, after snuffing up the whole, did not suffer the same change and become afraid of the hares ?

Other experiments, we are told, were tried with analogous results. Thus a glyceric extract of courage was obtained from a young lion, the olfactory nerves of a dog being again used as the collecting medium.

A difficulty which must make us hesitate before ascribing animal antipathies to some disagreement in their souls, making itself known by their specific emanations, is the following : the animals of uninhabited islands when they first come in contact with man entertain no antipathy for him, until his propensity for indiscriminate slaughter is learnt by experience. Can we assume that his emanations have changed in the meantime ? Again, a colony of mice had established themselves at the bottom of a deep mine, doubtless in order to prey upon the provisions, candles, &c., of the workmen, and had flourished there for many generations. One of them, being captured, was brought up, placed in a cage, and shown to a cat. The cat prowled around and tried to get at its prey, but the mouse gave not the least sign of alarm. Why should the emanations of a cat be less alarming to this mouse than to any other ? Is the tiger, our natural enemy,—which, according to Prof. Jäger, bears the same relation to us which a cat does to a mouse,—any more offensive to us than certain animals which never prey upon man at all, such as the polecat or the skunk ? If the timid man tempts the dog or the ox to attack him, on what principle does he diffuse panic among his fellow-men ?

In short, Prof. Jäger's theory is beset with many and serious difficulties. Nevertheless, or rather the more we consider it entitled to a careful examination, both as regards its conclusion and the phenomena upon which it is based; the science of odours has yet to be constituted, and we are convinced that it will amply repay the needful trouble.

One of our author's favourite ideas is that the social split between Jews and Christians, between Aryans and Negroes, &c., depends on the want of harmony in their specific emanations. The conflicting odours of races and nations play a great part in the history of mankind.

III. INTERNAL ENEMIES.

WE hear from time to time eloquent and perhaps over-charged lamentations over the danger to which public health is exposed from the consumption of articles of diet sophisticated with unwholesome ingredients. But we too often forget that we are liable to greater peril from the use of articles which are incapable of adulteration, but which carry in themselves seeds of disease far deadlier and more difficult to overcome than the mineral poisons. These seeds, or germs of disease, as we have ventured to call them, differ widely among themselves. There are, on the one hand, ferments or poisons—organic no doubt, and in all probability organised—which, when introduced into the human body, set up morbid and often fatal action, such as cholera, typhus, typhoid, scarlatina, diphtheria, and many other of the ills that flesh is heir to. But concerning the nature of these germs we are still very much in the dark. We cannot say whether each such disease springs from a distinct class of germs, which, under all circumstances, if only swallowed, inhaled, or otherwise introduced into the body, produce that disease alone and no other; or whether one and the same kind of germs may not, under different conditions, excite affections which we classify as distinct. Thus doubts are, we believe, entertained by eminent medical practitioners whether one identical poison may not occasion, according to circumstances, either scarlatina or diphtheria.

There are, on the other hand, organisms which we may swallow in our food, of a much larger size, and whose action and nature are much better understood. These creatures, when they enter our stomachs, instead of submitting, as does lifeless matter, to the digestive process, not merely retain their vitality, but undergo their normal development reproduce and multiply, and devour us in an almost litera sense of the word.

Amongst the most dangerous of these parasites is one which has only been known for about thirty years—the *Trichina spiralis.* This pest is at present far from uncommon in the flesh of swine, and now constitutes a very definite danger for the consumers of pork and bacon. It must not be inferred, as it is by many persons, that the *Trichina* suddenly sprang into being about the year 1850 from what was previously lifeless matter. It is perfectly possible that it may have existed in the swine for untold centuries, but that it was not detected and characterised as a distinct parasitic species. It is again conceivable that, though not recently called into existence, it may only of late years have found its way into the body of the pig. Some authorities, indeed, contend that its original home is that animal which Waterton always called the " Hanoverian," *i.e.*, the rat, but which, as it appears to have entered Western Europe from the plains of the Wolga, may claim another nationality. Rats, as is well known, will visit pigsties in the hope of plunder, and may no doubt be occasionally snapped up and devoured by pigs, which are semi-carnivorous animals. Whether every trichinised pig must at some period of his life have devoured a rat, or whether the parasite can be introduced into swine by some other channel, passing for instance from the mother to her young, is not quite decided, though the latter supposition is by no means out of the question.

A young hippopotamus which died in captivity has been found to be infested with *Trichinæ.* Hence the question has been raised whether the rat theory is correct, and whether the Pachydermata as a family may not be the original home of the *Trichinæ.*

It may here be remarked that rats have been gravely recommended as an important article of human food. A writer in a daily paper, not very long ago, gravely declared that the man who died of starvation so long as rats were to be had for nothing deserved severe punishment. How to escape the danger of swallowing *Trichinæ* was a thing not dreamt of in the philosophy of the humane enthusiast.

But *revenons a nos cochons,* if we may so far parody the proverb.

Suppose the *Trichinæ* have safely landed in the stomach of the pig, their career is somewhat remarkable. In the course of three or four days they find their way into the smaller intestines, where, when met with, it is known as an intestinal trichina. Here the female, which is about 0·04 to 0·10 of an inch in length, produces living young to the number of 500 to 1000 or more, a process which occupies sometimes three weeks, and then dies. The young parasites, which are scarcely one-twentieth of the size of their mother, do not remain where they were born, but perforate the coats of the intestines, and in the space of about six days distribute themselves through the entire body of their host, especially the anterior portion. They seem especially to haunt the muscles which lie between the ribs, and those of the neck, the jaws, and the eyes. When they have found quarters to their satisfaction their wanderings cease, they attain the full stature of their parent, and are now known as muscle-trichinæ. Their development occupies a space of about two weeks. Having thus reached maturity we might expect that their new task would be the perpetuation of their species; but nothing of the sort takes place: each trichina becomes encysted, coiling itself up something after the manner of an ammonite, and covering itself with a calcareous layer or capsule. Here, then, it remains possibly for years, motionless but yet alive. In this state it is found in the flesh of the pig when slaughtered. There are yet two remarkable points to be noticed in connection with trichinised swine. The animals as a rule do not appear to suffer in any marked degree from the presence of these parasites. So much at least may be said, that they do not present a diseased appearance. This circumstance greatly increases the danger of the *Trichinæ* finding their way into human food. If the pigs when thus attacked became sickly, emaciated, &c., buyers and market inspectors might have their suspicions excited. But, save by microscopical examination, there is no certain way of distinguishing the flesh of a trichinised pig from that of one that is perfectly sound. By what we might almost call a Satanic instinct they never attack the heart of their victim, in this respect resembling the ichneumon larvæ, which, while devouring a caterpillar, leave its vitals unattacked.

Let us now suppose that the flesh of a trichinised pig is served up at table, and is eaten by some unfortunate human being. The first step is that the calcareous cyst or capsule is dissolved by the acid juices of the stomach, and its inmate is set at liberty to repeat the career of its parents. But unfortunately its presence in man is always attended with great pain and serious danger, and in the majority of cases with

death. Of course the greater the number of *Trichinæ* swallowed the more violent is the resulting affeċtion. Even on
the very day when the fatal morsel has been consumed loss
of appetite, nausea, and diarrhœa occcasionally set in, and,
as a matter of course, resist all medical treatment.

Sometimes, however, the mischief only makes itself known
from about the seventh day to the end of the second or third
week. By this time the *Trichinæ* have perforated the intestines, and made their way into the muscular system.

In this second stage of the disease the sufferings of the
patient are often extreme. The face takes a peculiarly
bloated appearance and the eyelids swell ; violent muscular
pain sets in ; the hands and feet swell ; respiration, swallowing, and the movement of the jaws become difficult ;
and lastly, profuse and exhaustive sweats, and the ordinary
symptoms of typhoid fever ensue, with which, indeed, trichinosis is sometimes confounded. After about the fifth
week, if the number of parasites introduced into the body is
small, they may become encysted as in the pig, when the
alarming symptoms gradually disappear, and the patient
returns to his ordinary condition. Death, however, is more
frequently the result. We should mention that in man, as
in swine, the *Trichinæ* never cut short the sufferings of their
viċtim by attacking the heart.

It may, perhaps, be thought that trichinised porkers are
very rarely met with. Such is by no means the case. We
learn, on good authority, that in North Germany alone
763 pigs, slaughtered between the years 1864 and 1874, have
been found suffering from this affeċtion, and that several
hundred human beings have died from partaking of their
flesh. A single trichinous pig may easily prove fatal to a
hundred men. In England the cases of this fearful disease
have hitherto been less numerous than in Germany, from a
difference in national habits, to which reference will be made
below. Still many of our readers may possibly possess
slices of a man who died in London, of trichinosis, some
ten years ago, and whose muscles, cut into fine seċtions,
have been duly preserved as microscopic objeċts.

Lately there has been a very serious outbreak of trichinosis on board the reformatory school-ship *Cornwall*, where
eighteen decided cases of illness and one death have occurred. On a microscopical examination of the muscles
and viscera of the body, *Trichinæ* were distinċtly recognised.

In America the evil and its consequences are spreading,
and have reached the Pacific Coast, a fatal case having been
observed in Oregon. So that trichinised pork and baçon,

and trichinosis in mankind, may now claim their due place among the established ills which we must take into account.

From what has been already said it will doubtless be understood that when a horde of *Trichinæ* have once established themselves in the flesh of some unfortunate man, there is no known method for their expulsion or destruction. There they are and there they must remain, and the physician can do little more than seek to alleviate some of the symptoms, and to keep up the strength of his patient till the intruders pass into the encysted stage and cease to occasion torment.

Hence, as cure is scarcely possible, the more weight must be laid upon prevention, and this fortunately is quite within human reach. The most highly trichinised pork may be eaten with safety if every part of it has been exposed, for a sufficient length of time, to a temperature of not less than 212° F. Indeed the *Trichinæ* are found to be killed at much lower temperatures, even at 134° to 140° F. But in practice the difficulty lies here,—that though a sufficient temperature may be reached on the outside of a large piece of meat, yet, on account of the very defective conducting-power of the material, the heat at a little distance from the surface falls far short of the mark. Boiling for two hours is not more than sufficient to produce a temperature of 140° F. at the depth of 2 inches from the surface. Dr. Gerlach boiled a piece of meat 4 inches in thickness for an hour, and at the end of that time he was able to find *Trichinæ* still living in the interior! Roasting is still less efficacious, as the brown outer layer is a yet worse conductor of heat than the meat in its original condition. Dr. Dietzsch, chemist to the Industrial Museum of Zurich, lays down the rule that so far as boiled or roast meat has a reddish colour, or emits a reddish juice when cut, the *Trichinæ*, if present, will still be living.

Everyone therefore must admit that the process of cooking, so called, as it is carried on in most hotels, restaurants, coffee-houses, &c., in the " 'am and beef-shops " of London, and even in many private families, is no safeguard whatever. If *Trichinæ* were found in the meat as it left the butcher's shop, they will still be found in the half-raw mass served up at table. The only method for the effectual destruction of these and other parasites would be to abandon our old semi-barbarous habit of cooking meat in such huge masses that the outside may be charred whilst the "unknown interior " is still raw. The admirers of such crude matter —" rare " they call it—can scarcely now plead superior wholesomeness for the disgusting condition in which they expect joints to be served up.

But if we in England submit our animal food to a mere nominal cooking, some of our Continental neighbours go a step further, and omit the process altogether. In Switzerland, Germany, and we believe in Holland, ham, bacon, sausages, and what is called "hack fleisch" (*i.e.*, meat chopped up as if for sausages, but not put into a skin), are consumed positively raw. This is the reason why deaths from trichinosis are more plentiful in Central Europe than in England or France, and why in Germany every pig slaughtered—or, we presume, imported as dead meat—is subject to an official microscopic examination before it may be lawfully offered for sale. This must be pronounced a very judicious regulation so long as any portion of swine's flesh is to be consumed in a raw or a half-raw condition. Uncooked sausages are exceptionally dangerous, because they are often made from those portions of meat which lie nearest to the bones and the tendons, *i.e.*, precisely the parts where the *Trichinæ* are most apt to take up their abode.

Not a few persons entertain the superstition—we can give it no better name—that smoking hams or bacon will kill *Trichinæ*, if present, and will altogether act as a substitute for cooking. It is possible that if creosote, the supposed active ingredient in wood-smoke, were placed in a concentrated state in actual contact with a *Trichina* that the life of the parasite would be cut short; but even if the outside of a ham is painted over with a solution of creosote the interior offers a safe shelter. Long before the *Trichinæ* were all killed the ham would be rendered utterly unfit for food.

Thorough drying, without any attempt at smoking, is fatal to *Trichinæ* as far as its action penetrates; but a little reflection will show us that this process can rarely penetrate much below the surface, and must be utterly ineffectual as far as the centre of a ham is concerned.

Very similar is the action of salting. The *Trichinæ* die near the surface, but to kill them throughout would require such a proportion of salt that the meat could not be eaten. After a week's exposure to brine living *Trichinæ* were found by Dietzsch at an inch from the surface.

The microscopic detection of *Trichinæ* is fortunately not a difficult task. Dietzsch directs that thin portions of the flesh, not larger than a pin's head, should be cut with a sharp razor in a direction parallel with the fibres. It is placed on a slip of glass, teased out a little with the needles, moistened with a drop of glycerin or water, and examined with a magnifying power of from 50 to 100 diameters. If no *Trichinæ* are visible, it is prudent to re-examine after moistening the specimen with a single drop of the solution of potassa (1 part in 15 parts of water), when the muscular fibre becomes transparent, and the *Trichinæ*, if present, are brought into full view. It is important, in making the examination, to select the sections from those parts where *Trichinæ* most do congregate,—*e.g.*, from the sinewy ends of the muscles, especially those of the eyes, the jaws, the neck, &c.

We are well aware that the existence of *Trichinæ* and other parasites is urged by the Vegetarians as a powerful argument against the use of animal food. Strange to say, however, like many other reformers, they indulge at times in a little inconsistency. They extend to milk an exceptional toleration. Now milk, to the best of our knowledge, never contains *Trichinæ*, but if drawn from a cow affected with tuberculosis—no very rare occurrence—it is able, as has been experimentally proved, to set up tubercular disease in animals by which it is consumed. This dangerous property, moreover, is not in the least removed by boiling. So that Vegetarianism, as generally preached and practised, is no safeguard.

IV. THE ANTIQUITY OF MANKIND.*

WHILST the first appearance of human life upon the globe has risen almost to the rank of a " burning question," no small share of its attraction and of the attention which it receives are derived, singularly enough, from an extraneous and even an illegitimate, source. The true man of Science, whilst eager to learn the exact truth on this subject, is utterly indifferent what that truth may be. Prove to him that our forefathers have lived and suffered and died upon this planet for seven million, or merely for some seven thousand years, and he will listen

---

* Early Man in Britain, and his Place in the Tertiary Period. By W. Boyd Dawkins, F.R.S., &c. London: Macmillan and Co.

Fossil Men and their Modern Representatives. By J. W. Dawson, F.R.S., &c. London: Hodder and Stoughton.

with judicial impartiality. In neither case are his hopes or his fears called into play any more than on learning the exact composition of a mineral, the locality of a plant, or the structural peculiarities of a caterpillar.

Unfortunately, however, two numerous and active sections of the public approach this question not with the feelings of a judge, but with those of an impassioned and unscrupulous advocate. Though respectively hostile, they have both been mistrained into a common error. Both conceive that the interests of revealed Religion are gravely compromised if it should appear that mankind came into being earlier than the "4004 years before the vulgar Christian era," as assumed by Archbishop Usher. Guided by this erroneous principle, Secularists and Materialists are eager to admit, and, on the other hand, orthodox theologians are no less anxious to deny, the high antiquity which has lately been assigned to our race. How utterly unfit such frames of mind are for the quest after truth, and how greatly the final solution of the difficulty must be hindered by such feelings, needs no demonstration. The position of those who, like ourselves, are determined to judge this question from a purely scientific point of view, craving neither to confirm nor to infirm any theological doctrines, and holding fast to the great principle of Galileo so often enforced in the " Journal of Science," is rendered peculiarly difficult. Each party confounds us with its opponents.

One of the works which have led to our selecting this subject is written in the very spirit which we are endeavouring to recommend. Mr. Dawkins writes not to uphold any foregone conclusion. He examines the evidence in favour of the pre-historical, and even pre-mythical, existence of man impartially, critically, even sceptically, just as he would deal with the facts advanced to show the presence of any particular animal or vegetable form at some given geological epoch. It is naturally impossible to treat such a subject in an intelligible manner without giving a general survey of past geological changes and of the three great phases of organic life upon the earth. In the course of this sketch the author declares himself an Evolutionist. He declares "the argument in favour of the theory of Evolution founded on the specialisation of mammalian life in its progress from the Eocene times down to the present day seems to me so strong as to be almost irresistible." In a diagram he shows that in the Eocene there appear families and orders which have still their living representatives, but none of the present genera. In the Miocene, a step nearer in point of time brings us also a step nearer in point of development, as we now find not merely orders and families, but genera which still exist. We pass on again to the Pleiocene, and here we have even living species, though few in number, and accompanied by a majority of extinct species. In the Pleistocene, again, we find living species now no longer few, but numerous, whilst the proportion of extinct species has decreased. In the Prehistoric we see—as far as Europe is concerned—living species accompanied by one only extinct form.

These remarks and this diagram are not given by the author as a biological confession of faith; they are the summary of certain phenomena which bear closely upon the question at issue. Was man found in Britain or in the world in the Eocene epoch ? We look over the mammalian remains of those days, and we certainly find members of the order Primates, to which man belongs; but they are exclusively lemurs, members of the order lowest in their structure, and most remote from man and from the anthropoid apes. None of them, further, belong to any existing genus. The lemurs of those days were still closely linked to the Ungulates, or hoofed quadrupeds, and the beasts of prey bear a marsupial stamp. Surely, then, in the absence of any direct trace of man or of his works, we are warranted in concluding, with the author, that in such a fauna man could find no place. To seek for so highly specialised a being as man " where no now living genus of placental mammal was present would be an idle and hopeless quest."

We come next to the Meiocene Age. Mr. Dawkins describes the climate, the fauna, and the flora, and admits that the climate was favourable and that food was most abundant. Further, representatives of the higher apes were now present in Central and Southern Europe. Still the Meiocene fauna affords not even a single instance of any land mammal which still survives. Is it, then, probable that man, " the most highly specialised of all creatures, had his place in a fauna which is conspicuous by the absence of all the Mammalia now associated with him ?" " If," the author adds, " we accept the evidence advanced in favour of Meiocene man, it is incredible that he alone, of all the Mammalia living in those times in Europe, should not have perished or changed into some other form in the long lapse of ages during which many Meiocene genera and all the Meiocene species have become extinct. Those who believe in the doctrine of Evolution will see the full force of this argument against the presence of man in the

Meiocene fauna, not merely of Europe, but of the whole world."

This argument, in the absence of direct evidence of the existence of man in the Meiocene epoch, seems incontrovertible. It will be at once, however, perceived that—however conclusive against the presence of man, in the strictest sense of the term—it does not disprove the existence of beings somewhat closely approaching him. Dr. Hamy, M. De Mortillet, and others assert, indeed, that man inhabited France in the middle of the Meiocene epoch. Splinters of flint and the notched rib of an extinct manatee (*Halitherium* have been found in Meiocene deposits. Granting that such deposits had been undisturbed, the author and Prof. Gandry venture to ask whether these relics can have been the work of the *Dryopithecus,* a huge extinct anthropoid ape, rather than of man. No present apes are known to use stones except for nut-cracking. But Mr. Dawkins thinks it not improbable that some of the extinct higher apes may have possessed qualities not now found in their living successors. We must remember that the doctrine of Evolution nowise supposes an universal advance throughout the animal kingdom. There is nothing absurd or contradictory in the supposition that some species little inferior to man, or to the direct ancestors of man, may have become extinct.

In the Pleiocene age the improbability of the existence of man is greatly lessened. Not merely genera, but in any case one mammalian species is found to have survived down to our own days. But unfortunately, in these very formations, we have one of the most annoying instances of the "imperfections of the geological record." In Britain, at least, the strata of this epoch are either marine or have been subjected to the action of the sea. In France and Italy, however, we find marmots, elephants, oxen, and dogs making their first appearance in the world. Still Mr. Dawkins, from considerations very similar to those advanced as regards former epochs, considers the advent of man in the Pleiocene as highly doubtful. Of twenty-one fossil Mammalia proved to have inhabited Tuscany in this Age, one only, the hippopotamus, still survives on this earth. Nor is there any decided affirmative evidence to overweigh this improbability.

In the next following, or Pleistocene epoch, the case is altered. We are now introduced to mammalian forms still inhabiting the earth. Though the mammoth, the cavebear, the Irish elk, and the sabre-toothed tiger still survived, such well-known modern species as the shrew, the mole, the

beaver, wolf, fox, stag, roe, the wild bull (*Urus*), the wild
boar, and the horse have made their appearance.   Still the
proof in favour of man's presence, in Europe at least, in
the early Pleistocene is not free from doubt, and the author
agrees with Sir John Lubbock in referring the evidence to a
"suspense account."   In the middle Pleistocene man existed in Britain, as is proved by a wrought flint-flake found
by the Rev. Osmond Fisher, in the author's presence, in the
lower brick-earths at Crayford.   Mr. Dawkins guarantees
that it was *in situ*.   Four years later a second specimen,
also *in situ*, was found in the same series of beds at Erith.
The discovery of these two implements shows that man
lived " in the valley of the lower Thames before the Arctic
Mammalia had taken full possession of the Thames Valley,
and before the big-nosed rhinoceros had become extinct."
Among his animal neighbours were grisly bears, enormous
lions, hyenas, and wolves.   Against these formidable competitors the "river-drift men " had to struggle, armed with
but poor appliances.   Flakes of flint, quartzite, or chert,
roughly chipped to a cutting edge, served them for tools and
weapons,—a slight advance beyond the condition of the ape
who employs unsharpened stones to break nuts or to pound
the head of a venomous serpent.

It may be well here to mention that Mr. Dawkins is by
no means hasty in accepting evidences of the existence of
pre-historic men.   Professors Rütimeyer and Schwendauer,
for instance, had detected in the lignite beds of Dürnten
and Ultznach, of the Mid Pleistocene Age, a something
which seemed to be the fragmentary remains of fossilised
basket-work.   Our author, however, after examining these
relics, views them as knots from a decayed fir tree, without
any marks of human handiwork.

As we descend to the later Pleistocene Age the signs of
man's presence and activity become more numerous.   The
valley of the Thames and the fluviatile gravels of Salisbury,
and a variety of spots in the south-eastern parts of England,
have yielded rude stone-implements of the type generally
known as palæolithic.   Nothing similar has, however, yet
been found to the north-west of a line drawn from Bristol
to the Wash.   On what is now the Continent these same
palæolithic men appear to have been widely, though not
thickly, spread.   In France, Spain, Italy, the north of
Africa, Syria, and India, traces of their former occurrence
are found in the shape of stone-implements of the same
type.   Skeletons of these river-drift men have also been
discovered, though in so very fragmentary and imperfect a

state that no very precise idea of their physique can be
formed.   Mr. Dawkins, however, considers that at this Age,
remote as it is, man was present in Europe as such, *i.e.*, in
a form not widely differing from those of more recent days.
Hence it is plain that the intermediate forms, connecting
man with the lower anthropoids, must be sought for in the
early Pleistocene, and possibly in the Pleiocene, formations.
These pristine men, however, if sufficiently distinct from any
known ape, fossil or living, are no less incapable of being
identified with any human type now existing.   They must
doubtless be considered as the undifferentiated stock from
which the various races of mankind have branched out.
One interesting question remains : What was the relation
of these earlier men to that mysterious phenomenon, the
Glacial epoch ?   Mr. Dawkins concludes that in the milder
regions of Europe man was probably not merely glacial, but
even pre-glacial.   He only penetrated, however, to the north
of the Thames when that evil period had come to its end.

Formations of a later date reveal to us the footsteps of a
different and a higher race, the so-called cave-dwellers.
Both in Britain and in France the river-drift men were followed by population much less widely distributed, who may
be identified with the Eskimos of the Labrador and the extreme north of America.   Among their relics there occur
stone tools and weapons of a much more perfect kind than
those used by the river-drift men.   We find arrow-heads,
harpoons, saws, and borers.   The first attempts at decorative
art have even made their appearance.   Bones and teeth have
been discovered engraved with representations—rude, but
capable of recognition—of hunting-scenes.   Clothing and
personal ornaments had also made their appearance.   These
cave-men, however, who must be distinguished from their
neolithic followers as well as from their predecessors, the
river-drift men, were not in possession of the art of pottery ;
they did not bury or otherwise dispose of the remains of the
dead, and they had not domesticated the dog.   It must here
be remarked that in every case the lower civilisation preceded the higher.   Mr. Dawkins considers that the river-
drift men had probably lived for countless ages in Europe
before the arrival of the cave-men.   The Pleistocene Age,
it must be remembered, was of vast length.

We are now landed in that somewhat indefinite epoch
known as the pre-historical, where the climatic conditions,
the fauna, and the flora of Europe, no longer show any
marked diversity from what obtain at the present day.   Here
the subject gradually passes from the jurisdiction of the

geologist and the biologist to that of the antiquary, the interpreter of sagas, and the historian.

Now, therefore, whilst regretting that we can no further accompany Mr. Boyd Dawkins in his interesting survey, we may fitly record our opinion of the manner in which he has executed his task. It has rarely been our good fortune to meet so difficult a question treated with such indisputable acuteness, breadth of view, learning, and impartiality. Rash credulity and equally rash scepticism are equally avoided, and the result is a manual which the student and the cultured man of the world may each take up with full assurance that he will not be led astray.

The other work which we have coupled with that of Mr. Dawkins is one of a diametrically opposite character. Professor Dawson, at the very outset of his work, gives no doubtful sound. He utters his accustomed defiance to Evolutionists; he takes care to assert his belief in the absolute distinction between man and the lower animals, above all; he claims for the Hebrew Scriptures, by implication at least, the character of a physical revelation, thus perpetuating that wholly gratuitous warfare between Religion and Science which less prejudiced minds are seeking to abolish. Finally, he ascribes to the human race an existence not dating farther back than from 6000 to 8000 years. In other words, he starts with the assumption that traditional chronology is correct, and then strains all the powers of his mind, and applies all the resources of his extensive learning, to interpret facts so as to harmonise with this foregone conclusion. This procedure is, we maintain, fundamentally illegitimate. The man who really writes, as Prof. Dawson claims to do, "from the point of view of the geologist and naturalist," sees in traditions and cosmogonies merely a reflection of the beliefs current in early times, and starting from the facts he seeks for them the most rational interpretation. In the work before us every fact that tells for the long duration of the human race is questioned, and, if not denied, explained away. Because men in a certain low stage of civilisation were found in Canada and Newfoundland, by Jacques Cartier, in 1534, we are asked to believe that the drift-men and cave-men of Europe have not long disappeared from the globe! With every respect for Prof. Dawson as an indefatigable and successful collector of facts, we cannot accept as a guide in generalisation one who shows so strong a bias and so strong a determination to be guided by what we can only regard as perfectly foreign considerations.

ON   WATER   AND   AIR.*

By John Tyndall, D.C.L., LL.D., F.R.S.,

Professor of Natural Philosophy at the Royal Institution
of Great Britain.

Lecture V.

I TRUST that you already know as well as I do that the vapour of the ocean is raised by the sun into the atmosphere, and that when chilled it becomes precipitated as cloud, and that when the chilling goes far enough it is precipitated as snow. One sample of cloud manufacture I should like to bring to your attention, because it is a very interesting one ; and I should like you when you go to the Alps in future years, as I have no doubt many of you will, to be able to notice this phenomenon, and also to understand its nature. You sometimes find a perfectly transparent air charged with aqueous vapour in the true invisible vaporous condition, blowing against a cold mountain-crest. On the side of the mountain upon which this air impinges all is perfectly clear and bright, but on the other side of the mountain you have a vast cloud banner drawn out, which is the aqueous vapour of this perfectly transparent air precipitated into cloud by the cold crest of the mountain. I have seen it on many peaks frequently. On the Matterhorn, for instance, the air charged with aqueous vapour comes up from the plains of Italy ; meeting with the mountain-crest it is chilled, and condenses on the farther side of the mountain, forming a large streaming cloud, which appears to be attached to the summit of the peak, with perfect steadiness, in spite of the strong wind that is blowing. This steadiness, however, is only in appearance, the cloud being constantly dissipated at its farther extremity, but as constantly renewed by the continuous supply of warm moist air which gives rise to it.

We have now to consider the subject of the formation of glaciers, and as an example of a simple glacier will take the

* Being a Course of Six Lectures adapted to a Juvenile Auditory, delivered at the Royal Institution of Great Britain, Christmas, 1879. Specially reported for " The Journal of Science."

celebrated Mer de Glace, of Chamounix. It is not by any
means the largest glacier in Switzerland, but historically it
is very celebrated.   Here is a sketch map (Fig. 30) of the
Mer de Glace, with its tributary glaciers.   Now the vast
gathering ground of snow which forms these glaciers is

Fig. 30.

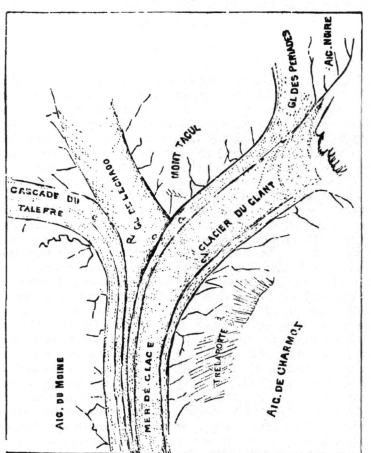

marked here : one is the great basin surrounded by lofty
mountains, called the basin of the Talèfre ; here a second
one, called the plateau of the Col de Lecleand ; and here
another great gathering ground is in the region of the Col
du Géant.   This great basin is where the snows collect and
contribute to the formation of the great glacier.

If I were to take a quantity of snow and subject it to pressure I should compress this snow into solid ice. Now the mountains surrounding this region are continually discharging their snows upon these great plateaux, and this continuous accumulation compresses the snow into solid ice; and rigid as this ice seems it is always moving downwards. It moves down almost like water, and fills the valley of the Col du Géant, forming the glacier of that name. This glacier is joined by another called the Glacier de Lechaud; and here another glacier comes down, falling apparently in broken fragments, in a kind of cascade, which is called the Cascade du Talèfre; and these three glaciers unite and form the celebrated Mer de Glace. You observe here the extraordinary fact that these great and wide glaciers squeeze themselves down through this narrow valley of the Mer de Glace at Trélaporte, as if the ice in point of fact were water. The quantity of squeezing endured at this place called Trélaporte is very great. The glacier coming down from the Col du Géant is 1134 yards across; the one from the Jorasse, the Glacier de Lechaud, is 825 yards across. The glacier coming down from the Talèfre is, at a certain place near the cascade, 638 yards across, and the sum of these three glaciers amounts to 2597 yards in width. All these three are squeezed together through that defile at Trélaporte, which is only 893 yards wide. This gives you an idea of the tremendous power with which these ice-masses are forced forward in order to form these glaciers. I say forced forward, and that implies, of course, that the glaciers are always in motion; and this is the case winter and summer. The summer motion of this Mer de Glace amounts to a certain number of inches per day; the winter motion is about half the amount. The summer motion is on the average about 30 inches a day; the winter motion is about 15 inches a day, the ice being more rigid and less yielding in winter than in summer.

You see the motion of the ice through this valley; it twists round and turns, and imitates, in all its movements, the motion of a river. Many distinguished men have worked at these glaciers, but perhaps none have contributed more to our knowledge of the matter than two men who are now no more—the celebrated Agassiz, who died some time ago in America, and the celebrated Principal Forbes of St. Andrews.

Here is a rough diagram of the Mer de Glace with certain lines drawn across it (Fig. 31). The velocity of the glacier where those lines are drawn has been determined. I will run briefly over the maximum, or what I call the greatest motion of the glacier. That motion is not always at the middle of the glacier, but near the middle, for the glacier in this respect resembles a river. Owing to the friction of the sides on the banks of the river, the water is held back, so that the place of maximum motion of a river is near the centre. Across the line A A' the maximum motion is about 33 inches in 24 hours; across the line B B' it is about 24 inches in 24 hours; across the line C C' it is about

FIG. 31.

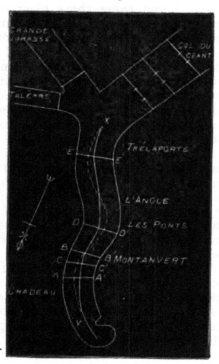

34 inches in 24 hours; across the line D D' it is about 23 inches in 24 hours; across the line E E' its maximum motion is about 20 inches in 24 hours. Also with the tributary glaciers the maximum motion across the line H H', on the Glacier du Géant, is 13 inches in 24 hours. The maximum motion across the line K K', on the Glacier de Lechaud, is 10 inches in 24 hours. Thus you see the velocity of the glacier changes. As I have said, the motion of a glacier resembles perfectly the motion of a river, in that

its motion augments from the banks to the centre, because at the centre it is free from the friction of the banks. Take the case of a river which flows through a sinuous valley. Suppose it to be coming down with considerable power, and having to make a bend. The most wondrous example of that kind in the world is perhaps the so-called whirlpool below the Falls of Niagara. There the River Niagara takes a sudden bend at right angles to its previous direction. The water comes down plunging with impetuous force, and impinges against the bank on the other side. It is thus caused to turn round and produce the wondrous whirlpool from which the water escapes on to Lake Ontario. But it is no matter how small the bend may be. The least bending, such as we see here exhibited in the Mer de Glace, has this effect in the case of a river. When the water moves downwards it is always carried on towards the concave bank of the river; and if the river bends in the other direction the greatest motion of the river will not follow the middle line, but will change in accordance with the flexures of the valley through which the river passes. This is the law of the motion of water; the law of the motion of ice is precisely similar. For instance, the continuous line x y in the diagram (Fig. 31) represents the centre of the glacier. The dotted line marks the maximum motion of the glacier; so you see that the swiftest motion is, as I have said, towards the concave side of the glacier. At Trélaporte the swiftest point is towards B on the line B B'. Then you have a point where the greatest motion coincides exactly with the middle of the glacier; then we come lower down, and here the point of greatest motion passes to the other side of the centre, and is towards D on the line D D'; and then afterwards it passes back again across the centre of the glacier, and its greatest motion is at c on the line c c'. Thus, as I have said, the motion of the glacier downwards resembles in every particular the motion of a river.

Let us turn again to our map (Fig. 30), and mark the wonderful power of the ice to mould itself to the valley through which it presses. See the extraordinary behaviour of the glacier in coming down that valley; and how the ice can accommodate itself to the flexures of the valley; and how the immense masses of ice which are tributaries of the Mer de Glace (the Cascade du Talèfre, the Glacier du Géant, the Glacier de Lechaud) weld together, and squeeze themselves into the extraordinary small space we find at the gorge at Trélaporte. Now it is possible for ice to be squeezed in this way, changing its form but not its volume, and appearing after the change as solid and as homogeneous as before.

A curious phenomenon was first observed by Faraday, who found that when two pieces of ice with moistened surfaces were placed in contact, they became cemented together by the freezing of a film of water between them. I have here two slabs of ice, the surfaces of which I will press together, and you see I have frozen the two slabs into one mass. Prof. Bottomley made an experiment recently which bears upon the subject. It is there in action before you. Upon the two uprights (Fig. 32) is placed a large block of clear ice (A B) with a loop of wire round it: twelve hours

FIG. 32.

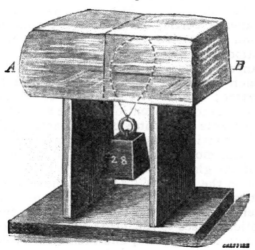

ago a 28 lbs. weight was attached to the wire. The wire immediately commenced to enter the ice, and you see by this time it has passed half-way through it, but the ice has remained undivided: except for a little opacity along the plane of passage, it shows no signs of ever having been divided. If we allow it to go on it will pass right through the ice, the severed surfaces re-freezing, and leaving the block entire as at the commencement.

I have here a number of moulds of different shapes, and if time permitted I could have shown you that by filling them with fragments of ice, and subjecting the ice to pressure, I could have moulded these fragments into cups, spheres, vases, &c., of solid ice; so by these illustrations you can readily understand how a substance which so readily changes its form under pressure, and so readily unites when broken, can be forced through narrow gorges, and can accommodate itself to the bendings of the valley through which it moves.

And now we have to dwell a little upon certain phenomena connected with this motion. And, first of all, with regard to these curious masses which are heaped up sometimes upon the glacier and sometimes at the side of the glacier, to which the name of "moraines" is given. You cannot have a better example of the origin of these moraines than you have here upon this Mer de Glace. The mountains bounding the glaciers are incessantly sending down stones and *débris* of all kinds which have been loosened by the frost or by the rain. These tumble down upon the sides of the glacier, and form what is called "lateral moraines." You sometimes find great heaps of matter thrown from the mountains, and the quantity depends upon the friability of the mountains. You find these great masses of stones and *débris* thrown down along the edges of the glacier. Well, now what follows? I will carry you up to that rock which is called the Grand Rognon, which is associated with the Glacier du Géant. You must imagine the surface of that Grand Rognon covered with this *débris*, which has fallen from this rock upon the ice. That ice is for ever moving forwards, and what occurs? Here, at the end of the Grand Rognon, you have a ridge of *débris* carried down by the moving ice, and you can trace it all along the Glacier du Géant down to Trélaporte. We have it marked *b* on the map (Fig. 30). Again, another moraine starts from a point of the Aiguille Noire (it is marked c on the map), and, coming down through the Glacier du Géant, you can trace it through the Mer de Glace. Another starts from that cascade just upon the Telèfre, and you can see it come down there. And when these moraines unite together upon the trunk glacier they form these lines (*e, d, e*) that you see here, not on the side, but in the middle of the glacier, and hence they are called "medial moraines," so that the side moraines of the tributary glaciers are converted into the medial moraines of the trunk glacier. A moment's reflection will make it clear to you that the number of medial moraines on the trunk glacier must be one less than the number of lateral moraines on the tributaries. Two lateral moraines form one medial moraine, three form two, four form three, and so on.

Well, so much for these moraines. When you come to visit the glaciers themselves, you may look down from a height upon a glacier and see these moraines perhaps like tracks or roads, in point of fact, coming down along the banks of the glaciers ; but when you come to examine them you will find that they are very different from what they appear to be. These moraines rise sometimes to a height of 50 or 60 feet above the level of the glacier. Looking at them you would imagine them to be heaps of stone and clay and dirt that have been brought together—a ridge of this *débris* heaped upon the glacier; but when you come to examine them more closely you find that there is simply a superficial covering of *débris* with a ridge of ice underneath. What is the reason of this? You will understand me immediately. Conceive the glacier covered with this *débris*. It is thereby protected from the melting action of the sun.

Fig. 33

The sun is free to play upon the ice right and left of the moraine, which it melts, and the consequence is that the protected part—that part on which the *débris* rests—is not melted, but remains as a great ridge of ice. In the same way are formed the glacier tables, as they are called. You sometimes find upon the glaciers rocks lifted up as if they had grown, as it were, out of the heart of the glacier : these are due entirely to an action of the same kind. There (Fig. 33) is the glacier table upon a surrounding of ice. That rock was thrown from the adjacent mountains upon the glacier, and has protected the ice beneath. The ice all round this protected ice melted, and by-and-bye this rock appeared to rise more and more, and finally you find it there as a so-called glacial table. It is not due to the lifting of the table or to the lifting of the rock, but to the melting of the unprotected ice all round the rock.

There are other cases in which protection sometimes comes into play in an extraordinary fashion, and there is nothing more interesting or more striking than what are called the sand cones of the glaciers. Rivers coming near moraines contract a certain amount of impurity, and carry sand along with them and distribute it over the ice; and the ice protected by this sand rises not in reality, but relatively. The surrounding ice is melted away, and these patches of sand appear to rise up, and you have upon the glaciers little mountains—little sand cones as they are called—which might be considered to be representative of the Alps themselves in miniature.

I now want to show you what occurs in consequence of the incessant action of the sun upon the glaciers. The rising of the moraines and the rising of the glacier tables prove that when the sun acts upon the glaciers it rapidly melts the ice. And what is the consequence? Rivers, rivulets, and streams, of great power and impetuosity, are sometimes formed in the glaciers upon those portions which are not very much crevassed (or cut across by open chasms in the ice). Those rivers which come upon a crevassed portion of a glacier plunge into the crevasses or chasms, and find their way to the bottom of the glacier and roll along the bottom as a sub-glacial river underneath the ice, and finally they issue as a river from a vault of ice at the end of the glacier. At the end of the Mer de Glace is a vault of this kind, which in the summer time is exceedingly dangerous to enter. I remember once standing and looking into that vault, and thinking whether I should go into it, it was of such magical beauty; and while I stood pondering upon the point of the prudence and possibility of entering it, the whole roof, weighing some 20 or 30 tons, fell down upon the bottom. In winter it is firm and fast, and can be safely entered and explored.

As I have said, from the termination or snout of the glaciers a river issues, the Mer de Glace being the source of the Arveiron, the Rhone glacier giving birth to the River Rhone, and the sources of the greatest rivers in Europe can all be traced back to the glaciers.

I have said that the centre of the glacier moves more rapidly than the sides, and I have referred to the crevasses, as they are called. That is the technical name

for those great chasms which are sometimes very deep indeed.
Now I want to propose a problem to you. There is a
drawing of a glacier (Fig. 34), and you observe that black

FIG. 34.

mark on the farther side. That is a great moraine of the
glacier cutting off the opposite side of the mountain.
Now I want you to look at the glacier itself, and I hope
you will answer the question that I shall propose. You see
the crevasses sweep round in this way to form curves. I
will ask the boys present to answer a question. The glacier
runs along there. Will anybody tell me what is the direction
in which the glacier is moving? Is it to the right or to the
left? [Voices : "To the left."] I venture to say that we shall
have in this assembly a majority of ten to one among the
boys saying that that glacier moves to the right. It is as if
the centre of the glacier pushed the ice forward, and the

FIG. 35.

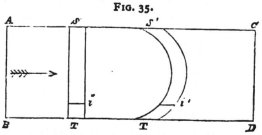

sides held it back. You will now see that the answer which
has been given is precisely the reverse of the truth. I

trust by means of the diagram (Fig. 35) to make the explanation evident to you. Let A C D B be a portion of the glacier, and, for the sake of simplicity, suppose the glacier flowing in the direction of the arrow. Suppose we take a transverse slice of the glacier (s T) and fix our attention upon a little square (T *i*) of this slice, that is a little square mass of ice, and I would ask the boys whether that slice of ice will remain straight after it has been two or three days subjected to the motion of the glacier. (Voices: No.) It will become curved, will it not (as at s' T)? The centre will move forward because of the quicker flow of the centre of the glacier; and therefore when that slice of ice (T *i*) comes down we shall have it transformed into ;the shape T *i'*, and our little square will be distorted to the shape which, as my young mathematical friends know, has usually applied to it the term rhombus, or lozenge. Now I want you to realise the exact mechanical state of matters. When the square (T *i*) is distorted to the shape of the lozenge (T *i'*) the diagonal T *i* of the square becomes the diagonal T *i'* of the lozenge, the diagonal of the lozenge being longer than the diagonal of the square. If the ice could freely stretch like treacle, one diagonal would stretch to the same length as the other, but the ice does not stretch in that way. It is strained, and it breaks across at right angles to the line of strain T *i'*, and forms a crevasse, or chasm, which, instead of being pointed downwards is pointed upwards, and thus you see that the motion of the glacier is exactly the reverse of what you supposed it to be.

Now I have to point out to you a few examples of violent crevassing. You have seen the real mechanical origin of those crevasses. They are due to the strain imposed upon the glacier by the different motions of its parts. When the crevassing is very violent, and when the glacier is pulled in different directions, you have the crevass converted into the most fantastic peaks and towers of ice. You hear these called in Chamouni and elsewhere *aiguilles* (needles). Now we pass on to a final result of the glaciers. There is the glacier opposite which I spend a couple of months of each year of my life—the great Aletch Glacier—and at a certain place that glacier at one period of its history turned down a lateral valley. It now only fills the great trunk valley. The glacier has retreated from this lateral valley, and in its place we have a little lake called the Margelin See, one of the most picturesque parts of the Alps, and upon this lake you have floating little blocks of ice which are rendered by the snow quite white; and they are, to all

intents and purposes, icebergs. The glaciers of Greenland
work their way down into the real ocean, and the ends of
these glaciers breaking away in large masses are sent floating out to sea; and it is these icebergs which constitute
some of the dangers of the Northern Atlantic. It was one
of these with which a vessel came into collision some time
ago. That iceberg against which the vessel impinged came
from one of these northern glaciers. It was a bit of one of
these northern glaciers, broken off and sent into the ocean
like the small masses of ice from the Swiss glaciers carried
into the Margelin See.

Now I must give a wrench to this topic, and pass on to
another subject.

We have to consider this atmosphere, and the physical
properties of the air which, in the long run, has sustained
the weight of all our glaciers and of all our rivers. It is a
constantly observed fact that when there is a great advance
in science there is always a general simmering and fermentation of scientific thought for some time previously, and
by-and-bye some particular individuals rise, so to say, out
of this general fermentation, and initiate the real principle
or law, or whatever it may be, about which people are
thinking during this intellectual simmering. And so it was
some three hundred years ago with regard to our atmosphere. People did not know that the atmosphere had
weight. It was shrewdly suspected, however, by that
great man, Galileo, and by Descartes, that the atmosphere
had weight. But this was not clearly made out even in
their day, and when we consider the knowledge that is now
open to every little boy and girl it is strange that this
knowledge was held back from those truly great men of
antiquity, men of the very greatest intellects. We have
now to deal with results which they would have given
anything to know, and which are now perfectly familiar to
us. Descartes realised that the air has weight. The weight
of the air in this room in which you are now assembled—
what would you guess it to amount to? Perhaps you
would say that it weighed a few pounds; but the weight
of 13 cubic feet of air is about a pound. There are perhaps nearly 80,000 cubic feet of air in this room; and if
you make a calculation you will find that the air in this
room at the present time would weigh something about
three tons. That, perhaps, you would not expect. Now
we have the air here with the whole of the atmosphere
above us. Some make the atmosphere of one height, some
another. There are various reasons why we should give

a kind of limit to the height of the atmosphere, but all give it a certain height ; some have held it to be about fifty miles in height, others one hundred, others two hundred ; but, at all events, whatever the height is these lower strata of the atmosphere have to bear the weight of all the atmosphere above them.

Now, upon my body at the present moment the entire weight of the atmosphere is pressing, and I wish you to clearly realise that this pressure is not only exerted upon my head, which has to bear all the weight of the atmosphere above it, but that it is just as much exerted upon my sides ; so that when you calculate the pressure exerted upon the human body, you must take the number of square inches upon the whole of the human body, and you must determine the pressure upon every square inch. What is that pressure ? The pressure upon every square inch is exactly equal to that bar of lead, which is a square inch thick and a yard in height. Now, how would you like to bear the weigh of a mass of lead surrounding your body? Well, what does that amount to in *its* totality ? It amounts to a pressure of no less than 14 pounds. Every man here bears a pressure of this kind—a pressure of 14 pounds upon every square inch of his body. The reason why we are not crushed by this pressure is that the liquids of our bodies are nearly incompressible. They are like water, and the air within our bodies is just as much compressed as the air without, so that there is really no difficulty in existing under these circumstances. Well, things went on in this way until one day, as it is said, some gardeners wanted to water their garden in Florence, and they came to Galileo, and told him that they found that the water refused to rise above a certain level in the pump. I will not vouch for the perfect accuracy of this account, but I think that it is very likely that it is correct. Here is a point worth noting. Great inventions are sometimes made before the reasons for them are known. In this way gunpowder was invented. The desire of man to kill his fellow man was so very strong and strenuous that it was the motive power to induce him to select the constituents of gunpowder in their proper proportions, although they knew not one iota about the chemistry of the thing. The chemist came afterwards. But it was not until afterwards that he explained that these men were correct, and that the proportions were remarkably just. And thus, as I have said, great inventions are made before the scientific reasons for these inventions are known. And thus it was with this discovery of the pump,

Well, these Florentine gardeners came to Galileo, and said : " We want to get up water 50 or 60 feet from our pump, but we find that it will not rise to a greater height than 32 feet." The explanation of the pump in those days was what ? No explanation at all. It is amazing how the words of a great man can bewilder the world ; and that not for a generation, or a century, but for thousands of years. And so it was with Aristotle. He gave an explanation, or, at least, his school gave an explanation of why water rises in a pump. The pump, as you know, is a cylinder with an attached valve, and into this cylinder is inserted a piston. Now, if I lift up this piston there is left behind it what is called a vacuum, that is, a space in which there is no air. And Aristotle's statement was that, as nature abhorred a vacuum, this empty space could not be allowed to exist, and the water ran up to fill the vacuum. Well, Galileo was rather soured by the treatment he had received at the hands of the world. He looked at these gardeners, and he said : " It is quite obvious that nature does not abhor a vacuum beyond a height of 30 or 32 feet." However, the problem was not solved by Galileo. It fell into the hands of his celebrated pupil, Torricelli. Torricelli reasoned thus :— " The atmosphere has weight. Here is a vessel, say, con- taining water. The atmosphere presses upon that water ; and if I, by means of my pump, remove the atmosphere, and so take away the pressure of the atmosphere above the water, the pressure of the atmosphere outside will cause the water to ascend." And what does that show ? A thought occurred. First of all, remember, it was only a thought ; and that is the way that scientific men advance. They gather up together the knowledge prevalent in their times, and they ask themselves what is likely to result from this knowledge. Torricelli divined the reason with his mind's eye before a human eye ever saw the fact. He said : " If I remove the atmospheric pressure from the interior of my pump, the pressure outside will force the water up ; and that column of 32 feet that those gardeners in Florence lifted by their pump answers to the outside pressure of the atmosphere. The 32 feet of water holds the weight of the whole atmosphere in equilibrium." This was his conjecture, but he did not content himself with that reason. He said : " If my reason be correct—if it be the fact that 32 feet of liquid water can hold the pressure or weight of the atmo- sphere in equilibrium—then, if I take a heavier liquid than water, I must have a shorter column supported. The weight of the atmosphere being a constant thing, will lift a shorter column of mercury, which is thirteen times heavier than water. If mercury is used instead of water, then, instead of a column of 32 feet being supported, I ought to have a column of only about 30 inches of mercury." I do not know any experiment in the whole course of science that must have carried more high expectation in its train than this experiment of Torricelli. Here is Torricelli's experiment. He filled with mercury a tall glass tube, closed at one end (C D, Fig. 36). He placed his finger, as I do, upon the open

FIG. 36.

end of the tube, and he turned it upside down. He then placed the end closed by the finger underneath the surface of some mercury contained in a vessel, and removed his finger. You understand the conditions of the experiment. When the tube of mercury is thus inverted in a vessel containing mercury, the mercury is exposed to the pressure of the atmosphere. If the pressure of the atmosphere can support the whole of that column, the tube will remain full; but if it can only support 30 inches, as Torricelli surmised, the mercury will descend in the tube. It will not descend quite down, but will stop at a certain point, and then we shall have a column of mercury that exactly balances the weight of the atmosphere. And there it is, as Torricelli foresaw with the eyes of his mind, the mercury falls until it reaches the point A, the column of mercury, A B (Fig. 36), being about 30 inches in length.

If I take a tube of water instead of mercury, when I invert the tube you will find no depression of the column. The tube will remain full, because the weight of the atmosphere is quite sufficient to keep it full. I invert a tube of water over a vessel of water in the same manner as I did with the mercury, and I withdraw my hand, and the tube remains full, as I anticipated. Here we have, then, this celebrated Torricellian experiment, on which our present barometers depend.

But another wise experiment followed this conjecture of Torricelli. There was a great man in those days named Pascal, and he, writing to his brother-in-law, M. Perrier, proposed an experiment to test the result of Torricelli. He had no doubt of the correctness of the result; but still as a philosopher he thought himself in duty bound to test it in all possible ways, for in the days to which I refer there was a contest between the philosophic spirit, which has now got such complete predominance in the world, and the old Aristotelian spirit, which explained these things in such an extraordinary and unscientific fashion; and hence it is that experiments like those proposed by Pascal had an extraordinary value, because it was the time of conflict between the era of scientific investigation and that era of loose speculation which had preceded it. Pascal wrote to his brother-in-law desiring him to make an experiment upon a celebrated mountain in Auvergne, " for," said Pascal, " if it be that the rising of a column of mercury is due to the pressure of the atmosphere, when we go higher into the atmosphere that column of mercury ought to diminish. The column of mercury supported at the top of the hill ought to be less in height than the column supported at the sea-level." Pascal's remarks at the time were very quaint, and at the same time very sarcastic, because it must be remembered that they were uttered in the midst of a controversy between the new and the preceding philosophy. Pascal writes thus to his brother-in-law :—" You see that if it happens that the weight of the mercury at the top of the hill is less than at the bottom (which I have many reasons to believe, though all those who have thought about it are of a different opinion),"—that is, the preceding philosophers,—" it will follow that the weight and pressure of the air are the sole cause of this suspension, and not the horror of a vacuum; since it is very certain," he adds, "that there is more air to weight on it at the bottom than at the top: while we cannot say that nature abhors a vacuum at the foot of a mountain more than on its summit." M. Perrier made the experiment. He found, as Pascal had predicted, that the mercury descended as he ascended the mountain, and as he came back again the column ascended, because of the augmented pressure of the atmosphere.

I will just say one or two words with regard to the next great step made in this field. It was a step made by that grand old fellow whom so many of our hearers have reason to know,—the burgomaster of the town of Magdeburg, in Prussia,—Otto von Guericke. This Otto von Guericke, who was the first to make the electric machine with a ball of sulphur which he rubbed in his hands, devised a means of exhausting the air, or taking the air out of a body. The original instrument invented by him for this purpose is still preserved in the city of Berlin.

Another illustration of the weight of the air I wish to give you here. Here is a very strong tin cylinder. I will extract the air from within that cylinder, and by-and-bye you will find that when the pressure within is withdrawn the pressure of the air outside will crush the vessel. [When the exhaustion of the air had proceeded for a short time the tin cylinder suddenly collapsed.]

Now, I want to direct your attention to an experiment bearing upon the result obtained by the gardeners at Florence. Here you see I have a tube 5 feet long. The pressure of the atmosphere lifts the water in that tube so as to fill it, although the same pressure is not competent to support a column of mercury of that height. This tube is prolonged by means of another tube to the base of the building, so that between the top, the glass tube, and the base of the building there is a distance of 33 feet. From the interior of the tube we exhaust the air, and the pressure of the atmosphere upon the surface of the liquid contained in the basin where this tube ends downstairs, forces the liquid up the tube as the vacuum is made. The water reaches a certain point, and then, according to the result enunciated by Galileo, beyond that point it cannot go. It reaches a point where the pressure of the atmosphere upon the basin down stairs is exactly equal to the weight of the column of water that we raise.

ANALYSES OF BOOKS.

Evolution and Involution. By George Thomson. London:
   Trübner and Co.

We opened this book in the hope of finding here the demonstration of some important and hitherto unrecognised law touching
the development of organic existence. We thought that the
author might have got sight of what has escaped the vision of
Darwin—to wit, the initial cause of those variations which
" Natural Selection " at the best can only preserve and accumulate. We were grievously disappointed. The "idea" of the
work is probably to test that supposed indefinite quantity, the
patience of unfortunate reviewers. We have here a repetition,
in its grossest form, of the threadbare and now utterly inexcus-
able error that Evolutionism is an attack upon the existence of
God. We quote the following most deplorable passage :—
" Messrs. Darwin, Spencer, Huxley, &c., in England, and a few
similarly inspired Germans, Frenchmen, and Americans, have at
the appointed time appeared on the stage of existence to enlighten
humanity, and to show what stupid fellows all former men have
been, and what more stupid fellows we are in adhering so long
to anything to which these stupid ancestors have given utterance.
It seems that the ancestors of the present races of men somehow stupidly took into their heads that there was a personal God
in being, the author and complement of all, and that we more
stupidly have been for ages worshipping this imaginary Being by
way of praying and singing psalms to Him, and after divers
other fashions, all attesting the folly and credulity of the human
race. Now, however, there is to be no more of this nonsense
and stupidity," &c.

This passage seems doubtless to Mr. George Thomson very
killing irony. We can pronounce it nothing but an entire mis-
apprehension of Evolutionist teachings,—a perversion simply
disgraceful to its author, who ought before writing on Evolution
at any rate to have acquired some acquaintance with its nature
and tendencies. Had he been conscientious enough to learn
before teaching, he would have found that the " few similarly inspired Germans, &c." comprise well nigh the total body of men
capable of understanding the question of the origin of the forms
of organic beings ; he would have found that the doctrine of
Evolution leaves the existence of a personal God precisely where
it was before, though it may lead us to different and worthier

conceptions of His mode of creation. He might have found that "Monsieur Comte," whom he curiously mixes up with the Evolutionists, was—like many other men who are neither Christians nor even Theists—a firm upholder of the permanence of species. He might perchance have learnt that the co-discoverer of the principle of "Natural Selection" is the author of the ablest refutation of Hume's argument against miracles, and that though certain believers in Evolution are, in common phraseology, "Infidels," yet their Infidelity and their Evolutionism stand in no causal connection.

To the biologist the work before us is simply something dead. It brings forward no facts either novel or which have escaped notice : it propounds no questions which may be solved by observation and experiment; it throws out no suggestions capable of being followed up. The author's law of "Evolution" and Involution is stated in the following terms :—"All beings in proportion as they assume personality and evolve out of the universe, in that proportion do they involve it within themselves and incorporate it, approaching at the same time absolutism in all its attributes." Concerning this collection of words the author modestly declares that, "like everything that is great, it is so simple and obvious that a child can comprehend it, yet upon this simple law hangs the revelation of existence and being to man." This seems to us one of the many laws which we have heard enunciated which, however simply, lead to nothing.

The Spectroscope in Medicine. By C. A. MacMunn, M.D. With three Chromo-lithographic Plates of Physiological and Pathological Spectra, and thirteen Woodcuts. London : J. and A. Churchill.

That the spectroscope, judiciously applied, is likely to become a valuable guide in Medicine scarcely requires demonstration. It has been already applied with success for ascertaining the presence of abnormal elements, such as the poisonous metals, in the body or its secretions,—a circumstance the more valuable as the quantity of material to be operated upon by the toxicologist may sometimes be too small for the ordinary processes of chemical analysis. According to Dr. Thudichum, morbid gases occurring in the animal economy may be recognised by its use. Dr. Bence Jones has employed the spectroscope for determining the time that certain salts take to reach any part of the body; for finding where diffusing substances go to, how long they are in passing from the stomach into the textures, how long they stay there, and how quickly they cease to appear in the excretions.

Still more instructive is the study of the absorption spectra of physiological and pathological fluids. The author not merely treats of spectroscopic examination as a means of detecting and recognising blood-stains, but gives an account of the changes produced in the blood by a variety of agencies. Thus we have a description of the spectrum of the blood in death from asphyxia; in death from the inhalation of nitrous oxide, and of carbonic oxide; the spectrum of blood treated with nitrogen dioxide, with sulphuretted hydrogen, with hydrocyanic acid and soluble cyanides. The author next explains the effect upon the blood of nitrites, of ammoniacal gas, of arsenic and antimonic hydrides. The use of such investigations in questions of medical and chemical jurisprudence may fairly be pronounced the smallest part of their value. Bile and urine are next dealt with in a similar manner.

It may doubtless be asked why the spectroscope has not already become one of the recognised auxiliaries of the physician, seeing that it will frequently be of signal use in the diagnosis of disease, in the comprehension of morbid changes, and in tracing them to their cause?

The author considers that many practitioners have been deterred from the use of this instrument by exaggerated ideas of the difficulties involved, and by the want of a simple treatise on its use and on its applications to Physiology and Pathology. Many persons still living will remember the time when the microscope was regarded with a similar distrust, which has only gradually—and we may say not entirely—been got rid of by the appearance of practical works on microscopic manipulation.

Such wants, as regards the spectroscope, the author makes it his business to supply, and in our opinion with very marked success. In the two introductory chapters he explains the essential parts of a spectroscope, the various kinds of spectra, the nature and use of the one- and two-prism chemical spectroscope, the art of mapping spectra, the nature of absorptive as distinguished from bright-line spectra, and the use of the microspectroscope. A supplement gives the bibliography of medical and physiological spectroscopy.

This work merits the attention not merely of the physician and the medical student, but of the toxicologist, and of all who engage in biological research. To all of these classes a familiarity with the use of the spectroscope is like the possession of an additional sense, and in acquiring such familiarity they will find Mr. MacMunn's treatise a needful guide. We hope it will enable many aspirants to qualify themselves for research in the fruitful field which Mr. Sorby has cultivated with such success. It must be distinctly remembered that the author is not a mere compiler, having no special and practical acquaintance with the subject of which he treats, but an actual and experienced observer.

A History of the British Marine Polyzoa. By THOMAS HINCKS,
  B.A., F.R.S.   London : John Van Voorst.

WE have here a most elaborate and admirable monograph, drawn
up by an author whose prolonged and successful researches in
this department of biology specially qualify him for the task.   In
the Introduction Mr. Hincks treats of the structure of the
Polyzoa, their embryology, distribution, and classification, and
discusses the meaning of some parts whose function is obscure.
Among these is the so-called " brown body " which has given
scope to much controversy.   Setting aside less probable views,
it has been regarded as a mere lifeless residuum of the dead
polypide, or as a special structure having a reproductive function.
In common with Prof. F. A. Smitt, the author has held the
former view, and has supported it by means of original observations.   He now, however, considers that the evidence tallies
better with the " residuary " theory of Joliet and Nitschke.
Still no very satisfactory explanation seems to have been given
of the observations of Prof. Smitt and of the author's own.

Another subject of dispute is the function of the " avicula-
rium," which in some genera—*Bugula* and *Bicellaria*—appears
as an " articulated assemblage attached to the zoœcium with a
formidable hooked beak and a mandible worked by powerful
muscles, perpetually snapping its jaws with monotonous energy,
and swaying to and fro with vigorous swing on its jointed base
—grotesque both in form and movement."   Mr. Hincks, who
traces the development of this " bird's head " through a long
series of gradational forms, considers that its function is defensive rather than subservient to the capture of nourishment.
Further observations will here be useful.   The passage of the
ova into the oœcium or ovicell has not been witnessed, nor is
the author aware of any observations showing how it is effected.
Mr. Hincks seems to believe not merely that the oœcium acts as
a marsupium, but that, under circumstances not explained, ova
are produced within it.

On the distribution of the Polyzoa the author remarks that
pending the publication of the results of the *Challenger* Expedition generalisation would be premature.

The remainder of the work is devoted to a systematic account
of the Polyzoa, and will prove a most useful guide to the student
of marine zoology.   The work is not merely illustrated with
woodcuts introduced in the text, but with eighty-three plates,
forming a companion volume.   The bibliography of the subject
is given in an Appendix.

Biologists will feel greatly indebted to Mr. Hincks for thus
placing before them the observations of a lifetime, and will not
be slow in acknowledging the merit of the publisher, already so
honourably known in the department of Natural History.

Science Primers: Introductory. Py Prof. HUXLEY, F.R.S. London : Macmillan and Co.

WE have here, in the compass of 94 pages, a most useful little book, which ought to circulate widely and to be extensively and carefully read by the young, and by not a few of their elders.

The first section, entitled " Nature and Science," may be likened to a dose of isinglass, as it is well fitted to clarify the popular mind. In it we find a plain account of sensations and things, of causes and effects, properties and powers, of Nature (so-called, accident or chance), of the laws and Nature, and of Science as merely the knowledge of the laws of Nature obtained by observation, experiment, and reasoning. The author easily and completely strips these subjects of the mystery in which they are always wrapped in the minds of the uneducated, and of that more numerous class the miseducated. One abuse of the term Science, very common in England, when, namely, it is confounded with Art, has escaped notice.

The remainder of the work deals in an equally lucid manner with the characteristic phenomena of material objects, mineral and organic, and with those of Mind.

The Geological Record for 1877. An Account of Works on Geology, Mineralogy, and Palæontology, published during the Year, with Supplements for 1874 to 1876. Edited by W. WHITAKER, B.A., F.G.S., of the Geological Survey of England. London : Taylor and Francis.

OUR readers are of course aware of the general character of this useful volume, and will remember that the term " Works " comprises not merely independent publications, but papers from the various scientific journals and from the Transactions of the learned societies. The compilation of such a " Record " involves an immense amount of labour, and such sins of omission and commission as Mr. Whitaker admits in his humorous Preface must be unavoidable. The delay in publication it seems has been caused by that haunting dread of anxious editors, the loss or delay of MS. in transmission. Some person, not named, appears to have got possession of a portion of the copy, and, in vulgar phraseology, to have " stuck to it " for some months.

The value of so complete a conspectus of the geological literature of each succeeding year needs no demonstration.

CORRESPONDENCE.

"INSTINCT AND MIND."

To the Editor of The Journal of Science.

SIR,—In an article in your last month's issue, bearing the above title, the writer, Mr. S. Billing, advances as a distinction between instinct and mind that the former is "tribal," *i.e.*, common to an entire species, variety, or breed. Thus, speaking of different kinds of dogs, he says, " The aptitude of each is tribal ; the gaze-hound has not the instinct of the pointer, or the shepherd's dog that of the retriever." Again—" The tribal characteristic is found in the habits of wild animals ; particular species of birds have the same constructive instinct ; and whenever a variation occurs in a particular species or variety, each member of the class practises the same peculiarities." Now if we are to under-stand from these passages that all the individuals of any given species or variety stand on the same level in point of intelligence, I must, in the name of all who have closely examined the doings of the lower animals, pronounce Mr. Billing's view utterly at variance with facts. Different breeds of dogs have, of course, peculiarities in common, just as have different races of men. But within each race there occur individual variations in saga-city, just as we find it to be the case in our own species. Thus all shepherd's dogs are to a certain extent similar as contrasted with pointers or retrievers. But to assert that one shepherd's dog is equal in intelligence to every other shepherd's dog is a statement not merely devoid of all foundation, but contrary to distinct observation.* These animals differ among themselves as decidedly, though not to the same extent, as do men. Every now and then we meet with a "collie" who far surpasses the common run of his kindred in perception of what is wanted, and in the variety and extent of his resources. On the other hand, there are stupid collies who are almost worthless to their em-ployers. Similar distinctions between individuals are daily traced in all species with which man comes so closely in contact that he is able to observe them with any approach to accuracy. It is well known that when monkeys of any particular species

* We took a very similar position in our critique on Mr. Billing's imterresting work " Scientific Materialism." See Journal of Science for 1879, p. 664.

accumulate at the Zoological Gardens, some of the surplus specimens are sold off. Now I understand that the purchasers often
declare themselves willing to pay a much higher price if they
might have the animal a few days on trial, so as to find whether
he was a clever monkey or a dull one. Whether we accept the
old doctrine of a distinction of kind between man and the lower
animals, or whether with the modern school we admit a mere
difference of degree, there is not, to my knowledge, a shadow of
foundation for the mental equality which Mr. Billing ascribes to
the members of each species.—I am, &c.,

R. N. M,

THE WILD BIRDS PROTECTION ACT.

To the Editor of the Journal of Science.

SIR,—I was much struck with the article in your last number
entitled " Our Friends, and how we Treat them." All that you
say about the practical nullity of the Act is true. As far at least
as the northern outskirts of London are concerned, the bird-
catchers are now daily and successfully at work, and the authorities, police, &c., take not the least notice. I hope that your
appeal both to the humanity and the interests of the nation will
not be ignored.—I am, &c.,

B. E.

THE ANCESTRY OF MAN.

To the Editor of the Journal of Science.

SIR,—In an article in your April number I find the following
words :—" the anthropomorphous apes, which our *savants* mark
out as the immediate ancestors of man." I was not aware that
any naturalist of standing considered man as descended from
any existing anthropoid. It has been expressly declared by the
leading Evolutionists that they do not place man on the same
ascending line as the gorilla, the chimpanzee, or the orang.—
I am, &c.,

ACCURACY.

CARNIVOROUS BEES.

To the Editor of the Journal of Science.

Sir,—In the Notes (No. LXXV. of your issue) you have an interesting observation upon bees eating butterflies, and observe that Fritz Müller states that in South Brazil he saw " bees licking the juice dropping from pieces of flesh which had been suspended to dry in the air." In England I have observed both wasps and bees sipping the juices which have exuded from a fresh steak when lying in a plate. I always supposed they did so for the purposes of drink. Mr. Darwin may probably be right, that they tear open the bodies of the moths to get the nectar contained in their stomachs. I think the more probable solution is, when we consider that they are not indifferent to the juice of meat, that the moisture, and not the nectar, is the object of their quest.—I am, &c.,

S. B.

FERTILITY OF HYBRIDS.

To the Editor of the Journal of Science.

Sir,—" Truth-seeker " might have found a more convincing proof of the fertility of hybrids than that he mentions from "the history of the American bison," in the statement of Haeckel, that the progeny of the hare and the rabbit are fertile, and have been so for several generations ; he has named the new species *Lepus Huxleyii.*—I am, &c.,

S. B.

[The hybrids which our correspondent mentions are well known in France as *Léporides,* and their fertility is established. But we fail to see that they furnish a more convincing proof than does the fertility of the hybrids between the American bison and the domestic cow.—Ed. J. S.]

PAUCITY OF ALMOND BLOSSOMS, &c.

To the Editor of the Journal of Science.

Sir,—The peculiarity of the season this year is curiously exemplified in the blossoming of the almond tree. Generally in February, or early in March, everyone knows that the tree is usually clothed with a profusion of light pink blossoms, with a total absence of green leaf. This year the leaves appear without the blossoms, or if there be any blossoms they are scarcely distinguishable. This is the more to be regretted, as the wall-fruit, peaches, nectarines, and apricots, follow in suit : there are scarcely any blossoms on them. The gardeners explain the phenomenon on the hypothesis of unripened wood.—I am, &c.,

OBSERVER.

Sunbury-on-Thames.

THE COLOUR-SENSE.

To the Editor of the Journal of Science.

Sir,—It has been maintained by Drs. Magnus and Geiger, and I believe by Mr. W. E. Gladstone, that the colour-sense in man is of recent origin. I rather incline to the opinion that it is decaying, if we may judge from the fondness for dull and impure colours now so conspicuous, and which perhaps may be due to the lowering of the tone of the nervous system which the conditions of modern life bring on.

A kindred fact is the love for bitter foods and beverages traceable to the increase of dyspepsia, and most striking among the most dyspeptic peoples.—I am, &c.,

SENEX.

THE MEANING OF BEAUTY.

To the Editor of the Journal of Science.

Sir,—When we see objects openly and conspicuously beautiful to the naked eye, such as certain birds, insects, flowers, &c., we

are apt to think that this wealth of colour and of design exists
with reference to man.   But suppose we take some such object
as the hoof of an ox or an ass, cut a thin section of it in certain
directions, and examine it under the microscope with the aid of
polarised light: what about the wonderful play of colour then
revealed to our vision ?   Will any teleologist maintain that the
minute structure of such hoofs was arranged with any reference
to the fact that after the lapse of many thousand years a few
microscopists should detect all this beauty ?—I am, &c.,

AN EVOLUTIONIST.

NOTES.

Prof. J. T. Short has published a work on the "North Americans of Antiquity," in which he contends that the mound-builders were not Red Indians, but were related to the Nahuas of Mexico; that man is not autochthonous in America, and has not been on that continent above 3000 years (?); that the ancient Americans were not a single race; that the Mayas and Nahuas possessed a very high degree of scientific and artistic knowledge, that they reached this continent from opposite directions, and that the civilisation developed by each people is indigenous.

The Influence of Light upon Animal Life.—We summarise here certain observations reported in the " Mittheilungen aus dem Embryologischen Institut zu Wien " (iv., p. 265), and " Untersuchungen zur Naturlehre des Menschen und der Thiere " (xii., p. 266):—It was established by Moleschott, as far back as 1855, that frogs in light exhaled more carbonic acid than in darkness. The question still remained whether a direct action of light upon the chemical process in the animal body was here concerned, or whether the light merely excited the animal to greater vivacity, and thus indirectly occasioned a greater evolution of carbonic acid. This point has been now experimentally decided by Moleschott and Fubini, who arrive at the following conclusions :—The action of light in promoting the metamorphosis of matter is exerted not merely through the eyes, but through the skin, and can be traced even in blind frogs, birds, and mammals. If the eye alone, or the skin alone, is stimulated by light, the increased escape of carbonic acid is smaller than when the entire body is exposed to light. The respiration of the tissues is increased by light in the same manner as the respiration of the entire body. Both in cold-blooded and warm-blooded animals, whether blind or seeing, the excretion of carbonic acid increases with the chemical activity of the light, the violet-blue ray being the most effectual.

Dr. S. L. Schenck has studied the development of tadpoles exposed respectively to red, blue, yellow, and green light, all other conditions being exactly alike. Under the red glass the embryos gave the first signs of movement, and were throughout the experiments the most active. Under the blue glass the exact contrary phenomena were observed. If the liquid was agitated many of them allowed themselves to drift about like lifeless matter. The tadpoles exposed to green and yellow light did not differ in their behaviour from such as were placed in full daylight. Those under the blue light appeared the most ravenous, but it could not be perceived that they were better nourished than the red. Frog-spawn illuminated from below only was materially retarded in its development. In tadpoles which received light only through a solution of potassic bichromate, cutting off all the rays from violet to yellow, the development of colouring-matter was very imperfect, and the creatures remained pale. In those exposed to light which had traversed an ammoniacal solution of copper oxide, the colouration was perfectly normal.

[We may remark that nothing is recorded of the intensified and increased growth which has been said to result from exposure to blue light. In our own experiments upon insects we have not observed the defective colouration which is here traced to the yellow rays.—Ed. J. S.]

Prof. E. D. Cope, in a lecture on Evolution delivered before the California Academy of Sciences, remarks, concerning the view that consciousness is a kind of force, " To the latter theory I cannot subscribe ; when it becomes possible to metamorphose music into potatoes, mathematics into mountains, and natural history into brown paper, then we can identify consciousness with force." He also gives the caution that " Lines of men in whom the sympathetic and generous qualities predominate over the self-preservative must inevitably become extinct."

We learn from the " American Naturalist " that the Chinese starling, the turtle dove, and the European sparrow are now very numerous in the Sandwich Islands, whilst the native birds, even in the interior, are becoming scarce, and are in danger of extinction.

M. Dieulafait, in a memoir recently presented to the Academy of Sciences, shows that copper exists in all plants growing upon rocks of the primary formation, and in such proportion that it may be detected in 1 grm. of their ash even by the ammonia-test. Plants flourishing upon relatively pure calcareous soils do not yield traces of copper. The author is still engaged with the question of the occurrence of this metal as a normal constituent of the animal system. He maintains that heat has taken no share in the formation of the dolomitic regions.

F. K., writing in the " Natur Forscher," points out that in certain geometrical spiders, belonging to the group Nephelinæ, the males are so minute as to have been overlooked or referred elsewhere. This is the case in the genus *Cærostris* of Africa, and *Celænia* of Australia.

A technological contemporary informs us that " of the sixty-two primary elements known in Nature, only eighteen are found in the human body, and of these seven are metallic. Iron is found in the blood, phosphorus in the brain, limestone in the bile, lime in the bones, dust and ashes in all." Our chemical readers will doubtless be puzzled at the three new elementary bodies, limestone, dust, and ashes.

In the " Scottish Naturalist " for April, Col. Drummond Hay shows, from his own observations and those of others, that fruit trees apparently disbudded by the bullfinches in early spring yield in summer luxuriant crops of fruit, the explanation being that the buds removed were infected with vermin, which were thus prevented from spreading. If this view is further confirmed the prejudice against the bullfinch as an enemy to our orchards must be pronounced unfounded.

Four species of *Helix*, including *H. studeriana* from the Seychelles, according to Vignier, bring forth living young.

According to the researches of Dr. Brush, the milk of ruminants, when swallowed, is coagulated by the acids of the stomach into a hard mass. Hence calves, lambs, kids, &c., which have taken no food but the milk of their mothers, always chew the cud. Animals which do not ruminate consequently find a difficulty in digesting the milk of a ruminant species. On the other hand, human milk, and that of mares, asses, and other non-ruminant animals, coagulates into small granular or flocculent masses which are easily digested. Hence, as was intimated in the " Journal of Science " (vi., p. 24), is not a natural food for man, nor for any non-ruminant creature.

According to J. Munk (" Virchow's Archiv.") glycerin does not rank among nutritive bodies. It is completely decomposed in the animal system, mere traces occurring in the urine. No formation of phospho-glyceric or sulpho-glyceric acid was observed.

Dr. Trouessart (" Annales des Sciences Naturelles ") has published an investigation of the geographical distribution of the bats in comparison with that of other land mammals. Only two families, the Vespertilionidæ and the Emballonuridæ, are common to both continents. The Pteropodidæ, Rhinolophidæ, and Nycteridæ are peculiar to the Eastern Hemisphere, and the Phyllostomidæ to the Western. The Vespertilionidæ have two cosmopolitan genera, *Vesperugo* and *Vespertilio*. The headquarters of the Pteropodidæ seem to be New Guinea. The bats do not belong in the neighbourhood of the Primates, but should rank below the Carnivora, with the Insectivora and the Rodents.

M. Flahault communicates to the " Bulletin de la Société Botanique de France " a paper on some exceptional cases where a green colouration appears in plants without the action of light. He finds that the green colouring-matter (chlorophyll) found within opaque seeds has been formed when the enveloping membranes of the seed are still transparent. Chlorophyll can be preserved for a long time in the absence of light. In the embryo of *Pinus* and *Thuja* chlorophyll is formed in the absence of light at the expense of the nutritive matter previously stored up.

Prof. E. D. Cope points out (" American Naturalist ") that, in accordance with what he calls the " Doctrine of the Unspecialised," the perfection produced by each successive age has not been the parent of future perfection. The largest and most perfectly armed animals (*e.g.*, machairodon) have been the first to succumb upon a change of circumstances. The lines of ancestry of the existing higher Mammalia commenced with types of small size.

In a memoir presented to the Academy of Sciences, M. H. Toussaint shows that tuberculosis is easily transmitted by the ingestion of tubercular matter, by heredity or by suckling, by inoculation, and even by simple cohabitation.

According to the experiments of Popoff, yeast proves rapidly fatal if introduced into the blood of vertebrate animals.

M. Giard (" Comptes Rendus," xc., p. 504) points out that the parasitic fungus *Entomophthora* has recently caused an epidemic among insects of the genus *Syrphus*, which are amongst our best destroyers of Aphides.

MM. J. Béchamp and E. Baltus, in a memoir read before the French Academy of Sciences, state that pancreatin, if injected into the veins of an animal, occasions serious functional derangement, and may prove fatal if the proportion reaches $0 \cdot 15$ grm. per kilo. of the weight of the subject operated upon.

M. Domeyko mentions (" Comptes Rendus," xc., p. 504) that the guano of Mejillones contains a considerable proportion of boric acid.

Dr. E. Erlenmeyer gives, in the " Bienen Zeitung," details of experiments showing that bees elaborate their wax from nonazotised matter. Carbo-hydrates serve also for the production of the fat found in the body of the bee. The food of these insects should not be very rich in nitrogenous matter.

Dr. Haberlandt, in the " Œster. Landwith. Wockenblatt," in treating of the cultivation of red clover, urges agriculturists to act upon the discoveries of Darwin, and protect the humble-bees necessary for the fertilisation of the blossoms of this plant.

Wernich, Baumann, and Nencki find it highly probable that bacteria are destroyed by certain products of the putrefaction to which they have themselves given rise. This conclusion seems to throw a light upon the course of certain epidemic diseases, the bacteria which have occasioned the affection being poisoned by the morbid products generated. When this has taken place the patient recovers, unless his vitality has been already exhausted.

According to the "Apotheker Zeitung" Dr. Offenberg has cured a woman who had been severely bitten by a mad dog, and who showed unmistakable symptoms of hydrophobia. He injected curare (woorali) under the skin, to the extent of 3 grains within five hours. The hydrophobic symptoms were overcome, but the specific action of the curare threatened to produce paralysis of the heart and respiratory organs, which was combatted by artificial respiration. Thus Waterton's anticipation of the remedial action of this drug has become a reality.

The Irish Anti-Vivisection Society is making capital out of a letter received from R. G. Butcher, an eminent surgeon, but who, as the "Medical Press" points out, has neither time nor disposition for biological research.

Dr. A. Vans Dunlop has bequeathed about £50,000 to the University of Edinburgh.

We learn from the "Medical Press" that the "Exchange and Mart" is opening a department for the sale of "charity votes."

Thomas Bell, F.R.S., well known for his treatises on "British Reptiles" and "British Quadrupeds," died on March 17th. He was at one time Professor of Zoology at King's College. From 1848 to 1853 he was Secretary of the Royal Society, and from that date to 1861 he held the presidency of the Linnean Society. The remainder of his life he spent at Selborne, in the very house once occupied by Gilbert White.

According to official returns three hundred persons are killed yearly by lightning in the island of Java.

M. Hospitalier, in the "Lumière Electrique," subjects the electric light of M. Tommasi to a very severe criticism, and pronounces its success hopeless.

Mr. D. Brooks concludes, from extended observations, that lightning is more rare in high mountain chains than in comparatively level districts.

According to Mr. Alexander Watt a vast amount of electricity is generated by railway locomotives, but is at present permitted to escape without utilisation.

We learn from the "Electro-Metallurgist" that two Americans, Messrs. Prescott, profess to have discovered that underground currents of electricity, flowing in all directions, form the true "earths" of lightning discharges. They assert that all houses, trees, &c., struck by lightning are underflowed by these currents, and that no houses, &c., standing on spots where there are no currents are ever struck. In protecting a house from lightning stroke, therefore, their method is to test the ground underneath, and if there are no earth-currents below to take no fu rther trouble; but if these currents are present, to earth the rod, which they erect in the parts of the ground below where they are strongest.

Mr. F. P. Perkins, in a paper read before the Society of Public Analysts, points out the universal presence of starch. In dust, whether suspended in the air or collected from the most varied places, this ingredient can always be recognised. The author therefore justly contended that, especially in chemico-legal investigations, no theories can be founded on its occurrence.

Two serviceable illuminators on the immersion principle have been produced : that by Messrs. Powell and Lealand consists of a combination of lenses capable of transmitting a pencil of light of 130° : it is connected optically with the lower surface of the slide by means of oil of cedar wood. The diaphragm arrangements are simple and effective ; the movement of a single lever permits of the employment of a central pencil, or of one or two marginal slots placed at right angles to each other. It is necessary, for the effective use of this condenser, that the objects should be mounted on the under side of the cover glass, and not on the slide. Like most illuminators that are of real utility, it requires some experience to obtain the best effects ; it is of great value in resolving difficult lined tests. The illuminator by Mr. J. W. Stephenson, F.R.A.S., consists of a plano-convex lens, worked on a 1-inch tool, and having a diameter of 1·2 inches, which is then " edged down " to 1 inch, as being more convenient in size, and as giving an aperture sufficient for the purpose. The upper or convex side of the lens is cut down or flattened, so as to give a surface four-tenths of an inch in diameter, with which the slide is to be connected, when in use, by a drop of oil or water. The upper curved surface of the lens is silvered, and beneath the lens a flat silvered plate, one-twentieth of an inch thick, and corresponding in size and position with the upper flattened surface, is balsamed. A suitable diaphragm can be placed one-eighth of an inch or less below the condenser. If used with a dry lens of the highest power on a balsam-mounted object, the light, unable to pass the upper surface of the covering glass, is thrown back on the object, giving opaque illumination : on the other hand, with dry objects adhering to the slide, the well-known dark ground illumination can be obtained with almost any objective. The lens should be accurately constructed, as any error in thickness is doubled by reflection.—W. T. Suffolk, F.R.M.S.

Prof. Owen has communicated to the Royal Society an account of the gigantic Australian lizard, *Megalenia prisca*, now extinct. The upper jaw appears to have been sheathed with horn, as in the tortoise. Upon the head were seven horns. Its closely now surviving representative is the diminutive Australian species, *Moloch horridus*.

According to Gautier, Scolosieboff, Caillol, and Liron, under the influence of arsenic an increased quantity of phosphoric acid is eliminated, and the glycero-phosphoric acid of the brain is replaced by glycero-arsenic acid.

Mr. J. Evans has presented to the Royal Society a report, by Mr. A. Hart Everett, on the exploration of the caves of Borneo. The results may be pronounced mainly of a negative character. Human and animal bones have been found in the caves, but no light has been thrown on the origin of the human race, the history of the development of the fauna characterising the Indo-Malayan sub-region has not been advanced, nor has any evidence been obtained showing what races of men inhabited Borneo previously to the immigration of the Malayan tribes. The north-west of Borneo has probably been elevated above the waters of the sea at too recent a date to give room to hopes of discovering cave-deposits of higher antiquity. Mr. Everett recommends explorations in the loftier portions of the island in the north-east.

The "Telegraphic Journal" of April 1st contains a description of a new form of lamp for the electric light, which has recently been invented by Mr. Charles Stewart, M.A. It consists of a number of square carbon rods placed radially upon a disk of wood or metal in such a manner that the inner ends of the carbon rods form a complete circle. There is a circular opening in the wooden disk through which the electric light is seen from underneath. The carbons, which are all forced towards the centre by a uniform pressure, move forward as they are consumed, and together form the positive electrode of the lamp. The negative electrode consists of a covered hemispherical cup of copper, which, before the current enters the lamp, rests upon the ring formed by the carbons. On the current entering the lamp an electro-magnet raises the metal electrode, and the electric arc is then formed between the circle of carbons and the metal electrode. There is a flow of water through the latter to keep it cool. The advantages which this lamp possesses are— (1.) It is automatic in its action. (2.) It is capable of burning for a very considerable period. (3.) It does not throw any shadows. (4.) It is of simple and comparatively inexpensive construction. (5.) The intensity of the light may be increased if so desired. This is the second lamp for the electric light which Mr. Stewart has recently invented.

ERRATUM.

Page 185, line 6 from bottom, *for* " oxygen and nitrogen " *read* " oxygen and hydrogen."

THE

JOURNAL OF SCIENCE.

JUNE, 1880.

I. INSANITY AND ITS DIFFICULTIES.*

DR. BEARD pronounces it "one of the paradoxes of psychology that we can best study the mysteries of the mind when that function is eclipsed by disease." Leaving to another opportunity, or perhaps to other hands, the discussion of the view here conveyed, that mind is merely a function, and not an entity, we see in the fact just mentioned, or rather in the exceptional facilities which it offers to the student, little that can be regarded as paradoxical, or even exceptional. For the successful investigation of any phenomenon whatever it is highly important that we should be able to view it under varying circumstances. But there is undoubtedly much that is paradoxical in connection with insanity. No subject, probably, of equal importance is so completely overlooked by the great mass of the educated public. Set on one side certain sections of the medical and the legal professions, a few philanthropists anxious that every practicable relief should be afforded to the victims of this scourge, and such private individuals as have relatives or friends in some asylum, and what cares the rest of the world about insanity? We no longer, indeed, like our powdered and periwigged forefathers, consider a visit to Bedlam as a holiday pastime. But little more than they do we in our turn recognise the terrible riddle which such institutions, Sphinx-like, propound for us to solve. We perhaps view madness merely as one of the horrors with which the sensational novelist of the day spices her† narratives, overlooking

---

* The Problems of Insanity: a Paper read before the New York Medico-Legal Society. By G. M. BEARD, A.M., M.D., &c. Reprint from the "Physician and Bulletin of the Medico-Legal Society."

† This department of literature is to such an extent in the hands of ladies that the feminine is here the more worthy gender.

the grim fact that among the poorer classes in England mental disease has during the last forty years been multiplied at the rate of 300 per cent, as against an increase of population of only 45 per cent. Nay, the late President of the British Psychological Association predicted that if the present growth of insanity continues at the same rate, we shall have, by the year 1912, a million and a quarter of lunatics in this our "highly favoured country." We believe that the estimate will rather fall short of than exceed the mark. We fear that even now the persons outside the pale of palpable insanity, but who are still guided by morbid emotions rather than by sound reason, and who are incapable of following a chain of causation of more than two links, may be counted by hundreds of thousands, and are consequently a serious and a formidable element in the community. What if the old saw "Quem Deus vult perdere" should prove to be literally true?

We do not, however, stand alone. Our kinsmen across the Atlantic are suffering, if less, only less than ourselves. France, Germany, Holland, Austria,—in fact Europe generally is moving in the same direction. So that not the British Empire alone, but modern civilisation as a whole, unlike that of classical antiquity, may perish, not from luxury, but from lunacy.

Surely, in face of such facts and of such prospects, not merely physicians and jurists, but society at large, might do well to seek out what are the causes of this growing plague, and what hope is there that it may yet be stayed.

Before asking after the cause of a phenomenon we shall do well to give it a definition, and here for all practical purposes we cannot do better than take the one which Dr. Beard offers :—"Insanity is a disease of the brain in which mental co-ordination is seriously impaired." This definition, it is true, does not stake out an absolute boundary line between sanity and insanity; but those who have studied Nature know full well that sharp antithetical classifications are in their essence not merely artificial, but delusive. Sanity shades gradually away into insanity, even as day into night and as each colour of the spectrum passes into its neighbour.

But the fact that we cannot run a hard and fast line between blue and green does not render us less able to distinguish them, and precisely similar is the case with mental health and mental disease.

Dr. Beard's definition of insanity has the further advantage that it makes no assumptions. Those who consider mind as a mere function or result of the brain, and those who, like the present writer, recognise in it an independent entity, must alike admit the correlation of insanity with cerebral disease.

Madness, then, as an affection of the brain, is only one member of a class all of which seem to be decidedly on the increase—the diseases of the nervous system. What fosters this class? Why, in spite of our improvements in the medical art, of our sanitary reforms, our advances in domestic comfort, our more wholesome food, and our freedom from some of the excesses of our forefathers, does our vitality thus appear to be withering?

Not a few writers of the present day seek for the cause of increasing insanity in drunkenness, or rather in alcoholism, with which we may couple the use of opium, of ether, chloral hydrate, perhaps of tobacco, and even of tea and coffee.

Thus Dr. W. S. Hallaran* considers the use of whiskey as one of the constant and growing causes of insanity. Dr. W. A. F. Browne, of Dumfries, who had devoted much care and attention to the question of insanity, considers that close on one-fifth of the cases of derangement must be traced to intemperance. Earl Shaftesbury, after acting as a Commissioner of Lunacy for twenty years, concluded that 60 per cent of all the cases of insanity in the United Kingdom and in America "arise from no other cause than from habits of intemperance." This conclusion, it is but right to add, was uttered nearly thirty years ago, and consequently before the recent progress of insanity.

Dr. F. R. Lees† even declares that "insanity in every country corresponds in the main to the use of intoxicating drinks." In support of this proposition he gives a tabular view of the population per each deranged person in certain countries, in juxtaposition with the annual average consumption of alcohol per head. This table at first glance seems to support the theory in question, since, starting from Cairo, where there is one idiot or lunatic in every 30,714 persons, and where the annual average consumption of alcohol is null, we proceed to England, where the insane rise to one in every 713 persons, with an average yearly ingestion of alcohol to the extent of 3 gallons. But the question at once arises—Are there no other points, save the consumption of alcohol, in which the social conditions of

* Observations on Insanity. Cork: 1818.
† An Argument for the Legislative Prohibition of the Liquor Traffic. London: Tweedie.

England differ from those of Cairo ?   On looking further we find that the average consumption of alcohol is alike in the following countries :—Ireland, New York State, England, and Holland with Belgium, viz., 3 gallons per annum ; yet the lunatics are in Ireland as 1 in 500 ; in New York State, 1 in 780 ; in England, 1 in 713 ; and in Holland with Belgium, 1 in 1046.   In Southern France the consumption of alcohol (3½ gallons) is somewhat greater than in the places above mentioned, but the proportion of insanity is 1 in 1500 —in other words, more alcohol than is taken in Ireland leads to exactly one-third the mental derangement.   Spain and Bretagne are alike in their small use of alcohol (1 gallon per head annually) ; but whilst Spain has only 1 lunatic in 7181 persons, Bretagne has one in 3500, or more than double ; more even than Italy (1 in 3785), which consumes exactly twice as much alcohol.

Whilst fully recognising alcohol and its associates as ranking among the causes of insanity, we fail to see that it plays the predominant part which is assigned it by the leaders of the Temperance Movement, or that it is in any especial manner responsible for the fearful increase of mental disease during the last forty years.

Dr. Beard, both from the observations of others and from his own investigations, concludes that excess in narcotics and stimulants, sensual indulgence, &c., are powerless to produce insanity in any considerable extent.   He has visited the so-called Sea Islands between Charleston and Savannah, —famous for their superior cotton,—and has there studied the Negro population, who have scarcely been brought into contact with civilisation, and who intellectually are little in advance of their African ancestors.   He declares that " All the exciting causes which philosophers have assigned as explanations of insanity, and of its increase in civilised countries, are operating there with constant and tremendous power.   These primitive people can go, when required, for weeks and months sleeping but one or two hours out of the twenty-four ; they can go for all day, or for two days, eating nothing or but little ; hog and hominy and fish, all the year round, they can eat without getting dyspepsia ; indulgence of passions, tenfold greater at least than is the habit of the whites, never injures them either permanently or temporarily ; alcohol, when they can get it, they drink with freedom, and become intoxicated like the whites, but rarely indeed manifest the symptoms of *delirium tremens,* and never of chronic alcoholism."   And what is the result of such lives ? Dr. Beard shall speak again ;—" There is almost no insanity

among these Negroes ; there is no functional nervous disease
or symptoms among them of any name or phase ; to suggest
spinal irritation, or hysteria, or hay-fever, or nervous dyspepsia among these people is but to joke. . . . Of nervous
diseases, from insanity down through all the grades, they
know little more or no more than their distant relatives on
the banks of the Congo." These people, it is added, are
"types of all Central Africa, of South America, of Australia,
of our not so very distant ancestors in Europe." So much
for the alcohol and general excess theory !

Another school of thinkers trace the modern increase of
mind-diseases to the decline of religious faith and the spread
of indifferentism, or even of what is conventionally known
as scepticism. It has even been said that insanity is not
so much disease as sin, and that no one who retains a firm
trust in God need fear becoming its victim. It has further
been declared that mental alienation is less abundant in
Catholic than in Protestant countries. The table of Dr. F.
R. Lees, referred to above, lends quite as much support to
these views as to the alcohol theory. Take the two most
decidedly Catholic countries in Europe, Spain and Italy.
The former has only 1 lunatic in 7181 persons, and the
latter—where scepticism has made some progress among
the upper classes—1 in 3785. The northern provinces of
France, where anti-clericalism and irreligion have their
stronghold, number 1 insane person per 1000 ; the more
Catholic south, 1 in 1500 ; and devout Bretagne, only 1 in
3500. But Ireland, also a Catholic country, with its 1 insane person in 500, is a fact in opposition which cannot be
overlooked. Nor can we ascribe to religious faith that immunity from mental disease which savage and semi-savage
races—such, for instance, as the Sea Island Negroes above
mentioned—evidently enjoy.

We come now to Dr. Beard's own theory, startling and
perhaps exaggerated, but containing at any rate such a basis
of truth as to constitute the most tremendous impeachment
of modern civilisation ever conceived. If we survey the
whole world we shall find that just in proportion as a nation,
instead of working to live, lives to work ; just in proportion
as life becomes a struggle, a race, a scurry ; as the margin
which separates the individual man from actual want is
narrowed, and as anxiety becomes the normal frame of mind
of the community, in that same proportion does insanity,
and indeed the entire class of nervous diseases, increase.
Dr. Beard asserts that "no climate, no institutions, no environment, can make insanity common except when united

with and reinforced by brain-work and in-door life." He ought, we think, to have here introduced a qualification— " brain work carried on under the influence of anxiety." No one has shown more fully and clearly than has Dr. Beard himself that intellectual work is in itself conducive to health and long life. Take the case of a man who can enter upon a course of scientific research, not depending upon its results for his livelihood, not compelled to complete his investigations by any fixed date; neither his bodily nor his mental health will be endangered; but compel a man to think, and discover, and invent, with the penalty of want hanging over him in case of failure, or merely delay, and we need not be surprised if his brain gives way. Of all the causes which debilitate the nervous system anxiety bears the palm.

Dr. Beard considers that there are " five features of the nineteenth century civilisation that are peculiar to it, unprecedented in history—the printing-press, the telegraph, steam-power, the sciences, and the mental activity of woman." He might have added, at least for England, " competitive examination." All these agencies, by increasing anxiety and worry, have become causes of insanity. " The telegraph alone has multiplied manifold the friction of life." Standing where we now do, in the last quarter of this nineteenth century, it is strangely suggestive to contrast the realities around us with the anticipations entertained fifty years ago as to the social results of mechanical inventions. They were to lighten toil, to give men leisure to think, in other words leisure to live; in short to make life easier. In the " Song of the Steam-Engine," written by a poet of the " Good Time Coming " school, this wonderful product of human ingenuity is made to say—

> " And soon I intend you may go and play
> While I manage the world myself !"

We need not waste ink and paper in showing how completely these expectations have been disappointed. But concerning the effects of the modern study of Science our author seems at times scarcely self-consistent. We find him here declaring that "the modern brain must carry and endure tenfold more than the ancient, and without a correlated increase of carrying and bearing force. Whence comes insanity, with its train of neuroses." Again, he recommends, among our prime needs for the arrest of the multiplication of the insane, " a partial reversion to the calmness and ignorance of our ancestors." To their calmness, say we, not merely partially, but totally; but to their ignorance ? Nay, without referring to what Dr. Beard has written in his " Longevity of Brain-Workers,"* we find him in the memoir before us ranking among the means for preventing nervous disease " the development of the intellect at the expense of the emotions, which is the tendency of the age in all highly civilised countries." Again, " We are made nervous and kept nervous through our emotions, and kept well by the activity of the intellect." If so, why is " the intellectual activity of the modern woman " pronounced to be " a potent and slightly considered cause of insanity." If, as Dr. Beard fully admits, barbarians and savages are much more emotional and less intellectual than ourselves, and at the same time free from insanity, why should we expect to be freed from mental disease by a further repression of emotion and cultivation of the intellect ? How does this prescription agree with the recommendation given a few pages back, to return to the ignorance of our fathers ?

Why, again, if insanity is so rare among the semi-savages and barbarians of inferior races, should it be—as we find it here stated—so prominently abundant among the poorest and most degraded classes of our great cities ? That their condition is more miserable than that of the savage we admit as decidedly as does Dr. Beard. Let us listen to his description of their lot :—" Civilisation grinds hardest on the poor, for it deprives them of most of the pleasures and delights and healthful influences of barbarism, without the compensations that the higher classes of civilisation enjoy ; it shuts them up in houses, gives them bad air and bad food and all phases of bad environment, crushes them, and keeps them crushed by rivalry and competition, corrodes them with envy, and gives them what many barbarians cannot get, *stimulants and narcotics in indefinite quantities at a moderate price and of the most possible easy access.*" We have not space to quote at greater length from this passage, which deserves to be pondered over by every philanthropist, clergyman, physician, and, most of all, by every legislator. But we must ask, where in the condition of these classes do we find the " brain-work," without which the author has declared that no other agencies can make insanity common. We call attention, too, to the words we have italicised, which seem to convey a half admission of the alcohol theory of insanity.

Dr. Beard's proposals for a revision of our school and college systems are well worthy of attention. He is as

* Quarterly Journal of Science, vol. v. (1875), p. 430.

hostile to cram as is the " Journal of Science," and in face of the overwhelming amount of the knowable he evidently leans, like ourselves, to Specialism, and would not examine the curator of a zoological museum in the higher mathematics and in dead languages.

II. THE HISTORY OF ANTOZONE AND PEROXIDE OF HYDROGEN.

By ALBERT R. LEEDS, Ph.D.

I. *Antozone.*

BY far the most important fact in the long and perplexing history of antozone is the recent discovery that there is no antozone. After giving rise to a very voluminous literature, filled with confused and contradictory statements, the mysterious body named by Schönbein " Antozone " has disappeared from the pages of chemistry, and been added to that daily increasing host of defunct chemical elements which, after a brief and troubled existence, have fallen into final oblivion. As it was, it never had a sturdy existence. It appeared to be a sort of chemical Will-o'-the-Wisp, a matter of exhalations, connecting its existence with the formation and disappearance of clouds and similar phenomena, and ever resisting the attempts of the experimenter to obtain it in some tangible form. The ghost of antozone, raised by Schönbein, and, together with its twin-brother, atmizone, expanded into great proportions by the labours of Meissner, was struck down by Von Babo (in his " Contributions towards a Knowledge of Ozone," 1863), and finally laid by the experiments of Nasse and Engler, on the phenomena attendant upon the action of oil of vitriol upon peroxide of barium (1870).

And when we consider for a moment the overwhelming host of acquisitions which are being yearly made to our stores of veritable chemical knowledge, the mind experiences a sensation of actual relief in seeing so many questionable statements expunged from the history of chemistry, and in getting hold, so to speak, of an unexpected *tabula rasa* on which to write discoveries of permanent value. Such being the case, it would certainly be an unremunerative toil to weigh and ponder the great bulk of conflicting data concerning antozone. The most we would feel willing to undertake would be to inquire into the grounds upon which such mistaken views were originally built; further, into the experiments which appeared to confirm these views, and eventually to win for them the credence of philosophers in general; and, finally, to examine narrowly into the validity of the experimental evidence which is regarded as demonstrating conclusively the non-existence of antozone.

The ground was prepared for the growth of a belief in the actual separate existence of antozone by the promulgation by Schönbein of his theory of Ozonides and Antozonides. Under the former class he included the peroxides which, in their action upon other bodies, manifested a strong likeness to ozone, the typical body of this class being the peroxide of lead.

Without enumerating all the features in their deportment towards other bodies, an enumeration which would serve only to confuse us, it will be sufficient for our present purpose to note that the properties of ozonides that Schönbein regarded as most characteristic were their power of liberating chlorine on contact with hydrochloric acid; of being reduced by peroxide of hydrogen to lower oxides (water and ordinary oxygen at the same time being generated); and of causing the tincture of the resin guaiacum to turn blue.

Antozonides, on the other hand, were those peroxides which, under the circumstances detailed above, behaved in quite contrary fashion,—under no circumstances liberating chlorine from a chloride, not decomposing peroxide of hydrogen, and not turning guaiacum tincture blue. The typical body of this class was peroxide of barium.

And inasmuch as Schönbein thought he had demonstrated that ozone is electro-negative oxygen, and that the ozonides were combinations of a lower oxide with ozone, he accordingly regarded the antozonides as combinations of a lower oxide with electro-positive oxygen. This electro-positive oxygen he appears to have named antozone to distinguish it from ozone, and to indicate the function it performed in antozonides, without claiming, at least at the outset, that it had been or could be isolated in a free condition. The fact that an ozonide and an antozonide could naturally decompose one another, and both at the same time undergo reduction to the state of lower oxides with liberation of ordinary oxygen, was regarded as lending great probability to the view that the oxygen in the two compounds existed in two opposed electro-chemical conditions.

This hypothesis of Schönbein was evolved at that epoch when the electro-chemical theories of Berzelius reigned paramount, and has the same general objection which is urged against the dualistic theory in general, that, instead of regarding a chemical compound as a new individual in which for the time being the specific identity of its components are lost, it assumes that these components, though unrecognisable for the time, nevertheless still exist. In other words, that in an ozonide there is ozone in combination with a lower oxide, and antozone in an antozonide. The validity of this reasoning is denied on the ground that a compound body may yield up its constituents in one form or in another form, according to the reagents, or according to the circumstances, &c., by which its decomposition is brought about. So with the bodies under consideration. It was pointed out by Brodie (1863) that the chemical differences in the deportment of the ozonides and antozonides were to be attributed to the nature of the substances with which in each class of bodies the oxygen was united, and to the nature of the substances taking part in the reactions, rather than to the existence in them of two different modifications of oxygen. For example, taking the evolution of chlorine when a chloride is brought into contact with an ozonide as the most characteristic of its properties, as was done by Schönbein, we certainly should not anticipate that peroxide of barium, which is the typical antozonide, in contact with chloride should evolve chlorine. But it was shown by Brodie (1861) that it did so or not according to circumstances—with concentrated hydrochloric acid, yielding chlorine ; with dilute, peroxide of hydrogen.

In the same direction tended the still earlier observations of Lenssen, that peroxide of hydrogen (an antozonide) could add oxygen to, or subtract oxygen from, an oxidisable body, according as the circumstances of the reaction, or as naturalists at the present time are fond of saying "the environment," are favourable to the formation of a higher or a lower stage of oxidation. Thus in alkaline solution it oxidises oxide of chromium to chromic acid, while it reduces chromic acid to oxide in the acid solution. The above facts are irreconcilable with the hypothesis that an ozonide contained ozone as such,—and an antozonide, antozone ; consequently the hypothesis and with it the terms employed have been abandoned.

But the existence or non-existence of ozone is not only

independent of the truth or falsity of any such hypothesis, but its properties have been studied with a minuteness and exactitude that render it in fact a much better known body than either sulphur or phosphorus. It is questionable whether or no sulphur and phosphorus are elementary bodies; but no one doubts that the substance-matter of ozone and ordinary oxygen is identical, and the relations existing between these allotropic conditions of one and the same elemental substance are clearly and sharply defined. How does the case stand with *antozone* ?

It is manifest that the theoretical speculations of Schönbein upon the existence of electro-negative and of electro-positive oxygen, in a state of combination with lower oxides in ozonides and antozonides respectively, would strongly incline him to the possibility of obtaining, in a free state, antozone, corresponding to the previously obtained modification of oxygen, ozone. Accordingly we find later that Schönbein thought that the gas set free by the action of oil of vitriol on barium peroxide contained antozone. He likewise formulated a number of characteristics by which the presence of antozone could be recognised. Without pausing to enumerate all of these, it will be of service to us, in obtaining a clear conception of Schönbein's conception of antozone, to specify the three most salient. They are— 1. That antozone, such as is made from barium peroxide, combines with water to form peroxide of hydrogen ; ozone, on the contrary, cannot oxidise water to the form of peroxide. 2. It does not turn manganous salts brown, while ozone does, a higher oxide of manganese in the latter case being formed. 3. It bleaches paper saturated with manganous and lead salts, after they have previously been turned brown by ozone. Unfortunately these marks of distinction were open to sources of mistake in their verification. But had the antozone been odourless, or incapable of turning iodopotassium starch-paper blue, Schönbein would have stated grounds of difference which would have rendered it possible readily to distinguish between it and ozone. On the contrary, in these two most striking points, according to Schönbein, antozone and ozone were nearly alike.

Perplexing as the subject was rendered by the numerous, and not unfrequently the contradictory, statements of Schönbein, it was enveloped in a far more disheartening nebulosity, and—it is hardly exaggeration to say—buried beneath a dense fog raised around it by the indefatigable and life-long labours of Meissner. Witness the following samples of Meissner's modes of conceiving and stating the nature of the problems under study, and ask yourselves whether, as he stated them, the problems were not too vague to admit of precise thinking or of crucial experimentation. Antozone, says Meissner, is identical with the gas which is set free by the action of sulphuric acid upon peroxide of barium, except in the two respects that, unlike this gas, it does not decompose iodide of potassium, and it does not smell. (In other words, it is identical with a gas from which it differs in two essential characters.) But (note how the accompanying qualification tends to clarify our ideas) this gas likewise loses its smell, clouds at the same time being formed, on coming into contact with moist air. (l)

According to Meissner ozone could not oxidise nitrogen, and probably antozone alone could not do so either, but both together could bring it about, in case moisture were present and other oxidisable bodies were absent. As the peculiarly distinguishing characteristic of antozone, Meissner rated its power of forming clouds in contact with water. When the water was abstracted from these clouds by contact with desiccating bodies, the dried antozone could form antozone again by transmission through water.

Finally, in opposition to Schönbein, Meissner held that antozone was not absorbed or acted upon by potassium iodide, so that if a mixture of ozone and antozone is passed through a solution of iodide of potassium the ozone is absorbed, while the antozone escapes and passes on free.

I have endeavoured to present above the views entertained by Schönbein, Meissner, and others concerning antozone, as lucidly as the contradictory and oftentimes vague statements made concerning it would allow, and have brought its history down to the time of the publication by Von Babo of the memoir before alluded to (1863), in which the weakness of the experimental evidence brought forth in support of a belief in its existence was for the first time clearly set forth. For Meissner, it will be recollected, saw, in its power of generating a cloud in contact with water, the distinguishing property of antozone. Von Babo discovered that the formation of a cloud is always to be noted when, in any manner whatsoever, ozone is decomposed, water being present. Meissner believed that the clouds could not be due to peroxide of hydrogen, because, according to him, the latter is not volatile. If, then, peroxide of hydrogen was not concerned in these phenomena, there was left—as the only other alternative under the circumstances—the hypothesis of a peculiar modification of oxygen capable of giving rise to them; and to this modification, which again was necessarily different from ozone, Meissner gave the name of Atmizone. Later he identified it with, and called it by, the same name as Schönbein's, Antozone. Von Babo, on the contrary, found that the clouds were only peroxide of hydrogen diffused through vapour of water, and capable of being transported along with it, and even passing with it through aqueous solutions for long distances, without being deposited or absorbed.

Unfortunately these results of Von Babo were encumbered with certain vague and doubtful speculations concerning the mode of genesis of the peroxide of hydrogen, through the interaction of ozone and water in the presence of an oxidisable substance. That they were in reality conclusive against the existence of the so-called antozone was not generally recognised until the labours of Nasse and Engler (1870), upon the gas set free by the action of sulphuric acid upon peroxide of barium, had confirmed their truth and illuminated their proper bearings and significance. Nasse and Engler, by simple and trenchant experiments, demonstrated that the gas evolved in this case was a mixture, containing not only ozone, but also water and peroxide of hydrogen. When the escaping gas was passed through a series of tubes surrounded with a freezing mixture, the latter underwent condensation, and the permanent gas which passed on was ozone. The condensed product, when subjected to appropriate tests, proved to be merely a solution of peroxide of hydrogen. Carry the simple explanation thus afforded with you, and see with what a flood of light it illuminates all the hitherto hopelessly obscure passages in the history of antozone, and enables one to give readily a natural explanation to phenomena which at the time of their original discovery perplexed mightily their discoverers, and led them to form many ingenious, but in the end harmful, hypotheses.

For instance, examine, with the aid thus given, Schönbein's first distinguishing characteristic of antozone, *i.e.*, as made from barium peroxide it combines with water to form peroxide of hydrogen. Since the gas given off in this reaction consists not only of ozone, but of peroxide of hydrogen, the peroxide of hydrogen which Schönbein thought was formed on its coming into contact with water really pre-existed. Consider his second test—that antozone does not turn manganous salts brown, while ozone does. This difference is likewise true of peroxide of hydrogen as compared with ozone. The same remark applies to his third test—that it bleaches papers saturated with manganous salts, after they

have been turned brown by ozone. The same effect precisely is produced by peroxide of hydrogen. Is there any
adequate explanation of these agreements short of con-
ceding that Schönbein's antozone is disguised peroxide of
hydrogen ?

But what shall we say of those numerous cases in which
Meissner thought that a mixture of ozone and antozone
was present, and that on removing the former, by passing
the mixture through a solution of iodide of potassium, the
latter went on alone, attended with its characteristic white
cloud ? The explanation is that afforded by the experiments
of Von Babo : when ozone decomposes potassium iodide
solution there is formed, in addition to free iodine, iodate of
potassium, and potassium peroxide, *peroxide of hydrogen.*
If anyone doubts the adequacy of this explanation, let him
try the following experiment :—Strongly ozonise some *dry*
oxygen by an electrical ozoniser; pass the ozonised gas
through a sulphuric acid wash-bottle, and then allow it to
descend upon a potassium iodide solution. The ozone will
undergo complete absorption, the solution becoming deeply
coloured by the liberated iodine. Resting upon the surface
will be seen a dense white cloud. This white cloud may
now be aspirated through many wash-bottles containing
water, and even a solution of chromic acid, and may stand
for hours over water before it completely disappears. But
on examining the waters used in washing it, they will be
found to contain peroxide of hydrogen. Apply the same
mode of solving the other statements made by Meissner,
remembering always that the peroxide of hydrogen which
is formed when ozone is decomposed by an aqueous solution
is attended by a white cloud, through which the peroxide of
hydrogen has diffused itself,—a white cloud of such permanence that it may be transmitted through many solutions
before undergoing absorption,—and their explanation will be
found both natural and easy.

In conclusion, why not make an end of the matter by
stating that antozone is peroxide of hydrogen ? The objection to so doing is, that along with the term antozone
there were attached many notions which are not true of
peroxide of hydrogen, such as its being electro-positive
oxygen, that it had the power of forming peroxide of hydrogen on coming into contact with water, &c. Finally, the
very name antozone implies a substance in its nature the
opposite of ozone, and supposes the existence of a theory
to account for the difference. For these reasons I deem it
more just to sum up the question by re-affirming the

affirmation made at the beginning, that *there is not, and never was, antozone.*

II. *Peroxide of Hydrogen.*

Though peroxide of hydrogen was discovered by Thénard more than half a century ago (1818), and has ever been a substance possessed of unusual interest in the eyes of chemists, yet the difficulties of its manufacture were so great that only recently has it ceased to be a chemical curiosity and come into use in the arts. Only a year ago a very dilute solution of the peroxide, imported from Europe, was sold in New York at the price of 16 dollars per gallon. But to-day a solution containing 8 to 10 per cent is retailed at about 1 dollar per lb. At this high price it is sold under fanciful names, and employed to bleach the hair, being used —as Dr. Warren would have styled it—as an Anticyano-chaitanthropopoion or Tetaragmenon Abracadabra, to change the dingy tresses of the titmice among the ladies to a ravishing tow. But there is much reason for believing that a most important future is before it ; and whether in the chemist's laboratory or in the arts, as a most powerful oxidising and reducing agent,—for it can act as both,—for bleaching purposes, &c., it is destined to play a great part. With its cheapening, many new uses will be found for it, and it is probable that before very long it will take its place, as Mr. G. E. Davis has strongly urged (" Chemical News," xxxix., p. 220), as an indispensable article upon the working table of every chemist.

But it is not these considerations which mainly interest us in connection with its scientific history. It is rather the accessions to our knowledge, which, more especially of late, have elucidated many obscure points connected with its sources and properties. That the method of preparation from peroxide of barium and hydrochloric acid (Thénard, 1818), or from the same oxide and carbonic acid (Duprey, 1862 ; Balard, 1862), is not used to obtain it on a commercial scale, is familar to many, the method of Pelouze, in which hydrofluoric or fluosilicic acid is employed to effect the decomposition, being that employed in the arts.

That peroxide of hydrogen was formed in the electrolysis of water strongly acidulated with sulphuric acid was stated by Meidinger (1853), and was apparently so well confirmed by the experiments of Bunsen (1854), C. Hoffmann (1867), and others, that until the researches of Berthelot (1875) were published the production of peroxide of hydrogen in electrolysis was looked upon as a fully established fact.

But the great French chemist showed that the body dissolved in the acid electrolyte did not exhibit the reactions characteristic of peroxide of hydrogen : it did not decompose potassium permanganate (Brodie's test), nor oxidise chromic to perchromic acid (Barreswil's test), nor convert calcium hydrate into an insoluble peroxide in alkaline solution (Berthelot's test ?). He demonstrated that it contained in solution the same oxide of sulphur which he had previously formed as a beautifully crystalline body by the long-continued exposure of dry ozone and dry sulphurous acid to the action of the silent electric discharge,—Berthelot's persulphuric anhydride, S_2O_7. Finally, during the course of the year just passed, Schöne has demonstrated, in his elaborate research upon the behaviour of peroxide of hydrogen towards the galvanic current (1879), that *in the electrolysis of water no hydrogen peroxide is formed.*

Will the same be found to be true of Schönbein's statement, that in the oxidation of phosphorus exposed to moist air, along with ozone, a by no means inconsiderable quantity of peroxide of hydrogen is formed ? This point was investigated by the author in the course of a research into the by-products obtained in the ozonation of air by phosphorus, with the result of confirming Schönbein's observation. The amount of hydrogen peroxide was determined by analysis of the water employed in washing the ozonised gas, the iodine liberated by the washed gas being attributed entirely to the decomposition effected in a neutral solution of potassium iodide by the ozone. The proportion of hydrogen peroxide to the ozone, as determined by this method, was only one to four hundred. But later the author has investigated the subject, estimating not only the hydrogen peroxide held back in solution, but the entire amount present in the ozonised gas, and has found that its proportion to that of the ozone may exceed one to three. The two substances, as Schöne has pointed out, may be present in the same vessel in quite a concentrated form, for a long interval, without effecting a complete mutual decomposition, and when highly dilute may coexist for hours.

One question of very great interest still remains :—*Is peroxide of hydrogen present in the atmosphere ?* As yet, except as an inference from other meteoric phenomena, there is no evidence that it is. Meissner (1863), Schönbein (1868), Struve and Schmid (1869), and Goppelsröder (1871), believed that they had succeeded in demonstrating the presence of peroxide of hydrogen in rain. Houzeau, whose authority in matters of chemical climatology no one would feel disposed to question, seeing that he gave a lifetime of arduous study to their elucidation, made very numerous analyses of the atmospheric precipitates at different seasons of the year occurring in the vicinity of Rouen (1868); but he did not succeed in finding peroxide of hydrogen either in snow- or rain-water, nor in natural or artificial dew.

But in the year 1874 Schöne made an elaborate investigation of the subject, and obtained results which established that—in that locality at least, and at the time his experiments were performed—hydrogen peroxide was present in certain atmospheric precipitates. Of 130 samples of rain-water, collected during the latter half of the year 1874, at Petrowskoje, near Moscow, he found only four in which hydrogen peroxide could not be detected. Of snow, of which twenty-nine samples were examined, there were twelve in which the presence of hydrogen peroxide could not be proven. As to the amount, Schöne found that it varied in rain-water between *one part in one million to one part in twenty-five millions.*

The problem how to detect with scientific certitude the presence of ozone, or peroxide of hydrogen, or both, in the excessively dilute condition in which—if ordinarily they exist at all—they must be present in the earth's atmosphere, is still unsolved. And while its importance, as a leading factor in chemical and medical climatology, is on all sides generally admitted, there appears to be scanty prospect of its speedy or satisfactory settlement.

III. THE ORIGIN OF FALLING MOTION.

By Charles Morris.

WHY do bodies fall? The attraction of gravitation may be the active cause of their passing from a state of rest into a state of motion. But attraction of gravitation does not create this motion. Nor can we well imagine gravitative energy to be a mode of motion convertible into other modes. However great the effect produced the force of gravitation remains unchanged. It is not transformed into motion of masses.

Whence, then, arises this motion ? It is a form of energy, and must be derived from some diverse form of energy which it replaces. If, for instance, a body begins to fall to the earth from a position of rest, we can safely assert that the motion it displays pre-existed, either in the earth, in the body, or in surrounding space. It was certainly not created for the occasion.

The theory of gravitation declares that the earth moves towards the falling body with a momentum equal to its own. If the body be supported above the earth, the support performs a double duty. It at once hinders the body from falling to the earth, and the earth from falling to the body. They compose parts of one rigid system. But if the support be removed the earth and the body at once become separate individuals, and they fall together, with equal momentums, until they again enter into rigid relations with each other.

The falling motion manifested by the descending body cannot, then, have been in some mysterious manner transferred to it from the earth ; for the earth's own motion is equally to be accounted for, and in that case we would have to look to the body for its source. No active motion could appear in such an equal mutual transfer of motive vigour. We must therefore look elsewhere for the source of the motive energy displayed.

Nor can it well have been derived from contiguous space. It is too instantaneous in its appearance, and too regular in its increase, to arise from any such transfer of moving energy.

It must therefore have had its origin in the moving bodies themselves. Not, however, as an ideal " potential energy " converted into a real " actual energy ;" but as a real motion, existing previously in some other form, and converted as needed into the form of mass motion.

Such motive energies exist as constituent forces of all matter. They present various modifications, and are named electrical, magnetic, chemical, cohesive, and temperate energies. These are partly modes of motion, partly modes of attraction : they are specialised manifestations of the general attractions and motions native to matter. The generalised form of attraction we possess in gravitative energy. The specialised forms are the organising attractions of substances, such as cohesion, chemism, and possibly magnetism. It is the same with motions. The generalised form is the free movement of gas particles ; the specialised forms are electricity, and heat as it exists in liquids and solids. But these two modes of motion are differently related to masses. Electricity is an organising energy. It only manifests itself through change in the organisation, or in the relations of bodies. Its only ready transformation is into heat.

Heat is a disrupting energy. It is the individual energy of the separate particles, and has nothing to do with the organisation of molecules into masses ; yet it is a generalised condition of motion only as it exists in gases. In liquids and solids it appears to be partly specialised ; most probably becoming some form of rotation in liquids and of vibration in solids.

Heat force is neither concerned in the organisation of the mass, nor is it closely related to the particle containing it. It is capable of ready transfer from particle to particle, and of ready change in direction.

We may look upon every separate molecule or distinct particle of a solid body as dwelling within a nest of attractions. The fixed organisation of the body most probably causes these attractions to become definite in direction, so that it is not improbable that the motion of each particle is confined to a fixed centre upon which these attractions converge, through which centre it must vibrate, or around which it must rotate.

But the forces acting upon the particle are not alone the attractive energies and the repulsive impacts of contiguous particles. The attractive or gravitative energy of the earth is also a powerful factor in the result. This energy must influence the direction in which the particle moves. It is therefore one of the various active forces to which this direction of motion must conform.

And gravitative energy is constant in vigour and direction. It does not vary as the forces of the surrounding particles may do. Thus every vibration or other movement of the particle has a vertical component, in response to gravitation, which must exercise a constant and unvarying influence upon the result.

Every particle, in fact, is incessantly falling. What we call a position of rest is really a position of constantly-arrested fall. If the surrounding attractions tend to force the particle towards a fixed point in space, the attraction of gravity tends to force it below this point. Thus it never moves to the exact point required by its contiguous attractions, but to a point nearer the earth, which forms a centre of all its attractions, that of gravitation included.

The distance between these two points is the distance to

which the particle falls during every vibration. It is arrested at this point by the surrounding attractions, the real and ultimate arresting force being the repelling impact of the particles of the supporting substance.

Every downward movement of the particle is thus aided by gravitative attraction. Every upward movement is retarded. These invigorated downstrokes become themselves an element in the problem; they add, by their impacting force, to the descending energy of the particles below them. Therefore the lower plane of particles manifests the combined gravitative energy of all the particles of the mass. This is what we call weight, this energy of impact, produced by gravity, of the particles of every substance upon its support. It constitutes an incessant rain of down-beating particles: they strike downwards with a vigour depending, primarily, upon their own response to gravity; secondarily, upon the gravitative pressure of the particles above them. The support must be strong enough to bear its load, or it will inevitably give way under this fierce and incessant rain. If the support be removed, what follows? The forces surrounding the particle remain the same. It descends in response to the gravitative component of these forces. This descent is not resisted by the surrounding energies, since all the particles descend at the same time from the same cause. The only real resistance to fall is the upward compact of particles occupying the space through which the fall must take place. If this resistance be removed or sufficiently decreased every particle of the mass must simultaneously descend in response to gravity, and the whole mass change its position.

Thus the heat movements of the particles are made to conform in direction by the attraction of the earth, this conformity constituting a movement of the mass as a whole. And this is a regularly increasing movement. As the mass moves its motion constitutes an energy. The motion caused in each instant by gravitation remains the same. But it has a separate effect in every separate instant, and these effects are persistent and constantly accumulate, producing a regularly increasing motion of descent.

Falling motion, then, appears to be a partial specialisation of the heat movements of the particles. These movements are made to conform in direction to a certain degree, under the influence of a fixed and persistent attraction. This descent is continuous, whether it be resisted or not. If resisted it cannot accumulate. Each momentary fall makes itself felt as weight by the resisting body; but these momentary falls are each obliterated by a reverse impact, and cannot be added together, constituting an increasing energy of fall. Only when the resistance is removed, and the particles are no longer driven back by impact, does the falling energy manifest itself in a downward movement of the mass.

Thus, when a body falls, part of its heat motion has been transformed, and has become motion of the mass as a whole. The generalised motion of heat has become partly specialised into motion of the mass. This is readily transformable again into heat; but it can only be so transformed by resistance. It is persistent as mass motion until some resisting energy overcomes it, when it again becomes heat.

And from this fact two conclusions necessarily arise. The first is, that a body whose mass motion is resisted must display an increase of temperature. The conformity in the motion of the particles is broken; they again move individually instead of collectively. Temperature effects appear in consequence.

The other conclusion is, that a body yielding to gravitation, in increasing its mass motion, must decrease in temperature. Its temperature is being converted into another form of force, and cannot continue to display its usual effects. The body grows colder in every direction except that of its mass motion, the movements of the particles being specialised in this direction, and their impacting force partly decreased in all other directions.

The heat thus lost, as heat, is probably regained from the radiations of the matter through which the body moves, so that its sum of forces is increased in consequence of a special transformation of a portion of them.

Where the motion of the body is decreased or increased by gravitation, without radiation of heat from other sources, certain interesting and perhaps important effects must ensue. If a mass be driven upward against gravitation its particles must continue to fall. The downstroke of their vertical component of motion, as caused by gravitation, is constantly more vigorous than the upstroke. The fall of the body is simply masked by its upward motion, and accumulates in the same manner as if the mass was descending. Thus the upward motion is more and more rapidly obliterated, and soon ceases to exist, the mass becoming momentarily at rest. What has become of the mass motion? Evidently there has been a simple change in the character of the motion of the particles. Instead of moving upwards more rapidly than they descended, they now move upward and downwards with the same vigour. The special mass motion has fallen back into the body, and has become vibratory movement of its particles. It has, in fact, become temperature, and the sum of temperature energy has increased through this loss of mass motion.

If now we apply this idea to the movement of the planetary bodies, some interesting deductions may be made. In the case of a comet moving from the sun we have an exact counterpart of that of a body thrown upward against gravity. The particles of the comet continue to fall towards the sun. These slight falls are masked in the mass motion of the comet, but they slowly consume this motion. They constantly accumulate, precisely as if the reverse motion did not exist. The comet is thus at once moving outward from the sun and falling inward to the sun, and its real motion is the difference between these opposite energies. Its mass motion is, in short, falling back into its substance, and becoming vibratory motion. Eventually the fall increases in vigour until it equals the outward motion of the mass. At this point the comet ceases to remove from the sun. The outward and inward movement of its particles have become equal ; the vertical component of their motion through space has become converted into a vibration about a fixed point in space. It has, in fact, become heat motion.

Thus the strange fact displays itself of a rapidly-increasing temperature in the comet, as a necessary consequence of its movement outward from the sun. In its return to the sun the opposite effect occurs. Its vibratory motion is gradually transformed into mass motion. Every new increment of mass motion thus gained is at the expense of the heat vibration, and the temperature necessarily decreases in consequence.

This effect is, of course, masked in its increased reception of radiant heat in approaching, and its rapid radiation into space while leaving the sun. It is in this like a falling body whose lost temperature is regained from the radiations of surrounding matter.

A precisely similar effect must occur in the case of every planet which has an elliptical orbit. The earth, for instance, after passing its perihelion point, begins to move outward from the sun, against gravitation ; but the fall of its particles towards the sun at once tends to consume this outward movement. The earth possesses really three movements, from whose composition its orbital movement results. One of these is a movement at a tangent to the radius of its orbit. This is resisted by a falling motion towards the sun, in response to gravitation. These two energies are exactly balanced; neither can accumulate at the expense of the other, and they result in a circular orbit. But there is a third motion, a vertical vibration in the line of the radius, a vibration of some three millions of miles in extent, each phase of which occupies six months. This vibratory movement has its full effect upon the resultant motion of the earth, changing its orbit from a circle into an ellipse.

But the vertical vibration is resisted by gravitation in its outward phase, and aided in its inward phase. The result is that a portion of the motive energy of the earth is consumed, by the resistance of solar gravitation, during its outward movement. This lost mass motion must fall back into the earth and become a vibration of particles, constituting an increase of temperature. Its inward movement is, on the contrary, aided by gravitation. The mass motion increases at the expense of the temperature energy.

The loss of mass motion in the earth, from this cause, between perihelion and aphelion, is about $1\frac{1}{4}$ miles, or 6600 feet, per second. It will consequently not be difficult to obtain an idea of the amount of variation in temperature from this cause. For we know that a mass of water, when arrested after a fall of 772 feet, gains 1° F. in temperature from a conversion of its mass motion into heat vibration. Now a fall of 772 feet yields a final velocity of about 220 feet per second. If the loss of this velocity yields water a temperature of 1°, the loss of 6600 feet per second of velocity by the earth should yield it an increased temperature of 30° F., supposing its mean specific heat to equal that of water. If the specific heat equalled that of iron the increased temperature would be about 270°, and if equal to mercury it would be 900°.

We have here a very marked result, but one that is not strikingly evident, from the fact that this lost motion is not an instantaneous arrest, but a gradual arrest extending over six months. The true result, then, is daily increase in temperature, for every particle of water in the earth of one-sixth of a degree, of $1\frac{1}{2}$° for every particle of iron, of 5 per cent for every particle of mercury, and a like result for every other substance in accordance with its specific heat. During the return movement of the earth, from aphelion to perihelion, the opposite effect results. Its mass motion increases, at the expense of its temperature, to an equal degree.

This variation in temperature cannot have any very evident effect at the surface of the earth, where it is lost in the much greater effect of the solar radiations. But in the earth's interior it may possibly produce important results. The variation in the earth's internal temperature, through loss by conduction, is exceedingly minute. But we have here a source of a considerable increase during six months of the year, and a like decrease during the succeeding six months.

These daily variations cannot be lost by radiation, but must accumulate, so that the temperature of internal water must vary 30° yearly, of mercury 900°, and of other substances in like manner. Although we do not know what results are likely to arise from such an annual variation in temperature, yet it is very possible that these results may be of an important character.

In the case of a planet of short period and great simplicity of orbit, such as we have in the planet Mercury, the effects resulting from this cause must be much greater than in the earth. It, indeed, must produce a marked effect on the surface temperature of Mercury, and an annual variation sufficient to partly neutralise the variations in the amount of solar heat upon this planet.

An interesting conclusion from the hypothesis here advanced is in regard to the simple and natural method in which one mode of motion becomes converted into another. The change from heat vibration into mass motion needs no special machinery and no difficult transfer of energy. Motion seems to be constantly at the command of attraction. The least definite pull in any fixed direction, if unresisted by opposing energy, at once converts heat motion into mass motion. This latter, in its turn, is persistent until resisted, when it immediately becomes converted into the independent movement of particles.

The change from electricity to heat is probably as simple in its nature. The impelling cause, in all cases, seems to be some variation in attractive conditions, to which the moving particles instantly respond, their modes of motion becoming special results of the modes of attraction.

And as there is but one motion, so there is, in all probability, but one attraction. Gravitation, chemism, and magnetism are probably modes of attraction, as heat, electricity, and mass movement are modes of motion. The different forms which these assume very likely result from the different relations of position assumed by the particles of matter. It is probable, also, that molecules have special relations of position between their constituent parts, and that their outward attractions become specialised in consequence. The relations of position between particles or masses at a distance from each other are general, and their attraction takes the generalised form of gravitation. The relations of position between particles in close contiguity are special, and their attractions become specialised. The modes of motion resulting are in direct response to the mode of attraction, and are readily convertible into each other at every variation in attraction.

As the generalised mode of attraction is gravitation, so the generalised mode of motion is the movement of the gas particle. This is so vigorous in its action as to resist the attracting energies of contiguous particles. Its motion is, therefore, influenced in vigour only through impact, and in direction only through impact and gravitative attraction. It is constantly falling in response to gravity, and constantly rebounding in response to impact. Wherever the resisting impacts are reduced in quantity the gas particles move in greater number, this movement constituting a wind, which increases in force as the resistance to the individual movements of the particles decreases in quantity.

Give the particles an opportunity to strike together with special ease in one direction, and a wind necessarily ensues. A fall, in response to gravitation, only ensues when the particles near the surface are separated by increased temperature, or through some other cause, so that their resistance to impact is decreased.

Attraction of gravitation, therefore, has no influence in increasing or decreasing the motive energy of matter. Its only influence is directive. It controls the direction of the motions of particles, so far as its control is not resisted by some other controlling attraction. The direction and mode of motion of the particle, at any instant, is a resultant of all the attractive and repulsive forces acting upon it at that instant, gravitation being simply a constant component of these forces.

The vigour of motion possessed by the particle can vary only in two ways. One of these is by impact, in which the energies of the two impacting particles may become changed, their sum remaining unchanged. The other is by the resistance of attraction. Here the particle loses motion, but gives its lost motion to the attracting particles, which it drags into swifter speed.

Motion cannot die nor be born. It can only be transferred in amount and changed in direction.

IV. THE AURORA.*

THIS is the day of *editions de luxe*. Even Science shares the decorative emotion, and appears in holiday attire. Of the many objects in Nature which attract alike the fancy of the artist and the more sober imagination of the scientific student, few possess the same kind of fascination as that afforded by the Northern Lights. Their mysterious and capricious uncertainty; their ever-changing aspect; their delicate *nuances* of tint; the weirdness of the surrounding objects in those regions where they are seen in greatest perfection; and, lastly, the completeness with which they baffle explanation—all these constitute a set of attributes possessed by no other natural phenomenon, not even by the rainbow.

The volume recently brought out by Mr. J. Rand Capron, entitled "Auroræ: their Characters and Spectra," is indeed worthy of its attractive title. Well printed, possessed of spacious margins, and illustrated with several fine chromolithographs and many uncoloured plates, the work now lying before us presents an inviting contrast to the many cramped and ill-printed treatises which enshrine so many precious chapters in Science.

All that can be gleaned from past or present literature on the subject of the Aurora, Mr. Capron has embodied in his treatise. While giving us epitomised notes of the latest researches of Lemström, Backhouse, and R. H. Procter, he does not forget the good work of Sir J. Franklin and of Parry, and the older observations and speculations of pre-scientific ages obtain their due meed of notice. The pages of Seneca, Aristotle, and Josephus alike bear witness to the interest excited by auroral displays; and from the first-mentioned author the passage may be recalled in which it is narrated how, under Tiberius Cæsar, the cohorts ran together in aid of the colony of Ostia, supposing it to be in flames, " when the glowing of the sky lasted through the great part of the night, shining dimly like a vast and smoking fire."

Almost every observer of Auroræ has formed his own impressions as to the character and origin of the phenomenon. It is hardly strange therefore, considering the immense

* Auroræ: their Characters and Spectra. By J. RAND CAPRON, F.R.A.S. London: E. and F. N. Spon. 1879.

variety of appearances which the Aurora may assume, that the different accounts differ greatly. One observer is struck by the steady pale gleam of the arch of light ; another by the ruddy streamers spread out fan-wise ; a third by the drifting coruscations which seem to sweep at a prodigious rate across the sky. What wonder if the accounts are occasionally tinged with details derived unconsciously from the observer's imagination, though set down in all sincerity as objective facts. Mr. Capron has impartially given all these marvellous details supplied by different observers ; and without pretending to sift their claims, or to reject, or criticise in any general manner, he lets each record stand upon its own merits. This method of treating the subject, while possessing many obvious advantages, has the one disadvantage of perplexing the reader, and compelling him to refer continually from the points in one set of observations to those in another. For many purposes the work would have been more valuable had something more been done towards guiding the reader towards a harmonious conception of the facts. But from the very nature of things this must be extremely difficult, and we should indeed be sorry to lose some of the precious details of isolated observations, which would run a risk of being shorn away in any such process of condensation. Isolated or exceptional facts, when really well established as facts, furnish indeed the most significant clues towards further phenomena : they are the finger-posts of scientific discovery. From amongst them we will select a few samples :—

On the 3rd of October, 1877, Herr Carl Bock witnessed an Aurora in Lapland sufficiently brilliant to enable him to sketch it in oil colours. A reduced facsimile of this painting is given opposite p. 25 of Mr. Capron's book. A significant point in Herr Bock's observation was that the auroral arch appeared to consist of two Auroræ, one behind the other, a quiescent arc in front, and a set of moving streamers behind.

Capt. McClintock makes the important observation that when Auroræ are present the atmosphere is never quite clear, there being usually a bank of low fog or cloud below the auroral streamers.

Prof. Lemström, in 1868, when accompanying the Swedish Polar Expedition, observed an auroral streak to burst forth suddenly during a fall of snow. The same author made some remarkable observations on the appearance of luminous beams around the tops of mountains, which the spectroscope proved to be of the same nature as the true Aurora.

Another precious and pregnant indication is to be found

on p. 38 of Mr. Capron's book, in the chapter on "Some Qualities of the Aurora," in which it is stated that Captain Parry with two companions saw a bright ray of the Aurora shoot down between him and the land, not more than two miles away. Accustomed as one is to regard the Aurora as a phenomenon of possibly the same magnitude as that of the zodiacal light surrounding the sun, this observation is of special and novel interest, particularly as it appears to be confirmed by similar phenomena noted by Sir W. Grove, Mr. Ladd, and Mr. Capron himself. The gleams of light witnessed by Sir W. Grove at Chester passed between himself and the houses, and, as they were continuous with the streamers above, " he seemed," as he says, "to be in the Aurora."

To obtain a knowledge of these and many other points of kindred interest, the student cannot possibly do better than consult Mr. Capron's exhaustive treatise : it teems with similar points ; it states them with perfect clearness of style, and it furnishes the enquirer with an extended bibliography of the whole subject. Amongst the more remote topics will be found a carefully-compiled chapter on the sounds declared by many observers to accompany the exhibition of the auroral lights, but concerning which there is much conflict of opinion. Another chapter deals with the curious reddish patches of light occasionally seen upon the surface of the moon, which are suspected—and not without some show of reason—to be of a kindred nature with the Aurora.

The general reader will find matter for pleasant thought in the brilliant and exciting descriptions of auroral displays quoted from the narratives of Arctic travellers, particularly from that of Lieut. Weyprecht, which is graphic in the extreme. We extract from p. 28 the following account of the Aurora Australis seen by Capt. Howes, of the *Southern Cross* :—

" Our ship was off Cape Horn in a violent gale, plunging furiously into a heavy sea, flooding her decks, and sometimes burying her whole bows beneath the waves. The heavens were black as death, not a star was to be seen, when the brilliant spectacle first appeared. I cannot describe the awful grandeur of the scene ; the heavens gradually changed from murky blackness till they became like vivid fire, reflecting a lurid glowing brilliancy over everything. The ocean appeared like a sea of vermilion lashed into fury by the storm ; the waves dashing furiously over our side ever and anon rushed to leeward in crimson torrents. Our whole

ship—sails, spars, and all—seemed to partake of the same
ruddy hues. They were as if lighted up by some terrible
conflagration. Taking all together—the howling, shrieking
storm; the noble ship plunging fearlessly beneath the
crimson-crested waves; the furious squalls of hail, snow,
and sleet, drifting over the vessel, and falling to leeward in
ruddy showers; the mysterious balls of electric fire resting
on our mast-heads, yard-arms, &c.; and, above all, the
awful sublimity of the heavens, through which coruscations
of auroral light would shoot in spiral streaks, and with meteoric brilliancy—there was presented a scene of grandeur
surpassing the wildest dreams of fancy."

A very considerable portion of Mr. Capron's work, and of
the illustrative plates, is devoted to the subject of the
Spectrum of the Aurora as noted by different observers, and
the various coincidences to be found between its bright lines
and bands, and the lines and bands exhibited by the spectra
of other known substances. Amongst more recent observers
Lemström and Vogel are disposed to the conclusion that the
spectrum of the Aurora agrees in the main with the spectrum
of the air as illuminated by the electric spark. Angström
concludes that the Aurora has two spectra, one of which
gives the very bright greenish line always present in the
Aurora, the other comprising the other and fainter lines.
The very important work of Lord Lindsay, Backhouse, H.
R. Procter, and Schuster in this department of research is
also mentioned. An observation by Lecoq de Boisbaudran,
cited à propos of the ruddy streaks so often seen in Auroræ,
that the red line in the spectrum of phosphoretted hydrogen
is increased in brilliancy by artificially cooling the flame,
deserves careful attention. Summing up the evidence of the
spectroscope as to the nature of the Aurora, Mr. Capron
concludes as follows:—"As the general result of spectrum
work on the Aurora up to the present time, we seem to have
quite failed in finding any spectrum which, as to position,
intensity, and general character of lines, well coincides with
that of the Aurora. Indeed, we may say we do not find any
spectrum so nearly allied to portions even of the Aurora
spectrum, as to lead us to conclude that we have discovered
the true nature of one spectrum of the Aurora (supposing
it to comprise, as some consider, two or more). The whole
subject may be characterised as still a scientific mystery,
which, however, we may hope some future observers, armed
with spectroscopes of large aperture and low dispersion, but
with sufficient means of measurement of line positions, and
possibly aided by photography, may help to solve."

On the general theory of the Aurora Mr. Capron is not very definite. He recounts a large number of experiments upon the action of powerful magnets upon the luminous discharges in vacuous tubes, undertaken with a view to elicit evidence on this question. His results show no important advance upon the researches of Plücker and De la Rive, and require, indeed, to be repeated by the light of the still more recent researches of Mr. Crookes, with which our readers are familiar.

Substantially Mr. Capron adopts Franklin's view, that the light of the Aurora is due to an electric discharge through the moistened air, the electricity concerned in the phenomenon being the product of evaporation in the region between the tropics, and passing at a great elevation into the Polar regions. This theory, it will be seen, almost precisely anticipates the theory of the Aurora put forward last autumn by Prof. Rowland. The theory of Lemström, and an account of the instrument devised by that physicist to illustrate the production of light by electric discharges as they pass into a stratum of rarefied air, are also given. Lemström takes Dalton's view, that the earth's magnetism acts in directing the position of the auroral discharge, but holds that the essential and unique cause of the formation of the light is the quiet discharge of positive electricity from the upper atmosphere to the earth—a discharge which has its counterpart in the lightning sparks of equatorial and temperate latitudes, but which differs from these simply by reason of the better conducting power of the colder and moister streams of air at the Poles.

Altogether the volume is, in spite of the drawback named above, a most valuable and admirable work, and we congratulate the author most sincerely on its timely appearance.

V.　FLIGHTY ASPIRATIONS.

By FRED. W. BREARBY, Hon. Sec. to the A.S.G.B.

ADVENTITIOUS circumstances sometimes place men at the helm, who, being ignorant of all duties save one, assume the possession of all, upon the faith of having steered wisely in times past.　So long as he retains the helm he will probably retain the confidence of his passengers; but let him act upon his assumption of knowledge in another sphere, and his ignorance may entail contempt. The human barnacles which fastened themselves upon the ship called " Progress "—sometimes the vehicle for the conveyance of such passengers as gas, lightning, steam on rail, &c.—have, it must be acknowledged, been nearly rubbed off, so rapidly has the " Progress " rushed through the waves of success.

So the helmsman has learned at last to stick to his tiller, and observe with respect, amongst other vessels, one freighted with such a cargo even as aërial navigation.

I have preserved an old barnacle.　It will be found in the " Quarterly Review " for the year 1819.　It was stuck to a ship that was being freighted with ideas for a railway :—

" We are not partisans of the fantastic projects relative to established institutions, and we cannot but laugh at an idea so impracticable as that of a road of iron upon which travel may be conducted by steam.　Can anything be more utterly absurd or more laughable than a steam-propelled waggon, moving twice as fast as our mail coaches?　It is much more possible to travel from Woolwich to the Arsenal by the aid of a Congreve rocket."

Don't you see that this barnacle was stuck upon a passing ship by the helmsman who quitted his tiller, and thereby manifested his intense ignorance.

This greatly dead and much-stained editor—as we may call him—may now be pictured laughing uproariously in presence of an enlightened audience, who look upon him with grave pity that so intelligent a man should be making such a humiliating exhibition of himself.

I am afraid that my remarks are not very respectful towards those editorial commentators who are apt to limit the aspirations of Science to their own conception of what is possible.　What, then, can be said of a barnacle stuck on with the authority of an acknowledged scientist such as Dr. Lardner, who, in his "Cyclopædia" of the edition of the year 1836, under the head of Hydrostatics,—which is too lengthy here to quote in full,—gives his reasons for asserting the impracticability of accomplishing, with any advantage, the then discussed employment of steam for ocean-going ships. He says—"But we have here supposed that the same means may be resorted to for propelling boats on a canal, and carriages on a railroad. It does not appear hitherto that this is practicable." He says again—"The friction of a carriage on a railroad moving 60 miles an hour would not be greater than if it moved but 1 mile an hour (!); while the resistance on a river or canal, were such a motion possible, would be multiplied 3600 times." By friction he means resistance, because in another place he says—"The resistance on the road, instead of increasing, as in the canal, in a faster proportion than the velocity, does not increase at all." So that we have it, upon the dictum of Dr. Lardner, that a wind blowing upon a surface at 60 miles an hour—the conditions are only reversed—produces no greater pressure than if it were blowing with a velocity of but 1 mile an hour. In each assertion of the rate of resistance Dr. Lardner was intensely wrong.*

The late Sir Wm. Fairbairn became a member of the Council of the Aëronautical Society of Great Britain. Now the supporters of this Society entertain two diverse opinions, but both parties aim at macadamising the aërial highway so as to make it subserve the purposes of transit. To the ordinary observer there is only one way. It is that which has been brought hitherto under his observation. Sir William was a balloonist. The balloon is a fact beyond dispute, so that all we have got to do, says he, is to propel it. It is given only to the man who has made its propulsion a study, and has been left gazing regretfully after the money which he has expended in his vain attempts after utility, to estimate rightly the opinions of that section which may be designated by the title "Gravitites," in opposition to that of "Levitites." The Gravitites contend that the object of aërial transit will be effected by opposing the resistance of the air to the action of gravity ; that whilst gravity is a constant force the resistance of the air is under control, so that it can be made subservient to the support of any weight, the surface of which is sufficiently extended, and propelled with the requisite

* A train of carriages and engine weighing 300 tons would meet with a resistance of 3870 lbs., at 10 miles an hour, which would be increased to 12,470 pounds at 60 miles an hour, irrespective of its advance against the air.

velocity. For instance, a sheet of stiff cardboard can be propelled horizontally by means of a finger loosely placed at the rear edge. By increasing the velocity beyond the necessary requirement the cardboard, or any plane surface, will depart from the horizontal in an upward direction, in obedience to the increased resistance of the air, and the rate of velocity being increased it will turn over towards the hand.

It is the desire of some workers to obtain support in the air by extending the area of such surface and propelling by screws : upon a small scale this has been proved practicable. Models of dimensions and weight capable of being launched from the hand are very effective ; but when those of a larger size, which cannot be thus manipulated, are attempted to be put to practical use, a preliminary run upon the ground is necessary, and hitherto the velocity under those conditions— being retarded by friction, although upon wheels—has not been attainable. This velocity is an absolute condition, so as to enable the apparatus to meet with that atmospheric resistance which would force it to leave the ground and continue its flight in the air. Certainly no rails have yet been laid down with the object of reducing friction, but the aid of an incline has been enlisted without effect. No experiment worthy of the object sought to be attained has yet been attempted by any one holding the opinion that eventual success lies in this direction.

My idea of a satisfactory trial would be the employment of great power, large and strong surface, and as frictionless a road as could be devised; for instance, upon a straight line of rail.

The interest which is attached to many scientific subjects is, however, absent in this, so far as respects the public and amongst scientific men generally. So little understood are the principles upon which the hope of flight is founded that it is well known that if a discussion is started in any scientific periodical there are scarcely any instructed minds to follow it up, and the subject dies away almost from its birth, eliciting nothing but worn-out ideas, and always drawing out the suggestion that gas should be used to take off the dead weight. This suggestion is as absurd as the converse one of using an aërial machine to propel a balloon.

Those who saw poor De Groof when he left Cremorne Gardens, in the hour of his death, dangling from the balloon in his comparatively fragile framework, will call to mind the diminutive appearance of the apparatus compared with the bulk of the balloon. To take off the dead weight would require as large a balloon as usual, but still of such a capacity as would dwarf any attached apparatus, and it is quite certain that if the apparatus had any power over the balloon it would not be exerted to propel it, but to drag it at the stern.

So the earnest workers and students are a very small minority. For want of guidance and the dissemination of fundamental facts, the result of experiment, many have been working in the dark, and doubtless, encouraged by the general ignorance, many pompous announcements have been made during the present century which have raised false hopes, and the reaction has had a most injurious effect upon the study of Aërology with a view to the sustentation of heavy bodies. The fact is that a triumph over the difficulties of aërial transport presents to the mind which can grasp the future such an Aladdin-lamp romance that the individual is inclined at once to self depreciation, and to say that "not for me is such a fate in store."

Some such effect has operated to produce apathy as is recorded by Stewart in his "History of the Early Days of the Steam-Engine," as follows :—"Every miscarriage thus added to the obstacles which at all times impeded the introduction of improvements, and the abortive attempts of ignorant and designing men were urged as reasons for disregarding the inventions of more honourable and meritorious individuals."

I cannot leave the subject of plane propulsion without reference to a late attempt by Mr. Lenfield, of Winchester, whose design was suspended from the skylight of the large room at the Society of Arts, at a general meeting of the Aëronautical Society of Great Britain, in 1879. This formidable-looking affair was 40 feet long by 18 feet wide, attached to a framework upon four wheels, the whole rising about 15 feet : 300 feet of canvas was stretched upon an upper frame, and below this, and upon the wheel-supported platform, the operator stood with his feet upon treadles, by which he worked two fan-blades, 9 ft. 6 ins. × 2 ft. 9 ins., in front of the apparatus. By this he obtained about seventy-five revolutions a minute, which enabled him, upon a macadamised road, to attain a speed of about 12 miles an hour,—totally insufficient, however, to enable him to obtain any fulcrum upon the air, for its weight including himself was 304 lbs. Nor was a subsequent attempt down an incline, by which he gained a speed of about 20 miles an hour, any more suggestive of aërial support.

In order to give some idea of the solid support which a body of air is calculated to afford to any surface passing over it in a horizontal direction, and which can be increased or diminished according to the angle in which it is propelled, I will quote some suggestive remarks of Mr. Glaisher, made at one of the Aëronautical Society's meetings, upon the subject of the captive balloon then lately exhibited at Chelsea by Mr. Giffard :—" That balloon has spoken to me trumpet-tongued. All my life I have been accustomed to weigh the air by grains. In winter I find the cubic foot to be about 570 grains, and in summer 20 or 30 grains less. When we took grains we thought the air light. When I saw the other day that the balloon would lift 16 tons or more, and consequently that the weight of air displaced by the balloon must be of greater weight, there must, I think, be something for members of the Society to work upon. When you see that balloon as a small ball only, yet know that air to so many tons weight was displaced by it, surely it held out the hope that some means would be found to solve the problem of aërial navigation."

The method to be adopted to attack this thin though weighty medium, so as to wrest from it the means of support in safety, and the mode of propulsion, is of course the subject of discussion and of some difference of opinion amongst experimenters.

Sir Wm. Fairbairn stuck another barnacle on the good ship " Progress " when he stated as his opinion, in a paper read at Stafford House, that " Man was never meant to fly ; that if the Almighty had intended him to do so He would have given him wings, and that the unalterable laws of Nature were against us." Now it is still a disputed point whether a man possesses the power to manipulate anything in the nature of wings, so as to afford him support and propulsion. The few experiments which have been made in this direction are not sufficiently authenticated for us to deduce any reliable data from them.

Without wishing to dogmatise, and especially without laughing like the writer in the " Quarterly Review " before referred to, I hold, with the Duke of Argyll, the opinion expressed in his own words when occupying the chair at one of our meetings :—" I think it quite certain that if the air is ever to be navigated it will not be by individual men flying ; but it is quite possible vessels may be invented which will carry a number of men, and the motive force of which will not be muscular action." I limit the application of these words to the action of wings by man's muscular efforts. I wish I could think that the late Sir Wm. Fairbairn had made the same reservation.

The laws of Nature, fortunately for us, are unalterable, and as often as they have been questioned as to their adaptability for aërial suppport they have returned a favourable reply.

It may be conceded that a properly constructed plane surface, propelled against the air, will meet with sufficient resistance to enable its course to be deflected upwards at an incline obedient to the angle at which such plane is driven, and sooner or later according to velocity.

An experiment is recorded in one of our annual reports which was made by M. de Louvrie. To a little carriage he fixed a thin plane surface, the angle of which he could alter at will. Placing this machine upon a level spot, he drew it along horizontally by means of a cord which was fastened to a dynamometer, and increased the speed until the machine left the ground, suspended by the pressure of the air on the plane.

In a large machine, such as Mr. Linfield's, where the balance cannot be adjusted with that facility which is readily attained with models launched from the hand, the difficulty will commence with the first tendency to rise. Supposing him to be unable to attain sufficient velocity with a new arrangement which he is constructing, his next step I presume would be to attach it to an engine on a railway, and repeat upon the largest scale the experiment just recorded. Now in such cases the laws of Nature are greatly in our favour, as proved by some experiments initiated by the Society to which I have the honour to act as Honorary Secretary. It remains as a condition—hitherto unfulfilled —that man must attain to perfection in his appliances before he can evoke the utmost effect which is capable of being wrested from Nature in her passive mood.

The experiments consisted in forcing a blast of air against various extents of surface presented to it at varying angles, in order to ascertain not only the force with which they would be driven back, but the weights which they would be able to lift by the air passing beneath the under surface of the various inclines.

Like the sheet of stiff cardboard propelled by the slight pressure of the finger against the posterior edge, supported by the pressure of the air underneath, it was required to know what that pressure was which tended to lift or tilt it up, because that knowledge would enable us to ascertain what weight it would bear to keep it from so tilting up, and also what amount of pressure forward, represented by the finger, would be required to propel it. And this is one of the results discovered, viz., that propelled against still air, at an angle of 15°, and at the rate of 25 miles an hour, a square foot of stiff plane surface—in this case it was steel plate—will support a weight of 1¼ lbs., whilst the resistance to its forward motion is only 5¼ ounces; so that it requires but little more than 5¼ ounces to propel it.

I must say that if we apply this calculation—the result of accurate experiment—to Mr. Linfield's arrangement of area, we do so to its great disadvantage, because he never contemplated advancing against the air at such an angle as 15°; and unfortunately, by some extraordinary oversight, the instrument employed in the experiment, and made expressly with that object, was unable to record any angle less than 15°.

Those very angles which most concern aërial experimenters were left out of the question. It may, however, readily be conceived that at less angles the resistance to the forward motion is less, consequently the power required to propel is less, except that greater velocity is required to keep the same weight in suspension.

But we will take the calculations at 25 miles an hour and 15°, and 300 square feet of plane surface. At that velocity a rigid plane surface would support, at the least, 450 lbs.: in reality much more than that, because the supporting effect of a plane increases in some yet undetermined ratio for each additional square foot. This, however, is not a rigid surface, and therefore some element of uncertainty exists.

However, it appears to be necessary for success to be able to propel such a surface at the rate of 25 miles an hour. We have it reported that down an incline Mr. Linfield succeeded in obtaining 20 miles an hour; but inasmuch as upon the level ground he did not travel more than 12 miles an hour, with the greatest number of revolutions which it was possible for him to impart by his muscular efforts to the screw-propeller, it is evident that the excess down the incline was due only to gravity. The additional air-pressure must have had the same effect upon his screw-blades as the wind produces upon the sails of a windmill, thereby accelerating the speed of the screw without producing an increase of propulsive force, because the air would pass with a greater velocity than that at which the screw was working, rendering it impossible for him to keep his feet upon the treadles. The aid of an engine and railway would therefore be no assistance to anyone trying a similar experiment. It is, I think, conclusive that man has not sufficient power to revolve a screw capable of giving him the necessary speed to leave the ground. Whether if he left the ground, and travelled only in the frictionless air, he would under such conditions possess the power to continue his flight, remains undetermined. He could only be introduced into such a position for determining the truth by first being flown as a kite, without any screw action, and then, having attained that position, he could set to work with his screw. The question of balance is a most important one for consideration. I have known repeated attempts to effect the flight of models entirely fail for want of a proper adjustment of weight. The attainment of that necessary condition for success would be not the least of the difficulties to be encountered by the plane propeller, where he himself is the adjustable weight, yet confined to his work at the treadle.

The conclusion is I think inevitable, that another and more powerful motor than that of a man is necessary to get the initial velocity. I think that it will have occurred to the reader of this paper that the screw-blades themselves are a source of hindrance to progression; that whilst al else lies along the plane of progression, the screw-blades offer a direct opposition to the air the moment that the operator travels by any extraneous aid *faster than he can revolve the screw.*

Here is unmistakably shown the enormous advantage of the wing-movement, where the means, both of support and propulsion, lie in the plane of progression, and only edge-resistance is offered to the air. The notion of wings, however, is nearly always ridiculed. Somehow we are inclined to contemplate them as we are accustomed to see them pourtrayed upon the backs of angels, where the muscular development is singularly wanting; and looking at a full-grown angel, as depicted by some of our artists, one does really doubt his ability to fly.

Had Sir Wm. Fairbairn such an imbecile in his mind? Possibly, as those who entertain such flighty aspirations are supposed to be devoid of common sense, it might be supposed that we relied upon these angelic prototypes as our authority. Well, I will at once disabuse their minds. We object to them strongly, not only because the proportions and malformation are unworthy of the object, but because they have n't any tails l

That, however, which painfully strikes me as wanting in any plane surface used merely as a fixed plane, is the apparent absence of stability. I should not care to trust myse to the possible vagaries of a screw-propelled plane. It seems to me to be a structure without any life in it, unable, as it were, to help itself. It represents the fag end of a bird's flight when he rests from his labours, and, by the aid of the impetus gained, continues his journey with motionless wings extended as a plane.

This plane he can also use as a propeller; and therein consists one of the grand instances of superiority over any other mode of aërial transit. But it is the manner of its use which demands from us the acknowledgment of its great advantage.

The action of a bird's wings is exerted in a space of *three dimensions*, measured by the length of the wings from tip to tip—the arc of vibration of wings—and the length of the body, to which might perhaps be added the effect, both in front and rear, of a reaction arising from their attack upon the air. The stability which by this process is attained, within the knowledge of every one, exists in spite of the fact that the wings are moved up and down, although alternately above and below the centre of gravity of the bird, and although the head and the tail are really acting as the scales of a balance of which the wings are the centre. Those who have watched the flight of wild ducks must have been struck with the peculiar contour which they present. The wings appear to be the cross centre of a cylindrical shaft. The dimensions of a tame duck which I have just measured are—Extreme length from toe to beak, 29¾ inches; from beak to root of wing, 14 inches; stretch of wing from tip to tip, 32 inches; so that the wings are only 2¼ inches longer than the whole body, which in flight would present this shape:

In all probability this bird flies in a stratum of air— taking 1 foot as the extent of vibration of wing—more than 6 cubic feet of which is put into a state of commotion in every sense conducive to its support and balance.

Whatever may be thought, therefore, as to the folly of obtaining flight by wing-vibration, nevertheless it is to that, or to some application of the principle, that we must come in our attempt to make flight serviceable for man.

It is very doubtful whether man has the power necessary for the manipulation of wings of dimensions and strength sufficient to afford him support, but I think it very possible and probable that he may construct an apparatus with surface sufficiently large to sustain him and the additional weight of a motor powerful enough to attack the air so as to obtain support and proportions from it.

We know that the long-winged bird, such as an albatross, has a very flexible wing, which hangs quite droopingly from the body, and which being vibrated produces a wave-action from the root of the wing to the tip.

It is not the lot of many who will read this paper to shake carpets, but to all it will be understood how the wave of air compressed in the downward shake is propelled underneath, so as to throw the carpet into waves, and how—if with sufficient power—those waves are bound to escape at the opposite side ; and if with rapidity of wave-action, how the carpet may be made to hover above the floor without contact. Suppose that we can communicate such an undulation to a fabric free to vibrate in the air ; then its progress would be exactly opposite to the direction of the wave, and it would entirely depend upon the power that we could employ to enable us to determine what weight of fabric could be thus thrown into wave-action, and thereby supported and propelled ; or, in other words, how light a fabric could be used, and what additional weight could be substituted. Had Mr. Linfield's arrangement of 300 square feet of fabric been provided with the power to impart this wave-action, a very different result might have been recorded.

All the experiments which I have made in this direction, with models set free in the air, have pointed to probable success when manufactured upon a scale of utility. This means the capability of sustaining and propelling 1 lb. of weight for every square foot of surface, and this capacity ought to be sufficient to provide for the sustentation of the necessary motive power. The action of such an apparatus may be thus described :—A kite-like structure, the two arms of which are not fixed like the kite, but are free to move, in the manner of the wings of a bird, with a sweep of perhaps 6 feet ; width of such arms from tip to tip, 20 feet ; the fabric attached to the arms, and extended backwards some 30 feet, affording with its triangular shape above 300 square feet ; from thence a tail, capable of elevation or depression ; the whole capable, by the aid of steam, of acting upon a stratum of air of perhaps 3000 cubic feet.

It may be disputed—as it has been—that a fabric shaken as described has any propelling tendency, because, say the

objectors, " If you leave your hold it will recede from you."
My prescription would be " Keep it well shaken." But although I have advanced this illustration of the carpet, the
cases are not exactly parallel. By the peculiar vibration
given to my arrangement a double wave-action is imparted,
—that is, along the wing-arms, and obliquely throughout
the fabric from front to back,—so that the air is propelled
laterally, the fabric being attached to the body or shaft
which runs down the centre like the back-bone of a kite.

For all answer the propelling power of such an arrangement has been proved to be effective. The sustaining property is not disputed. Moreover, owing to the " bellying "
of a large surface, from the great resistance offered by the
air, a gradual and safe descent is ensured in the direction of
advance, upon cessation of the motive power.

This is my " Flighty Aspiration," irrespective of details,
and I contemplate its construction.

ANALYSES OF BOOKS.

The Theory of Colour in its Relation to Art and Art-Industry.
By Dr. WILHELM VON BEZOLD, Professor of Physics at the
Royal Polytechnic School of Munich, and Member of the
Royal Bavarian Academy of Sciences. Translated from the
German by S. R. KOEHLER, with an Introduction and Notes
by E. C. PICKERING. Boston: L. Prang and Co. London:
Trübner and Co.

WE have here a thorough and masterly exposition of the laws of
colour as applied to industrial and artistic purposes. The author
has most important lessons to convey, not merely to the painter,
but to the tissue printer, the dyer, the paper-stainer, the uphol-
sterer, the manufacturer of glass and porcelain, the worker in
mosaic, and the jeweller, all of whom will find here valuable
principles laid down for their guidance. Natural good taste and
what is called a quick eye for colour will doubtless do much, but
without distinct and systematic knowledge they will frequently
fail to save their predecessor going astray. It may even be
suggested that the propensity for sombre, dull, and impure colouration in modern costumes is to be traced to ignorance.
Fearful of the wretched effects admittedly produced by the improper combination of pure colours, we take refuge in a general
murkiness, which at any rate makes our failures less conspicuous.
A work like the present may therefore justly claim a very high
industrial, artistic, and social-æsthetic importance.

The author naturally opens with an examination of the physical basis of the theory of colour, involving, of course, a description of the spectrum. Here he explains very clearly that great
stumbling-block to numbers of even well educated persons—the
essentially subjective character of colour. He also takes an
early opportunity of pointing out the distinction which exists between colours and pigments. If we mix together a yellow and
a blue pigment,—say gamboge and ultramarine,—or if we dye a
piece of silk with indigo and then top it with weld, we obtain, as
all the world knows, a green. On the faith of this and similar
experiments it was decided that green is a compound colour, and
that the only simple primary colours are yellow, red, and blue.
The reason, however, why green is produced by the mixture of a
yellow and a blue pigment is not difficult to trace. The light
passing through them undergoes what may well be called a process of subtraction. Suppose we pass light through Prussian

blue and through gamboge, and allow it to fall upon a prism. We shall find the spectra thus obtained very different from that of white light. In that of gamboge, the violet, ultramarine, and a portion of the blue are cut off. In the spectrum of the Prussian blue, the red, orange, and yellow are cut off. Now every ray of light that falls upon a mixture of these two pigments undergoes this double process in the particles from which it is transmitted and reflected, and the result of the subtraction process is that green remains. But if we throw the blue portion of one spectrum of white light upon the yellow portion of another, we have a process of addition, not subtraction, and the result is a grayish white. From modern researches it appears that if we leave pigments out of the question, and attend to colours in the simple sense of the word, the three fundamentals are red, green, and a blue-violet.

The author's classification of colours is different from the celebrated circles of Prof. Chevreul. His arrangement is represented by a cone, on the base of which are arranged in a circle the following colours :—Carmine, vermillion, orange, yellow, yellowish green, green, bluish green, turquoise, ultramarine, bluish violet, purplish violet, and purple. All these colours, it will be observed, are prismatic except " purple." This name the author gives to a colour which is not found in the spectrum, but which would form the intermediate gradations between carmine and violet. It coincides tolerably with that of the dye-ware commonly known as magenta, but technically called rosanilin acetate. If we then place white in the centre of this circle, all the gradations between it and the above twelve colours can be arranged in lines running from the centre to the circumference. Here therefore will be the places of such colours as rose, pink, peach, lilac, and lavender, which are respectively dilutions of carmine and of the various grades of purple and violet.

Again, at the apex of the cone we place black. On lines ranging along the exterior surface of the cone, from the base to the apex, are arranged such colours as olive (a green saddened with black), the browns (which according to their various nature may be either yellowish green, yellow, or red saddened in a similar manner), wine-colour (which may be called a blackened purple), &c. The true greys—*i.e.*, the gradations from white to black—fall along the line passing from the centre of the base to the apex.

In the fourth chapter the author deals with the theory of contrast—a subject based almost entirely upon physiological considerations. A number of easy and familiar experiments prove that any coloured object, if regarded fixedly for some little time, causes the observer, on turning away, to see its image, but changed to the complementary colour. Thus after looking at a red wafer placed upon a sheet of white or grey paper, we fancy we see a bluish green circle of a similar size to the wafer. But contrast is not merely successive, but also simultaneous. The author formulates the following general law :—" If by the side of a given colour we place any other colour, the first will apparently be changed, as if some of the complementary of the second colour had been mixed with it."

In the fifth chapter Prof. von Bezold enters upon the practical application of the preceding considerations,—the uses and combinations of colour in the decorative arts and in painting. Here he refutes the theory of Field, reproduced by Owen Jones, that the most favourable impression is obtained by employing colours in quantities so proportioned that their mixture would produce pure grey. He shows that a pattern of warm colours upon a cold ground produces a better effect than the opposite arrangement. The number of useful hints scattered through the book must, we feel sure, both surprise and please the practical man, who will here see sound reasons for much that he has hitherto done instinctively or arrived at by a process of groping. He will see why yellow, yellowish green, and turquoise never produce a happy effect upon velvet, a material which is best adapted for violets, purples, reds, and greens; in a word, for full colours. Where the surface reflection is great, and where much white is necessarily mixed with the colours, the very reverse is the case. Violets on satin are apt to take a faded or washed-out appearance. Turning to a totally different region we find the reason why the inside of a golden cup appears of a deeper and richer colour than the outside,—*i.e.*, because the inside is, by reflection, illuminated with yellow light. Hence the inside of silver vessels is very commonly gilt, but the outside rarely.

An important feature of the work consists of its beautifully coloured illustrations, which enable the reader to make observations, and we may almost say experiments, confirmatory of the doctrines taught in the text. For instance, we have certain words, in a bold heavy type, printed with ordinary black ink upon a green paper. If we look at them through the appended leaf of thin tissue paper the black letters become purple.

We consider that this work can scarcely be over-estimated; and whilst regretting that our examination has necessarily been brief, we recommend it most warmly to all who are concerned with the use of colours.

The language of the translator is not incapable of improvement. Thus, p. 188, we read that "mouldings which ought to be horizontal *may be given* a slight inclination." We were not aware that this modern solecism had spread to America.

The Constitution of the Earth. Being an Interpretation of the Laws of God in Nature, by which the Earth and its Organic Life have been derived from the Sun by a Progressive Development. By ROBERT WARD. London: George Bell and Sons.

WE have here an exceedingly interesting work, which, though by no means free from errors, places before us bold and startling conclusions, and thus prompts its readers to research. The author propounds the following four laws :—

1. Circumstances govern the creation of things, and therefore all things exist by virtue of their circumstances.
2. As no two things can occupy the same place in the universe, therefore no two things can be exactly alike.
3. The differences or similitudes between two things must be in proportion to the differences or similitudes in the circumstances under which they have come into existence, or by which they are sustained.
4. A change in the circumstances of things necessarily involves a corresponding change in the things themselves.

By means of these laws, applied to a variety of phenomena, the author arrives at conclusions strangely at variance with the teachings of our received scientific text-books. He holds that the earth and all the planets have been derived from the sun, and in like manner the satellites from their primaries. Each planet in the course of ages gradually recedes from the sun, increasing at the same time in size. The rotatory motion of each increases in a ratio having a relation to the distance. As the planets grow older and become more developed, they become more independent of the sun for both light and heat. "We are justified in assuming that magnificent worlds, like Jupiter, are the abodes of life of corresponding importance, and that their quicker motions in some way compensate for the weaker rays of the sun, which they will ultimately dispense with altogether." The southern portion of the globe, which is more immediately presented to the sun when the earth is nearest thereto, is in an inferior state of development to the northern. The moon has no rotatory movement; it is a child of the earth, created long ages after the earth, and its want of air and water are not marks of decay, but of a development as yet imperfect. Our world, instead of being united to the sun, "is destined to a glorious future, in which it will expand in magnitude, increase in created splendour, evolve new properties, new laws, new elements, add satellite within satellite to its train, and ultimately become the centre of a system as magnificent as the sun's!" When the earth was no older than the moon now is, it would be in a similarly undeveloped condition.

The author denies the circulation of matter. He contends that there is no evidence showing that the exact amount of water which falls from the atmosphere is again returned to the atmosphere, and that the ether in which the earth moves affords an obvious and boundless material for the creation of clouds.

We must here interrupt our exposition of the author's views. When we have full proof, as Mr. Ward admits, that portions of water exposed to the air disappear by evaporation, it seems to us unwarrantable to call in another source for clouds and rainfall of which we have no proof whatever. The experiments of Dr. Wells do not demonstrate that the dew-forming vapours have any other origin than the moisture exhaled from the earth.

The author then contends that entire oceans of water have been converted into solid rock, "in opposition to this theory of the mechanical circulation of matter." The lime, *e.g.*, of which coral-reefs are formed he supposes not to be matter which the water had dissolved out of the crust of the earth, and which the coral animalcules merely separate out in a solid form, but to be generated by the water. Thus "unless replenished by condensations from the ether, the entire ocean seems destined to be transformed into stone." Water, we are told, begins in vapour and ends in salt. The saltness of the ocean is the beginning of its solidification. But where is the actual evidence that pure water, preserved in vessels which it cannot dissolve, will ever become anything but water, or that we can obtain from it anything save the elements of water—oxygen and hydrogen? Let Mr. Ward obtain from water one grain of salt or lime which is not derived from some extraneous source, and he will at the first vacancy be elected President of the Royal Society. Salt is the only mineral body which we eat as such ; but does it therefore follow that the lime, phosphorus, iron, sulphur, &c., found in our bodies are due to its transformations? All these substances can be detected in our food.

But we must hasten on, for Mr. Ward has yet much to tell us. The chinks and fissures in the earth's surface, the ravines, canons, and river valleys are the results of the earth's expansion. Her outer layers crack like the bark of a growing tree. As regards the hypothetical growth of the earth, the author's views coincide with those expressed by Capt. A. W. Drayson, R.A., in a work published as far back as 1859, under the title "The Earth we Inhabit." This writer states that "whilst the most perfect accuracy was supposed to have been attained in Astronomy and Surveying, still, when the results obtained by the two sciences were compared, the most alarming differences were almost invariably found to exist. The more perfect the instruments and the more skilful the observer, the more surely was a difference found." Both Capt. Drayson and Mr. Ward ask, essentially, "Is it a fact that, whenever distances have been carefully measured after the lapse of any considerable number of years, differences have always been found to exist, and in every instance the later measurement has shown a greater number of feet and inches than the former?" It may here be mentioned that some years ago we were informed by a solicitor who had a very large practice in the transfer of real property that he had always found the size of an estate, as given in old title-deeds, fall most decidedly short of the results as ascertained by a recent survey. He ascribed this discrepancy to the inaccuracy of the old measurements, though it must be conceded as strange that the error should always fall in the same direction. It is asserted, in Capt. Drayson's work, that since 1831 the equatorial diameter of the earth has grown 5574 feet and the polar diameter 3180 feet. The latitude of Edinburgh Observatory, as given in 1827 by Francis Baily, then President of the Astronomical Society, when compared with 1858, shows an apparent movement of 1373 yards. The Observatory of Berlin in 1845, as compared with 1858, differs to the extent of a mile and a hundred yards. The general public, and even men of science whose speciality lies elsewhere, must stand aghast at such statements, and be at a loss which is the less incredible supposition—that eminent astronomers should be in error on such a point as the latitude of an observatory, or that buildings should be capable of undergoing a movement of translation, amounting, as in the case of Edinburgh, to 40 yards per year.

It will be remembered that the distance of the earth from the sun, as deduced in 1874 from the observations made on the transit of Venus, was found greater than had been previously calculated. This Mr. Proctor refers to the untrustworthy nature of Delisle's method, whilst our author takes it as a confirmation of one point in his theory—the recession of the earth from the sun.

It must be admitted that the work contains not a few errors in points of fact. Thus we are told that aniline is the base of all the colouring-matters obtained from coal-tar; and again that picric acid, peonine, azuline, &c., are aniline-colours; that lithium is ranged among the non-metallic bodies, &c. We find also reasoning which to us appears strangely inconclusive. The modern doctrine that colour is nothing inherent in bodies themselves, but is merely a form of animal consciousness, is declared "open to the somewhat startling objection that if true it would follow that there would be no rainbow if there were no human beings." To us this seems no objection at all. Still the author's views, if novel, are tangible; they admit of being verified or refuted by observation and experiment, and to these he appeals. Hence we must consider his work deserving of serious attention.

As a defect, we feel compelled to notice the number of typographical errors. Thus we read of albumious, oeline, naphthalian, corana, &c. A modern work is quoted as "Kingzett on the Alki Trade."

Six Lectures on Physical Geography. By the Rev. S. HAUGHTON, M.D., D.C.L., F.R.S., &c. Dublin: Hodges, Foster, and Figgis. London: Longmans, Green, and Co.

THE work before us is one which for many reasons deserves our most hearty welcome. It is an exposition of some of the latest results of modern science, and though popular, in the sense of being intelligible to the educated lay public, it is not the less accurate, comprehensive, and thoughtful. To many well-meaning but timorous souls the author's attitude with reference to certain burning questions will prove not a little salutary. Dr. Haughton is no rash and shallow theorist, but a writer whose sound learning and sobriety of judgment are beyond dispute. He is, further, a believer in a personal God, and in the authority of the Scriptures as a moral and spiritual revelation. But he finds no difficulty, on this account, in accepting the doctrine of Evolution, whether as regards the heavenly bodies or their organic inhabitants, and in admitting a minimum of two hundred million years for the past life-time of our planet.

Among the many interesting subjects dealt with in this volume we will first glance at one which has often been discussed in the "Journal of Science,"—to wit, the Glacial epoch. Dr. Haughton rejects "with contempt" the theories which refer this dreadful visitation to a change in the position of the earth's axis of rotation. He also rejects all solutions based upon the secular cooling of the earth. It is of course obvious that such a gradual cooling down, if it might serve to explain the advent of the Glacial epoch, could not possibly account for the return of a better temperature. The changes in geological climate he refers to the gradual cooling of the sun, involving a corresponding refrigeration of the earth's surface. Glaciation he thinks due to a temporarily diminished rate of heat-radiation from the sun. A natural consequence of such a change would be the precipitation of the aqueous vapour suspended in the atmosphere. Now as this same watery vapour is one of the main agencies which prevent the earth from radiating into space most of the heat which it receives from the sun, and which thus moderate the cold of the night and of the winter, it follows that such a precipitation would be "followed by an increased radiation of non-luminous heat into space from the earth's surface." Now, though we have no positive proof of a transient decline in the sun's heating power during the Glacial epoch, we still know that bodies like our sun do not decrease in radiance at one uniform rate. They have been observed to fade down, and again to blaze forth with increased splendour. In supposing that the sun has once, or more than once, been thus dimmed, we merely assume in him such changes as have occurred elsewhere in the universe. As a necessary consequence, if glaciation was due to a temporary decrease in the sun's radiant heat, it must have occurred simultaneously in both hemispheres, as was concluded by our late friend Mr. Belt, from his geological researches. It need scarcely be said, however, that Dr. Haughton utterly disagrees with Mr. Belt as concerning the cause of the Ice Age. Our author, in passing, notes the absence of all evidence of glaciation in Siberia. Whether this absence extends to all parts of that vast region, and whether future research may not somewhat modify our conclusions in this respect, are perhaps open questions. But what if, during the Glacial epoch, north-eastern Asia was mainly under water? It seems, even since what may be called historical times, to have been undergoing a gradual elevation.

As factors in the climatic peculiarities of earlier geological ages, Dr. Haughton calls attention to two points often overlooked, viz., the large quantity of carbonic acid in the atmosphere during Palæozoic times, and the excess of watery vapour in the Miocene period. Both these bodies, especially the former, have a much greater power than oxygen and nitrogen in arresting the passage of non-luminous heat. It is scarcely too much to say that if the atmosphere were equal in this respect to carbonic acid, frost would be unknown outside the Polar regions.

As regards the origin of life upon our globe Dr. Haughton rejects, with well-merited ridicule, the wild hypothesis which seeks to derive organic existence from an aërolite,—the " moss-covered fragment " of some disrupted world. As we pointed out on a former occasion, this assumption merely postpones the difficulty,—" lengthens out the disease," like Falstaff's loans, unless we can explain how life took its rise in that other world. Our author sees nothing unphilosophical in the assumption that organisms may have been produced from lifeless matter, under peculiar, Divinely-arranged, conditions.

Dr. Haughton's views on the future of our globe do not agree with the expectations of poets and orators. He shows that ultimately Mother Earth will become an airless and waterless orb, revolving on its own axis in the same length of time as she revolves round the sun, one hemisphere being scorched with intolerable heat and the other pinched with unceasing frost,—a counterpart, in fact, of the moon. Indeed the author is not, in the strict sense of the term, a Uniformitarian. He contends— and what geologist can contradict him ?—that the earth has already experienced many marked changes of climate. But such changes will scarcely be denied by the ordinary Uniformitarian, who simply holds, in opposition to the Catastrophist, that the organic population of the globe has not been completely extirpated some eight or ten times, being then, after a " period of repose," succeeded by a fresh creation. We find in the work before us no evidence to connect Dr. Haughton with this latter school.

To the following passage in the opening Lecture some exception may perhaps be taken. The author says:—" There have been from the earliest times two classes of speculation on this subject,—two distinct poles or forms of thought. In the one case we have a race of people distinctly imagining to themselves a personal Creator of themselves and of the universe; a tone of thought leading to an elevation of man above surrounding things, and to a contempt for Nature. In the other case we have a belief in blind formative forces animating matter, involving with it a contempt for man as a miserable unit, as the sport of circumstances, and a corresponding high view of the dignity of Nature." We must confess ourselves unable to see why the conception of a personal Creator should lead to a " contempt for Nature," which is His work. Nay, the entire antithesis between Man and Nature seems to us not less vicious than the old contrast between the earth and the heavenly bodies, —which now survives merely as a rhetorical flourish.

In conclusion we most warmly recommend these Lectures to our readers, and we salute the author as one who, like ourselves, recognises in Evolution God's method of creation.

The Art of Bookbinding. By JOSEPH W. ZÆHNSDORF. London: George Bell and Sons. 1880.

THE review of a work giving an account of the processes carried on in a mechanical trade may seem out of place in a journal devoted to a record of scientific progress; but books are a necessity to everyone, whatever may be the nature of his pursuits: therefore a practical treatise on a subject so important as the preservation of books must be welcome to a very large number of readers.

In the Introduction the author gives an account of the early modes used for preserving and keeping documents together, noticing the fastening of inscribed metal plates with rings, the joining of strips of vellum into a continuous roll, as in the sacred writings of the Jews, a practice continued in the synagogues at the present time. Also the most ancient form of book composed of separate leaves,—the sacred books of Ceylon, written upon palm leaves, and tied together with a silken cord passed through one end. The history is continued by describing the various ways in which vellum and paper manuscripts were placed in libraries: the bindings of some of these, as time progressed, were real works of art.

After the invention of printing books were always bound in volumes, as we are at present accustomed to see them, and in the sixteenth century some of the finest specimens of art binding were executed. To give the reader an idea of the beauty of design of these medieval and renaissance bindings, ten photo-lithographs, from specimens in the possession of the author, are given in illustration. Impressions of tools of various periods serve as vignettes and tail-pieces, and woodcuts of machines and implements occur frequently throughout the volume.

The twenty-four chapters into which the remainder of the work is divided give a very detailed account of the various processes pursued in the binding of a book; and so minutely and carefully has this been done that a clever amateur would find no difficulty in binding books for himself. The author draws attention to a fact little known to the public, of the existence of two distinct kinds of binding, viz., "common" and "extra work." It is with the latter, or binding by hand, rather than the machine pro-cesses or "common" binding, that the writer treats upon at the greatest length.

Even those who do not wish to bind books for themselves may read Mr. Zæhnsdorf's interesting work with profit, as it shows how to distinguish good binding from bad. The remarks on the care of books and their preservation in libraries are of great value, as are also the processes for the restoration of old and injured books. A glossary of technical terms and implements concludes the work.

The author must be congratulated upon adding one to a class of books of which we have but too few; practical descriptions of any manipulative processes are far from common, and persons engaged in researches requiring apparatus of any kind must often have found difficulties in their way, and how much they needed some book of reference on working in metal, wood, glass, or other material. It would be well if those who possess me-chanical skill, combined with the power of description, would follow Mr. Zæhnsdorf's example, and publish the results of their experience.

The Cobham Journals. Abstracts and Summaries of Meteor-ological and Phenological Observations made by Miss CAROLINE MOLESWORTH, at Cobham, Surrey, in the Years 1825 to 1850. With Introduction, Tables, &c., by ELEANOR A. ORMEROD, F.M.S. London : Stanford.

MISS ORMEROD has here placed at the disposal of the scientific world a series of observations of very considerable value. For twenty-five years Miss Molesworth made regular observations of the weather, recording maximum and minimum temperature, direction of wind, occurrence or absence of rain, and general character of weather. The barometer and the wind-gauge were not consulted. At the end of each month the mean temperature and its relation to the average for the month, and the total rainfall are recorded. The purpose of these observations was to ascertain the influence of variation in the character of the seasons upon organic life. The leafing, flowering, fruiting, &c., of plants are recorded, and the appearances, migrations, &c., of birds, insects, and Mollusca. Of these observations many are sufficiently numerous to admit of tabulation, a task which has been performed by Miss Ormerod.

The work will be a useful manual of reference to observers of animal and vegetable life, and Miss Ormerod has done a useful and laborious task in rendering it available. She points out, however, certain shortcomings in these "Journals." Many of the native plants are mentioned merely by their local names; and as for the garden plants, which figure in the list to a considerable extent, their time of flowering, leafing, &c., must be influenced to no small extent by the manner of cultivation to which they are submitted. The insects mentioned are, as a rule, not specifically named, but merely characterised as "Aphides," "gnats," "two species of butterfly," &c.

These tables may be usefully compared with those of the Rev. Leonard Jenyns, which are based upon a much shorter series of observations.

Science Lectures for the People. Science Lectures delivered in Manchester, 1879-80. Eleventh Series. Manchester and London : John Heywood.

THIS series of the Manchester Science Lectures must, we regret to say, be regarded as the last. In the Preface Prof. Roscoe remarks that, in spite of the character of these discourses and the eminence of the men by whom they have been delivered, public interest, as measured by the number of persons attending, "has so far declined that the Committee have had no alternative but to discontinue the Lectures." This fact may be partly explained by the peculiar complexion of the times and the withdrawal of public interest to other spheres. Not the less is it a humiliating result.

The first lecture in this series is on "Islands as illustrating the Laws of the Geographical Distribution of Animals," by Mr. Wallace, whose name is here made to appear as "Arthur R. Wallace." The lecture contains a brief but clear and comprehensive summary of the main facts laid down in the lecturer's classical work on the distribution of animals.

Mr. Waterhouse Hawkins discoursed on the "Age of Dragons."

He explained the legendary accounts of these monsters as exaggerated descriptions of the pterodactyls. It follows as a matter of course that these creatures, if they were the archetypes of the dragons of mythology, must have still inhabited the world after the appearance of man upon the scene.

The "Physical Aspects of Palestine" were ably expounded by the Rev. Canon Tristram, F.R.S. It was pointed out that the lower valley of the Jordan and the basin of the Dead Sea, depressed as they are 1300 feet below the level of the Mediteranean, have retained, as it were, a fragment of the Miocene climate, and may have served as a refuge for certain tropical forms of life during the Glacial period. Certain it is that this district contains species quite distinct from those of North Africa, Egypt, or Arabia, and decidedly resembling South Indian and South African forms. Thus the author found in the Jordan valley a sun-bird (*Nectarinia oseæ*) whose nearest representatives are in the Deccan and a kingfisher (*Halcyon Smyrnensis*) not found elsewhere nearer than Madras. Many of the butterflies, locusts, and beetles of the region belong not to the Mediterranean district of the Palæarctic fauna, but to the Oriental or the Ethiopean. The lecturer very acutely argues that the Arabian Desert is older than the Sahara, because the fauna of the former has a more pronounced desert type.

Captain W. de W. Abney's "Traps to catch Sunbeams," the final lecture, deals with the action of the sun upon the earth.

Eleventh Annual Report of the United States Geological and Geographical Survey of the Territories. Being a Report of Progress for the Year 1877. By F. V. HAYDEN, United States Geologist. Washington: Government Printing-Office. 1879.

THIS volume is devoted to the territories of Idaho and Wyoming. In Fossil Entomology a rich harvest has been reaped in the Tertiary basin of Florissant, from whence 6000 to 7000 insects and 3000 plants have been already received, whilst as many more were expected by the close of the year (1877). There is every reason to believe that the study of these specimens must throw a new light on the origin and early history of insects.

Prof. Leidy has been engaged with the study of the Fossil Foraminifera of the Uintah Mountains and the Salt Lake Basin. The botanical department was conducted by two of the highest authorities in that Science, Prof. Asa Gray and Sir J. D. Hooker. Their valuable report is already before the English public.

The following remark on the practical value of Palæontology is worth recording :—" The sums which have been expended by private enterprise in the search for coal in the different States of the Union, in places where anyone possessing the merest rudiments of palæontological knowledge would have known better, is enormous, and this waste of labour and capital can be stopped only by a proper diffusion of the knowledge referred to."

We find an account of the so-called " spring " of the rattlesnake, which has been much exaggerated :—" The head, neck, and upper portion of the body, which may be raised to the height of more than a foot from the ground, are curved slightly backward, and then thrown forward with great violence. So far as we observed, the distance of the spring may amount to about two thirds the length of the snake."

A valuable feature of the volume is the Report on the Cretaceous Fossils of the Western States and Territories, by Dr. C. A. White.

A Treatise on Statics, containing the Fundamental Principles of Electrostatics and Elasticity. By G. M. MINCHIN, M.A. Second Edition. (Clarendon Press Series.) London and Oxford: Macmillan.

THE work before us, as compared with the former edition contains several important improvements. Not only has it been largely, if not entirely, freed from those misprints and clerical errors which in a mathematical book are at once so serious and so difficult to avoid, but the examples have been re-arranged in the order of their respective difficulty. Further, the demonstration of the parallelogram of forces has been based entirely on Newton's definition of force. The work is further enriched with the principal propositions of graphic statics.

The section dealing with electrostatics has been enlarged, and a chapter on Strains and Stresses has been introduced. On this subject the author remarks :—" In view of the enormous development of mathematical physics, and the wonderful inventions depending on the small strains and vibrations of natural solids which have been made within the last few years, the study of the equilibrium and motion of bodies as they are, and not as they appear in abstraction, is surely a subject of which it is impossible to exaggerate the importance. We may well ask whether, in this country, too much time is not spent in the discussion of neat mathematical unrealities—in the calculation of the behaviour of impossible bodies under impossible circumstances."

It may be safely said that these additions are decided improvements, and that this second edition will be found a decided advance on the foregoing.

Geodesy. By Colonel A. R. CLARKE, C.B., F.R.S. (Clarendon Press Series.)   London : Macmillan and Co.

THE author commences with a sketch of geodetical surveys from Snellius, Picard, and Cassini, down to the operations of Everest and Walker in India, of Struve in Russia, and of Maclear at the Cape of Good Hope. After two chapters devoted to Spherical Trigonometry and the Method of Least Squares, he proceeds to the theory of the Earth's figure. Here it is remarked that, as regards the attraction of mountains, there is little correspondence between theory and observation, since the attraction even of the Himalayas is only perceptible at places quite close to them. Archdeacon Pratt proposes the theory that the elevations and depressions of the earth's surface have arisen from the mass having contracted unequally in solidifying, and that under mountains and plains there is a deficiency of matter approximately equal to the mass above the sea-level, whilst below ocean-beds there is an excess of matter equal to the deficiency in the ocean as compared with an equal volume of rock. On this supposition the amount of matter in any vertical column drawn from the surface to a level surface below the crust would be approximately the same in every part of the earth. We learn that the meridian of the greater equatorial diameter passes through Ireland and Portugal, cutting off a small bit of the north-west corner of Africa : in the opposite hemisphere this meridian cuts the north-east corner of Asia, and passes through the southern island of New Zealand. The meridian containing the smaller diameter of the earth passes through Ceylon on the one side of the earth, and bisects North America on the other. This position of the axes, it is said, corresponds remarkably with the distribution of land and water on the surface of the globe.

In days when such hypotheses as the growth of the earth and the secular transfer of the ocean from the southern to the northern hemisphere, and *vice versa*, are brought forward, Col. Clarke's work must be of especial value.

CORRESPONDENCE.

SOME FACTS ON ANTHROPOLOGY.

To the Editor of The Journal of Science.

Sir,—The fertility and durability of crosses between diverse human races has been, and continues to be, the subject of debate among anthropologists, and thus any little fact, the result of observation, is interesting to its professors.

During a sojourn of three years on the West Coast of Africa, the subject being interesting to me I made it a matter of observation and enquiry. In result I found the intercourse between a white man and negro woman was generally fruitful. The issue intermarrying with the issue of a similar character, generally, I may say was always productive and apparently healthy: in the second generation, on such intermarriage, the issue were weakly and unhealthy; more so in a third generation, when usually they died out. If a mulatto coon of the second generation had issue by a white or a black man, the virility of the descendants was restored. If the issue was again crossed by a white or a black man, so far as vigour and appearance went they were almost indistinguishable from the stock of the male parentage. What the result would be between a white mother and a black father I was unable to ascertain. Probably on the coast such an alliance was never formed.

Since my return to England I met with such an instance. A negro married a white woman, and had issue a son, who intermarried with a niece of the wife, and had issue three sons: the second son married a woman of the same stock as his mother, and has seven children (five girls and two boys), who are as healthy as the children of the white poor usually are; the girls have less of the negro type than the boys. The father of the children is very swarthy, and has fine curly silky hair, and a slight figure. His negro descent is unmistakable; although he has no thick lips, yet there is something about him, to those who are acquainted with the class, which would indicate the negro descent. The elder brother is a short burly figure, more negro-like than the brother just described. He married a white woman, but of an alien stock; his issue are sturdy and strong. The third son would anywhere be taken for a man without the negro cross. He married also a woman unconnected by blood; the marriage is unfruitful.

The latter case is interesting as showing the issue of three generations through pure white mothers. The first family are not so robust as the second family: this probably is due to the intermarrying with women of the same stock.—I am, &c.,

S. B.

THE WILD BIRDS' PROTECTION ACT.

To the Editor of the Journal of Science.

SIR,—Your statements concerning the failure of this Act are, I believe, correct; but the reason is apparently because the penalty inflicted, in the few cases which have been brought before the magistrate, is too low to be of any effect. A fine of half-a-crown will never prevent such offences. Moreover, it is rumoured that the men who actually ensnare birds are merely the servants of dealers, who agree to hold them harmless in case of a penalty. The Act requires sharpening in several respects. I do not, however, think that you need expect the Society for the Prevention of Cruelty to Animals to take action.—I am, &c.,

H. T., of B.

FROST AS A VERMIN-DESTROYER.

To the Editor of the Journal of Science.

SIR,—Well-meaning people sometimes tell us that severe winters are very serviceable, by destroying noxious insects. Such, however, is not the evidence of facts. The prolonged and deadly cold of the winter of 1878-79 has not prevented the ravages of certain caterpillars from being unusually extensive. *Abraxas grossulariata* and *Plusia gamma* were exceptionally plentiful, in every stage of their existence, during the *quasi*-summer. The last winter, again, was much more severe than ordinary, but it has again proved utterly unavailing against the former of these two species, which is literally swarming.—I am, &c.,

AN OLD NATURALIST.

THE PROTECTION OF RESEARCH.

To the Editor of the Journal of Science.

SIR,—Unless I am greatly mistaken you once suggested the formation of a Biological Defence League, to resist any further restrictions upon research. May I ask whether you have, or has any one, taken any steps in that direction? It must not be supposed that the anti-biological agitation of the last few years has died out; and probably the next Session of Parliament will not pass without the introduction of some Bill of a still more sweeping character than that which has unfortunately become the law of the land, and against which you so ably and earnestly protested.—I am, &c.,

M. D.

[We have found a strange amount of apathy on the question. We shall be very glad, however, to receive communications on this important question. Our correspondent may feel certain that nothing shall be wanting on our part to baffle the machinations of organised ignorance.—ED. J. S.]

NOTES.

PROF. OWEN has communicated to the Royal Society a paper on the Ova of *Echidna hystrix*. In most respects they correspond closely with those of the *Ornithorhynchus*. The fission of the germ-mass, corresponding to that described by Barry and Bischoff in the rabbit's ovum, strengthens the conclusion that the *Monotremata* are viviparous. The functional equality of both uteri in the genus *Echidna* corresponds with the equal development of the right with the left female organs, in which it differs anatomically from the *Ornithorhynchus*.

Messrs. H. T. Brown and J. Heron have laid before the Royal Society a memoir on the "Hydrolytic Ferments of the Pancreas and Small Intestine." They conclude that the action of artificial pancreatic juice upon starch paste at 40° C. is similar to that of unheated malt-extract acting at 60° or under, the composition of the starch-products becoming comparatively stationary when 80·8 per cent of maltose has been produced. Neither artificial pancreatic juice nor the tissue of the gland itself contains any ferment capable of inverting cane-sugar. The small intestine is capable of hydrolising maltose, inverting cane-sugar, and acting feebly as an amylolytic ferment. The action of the tissue of the small intestine in bringing about these changes is far greater than that of its mere aqueous infusion, and differs materially in different regions of the intestine. This variability is independent of the relative frequency either of the glands of Lieberkühn or of those of Brunner, but appears to be correlative with the distribution of Peyer's glands. In the transition from colloidal starch to readily diffusible and easily assimilated dextrose, the actions of the pancreas and Peyer's glands are mutually dependent and complementary. The pancreas readily converts the starch to maltose, but is capable only of a very slow transformation of the resulting maltose to dextrose. Peyer's glands, almost powerless upon starch, take up the work when the pancreatic juice almost ceases to act, and complete the conversion into dextrose.

Dr. J. Burdon-Sanderson, F.R.S., and Mr. F. J. M. Page, B.Sc., have submitted to the Royal Society the results of experimental researches on the time-relations of the excitatory process in the ventricle of the heart of the frog. The facts observed agree with the following theories :—Every excited part is negative to every unexcited part as long as the state of excitation lasts,—*i.e.*, the time it lasts in each structural element is measured by the time-interval between the beginning of the initial and the beginning of the terminal phase of the variation.

Dr. Burdon-Sanderson has also submitted to the Royal Society an instrument for investigating the successive phases of the electrical change which takes place in the excitable parts of plants and animals in consequence of excitation.

Dr. Theobald Fischer considers that the climate of North Africa might be better amended by the planting of forests than by the proposed inland sea.

General D. Ruggles, of Virginia, has taken out a patent for the artificial production of rain. It need scarcely be said that in England a process for the prevention of rain and the dispersal of clouds would be very much more useful.

Prof. J. Milne, F.G.S., of the Japanese Imperial College of Engineering, in an article in the " Geological Magazine," shows that volcanoes are chiefly distributed along the borders of land which slopes steeply beneath the sea, and gives reasons why this should be the case.

According to the comparative observations made by M. Alluard at Clermont and on the Puy-de-Dôme during the past winter, the general rule may be deduced that whenever a zone of high pressures covers Central Europe, and especially France, there is, in our climate, an interversion of the temperature with the altitude, the cold being greater in the low grounds than on the heights.

[In the terrible frost of the early morning of Christmas-day, 1860, the same phenomenon was observed in various localities in England. We learn that at Ainley Top, near Halifax, a lofty and generally cold situation, the thermometer stood 4° F. higher than at Elland in the valley below.—ED. J. S.]

M. Inostranzeff describes a mineral carbon, from the banks of Lake Orega, much harder than anthracite, which it also surpasses in electric conductivity. If we arrange the forms of coal in the following series, lignite, bituminous coal, and anthracite, the new mineral follows after anthracite, and is still more divested of all organic character. It is not readily combustible.

Mr. Searles V. Wood, in the " Geological Magazine," asks how the glaciation of North America, as compared with that of Europe, is to be explained on Dr. Croll's well-known eccentricity theory? The superiority of temperature in Western Europe as compared with Eastern America, in corresponding latitudes, is generally admitted to be due to the existing oceanic currents. How, then, could a similar relation of temperatures prevail when these warm currents were supposed to be completely diverted from the glaciated ocean?

Dr. Phipson, in his researches upon palmelline, describes a simple method of preservation applicable to many other substances, and which he describes in the "Chemical News." It is sufficient to add a small quantity of ether to the solution of palmelline in a test-tube, cork it, and turn it over once or twice so as to dissolve as much ether as possible in the liquid, in order to preserve it with all its properties for several months. As long as the contents of the tube have a strong odour of ether, no decomposition sets in, and the optical properties of the palmelline remain intact. When decomposition occurs, the beautiful rose and yellow dichroic tint of the solution fades away, the liquid gives off an odour of ammonia, and swarms with *Bucterium*, *Vibrio*, and *Spirillium*, the latter not readily to be distinguished, save by their small size and more rapid movements, from the *Spirillium* found in the blood in cases of relapsing fever.

According to M. Tayon, ewes found in the neighbourhood of Montpellier have four teats, all yielding milk. A similar anomaly has been pointed out by Mr. Darwin, on the authority of Mr. Hodgson, in the Agia breed of sheep, at the foot of the Himalayas.

An American contemporary enlarges on the *vis medicatrix. naturæ.* Would it not be well to mind the *vis deletrix naturæ* to which the late G. H. Lewes drew attention?

The McGill University, of Montreal, will shortly be provided with a large library, reading-room, and museum. Mr. Redpath, a member of the governing board, has promised to provide the necessary funds.

Mr. R. Barrett, writing in the "Victorian Review," asserts that nine-tenths of the black fellows in Australia die of consumption—a curious commentary on the practice of sending consumptive patients to Australia.

M. A. Veeder finds, from careful microscopic observation, that freezing does not free water from filth due to the presence of sewage or decomposing vegetable matter.—*Amer. Naturalist.*

Dr. Aitcheson reports that in the Kurum Valley, in Affghanistan, there is an intermingling of three very distinct floras,— those of India, Thibet, and Western Asia.

Wickersheimer's preservative fluid for animal and vegetable tissues is composed as follows :—

Alum	100 parts.
Common salt	25 „
Saltpetre	12 „
Potash	60 „
Arsenious acid	10 „

Dissolve in 3000 parts of boiling water. After cooling and filtering, add to every 10 pints of this solution 4 pints of glycerin and 1 pint of methyl alcohol.

According to Dr. Chapman, of Philadelphia, the placentation of the elephant is non-deciduate, diffuse in character, with a zonary form.

The management of the Philadelphia Academy is still a burning question in the American scientific press. The "American Naturalist" quotes a statement of the "leading officer of the Academy" to the effect that "original research was not the sole object of the Society;" and again, that "no part of the museum or library can be held in reserve for the exclusive use of any class of specialists." On these utterances our esteemed contemporary remarks that research requires the "exclusive use" of material as long as the research may last.

Mr. W. Trelease, writing in the "American Naturalist," quotes from Kirby and Spence instances of the carnivorous habits of the honey-bee.

Referring to the fertilisation of flowers by humming-birds, the same writer states that the ruby-throat visits a very large proportion of the plants of North America.

The total quantity of gold got in Victoria during the last three months of 1879 was 209,411 ounces.

Mr. F. W. Rudler announces, in the "Mineralogical Magazine," the occurrence of celestine (strontium sulphate) in the New Red Marl at Sidmouth. He describes the crystals as the finest he has ever seen in form, lustre, and transparence.

After much controversy the *Hyœnodon* has been restored by Prof. Gaudry to the situation originally assigned it in 1838 by its discoverers, Laizer and Parieu, *i.e.*, among marsupial carnivores such as *Thylacoleo* and *Dasyurus*.

Dr. H. Trautschold, of Moscow, maintains that the level of the ocean is getting lower, and the total quantity of water on the surface of the globe diminishing.

It appears that, with few exceptions, Australian trees flourish as well in California as in their native country. It is hence, conversely, to be expected that the native vegetation of Calfornia can be successfully acclimatised in Australia, and that the crops which succeed in the one country are likely to do well in the other.

G. Boericke recommends perosmic acid along with oxalic acid for colouring microscopic objects. The sections or pieces of tissue are steeped for an hour in a 1 per cent solution of perosmic acid, carefully washed, and then immersed for twenty four hours in a saturated solution of oxalic acid; peculiar colourations are observed when examined microscopically in water or glycerin.

According to the "American Journal of Microscopy," silver-wire, in which the most delicate test could detect no difference of diameter, has been run through plates of rubies to the length of 170 miles.

An Exhibition illustrative of Prehistoric German Anthropology will be opened at Berlin in August. Prof. Virchow is President of the Committee of Arrangements.

The new Royal Irish University does not promise to become of great importance in Science. The Senate consists of ecclesiastics, politicians, and lawyers,—classes of men seldom favourable to the investigation of Nature.

A shower of dust was observed from April 21st to 25th, in the Departments of Basses Alpes, Isère, and Ain. It consisted of fragments of mica, garnet, and orthose, accompanied by starch grains, diatoms, and fragments of the integuments of Infusoria. There is neither native iron nor magnetic oxide, whence the dust cannot be of cosmic origin.

According to M. Des Cloiseaux the crystalline form of magnesium is a regular hexagonal prism.

A hemipterous insect (*Hysteropterum apterum*) has appeared upon the vines in the Gironde, and is occasioning injury.

Mr. J. W. Mallet, F.R.S., has communicated to the Royal Society a memoir on the "Atomic Weight and Valence of Aluminium." As the mean of twenty-five determinations he takes $Al=27.019$, with a probable error of ± 0.0030. Further, taking the eighteen elements whose atomic weights have been determined with the greatest attainable precision, he calculates the probability that nine of these numbers should lie, as they actually do, within 0.1 of integers, supposing the value of the true numbers to be determined only by chance, and finds it only as 1 to 235.2. This example seems to show that Prout's law is not yet absolutely overturned, but that there exists a heavy and seemingly increasing probability in favour of it, or of some modification of it.

According to the Registrar-General's returns males are increasing in numbers and females decreasing. In 1879 the boys born amounted to 442,289 as against 433,577 girls. The number of females who died exceeded the males by above 30,000.

We have from time to time referred to the indiscreet dabbling of political organs in scientific matters. We find that the "Times," noticing the successful experiments of Profs. Golgi and Raggi, of Pavia, on the transfusion of blood for anæmia in the case of a lunatic, gravely states that the blood was injected into the "peritonitis"!

Dr. Huggins most truly declares that one of the great charms of the study of Nature lies in the circumstance that no new advance, however small, is ever final. There are no blind alleys in scientific investigation. Every new fact is the opening of a new path.

Mr. Joseph Beck has recently added to the capabilities of the large model stand produced by his firm. The stage is capable of being inclined to any extent, the amount of rotation being recorded on a divided plate ; or it can be turned completely over, in case it is desired to view objects without the interference with oblique illumination caused by the thickness of the stage. The swinging sub-stage bar moves independently of the mirror, or the mirror can at pleasure be attached to it, and then partakes of its motion. After adjustment of the angular direction of the light, a lever permits of its focal adjustment in the optical axis of the instrument. The whole of the illuminating apparatus, including the mirror, can be brought above the stage if desired. All the movements are registered on divided circles or scales. Provision is made for attaching a lamp to the stand of the microscope, and giving a power of motion which can be turned to good account in the examination of opaque objects. Although the arrangements of the microscope are necessarily somewhat complex, it offers unusual facilities for using illuminators of various kinds, and in a way not easily attainable with any other instrument.

Messrs. R. and J. Beck have published a description of the structures to be seen in a slide issued by them, of a section of the stem of a lime tree (*Tilia Europœa*). The section is transverse, but slightly oblique, at the junction of a branch, and double-stained, the period of growth being two and a half years. The preparation is of great beauty, and is so carefully made that almost the whole structure of the stem is capable of demonstration. The page of explanation will prove of great value to anyone who will carefully go through it, and endeavour to make out the structures described with the microscope. It is to be hoped that Messrs. Beck will continue to publish such descriptions whenever they obtain an object of educational value : such a course will do much to instruct those who would otherwise be contented merely to look at an object for its prettiness, and allow the microscope to become only a toy in their hands.

THE

JOURNAL OF SCIENCE.

JULY, 1880.

.

I. THE EVOLUTION OF SCIENTIFIC KNOWLEDGE.

By C. Lloyd Morgan.

IN the " Westminster Review " for April, 1857, appeared an Essay on " Progress: its Law and Cause," by Mr. Herbert Spencer, in which was first sketched the *Law of Evolution* more fully developed in the First Principles. Of that law the late Prof. Clifford said, some twelve years ago, that " it is to the ideas which preceded it even more than the theory of gravitation was to the guesses of Hooke and the facts of Kepler." And now it seems almost an impertinence to the reader to occupy any space here with a definition of this Law of Evolution. Still, impertinence though it be, I venture to preface what I have to say on the Evolution of Scientific Knowledge with a few words on the general law, by way of reminder. The processes of Evolution, according to Mr. Herbert Spencer, are processes of *differentiation* and *integration ;* that is to say, they are processes by which the parts of that which is being evolved become more *different*, and by which those parts, at the same time, become more *dependent upon each other*, and are bound together into a more *definite, complex whole.* Where, as in most cases, Evolution is accompanied by growth, the differentiation and integration are not confined to the parts which originally existed in the system, but are extended to the material which is gradually incorporated with the original matter. And where, as is always more or less the case, the parts of the evolving system are in motion, this motion also is subject to the processes of integration and differentiation.

A hypothetical example of the simplest possible kind will not here be out of place. Take a tribe of savages in the pre-social stage. Each individual does for himself all that he wants done: he cuts his own bow, makes his own arrows, and shoots for himself the wild creatures from which he has himself to prepare food, and perhaps clothing. Now let Evolution come into play. Differentiation sets in. There is an incipient division of labour. One individual devotes himself to the making of bows and arrows on the understanding that they whom he supplies shall procure him food and clothing; another dresses the food or the skins on similar terms; and so on. It is needless to elaborate this example; for without elaboration it is clear that the individuals of a tribe so far evolved have become more different, and at the same time more dependent on each other; while the tribe itself has been converted into a more definite, complex whole.

Such being, therefore, the law of Evolution, I propose in this paper to apply it to the history of Science. I shall endeavour to show that the advance in complexity, definiteness, and integration, which constitutes Evolution, is seen not less clearly in our *knowledge of phenomena* than in the phenomena themselves. It matters not, I believe, to which branch of Science we turn our attention; all tell the same tale. Definiteness of observation, complexity of subject matter, interdependence of phenomena, comprehensiveness of generalisation—all these advance hand in hand. The history of science is the history of an evolution, and the Law of Evolution is the outcome of that evolution.

At the outset, however, an objection may be raised to this application of a physical law to the products and processes of the mind. And to those who see no connection between consciousness and the vibration of brain-molecules, who recognise no physical basis of mind, the objection is probably insuperable. But while the fact cannot be too frequently insisted upon, that we are utterly and completely unable to conceive *how* the vibrations, decompositions, or isomeric changes of grey nitrogenous matter may be accompanied by the phenomena of conscious thought; while we must honestly confess that this is, and probably ever must be, an inscrutable mystery; still we seem forced, by the balance of scientific evidence, to infer that there *does* exist a definite connection between what we characteristically call *brain power* and thought. And if this be so, then it is evident that the evolution of scientific knowledge is but the sign of the evolution of one portion of the individual and social organisation.

For just as the varying sound of the voice from feeble treble to resonant bass testifies to the gradual development of the vocal organs, so does the evolution of scientific knowledge testify to the evolution of the brain power and brain complexity of which that knowledge is one of the products.

Before we proceed to the consideration of some special instances of this evolution as exemplified by the sciences of astronomy, geology, and chemistry, it will be well to devote a short space to the broader question, From what has Science in general been evolved ?   To this question there can, I think, be but one answer. The more or less organised knowledge which we call Science has sprung from the unorganised general knowledge of our uncultured ancestors. In an Essay on the " Genesis of Science " Mr. Herbert Spencer has treated this subject with masterly clearness. The first step towards knowledge of any kind is classification ; and classification is based on the recognition of likeness and unlikeness. We only know an object when we recognise that it is like something we have before met with. If it be like nothing that we have before seen or heard of, we say that we do not know what it is. And just as classification is the grouping together of like things, so is reasoning the grouping together of like relations among things. Now it is by the extension of these processes of grouping together like things and like relations among things that Science arises. But so far the science is only *qualitative*. It is only when the recognition of *likeness* grows into the recognition of *equality* that science becomes quantitative ; for equality—equality between things and equality between relations—is the fundamental conception which underlies all mathematics. Out of this conception of equality therefore springs *exact science*, endowed with the power of quantitative prevision. We may, in fact, from one point of view, divide Science into three stages—that of merely qualitative prevision, that of vaguely quantitative prevision, and that of exact quantitative prevision.   Let us take an example of each. From a study of the appearance of the sky, and from observations of the barometer and hygrometer, I may be able to foretell a storm ; but my prediction cannot be in any sense quantitative, for the storm may last a couple of hours or continue for as many days. After careful diagnosis the skilful physician knows that the administration of a certain drug will have a certain effect upon his patient ; but his prevision is only vaguely, not exactly, quantitative. On the other hand, after taking into consideration certain perturbations of the planetary system, M. Leverrier, in 1846, inferred the presence of a hitherto unrecognised planet, and recommended M. Galli, of the Berlin Observatory, to direct his telescope, on the 23rd September, to a definite point in the heavens : he did so, and beheld the planet Neptune. Here, then, was exact quantitative prevision. In the last of these examples we see science *completely* differentiated from common knowledge. In the first we see science *scarcely* differentiated from common knowledge. The child who having been burnt dreads fire exhibits qualitative prevision, but it can scarcely be said that his knowledge is science. And this serves to show how difficult it is to separate common knowledge from incipient science. Looking at the question then in this general way, there can, I think, be no doubt that the passage of the general mass of knowledge into science exemplifies evolution. In the separation of the special sciences from each other we have a clear case of differentiation ; in the growing inter-dependencies of the sciences (to be more particularly noticed presently) we have a further trait of evolution ; while daily integrations of our knowledge tend more completely to fuse the body of scientific knowledge into a definite complex whole.

Let us now turn to more special examples.

It is probable that in very early times men were led to speculate on the constitution of that starry firmament which night after night met their gaze. What may have been the exact nature of the views which were the outcome of this primitive guessing we shall probably never know, but we may form some idea by enquiring what are the views of the uncivilised to-day. We find, on doing so, that the South Australians think "the constellations are groups of children," and " three stars in one of the constellations are said to have been formerly on the earth : one is the man, another his wife, and the smaller one their dog ; and their employment is that of hunting opossums through the sky." We find, too, that the Esquimaux think the sun, moon, and stars are " spirits of departed Esquimaux, or of some of the lower animals." We find again—but we need go no further. Little is to be learnt from these guesses ; but we may fairly suppose that it was long ere vague mythological ideas gave place to conceptions based on physical analogies. Among the early Greeks, however, the phenomena of the heavens were, according to Whewell, explained on the supposition that the sky is a concave sphere or dome, to which the stars are fixed, and that the celestial sphere revolves perpetually and uniformly about the pole or fixed point. Here, then, we have an explanation in some sort physical, but one of extreme simplicity and generality, and one, therefore, which exemplifies an early stage in evolution. But ere long this simplicity gave way to incipient complexity. To account for the way in which the appearances of different nights succeed each other, the sun also was supposed to move round among the stars on the surface of the concave sphere; and, if we are to believe Pliny, Anaximander was the first to point out that the circle in which the sun moves is oblique to the circles in which the stars move about the poles. Other irregularities were in process of time discovered, and the celestial mechanism by which they were explained grew proportionally in complexity. The wandering planets changed their course, moving now forwards and now backwards; and to account for this motion each was supposed to be placed on the rim of an invisible wheel, which revolves on its centre while it moves around the sphere. Such a wheel was called an epicycle. Then it was discovered that the motions of the sun and moon also were irregular, so that they too were placed on the rims of imaginary epicycles; while it was found that, for purposes of calculation, the same results were reached if—abandoning the epicycle—the sun were supposed to revolve in a circular orbit in which the earth does not occupy a central position, but is placed rather nearer to one side. Such an orbit was called an eccentric. Finally, as further anomalies and irregularities were discovered in the motions of the sun, moon, and planets, further extensions of the hypothesis of eccentrics and epicycles were rendered necessary until the master-mind of Hipparchus formulated and organised this system, the essence of which consists in the resolution of the apparently regular motions of the heavenly bodies into an assemblage of circular and uniform motions.

In those days, it must be remembered, circular motions were the only motions admissible; the idea of "such disorder among divine and eternal things as that they should sometimes move quicker, and sometimes slower, and sometimes stand still," was considered impious, "for no one," it was said, "would tolerate such anomaly in the movements even of a man who was decent and orderly." And thus there sprang up that complex system which gave rise to the celebrated saying of Alphonso X., of Castille, that "if God had consulted him at the creation, the universe should have been on a better and simpler plan."

That the history of early astronomic thought above sketched exhibits an advance from the vague to the definite, and from the simple to the complex, while it shows also an increase in integration and dependence of parts, cannot, I think, be for a moment doubted. And the views of Hipparchus, as developed by Ptolemy, may perhaps be looked upon as the culmination of the evolution of the geocentric idea, while the later history of Astronomy exhibits the evolution of the heliocentric idea.

A scientific theory in some respects resembles an organism, and especially in this—that it must harmonise with its environment or die. The environment of theory is fact. So long as a theory is in harmony with the known facts of Nature it can exist, the development of a theory being its modification in accordance with newly-discovered facts. But when the plasticity of a theory ceases,—when it refuses to accommodate itself to fact,—its days are numbered, and it must give place to a more fortunate rival in the struggle for existence. In this way the Earth-centre theory of our system, organised by Hipparchus and developed by Ptolemy, had to give place to the Sun-centre theory, foreshadowed by Pythagoras and worked out by Copernicus. As Whewell truly remarks, so long as the *positions* only of the heavenly bodies were considered, the hypothesis of Hipparchus is a close representation of the truth ; but when once the processes of measurement gave sufficiently accurate results with respect to the *distances* of these bodies, the theory and the environing facts were out of harmony, and the theory was doomed. And when Galileo discovered, with the newly-invented telescope, in the system of Jupiter and his moons, a model of the Solar System ; when he found that Venus, in the course of her revolution, assumes the same succession of phases which the moon exhibits in the course of a month ; and when Kepler observed the transit of Mercury, and Horrox the transit of Venus ; then the fate of the old hypothesis was sealed, and the success of the new theory was secured.

Copernicus, however, retained the conception of circular motion, and the consequent existence of epicycles. But the idea of epicycles, like the geocentric idea, ere long ceased to be in harmony with the environment of fact. Kepler, we are told, attempted to reconcile the theory of Mars to the theory of eccentrics and epicycles, the event of which was the complete overthrow of that hypothesis, and the proposition in its stead of the theory the central truth of which has long since been abundantly established, that the planets move in ellipses. And this, be it noted, was a substitution of a more complex and integrated kind of motion for a combination of more simple kinds of motion.

As we now know, indeed, the motion substituted is even more complicated than Kepler supposed. For not only does the ellipse itself revolve slowly round the sun, but its shape undergoes change, being sometimes more nearly circular than at others, while, at the same time, the plane of the planet's motion oscillates about a mean position. Thus it comes about that the ellipse which accurately represents the planet's motion at one part of its course does not accurately represent that motion at another part of its course.

But the indefatigable industry of Kepler, besides establishing the elliptical theory, led him to the discovery of two other fundamental laws of formal astronomy,—that the line drawn from sun to planet sweeps over equal areas in equal times; and that the square of the time taken to describe a planet's orbit, divided by the cube of its mean distance from the sun, is a fraction which is the same for every planet of the system. And these formed the bases for the grand generalisation which was inevitably to follow.

For the next great step in astronomy was Newton's splendid induction. So gigantic was the onward stride then made—a stride without parallel in the history of Science—that it seems at first sight impossible to reconcile it with the gradual advance implied in a development by Evolution.

But a closer study of history makes evident the parallel but imperfect generalisations which were simultaneous with this more perfect and exact generalisation. While Newton at Oxford was pondering on cosmical gravitation, Borelli in Florence was publishing his theory of the " balancing of the planets," arising, as he conjectured, from the equality of an " appetite for uniting themselves with the globe round which they revolve," and the " tendency to recede from the centre of revolution." While Newton was preparing his " Principia," Huyghens, Wren, Halley, and others seem to have possessed a general idea that the attractive force exercised by the sun varies inversely as the square of the distance from the centre. Hooke even went so far as to claim priority in publication to Newton himself. And it is undoubtedly true, as Whewell points out, that Hooke's *assertion* was prior to Newton's demonstration. Francis Bacon, again, had not only speculated on the mutual attraction of the particles of matter, but devised an experiment to ascertain " whether the gravity of bodies to the earth arose from an attraction of the parts of matter towards each other, or was a tendency towards the centre of the earth." But these were but foreshadowings of the truth. It remained for Newton to demonstrate that the same law—that the attraction is directly as the joint mass of the attracting and attracted bodies, and inversely as the square of their distance asunder—holds good for the sun's attractive influence on different planets and on the same planet in different parts of its orbit ; for the earth's attractive influence on the moon, and on bodies near the earth's surface ; for the mutual attraction of sun, moon, earth, planets, and satellites on each other ; and for the attractions of the individual particles of which these masses are composed.

The bearing of these well-known facts on the theory of Evolution will now be evident. Not only does the conception of Universal Gravitation exhibit a great advance in the ideas of the inter-dependence of the members of the solar system,--not only does it show an onward stride in the integration of our knowledge, but it displays also a vast increase in the orderly complexity of our views of that system by introducing definite conceptions of matter and force in addition to those of motion and distance. And here we might well leave the history of Astronomy, satisfied that it has afforded ample illustration of the law under consideration. But for the sake of rounding off the argument attention may be drawn to the advance in the traits which characterise evolution implied in the nebular hypothesis of Kant and Laplace, which affords a not improbable conception of the mode of development of our system, and implied also in the results of modern spectroscopic researches, which teach us that the chemistry of the sun and stars is not dissimilar to the chemistry of the earth, and which have raised the physics of the sun to the rank of an independent science. When we take into consideration, too, the conceptions concerning stellar distribution, started by Wright, developed by William Herschel, and now criticised and opposed by Proctor, and add to these the results concerning the motion of the sun through space, the velocities of motion of certain stars determined by Huggins, and the "drifting" of star groups inferred by Proctor, we shall not lack instances of advance in definiteness of knowledge, in inter-dependence of ideas, and in complexity of our total conception of the phenomena of the heavens ; while the labours of Schiaporelli, Huggins, Donati, Lockyer, and Sir William Thomson, the results of which point to an intimate connection between nebulæ, comets, meteorites, and falling stars, bring into view a proportionate advance in the integration of that knowledge.

If now we pass from our knowledge of the solar and sidereal system to that of the crust of the earth, we shall find many facts which point clearly in the same direction.

A few of these may be noticed. When Lister, in 1683, proposed to the Royal Society that maps of the soils and minerals of England should be made, he probably was not aware of the existence of continuous layers of rock or *strata.* It was perhaps Hooke who first conceived the idea of " *raising a chronology* " out of the records of the rocks, while to Woodward has been attributed the earliest definite enunciation of the continuity of strata. We shall not therefore be far wrong in saying that, previously to the middle of the seventeenth century, the prevalent ideas of the structure of the earth's crust were altogether general, disconnected, and indefinite. From such views have been evolved, by gradual stages, the special, connected, and definite conceptions of modern geologists ; for whatever may be said of geological theory, there can be no doubt that the tabulation of British strata may be thus characterised. And it is only necessary to point to the recently-established connection between sedimentary, metamorphic, plutonic, and volcanic rocks, to show clearly the marked integration which has gone on in our knowledge of the crust of the earth. Nor do we find a different result when we turn from descriptive geology to causal geology. Although Ray, in 1692, " enlarged upon the effects of running water upon the land, and of the encroachment of the sea upon the shores," yet when Buffon, half a century later, maintained that " the waters of the sea have produced the mountains and valleys of the land," his attempts at a natural explanation of the origin of the features of the earth were so far out of harmony with the accepted tenets of his age that he was politely requested, by the Faculty of Theology in Paris, to recant. When now we contrast the total ignorance of, or the wilful blindness to, the action of natural causes implied by this act, with the view generally accepted to-day, owing to the strenuous advocacy of Hutton and Lyell and their successors, that every feature of the earth's surface, and every record buried in the rocks, is the result of causes similar in the main to those now in action,—when we compare the views of Woodward or Whiston (dealing as they do with universal deluges and comet-tails) with the views of Ramsay or of Judd,—we shall not fail to see a marked advance in all those special traits which characterise evolution. And now the broad generalisation that, with trifling exceptions, all geological action is due to the antagonism of sun-heat and earth-heat shows an integration of knowledge that can scarcely be carried further.

If now we turn from the history of our knowledge of the

earth's crust to the history of our knowledge concerning the constitution of the matter of which that crust is composed, we shall obtain further exemplification of the Law of Evolution. Among the early Greeks we find prevalent the doctrine of the four elements, Earth, Water, Air or Steam, and Fire, of which all bodies were supposed to be composed. But, as Dr. Roscoe points out, these terms denoted rather general properties than particular substances. " Thus earth implied the properties of dryness and coldness; water, those of coldness and wetness; air or steam, wetness and heat; fire, dryness and heat. All matter was, indeed, supposed to be of one kind, the variety which we observe being accounted for by the greater or less abundance of these four *conditions*." Here, then, we have a theory which is sufficiently vague, general, and indefinite. Somewhat similar, but more advanced, are the views of the Arabian alchemist, Geber, according to whom " the essential differences between the metals are due to the preponderance of one of two principles, mercury and sulphur; of which all the metals are composed." According to him " the noble metals contained a very pure mercury, and were therefore unalterable by heat, whilst the base metals contained so much sulphur that they lost their metallic properties in the fire." To these two principles Basil Valentine subsequently added a third, which he called Salt: while Lemery, calling these the active principles, added two more, which he termed passive, namely, water or phlegm and earth. The first philosopher, however, who seems to have grasped the idea of the distinction between an elementary and a compound body, was Robert Boyle; and this conception cannot but be regarded as an immense advance in descriptive chemistry. But the theory which had the most marked influence on the history of chemistry in the seventeenth century was that which was started by Becher and developed by Stahl, and which is known as the Phlogistic theory. When magnesium wire is burnt in the air a white solid is formed. We now know that this is the result of the combination of the metal with the oxygen of the air; but, according to the Phlogistic theory, this white substance results from the fact that the metal has lost a "combustible principle" which Stahl termed *phlogiston*. It was upon this hypothesis that Black, Priestley, and Cavendish—the founders of exact chemical science—worked, and it was their labours which afforded the observations which resulted in the complete overthrow of that theory. For the Phlogistic hypothesis, like the geocentric conception of the Solar System, was ere long found to be out of harmony with the facts, and was therefore worsted in the struggle for the Survival of the Fittest. It was found that chemical substances during combustion, instead of losing weight by the abstraction of phlogiston, gained in weight, the products of combustion being heavier than the body burnt. It was in vain to try and " bolster up the old theory " by the hypothesis that phlogiston is so light in its nature that it makes the bodies with which it combines lighter than they were before, though this view seems seriously to have been entertained. When Lavoisier found that mercury, on being heated for a long time in the air, absorbed oxygen and increased in weight, and that when the material formed was subsequently heated it lost weight and gave up just so much oxygen as it had before absorbed, the Phlogiston theory was doomed, and the science of chemistry was placed on its true basis; while at the same time, and by the same man, the foundation of quantitative chemistry was laid by the distinct assertion of the indestructibility of matter.

But we cannot say that a true quantitative theory of the chemical constitution of matter was in existence before Dalton, in 1803, published his first table of atomic weights, and showed "that elements combine in *definite* proportions; that these determining proportions operate reciprocally; and that when, between the same elements, several combining proportions occur, they are related as multiples." This conception of definite combination by *weight*, which has been termed " the pole star round which all other chemical phenomena revolve," was not long afterwards supplemented by Gay-Lussac's law of definite combination by *volume*, which received an explanation in 1811 at the hands of Avogadro, who assumed that equal volumes of all substances, when in the state of gas and under like conditions, contain the same number of molecules,—an hypothesis which is borne out independently by both chemical and physical considerations.

When we add to the advances in chemical science already noticed Sir Humphry Davy's discovery of the compound nature of the alkalies; the determination, by Dulong and Petit, that the atoms of many of the elements have the same capacity for heat; the proof, by Berzelius, that organic bodies obey the same laws as inorganic substances; followed by the artificial production, by Wöhler, of urea, a material before supposed to be exclusively the product of life: when we consider the overthrow of the dualistic theory, that salts result from the union of an acid oxide with a basic oxide, by the modern view that they result from the displacement of the hydrogen in an acid by a metallic or other radical ; and when the reader is reminded of the advances implied in the terms atomicity, isomorphism, allotropy, isomerism, diffusion, dialysis, crystalloid and colloid, and many others, he will admit that, though chemical science yet awaits its Newton, it exemplifies a continuous advance in definiteness, dependence of parts, orderly complexity, and integration.

Without staying to point out that our knowledge of the physical constitution of matter exemplifies evolution not less clearly than that of the chemical constitution of matter, attention may now be very briefly drawn to the conceptions of Force. To illustrate the indistinctness of ideas on this subject which prevailed in comparatively early times, I may give a quotation from Pliny, cited by Whewell :—" What," he cries, " is more violent than the sea and the winds ? what a greater work of art than a ship ? yet one little fish (the *Echineis*) can hold back all these when they strain the same way. The winds may blow, the waves may rage; but this small creature controls their fury, and stops a vessel when chains and an anchor would not hold it ; and this it does not by hard labour, but merely by adhering to it." " In our own memory," he continues, " one of these animals held fast the ship of Caius, the Emperor, when he was sailing from Astura to Antium." Now although it may be objected that the Greeks, long before Pliny wrote, had some definite mechanical notions, and that this quotation gives an over-estimate of the amount of indefiniteness of early knowledge on this subject, still the fact remains that so indefinite was this knowledge before the time of Galileo that the first law of motion—that a moving body, if entirely left to itself, will continue moving in the same direction and with the same velocity—had not been distinctly grasped. And even when this highly abstract law *was* at length grasped,—when it was seen that under ordinary circumstances (on our earth, for example) the onward motion of a body is stopped by the resistance afforded by the air, or by solid or liquid bodies,—even then for a long time no account was taken of the motion thus apparently lost ; and it was reserved for modern times to show that the force which this motion implies is not destroyed by the resistance, but only takes on a new form in the molecular motion of heat, developed in the resisting and resisted bodies. And this conversion of the motion of a mass of matter into molecular motion was subsequently found to be only a particular instance of the conservation of energy, the discovery of which law, considered as it is by some the chief glory of modern science, exemplifies clearly the results of an evolution. As a quantitative law it has increased the definiteness of our knowledge ; as a generalisation which comprehends *all* forms of energy, it has increased the integration of our knowledge ; and, implying as it does the transformations of the various forms of energy into each other, it exemplifies also an increase in dependence and orderly harmony.

Having now given some examples of the evolution of certain branches of scientific knowledge (lack of space precluding further exemplification from the biological, sociological, and psychological sciences), the fact must be pointed out that in no case is the evolution of one branch of such knowledge independent of the evolution of other branches. The term *branch*, indeed, suggests the conception of a general tree of knowledge of which they are the offshoots. But the special sciences might perhaps be more profitably likened to the organs of the animal organism.   For not only do the several organs progress together with the progress of the whole of which they are the parts, but each organ ministers in its own fashion to the welfare of the whole and of the other organs.   Mr. Herbert Spencer has given many examples of the dependence of the advance of one branch of science upon the advance of other branches. It is therefore unnecessary here to do more than indicate the nature of the evidence.   Where would Astronomy be now, it may be asked, without the advance in Optics implied in achromatic telescopes ; without the discoveries in Mechanics of the laws of motion, and of the anachronism of the pendulum ; without the determination of the specific gravity of the earth, and the measurement of a degree on the earth's surface ?   There is indeed scarcely a branch of Science upon which Astronomy does not call for aid.   In addition to those just mentioned she relies on Atmospheric Physics for tables of atmospheric refraction ; upon Chemistry for photographic processes ; upon Electricity for various recording instruments ; and upon Psychology for the personal equation,—the time which elapses between seeing and registering which varies in different individuals.   These facts are sufficient to exemplify the inter-dependence of the sciences, and they form not a tithe of the number which could be adduced.   We have only to trace the interaction of terrestrial and astronomical physics on Mathematics, to watch how new problems in physics called forth new mathematical processes, which processes enabled further physical advance, and thus led to fresh problems ; we have only to consider how Chemistry has aided electrical science, and been aided in turn by that science; we have only to observe how Geology has profited by the advance of Biology, and at the same time has aided in solving important biological problems; we have only, in a word, to study the history of Science in a scientific spirit, to see how completely the organs of the body scientific are dependent on each other, and are bound together into a definite complex whole.

Let it then be clearly noted that the advance in scientific knowledge is not merely, as some suppose, an increase of the mass of accumulated facts. It is this, but it is much more besides. Just as the more extended view which the mountaineer obtains as he rises above the valley does not charm so much by the multiplicity of objects as by the connection which is disclosed among them, so, too, the more extended view which the philosopher obtains, as he climbs the hill of Science, does not owe its value so much to the number of facts accumulated as to their definite organisation.

In conclusion, one or two general facts may be pointed out. There is no more striking trait in the evolution of knowledge than the fact that not unfrequently the same discovery is made almost simultaneously by different workers labouring altogether independently of each other. Instance the theory of Natural Selection elaborated simultaneously by Darwin and Wallace; instance, again, the independent liquefaction of oxygen by MM. Pictet and Cailletet. Or again, perhaps, we find that the same law or theory is advanced by several men as a speculation, and by one master mind as a demonstration. This may be said to have been the case in the discovery of the Law of Universal Gravitation. Now these facts are in full accordance with the theory that the development of our knowledge is an evolution. When the environment of ascertained fact has reached a certain definite state, its influence inevitably calls forth the development of a new theory which shall be in harmony with all the conditions. In the minds of Newton's contemporaries the environment of accumulated facts called forth general conceptions more or less in harmony with these facts; but in the master mind of Newton not only was the environment of fact more extended, from his powerful grasp of intellect, —not only was that environment more pressing, from his constant habit of earnest thought,—but the conceptions were more definite, from the extraordinary depth of his mathematical insight. The result was the production of a law and a book which have been the wonder of all after-time.

Newton's contemporaries were at a high enough level of thought to accept his generalisation, which thereupon became part of the environment which induced subsequent though minor generalisations. But this has not always been so. Sometimes the conception of a master mind has fallen amid conceptions having so little in common with it, that its influence has not been felt until the subsequent advances of knowledge have caused its re-development, and called the attention of the world to a genius who lived before his time. Such was the case with some of the generalisations of Archimedes, and of Roger Bacon, to mention no other names.

Finally, we may notice the fact that just as in the organic world we have the highly organised human being existing side by side with the lowly organised *entozoon*, so too we have, in the world of thought, conceptions in some sort in harmony with the grandeur of the Cosmos side by side with conceptions moulded to the meanest and most trivial facts. But there is nothing here at variance with the theory of Evolution. The human being and the entozoon are each more or less in harmony with their several environments, and any advance in the development of each is such as to bring it more clearly in harmony with all the conditions. So, too, the conceptions of the philosopher and the clown are each more or less in harmony with the environing facts by which they are respectively surrounded; and here too, in each case, any advance in development is such as to bring the conceptions more closely in harmony with the surrounding facts.

To this parallel between the organic world and the world of thought we may add another. It is now generally admitted that the evolution of the individual is a condensed epitome of the evolution of the species to which that individual belongs. The evolution, for example, of an individual frog from the undifferentiated egg to the complex adult epitomises the evolution, through long ages, of frogs from simpler forms of life. So, too, does the evolution of the conceptions of the individual philosopher epitomise the evolution of philosophic thought in general. Both in the individual and in the race the discovery of law is itself subject to law, and, if there be any truth in the views above advocated, the law to which it is subject is the Law of Evolution.

II. HABITS AND ANATOMY OF THE HONEY-BEARING ANT.

By Charles Morris.

IN the February number of the "Journal of Science" were published certain results of Dr. McCook's investigations on the habits of the honey-bearing ants. Since the period in which these observations were made he has had under his constant supervision an artificial formicary of these ants, and has made some further interesting communications in regard to them, of which the following is a condensed sketch.

The most striking points of these communications relate to two particulars, one bearing on the sympathy, or spirit of beneficence, in the ants ; the other relating to their anatomy. The first of these particulars is in the same line of research as that followed by Sir John Lubbock, in his observations on the behaviour of ants towards captive friends and enemies, in which he discovered that, while the ants were full of hostility against individual foes, they showed no evidence of sympathy for friends in trouble.

Dr. McCook's observations lead to the same result. In making his artificial formicaries he simply pressed a quantity of earth compactly into a glass bottle, placed his captured colony upon the surface of this earth, and left them to establish their new home in their own way. It was not long before the workers were busily engaged in the business of excavating ; galleries were speedily formed in the earth, and the mined-out materials deposited upon the surface. But in this active labour the comfort of the poor honey-bearers was utterly ignored. They lay helplessly where they had been dropped, and were treated by the other ants as if they had been so many lifeless impediments to their work. For, instead of making a detour around them, the workers went straight forward, clambering over any of these helpless magazines of sweets that lay in their path, and even dropping the pellets of earth which they brought out from the excavations upon and around them, until some of the honey-bearers were almost buried by their own heedless friends.

There seemed here, indeed, a double lack—a lack both of

sympathy and of intelligence. In the new circumstances under which these ants were placed, their strongest instinct —that of excavation—at once displayed itself; but they seemed quite incapable of handling the question of the proper disposal of their food-magazines, under such unusual conditions. Fortunately for the honey-bearers they were not quite helpless, despite the great weight which they bore. They were able to drag themselves slowly forward, and in time they seem to have all managed to reach the underground habitation without aid from their friends. At least such was Dr. McCook's belief, although he did not observe the complete process. There was no evidence of any assistance from the workers.

Observations made in the natural nest of these ants gave different results. When they were dug into for the purpose of examination, and ants of every age and caste exposed to the light of day, the workers made the most vigorous efforts to carry all their helpless charge into the unbroken galleries of the underground city. The eggs, the young, and the honey-bearers were alike carried to this place of safety, with the most devoted energy. Whether the honey-ants were replaced in their favourite position on the ceiling of their chambers is not so certain. In our former article the workers were spoken of as so replacing them. But this statement was not derived from actual observation, but from the fact of their being often found in this position on further excavation.

There is some reason, however, to believe that they may have regained this position by their own efforts, to judge from facts observed in the artificial formicary. As already mentioned, the honey-bearers are not quite helpless: they have the full use of their legs, though their movements are necessarily made at a disadvantage, from the angle into which the head and thorax are thrown by the swollen condition of the abdomen; yet they have been observed to move by their own efforts, and it is not impossible that they themselves regain their favourite position on the ceiling of the nest. The reason of their preferring this position may be from the uncomfortable attitude which they are forced to assume on the floor of the nest.

It may seem that an intense muscular effort would be required to sustain their great weight in this position. That ants, and insects generally, are excessively muscular, as compared with the larger animals, is well known. And the honey-bearers are more muscular than ants generally, their legs being simply bundles of powerful muscles. But it is

rather difficult to conceive how muscular effort can be
brought to bear to overcome the action of gravity in this
position, unless by some clasping of the terminal hooks of
the feet around the excrescences of the rough ceiling. It
seems more probable that support is gained by the action of
the pulvillus, or sucking-disk,—that cushion-like organ
which ants possess in common with so many other insects,
and which enables this class of animals to walk on smooth
surfaces in a vertical position. If the honey-bearer thus
brings suction to bear to aid in its support, it simply avails
itself of the natural force of atmospheric pressure to overcome the force of gravity, and has no occasion for muscular
exertion as an aid in sustaining it.

The main argument against any aid from the working
ants in replacing fallen honey-bearers comes from observations of their behaviour in the artificial formicary. From
various causes these ants occasionally fell to the floor of
their chambers, often alighting in the most uncomfortable
positions. In some instances they lay on the base of the
rounded abdomen, with the head and thorax upward, looking
not unlike the old-fashioned flat-bottomed Dutch dolls; in
other cases they lay head downward, quite buried under
their load of sweets. Yet, uncomfortable as these positions
must have been, the working ants made no efforts to replace
them in a more desirable position, much less to carry them
back to their elevated dormitory. They would attend to
their wants so far as the need of cleansing them was concerned, but no further.

To discover if the honey-bearers were satisfied in these
unnatural postures, Dr. McCook tried the expedient of introducing a straw or broom splint to such as were within his
reach. In every instance he found them to vigorously grasp
this substance, and to cling to it until they were drawn into
a more natural posture.

This evidence of lack of sympathy with friends in trouble
was strengthened by another observation. On one occasion
a large piece of compact earth fell upon one of the honey
ants, burying and compressing its abdomen, and leaving
only the anterior portions of the insect visible; yet the
other ants made no effort to relieve it from this difficulty,
which they could readily have done, but moved about it
with the most supreme indifference to its unfortunate condition.

An observation of some importance in respect to the
question of ant intelligence is that regarding the demeanour
of the ants towards dead honey-bearers. In this case it is

their habit to separate the head and thorax from the honey-bag, burying the former in the fixed cemetery which these ants usually establish in the earth outside their nests. But though the honey-bag remains, full of its sweet contents, the ants—either from respect for the dead or from lack of mental power to devise a new means of getting at its honeyed freight—seem to make no effort to penetrate its transparent wall. This is singular, in view of the avidity with which they will lick up the smallest portion of sweet food offered them in any uncovered condition.

We may supplement our remarks on these evidences of lack of personal sympathy and of benevolence in the honey-bearing ants with some description of their anatomy. In our former article it was said that the whole abdomen appeared to be occupied by the honey, its organs seeming to be obliterated, so that only a thin transparent skin remained. But anatomical observation shows that this external appearance does not give the true facts of the case. All the abdominal organs remain, but so strangely distorted and compressed as to be almost imperceptible. The fact is that any of these ants may, if necessary, be converted into a honey-bearer, and that the worker, when on her way home with her abdomen distended with the fruits of her nocturnal labour, has made a step towards the condition of the fully-developed honey-bearer.

It need scarcely be repeated that the intestinal tract of the abdomen of the ant is possessed of three special expansions, named respectively the crop, the gizzard, and the stomach, from their relations in function to the similarly-named organs in birds. Of these the crop, into which the œsophagus immediately opens, is the recipient of the honey. As its stores increase, by continual additions, it expands more and more, pressing outward the extensible walls of the abdomen, and compressing the remaining portions of the intestine into a smaller and smaller space.

In a fully-laden honey-bearer the crop has become so expanded that it fills nearly the whole interior of the greatly dilated abdomen; the dorsal vessel, or heart, being compressed and flattened against its upper wall; while the gizzard, stomach, and intestine are similarly compressed against the posterior wall. The compression of these organs is so great as seemingly to preclude their functional action, the stomach appearing to be quite incapacitated for its normal office of digestion. Yet the continued vitality of the ant is sufficient evidence that alimentation must still exist; and as it is not at all probable that the crop could assume the function of a digesting organ without injury to its stores, it seems as if some of the liquid food must make its way into the stomach and intestine, despite their extreme compression, and be there prepared for aliment.

It is, in fact, a puzzling question. Dr. McCook is inclined to think stomach digestion, in some instances, impossible. But the continued vitality of the ant seems to render it necessary, despite its apparent impossibility.

III. THE "LAWS OF EMPHASIS AND SYMMETRY."

By J. W. SLATER.

THE presence and the distribution of colour in the organic world have not passed unnoticed in the "Journal of Science." Sexual selection, the need of concealment, the influence of light, of temperature and diet, and even certain "waves of beauty," have all been put forward as affording partial or complete solutions of the questions involved. No one, however, will contend that the explanations given are fully satisfactory, or indeed cover more than a corner of the subject. The general rule with those naturalists—and unfortunately also un-naturalists— who have attempted to show why an animal displays some particular colours in some particular design is to overlook all phases of the subject save the one which has first attracted their attention. But are the chemico-physical, the vital, and the utilitarian theories of animal colouration necessarily contradictory and mutually destructive? I think not. Suppose it is shown that an animal of some given colour harmonises better with its surroundings than if its hues were modified; it may then be argued that, on the principle of the Survival of the Fittest, individuals of the species in question will have been able to escape the notice of their enemies, and to leave posterity the more successfully, the more decidedly they possessed such colour. But it is surely not contradictory to this explanation if we point out some constituent in the food of the species from which the colour may have been derived. On the contrary, the two principles seem to support each other. To take an especial instance :—It has been explained that larvæ which feed upon the leaves and stems of plants in open daylight assume a green colour, like the objects by which they are surrounded; whilst those which prey upon roots, bark, and wood, and especially such as pass their existence in the dark,—whether under ground or in the interior of fruits, seeds, or trunks of trees,—are of a pale grey or livid colour. All these facts are doubtless intelligible on the hypothesis of " protective colouration." But they are also by no means antagonistic to a chemical explanation. The present writer has pointed out, on a former occasion,* that larvæ feeding upon leaves may readily be coloured green by retaining in their tissues, in an undecomposed state, the chlorophyll con. tained in their diet. On the other hand, larvæ which feed upon substances devoid of chlorophyll may be considered much less likely to display a green colouration. Thus the caterpillars of *Cossus ligniperda* and of *Zeuzera æsculi* are not green ; but they have not been consuming a green pigment. The same rule holds good with the larvæ of the Elateridæ, . the Buprestidæ, Lamellicornes, Longicornes, &c. It may surely be submitted that before a mimetic or otherwise protective colouration can be developed by the agency of Natural Selection the material for the production of such colour must be present. It is, of course, admitted that many caterpillars consuming chlorophyll are not green ; but in these cases their colours are such as may be formed by the decomposition or modification of chlorophyll,—colours such as we see actually produced in the leaf when fading. The difficulty remaining is to know why in some species the green pigment remains unchanged till the insect reaches maturity, and why in others it is immediately decomposed, or at least disappears after some time. Why is the green colour in *Chærocampa elpenor* sometimes absent in the larva, and why does it appear in the perfect insect ? We may say that the chlorophyll is altered by oxidation, and restored to its original condition by a process of reduction. But why do these changes take place in some species and not in others whose vital conditions are, as far as we can judge, essentially identical ? At the same time, when we seek to account for these peculiarities of colour and design on the principles of mimetism and protection, the same questions still lurk in the back-ground : how are the colours requisite

* Quarterly Journal of Science, vol. viii. (1878), p. 48.

to make a species inconspicuous elaborated, and how are they despatched to the parts where we find them?

Certain interesting and novel ideas on these subjects have been recently brought forward by Mr. A. Tylor, in a Paper read before the Anthropological Society.* He considers that the forms and decorations of organised beings seem to be regulated by laws which he calls those of Emphasis and of Symmetry.

Symmetry, of course, calls for no explanation, as it is generally known that the two lateral halves of vertebrate and articulate animals are substantially alike, both in form and colouration, exceptions in the latter respect being scarcely more common than in the former, save among certain domestic mammals, such as horses, cows, dogs, and cats, which, when not concolorous, frequently show a different design on the left and the right side of the body.

Emphasis, Mr. Tylor defines as the marking out by form or decoration of the important parts or organs. He considers that the emphasised functional decorations group themselves into two classes, and that these classes coincide in their occurrence with the two great divisions of the Vertebrata and the Invertebrata. In the former class the emphasised ornamentation is axial, being the outward expression of the vertebral column with its appendages. In the Invertebrata the decoration tends to follow the outline of the animal, and so developes borders.

These generalisations deserve a careful examination. In the Invertebrata, or at least in the Articulata, borders marking the outline of the body, or of some particular part, or running parallel with such outline, are of very frequent occurrence. The margins of the wings of butterflies are very often coloured differently from the disk of the wing, and there are often repetitions of the outline in the form of bands and lines of spots as we proceed inwards. Such patterns occur to a great extent in genera and families by no means closely connected, and have often excited the attention and the comments of naturalists.

In considering these designs, which seem, so to speak, based upon the outline of the wings, the idea at once suggests itself that they may be—in part at least—explained by the well-known phenomena of capillarity and of the varying

* "On a New Method of expressing the Law of Specific Changes and Typical Differences of Species and Genera in the Organic World, and especially the Cause of the Particular Form of Man."

diffusive power of different colouring-matters. Everyone must have observed that if a coloured liquid, not too intense in its shade,—*e.g.*, wine or tea,—is allowed to fall upon some white fibrous material, the spot produced will be darkest at the edges, whilst in the centre the linen or paper will scarcely appear stained. If, again, several colours differing in diffusibility are dissolved in the same liquid, they may be to a certain extent separated from each other on the same principle. Thus in the earlier days of the manufacture of the aniline dyes, samples were often met with which— from inattention or from imperfect purification—consisted of two or more colouring-matters mixed together. To detect such mixtures it was usual to place a drop of the solution on a sheet of white blotting-paper, when the different colours appeared as concentric rings. If we suppose the tinctorial matters present in the fluids of the butterfly towards the close of its pupa-life being liberated in the porous tissues of the wings, we can form some remote conception of the manner in which these patterns are produced.

It will perhaps, however, be asked by some naturalists whether the parts where these borderings chiefly occur— *i.e.*, wings and elytra—can fairly be pronounced " important parts " in the same sense as is the central axis in vertebrate animals ? Mr. Wallace* points out as a remarkable fact that black spots, ocelli, and bright patches of colour are generally on the tips, margins, or disks of the wings, at a distance from the vital organs, and considers that this position of the more conspicuous parts may be a protection to the insects.

In vertebrate animals Mr. Tylor considers that empha- sised ornamentation has what he calls an axial character, " being the outward expression of the central axis or verte- bral column with its appendages." It is perfectly correct that in a vast number of vertebrates the head is prominently decorated, and in many mammals and reptiles the back is adorned with stripes, or chains of spots, from which other stripes branch off at right angles, or nearly so, to the main axis of the body. In birds I believe a dorsal stripe, if it occurs at all, is very rare. This design, in which the con- trasting colours are often very striking, is referred to in the " Journal of Science,"† and is there traced much more widely among Articulata than among vertebrate animals.

* " Colours of Animals and Plants," in Macmillan's Magazine, No. 215, p. 402.
† February, 1879, p. 196, and July, 1879, p. 496.

The central line, indeed, is often wanting, but the lateral
bands take their origin where such a line would run, or
intersecting it from one side to the other. Such lateral
bands, with or without a distinctly marked axial line, occur
on the abdomen, are met with in many Homoptera, and in
certain Lepidoptera, such as Sphingidæ, Chelonidæ, &c.
Among Coleoptera, where the transverse stripes occur in
groups too numerous to mention here, they run across both
elytra. It would seem, therefore, that among invertebrate
animals both these kinds of decoration prevail, whilst among
vertebrates we find the axial designs alone.

Mr. Tylor's views, however, call for further investigation.

IV. INSTINCT AND MIND.*

By S. Billing.

AT page 339, in the May issue, appear some observations by R. N. M., on the article "Instinct and
Mind" in the April issue (p. 288).

My object in framing the paper was to set forth distinct-
ively the views I hold of the differences, so to speak, of the
mentality apparent in men and animals. I do not object to
a fair discussion of the propositions which therein appear,
as my sole object was to elicit the truth, and in a measure
to answer the unflinching materialism with which the *savans*
of the day envelope the subject. On the assumption that
Descartes had said so, animals are pronounced to be automata. This authorisation of Descartes for the automatic
theory the late George Henry Lewes wholly denied (*vide*
"Physical Basis of Mind"). The halt is not made with
animals; men also are pronounced to be automata. The
logical corollary certainly would be that if animals are
automata, men are also. All the facts, physical and psychical, set forth on this subject lead directly to a contrary
inference. The objection I have to R. N. M.'s note, to say
the least, is that his interpretation of my views is wholly

* Sequel to article at page 228,

erroneous, and that the only answer to the phantom he has set up should be one of negation.

When a subject is treated upon general principles, particularisations, unless for the purposes of illustration, are generally avoided ; and had I said, when speaking in illustration of the particular instincts of animals, as dogs, &c., that of course I do not mean that each individual has the exact sagacity of every other individual of the species,—had I adopted this course I should have felt that I was needlessly ignoring the common intelligence of your readers : but as there appears to be one among them who cannot, or will not, understand plain and terse reasoning, I will endeavour to particularise.

When I speak of the aptitude of varieties or species to work in a given direction, which I term tribal instinct, I neither say, infer, or mean to say, that there are not different exemplifications of the particular potence or sagacity. If it were not so, what would be the use to animals of the discriminative capacity (instinctive mentality) I ascribe to them ? Dogs and other creatures possess character in expression and in act. I do not say, or mean to say, that all have the same aptness, *i.e.*, that the characteristic which is applicable to the variety has received the same development in each individual of the variety or species. Dogs and other creatures vary as much as men do, the difference to be found with them being one of degree ; *e.g.*, all men possess the power of abstract reasoning, but all men do not exemplify it in the same degree. The varieties of birds build their nest in the same pattern, but all are not equally neat in the display of the constructive faculty.

In the article in question, when speaking of tribal instincts, I generalised, and it seemed to me the cultural power which makes the difference between individuals of the same class would have been understood ; therefore, in the face of the unfounded assumptions of R. N. M., one of two things must be assumed,—either that I am wanting in common observation, or that he is incapable of comprehending the argument he attacks. As to dogs, every person, not sportsmen only, knows that there is a broad distinction between the sagacity of one sporting dog and another ; but this does not imply a variation in the characteristics of the particular species. One dog, according to its adaptability for its particular purpose, may be worth a large sum of money, whilst another dog with the same characteristics, but without the same development, may for the particular purpose be worthless. It is the same with shepherd's dogs

(collies). I neither asserted nor inferred that there was no
distinction between them in capacity; nor do I say or infer
"that every shepherd's dog is equal in intelligence to every
other shepherd's dog;" but I do say that one shepherd's dog
displays the same instincts as every other shepherd's dog.
All this is very puerile, and I am almost ashamed in being
compelled to write it.

The broad distinction I draw is that instinctive mentality
(the animal attribute) and abstract intellectuality (the attribute of man) are not the same, *not even in origin.* The one,
being derived from the senses, is shared in common with the
animal and animal man; the other, the intellectuality which
deals in abstractions, has no place in the sensuous expressions, and is possessed by man alone. Animal instincts
are developed in particular directions, and the characteristics
they exemplify run through the whole race, tribe, or species.
If it were not for this particular tribal development there
would be no distinction as class, variety, race, or species,
and all animation would be one indistinguishable mass or
confusion. I do not suppose that the peculiar characteristic running through the tribe has its expression or development in an equal degree in each individual composing it, as
I do not suppose that all men have the critical acumen in
an equal degree; but I do suppose that all men have a
critical acumen.

As R. N. M. appears wholly to have mistaken the drift of
the argument,—for in no way do I infer that the instinctive
mentality is equal in each member of the species,—perhaps
he will trouble himself by reading the article again, when
probably he may take a new view of the subject, and assure
himself that the author does not ascribe a mental equality
to the members of each species.

V. TUBERCULOSIS TRANSMISSIBLE THROUGH THE MEAT AND MILK OF THE ANIMALS AFFECTED WITH IT, WHEN CONSUMED BY YOUNG CHILDREN.

THE following Memoir, read before the Farmers' Club of the American Institute, by the President, A. S. Heath, M.D., has been courteously forwarded to us at the author's request :—

In 1865 Villemin proved, by repeated experiments, that it was possible to produce consumption in previously healthy animals. He found that finely-divided tuberculous matter, when introduced under the skin of rabbits and guinea-pigs, produced tubercles in three weeks in their lungs; thus proving from these experiments that *tuberculosis* should be classed as a specific infective disease, capable of being conveyed by inoculation, like small-pox or syphilis. Numerous pathologists have verified Villemin's experiment. It was also found, by Dr. Wilson Fox and Dr. Sanderson, that pneumonic matter, pus, putrid matter, &c., would produce disease in healthy animals, and transmit it through their meat and milk, to dogs, cats, hogs, and through milk to young children and animals to whom it had been fed.

Cows living under bad hygienic conditions, as in man, under similar conditions predispose to tuberculosis in themselves, and render their milk poisonous to children. In New York city most of the cows are diseased from this cause, and by being fed on unsound food.

In a future *paper* I shall pay my special respects to city cows and *cow stables*, and show, I trust, that the milk from diseased cows poisons thousands of city children, who are supposed to die from *cholera infantum*, when, in fact, they die from tubercles of the intestines, resulting in wasting diarrhœa. Consumption is infinitely more common in city-kept cows than it is believed to be, even by physicians. I recently called Prof. Chandler's attention to the unusual number of cows crowded into city stables; and from his taking copies of notes sent to me by Dr. James D. Hopkins, I feel assured that so efficient a sanitarian as he is will have these nuisances abated, and promptly too.

Tuberculosis prevails extensively among domestic animals over the entire globe, and especially in populous and crowded localities.   In Mexico 34 per cent of slaughtered animals supply tuberculous meat, and it is probable that the milk cows are affected to the extent of 50 per cent in the large towns.

Van Hertsen, of Belgium, found tubercles in all the tissues of an apparently healthy bull seven years old.   From these facts it is apparent that there is great danger in eating uncooked beef, for fear of contracting consumption.   The sources from which consumption are derived is now known to be infinitely more numerous than former pathologists supposed.

It is more dangerous to eat the milk of tuberculous animals than to eat the meat ; for the milk is seldom cooked, while the meat is almost always cooked.   Cooking is a most valuable sanitary measure.   Cows confined in dark, damp, unventilated city stables, become tuberculous eventually to the extent of 75 per cent.   Fleming says :—" For it must be borne in mind that there are few animals which have been kept for any length of time in cow-sheds, and fed and milked in the usual manner, which are not more or less phthisical : more particularly is this the case if the dwellings are bad."

The milk of tuberculous cows is of a poor quality, besides being liable to produce the disease.

Klebs has produced tubercles in rabbits, guinea-pigs, and dogs, by giving them the milk of diseased cows.

This milk given to young children produces catarrh of the intestines before the tubercles are deposited in the lungs. It is not often that the intestines of young children who die from what is supposed to have been *cholera infantum* are examined after death ; but doubtless tubercle of the intestines would be frequently found.

Garlach and others have demonstrated that the milk of tuberculous cattle will produce phthisis in creatures fed with it.

Fleming says :—" It is certain that *tuberculosis* is a somewhat common and a very destructive disease, among dairy cattle especially, and more especially those of towns." And *consumption* is one of the most fatal diseases of large cities, and doubtless from this cause.   Marasmus is undoubtedly largely attributable to diseased milk, and many thousands of children perish from *tuberculosis*.   The excessively hot weather of parts of July and August is productive of an irritable condition of the stomach and bowels of young and teething children, which condition acid, impure, or tuberculous milk greatly aggravates, and renders poisoning with diseased milk from unhealthy cows more common than it is popularly known.

Neimeyer says that the predisposition to consumption is strongest in persons of feeble and delicate constitution, and especially that children poorly nourished are most subject to the disease. The children fed on the milk of tuberculous cows must, of necessity, suffer in a twofold sense—from bad food, and poisonous food also. From a seventh to a fifth of all deaths are caused by consumption, and nearly half of the *post mortems* show the traces of nutritive disorders from which pulmonary consumption proceeds, and " consumption of the bowels " is the more frequent form of the disease in children, as a result of bad food and diseased milk.

It has always been my aim to be suggestive in my papers rather than exhaustive, and, as lawyers say, " I here rest my case."

VI. NATURAL SCIENCE AND MORALITY.

By S. Tolver Preston.

" I say that Natural Knowledge, in desiring to ascertain the laws of comfort, has been driven to discover those of conduct, and to lay the foundations of a new morality."—Huxley, *on the Advisableness of Improving Natural Knowledge.*

THE view that happiness must be the standard of mo-rality has recurred again and again, as if by inevitable logical sequence, to the leaders of thought in all time ; and this doctrine is so well in accordance with the most advanced modern ideas that it will not be our task to inculcate this maxim here, but rather to attempt to reconcile some of the difficulties which appear to beset its universal adoption as a standard of morality.

The grand difficulty that has stood in the way of this has been the opinion that the pursuit by the individual of his own happiness, or a regard to his own interests, clashes with the interests of others, tends to make the individual prey upon the rest of society, and is subversive of all harmony and concord; in fact that self-interest and selfishness are synonymous, and that by such a moral standard that desirable consummation, the greatest happiness of the greatest number, would be rendered an impossibility. The late John Stuart Mill, in his celebrated work "Utilitarianism," while fully recognising the worth of happiness as a standard of morality, was nevertheless probably led by the above-mentioned difficulty to advocate the maxim that each one was to make his own happiness *subservient* to that of the greatest number,—a dogma that must fail in practice, owing to the absence of logical incentive to carrying it out. Our task will be to show that, so far from the greatest happiness of the greatest number being inconsistent with each individual consulting his own happiness, this desirable consummation can only be attained by that means.

One hundred and sixty-six years have elapsed since Bernard de Mandeville argued that self-interest being the guide of action, " those creatures would flourish most which are least possessed of understanding; for the more they know, the more would their appetites to be satisfied *at each other's expense* be increased, and therefore the more would they war with and exterminate each other." Whence man, by reason of his understanding, would be least fitted to agree long together in multitudes.*

The Grand Jury of Middlesex of that day were seized with a panic: they seem to have feared that De Mandeville's theory, that society rested upon a fiction, was true, and therefore to have burned the book in which that fiction was exposed. The panic has not yet subsided. Many worthy people dread the theory of the Survival of the Fittest, because, while they recognise that the fittest are those who can best provide for themselves, they are still chained to the old error which supposed selfishness to be the ideal practice of self-interest.

Before the dawn of Political Economy there was some plausibility in the theory that the wealth of the individual could only be increased at the expense of his neighbour, and that consequently true happiness was only to be found in a

---

* " An Enquiry into the Origin of Moral Virtue," appended to the republication (1714) of " The Grumbling Hive." It is interesting to notice that although self-interest is recognised here as the incentive to conduct, there is a failure to reconcile it with order and stability in society, or only *half* the truth is recognised. This was the same with Hobbes (as related in the " Leviathan "), and with many others. For an admirable and lucid sketch of some of the more important systems of morality, the reader may be referred to LANGE's notable historical work " Geschichte des Materialismus " (of which we believe an English translation is now published).

" small and peaceable society, in which men, neither envied nor esteemed by their neighbours, should be contented to live upon the natural products of the spot they inhabit."[*] But now that we have to face the fact that a savage who lives solely on the produce of the chase is tolerably reckless of the life which requires some 78 square miles for its sustenance, while a Belgian clings to that which is supported on 2 acres,[†] we are driven to the inference that there must be some flaw in Bernard de Mandeville's conclusion. For the purpose of his argument De Mandeville, in analogy with Hobbs and others, took the wealth to be extracted from a given area as a constant quantity, left out of account man's labour, and estimated the happiness of the individual in any country by dividing the uncultivated (or natural) products of the country by the number of inhabitants. Political economists have reversed all this; they recognise labour as a source of wealth whose value varies with the intelligence and sociability of the labourer; so that the wealth of *each* individual may be greatly increased by co-operation of numbers.

If De Mandeville had been right in assuming the total wealth to be constant and independent of man, he would, of course, have been correct in the deduction that through self-interest each individual would increase his own wealth at the expense of his neighbour; but when the facts are known that in the most wealthy countries the proportion derived from natural or uncultivated products is almost insignificant compared with that which can only be obtained by the co-operation of numbers of individuals, it is certainly remarkable that some Utilitarians of the nineteenth century should fall into the error that the pursuit by man of his own self-interest would be synonymous with selfishness, or would tend to make him isolate himself from his neighbours, and prey upon them; whereas we may see, from the above considerations, that precisely the reverse of this may be true, or that sociability and co-operation may be in reality the highest forms of self-interest.

Nor can that purely passive selfishness which stops short of actual dishonesty (the ordinary selfishness of private life) be carried out in an intelligent society without great loss to the individuals who practise it. For it is an every-day occurrence for A, by relinquishing a small pleasure, to be able to render a large service to B; and when under such circumstances A does so sacrifice his own immediate smaller

---

[*] BERNARD DE MANDEVILLE, Fable of the Bees.
[†] LUBBOCK, Prehistoric Times, 4th ed., p. 607.

happiness, his action should be determined not by the dogma that, "between his own happiness and that of others, Utilitarianism requires him to be as strictly impartial as a disinterested spectator,"* but because he clearly perceives that if all agree to act similarly, all, including therefore himself, will be benefited. To-day A relinquishes a small pleasure, and B gains a great one : to-morrow B may do the same for C, C for D, and so on, until ultimately Z may sacrifice his own immediate smaller happiness for the greater happiness of A. Seeing, then, that by the practice of unselfishness each individual in our mutual-benefit society has succeeded in exchanging a smaller pleasure for a larger one, it seems but natural to describe unselfishness as self-interest, and it appears to be only by a most unfortunate oversight that the late John Stuart Mill persisted in repudiating the idea that it was desirable for individuals to act each for his own interest.†

But if selfishness is the opposite of self-interest, by the practice of which civilised man would quickly reduce himself to the condition of brute beasts, it becomes easy for the naturalist to conclude that man may have evolved himself from some lower form simply by virtue of improvement in power to detect his own self-interest. Dimly perceiving the advantages of association, mankind has in this view gradually drifted by the rude method of trial and error into codes of written and unwritten laws, which, less or more efficiently, make selfishness immediately disadvantageous to the individual who practises it, so that the simple guide of action for each and all may be self-interest. Thus, to take a simple case for mere sake of illustration : if an intelligent man, influenced solely by the desire to obtain certain goods

---

* Utilitarianism, 1863, pp. 24, 25.

† " I must again repeat what assailants of Utilitarianism seldom have the justice to acknowledge, that the happiness which forms the Utilitarian standard of what is right is not the agent's own happiness, *but* that of all concerned."— MILL, *Utilitarianism*, 1863, p. 24. The important word " but " is italicised by ourselves : it implies the incompatibility of two interests which, as we contend, actually coincide. Mill also remarks that there is " happiness in absolutely sacrificing one's own happiness to the happiness of the greatest number."— (Pp. 24, 25.) This apparent contradiction, or seemingly irreconcilable statement, can be due to nothing else than the failure to realise that the happiness or interest of the individual need not be incompatible with that of the greatest number ; but that it may be to the interest of the individual to forego certain benefits for the sake of gaining the esteem and friendship of his fellows, the reward of whose esteem would more than compensate the privation undergone, so that no absolute " sacrifice " of happiness would occur. Indeed, no doubt one of the principal rewards of the labours of unselfish people is to be found in the inestimable prize of the real and cordial friendship of their neighbours and companions.

at the least possible expenditure of labour, could earn them in ten days, acquire them in an underhand manner in one, the risk of detection in the latter case being, say, one to ten —then a penalty (fixed by society) of something more than 100 days' labour, in the event of detection, would be sufficient to make him see that it was to his interest to adopt that method of obtaining the said goods which was concurrent with the interests of his neighbours.' Or again, if one man, observing that his neighbour never dreamed of going a step out of his way to help anyone else, himself resolutely determined not to move so much as a little finger to his neighbour's assistance, it would not be long before passive selfishness died out. For not to retaliate would be to offer a premium on selfishness, just as not to punish a theft would be to offer a premium on thieving, or to encourage it. Selfishness is analogous to thieving (in kind at least), since by it an individual obtains an unfair advantage at the expense of his neighbour. The course taken by society must obviously be to act so towards selfish persons that selfishness (like thieving) is rendered unprofitable, or against the interests of the individual who practises it. This is tacitly done ; but unfortunately, as regards *doctrine,* the contrary maxim is commonly preached, though in practical life it never can be and never is acted upon, as indeed it would be highly undesirable if it were. The apparently amiable doctrine that one should return good for evil, love one's enemies, &c. [like some other maxims that may recommend themselves on a superficial view], shows itself on analysis to be highly dangerous, constituting the strongest possible incentive to selfishness, and consequently the general practice of which would ruin society. From the very fact, however, that the ideal aimed at in this kind of doctrine is unattainable [on account of its inherent defects], it unfortunately comes on that account to be looked upon as something nobler and above this world, and forms a never-ending resource for sermonising and for characterising mankind as " miserable sinners." It may be safely concluded that the larger proportion of the asserted wickedness of this world is of clerical imagination. Without inquiring too closely into (perhaps unconscious) motives, it is none the less obvious that the more degrading the picture drawn of humanity, or the blacker the colours in which this world is painted, the brighter must the painters inevitably appear by contrast, and the stronger must seem the motive for their *raison d'etre.* This is unavoidable, and it must at least be admitted that the colours selected to paint humanity are of sufficiently sable hue.

If, instead of preaching the " wickedness " attendant on breaches of the law, society were to take care to inculcate on its members the advantage which accrues to each from the general practice of honesty, and to point out the efficacy of the arrangements it has made accordingly to prevent an occasional selfish man, residing among unselfish neighbours, from advancing himself at their expense,—in short, if it were taught that the question of honesty and dishonesty was one of profit and loss (or that knaves in the long run *are* fools), —then probably less machinery would suffice for the repression of crime. Unfortunately, however, popular religious doctrine seems to dissociate rather than to identify the path of virtue with that of self-interest ; as, for example, we have the saying about the " thorny and difficult path of Virtue," and the " broad and easy path of Vice,"—which, of course, is tantamount to setting a premium on vice. And yet what could be more contrary to the truth than the spirit of this kind of doctrine ? Also there can be no doubt that the holding out of rewards and punishments in a future world is a strong incentive to crime ; for it is justly argued by the would-be delinquent that if virtue require a future reward, it cannot therefore be remunerative in this life, or the practice of virtue cannot be consistent with self-interest. Since, therefore, the belief in a future world is necessarily very shadowy, the criminal naturally infers that it is desirable to tread the " broad and easy " path of Vice. Moreover, when it is commonly taught that such and such a course is " wicked," one may be inevitably led to conclude that in the absence of any more tangible reason than this against it the course must be advantageous.*

* There can be little doubt that one of the main causes for war may be reasonably traced to the continual preaching that it is " wicked," a vague phrase whereby a sort of attraction is attached to it, and the fact pushed out of sight that self-interest is the principle to appeal to here (or that it is futile to attempt to dissociate right from self-interest). For since the principle of co-operation and association is the very essence of the morality of self-interest, it would be seen that, from the fact that war strikes directly at the root of this principle, it violates the fundamental groundwork of the self-interest morality. Indeed men have already learnt this fact in their individual relations, and its influence has always been spreading wider and wider. We know that formerly, in feudal times, people inhabiting small tracts of country, or even families and near neighbours, used to arm themselves and be in continual war with each other ; and even the croakers [or " parrots of society," as the late Charles Dickens called them], who think the world goes backwards, and say that disarmament is impossible, laugh at the folly of the feudal times. It can only be a question of lapse of time for an appreciation of this folly to extend to larger tracts of country (or nations). Certainly the self-interest morality will have a great field here. The total violation of interests indicated by the self-inflicted punishment of the crushing armaments under which nations groan at present, affords a pitiable instance of the absence of self-help in reasoning beings (or

A little power of penetration should suffice to discover that a civilised man (or a man of high intellect) is capable of more happiness than a savage, and that in a civilised society an individual is more or less directly dependent upon the goodwill of his neighbours for almost all his pleasure in life ; whence it follows that desire for happiness on the Earth alone would itself induce all intelligent people so to conduct their lives as to secure the friendship of their fellows. For this purpose the strictest honesty and sincerity, practised as an undeviating *principle*, is obviously indispensable. For where would be friendship without sincerity ? When one considers that a man must have a *character* for honesty and sincerity in order to secure the pleasure of the real esteem and goodwill of his fellows, and that a *single* act of dishonesty or deceit may destroy his entire character (or reputation), one may see how utterly insignificant the temporary gain due to such an act would be compared with the prospective loss attendant thereon, and therefore on how firm and impregnable a basis stands the morality of self-interest.*

It is only to the absence of adequate appreciation of this fact, and the sort of fear that society rests upon a fiction, that some of those monstrous and terrifying doctrines unfortunately identified with religion can be attributed. To select a single example as a representative case : could any greater incongruity be imagined than the coupling of a God of Love with *eternal* punishment ? *i.e.*, a punishment which (measured by its duration) is *infinitely* greater in amount than that which the most implacable hatred could devise or

---

the absence of power to come to an agreement for their own advantage). The time may not be far distant when such a state of things will come to be looked back upon with something like contempt.

* It is a noteworthy fact that if the lives of those men who have accumulated such *exceptional* fortunes as to call for biographies be examined, it will be found that *exceptional* integrity and honesty were the main characteristics of all their transactions, which was the secret of the unbounded confidence inspired in their business relations. These men possessed sufficient power of penetration to see that that superficial sharpness which imagines an advantage in a little deceit or duplicity is in reality no more than stupidity. The parable of the unjust steward, who attempted to deceive by inducing his lord's debtors to falsify their accounts, contains exactly that exterior of superficial shrewdness which may be well adapted to mislead the unthinking ; but it will scarcely injure a man of intellect. He will see plainly enough that, so far from unjust stewards being "*wiser in their generation*," they are in reality fools (irrespective of any time or epoch). Nevertheless, can we wonder that dishonesty and underhand dealings are still so rife, when doctrine of this kind is actually included in the code of moral instruction.

the most outrageous injustice invent.* Dogmas of this
nature are simply illustrations of the lengths to which doctrines may go without universal repudiation, when society
has an instinctive dread that to disclaim them openly would
affect injuriously its own stability. It is the old instance of
the baneful influence of the false idea that good can ever
come out of error. In short, it is only necessary for anyone
to reflect impartially on the subject (especially from the
points of view indicated) in order that the evil which has
resulted from such doctrines may make itself plainly apparent; and their retention after the true groundwork of
morality is recognised could not be palliated by even the
semblance of an excuse. Indeed such dogmas constitute an
evident insult to the justice and goodness of the Deity : that
they are precisely of that character which is calculated to
allow Clericalism to predominate over the rest of society is
unquestionable, whether that motive had any part in their
original invention or not.†

At the same time, is it not a melancholy consideration
that doctrines of the above character (though happily excluded from our Board Schools) are still taught to young
children, at the very dawning of their mental faculties,
before they are sufficiently matured to distinguish truth from
error, and without experience to know representative
opinions (especially those of the unbiassed and intellectual
few) on these subjects. Thus the child imagines himself
isolated in his opposition to these doctrines, and years may be
miserably spent and intellectual energy wasted in fruitless

---

* According to this dogma of infinitely lasting punishment, the punishment
for vice in this life would be *infinitely* inadequate, which is practically tantamount to teaching that the pursuit of vice must be infinitely profitable in this
life.

† Putting the case as an *à priori* problem (as is sometimes usefully done in
physics),—then, in order to predominate over the rest of society, the condition
fundamentally required is to appeal in the strongest possible way to the interests of mankind, by inventing some startling and terrifying danger, together
with some remedy or means for escape equally startling and exceptional (if
possible) ; when the rest of society will naturally run after those offering the
remedy as their rescuers and benefactors. [It almost unavoidably reminds one
of the story of the bees who offered their comrades Heaven, and took the
honey.] If, in the attempt to strain the magnitude of the danger to an extreme
pitch (illustrated by making the punishment *eternal*), the limits of justice be
passed, this is a matter of secondary consequence ; since the very incongruity
of the doctrine, especially if coupled with an affectation of mystery (which
applies equally to the scheme of escape proposed), may tend to make it fasci-
nate all the more, from its seeming originality. It is a well-known fact that it
is the policy of men who exercise an ascendancy over others never to be sparing
on the side of boldness, since the very audacity of the incongruities indulged in
may tend to cause additional cringing rather than revolt. The more intelligent
portion of mankind may not be disposed to examine the error too closely, from
a sort of undefined idea that it may conduce to the stability of society.

efforts to reconcile the impossible, in the vain attempt to put in practice unnatural and ridiculous moral maxims,[*] or in the struggle with beliefs that disintegrate the mind. Of this one of the authors of this essay[†] can speak from personal experience (as no doubt many others could); and all this is assumed to be necessary to benefit society in general, as if society rested upon an unstable basis that required fictions to support it. Fortunately the majority escape this evil of young days, simply because they do not inquire into or realise what is taught them ; and it would be all the same, in their case, which of the thousand and one creeds of the world were inculcated.[‡] Youthful minds of an exceptionally penetrating and inquiring character run the greatest risk of becoming hopelessly entangled here, or it is reserved for the most inquiring and thoughtful, and therefore probably those who would have exceptional capacities for becoming useful members of society, to bear the brunt of this, in order to sustain a system for the fictitious benefit of the many.

That there is no limit to the depths of absurdity and superstition to which even men of education will descend (and in this nineteenth century) in matters wrongly termed "religion,"—especially where sectarian interests are involved, —is fortunately not without such instructive illustration as will serve to keep the thoughtful on their guard. The never-

* Mill makes a remark bearing on this point ("Utilitarianism," p. 44) :—"Unhappily it [the moral faculty] is also susceptible, by a sufficient use of the external sanctions [*i.e.*, eternal punishment, &c.] and of the force of early impressions, of being cultivated in almost any direction ; so that there is hardly anything so absurd, or so mischievous, that it may not, by means of these influences be made to act on the human mind with all the authority of conscience."

† The other author, a friend formerly largely associated in the thought and preparation of the scheme of this essay, and who had an equal (perhaps greater) share in the development of the main principle, has reasons for remaining anonymous for the present. The work and study connected with the essay has extended, from time to time, over some three years. This is mentioned to avoid any idea of the publication having been entered on prematurely.

‡ The fact that religious belief is a mere unrealised dead letter (or profession) with the majority, so as to have no practical effect on their lives, is well illustrated by Mill in his celebrated essay on "Liberty." At the same time, is it not a sad thought, in view of the enormous number of diverse creeds in the world (each sect maintaining its own to be the only *true* one), to contemplate the means for the brain-poisoning of youth that the prevalence of so much error must afford ? It would not be of so much consequence if these doctrines were not instilled before the intellect is sufficiently developed to distinguish truth from error. For if ignorance be a great evil, how much greater must be the scourge of false doctrine ! For it is indeed far more difficult to un-learn than to learn. Moreover, does not the prevalence of so many diverse creeds in the world afford a signal illustration of the recklessness of invention on the one side, and of credulity on the other ?

to-be-forgotten spectacle (only a few years since) of the
Cardinals—men who had passed through Universities—
sitting in numbers in solemn Council on the Infallibility of
the Pope, remains a standing warning that there is no fable,
however wild and absurd, no superstition, however monstrous
and incredible, which, under the guise of "religion," will
not gain masses of adherents; and therefore this shows,
with incontrovertible logic, how necessary it is to inquire
into everything and be on the alert if we would keep clear
of error. No one could say that this is not a fair illustrative
case, or that the warning it contains in regard to the doctrines of Clericalism may not be as applicable to one country
as to another.

One would not desire to prohibit speculation on so-called
"religious topics," but let us take especial care that speculations are not at any time made up into a book and taught
as *truths*, and above all let us be on our guard that the
speculations are not irreconcilable with each other, or
directly opposed to the attributes of goodness and justice
that are ascribed to the Deity.

Let speculations and scientific inductions be carefully
distinguished from each other. While a relic of barbarous tradition tells the degrading narrative of the
Fall of Man, inductive science points to the ennobling
view of his Rise, thus opening out a practically limitless
field for a greater rise in the future, progress in the past
being the best guarantee and incentive to progress in the
future.

Surely there could be no nobler doctrine than that inculcated by the self-interest or individual happiness morality,
viz., that man's interests and happiness lie in the practice
of virtue, or that the path of virtue and that of self-interest
are identical with each other. What higher incentive could
there be to an upright life than this? Those who oppose
this doctrine must be prepared to contend [as some superficial people, who imagine they are sharp, do] that virtue or
strict integrity is *not* its own reward.* There would seem
to be a sort of cringing or slavish disposition to some extent
prevalent which thinks that virtue can only be attached to
privation and absence of freedom, as if it were thought that
the Deity took a pleasure in seeing his creatures practise

---

* The follower of the morality of self-interest is contented with the reward
that virtue brings with it; not looking to an enormous (infinite) reward in the
future. He also does his best to lead mankind by teaching them that right
conduct is in accordance with self-interest, not to coerce them by a degrading
system of terrorism.

a course of conduct that made them miserable.* Instead of ennobling virtue by regarding its practice as the privilege and interest of a free man, there is often rather a tendency to degrade it by identification with the abject "*duty*" of a slave.

The natural or un-sectarian morality (grounded on Natural Science) constitutes the very ideal of liberty, the freedom of contributing to one's happiness. This morality might therefore be termed with equal propriety the morality of liberty; and the very fact of its constituting the perfection of liberty might be viewed as an additional confirming illustration of its truth, in so far as the complete achievement of liberty is justly regarded as one of the last conquests of human progress.

Morality in Relation to Evolution.

If the morality of self-interest be brought under the test of the theory of Evolution, we think that it will not fail to become clearly apparent that the two harmonise in a remarkable manner. For Natural Selection has been recognised to adapt a living being to the conditions of life, and accordingly tends to produce in animals such "instincts" as are adapted to protect them from danger. "Sociability" (by which animals congregate in troops) is one of these instincts. Natural Selection may therefore be said to tend to develop such instincts in animated creatures as to cause them to act in a way conducive to their *interests* (which is the self-interest morality).

It becomes evident, therefore, that in the case of man— if the power of reason (attendant on brain development) be sufficiently augmented—this may largely replace (in regard to conduct) the "instincts" formerly established by the rough drill of Natural Selection. While the lower animals blunder, and Natural Selection corrects their errors by working upon the brain to develop instincts which check the repetition of errors ; man, on the other hand, by using his reason aright, may avoid blundering, and thus may emancipate himself, to a great extent, from the rough discipline of Natural Selection.

There is evidently a great difference (in degree) in this respect between man and the lower animals. For self-interest being the guide of conduct, one of the highest

* The animated coffin-like types of humanity, immured in the cloisters of the Jesuits, may serve as instructive illustrations of this principle carried out to an extreme degree.

attainments of knowledge must therefore be *to know one's interests.** It could not therefore be expected that the lower animals would have advanced anything like so far as man in this respect. That most important of interests, "Sociability," which requires some penetration and thought to appreciate its value and consequences beforehand, is where the lower animals notably fail; and it is a significant fact that the higher the animals stand in the scale of intelligence the more do they appreciate the value of sociability. Thus the ant, various mammals, the higher apes, &c., associate in communities, and are known to be distinguished for their exceptional brain development. Man therefore has progressed in proportion as he has discovered the value (interest) of sociability; *i.e.*, he has advanced in the same ratio as he has gained a knowledge of his own interests (all blunders being errors against one's interests).†

It forms a noteworthy confirmation of this to consider the progress of any civilised nation in the past. At first we may observe that the knowledge of self-interest had only developed so far as to cause small communities or tribes to associate together, who, however, were in continual war with neighbouring tribes. If we thence look at the feudal times, then the knowledge of self-interest had spread further, and there was much greater harmony and association; but still the parts of a single nation were in frequent strife and contention. At the present day the harmony has extended itself to whole nations; but still these are occasionally at war: nevertheless the violation of mutual interests here involved is becoming every year more and more clearly seen. Thus we may perceive that the advance in intelligence, by affording a clearer appreciation of self-interest, has always coincided with the development of association,

---

* This fact may make it cease to be surprising that people may run after the most pernicious sectarian delusions and imagine them to be to their interests.

† As in some respects an instructive illustration of the opposition of selfishness and self-interest, Free Trade might be mentioned. At one time, when intellect was, perhaps, not quite so highly developed as now, a species of commercial suicide (called by irony "Protection") is known to have been largely practised, by which it was sought to derive benefit at the expense of one's neighbour by taxing his goods. The strangulation of trade and violation of self-interest thus resulting might be compared to the condition of the selfish individual who isolates himself from his neighbours, and who makes no true friends, and, nevertheless, whose intellect is often of so low an order as to be unable to discover the cause of his unhappiness. It is notorious that selfish individuals are generally of inferior intellectual capacity. No one will probably doubt for one moment that selfish persons who isolate themselves, and miss the great benefits of sympathy and friendship, are unhappy. If so, the morality of self-interest (as the opposite of selfishness) must commend itself as an irrefragible truth.

in order to gain the inestimable benefits of co-operation and friendship among mankind.

Mr. Darwin, in his work "The Descent of Man," appears to consider that man (in relation to the principle of Evolution) may be still influenced to a certain extent by some of the more important "instincts" which formerly belonged to a lower state, such as "Sociability" for instance. But it would not seem to follow necessarily from this (and perhaps this would not be essentially implied) that man might not now, by the light of his reason, test these instincts, in order to see whether they are desirable or not. At least it may probably be conceded that it would not be a thing to be wished that man should be dominated by " instinct " without the control of reason, or this would surely be a somewhat low (and undefined) basis on which to rest morality. It becomes only necessary, therefore, to trace an " instinct " [shown to be dependent on Natural Selection] up to its rational basis, in order to see that this is self-interest. This amounts to no more than taking the final step of advancing the undefined "*instinct*" up to its definition. We must not shrink from this through fear of discovering the bogey Selfishness behind it. The worst of this confounding of self-interest with selfishness is that it has caused inquirers to fear reason. It would be a pitiful state of things if we were afraid to look Reason in the face. Morality thus loses all its dignity. While a lower animal may act by such and such an " instinct " (" sociability " for instance) without being able to appreciate the *cause* of the "instinct," man, on the other hand, may be able to define the reasons for it, and even to say beforehand whether a given course of action is desirable or not.

There cannot, we think, be a shadow of a doubt, on analysing the question impartially, that the extraordinary fact of no generally recognised standard of conduct existing—in spite of the immense advance of the other sciences—is mainly due to the mistaking of self-interest for selfishness (its opposite). For it is a notorious fact that the self-interest morality has been driven home by hard logic again and again, by the ablest minds from the Greeks downwards, but its fitness or suitability has escaped appreciation, or the bugbear of Selfishness has always intruded itself and prevented its adoption. Nevertheless, it may be observed that the only escape from selfishness is by recourse to the morality of self-interest. If a man by practising unselfishness earns the immense benefit of the high esteem and friendship of his neighbours, is he to forego this benefit and become selfish in order to avoid following his interests ? What other course would be possible ? Herein lies surely the impossibility of escape from the self-interest morality, and (may we not add) the absurdity of the attempt to do so. This unfortunately perverted tendency to exterminate self (owing to the mixing up of selfishness with self-interest) has no doubt tended to stunt and wither some of the best impulses of our nature, viz., those which urge us to earn the praise and esteem of our fellows by good actions. Where, indeed, we might ask, would even love or friendship be without self ? Abolish self (in the form of the happiness enjoyed) on each side, and where would be the friendship or the love ?

It may be safely concluded that all great systems have a *simple* principle on their basis, and morality makes no exception to this. The turgid or diffuse discussions that one sometimes finds on this question may be doubtless the not unnatural result of the immense difficulties inevitably encountered in wandering from the truth, under the frightening influence of the bugbear of Selfishness. The very ingenuity (sometimes almost desperate) of the attempts made to avoid basing morality on self-interest are surely themselves among the best illustrations of its validity. It is hardly likely that so fundamental a truth could have eluded general recognition, had it not been for this peculiar oversight. Indeed it has been ably argued, by many reasoners of admitted ability, that a man *cannot* act excepting by something which affects his interests, or touches his individuality in some way ; for that which does not affect *him* cannot make *him* act (or is not a rational *motive*). Hence it would result that the morality of self-interest (or individual happiness) always *is* —tacitly, or even unconsciously—followed. It only therefore remains to recognise its *fitness*, officially and openly, in order to derive that benefit which attends the appreciation of any great truth.

It might possibly be thought by some that we have criticised unnecessarily some dogmas and (so-called) " moral " precepts, which are sometimes unfortunately taught as part of education. But it should be noticed that truth cannot be effectively illustrated excepting by contrast with error ; and it will be sufficiently clear that some of the dogmas and points of doctrine referred to are, beyond question, highly dangerous. Moreover, only a few instances have been selected for analysis, where many might have been noticed ; and we have every ground for confidence that a good purpose will be served thereby.

The more the question is examined, the more apparent

will it surely become that the neglect to identify morality with self-interest has caused great evils. It has acted as the strongest discouragement to virtue, by making it appear *against* one's interests, and has given rise to the invention of those pernicious dogmas (above referred to) which are worse than vain attempts at terrorism. If in the general system of education it were invariably taught that the path of virtue, or strict honesty and sobriety, was absolutely in acccordance with self-interest (in fact that virtue is its own reward), and that such practices as intemperance, thieving, or deceit were to be avoided, because they were *against one's interests*,—instead of the absurd statement that they are " wicked " (which only makes them more attractive, from the intangible nature of the reason), there can be no question that immense good would result. In fact it would be doing more than making morality stand upon reason,—its only sure basis.

Responsibility and Physical Causation.

It has been argued by some that from the fact of the original formation of man's character having been determined by causes beyond his control (or because a person is not reponsible for his inherited brain structure), that therefore he cannot be made accountable for his actions. Mill notices this view in his "Utilitarianism," p. 83, viz.: "The Owenite invokes the admitted principle that it is unjust to punish anyone for what he cannot help." But we think it may be made clear that the supposed absence of responsibility under the above conditions is a fallacy, and that in addition to this, the doctrine of strict causal sequence in nature may enable us to arrive at what might be capable of forming a rough basis for a scientific penal code. We will endeavour to point this out in as clear terms as possible.

When any crime is contemplated, the eventuality of punishment is taken into consideration beforehand, and balanced against the direct material gain that would ensue from the crime, the chances of escape being duly allowed for, and it is this balancing process that accompanies the decision of the wrong doer to commit the offence. The additional punishment, consisting in forfeiture of position in society (which would probably be of itself more than sufficient to deter any respectable member) does not of course influence the habitual criminal. If therefore the punishment fixed beforehand by society (*i.e.*, by the penal code) be such that when the criminal has duly allowed for the chances of escape, the amount of punishment (as a contingency) seems in his judgment to be less than the direct material gain derivable from the crime, then he is led to commit the offence.   He therefore, of course, *deserves* the punishment if detected, because this was precisely what he contemplated beforehand, and which entered in as a factor in determining his decision.   To remit the punishment would be exactly like remitting afterwards the loss sustained in a lottery which was contemplated as an eventuality beforehand (the injustice of which would, of course, be self-evident).   Since crime is committed for the sake of the material gain that attends it, to repeal the punishment would be to offer a reward for wrong doing.   The absurdity of society offering a premium for misdemeanours is too evident to need further comment.   Indeed, the removal of penalties for crime would precisely resemble (in principle) the cancelling of prizes in an honest contest, the prospect of earning which had induced the competitors to contend.

No doubt the criminal (like the case of the lottery) may miscalculate somewhat beforehand the value of the material gain attendant on an evil action, when balanced against the contingent loss (represented by the punishment), and, doubtless, society is obliged to fix beforehand the punishment somewhat higher than the value of the prospective gain accompanying the misdeed, in order to deter from evil actions.   But on this account the criminal is by no means a subject for unmitigated pity.   At the very outside (even if this concession were perfectly above suspicion) he could only deserve the relatively insignificant amount of pity due to the *surplus* of punishment over its true contingent value, which society is obliged to put on in order to make dishonesty unprofitable—and the existence of which *surplus* (in the penalties) the criminal has failed to see beforehand, either from imperfect reasoning faculties or a neglected education. He may be compared to a foolish gambler who goes on playing when the value of the chances of the table is calculated against him.

That a principle, mathematical in its nature, underlies the system of punishments, so as to be capable of forming a rough basis to a scientific code, will probably have become tolerably evident from the above considerations.   For there is clearly for every crime a certain amount of punishment which is merely the exact equivalent of the material advantage gained by the commission of that crime.   The probability of escape must, of course, be allowed for, so that, for instance, when the chance of detection (derivable from statistics) is in the ratio of one to a hundred, the punishment equivalent to the stealing of as much as a thief could earn by honest work in a day, would be a hundred days' labour. This would be the *minimum* mathematical value of the penalty under the above conditions, and if society did not counteract the advantage gained in the theft by at least this amount of .punishment, it would be· absolutely offering a reward for stealing. But it is necessary that a flourishing society should do more (or it must fix the penalty somewhat in excess of its true *minimum* value) in order to make the unfair method of attaining definite ends positively disadvantageous, so that it may not be adopted except by members of inferior reasoning power. No doubt special considerations may influence the administration of the code in special cases, but the recognition of a broad or general principle underlying the penalties is not on this account of less value or importance.

It may, perhaps, assist in appreciating that the above principle is a just one (in regard to the *minimum* value of the penalty) to observe that if detection were *certain* (in the case of a sum stolen, for instance), then the mere deprivation of the sum afterwards would be sufficient as a penalty to check thieving (as it would destroy all profit). It must follow logically from this, therefore (on the same principle), that when detection is *not* certain, a fine equal to the chance of escape multiplied by the value of the sum abstracted, would also be a sufficient penalty, because all means of gain would thus be entirely extinguished (and a margin of loss remain in the trouble of abstracting the sum). This is evidently merely an instance of varying the punishment by inflicting fines instead of the equivalent labour.

The above analysis may perhaps serve to make it sufficiently clear that the feelings of responsibility, praise, and blame (originally formed probably as "instincts" through natural selection), have a distinct rational foundation, and are in harmony with the doctrine of strict causal sequence in nature. The penal code may be regarded as merely a more emphatic method of awarding blame, or of teaching people that selfishness is the opposite of self-interest. It may be added that those who are interested in the related question of strict causation in physical events, may be referred to a recent letter by one of the authors in "Nature," May 13th, p. 29, "On a Point Relating to Brain Dynamics." It should be remarked, however, that we have since learnt through Mr. George Romanes ("Nature, May 27th, p. 75), that the mode of reconciliation of the rival views on Free Will *v.*

Necessity suggested in that letter, was very analogous to a means proposed by the late Prof. Clifford in an oral lecture. This independent deduction of the same result by different minds may perhaps be regarded rather as a confirmation of its truth than not.   Mr. Romanes, who apparently accepts the reasoning given in the letter (on "Brain Dynamics") "as far as it goes," nevertheless remarks that both there and in Prof. Clifford's lecture, "The Prince of Denmark," responsibility had been omitted; and he seems to hold the view that the feelings of responsibility, praise, and blame, cannot be reconciled with the doctrine of strict physical causation, and suggests at the end of his letter that these feelings may be destitute of any rational basis.   The follow-ing is the passage by Mr. Romanes:—

"What then, it cannot but be asked, is the psychological explanation of these deeply-rooted feelings of responsibility, praise, and blame, which can never be eradicated by any evidence of their irrationality?   To me it appears the only answer is that these feelings have been gradually formed as instincts, which, while undoubtedly of much benefit to the race, are destitute of any rational justification."—("Nature," p. 76).

This is the only point where we would venture to differ with Mr. Romanes (while otherwise fully endorsing his letter).   Possibly the above carefully considered conclusions may serve as some help out of this difficulty, which has always been regarded as a formidable one.   It would at all events seem to us *à priori* more probable that the function of science should rather be to *explain* the "instincts" de-veloped in man, than to show them to be devoid of rational foundation.   Precisely on account of the beneficial light that science may be expected to shed on matters of this kind, does it become all the more difficult to understand the half-expressed repugnance of some to scientific inquiry on subjects of this class—almost as if it were imagined that the discovery of truth was a thing more to be dreaded than the persistence of error.

Conclusion.—Since life is valuable only in proportion to its happiness, or happiness is the object of existence, the struggle for life therefore becomes synonymous with the struggle for happiness, and the practice of conduct favour-able to happiness constitutes morality.   Just as the life of the individual receives important aid from the community (to whom the individual owes some of the essential condi-tions for his continued existence); so in the same way the happiness of the individual is promoted by the community in many important respects.

There cannot be the slightest fear of any principle here by which the pursuit by each individual of his own happiness could take place at the expense of that of the community; for since one of the most important elements in the happiness of the individual is the good will or friendship of his fellow men, he could not be said to be "pursuing his happiness" in forfeiting this; and since any ill-considered attempt to further one's happiness at the expense of others, is instantly felt by them and retaliated upon by the withdrawal of friendship (or the more active reproof of the penal code); this, therefore, by infallibly teaching the individual that the attempted pursuit of happiness *at the expense* of the community is in reality a violation of self-interest (or opposed to his happiness), would infallibly bring his proceedings to a check. Thus a self-righting principle in the moral world (much in analogy to the self-correction of the equilibrium of the moving parts of a system under the great kinetic theory) exists, by which the individual happiness is made alone consistent with the greatest happiness of the greatest number.*

In fact, morality is seen to contain that essential element

---

* How, indeed, could the greatest happiness of the greatest number be secured, if each of the units of that number (the individual) neglected the pursuit of his own happiness? In fact, since the more an individual is happy, the greater is the happiness he inevitably sheds around him; so in this sense it may be considered almost a "duty" for the individual to be happy. It is certain that the highest ideal of morality can never be reached without. It may be observed, that the energy of the automatic correction in the moral world, is always proportional to the disturbance (as in the physical world under the kinetic theory). Thus the more an individual attempts to further his own happiness *at the expense* of others, the more violent is the correction or recoil which acts to diminish his happiness—so keeping him in check. In an analogous way, the more the equilibrium of a gaseous body is disturbed by some molecules acquiring excessive velocities, the greater is the tendency of the surrounding molecules to check the disturbance (or to restore equilibrium). At the same time it is well to keep distinctly in view, that the existence of the community does not, on the whole, tend to diminish individual happiness, but (on the contrary) distinctly to increase it: since the pleasures of sociability are among the greatest. There is, therefore, no restriction of liberty here; for an individual, even if he had his choice, would not wish to exist entirely alone (indeed, solitary confinement is considered one of the worst of punishments). The community increase the happiness of the individual, and (inversely) the happiness of the individual diffuses itself around him. Thus the conditions for a perfect harmony are seen to exist. It is only the blundering against self-interest, owing to ignorance and false sectarian doctrine, that causes the occasional discord. The knowledge of self-interest—or of the conditions for individual happiness—being the highest achievement of knowledge (as the final end of morality); it is scarcely to be expected that this should be reached yet, though signs of a rapid advance are not wanting. For the progress of science, by enlightening ignorance, will thereby remove the main cause of unhappiness. This may be still further facilitated as the public gradually come to have a juster appreciation of their true friends.

of stability within itself which is the very condition for the existence of self-evolved systems. It will be, of course, understood here that one of the most important elements in the pursuit of individual happiness, is the cultivation of the esteem of one's fellow men by the performance of kind offices, since friendship and sociability are among the mainsprings of individual happiness.

ANALYSES OF BOOKS.

Catalogue of Books and Papers relating to Electricity, Magnetism, the Electric Telegraph, &c., including the Ronalds Library. Compiled by Sir FRANCIS RONALDS, F.R.S. With a Biographical Memoir. Edited by ALFRED J. FROST. London : E. and F. N. Spon.

WE have here a very valuable compilation issued to the world under the auspices of the Society of Telegraph Engineers. After Sir Francis Ronalds retired from the direction of the Kew Observatory, he made the collection of a library of works on Electricity and the allied branches of Science, and the compilation of this Catalogue, which contains 13,000 entries, the object of his life. On his death, in 1873, he bequeathed the library to his brother-in-law, the late Mr. Samuel Carter, with the stipulation that the books should not be dispersed, but preserved so as to be accessible to persons engaged in the study of electrical science. This condition Mr. Carter very wisely fulfilled by handing over the library to the Society of Telegraphic Engineers, represented by a special body of trustees. The Society undertook the publication of the Catalogue and the binding of the books. When this latter task is completed the library will be open not merely to the members of the Society, but, under the necessary regulations, to the public.

It is doubtless forgotten by many that as early as 1816 Ronalds demonstrated, by actual experiment, the possibility of an electric telegraph, and attempted to bring his invention under the notice of the Government. The following communication, which he received from the Secretary of the Admiralty, Mr. Barrow, afterwards Sir John Barrow, and the author of the article "Telegraph" in the seventh edition of the "Encyclopædia Britannica," is a treasure worth preserving :—

"Mr. Barrow presents his compliments to Mr. Ronalds, and acquaints him, with reference to his note of the 3rd inst., that telegraphs of any kind are now wholly unnecessary, and that no other than the one now in use will be adopted.—*Admiralty Office, August* 5, 1816."

The "one now in use" was the old semaphore, useless in the night and in foggy weather. Long afterwards, *at the age of eighty-three*, he received the honour of knighthood !

Heat, a Mode of Motion. By JOHN TYNDALL, D.C.L., LL.D., F.R.S. Sixth Edition. London : Longmans and Co.

WHEN a scientific book—and especially one which has not earned popularity by appealing to unscientific or anti-scientific prejudice —reaches its sixth edition the ground for the reviewer is much narrowed. If he censure, the majority is against him ; if he praise, or merely expound, his readers, if men of high culture, call his critique an *Ilias post Homerum,* or, if of a more vulgar stamp, they long to confer upon him that vindictive pinch which in some rural districts is still bestowed upon the bringer of stale news.

The volume before us is not a reprint ; it has been modified, improved, and extended. Upon one remark in the Preface we must pause for a moment. Says the author—" Far be it from me to claim for Science a position which would exclude other forms of culture. A distinguished friend of mine may count on an ally in the scientific ranks when he opposes, on behalf of literature, every attempt to render science the intellectual all in all. Ours would be a grey world if illuminated solely by the dry light of the understanding." This is well said ; but as there is no earthly prospect of Science ever becoming the " intellectual all in all,"—would not " mental " be the better word ?—as there is no possibility of " other forms of culture " being excluded, such concessions seem scarcely needful. But at least every man should have his free choice, and not be forcibly restrained from studying things without he has first spent a serious portion of his life upon words.

In the Preface we find an interesting notice of a work of Carnot's, written and published as far back as 1824, in which he developes the relation between heat and work which has since been independently discovered by Mayer and Joule.

It is somewhat remarkable that whilst the modern view of heat as a mode of motion was generally held by scientific men and philosophers in the seventeenth and the earlier part of the eighteenth centuries,—*e.g.,* Locke, Boyle, Euler, &c.,—the opposite or material theory seems to have gained ground in the first half of the present century, especially in France, where even such a man as Berthollet came forward on behalf of the " received theory of caloric "—a word still used by leader-writers in their often indiscreet " meddling and muddling " in scientific questions.

A very interesting fact is that established by Prof. Frankland, that, by merely condensing the air around it, the pale flame of a spirit-lamp may be rendered luminous, and even smoky, " the oxygen being by the compression rendered too sluggish to effect the complete combustion of the carbon."

We find it mentioned that the liquefaction of chlorine was effected by Northmore as early as 1805, and it is at least probable that sulphurous acid was liquefied by Monge and Clouet even prior to 1800. Faraday, in 1824, openly admitted that the merit of first liquefying the gases belonged neither to Davy nor to himself.

Touching on the Glacial epoch of geologists, Prof. Tyndall rejects the explanations generally given, such as the reduced radiant power of the sun, the passage of the solar system through a colder region of space, or a re-distribution of land and water, so contrived as to lower the temperature of the globe. He declares that to produce glaciation " we cannot afford to lose an iota of solar action," but want an " improved condenser." How this condensation was effected, nor what it could have been other than a reduction of temperature, we do not find stated.

In treating of the " limits of science " we are gratified to note that Prof. Tyndall takes substantially the same view as Prof. Du Bois-Reymond. He fully admits that, in passing from the region of physics to that of thought, " we meet a problem not only beyond our present powers, but transcending any conceivable expansion of the powers we now possess." Here, then, is a distinct recognition that mental phenomena cannot be explained by the properties of matter and energy. Prof. Tyndall, therefore, cannot belong to the " Kraft-stoff " school, and those who accuse him of materialism would do well to consider with due care the words we have just quoted.

" Heat, a Mode of Motion " must be regarded as a most comprehensive and accurate survey of an important branch of Science, and all the more valuable because not encumbered with that parade of mathematical formulæ which renders so many works on physics unintelligible save to one particular order of minds.

Degeneration. A Chapter on Darwinism. By Professor E. RAY LANKESTER, F.R.S. London : Macmillan and Co.

THIS interesting and valuable little work is in substance a discourse delivered by the author at the Sheffield meeting of the British Association. In the introductory portion we meet with the suggestion—humorous, but not the less appropriate—that the Association should plainly state *what* it is seeking to advance. There is scarcely a word in the English language so strangely and vaguely applied—or rather misapplied—as " Science." In vain did Prof. Whewell point out clearly and succinctly the distinction between Science and Art, and between Science and unsystematised knowledge of facts, objects, events. We still hear the British workman use the term as a synonym for pugilism. Persons of culture set our teeth on edge by speaking of the feats of " Farini's Zazel," or of Maskelyne and Cook, or of the " eagle-swoop " of Maraz, as "triumphs of Science."

Prof. Lankester considers that the "most frequent and ob-jectionable misuse of the word ' Science ' is that which consists in confounding science with invention—[we think ' industrial art ' would have been a happier word]—in applying the term which should be reserved for a particular kind of knowledge to the practical applications of that knowledge. Such things as electric lighting and telegraphs, the steam-engine, gas, and the smoky chimneys of factories, are by a certain school of public teachers, foremost among whom is the late Oxford Professor of Fine Art, persistently ascribed to Science, or gravely pointed out as the pestilential products of a scientific spirit."

Scientific men, in fact, find themselves placed between two fires. Men of business ridicule us for seeking after the cause of the Aurora Borealis, or the laws which govern the distribution of colour in the plumage of a bird or the wing of a butterfly, and consider such knowledge utterly useless and frivolous. On the other hand, poets, parsons, artists, orators, metaphysicians sneer at us because the facts we discover or the generalisations we establish admit of, and sometimes receive, practical applications, possibly of an unæsthetic character. How completely these two kinds of objections refute each other is completely overlooked.

Passing from these introductory considerations to a sketch of the doctrine of organic development as regarded from the Darwinian point of view, the author shows that the action of external forces upon a living being is not necessarily and invari-ably in one sole direction. Three results are possible : the organism may either remain *in statu quo*, or it may increase in complexity of structure, or, again, it may decrease in complexity of structure. These three possible results Prof. Lankester calls respectively balance, elaboration, and degeneration. The two former of these cases have been fully recognised : it is admitted that certain animal species have remained from remote geolo-gical ages unmodified,—a fact which certain writers, miscon-struing the very essence of the doctrine of development, have wrested into an argument for the separate, and we might say the desultory, creation of every species. But degeneration, or degradation in structure, the descendant being lower in organisa-tion than its ancestors, has not been admitted by naturalists save in a few exceptional animals,—parasites,—and great credit is due to Dr. Dohrn, of the Naples Aquarium, for applying it in explanation of some animal relationships hitherto perplexing and mysterious. Degeneration is defined as a gradual change of structure in which the organism becomes adapted to less varied and less complex conditions of life. Elaboration, on the other hand, is a gradual structural change in which the organism be-comes adapted to more and more varied and complex conditions of existence. " In Elaboration there is a new *expression* of form, corresponding to new perfection of work in the animal machine. In Degeneration there is *suppression* of form, corresponding to the cessation of work." The author recognises that elaboration of one organ may be, and often is, the accompaniment of degeneration in others. An animal is only to be viewed as an instance of degeneration when it is, as a whole, left in a lower condition than that of its ancestors. Prof. Lankester considers that any new set of conditions which secure the food and the safety of any animal without effort lead as a rule to degeneration.

Some curious instances of degeneration are given. Among others the barnacle, a degraded crustacean, classed by Cuvier and others among molluscs, and only restored to its rightful rank in consequence of the discovery of its juvenile stages of form. The Ascidians the author regards as degenerate Vertebrates, a conclusion justified by the researches of Kowalewsky on their development. They pass through a tadpole state almost identical with that of the common frog, but instead of being elaborated they degenerate. The Convoluta worms, which contain chlorophyll, and have the power of nourishing themselves, like plants, upon the carbonic acid dissolved in the water around them, are also adduced as an instance of degeneration. As the chief causes of such structural degradation the author enumerates parasitism, immobility, vegetative nutrition, and excessive reduction in size. Where these conditions are present degeneration may be suspected, even in the absence of any embryological evidence.

Finally, it is shown that degeneration is not a mere geological question. In vegetable life it plays, as might be expected, a decided part. In language it has long been recognised, and it must be taken into account in anthropology and in social science. Prof. Lankester hints at the possibility of a degeneration in what are at present the leading types of mankind. Who can venture to pronounce such a decay out of the question?

The hypothesis of degeneration is worthy of very careful study, and we must consequently recommend this thoughtful and suggestive treatise to biologists.

Letter of the Commissioner of Agriculture on Sorghum Sugar. Washington: Government Printing-Office.

A PAMPHLET recommending the cultivation of sorghum in the United States as a source of sugar, and giving descriptions and figures of the machinery required for extraction.

*Alphabetical Manual of Blowpipe Analysis, showing all Known
    Methods, Old and New.*   By Lieut.-Colonel W. A. Ross,
    F.C.S. (Berlin).   London : Trübner and Co.

FOUR years ago we had the pleasure of examining a work by
Col. Ross, entitled " Pyrology, or Fire Chemistry." We could not,
in common fairness, pronounce it other than valuable and suggestive, and we felt bound to recommend it to chemists and
mineralogists as well deserving practical criticism. We declared
that it would be well worth the while of any student possessing
the necessary time, patience, and skill, to work through the book
and verify the phenomena described. How far the chemical
public has condescended to take our advice is doubtful. Had it
done so to any appreciable extent we should certainly have read
of confirmations of some of the observations made, and certain
interesting facts at which the author glances in passing would
have been followed up.

Meantime Colonel Ross has not been idle. In the little volume
before us he has embodied, in a convenient and portable form,
the results of his twenty years' experience in blowpipe analysis.
We find here, within the compass of 150 pages, a description of
the necessary apparatus and reagents, and of the methods for
their application,—both those discovered by the author and those
of earlier experimentalists which he has verified.

A curious feature of the work is a process for the quantitative
determination of cobalt in ores and furnace-products, based on
the colours produced by the oxide when dissolved in vitreous reagents. The mechanical details of the operation, which, as it is
remarked, must be conducted with very great care, are not given.

Methods for the detection of some of the rarer metals are not
wanting, and here the author frequently resorts to the spectroscopic examination of the pyrocones and pyrochromes which are
produced. He has devised for this purpose a small spectroscope
which can be worn like a pair of spectacles, thus leaving the
hands at liberty. He cautions the beginner never to attempt to
learn the reactions of any substance from native minerals, however pure they may be supposed, but always to work upon pure
oxides.

A passage which puzzles us, and some scientific friends whose
attention we have called to it, is the following :—

" *Chemical Water.* A combustible but not vaporisable compound, composed apparently of hydrogen and oxygen, present in
every natural and in almost every artificially prepared oxide. It
ignites at apparently red-heat, and burns with an orange pyro-
chrome, *i.e.*, a non-luminous flame tinged with colour."

The detection of calcium phosphate in the body of a fly is a
curious observation. The author remarks that " volatile metals
may also be thus detected ; for if an ordinary healthy fly be consumed as above, and the resulting bead compared with another

in which a fly poisoned on arsenical paper has been burned, a very marked difference can be observed between the two, even with an ordinary lens." It is quite possible that the number of elements which enter into the systems of plants and animals may prove, on careful examination, much more numerous than has been hitherto supposed.

A paragraph headed "Diamonds, Artificial, to make" is capable of being misunderstood. The substance produced, consisting exclusively of lime and boric acid, has little chemical analogy with the diamond. The author states that he has discovered the colouring principle of the sapphire, and can now make stones, chiefly of alumina, a deep blue colour without using any colouring oxide whatever. What this colouring principle is the reader is not informed.

To those chemists—unfortunately too few, in England at least —who study the reactions of substances in the dry way, as well as to mineralogists, geologists, mining engineers, &c., this book may be safely recommended, and from its portable character it will prove a useful travelling companion on exploring expeditions.

We wish Colonel Ross continued success in that department of chemistry to which he has devoted himself with such ardour and perseverance, and a wider appreciation than has hitherto been his lot.

Journal of the Society of Telegraph Engineers. Vol. ix., No. 32. May, 1880.

THIS number contains a summary of the present position of electric lighting on the Jablochkoff system. The author considers that if we take it for granted that the question of electric lighting does not progress more rapidly than the other great discoveries and inventions of the century, one may fairly expect that it will, within a very reasonable time, become of general and every-day use.

The desirability of getting rid of the term "candle-power," in the measurement of light, was suggested.

Prof. Hughes read a paper on some effects produced by the immersion of iron and steel wire in acidulated water. Whatever the quality of the metal, or the nature and proportions of the acid, the wire is rendered brittle, whilst similar treatment produces no such effects upon copper and brass. The author adopts the view suggested by Prof. W. Chandler Roberts, that the result is due to the absorption of hydrogen.

*Results of Astronomical Observations made at the Melbourne
Observatory in the Years 1871 to 1875 inclusive.*   Under the
direction of R. L. J. ELLERY, Government Astronomer to
the Colony of Victoria.   Melbourne : J. Ferres.

WE have here a collection of observations which are of course
valuable data for the astronomer, but which are incapable of
being criticised.   The volume opens with a notice of the exact
position of the Observatory, upon which follow a list of the staff
and an account of the instruments.

─────────

The American Naturalist.   May, 1880.

AMONG the valuable matter found in this excellent journal we are
particularly struck with an article by J. S. Lippincott, entitled
" The Critics of Evolution."   We have never seen a happier or
more complete exposure and refutation of the calumnies and
personal imputations with which some authors and orators have
sought to encounter the doctrine of Evolution, especially in that
phase commonly known as Darwinism.   If anything could make
such men as Joseph Cook, C. Hodge, and F. O. Morris feel
intensely ashamed of themselves it would be a careful perusal of
Mr. Lippincott's statements.   He shows that, so far from having
undertaken an atheistic propagandism, as the Rev. F. O. Morris
·has dared to assert,* Mr, Darwin admits a Divine agency and
Divine supervision in forming and peopling the world.   It is
perfectly true that in the earlier days of the Darwinian controversy the new theory—as it was supposed—was eagerly accepted
by the so-called free-thinkers, and was advocated from heterodox
platforms and in heterodox journals, a championship which
official " free-thought " now repents of, and is very little likely
to continue.   But wherefore was Evolution thus welcomed ?
Simply because ignorant or jealous critics had pronounced it
atheistic.   Mr. Darwin's great sin, in the eyes of the modern
Pharisees, was that he had shed some light on God's ways of
working !   It is strange, but instructive, to learn that " of all the
younger brood of naturalists whom Agassiz educated, every one
—Morse, Shaler, Verrill, Niles, Hyatt, Scudder, Putnam, even
his own son—has accepted Evolution."

* " I do my little best or worst to shake their faith."   *See* Journal of Science,
1878, p. 468.

CORRESPONDENCE.

CURIOUS PHENOMENON IN A FROG.

To the Editor of The Journal of Science.

SIR,—Being on an occasion in my greenhouse I observed a cat playing with a small frog—patting it with *velvet* paw, with the evident purpose, for its own amusement, of making it jump: this continued some time. Whilst absorbed in the contemplation of a plant my attention was arrested by a very peculiar noise—a kind of suppressed shriek, seemingly expressive of pain. I looked about to ascertain the cause : the cat was still playing with the frog, and from the frog I found the sound proceeded. When the cat patted the frog the cry was emitted, evidently from an excitation of fear, or from weariness excited by the forced activity induced by the cat's undesired attentions. I was the more surprised at the sound because I was under the impression that all the vocal expressions of frogs was the croak. Naturalists are silent on this phenomenon. On the gardener coming into the house I called his attention to the peculiarity of the cry emitted by the frog. He treated the matter as a very commonplace incident, and said it was common with frogs when in terror; that he had frequently heard it, and that it was always uttered by frogs when pursued by a snake or a viper.—I am, &c.,

S. B.

[Our correspondent's gardener is perfectly correct ; frogs do utter shrieks if pursued by serpents. We cannot at the moment refer to any text-book of natural history in which this fact is mentioned.—ED. J. S.]

ROOKS SETTLING.

To the Editor of the Journal of Science.

SIR,—In my neighbourhood are many rooks. Observing them, I saw a rook (they generally, if not always, fly with the wind) about to settle, when it made a semicircle before alighting. I thought this movement was exceptional, but, continuing my observations, I found that all the rooks made a half circle *against the wind* before settling. I thought it a curious instinct. Repeated observations have confirmed me in the idea that it is their universal habit, occasioned, I suppose, by the necessity of breaking the impetus of the momentum acquired by their downward flight on the wind. The change in direction is probably occasioned by the interstices to be observed in the spreaded wing of a rook, disabling them by any other means to arrest the acquired velocity. The turn made is always against the wind, and even then on settling they appear to strike the ground with a heavy thud. Doubtless was the ground reached in direct downward flight, the weight of the bird and the inability of control—occasioned, most likely, by the interstices between the feathers of the wings—would cause fracture or other serious inconvenience.—I am, &c.,

S. B.

ORIGIN OF THE PERCEPTION OF NUMBER.

To the Editor of the Journal of Science.

Sir,—A little boy, the son of a friend of mine living at Aylesbury, seeing three rooks perched on the branch of a tree, exclaimed, "Quo, quo, quo!" It struck me that these words explain the manner in which the number of objects is apprehended by young children and by the lower animals.—I am, &c.

C. S.

NOTES.

Mr. J. W. Redhouse, in a communication to the Royal Society of Literature, combats the ordinary theory that the Aryan race of mankind originated in the Pamir highlands of Central Asia, spreading thence in a north-westerly direction into Europe, and south-easterly into India. He considers that the Polar regions, which at one time possessed a tropical temperature, were the original home of man.

Signor Beccari has observed that the visits of a certain ant (*Pheidola Javana ?*) seem essential to the health, and even the existence, of certain plants of the genus *Myrmecodia*, from Borneo.

According to a communication made to the Linnean Society, on behalf of Mrs. Bunbury, of Western Australia, it appears that the once common flowers of that colony are becoming rapidly scarce in the pasture lands, and that it is even difficult to propagate them by culture.

The "Revue Internationale des Sciences Biologiques" gives an extensive memoir by Dr. Carl Hoberland, on "Infanticide among Ancient and Modern Nations." He traces the origin of this custom to the difficulty of subsistence, the sacrifice being in the outset urged by the male parent, and opposed or reluctantly submitted to by the mother. (We observe a parallel case among certain of the lower animals, where the young are often destroyed by the father, and are defended against him and concealed by the mother.)

The invention of the diaphote, if all that is alleged concerning its powers proves correct, recals the magic mirrors in which, according to tradition, mediæval necromancers were able to show their clients persons, objects, and events in distant countries.

According to M. Corder, the groups of shooting stars proceeding from different centres have distinctive characters as regards colouration.

The Chinese consider the use of cows' milk, as an article of human diet, unnatural and immoral. It does not appear that this opinion is based on the comparative indigestibility of the milk of ruminant animals.

Dr. Manson has observed that in persons affected by *Filariæ* the blood is comparatively free from these parasites during the day, but swarms with them at night, especially during those hours when mosquitoes bite most freely. He has also described the life-history of the *Filaria* during its development in the body of the mosquito.

Mr. F. Galton, in a paper read before the Anthropological Society, described a curious psychological phenomenon. About one in every thirty adult males, or fifteen females, whenever they think of numerals, see them in a vivid mental picture, and each number always occupies the same relative position in their field of view.

The following incident was communicated to the " Popular Science Monthly " by a valued correspondent :—" A bull dog and a Newfoundland came into collision in Federal Street. The Newfoundland took to his heels for safety, and was closely pursued. Seeing that he was likely to be overtaken he caught up a bit of dirt from the street, and at the critical moment dropped it as if it were something of value he was obliged to give up. The *ruse* succeeded ; for the bull-dog stopped to pick up the supposed tit-bit, arid the Newfoundland escaped. The disgust manifested by the vicious brute when he found how he had been outwitted is said to have been very comical."

An " Anti-Vivisectionist Conference " was held in Edinburgh on the 22nd of May, under the presidency of the Earl of Kintore. Balaam's ass was brought forward as evidence in favour of the agitators.

The " Medical Press " reports a fatal case of hydrophobia resulting from the bite of a cat. No light is thrown upon the manner in which the cat became affected.

According to the " Medical Press " the death-rate in St. Petersburg is 59 per 1000, or more than three times that of London.

With reference to Dr. Croll's " eccentricity theory " of glaciation, Mr. A. R. Wallace writes to the " Geological Magazine " to show that, independent of ocean currents, Europe and North America are subject to different climatic conditions, which explain why the ice-fields should extend farther south in the western continent. The author is about to publish an important modification of Dr. Croll's theory, which we shall await with much interest.

Mr. P. M. Bose, in a valuable paper on Extinct Carnivora, communicated to the " Geological Magazine," shows that, starting from *Arctocyon,* the most primitive Carnivora known, there are two divergent series, one comprising *Palæonictis, Amphicyon,* and *Cynodon,* and the other *Proviverra, Hyænodon, Pterodon, Amblectonus, Oxyæna,* and probably also *Synoplotherium, Mesonyx, Patriofelis,* and *Sinopa.* The former series approach, in the form of their teeth, the typical Carnivora of which they were doubtless the ancestors. The second series formed an exceptional group, of which *Hyænodon* was the last and most highly organised form.

M. Ch. Richet has laid before the Academy of Sciences an account of researches on the action of alkaline and acid mediums upon the life of Crustaceans. He finds that such liquids are not poisonous in the direct ratio of their acidity and alkalinity. Nitric acid is five times more poisonous than sulphuric, and twenty-five times more poisonous than acetic and tartaric acids, for equal weights. Ammonia is, in equal weights, thirty times more poisonous than baryta, and fifteen times more poisonous than soda, and even more deadly than strychnine. As all these substances, if injected into the circulatory system, would prove fatal, the difference is probably due to the degree in which they are absorbed.

M. Viallanes, writing in the " Comptes Rendus," demonstrates that the heart of insects is at first a simple tube open merely at its two extremities. As long as there are no lateral orifices the heart is completely arterial. The author has also indicated the mode of the formation of the lateral orifices and of the pericardic sinus.

M. Thibaut finds that the production of urea in the animal system is not exclusively confined to the liver, but extends in a slight degree through the whole organism.

The " Bulletin de la Société Botanique de France " records a case of morbid alcoholic fermentation observed in the roots of apple-trees in Normandy. No yeast-cells or other microphytes were found in the affected roots.

We perceive with regret that the practice of stocking uninhabited islands with goats and rabbits still continues. The consequence is the destruction of the native flora, and, secondarily, of the fauna also.

Dr. W. K. Parker, F.R.S., has communicated to the Royal Society some very important researches on the structure and development of the skull in the Batrachia. Concerning the geographical distribution of frogs, he remarks that there is a sort of *facies* or character about the allied types of any great geographical region which indicates that certain external characters repeat themselves in different parts of the world. The European and Indian regions yield the highest kinds ; Australia and South America the lowest and most generalised.

Mr. J. Norman Lockyer, F.R.S., has laid before the Royal Society a further note on the spectrum of carbon. The additional phenomena described are a blue line with a wave-length of 4266, a set of blue flutings from 4215 to 4151, and another set of ultra-violet flutings from 3885 to 3843. The blue flutings correspond to the lowest temperature, the violet flutings to an intermediate, and the blue line to the highest temperature. The blue line doubtless represents the most simple molecular grouping.

Mr. W. H. Preece has communicated to the Royal Society a paper on some "Thermal Effects of Electric Currents." The facts elicited are the extreme rapidity with which fine wires acquire and lose their increased temperatures, and the excessive sensibility to linear expansion which fine wires of high resistance evince.

Dr. W. J. Russell, F.R.S., and S. West, M.B., have communicated to the Royal Society a memoir on the relation of the urea to the total nitrogen of the urine in disease. The authors conclude that the chemistry of the urine remains the same in disease as in health, the urea forming 90 per cent of the total nitrogen.

The same authors have also contributed a paper on the determination of the "vital nitrogen" in man. They give the average as 8 grms. = 17·5 grms., or 260 grains of urea.

Mr. J. Costerns has examined the action of saline solutions upon the life of vegetable protoplasm. He observes that the cellules of the red beet-root, if the air has free access, are injuriously affected by solutions of common salt and saltpetre. In presence of very small quantities of air these same liquids prolong the life of the cellules.—*Archives Néerlandaises.*

M. J. A. Roorda Smit concludes that in South Africa the diamond is found in a primitive gangue which is of volcanic origin and has experienced merely secondary modifications. The mines are extinct volcanic craters, and the diamonds have been formed at the expense of organic matter under the joint influence of great pressure and strong heat.—*Archives Néerlandaises.*

We have the pleasure of announcing that the New York Anti-Vivisection Bill has fallen through.

Mr. T. Wrigley, of Timberhurst, near Bury, has bequeathed £10,000 to Owens College.

M. Selzer communicates to the Imperial Academy of Sciences at Vienna that he estimates the optic fibres in the human eye at 438,000, and the retinal cones at 3,360,000.

The "Pharmaceutische Central Halle" recommends the following mixture for the destruction of parasites on plants :—

Boracic acid	...	...	...	...	...	10 parts.
Salicylic acid	...	...	...	...	...	5 ,,
Rectified spirit	,,.	...	...	...	20 ,,
Water	...	...	...	...	...	200 ,,

The liquid is applied to the plants by means of a spray-producer.

M. Ekunina ("Journal für Praktische Chemie") has investigated the acid reaction which appears in the animal tissues after death, and finds it due to a decomposition of the juices produced by *Schizomycetes*. Volatile fatty acids first appear, generated by the incipient decomposition of the albumen, soon followed by the two forms of lactic acid derived from glycogen. The more carbo-hydrates are present in any tissue the longer its acid reaction remains after death. In from twenty to forty hours the lactic acids disappear, and are succeeded by succinic acid. Finally the reaction becomes alkaline, in consequence of the ammonia formed by the decomposing albumen.

Prof. Huxley, in his lecture on the "Coming of Age of the Origin of Species," delivered at the Royal Institution, says—"It was gravely maintained and taught that the end of every geological epoch was signalised by a cataclysm, by which every living thing on the globe was swept away, to be replaced by a brand-new creation when the world was restored to quietness. . . . I may be wrong, but I doubt if at the present time there is a single responsible representative of these opinions left." In a work entitled "Outlines of Geology and Geological Notes of Ireland, by Mr. W. Hughes, and which has reached its third edition, the ultra-catastrophist view is stated in its most unmitigated form. The author declares that "this earth was subjected to repeated changes, that thousands of years intervened between them, and that each completely destroyed all animal and vegetable life." The progress of sound Science has not been so rapid as Prof. Huxley supposes.

Dr. G. Thin has laid before the Royal Society a memoir on *Bacterium fœtidum*, an organism associated with profuse sweating from the soles of the feet. The exudation in question has no offensive smell as it escapes from the skin, but acquires its characteristic odour when absorbed by the stocking. It is not pure sweat, but contains an admixture of blood-serum, and has an alkaline reaction. The author describes and figures the stages of development of the *Bacterium*.

Mr. J. B. Hannay communicates to the Royal Society a memoir on the state of fluids at their critical temperatures. He contends that the difference between the fluid and the gaseous states is not entirely dependent upon the length of the mean free path, but also upon the mean velocity of the molecule.

According to a memoir submitted to the Academy of Sciences by M. Marangoni, the swim-bladder is the organ which regulates the migrations of fishes. Those species which frequent shallow and warm waters, and are always found at the bottom, are without swim-bladder, and do not migrate. Those which have a swim-bladder generally inhabit deep waters, but rise to the surface to spawn. Fishes do not rise spontaneously, and have to struggle with their fins against the influence of their swim-bladder. This organ renders them instable, both as to level and position. This, the author considers, helps to render them agile, as the most active of terrestrial animals are those which have the least stability.

G. D. Liveing, F.R.S., and J. Dewar, F.R.S., communicate to the Royal Society a memoir on the history of the carbon spectrum, in reply to Mr. Lockyer's paper on the same subject, read before the Society April 29th, 1880. In a subsequent paper, on the spectra of the compounds of carbon with hydrogen and nitrogen, the same authors remark—"As we have now demonstrated the utterly unsatisfactory character of the crucial experiment on the strength of which Mr. Lockyer condemns in so sweeping a manner our conclusions, it follows that the whole fabric of his argument is utterly futile, and it seems unnecessary to add any further comments.

J. B. Ramsay, F.R.S.E., has forwarded to the Royal Society a paper on the solubility of solids in gases. He concludes that a gas must have a certain density before it will act as a solvent, and when its volume is increased to more than twice its liquid volume its solvent action is almost destroyed. The volume remaining the same, the solvent power rises with the temperature.

R. C. Rowe, F.I.C. ,Cambridge, has communicated to the Royal Society a simplification of Abel's theorem.

Mr. W. Crookes, F.R.S., has communicated to the Royal Society, in the form of a letter to the Secretary, Prof. Stokes, a condensed summary of the evidence in proof of the existence of a fourth state of matter. In conclusion he says, "That which we call matter is nothing more than the effect upon our senses of the movements of molecules. The space covered by the motion of molecules has no more right to be called matter than the air traversed by a rifle bullet has to be called lead. From this point of view, then, matter is but a mode of motion ; at the absolute zero of temperature the inter-molecular movement would stop, and, although *something* retaining the properties of inertia and weight would remain, *matter*, as we know it, would cease to exist."

According to Mr. A. Wilcocks, the shadows thrown by the planet Venus are distinguished from those thrown by the sun and the moon by their sharpness, the penumbra being totally wanting. This is due to the fact that to our eyes Venus is merely a luminous point, whilst the light of the sun and of the moon emanates from broad disks.—*Proc. Amer. Phil. Soc.*, xvii., p. 705.

THE

JOURNAL OF SCIENCE.

AUGUST, 1880.

I. REPORT ON SCIENTIFIC SOCIETIES.

By Dr. C. K. AKIN.

OUR readers are doubtless aware that some years ago there was a formal inquiry as to the most efficient means for the advancement of Science in England. The following letters on this important subject were addressed to Professor Stokes, Secretary of the Royal Society, by a gentleman well acquainted with the universities and learned societies of most parts of the Continent. They have never been published, and as they have been kindly placed at our disposal by the author we gladly embrace the opportunity of laying them before our readers.—ED. J. S.

Pesth, June 24, 1870.

MY DEAR SIR,

As the terms of appointment of the Royal Commission of which you are a member make reference to England only, perhaps you were astonished at my wishing to tender it any advice. But, on consideration, you will no doubt agree with me that, Science being in its essence cosmopolitan, whatever concerns the science of any nation cannot but awaken the interest of scientific men at large, of whatever country. You will also recollect that, in the course of a more than three years' residence, in the aggregate, at Cambridge, Oxford, and London, I have had some experience of the state of Science in England; which I have been able to compare with the corresponding state of things in Germany and in France, having spent, both previous and subsequent to my stay in England, seven or eight years on the whole at

Berlin, Göttingen, Heidelberg, Königsberg, and Paris. My
motive for stepping thus forward is the hope of being possibly instrumental in effecting some good ; but I freely confess
also to the desire, which I am apt to entertain like many
other persons, of discharging my mind upon a fitting occasion of thoughts which have weighed for years upon it ; and,
with regard to the matter on hand, I may perhaps mention
that my meditations concerning it date as far back as my
experiences.

I intend to divide the remarks I shall make under four
heads, having reference to—1, the composition and functions
of scientific societies ; 2, the efficiency of the universities
and kindred institutions with regard to scientific instruction
and research ; 3, the services of scientific literature of a
periodic and non-periodic character; and 4, the recognition
of Science by Society and the State.

In this present letter I shall confine myself to a consideration, very summary, of the subject first mentioned, viz.:—

Of Scientific Societies.—The scientific societies extant in
England may be distributed among four classes : the *first*,
containing the Royal Society, so-called κατ' ἐξοχην; the
second, the remaining London Societies ; the *third*, the
British Association ; and the *fourth*, the Provincial Societies. If it be now asked, what are the functions of these
several associations, and in what do their aims differ or
agree ? the answer will prove a little embarrassing, and, if
candidly given, not over satisfactory.

As regards the Royal Society, there was undoubtedly a
time when it proved of great utility to Science. Its Transactions were the *Acta eruditorum* or *Journal des Savants* of
England ; its Secretary corresponded with many of the
foreign *virtuosi*, acting thus as a connecting-link between
English and continental philosophers ; finally, its Curator
repeated and made known all the newest facts and experiments at the Society's meetings. That was in the early
stages of the Society's existence, when it also kept a laboratory of its own ; but now all this is changed. The Society
still publishes its Transactions, containing highly valuable
papers; but it thereby confers no boon upon Science, as
many of the papers are reprinted, and in fact all might be
published, with greater despatch and in a more accessible
form, through the means afforded by the scientific periodicals.
The office of Secretary still exists, and there are even three
secretaries acting at present, one of whom is specially
designated as the Foreign Secretary ; but their functions bear
no trace of what they originally had been. The curatorship,

on the other hand, as well as the laboratory, have ceased to exist.

In consequence of the great subdivision of Science the meetings of the Royal Society are attended by few members, still fewer of whom generally take any interest in the papers that are read. As for the medals awarded by the Society they can be considered only as badges, but scarcely as recompenses, still less as incentives to work. Similarly, the sums annually distributed out of the donation fund, and the grant obtained from Government, have been hitherto productive of little good, at least in so far as they have mostly conduced to such researches only as without such aid would have been executed as well; and strange to say, the fund and grant, however unimportant, are not even annually exhausted. The greatest service that Science has recently had to thank the Society for consists in the publication of the Catalogue of Memoirs which has been undertaken by the Royal Society, partly, upon the initiative of the British Association, and at the expense (so far as printing, paper, and binding) of Government.

I trust that neither yourself as Secretary, nor anyone else, will consider the above remarks as conveying anything disrespectful to the Royal Society, which, from the great number of highly eminent persons it includes, cannot but command, even if it had no other titles, everybody's esteem. But if I may quote words of my own, "a Society may be said to fulfil a useful purpose if its members combined produce a greater amount of valuable results than they would in the aggregate if in a state of isolation; or, still better, if the Society in its corporate capacity performs services which no individual or individuals unconnected with each other could render." Now, I may be permitted to ask, has such been the Royal Society's case of late years? In my opinion it is more of an ornament than a utility to the State; and the tendency to make it such is apparent also in the comparatively new practice, according to which the number of Fellows annually elected has been limited, and that at quite an inconsiderable figure. Such a proceeding is intelligible enough, although detrimental likewise, in the case of the foreign academies, the members of which are paid, and considered somewhat as functionaries of State; but it must appear strange on the part of a voluntary association, the members of which pay an annual contribution, destined for the " promotion of natural knowledge."

The remaining London Societies—being formed on the model of the Royal Society, but less exclusive in their elec-

tions, although more narrow in the circle they embrace—
enjoy upon the whole a greater popularity and show more
signs of activity. Still, although at times lively, their meetings produce in the end nothing, or scarcely anything, but
papers, the publication of which is the only observable sign
of their existence. To this latter subject, however, I propose to revert again when I shall come to discuss the state
of scientific literature.

The British Association is in part like to the Royal
Society from the number of sciences it embraces, but in part
also it is similar to the other scientific societies, admitting,
as it does, all and any persons who are willing to pay.

It also resembles the Provincial Societies—concerning
which nothing in particular need be said—from its holding
its Sessions in provincial towns; and in sowing thus broad-
cast the seeds of Science all over the country may be said
to consist of the principal service rendered by the British
Association. For as regards its reports and grants, the
same remarks might be applied that I have made above;
and as for the Observatory kept at the Association's expense
at Kew, it is not clear to me for what reasons its objects
could not be attained at Greenwich as well.

The defects I have noticed in the working of the scientific societies arise, in my opinion, from the want of a proper
organisation of each, and the absence of any co-operation
among themselves. As a remedy, I propose that some kind
of connexion should be established between the several
societies and the Royal Society, on the footing of the affili-
ation that subsists between the colleges and the university.
Every chartered society should be represented on the council of the Royal Society, which would thus become a kind
of metropolitan board of science in permanency. In the
same way the London and the provincial societies should be
represented by delegates on the council of the British
Association, which, thus constituted, might be considered as
a kind of national synod of science, holding an annual
session of short duration. I do not propose this organisation, of course, for its own sake, but as a means to several
ends, the nature of which I shall explain in subsequent
letters. For the present it will be sufficient for me to observe that the scheme I propose will scarcely require any
change, although it implies considerable innovation ; and I
apprehend there will be little difficulty in carrying it out, if
once it be recognised as useful, from the fact that both the
Royal Society and the British Association contain already
the greater number of the more influential members of the

several London and provincial scientific societies. But should there any difficulties arise,—such as impeded the union that it was once intended to establish between the Geological Society, I believe, and the Royal Society,—then it will be a case for the Royal Commissioners to step in, as was done in the instance of the colleges on the occasion of the university reform.

Requesting of you the favour to communicate this letter to your fellow Commissioners,

I remain, my dear sir,
Yours very sincerely,
C. K. Akin.

To G. G. Stokes, Esq., M.A., Sec. R.S., one of the Royal Commissioners on Scientific Instruction and the Advancement of Science, London.

———

Pesth, June 26, 1870.

My Dear Sir,

In the present letter I shall consider the efficiency of the instruction and aids to research derived—

Of the Universities and kindred Institutions.—As for the universities, my observations have reference only to Oxford and Cambridge, which I know by personal experience; and as for the other institutions, I shall limit my remarks to the Royal and the London Institutions.

To the minds of the uninitiated, especially of the uninitiated foreigners, an account of the English Universities, of their colleges and foundations for professorships, fellowships, scholarships, and the like, must appear as the realisation of the ideal institution which is to ensure the perpetual advance of science and learning; yet when we come to look to facts, however useful colleges may have proved with regard to education, their tendency has not been to foster any such scientific activity as that of which their several foundations seem to hold out the prospect. Perhaps it might not be uninstructive to count up the number of modern English scholars, philosophers, and thinkers who have lived, or still live, without the pale of the universities; and it will be contested by no one that, in Germany, at least, the universities *are* in a much higher degree centres both of national and European thought than the English Universities were ever *asserted* to be by their best friends or warmest admirers.

And if England be still foremost among nations in most branches of Science, it is owing mainly to that private and individual energy of mind which distinguishes the English race beyond any other, and in a much less degree only to the corporate action of the universities or to their system. In Germany there are no foundations to secure " a learned leisure " to the successful student, and no mitres to crown the careers of eminent scholars and mathematicians ; moreover, within their juridical, medical, and theological faculties, they afford all that professional instruction of which the English Universities—in my opinion, properly so—are absolved, either wholly or in part, by the inns of court, the hospitals, and certain theological colleges. Yet, notwithstanding that defect on the one hand, and the practical bias thus imparted on the other hand, the eminence in, and devotion to, abstract science and pure learning of the German Universities is actually in Europe unsurpassed.

The causes of the differences noticed in the comparative working of the English and German University systems appear to me to be *two* in number. In the *first* place, the English Universities are much less considered as " seats of learning " than as " seats of education," if I may use an expression actually applied by Mr. Gladstone to them. In Germany the *general education* of youths is supposed to be finished at those establishments which, by the age of their pupils, if not by their constitution or systems of teaching, correspond to the English Grammar and Public Schools. At the University the *instruction* is either *special* or *professional;* special in the case of those who make Science their calling, and who frequent principally the so-called philosophic faculty ; and professional in the case of others (the majority) who wish to become doctors, barristers, lawyers, judges, or priests. Teachers at the German Universities, instead of doing the work of schoolmasters,—which is rarely compatible with the prosecution of independent research,— form " schools," in the higher sense of the word, in which a tradition is handed down from individual to individual and from generation to generation.

All this is different in, or absent from, the English University system. Even at Cambridge, the mother university of so many famous mathematicians, every fresh mathematical inquirer has to set up for himself and begin *de novo ;* for the tradition which there exists relates rather to the modes of tuition than to any methods of investigation. And this brings me to the *second* point I wished to notice, as one of the main causes of the comparatively small influence the universities have exerted on the growth of Science in England. In consequence of the competitive system adopted at the examinations, tuition at the universities has developed into a species of " cramming " *of a very high order*, or, as it may perhaps better be called, a species of " drilling" rather than of training of the mind. As a further consequence, the mental grasp of a high wrangler is generally quite astonishing, and he is capable of solving, so to speak, any problems that may be set him ; but he has mostly lost that inventive faculty which busies itself about the problems implied by Nature, and the very existence of which must be divined. Mostly, great philosophers have worked at questions which *none* had formulated before them ; and many have declared their inability to work at others, or shown their incompetency with regard to such subjects as had not been objects of their predilection. To force a person to inquiries of *any* kind may be a commendable proceeding in the case of individuals whose callings will bring them in contact with the problems of the practical arts, or generally of life ; but neither discoverers nor inventors of a high order can be bred in such fashion. And, as long as fellowships shall be considered as prizes to be obtained only at competitive examinations, it is not to be expected, in my opinion, that the rich foundations of the universities will bear that fruit to Science which they might or ought to; although "Science," in the more limited sense, be admitted in future on the same footing at the examinations as " classics " or " mathematics." The exhaustive strain on the mind produced by these examinations, or the preparation for them, is another element to be considered, and which induces Fellows to repose rather on their laurels than to exert themselves in any further fatiguing work.

Probably a greater part than to the universities is to be ascribed in the spread and development of modern science to the Royal Institution, which has been the scene of the teaching and labours of the three by far greatest philosophers of our century,—of Young, of Davy, and of Faraday,—for neither of whom was there, nor for the two latter of whom could there be, any room at the universities. But, however valuable and important the results achieved by these extraordinary men may be, and although other distinguished persons have worked by their side or may have acted, in one or other case, as the successors of either, the Royal Institution cannot be considered as a school or seminary of philosophers or scientific men. Such it was not even its original object to be; that being, if memory deceive me not, very different, as conceived by Sir Benjamin Thompson (Count Rumford) and the other founders, from what it has actually turned out, and still more different from any such design as has above been mentioned.

The object chiefly aimed at at present is the diffusion of scientific knowledge, which, if I mistake not, is also the avowed, and almost only, object of the London Institution. Perhaps it should be stated that, as far as Science is concerned, the beneficial effects of this popularising tendency, chiefly promoted by the establishments under consideration, are not always unmixed with evil ; for, as was once remarked to me by a person whose name I need not recall to you, " Popular lectures are more frequently intended to make people believe they understand Science than to make them actually understand it." And if once a lecturer, instead of aiming to elevate and expand the minds of his hearers, contract the habit of narrowing Science itself to the degree of exiguity supposed to correspond to the general hearer's intellect, and thereby disfigure and debase it, Science frequently becomes with him, also in his capacity of inquirer, an object for show and exhibition, rather than for disinterested research and truthful exposition.

In the proposals which I shall now submit I am chiefly guided by the following considerations, which I have already had occasion to state in print :—" It is especially by encouraging those who travel along some other than the beaten track of ordinary men of science that corporations representing the interests of science might tend to promote it. . . . It is true that there are not many persons answering to this description to be found at any particular time [many more, however, than one naturally becomes aware of]. It is the exclusive glory of this country to have produced men like Davy and Faraday, for instance, for whom there would have been absolutely no room abroad to develope their talents; yet a little less constancy on the part of Mr. Faraday, if biographical notices inform us aright, would have deprived the world of a discoverer as unrivalled and prolific in his line as Newton was in his; and who knows but that the cold shade of neglect, the want of due encouragement extended to merit exhibited in some unusual way, may not be nipping every now and then some great discoverer in the bud ? - . . A German physician still living [and *now* heaped with useless glory], who is now acknowledged to have been among the first to recognise one of the most important laws in natural philosophy, encountered so much annoyance and vexation whilst endeavouring to bring his discovery before the public in a manner to ensure the attention of men of science, that he absolutely went out of his mind. . . . In the same way many a promising suggestion happens to be advanced by persons who have either not the leisure, the convenience, or the instruments to verify it by experiment, and is therefore totally disregarded "—or else dishonestly appropriated by unscrupulous persons, to the detriment of the discoverer, and thus indirectly of Science itself.

Starting from these principles, and considering that the universities are chiefly engaged in the *general tuition* of Science, while the kindred institutions mainly aim at its *popular diffusion*, I believe there is room for the creation of an establishment the object of which shall be the *special advancement* of Science, both by instruction and research. The inner organisation and outer relations of this establishment, which, for brevity's sake, I will call the Royal College, I think should be based upon the following foundations and governed by the following rules:—The members of the college should consist of probationers, fellows, deans, and president or master. As *probationer* any person should be admitted who could satisfy the council of any chartered scientific society, under the restriction to be later apparent, of his willingness and capacity for scientific work in the department being of such society's competency. After the lapse of one year, and on presentation of a report by the competent dean to the council of the Royal Society, which shall be judged by it satisfactory, the probationer will be admitted as a *fellow* into the college. During his stage of probation, the person introduced to the college by any society shall have his expenses for apparatus, &c., defrayed by such society (as also his personal expenses, if the society be so inclined) ; on being, however, admitted a fellow, he shall be both personally maintained and furnished with all necessary requisites for work at the cost of the college. As *deans* shall be elected such of the fellows of at least three years' standing as by their labours in the latter capacity shall have given most satisfaction to the council of the British Association, that shall be enabled to take cognisance of each fellow's work ; it being understood that both the councils, that of the Royal Society above mentioned as well as that of the British Association, shall have been so constituted as stated in my former letter. Fellows should be nominated for three years, but capable of re-appointment ; and the deans also, each of whom shall represent, singly, one distinct department of science, should be elected for the same term, but not be capable of re-election, the functions

of dean, as to be explained below, requiring a constant infusion of fresh blood. The president or *master*, being the
administrator and governor of the college, shall be nominated by the Crown, upon a joint presentation of the two
councils of the Royal Society and the British Association.

The duties of probationers and fellows shall be simply to
work, the former in order to be admitted as fellows, the latter
in order to be re-appointed; those of deans, on the other
hand, will be of a twofold nature. In the first place, it will
be their office to aid and supervise in their work both the
probationers and fellows, each of his own department; and,
secondly, each dean will give in each year or session one
course of lectures, the subject of which shall be chosen by
himself, but varied from year to year. It is supposed, moreover, as evident of itself, that it will become the dean to
work also on his own account in order to be re-elected in
his capacity of fellow, which shall remain independent of
that of dean. The duties, lastly, of the master, as already
referred to, shall consist mainly in the government and
administration of the college, as well as in its representation
in all external relations.

I now come to the chief question, How shall the expense
of this establishment be defrayed? According to my views
the British Association, the Royal Society, and the principal
chartered societies, firstly, should tax themselves to a certain
amount annually, in order to realise a scheme which I cannot but believe is in harmony with and will greatly promote
the tendency of them all, viz., the advancement of Science.
In what manner, next, the Government certainly, and perhaps also the universities, ought to be induced to contribute,
both to its maintenance and foundation, is a subject for the
Commission to deliberate upon. I will only further state
that I think the college might perhaps be suitably connected
with the British Association's establishment at Kew; and
that, besides laboratories, it should contain all the necessary
adjuncts for living furnished by colleges at Cambridge or
Oxford. Should the universities be willing to partake in the
erection and maintenance of the Royal College, of course
they ought to enjoy the same rights of presentation as
awarded to the chartered societies, as also some influence in
other appointments and elections.

Believe me, my dear sir,

Yours sincerely,

C. K. AKIN.

Prof. STOKES, Sec. R.S.

PS.—Of a not quite unimportant function the Royal College, by its members, will be able, and made, to fulfil with regard to scientific literature, I think it preferable to speak in my next letter, in which the latter subject shall be treated as a whole.

C. K. A.

Pesth, June 28, 1870.

MY DEAR SIR,

Conforming to the order adopted in my first letter, in enumerating the subject matters to be discussed, I shall treat in this present letter—

Of Scientific Literature, Periodic and Non-periodic.—The class of works which comes here under consideration is composed of *three* different categories: the first comprising the works intended for specialists or scientific men exclusively; the *second*, those designed for amateurs, if not wholly yet principally; and, finally, the *third*, those written simply for popular readers.

To the category *first* mentioned belong, independently of treatises and similar works, the manifold scientific journals, periodicals, proceedings, and transactions, published partly by booksellers and partly by societies. Now nothing can exceed the unsatisfactory state of this latter species of literature. A physicist, for instance, in order to become familiar with the current progress of his science, has to consult some twenty different works published periodically, yet at times more or less uncertain; but if he be a little more scrupulous or comprehensive in his tastes, and have a mastery of other languages besides English, German, and French, which in the above estimate I have alone taken into account—then, instead of twenty, there will be even two or three times twenty such works to call for his attention. The statement holds as good for most other sciences as it does for natural philosophy; and in one or two branches only has there recently been any attempt made to *condense,* rather than to concentrate or *centralise,* all new publications scattered periodically through a great variety of works and bearing upon some distinct portion of science. But, without wishing to detract from the usefulness of these undertakings, it may be asserted that they rather indicate than supply a want.

As far as the inconvenience just referred to goes, men of science are the only sufferers; but there is another perplexiity to which they are liable in common with the important class of amateurs, and which is owing to deficiencies in the *second* species of literature before adverted to. Next to the desire for complete information concerning his special department, every man of science is, or should be, animated also by the wish for some such knowledge of the progress achieved in all kindred departments as would satisfy, ordinarily, the simple amateur. More than one attempt has been made already to supply the want here stated, but each has hitherto ended in failure, excepting the one last made, which is highly to be commended, but is yet not, perhaps, quite suffi-cient. In fairness it should be stated that the task of carrying on a publication of this kind, so as to be successful and complete, may almost be called formidable.

The *third* and last category of books may here deserve also some words of notice. In the introduction to his work on natural philosophy, Dr. Young has made the following observation :—" I shall esteem it better to seek substantial utility than temporary amusement, for if we fail to be useful for want of being sufficiently entertaining we remain at least respectable ; but if we are unsuccessful in our attempts to amuse, we immediately appear trifling and contemptible." Such are not the views acted on or adopted by many modern popular writers. While the most abstruse questions of politics, for instance, are daily discussed, not in treatises only, but in newspapers even, in a becoming manner, popular authors on Science, both in periodicals and independent works, generally address the public in a style befitting rather the intellect of children than that of grown-up people, for whom those yet are designed. Even in books calculated for well-educated persons, the matter is frequently dressed up in the fashion generally deemed suitable only for "the million " in other departments of literature ; and it is im-possible not to be astonished, if we once come to reflect upon it, at the amount of conceit shown in this way by authors and tolerated by the public. It seems almost as if it were agreed upon all hands that Science is as much above the comprehension of ordinary men as Politics are supposed to be above that of ordinary women ; and as if men in the one case, like women in the other, concurred themselves in the verdict.

In order to remedy the defects noticed in the first two species of literature (for with the third I am here, so far, not concerned), I propose the following scheme :—

In the *first* place, all scientific societies of any importance, in England as well as in foreign countries, should be invited to publish their transactions in such form as to make each single paper contained in them purchaseable *separate*. From all such societies should be obtained also a limited number of copies of their complete Transactions, the several papers published in which shall be collected and bound up in volumes corresponding, as far as their *matter* goes, to the various branches of Science. Such volumes might be issued annually, one for every department, and bear the title " Contributions to Knowledge," with indication of the year and the name of the science.

The measure just explained takes into account current literature only ; but even greater difficulties attach to the accumulated store of works published in the course of centuries or ages, if information concerning their contents be desired. As you stated last year at Exeter, " To make oneself, without assistance, well acquainted with what has been done, it is requisite to have access to an extensive library, to be able to read with facility several modern languages, and to have leisure to hunt through the tables of contents, or at least the indices of a number of serial works." And as you went on to say, " With a view to meet this difficulty, the British Association has requested individuals who were more specially conversant with particular departments of science to draw up reports on the present state of our knowledge in, or the recent progress of, special branches." Such, however, is the labour involved in the preparation of such reports that all the influence of the Association has *not* been sufficient to induce a number of scientific men to *regularly* undertake it ; and as it is found in practice that mostly young persons only have sufficient enthusiasm for labour of this kind, so the British Association has shown only in the early stages of its existence much activity in this direction. A most valuable surrogate for reports of this kind will be furnished by the great Catalogue of Memoirs published by the Royal Society at the British Association's instigation ; but more than either reports or a mere catalogue is required in order to satisfy our want for some synopsis of the work done in the various branches of science.

I therefore propose as a *second* measure, and co-ordinate with the above, the publication at times of volumes comprising collections of all papers published within a stated period, and having reference to some such branch or question in science as has prominently sprung into notice or is rapidly progressing; for concerning such, as you observed, it is by no means easy now to get acquainted with its actual state, nor, as I will add, with its history. The papers should be

communicated either as a whole, or in such abridgments as
will bring out their importance in respect of the matter
treated. In all cases the author's own words ought to be
used, and perhaps they should be given also in the original
language; for it is not so much the diversity of languages
as the multiplicity of works familiarity with and access to
which is required that is the great stumbling-block in the
way of getting up any question. This publication might
bear the name of "Archives of Science," with sub-titles
corresponding to the subject-matter of the volumes. As an
annex or supplement to this collection might be published
also, from time to time, volumes comprising the collected
works of those philosophers whose merits in science would
deserve such a memorial of their work. As belonging to
this part of the subject, and preparatory to the above serials,
I would suggest the speedy publication of indexes to such
works, periodically issued, as now lack them; and among
which even those important collections the "Philosophical
Transactions" and "Philosophical Magazine" must be
numbered, at least in respect of their latest series.

In the *third* instance I propose the publication of periodicals to be issued monthly, each devoted to a distinct branch
of science, and the objects of which I can most shortly
explain by a statement of what should be their uniform
table of contents, viz., (1) original communications on new
work; (2) reprints or abstracts of papers published in
journals or proceedings; (3) notices of books and bibliography; (4) correspondence and discussions; and (5) mis-
cellaneous. These publications, which I will call *bulletins*,
should be like the Archives polyglot, for the reasons above
stated and in the manner there explained; actually the
existence of a periodical of this kind (the "Astronomishe
Nachrichten"), which stands high in public estimation and
usefulness, proves the practicability of the principle. But
in connexion with the *bulletins* should be issued another
serial, and this *wholly* in English, to be published weekly,
and giving a summary of progress in science and other
intelligence, culled from the *bulletins*, and adapted for the
amateur or cultivated general reader who takes some interest
in science. This publication should contain (1) leaders on
scientific questions of the day; (2) abstracts, reprints, or
translations from other journals; (3) proceedings of scientific societies; (4) reviews of books and bibliography;
(5) news and intelligence; (6) complete lists of scientific
papers published in periodicals. This publication, which
has its near prototype in an extant periodical, might be
designated as the "Journal of Science."

With regard to the realisation of these several projects, I believe that the British Association might well take the initiative in respect of the *first*, or the formation and issuing of the contributions; while the *second*, relating to the collection and publication of the Archives, might be carried on under the auspices of the Royal Society. The composition and editing of the Bulletins and Journal, *finally*, should form part of the duties of the Royal College. The expense entailed by these publications, in so far as it may not be covered by their sale, should be borne by the corporations mentioned and the chartered societies jointly, or eventually by the State, as is done in the case of certain publications now issued in the interests of history. The universities also might help by lending the aid of their presses.

Should the suggestions here made be adopted and acted on, England would confer benefits on science and scientific men all over the world; but at the same time it would set an example which other countries would be sure to follow, and would start enterprises in the costs and glory of which other nations, no doubt, would hasten to share, if the stimulus were once given. Actually, international societies have been formed for various purposes, some even for strictly scientific objects—as in the case of the International Association of Astronomers; and there is no reason why those beneficial institutions should not be extended and more largely adapted to science, so as to make all nations work, either by congresses, conferences, or associations, for objects which are common to them all.

In conclusion, I will only state that, of the various proposals made in these letters, those above developed I have perhaps most maturely considered, and with their importance I am most deeply impressed, as they have engaged my attention and thoughts earliest and oftenest. Without a knowledge of previous researches, as you said last year at Exeter, " there is always the risk that a scientific man may spend his strength in doing over again what has been done already, whereas with better direction the same expenditure of time and labour might have resulted in some substantial addition to our knowledge." And, if it be the primary object of Science to seek and to find out truth, to preserve and rescue truth from oblivion is to her progress of no less vital importance, as all must agree.

Mentioning, finally, for your information, that my next letter will be the last of the series,

I remain, my dear sir,
Yours very sincerely,

Prof. STOKES, Sec. R.S.   C. K. AKIN.

(To be continued.)

II. THE CONSTITUTION OF THE EARTH.

By ROBERT WARD.

THE notice which was given, in the " Journal of Science" for June, of my book, "The Constitution of the Earth," well deserves my thanks as an honest attempt to give a fair idea of its contents. A book which, as the reviewer observes, is "strangely at variance" with the teachings of our received scientific text-books," might have been safely described in garbled language which appealed to the prejudices of the reader, and the writer might have triumphantly retired from his work with the proud consciousness of having annihilated the humble individual who had presumed to disturb the orthodox scientific teaching of this enlightened age. It is undoubtedly the fault of my book that it apparently comes into conflict with many ideas which have gained a place in our scientific text-books; but I believe that I am entitled to plead, in mitigation of my offence, that the various departments of science are often greatly in conflict with each other. In his " Physiology of Common Life," the late G. H. Lewes tells us that "we must never attempt to solve the problems of one science by the order of conceptions peculiar to another." I believe that the observation is pregnant with truth, and yet we would not be justified in concluding that there is some actual discordance in Nature which thus results in conclusions which are inharmonious, if not actually antagonistic. Rather we must infer that the seeming incongruity arises from the relativity of human knowledge, which is therefore in every direction imperfect. If I understand aright the view of creation which I have attempted to unfold in my book, instead of being in actual conflict with the great truths which are taught in our text-books, it goes far to reconcile many statements in the different departments of science which are more or less inconsistent with each other; and my faith in it is all the more because, as I believe, it also reconciles the fundamental doctrines of Religion with the great truths of Science.

My book might, appropriately enough, have been entitled " The Evolution of the Earth," but, as explained in the introductory chapter, its present title was chosen thirty-six years ago, when I published "An Essay on the Constitution of the Earth," which contained, in a more or less crude and imperfect shape, the ideas which it is the object of the later work to elucidate, and when the term " Evolution " was not so scientifically popular as it is at the present time. Instead of being the result of some occult or mechanical disposition of things to change, I maintain that the evolution of species has been the result of a great scheme of creation, by which the Earth itself has undergone a progressive development. It seems to me that what are usually described as the " Laws of Nature " are simply *man's recognition of the order, harmony, or consistency, with which the Creator operates in the Universe.* I deny the existence of "immutable elements " or of "eternal and imperishable" matter in any shape. Matter, as it is known to our consciousness through the senses, is continually coming into existence and passing away. Only in the abstract sense can it be said to be unchangeable, for the simple reason that matter devoid of every quality by which it is known to us is only an abstraction, and has no more sensible existence than any other metaphysical conception. We recognise the continual activity of Nature when we observe that all things grow older. Time is, in fact, the symbol or abstract representative of the boundless succession of created things which have existed in the past or may exist in the future.

The order and consistency of the action of God in creation may be formulated as follows :—

1. *The Law of Existence.*—Circumstances govern the creation of things, and therefore all things exist by virtue of their circumstances.

2. *The Law of Diversity.*—As no two things can occupy the same place in the universe, therefore no two things can be exactly alike.

3. *The Law of Organisation, Variation, or Development.*— The differences or similitudes between two things must be in proportion to the differences or similitudes in the circumstances under which they have come into existence or by which they are sustained.

4. *The Law of Motion.*—A change in the circumstances of things necessarily involves a corresponding change in the things themselves.

Except for the theories of "immutable elements " and "imperishable matter," no one would think of disputing that no form of matter of which we have any sensible knowledge can exist independent of the conditions of its existence. As Paley, in his Natural Theology, has declared—"The bodies of animals hold, in their constitution and properties, a close and important relation to natures altogether external to their own —to inanimate substances, and to the specific qualities of these ; *e.g.*, they hold a strict relation to the elements by which they are surrounded." And Chemistry is essentially the science which notes the circumstances under which the various forms of inorganic matter exist, or by which they are changed.

The Law of Diversity follows the Law of Existence as an inevitable consequence, and the third law is also a necessary consequence of the first and second. If circumstances govern the creation of things, it follows that the differences or similitudes between two things must be in proportion to the differences or similitudes in the circumstances under which they have come into existence or by which they are sustained, which comprehend all the circumstances which have contributed to their actual present state. Hence each individual existence, or form of matter, may be described as an embodiment of all the circumstances through which it has passed from the remotest beginning.

Nor does the Law of Motion less strictly follow from the first law. If we admit that all things exist by virtue of their surroundings, a change in the circumstances of things must necessarily involve a corresponding change in the things themselves. It is obvious, however, that the effect of existing circumstances must be governed by the actual creation of the circumstances which preceded. Hence what I have described as Individuality comes into conflict with the existing conditions, and motion of some kind (chemical, physiological, or mechanical ?) is the result.

The belief that creation is a thing only of the past is utterly inconsistent with the religious idea of an ever-present and sustaining God. Evolution is, in fact, only another name for creation by the orderly and consistent action of an Almighty Creator. The same laws which have been observed in the evolution of species have been concerned in the origin and development of the globe upon which they are found. If plants and animals exist by virtue of the earth's existence, it follows that *they must be constituted according to the constitution of the earth.* The earth is constitutionally related to the sun ; the condition of the air, the land, and the water, of animals and plants, is dependent on the earth's relationship to that body ; the earth's relationship to, or dependency on, the sun is paramount to all relationship in respect of other bodies. If we assume that the earth has derived its existence from the sun, and been constituted under its influence, we are presented with a reason of relationship as clear as that every effect must be consistent with its cause.

Thus, by an inductive process of reasoning, we find that the earth is an emanation from the sun. All the undoubted truths of Science are consistent with the conclusion that the Earth and all the other planets of our system have been derived from the sun by a natural process of progressive development; they are, in fact, the children of the Sun, born in the order in which we find them placed in our planetary system, the most remote being the oldest, and intervals of millions of years having transpired between the times of their first foundation. Thus there is a relation between the sizes of the planets and their distances from the sun; between the number of satellites of a planet and its distance from the sun; between the revolutionary motion of a planet and its distance from the sun; and between the rotary motion of a planet and its distance from the sun. Such relationship suggests a higher development in the planets which are most distant. The more distant planets are usually larger, have slower revolutionary and quicker rotary motions, and have more satellites, though such characteristics do not follow with that mathematical accuracy which we might expect if the relationship was of a purely mechanical instead of a physiological nature, as now propounded. The Moon, which I take to be the child of the Earth, has no rotary motion in relation to its primary; it has no atmosphere; it is destitute of water; and no signs have yet been discovered of vegetation or inhabitants. Born, as I assume, long ages after the Earth, it is altogether in an inferior state of development.

Geology admittedly points to a time when there was neither animal nor vegetable life on the earth. It is supposed to be an established fact that organic life originated in the simplest forms, which were slowly followed by others of a higher type; and all the changes which have taken place are believed to be due to the present "order of Nature."

Such conclusions are inconsistent with the idea of an accidental mechanical arrangement of the stratified rocks which constitute the outward covering of our globe, and which, for the most part, present the appearance of aggregations of matter consequent upon chemical or physiological processes. Moreover, they suggest the idea of being actual *additions* to the original matter of which the earth was formed before organic life began to be developed. That they

are, in the case of coal and lime, at least temporary additions is beyond dispute. The coral zoophyte has the power
of secreting lime from the waters of the ocean, with which
it builds up reefs of vast thickness, and entire islands are
thus created. As the result of the *Challenger* Expedition, it
has been discovered that the minute shells of microscopic
animals which live at or near the surface are constantly
being showered down upon the bottom of the ocean, where
they are ultimately transformed into limestone and iron-
stone. An estimate of the enormous extent of this deposit
may be formed by reference to the fact that the ocean covers
three-fourths of the entire globe. Upon the mechanical
theory of the formation of the earth, it is assumed that the
lime thus deposited was previously held in suspension in the
water; but the chemist informs us that salt, not lime, is
the chief ingredient of sea-water. This deposit of lime has
been going on for millions of ages, and there is no reason
to believe that the rate of increase is at all diminishing.
Moreover, Sir Wyville Thomson concludes that the " 'red
clay' is not an additional substance introduced from without," but " is produced by the removal, by some means or
other, over these [deep] areas of the carbonate of lime,
which forms probably about 98 per cent of the material of
the 'globigerina-ooze.' "

The natural inference is that water, like every other form
of matter of which we have any *real* (as contrasted with
theoretical) knowledge, is undergoing a constant change, by
which lime, iron, salt, and numerous other substances, are
constantly being developed or evolved; such evolution being
the result of general or special changes of circumstances to
which water is continually being subjected. The theory of
the mechanical circulation of water, instead of being founded
upon facts, is largely a creation of the imagination. In
place of being immutable or unchangeable, water is one of
the most unstable substances of which we have any knowledge. Under mere changes of temperature, it (that is, the
abstract article) is alternately a vapour, a liquid, or a solid;
under some conditions it becomes putrid, whilst under others
it forms the chief substance of the growing plant or the
living animal. In the human body it is an important element in the various processes by which upwards of fifty
more or less stable substances are formed. We have,
indeed, no reason to believe that water is an exception to
the other various forms of matter of which, through the
senses, we have any conscious knowledge. *As a fact*, it
falls upon the earth to a depth which has been estimated at

from 20 to as much as 250 inches in a year ; and, if we take the average fall at 36 inches, the ponderable matter which thus falls upon every square mile of surface amounts, in round numbers, to no less than 2,326,000 tons. That, in certain dry and warm conditions of the air, some portions of this water rise again in the form of vapour is undoubtedly true, but even such comparatively small quantity probably all falls once more in the form of dew. Certainly it is a *pure theory* that all the water which falls is lifted up again to produce those beautiful creations of the atmosphere which, beginning in the light and fleecy curl-cloud, pass, by a process of growth, to the dark grey rain-cloud, and finally fall in rain, snow, or hail.

True, we cannot take pure water, which is said to be a compound of oxygen and hydrogen, and transform it at pleasure into salt, or lime, or iron. Man can only work successfully in the ways of God. We cannot transform, by any mere effort of our will, the acorn into the oak tree, and yet, if the acorn is planted in the soil and subjected to all the other necessary conditions, it will grow into an oak tree. Man has discovered the means by which a mixture of two light gases can be transformed into water,—why should water not be changeable into salt ? We certainly know how to reverse the process. By means of heat, salt can be fused or liquefied, and so transformed into vapour or gas, in which condition it will return to the air from whence it descended, modified, however, by the circumstances through which it has passed.

Nor is this a view of creation essentially at variance with the accepted theory of scientists as to the formation of the earth. The theory of Immanuel Kant assumes that every form of matter was created out of gas, or the ethereal substance which occupied boundless space. I assume the same, with this difference, that, whilst Kant propounded an entirely fanciful process, I maintain that all the elementary bodies, and every form of matter of which the senses can take cognizance, have been the result of natural processes which are now going on. He set up an idol in the name of gravity ; I worship only the " living and true God," whose power and presence I recognise in all created things.

If the earth be not an emanation from the sun by means of the laws of God in creation to which I have briefly referred in this article, but which are described more at length in my book, how have all the several kinds of " immutable atoms " of which the elementary bodies are composed, and which it is assumed once existed in the form of gas, been collected together and formed into the solid substances which now exist. The theory of gravitation, as now taught, returns no answer to this question ; but if, as Newton suggested, the cause of gravity is to be found " in the general laws of Nature by which the things themselves are formed," the solution is consistent with the constitution of the earth as now expounded.

The space at my disposal prevents me from entering at length into the physical evidences of the earth's expansion. These, however, exist in abundance in the cracks, veins, fissures, ravines, and valleys which are to be found in all directions on the earth's surface, and which present in a large way very much the appearance of the cracks in the bark of growing trees. Assume that the crust of the earth is an addition to its surface, and that, besides the mechanical and other movements described by geologists, we have to add the important effects of growth, and all the difficulties and disputes which prevail, as to the meaning of existing phenomena, are satisfactorily solved.

The sun is often described as the source of life, and its importance in connexion with all the physiological processes which take place on the earth demonstrate the existence of a relationship the reverse of mechanical. It is the active cause of light and heat, and the fossil sunbeams of millions of ages ago are now used to light up our factories and workshops when, by the rotation of the earth on its axis, we cease to enjoy the direct effect of its life-creating influence. Even the colours of the ancient sunshine have not been lost to us, but are reproduced and utilised in the shape of aniline dyes. The spectroscope, as I have endeavoured to demonstrate in my book, not only decomposes the light of the sun into all the colours of the rainbow, but also exhibits evidence of the presence of numerous embryonic elements, which may be capable, under the necessary conditions, of being converted into the solid state. Nor is the proof by measurement of the earth's growth absent, Capt. Drayson having shown that, whenever a base line has been measured a second time, after the absence of years, the later measurement was invariably found to be longer than the former. And the latest measurement of the earth's distance from the sun is also confirmatory of the view of creation now propounded ; the most careful calculations of eminent astronomers, in connexion with the transit of Venus, showing that the distance between these two bodies is increasing, former measurements having also supported the same conclusion.

III. THE PRESENT STATE OF THE VIVI-SECTION QUESTION.*

WHEN the Anti-Vivisection outcry had become so far successful that a Royal Commission was appointed to investigate the charges brought against medical men and experimental physiologists, and when a Bill for official interference was introduced under the auspices of the late Ministry, a professional contemporary expressed the amiable hope that the measure would not " in any material way " hinder the progress of research, and would " calm the needless apprehension and put an end to the odious mis-representations which have been recently rife concerning this subject, and which have been in ignorance adopted by persons of consideration, who will probably in future take more pains to be correctly informed." We ventured at the time to express our fears that the result would be very different.† Our misgivings have been more than verified. Research has been impeded to a very serious degree. In proof we need only quote the instance narrated by Dr. Pye-Smith, F.R.S., in an Address delivered at the Sheffield Meeting of the British Association. It appears that " two thoroughly qualified men were anxious to carry out an important investigation on the treatment of snake-bites. They procured venomous snakes from a distance, and applied for the special certificates necessary. Considerable delay ensued ; various objections were raised and set at rest, and at last all the certificates were obtained ; but meantime the snakes had died !" Dr. Ogle‡ informs us that " at a certain hospital experiments on a most important investigation had not been carried out, owing to the risk of bringing that hospital into public notoriety." Thus between the law and Mrs. Grundy, research is interfered with in a most mate-rial way.

Yet, although biological investigations requiring experiments upon living animals are now rendered all but impossible, it must not be thought that the agitation has decreased,

---

* Ueber den Wissenschaftlichen Missbrauch der Vivisection. Von FRIED-RICH ZÖLLNER, Professor der Astrophysik an der Universität Leipzig. Leipzig : L. Staackman.
† Quarterly Journal of Science, vi. (1876), p. 333.
‡ Harveian Oration for 1880, reported in Medical Press and Circular, June 30, 1880.

that the enormous concessions made are accepted as final, and that the odious misrepresentations have been withdrawn. According to the Report for 1878 of the Inspector under the Vivisection Act, only ten licenses have been applied for in Ireland; only five of these have been acted on, and "no pain has been inflicted even in one of the twenty-four experiments performed." For all this the "Irish Anti-Vivisection Society" still exists and continues its labours! Indeed it seems to us that since the passing of the Act 39 and 40 Vict. the agitation has become even more unscrupulous in its character. We know that the most questionable means have been used to obtain signatures to the petitions for the abolition of vivisection. Persons, children included, who have never given the subject a moment's thought, and who scarcely know what "vivisection" means, are besieged to "just sign their names," and for the sake of quietness they often consent. We know a young lady, the daughter of an eminent artist, who, whilst copying a picture in one of our public art-galleries, was accosted by two strange ladies, who pressed very hard for her signature, and when she declined, on the rational ground that she was totally ignorant of the matter, went off quite exasperated. Is it likely that she was the only young student thus solicited? Perhaps if Prof. Zöllner were aware of the manner in which agitations are got up and conducted in England, he would lay much less weight on the number of signatures to the petitions against vivisection.

At one time it was contended by the anti-vivisectionists that Harvey's great discovery was not due to experiments upon living animals. Finding that position untenable, they now seek to throw discredit upon Harvey, and insinuate that he had been anticipated by Servetus. We have seen this charge against our great English physiologist stated in an advertisement which, if not carefully read, may lead to the belief that Sir W. W. Gull, M.D., F.R.S., and other heads of the profession, are connected with the Anti-vivisection agitation.

It must be distinctly understood, however, that the rival societies for the suppression of research—bodies not too amicable among themselves—have not succeeded in bringing forward any cogent or novel arguments on their side, or in meeting those advanced by the defenders of research. The great charge of inconsistency brought against the agitators[*] has never been met. A peer of the realm did, indeed,

---

[*] *See* Quarterly Journal of Science, vi. (1876), p. 318.

indulge in the hacknied sophism that " one wrong does not excuse another." We bid him first prove that the infliction of pain and death upon animals for important purposes is a wrong at all. A certain literary organ tells us that if we can only learn how to escape diseases at the cost of the lower animals, then *noblesse oblige* that we should be content to suffer! It might be thought that even the editor of a " Society journal " could scarcely fail to see that the same principle, if it holds good here, must *à fortiori* forbid the infliction of pain and death upon animals for amusement, for convenience, or for luxury.

There is another inconsistency in the position. As the law at present stands we may, if we think proper, keep a number of live cobras, rattlesnakes, &c. We may feed them with rats, mice, guinea-pigs, rabbits, and we may even take pleasure, if so disposed, in the sufferings of the victims. In all this we neither break the law nor are our proceedings denounced as " orgies of diabolism." But suppose, after a rabbit has been bitten, we take him out and try to cure him, we are then performing a painful and cruel experiment. Unless we have a license from the Home Secretary we are in peril of fine and imprisonment, and whether we hold such certificate or not we shall receive anonymous post-cards fraught with abuse. Nay, if we even take the wounded rabbit, and after his death submit his fluids and tissues to chemical or microscopical examination, we are probably liable to the same doom. Or, again, we may poison rats and mice, if we think proper, with strychnine, phosphorus, arsenic, or baryta, without fear of legal consequences. If we have a dog or a cat which we wish to get rid of, no one questions our right to give it a dose of potassium cyanide. In short, we are allowed to poison animals at pleasure so long as they are our own property. But suppose that if we mix with the food of such rats, dogs, or cats, some unknown drug or some novel chemical preparation, and mark the results; or suppose we administer to such creatures an extract obtained from the viscera of an animal or man supposed to have been poisoned? We have then broken the law, and outraged public opinion. Thus we see that actions which are inno- cent as long as performed from wantonness, from the sheer love of killing, or at best to get rid of a nuisance, become criminal if done for the solution of an important physiolo- gical question : the higher the motive, the worse the action! This, then, is the outcome of humanitarian meddling and muddling and of governmental submission to the demands of organised ignorance.

We are not aware that our British anti-vivisectionists have ever attempted to justify the contradictions and inconsistencies which have just been pointed out. But as the agitation has become "international,"—or epidemic, which is much the same thing,—we felt bound to hear what foreign denouncers of biological experimentation might have to advance. One work especially attracted our attention. Its author, Prof. Zöllner, is a man of totally different calibre from the English anti-vivisectionist oracles. He is not a hysterical novelist nor a sensational philosopher. He is neither a sportsman quite ready to practise and uphold killing for killing's sake, nor a worshipper of Anubis, prepared to inflict any inconvenience or injury upon his fellowmen so that dogs may be made more comfortable. Nor, lastly, is he a physician or surgeon, careless for the advance of Science so long as his own practice may be increased. On the contrary, he is a thinker whose erudition is no less wonderful than his acuteness and profundity, who has "won his spurs" in physical and astronomical research, and whose passionate love for truth and fearlessness in its utterance are universally recognised.

It might naturally be awaited that such a man, if he entered upon this question at all, would do it fundamental justice. We expected a philosophic discussion of the nature and limits of man's sovereignty over the lower animals, and of his rights to inflict upon them pain or death. We trusted that he would have shown whether, in how far, and under what circumstances vivisection is in excess of such rights, and a departure from man's duties towards his fellow tenants of the globe. We thought he would show, or at least attempt to show, when the conclusions drawn from vivisectory experiments were trustworthy, and when they might be regarded as illusive. This we felt the more inclined to expect since he states that "the right of judgment on the scientific-physiological value of vivisection in its previous extent belongs in the first place to physicists, and not to physiologists and physicians." This is a paradox which we are not prepared to accept. We conceived it quite possible that we might meet in this book arguments against experimentation such as we have never before encountered, and which might seriously alter our convictions.

We must first note that Prof. Zöllner is not an advocate of the "total abolition and utter suppression" of vivisection. He denounces its abuse (*missbrauch*). Now it would be utterly illogical to speak of the abuse of a "grievous sin "*

---

* See bills displayed at railway-stations and elsewhere.

of an evil which it is sought utterly to abolish. The abuse of anything implies that there is a legitimate use. Hence Prof. Zöllner is by no means to be regarded as one of those advocates, blind to facts and deaf to reason, who seem to spring up in connection with every agitation and movement. Yet we scarcely see that he draws the line between the use and the abuse of vivisection with the clearness which would be desirable, and his language at times seems to savour of the most violent English school. Still, taking the title of the book into consideration, we are bound to put the most favourable construction upon his words. On our part we are free to admit that vivisection has been abused, and that some writers—not content with defending its legitimate and necessary employment—have weakened their position by denying or justifying such abuses. We fully grant that painful experiments upon living animals should be undertaken simply for purposes of research, for the solution of novel or undecided questions. We would neither practise nor countenance vivisection for mere demonstration, or for the acquirement or display of manipulative skill. We hold that every animal irremediably injured should be at once put to death after the operation or experiment is at an end. We deplore that physiologists should in any degree have departed from these conditions, and should thus have given a foothold for agitators and sentimentalists, and have gravely compromised the future of biological research. But whilst admitting and regretting such aberrations and excesses, we must not forget that nothing of the sort is chargeable to English men of Science in modern days. Even the Secretary of the Society for the Prevention of Cruelty to Animals, when giving evidence before the Royal Commission on Vivisection, acknowledged that he did not know of a single case of wanton cruelty.* Such being the facts, and due regard being had to the ultra-humanitarian tendencies of the nation which—save when sport or fashion is concerned—is so exceedingly averse to give pain, it might have been thought that special legislation was unnecessary, and that public opinion would have proved amply sufficient to restrain the abuses of vivisection.

But whilst finding in the fundamental idea of Professor Zöllner's work much that we can accept, we regret to note a striking absence of those features which we considered ourselves warranted to expect. The vivisection question is

---

* It would have been well had Prof. Zöllner and others noticed how much the Report of the Royal Commission goes beyond the facts elicited, and how far it is, in turn, outstripped by the Bill as proposed, and even as carried.

here so much mixed up with other matter that were the title, the table of contents, and the heads of chapters struck out, few persons, if any, would consider the work as an anti-viviseĉionist treatise. The author's attention is continually drawn off from his ostensible subjeĉt to the influence of the Jews in Germany, to their conduĉt in the Reichstag, in the Universities (where they appear to hold a number of professorships very high in proportion to the percentage which they form of the German nation), and in the political and literary press. It appears that there are in the German universities no fewer than seventy professors of unmixed Hebrew race—more than three times the number which might be expeĉted from the proportion of Jews in the empire. Concerning this faĉt Prof. Zöllner, like many Germans in all ranks of life, appears greatly exercised. To this subjeĉt he returns again and again, and no small portion of the work is taken up with remarks on Herr Lasker, a Jewish member of the German parliament, and of. Lasalle, also a Jew, and the founder of social democracy in Germany. The space devoted to these two men seems—in view of their very remote and doubtful conneĉtion with the subjeĉt of the work—surprising. Social democracy and Russian nihilism are also introduced, apparently in virtue of the circumstance that the would-be regicide, Nobiling, had enjoyed a medical education, and had possibly been a speĉtator of viviseĉtions.

Prof. Zöllner is no admirer of the modern mania for the " higher education " of women and for the admission of female students to the medical schools. He relates, with justifiable indignation, that at the University of Geneva two young ladies were busy with the disseĉtion of the lower extremities of a male subjeĉt, whilst two young men were at the same time engaged on the head and breast. There is also a notice of the scandalous career of the female students of the University of St. Petersburg, which we do not care to quote. The author returns to the subjeĉt in a subsequent portion of his work, and grows somewhat Rabelaisian thereon. We fully share his displeasure, but cannot reproduce his remarks, humorous though they be.

We find, also, certain interesting statistical results concerning the kingdom of Saxony, the region where it appears that the anti-viviseĉtionist agitation is more rampant than in any other part of Germany. It appears that the total yearly number of condemned criminals has increased from 11,001 in 1871 to 21,319 in 1877, or a total increase of 93·79 per cent, as against an increase in population of 7 per cent. Murders and other crimes of violence have multiplied at the rate of 473 per cent, and indecent assaults upon children under 14 years of age in the proportion of 825 per cent. The author refuses to attribute this moral decline to the consequences of the war, but ascribes it to the teachings of the schools, the universities, and of the press.

Another subject which plays here no small part is Spiritualism. By the prosecution of Slade we are told that England " set the seal on its intellectual decline." On this subject we wish to give no uncertain sound : much as we honour Prof. E. Ray Lankester for his achievements in biological science, we hold that in the Slade prosecution he committed a fearful mistake. To hand over to solicitors and counsel, to police magistrates and quarter sessions, a question which, if capable of solution at all, can only be decided by men of Science, was a piece of renunciation or self-abnegation which cannot be too deeply deplored, and which is doubly to be regretted in a country where Science is so little honoured as in England. However much Spiritualism may have been complicated by deceptions or delusions, it is the duty of scientific men to *make sure* that there is in the phenomena produced nothing more than is fairly referrible to jugglery or "unconscious cerebration." Till this has been done, to call in the aid of such rough and ready tests of truth as courts of justice can supply is nothing short of a formal abdication and a confession of impotence. At the same time, had Prof. Zöllner been better acquainted with our institutions he would scarcely have found, in this prosecution, grounds of accusation against either Government or people. He is doubtless not aware that so long as a statute—however obsolete—remains unrepealed the authorities cannot prevent any person from applying it to his purposes. In connection with this subject it must not be forgotten that Prof. Zöllner, unlike some of his English colleagues, proposes, or at least hints at, a substitute for vivisection,—to wit, "bio-magnetism." This idea surely had better have been developed in full, even if some of the personalities here introduced had been omitted. Such personalities, indeed, form a most strange—and we do not hesitate to say a most deplorable—feature in Prof. Zöllner's work. Virchow, Helmholtz, Du Bois-Reymond, Vogt, Pagenstecher, Bernard,—every physiologist of note who has either practised or defended vivisection,—is denounced in a manner which is very far from carrying conviction to the mind of the reader. Let us take a specimen : Professor Tyndall is no biologist, and has probably never dissected any animal, living or dead ; but he publicly declared, in 1875, that he should regard any interference with experimental* vivisection as a great misfortune to Science. He is therefore treated to four pages of irrelevant personality. Our author writes : " Prof. Tyndall decided a few years ago, at the age of 57, to enter into the holy estate of matrimony. The sacrifice (*opfer*) of his ardent affection was a Lady Hamilton, a wealthy lady belonging to the highest English aristocracy. I cannot say with certainty whether the happy or unhappy spouse of Tyndall belongs to that eccentric family of Hamiltons from which the hereditary Prince Albert of Monaco selected a wife, from whom he is now being divorced." The Monaco scandal is then given in detail in a note, and the author proceeds to quote a letter which Prof. Tyndall is said to have written during his courtship, as a model for " Darwinians and Materialists." We must here ask—Is this scientific criticism ? Is it in any way relevant to vivisection ? Is it worthy of an author of Prof. Zöllner's merit ? This style of controversy is the more to be regretted since he appears highly indignant at any personalities on the part of his opponents. What, again, shall we say of the following passage ?—"At that time Berlin, the learned and cultivated, was so completely chloroformised by the *fœtor Judaicus* that it listened as devoutly to the Hebrew melodies of the lyric poet Heine, and to the oracles of the naturalist Alexander von Humboldt, as it now does to the songs of Rudolf Löwenstein and the theories of the academicians Virchow and E. du Bois-Reymond." Humboldt died before the vivisection question broke out, but he is repeatedly made here the object of unfriendly remarks. Prof. Zöllner seems to have many enemies, living and dead, and he may have the best causes for his enmity. But what light do all these disputes and grudges throw upon the subject in hand ? Do they prove vivisection wrong ? Do they mark out the limits, if any, within which it ought to be tolerated ? An interesting and instructive fact here mentioned might well be taken to heart by English biologists, in view of the restrictions under which they already labour and of the further attacks with which they are menaced. The medical faculties of the following universities have formally protested against the anti-vivisection agitation :—Bâle, Bern, Bonn, Dorpat, Erlangen, Freiburg, Graetz, Greifswalde, Halle, Heidelberg, Kiel, Königsberg, Leipzig, Marburg, Munich, Prague, Vienna, and Zürich. Could no similar expression of qualified opinion be produced in the United Kingdom ?

* We call attention to the word " experimental." Prof. Tyndall does not appear to have advocated vivisection for mere demonstration.

The main points of Prof. Zöllner's work, stripped from all irrelevancies, are—First, his approval of restrictions and limitations on vivisection on behalf of Government. Such regulations, we admit, might work much better in Germany, where the authorities respect Science more than in England and interfere less clumsily. If licenses to perform painful experiments are required in any continental country, they will be issued by some competent and qualified official, and not by a Secretary of State whose time is engrossed with party affairs, and who knows and cares as much about biology as does an average bricklayer. Secondly, the proposal of a substitute for vivisection, most welcome if practicable, but on which the learned author might have been more explicit. Thirdly, the claim that physicists are better judges than physiologists of the value and necessity of vivisection,—a doctrine which we shall be prepared to accept when it is proved that organisms are mere machines or automata, when the chasm between animation and inanimation has been bridged over, and when irritability is shown to be merely a special case of attraction and repulsion. But, if we mistake not, Prof. Zöllner is as far removed as ourselves from believing in the probability of such a consummation.

ON WATER AND AIR.*

By JOHN TYNDALL, D.C.L., LL.D., F.R.S.,
Professor of Natural Philosophy at the Royal Institution
of Great Britain.

LECTURE VI.

IN our last lecture we entered upon the examination of air, and I confined myself in that examination to what we call the physical properties of air, not going into its chemical qualities. We learnt how the pressure of the atmosphere was determined,—how that immortal experiment of Torricelli was made ; and, as I said at the time, I think that no more impressive experiment ever could have been made, or at least that there was no greater example of that outgoing, so to say, of the human mind in advance of experiment, than in the case of Torricelli. It was an instance of pure reasoning by which he inferred that a mercury column 30 inches high would be supported by the pressure of the atmosphere. His seeing this beforehand, and his bringing his foresight to the test of fact, and seeing that fact realised and experienced, must have given him a joy and pleasure not frequently experienced.

Well, this was in 1654; but immediately afterwards the discovery of the air-pump placed in the hands of investigators a means of pursuing this subject farther, and one great man—a man whose name is for ever memorable—threw himself with ardour at the time into the investigation of this subject. This was the illustrious Robert Boyle. He improved the air-pump, and he made a great number of experiments with his improved " pneumatic machine," as it was called sometimes. He was a man of great thought and great power of experiment, and great reasoning power. Herein consists the power of a philosopher in examining Nature—that he shall have power not only to investigate the facts of Nature by inventing experiments, but the power of reasoning out those experiments to their legitimate results. Boyle possessed both these qualities in a high

* Being a Course of Six Lectures adapted to a Juvenile Auditory, delivered at the Royal Institution of Great Britain, Christmas, 1879. Specially reported for " The Journal of Science."

degree, and it was soon found, after the experiment of Torricelli, that the height of the column of mercury varied from day to day, and varied sometimes very rapidly indeed ; and that proved to people in those days that the pressure of the atmosphere was by no means a constant thing. Sometimes the atmosphere was lighter, and sometimes it was heavier than at other times ; and Boyle observed for five weeks the oscillations of the height of this barometric column, and in that way saw clearly the variations in the pressure of the atmosphere. It is these pressures, or at least these variations of pressure, that now enable us more or less to predict the advent of certain kinds of weather. Here in London, for instance, when the north-east wind blows, you would usually find a high barometer as it is called,—that is to say, the atmosphere, being chilled by the north-east wind, is denser or heavier, and it is able to support a higher mercurial column than when, for instance, the west wind blows, which comes to us charged with light aqueous vapour from the Atlantic, warmed more or less, and the consequence is that this lighter air is not able to sustain the same height of the mercurial column. Well, this was one of the early observations made by Robert Boyle, and he made various other observations with his air-pump. He made observations upon the influence of air upon respiration, and we know a great deal about these matters now ; but it was of infinite importance to break ground in this region—to get some experimental grasp of this region which was not at all known before. The proper action of the air upon animals was not known, and he put snails and caterpillars, and mice and rats, and I think even a dog, under the receiver of his air-pump, to observe the effect of withdrawing the air from those animals. He produced very wonderful results, which caused him to lift his heart in gratitude to the Creator for making the air so essential to animal life and enjoyment.

I dwell upon Boyle particularly, because he was a scientific character and a philosopher worthy of all imitation by the younger scientific philosopher. He was a man who not only possessed the highest intellectual culture of his age, and who made important discoveries, but he united to this power of a scientific man those characters and qualities that we essentially ascribe to the human heart—the qualities of tenderness, and courage, and courtesy. And let me say to you, my young friends, once for all, that scientific knowledge unless backed up by those other qualities leaves you not rounded men, but only lopsided men.

Well, this I have thought it my duty to say with regard

to the illustrious Robert Boyle. One experiment which has been made has an immortal interest, and is of the utmost importance in relation to this subject. I have here (Fig. 37) a simple instrument, similar to the one with which Boyle operated. The long arm (A) of the bent tube (A B) is open at the top. The short arm is closed at B. We have introduced some mercury into the bent part of the tube, and the fluid metal now stands at the same level (at o) in both the

Fig. 37.

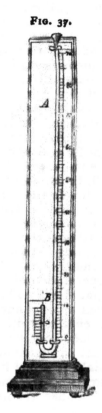

long and short arms. The space above the mercury in the closed tube contains a certain bulk of atmospheric air, submitted to the ordinary pressure of the air through the open tube A. And now I will ask Mr. Cottrell to pour carefully into this longer arm a quantity of mercury. What will occur? The mercury falling into the tube will press the mercury which is in the bent part up into the shorter arm. The air in this short arm will be squeezed more forcibly together, and the greater the amount of mercury which we pour into this longer tube the greater will be the amount of squeezing. Now, what Boyle wanted to ascertain is this: what is the relation between the pressure of the mercury brought to bear upon this air and the volume of the air; and I will tell you what his result was. He found that when the pressure was doubled the volume was halved, and when the pressure was quadrupled the volume was reduced to one-quarter, so that as the pressure augmented the volume decreased in precisely the same proportion. Or, turning it into other language, instead of "volume" let me say "density." The density means the quantity of matter crowded into a certain space. If we use the term density Boyle's law would be this, that the density of the air is directly proportional to the pressure exerted upon it.

And now the mercury has risen in the long tube, and has reduced the volume of the air enclosed in the short arm to one-half its previous volume. If we now measure the height of the two columns, we shall find that the level of the mercury in the long arm stands very nearly at 30 inches above that in the closed portion. Thus the pressure of an additional atmosphere has reduced the bulk of the contained air to one-half. In this way, then, by doubling the pressure we halve the volume of the air, and by quadrupling it we should render the volume of the air one-fourth. In this way Boyle proved that the density of the air was exactly proportional to the pressure brought to bear upon it. I could not possibly dwell upon the subject were it not that I think that honour ought to be given where it is due; and for a long series of generations, I may say, the discovery of this law has been ascribed to another person,—a most meritorious man,—a philosopher named Mariotte; and you will find that in books, particularly in Continental works, this discovery is always referred to as the law of Mariotte, but Boyle preceded Mariotte. He took the greatest care in his experiments, and he established with the utmost rigour the existence of this law anterior to Mariotte, and therefore the law ought undoubtedly to be called the law of Boyle instead of the law of Mariotte. Boyle, as I mentioned in our last lecture, had very clear notions regarding pressure of the atmosphere, and regarding what he described in beautiful, or, I may say, almost poetic language; for men like Boyle have a strong poetic instinct,—that is to say, they can see resemblances between utterly remote things. They can pierce, as it were, into the nature of things, and see things resembling each other where the common eye sees no resemblance whatever; and thus, as I expressed in my last

lecture, he compared the particles of air to little springs
pressing against each other. When the air was pressed
these springs yielded, and the air was forced into a smaller
space. When the pressure was relaxed these little spiral
springs, as it were, expanded, and caused the air to expand.
In this way he figured the mechanical action of the air, and
he wrote some celebrated papers on what he called the
weight and the spring of the air. There are one or two
illustrations of this spring of air which I think will impress
it upon your mind. One which I should like to mention to
you occurs to me just at the moment. Sometimes in going
up a mountain you notice a curious pain in the ear; sometimes, at intervals, you feel as if a kind of bubble of gas
got into your ear: some persons are subject to this feeling,
and some are not. The cause of this effect is this:—Here
we have the passage of the ear, through which all sound, or
the great body of sound, passes. That passage is stopped
at a certain place by a membrane called the tympanic membrane, or, to use the common term, the drum of the ear.
Beyond that there is another membrane. When the sound
vibrations come into the ear they strike upon the drum of
the ear, and the vibrations are transmitted to the membrane
on the other side, and sent on to the auditory nerve, and
from the auditory nerve to the brain, where they are translated into sound. It is necessary that the same pressure
should exist within the drum of the ear and without it. If
the pressure were greater within than without it would bulge
the tympanic membrane or drum-head outwards, and cause
pain. If the pressure within were less than without, the
excess of the pressure of the atmosphere outside would force
the drum inwards, and cause pain; and this is very frequently the case. I do not experience the pain myself, but
I have often walked beside others who have experienced it.
How is this pain to be avoided? There is, among the wonderful adaptations that we see here, a tube from the mouth
to the drum of the ear: that is the only passage to the ear.
This tube is called the Eustachian tube. When you are
climbing a mountain, and feel this inconvenience, simply
go through the process of swallowing. The act of swallow-
·ing opens the Eustachian tube, and allows the air to get in
or to get out, as the case may be, and so establishes an
equilibrium between the air within and the air without, and
thus you have all the inconvenience removed. This is due
to what Boyle called the spring of the air.

There is another instance of the same spring which I
dare say you are all well acquainted with. It is called the

Cartesian Diver. The diver (Fig. 38) consists of a glass figure having attached to its head a hollow ball. The figure and ball are entirely closed, except a small opening in the curled tail, D, which is affixed to the body of the figure. A quantity of water has been introduced into the figure and ball, so that at the present time it is slightly lighter than the water in which it is contained. In the jar, A B, the proportions between the air and the water are such that the diver floats; but if I press upon this sheet india-rubber

Fig. 38.

with which the top of the jar is closed, what do I do? I squeeze the air at the top of the water, there being a little air underneath this piece of india-rubber. The compression of the air is transmitted to the water; and this is the point that I want to make clear to you. The water has the power of transmitting the pressure from the air above it, and that pressure is transmitted to the little diver below. What is the effect? It is that the pressure squeezes a certain amount of water into the diver, and the consequence is that the air within the hollow ball is compressed, and is replaced

to a certain extent with water; and this makes the little
figure heavier, and down it goes; it sinks to the bottom of
the jar.   I now release the pressure, and the spring of the
air within the ball ejects the water that was forced into it,
and the figure rises to the surface.

Now let us make another experiment.   I will press this
covering, and at a certain point I will suddenly withdraw
the pressure.   What will occur?   Boyle's spring of the air
will again come into play.   The air that is compressed in
the hollow ball will suddenly relax itself, will suddenly expand like a compressed spring, and will drive out the water

Fig. 39.

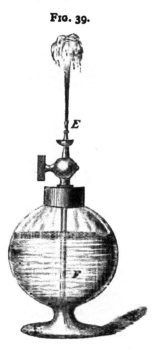

in the tail as before, and, inasmuch as the tail is curled
round, the spring of the air will cause the little creature to
pirouette.   The spring of the air might be illustrated in
fifty ways.   For instance, here (Fig. 39) I have a small
fountain.   It is a spherical vessel, which has inserted into
it a metal tube, B F, provided with a stopcock.   It is about
two-thirds filled with water.   I have here a condensing
syringe, which I will attach to the end of the tube at B, and
by means of it will compress air into the fountain.   The
air will pass down the tube B F, and collect in the space

above the water. Now, having forced sufficient into the fountain, I will shut the stopcock and remove the syringe. Many fountains play upon this principle, but it is simply an illustration of the spring of air which was so completely investigated by Robert Boyle. At the present time you have compression of the air by means of this spring to which Robert Boyle refers. This spring of the air which has been forced upon the surface of the water will drive out the water. I will now open the stopcock, and you see that the water is ejected through the tube B F to a considerable height, and continues to play until the vessel is emptied.

There is another experiment referred to by Boyle to which I will draw your attention. Here are two plates of marble that have been in this institution for fifty or sixty years; at all events they were here long before my arrival, and they were obviously made for the purpose of illustrating the experiment of Boyle. We take these two plates of marble, which are intended to be very smooth, and slide their flat surfaces over each other : there is a certain definite way of causing them to cling together. I feel the clinging force very much. They ought to cling and hold fast together. Now Boyle entertained the idea—and the idea has come down to the present time—that when you pass one of these plates over the other, and cause them to come into close contact with each other, you squeeze out the air, and hence the atmospheric pressure outside comes into play and keeps the plates of marble together. I am indebted to one of the greatest mechanical geniuses of our age, Sir Joseph Whitworth, for the two metal planes that are now before you. I want to show you their action. These two planes of iron were produced by Sir Joseph Whitworth, who has done this in a manner that really is perfectly wonderful. He has produced what are called true planes. He started in life with the idea of making mechanism as perfect as possible, and his first idea was to obtain perfectly flat surfaces. I wish I had time to go fully into the manner in which he has done it. He has done it in the simplest manner possible, by first of all making them approximate to smooth surfaces. He then placed a coating of rouge on one of the surfaces, and, placing this upon the other surface, pressed them together. If they were perfectly smooth the rouge would cover both surfaces uniformly; but there are always little hollows, and hills, and eminences, which left the rouge in spots and patches, and by scraping away these eminences he arrives at his true planes. When I place the two together there is a most extraordinary effect. The upper one appears to be floating upon a kind of oil, and that is due undoubtedly to the air beneath; and it was supposed, even by Whitworth himself, that when you press down this plate you by so doing squeezed out the air, and that this one was firmly clasped to the other, and that this firm clasping of the under plate to the upper plate was due to the air being squeezed out. There are many who thought that this was not the case, and that it was a case of firm molecular cohesion. To experimentally test this question the following

FIG. 40.

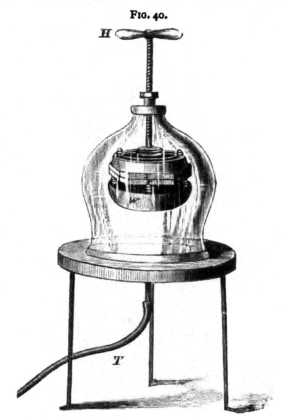

arrangement was made :—Here are two Whitworth planes, P, P (Fig. 40); they are placed under the receiver of an air-pump. The planes have been pressed together, and are firmly adhering. The under plane has attached to it a weight of lead weighing 20 lbs. The upper plane is attached to a rod which slides air-tight through the upper part of the receiver, and by means of the handle H I am able to lift the planes with the lead weight attached. The receiver is placed air-tight upon a circular plate, and connected to an air-pump by the tube T. By means of the pump we have exhausted the receiver of air, and the lower plane, with its lead weight attached, is hanging there, adhering firmly to the upper plane. Had a weight of 100 lbs., instead of 20 lbs., been attached to the lower plane, it would have been sustained by the powerful attraction of the two surfaces, so that this clinging force, whatever it is, is independent of the pressure of the atmosphere. It is, in point of fact, part and parcel of the force which causes the individual molecules of the air itself to cling together. If you bring the two surfaces closely together the molecular cohesion comes into play, and holds the plates together. This experiment shows that in this particular, at all events, Boyle was not correct in his conclusions; but the number of things in which he was correct is simply wonderful. He observed, as I have said, those changes in the height of the mercurial barometer, and he observed many other things. Usually the credit of proving that sound cannot pass through a vacuum was ascribed to a most meritorious man, and far be it from me to say a word against him,—to Hawksbee, in 1709,—but nearly half a century previously Boyle actually put his watch under the receiver of his air-pump, and he found that when he exhausted the receiver the ticking of his watch ceased to be heard; so that nearly half a century before the time of Hawksbee, Boyle had proved this fact. This error on his part, in his conclusion as to the cause of the clinging of the plane surfaces, is a mere trivial thing compared with the amount of truth which he established.

Now I have done referring to the transmission of the pressure through the water towards this little Cartesian diver. I do not know whether the experiment which I am now going to make is destined to succeed, because I had not the opportunity of making the experiment beforehand, having only one of these bottles at hand; but I want to see whether the transmission of a shock through this water will be sufficient to produce a striking mechanical effect which will be the means of showing the transmission. Here is one of what are called Rupert's drops. It is a drop of glass cooled quickly, and drawn out into a long tail. This glass is in such a state that the slightest crack or rupture, such as the breaking off of the long tail, causes the whole thing to fly into powder. As I said, this rupture of the Rupert's drop is accompanied with a powerful mechanical shock. Here is a bottle of water (B, Fig. 41); I do not know whether it is calculated to produce the effect. I want to show you not only the transmission of gentle pressure, but the transmission of a shock through water; and if I hold this Rupert's drop (D) in the bottle of water, and break off the tail, the

shock ought to be transmitted through the water, and if the experiment is successful the bottle will be broken. I will now break off the tail of the drop, and the bottle is broken to pieces. The breaking of the bottle is entirely due to the transmission of the force through the water.

I introduce that experiment to your notice in order to make you understand another of far greater practical importance, and that is an experiment for which I am indebted to the obliging kindness of my friend Prof. Abel, of Woolwich. I have no doubt that it will have an important future. Here is a bombshell, and I will pour water into that bombshell: it is very thick. Here is a small charge of powder, and we will explode that charge of powder in the middle of that bombshell filled with water. We place the shell for protection in a strong metal case. The force will be sufficient to break the bombshell, and I should not like even that small quantity of powder to have its way in this room without carefully covering it up. Two electric wires lead from a small fuze which is placed in the centre of the powder, and we will fire the charge by means of the electric current, and if the experiment is successful the shock will be transmitted to the water and burst the bomb. Now I will press

that button and make contact with the battery, and you
hear the noise of the explosion, and there you have the
result.   The bombshell is shattered to fragments, and reduced to that state by the force of the shock that has been
transmitted to the water.   I do not know what is to be the
future of this form of bombshell, but I can see that it is of
very great importance.   Now calamities in coal-mines often
occur by the use of gunpowder, and it is possible—I will
not say that it is certain—that by surrounding our charges
of gunpowder with water in this way, we can do away with
the danger that now is associated with the use of gunpowder
in mines.

I want to refer for a moment to the condensation of gases
in so far as it bears relationship to that of vapour.   You
know that aqueous vapour is very easily condensed.   It was
long ago conjectured that the very air we breathe, and that
oxygen, hydrogen, and nitrogen were the vapours of bodies,
and that they could be congealed to a certain extent and
reduced to a liquid condition.   Various attempts were made,
and various easily condensible gases were liquefied, but
quite recently some of the more refractory gases have been
condensed and rendered liquid by pressure.   This has been
done by two experimenters, Messrs. Cailletet and Pictet,
working almost simultaneously.   It is one of the cases in
which we find the same thoughts on the same subject occurring to two different minds.   Oxygen has been liquefied and
nitrogen has been liquefied, and you know that the air we
breathe is a mixture of oxygen and nitrogen.   In 100 cubic
metres of air there are 20 cubic metres of oxygen and
80 cubic metres of nitrogen.   I will not take the atmosphere,
or oxygen, or nitrogen, because these gases require for their
liquefaction appliances which we have not the means of
using here ; but you remember the gas I brought before you
in our first lecture,—the gas we obtained by the action of an
acid upon fragments of marble,—carbonic acid gas.   This
gas I intend to liquefy in the apparatus which is now before
you (Fig. 42).   B is a strong iron bottle containing mercury,
into which is fixed a thick glass tube, P T.   The end of the
tube at P is closed, the lower end being open.   This tube is
held firmly in position by a strong iron nut, which is
screwed into the bottle at $n\,n'$.   Before introduction into the
iron bottle, the glass tube was filled with pure carbonic acid
gas,   The tube W, which opens into the iron bottle, is connected to a powerful hydraulic pump.   We will now urge
water, by means of this pump, through the tube W, into the
iron bottle, and force the mercury into the glass tube

containing the gas. You see the mercury rising higher and higher as the pressure is increased, and it is now within a couple of inches of the top of the tube, at g, the whole of the gas contained in the lower portion of the tube being

FIG. 42.

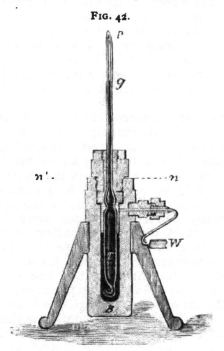

compressed into the small space, g P. Observe the top of the column of mercury, and you can see collected there a quantity of liquid, clear and colourless; that is liquid carbonic acid gas. By this compression we have actually changed the gas into a liquid. You see that it is actually boiling before you, even at this temperature. If I lower the pressure it will boil still more violently; the liquid will disappear, and again become a gas. If I increase the pressure the liquid gas is again formed. Now what I want you to observe is the effect of the sudden expansion of this atmosphere above the column of liquid carbonic acid. If you take a champagne bottle, say half empty, allowing the carbonic acid to fill the space above the champagne, and suddenly open it, you hear an explosion, and you find the space which was formerly perfectly clear above the liquid filled with cloud. That is due to the chilling by expansion which I illustrated a lecture or two ago. Now I will try to produce that cloud in this apparatus; although it is composed of transparent carbonic acid, still it will appear almost dark for the time being. To produce this effect we must lower the pressure very suddenly; and observe, the instant I do so the space above the liquid is filled with a dense cloud.

And now I have to refer to an instrument which we employ in this experiment. It is what we call the hydraulic press. Its construction you will readily understand from this diagram (Fig. 43). A strong iron cylinder, C, with very thick sides, has inserted into it an iron ram, P, working water-tight in the collar of the cylinder. On the ram, P,

FIG. 43.

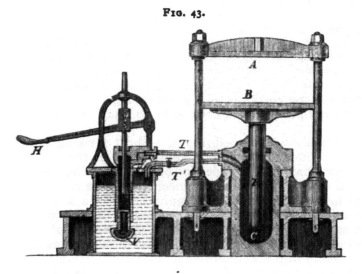

there is an iron table, B, on which the substance to be pressed is placed. Strong iron columns support an iron plate, A, fixed directly over the table, B. By means of a small pipe, T, the cylinder, C, which is filled with water, communicates with a small force-pump, P′, which is worked by means of the lever, H. When the lever H is lifted, the piston P′ ascends, the valve V opens, and the cylinder of the pump is filled with water. When the lever is forced down again, the valve V closes, and the water is forced through a valve in the pipe T into the cylinder C, and the piston P, carrying the table B, is forced upwards. By a repetition of this action enormous pressure can be obtained. When it is necessary to relieve the pressure, the screw plug inserted into the tube T′ is opened, and the water flows from the cylinder C, through the tube T′, into the cistern, and the

piston P descends to its original position. The principle of
the hydraulic press is simply the forcing of water from a
narrow cylinder into a wide one. We have here a small
sample of the hydraulic press. It is the pressure of that
which we have been working here which has converted the
carbonic acid gas into the liquid form, and that without the
exercise of very particular force by the human arm.

Now I want to try whether we cannot break a bar of iron
by the hydraulic press. You have here no creation of
power, but you have the means of applying power. You
have the means of adding power to power, until, by the
addition of small fractions of power, you are able to break
that bar. We will place the arrangement for breaking this
thick iron bar between the fixed plate A and the movable
table B, and work the lever H as before described; the
piston P rises, and now I have half a ton of pressure upon
that bar, and you see it has snapped asunder. It has been
placed across two supports, and a point pressed upon the
centre; and by certainly not a very inordinate amount of
effort on my part, I have been able to snap that thick bar
asunder.

Now I wish to bring before you a most striking example
—perhaps the most striking application of the hydraulic
press that has been made in our time. The world is
indebted to that illustrious engineer Sir William Armstrong
for the introduction of the hydraulic press into a variety of
engineering operations. This occurred to Sir Joseph Whitworth some years ago, and I had an opportunity of talking
to him at the time, and I saw that the problem that he had
placed before him was enormously difficult. Here I have a
number of specimens of steel. This specimen has been
cast as steel used to be formerly cast. It has been cut
across, and you see that it is honeycombed,—full of bubbles
of air, so that you never could trust steel of that kind. It
might be very hard, very strong, but still you never could
trust to it, because you never knew where these bubbles
might not exist in the steel. Here is another sample of
steel which is perfectly close, without the trace of a bubble.
There is not a bubble large enough to admit a pin's head or
a pin's point. It is perfectly compact. Now the problem
which Sir Joseph Whitworth set before him was to convert
steel of the old kind into steel of this close compact character. And how did he do it? He devised a mould of a
certain quality,—a mould that would not allow the molten
metal to pass through it, but would allow all these gases to
pass through,—and he brought enormous pressure to bear

upon the steel when it was in a molten condition, and he produced what is certainly one of the greatest achievements of the time—this compact quality of compressed steel.

But here I am going to refer to one of the achievements of the hydraulic press. Whitworth places his steel under pressure and squeezes out every bubble, and he converts the steel from a material which is full of holes and unreliable into a steel of this compact character, which can be thoroughly depended upon. Now, thanks to Sir Joseph Whitworth, I am able to show you the manner in which he tests this steel. There is a hydraulic press, of very great power, invented by Sir Joseph Whitworth himself, and we will place a piece of Whitworth iron within the apparatus. The reason I wished to bring this before you was that I saw some time ago, in Great George Street, a pressure of 40 or 50, or 60, tons per square inch applied to the cylinder of iron that I now hold in my hand, and it was elongated and ultimately drawn asunder. The hydraulic apparatus is so arranged as to work by traction and exert a pull upon this cylinder of iron we are testing. This cylinder is exactly a square inch in area, and we will work the press, and exert an enormous strain upon it. It is worth while remembering what occurs. For a time it will resist the strain, and if we release the pressure it will go back again ; but when the pressure goes beyond a certain point the iron passes what is called its limit of elasticity, and it begins to be elongated, and then after a time it snaps asunder. By the continuation of the strain on the iron cylinder we have reached a pressure of 20 tons to the square inch. This point is the limit of elasticity of the iron, and you see the cylinder begins to elongate, and is reduced in diameter from one inch to nearly a half inch ; and now the cylinder has broken asunder. That has required a force of 32 tons to the square inch. The end of this cylinder is a square inch in area, and therefore it is as if you had screwed it up and attached to it a weight of 32 tons, which would be just sufficient to break the cylinder. That is called the strength of the cylinder ; and here another point comes into play which is of very great importance, and that is the ductility of the cylinder. The ductility is measured in this way :—The power necessary to break a cylinder is measured by the number of tons required to cause it to give way : the ductility is measured carefully by placing the two broken halves together, and measuring the elongation of those two halves. That gives the ductility. The specimen we have just broken is a very ductile metal, but it is not of great strength. Here, however, is a cylinder which is composed of Whitworth steel, which is enormously strong : we will place it in the hydraulic press, and subject it to the straining action. It may not at all succeed in the form of showing you the breaking of the metal, because the steel is so good that it may resist a force of 100 tons.   There is a force of 10 tons on it now, and we shall go on augmenting the pressure : 80 tons are now working upon that cylinder ; the gauge tells the story. You see what the heaping of small impulses will do—small increments of power.   There is now a pressure of 100 tons to the square inch, and the steel has stood this test. That steel is not to be broken by that force.

And now, having said so much, I have only to offer you my best thanks for the attention you have given me throughout this course of lectures. I do not know that I have ever given a course of lectures in this place that has gratified me more ; for I do not know that I have ever seen the young folks—the lads or the girls—more attentive than they have been during this course of lectures. I have to thank my excellent friend, Prof. Dewar ; I have to thank Sir Joseph Whitworth ; I have to thank Prof. Abel ; and last, not least, I have to thank my excellent assistant, Mr. Cottrell.

ANALYSES OF BOOKS.

The Disestablishment of the Sun. By JOHN BLAND. London : Sprague and Co.

IT has been asserted, perhaps in too boastful a strain, that Science knows neither sects nor heresies, and that her revelations are based upon such evidence as to render dissent altogether impossible. What, then, will be the sensations of our leading astronomers, physicists, meteorologists, and of the educated public in general, to find it boldly declared that there is no heat in the sun's rays ? Such an assertion and others scarcely less startling are made in the pamphlet before us. Now we do not belong to that numerous and influential class who seek to establish a scientific orthodoxy, and who treat all dissidents with sublime contempt. We hold that he who looks at any question from a novel point of view may possibly see something worth recording, and that he who detects a flaw in our received theories gives us an opportunity not to be neglected. Still we must confess ourselves disappointed in this little work. We expected to have our attention called to facts hitherto overlooked, which the author had possibly discovered. We looked for experimental evidence. But our expectations have not been fulfilled. We must even doubt whether the author has in all cases taken the trouble thoroughly to apprehend the received theories which he is assailing, and the facts upon which such theories repose.

Mr. Bland's objects are—" To show the fallacy of the popular belief in the heat of the sun. To prove that the only heat we have on the surface of the earth comes from an internal source. To explain the primary function of vegetation. To explain the law under which trees develop. To show that at one time it is probable there was no water on the surface of the earth. To account for the formation of mountains. To show that it is probable that the quantity of water on the surface of the earth is still increasing. To account for the deposit of marine shells on hills. To account for the formation of coal strata. To show that it is probable that at one time no air existed on the surface of the earth. To explain the existence of heat under a burning-glass held in the rays of the sun, &c., &c."

Many of these explanations are of course perfectly unnecessary. The formation of mountains, the presence of shells not on but in hills, the formation of the coal strata, have long since passed out of the range of problems to be discussed. To say that at one time there was no water on the surface of the earth is a needless truism, if, as Mr. Bland holds, in perfect accordance with the commonly received theory, the temperature of our planet was greatly in excess of that at which water can exist. To deny the existence of air is another question. Surely the higher the temperature the more matter must have been present in a gaseous state. Our author forgets, also, that no one seeks to account for the lower temperature of the poles than that of the Equator by their greater distance. The theory of direct and oblique rays, which he pronounces a " myth," is perfectly satisfactory. Suppose a number of parallel rays falling upon a given surface : the more such surface is deflected from a direction at right angles to the rays, the fewer of such rays fall upon each square foot of the surface. The author argues that if the poles are cold because the heat-rays fall upon them obliquely, a spot on the Equator at midnight ought to be colder still, because it is out of the sun's rays altogether ! He forgets that at the poles the sun's rays are absent for one-half the year entirely, and fall during the other half very obliquely and scantily upon the soil ; whereas at the Equator the surface of the soil has not time to cool down during a single night. This is, indeed, Mr. Bland's main error : he seems to think that if the sun is the source of heat, the temperature of any place should depend solely upon its being above the horizon, or not, at any given moment. In consequence of this mistaken assumption he asks—How is it that the temperature often gets higher in the night ? and how is it there is not a great difference in the temperature when there is a total eclipse ? The first of these questions is solved by reference to a change in the wind : as to the second, a chilly feeling has been observed during an eclipse ; but the occultation is too brief, and extends over too small a part of the earth's surface, to bring on, *e. g.*, a frost. He asks, further, How is it that rivers become frozen on the surface rather than at the bottom ? We thought it was now generally known that water expands in freezing, and the ice formed, being specifically lighter, floats upon the surface. Again, " How is it that the frost never penetrates more than a comparatively few feet into the earth ?" Because earth is a very poor conductor. But if any appreciable quantity of heat were emitted from the interior of the earth, how is it that ice buried a few yards in the ground may be preserved during a whole summer, whilst if laid on the surface it melts in the first mild days of spring ?

The author even asserts that the sun, so far from imparting, robs us of heat. He makes use of this hypothesis to explain away the effects of a burning-glass. He says—" If a burning-glass be held in the shade no change in the temperature is observed in its neighbourhood. If the shadow be removed an entirely different state of things is observed. The rays of heat that were lazily ascending *straighten out* and quicken, as if endowed with life. The evidence that they straighten may be stated in a few words. It is known that the rays of light come straight from the sun. If a burning-glass be held so that it fairly intercepts the sun's rays, the foci of the concentrated rays of light and heat are identical under the glass. It is known that glass obstructs the passage of heat [!] and admits the passage of light. As the rays of heat and light must be reflected [should be, refracted] at the same angle to have the same foci, it is assumed [*i.e.*, by the author] that they are moving in the same straight line, but in opposite directions." Here is, in the first place, a grave error in fact. Glass does not obstruct the passage of radiant heat from incandescent bodies. Further, if Mr. Bland will take a powerful burning lens he will find that the intensity both of light and heat decreases *pari passu*, going from the focus towards the lens. This is what should happen if the light and heat move in the same direction, according to the common theory, and what should *not* happen if the light only passes through and is concentrated by the lens, whilst the heat arrives at the focus from the opposite side and is refracted and concentrated by nobody knows what.

We have not space to make any further extracts from this pamphlet. We think if the author would work for a few months in a good physical laboratory he would greatly alter his views. His manner of dealing with heat reminds us of Goethe's researches in optics.

Reports of the Mining Surveyors and Registrars of Victoria. For the Quarter ending September 30, 1879. Melbourne: J. Ferres.

THE total quantity of gold raised in the colony during the quarter has been 189,648 ozs. 14 dwts.

An Appendix by the Government botanist, Baron F. von Mueller, consists of observations on new vegetable fossils of the auriferous drifts.

Guide to the Geology of London and the Neighbourhood. By W. WHITTAKER, B.A., F.G.S. London: Longmans and Co., and E. Stanford.

THIS pamphlet is an explanation of the Geological Survey-Map of " London and its Environs," and of the Geological Model of London in the Museum of Practical Geology. The author points out two errors: that the " London Basin "—so-called—is deep in proportion to its length and breadth, and that the Tertiary beds have been deposited in a hollow cut out in the chalk. Mr. Godwin-Austen inferred from the exposures of older rocks in the Ardennes (Belgium) and in the Mendips (Somerset) that these are parts of one grand line of elevation, connected underground nearly along the valley of the Thames. His conclusion has been since verified by borings, but the direction of the dip of these strata has not yet been ascertained.

The author considers that no important supply of water need be expected on tapping the Lower Greensand, on account of its thinness and its very limited outcrop.

Records of the Geological Survey of India. Vol. xiii., Part 1. 1880.

THIS issue contains the annual summary of the Survey and of the Geological Museum at Calcutta, for the year 1879. It is remarked that Mr. Griesbach, in examining the higher Himalaya of Kumann and Hundes, has discovered and brought in a good harvest of fossils.

We note with interest that two natives have received permanent appointments as Sub-assistants. The author seems, however, to doubt whether they can ever prove competent for independent field-work, which requires "the very quality which more than any other makes the western man differ from the eastern."

In the list of societies and institutions with which the Survey exchanges publications, we find the "Geological Survey of New Zealand" and the "New Zealand Institute" described as at Washington!

The memoirs comprised in this number are—"Additional Notes on the Geology of the Upper Godavari Basin," by W. King; "Geology of Ladak," by R. Lydekker; "Teeth of Fossil Fishes from Ramri Island and the Punjab," by R. Lydekker; "Notes on certain Fossil Genera in the Palæozoic and Secondary Rocks of Europe, Asia, and Australia," by Ottokar Feistmantel; "Notes on Fossil Plants from Kattywar, Shekh Budin, and Sirgujah," by O. Feistmantel; "On Volcanic Foci of Eruption in the Konkan," by G. T. Clark.

Geological Survey of Canada. Report of Progress for 1877-78. Published by authority of Parliament. Montreal : Dawson Brothers.

THIS report, as usual, contains a goodly store of interesting matter. Considerable attention has been paid to the apatite beds, and a very large number of samples have been analysed. The mineral is decidedly rich in phosphoric acid, averaging about 40 per cent. The carbonates are trifling ; iron is in many cases absent altogether, and alumina ranges from 0·267 to 1·969 per cent. The only drawback is the presence of fluoride of calcium, averaging 7 per cent. This constituent requires special precautions to prevent nuisance when the apatite is used in the manufacture of superphosphate. It may be pronounced decidedly superior to the Nassau apatite, which contains less phosphoric acid, more iron and silica, about the same fluorides, and alumina. To what extent these samples are fairly representative of the mineral veins seems, however, an open question. The apatite occurs in Ottawa county, and is accompanied by a great variety of mineralogical species, mainly similar to those found associated with apatite in Norway.

The gold deposits of British Columbia are described at some length. The precious metal is found both in alluvial deposits and as auriferous quartz. Near Kelley's lake a quartz vein has been discovered yielding 1·21 ounces of gold and 2·43 ounces of silver to the ton. Pellets of native silver have been found on the Similkameen, and the ore at Cherry Creek has been found to contain 658 ounces of silver per ton. On the South Similkameen the gold is mixed with coarse scales of platinum. Rich copper-ores and native copper have also been met with, but pending the possible discovery of deposits yet unnoticed the mineral resources of British Columbia do not seem of capital importance.

Cape Breton is rich in fine marble, which at present lacks a market, being excluded from the United States by a protectionist duty of 50 cents per cubic foot. Many of the beds are perfectly fit for statuary purposes.

The palæontological observations are not remarkably important. The Tertiary strata of British Columbia have yielded fossil insects in good preservation, but presenting few points of interest.

The Dominion must undoubtedly have many secrets, geological and biological, which are yet to be brought to light, and we must hope that the Survey will not on any account be cut short in its operations.

Several important books stand over for review, on account of the want of space.

CORRESPONDENCE.

. The Editor does not hold himself responsible for statements of facts or
opinions expressed in Correspondence, or in Articles bearing the signature
of their respective authors.

ON A POINT RELATING TO BRAIN DYNAMICS.

To the Editor of The Journal of Science.

SIR,—The following letter "On a Point relating to Brain Dynamics" (referred to by me in the article on " Natural Science and
Morality," in your last number) might possibly have some interest as a mode of reconciling *Free Will* v. *Necessity,* said to be
similar to that proposed by the late Prof. Clifford in an oral
lecture, at St. George's Hall, on "Body and Mind." I am not
aware if this lecture has been published, and the views in the
subjoined letter were of course arrived at quite independently.—
I am, &c.,

S. TOLVER PRESTON.

From " Nature," May 13, 1880.

ANY attempt to grapple with the doctrine of Free Will *v.*
Necessity on the old lines would probably (and deservedly so)
not attract much attention. The object of this paper is to place
a consideration of extreme simplicity under critical notice, which
would seem to be capable of affording a key to the complete
reconciliation of the divergent views on a common basis ; and
since the matter to be dealt with will be strictly within the
domain of natural science, a clear analysis will be rendered
possible.

It is well known that the only attempt to harmonise the doctrine of Free Will with the principle of the Conservation of
Energy consists in supposing that living creatures have a power,
by the mere exercise of their " will," of deflecting particles of
matter within their bodies from their natural paths, without
thereby altering the total energy of the particles.* This, there-

---

* The necessity for this special assumption, in order to prevent Free Will
from coming into direct collision with the principle of the Conservation of

fore, it will be observed in the first instance, assumes a peculiar physical state of things to exist within the body of an animal which does not prevail elsewhere, or it supposes that the laws of Nature have not a general application, but that the animal world must be made an exception. This at the very outset evidently involves a very questionable admission. My purpose is simply to point out that by taking into account a special consideration based on the evidence of modern physiology as to the functions of the brain, such an assumption as the above is rendered entirely superfluous, and that even if it could be supported it would still miss the main object in view.

Whatever room for speculation there may be as to the exact nature of the mental faculties, it is at least very generally admitted that these faculties are most intimately *connected* with or dependent on brain structure. Modern physiological research has at least placed this fact beyond question, or it is allowed that the mental faculties have at all events *a physical side.* From this it must follow, therefore, that what we call "identity," character, or individuality (as involved in " mind ") must be dependent on the special structure of the brain; indeed this view is so widely prevalent that it becomes almost superfluous to insist upon it. Now it may be safely assumed that no upholder of Free Will would wish for more than that a person should act in strict accordance with his identity or individuality, for the object of Free Will certainly is not to annihilate individuality (or those personal *traits* which constitute character). But is not this precisely what would occur if this contention for a mysterious power of deflecting particles within the body could be carried out ? for the effect of this contention would be to make the brain superfluous as a directing mechanism, which would be tantamount to abolishing it (together with the individuality, of which it is the seat). For where would be the use of the elaborate mechanism of the brain for directing the movements of the body if we are to have power of carrying out this same object by deflecting particles by " volition " (whatever that may mean) ? This would be to substitute for the brain, with which the identity is bound up, the empty nothing " volition." In that the brain directs the corporeal movements; the identity, or that which constitutes the very essence of individuality, thereby directs. What more would we have ? Attempt to supplant the brain by the vague notion " volition," and the individuality ceases to exist ; or that very end is attained which those who support Free Will most wish to avoid.

From the very fact that the brain is *known* to exist, it therefore should be perfectly conceivable (if not even *à priori* a natural

---

Energy, is so obvious that it will probably be regarded as superfluous to give references to particular authors.

conclusion) that the brain might be a mechanism competent to regulate all* the motions of the corporeal system (for a set of dynamical conditions adapted to *any* effect is conceivable). In view of this, does not the assumption of this mysterious "deflecting power" seem all the more unwarrantable, or even absurd; as if it were imagined that the brain, being already there to direct the corporeal movements, something additional were necessary to direct the brain, or as if it were supposed that [the brain being the seat of the identity] something besides the identity were required to direct the actions of the body? This would seem to be no more than a specimen of the kind of incongruities which may be expected to present themselves by any attempt to evade physical principles.

It could not, however, be said that the opposite party were entirely free from error. For there appears to have been a notable oversight on the side of those who uphold strict Causal Sequence in Nature (sometimes called " Necessity ") in failing to appreciate adequately the important influence (on the question of Free Will) of the fact that the brain is the seat of individuality, as above insisted on. For the omission to give due import to this fact has naturally made strict Causal Sequence to appear as a sort of grinding process, whereby man's actions are determined *independently* of his individuality; a view which is no doubt repulsive, and may have served as some excuse for the invention of the curious device of deflecting particles by the " mind " or " will." It will be observed, however, that by simply substituting the word " brain " (which includes " mind ") for the word " mind " in the foregoing sentence, a deflection of particles of matter (represented by the direction of material operations by the brain) then can take place in accordance with, and not in opposition to, the laws of Nature. For from the very fact of the brain substance forming part of the material universe, it must of course influence and direct material operations in conformity with natural causes.

Could it be justly said that there is any *compulsion* in this? Can there be compulsion in being obliged to act in accordance with one's individuality or identity (determined by brain structure), since the only conceivable escape from this would be to act *in opposition* to one's identity (scarcely a desirable end)? But, it may be argued, there is still some coercion left here,

---

* Does it not seem a violation of principle, or a kind of inconsistency, to recognise that the brain does, in fact, direct certain motions of the corporeal system (and even those of a complex character, such as the digestion of the food, the circulation, &c.), and yet to assume that the brain would be incompetent to direct *all* the motions of the body? It may be said that a reasoning process accompanies the direction of some of these motions, but not others. But then is not reasoning itself a brain process, or is it not universally admitted that the reasoning faculty (whatever its exact nature) is at least *connected with* the brain, or has a physical side, just as, indeed, the mental faculties generally (or " mind ") could not exist without brain?

because, although brain structure may be the seat of individuality or "mind," nevertheless, since our brains were originally formed by the operation of causes beyond our control, there is coercion in this aspect of the case. But then do even the most ardent supporters of Free Will ever dream of upholding the expectation that an individual should have a control in the original formation of his brain? or do they not concede (and rightly) that the ideal of Free Will is that an individual should act in strict accordance with (and not in opposition to) his own identity? Yet this is precisely what the believer in strict Causal Sequence, who has a just appreciation of the functions of the brain, will maintain must necessarily occur. Solely in virtue of the fact that there is strict Causal Sequence in Nature are the actions brought into strict conformity with individual brain structures (or with identity). If the principles of dynamics were not rigid, or if the laws of Nature were liable to alteration, a man's actions might sometimes be in harmony with his brain structure, sometimes in discord with it; or any number of persons, though possessing totally different brain structures, might act identically. The questionable expediency of the proceeding of those who are disposed to grumble at what they term the "iron" laws of Nature becomes apparent here.

But is it not, after all, more satisfactory to look to a definite physical basis for identity or individuality, as dependent on the magnificent mechanism of the brain, in preference to the superficial view of ignoring all this? No doubt there have been misunderstandings on both sides of this Free Will *v.* Necessity question,—the Free Will party, failing to appreciate justly the sequence of cause and effect; the Necessitarians, on the other hand, omitting to realise fully the important bearing of the relation of individuality to brain-structure on this question. No logical ground could be given why a complete agreement should not be possible on this subject; for there can evidently be but one correct view on any subject or question whatever. Moreover, from the very fact of the fundamental character of this question, it would follow necessarily that the wrong view on this subject must involve a great error, which therefore could hardly escape detection under a careful analysis. The divergence of views here is, however, no doubt more apparent than real; for if Free Will may be justly regarded as the freedom to act in accordance with identity (or as the assertion of individuality), then such freedom of will actually exists, and, moreover, the very condition for its existence is seen to be the prevalence of that strict Causal Sequence in Nature demanded by the Necessitarians. Thus the two views would show themselves capable of reconciliation on a common basis. That this fact should have apparently hitherto escaped appreciation may possibly be to some extent due to that spirit of partisanship which has so largely entered into this question, whereby the judgment may be allowed to be unconsciously biassed, so that in some cases, instead of searching impartially as to what *is* truth, the inquiry has perhaps rather been as to what *ought* to be truth.

S. TOLVER PRESTON.

STRANGE PROPENSITY OF CATS.

To the Editor of the Journal of Science.

SIR,—Perhaps you, or some one of your correspondents, may be able to throw a light on the attraction which the pretty blue-flowering annual *Nemophila insignis* seems to exert upon cats. They roll on it, nibble it, and in short maltreat it in such a way that its cultivation in a suburban garden is, according to my experience, impossible. I should like to know whether this circumstance has been observed by others, and whether any experiments have ever been made with a view of explaining the matter.—I am, &c.,

EXETER.

[The fact noticed by our correspondent was discussed some years ago in "The Zoologist," and it has repeatedly come under our personal observation and been mentioned to us by friends in various parts. We are not aware that the plant has ever been chemically examined.—ED. J. S.]

INSTINCT AND MIND.

To the Editor of the Journal of Science.

SIR,—At an early opportunity I shall crave your kind permission to make a few remarks on Mr. S. Billing's interesting article, which appears in your July number.—I am, &c.,

R. M. M.

NOTES.

Mr. J. W. Mallet, F.R.S., in a communication to the Royal Society, shows that although hydrogen is occluded by aluminium its quantity is too small to vitiate the determination of the atomic weight of that metal.

Among other papers recently read before the Royal Society are—" Notes of Observations of Musical Beats," by A. J. Ellis, F.R.S., &c. ; " The Aluminium Iodine Reaction," by Dr. J. H. Gladstone, F.R.S. ; " On the Critical Point of Mixed Vapours," by James Dewar, F.R.S.; " Experimental Researches on the Electric Discharge with the Chloride of Silver Battery," by W. De la Rue, D.C.L., F.R.S., and Dr. H. W. Müller, F.R.S. ; and on the Lowering of the Freezing-point of Water by Pressure," by James Dewar, F.R.S.

Prof. E. Ray Lankester has communicated to the Royal Society a note on the discovery of a fresh-water Medusa, of the Order Trachomedusæ, in a tank in the water-lily house of the Royal Botanical Society. It has received the name of *Peregrinella Sowerbii*.

Prof. E. A. Schäfer, F.R.S., and Mr. F. A. Dixey have laid before the Royal Society a preliminary note on the ossification of the terminal phalanges fo the digits. The process begins at the tip, and not at the centre.

Prof. W. C. Williamson, F.R.S., in a paper presented to the Royal Society, gives a very interesting account of the organisation of the fossil plants of the coal-measures.

Prof. Hoppe-Seyler has obtained from the alcoholic solution of chlorophyll crystals of a body which he names chlorophyllon, and to which he ascribes the decomposition of carbonic acid under the influence of light, and the peculiar fluorescence of chlorophyll. On treatment with caustic alkalies at elevated temperatures a purple-red compound is obtained, dichromatic acid, which possesses very remarkable optical properties. — *Zeitschrift Physiol. Chemie*, iv., p, 193.

According to the " American Naturalist " the English sparrow has become such a nuisance in Iowa that the State legislature is considering plans for its extermination.

From the same journal it appears that the arrival of migratory birds from the south is determined by hot winds lasting about three days.

Among eminent scientific men recently deceased we must mention Prof. D. T. Ansted, F.R.S., and Count de Castelnau.

According to a memoir communicated to the Academy of Sciences, the equatorial and polar diameters of the planet Mars agree with the hypothesis of an anterior fluidity.

M. Schützenberger, ex-Professor of the University of Strasburg, gives, in the " Revue d'Anthropologie," a curious account of a dog which was in the habit of stealing carrots, not of course for its own use, but to give to a horse—*un grand diable de cheval* —for which he had a particular affection.

Since the commencement of the present year the deaths at Paris have exceeded the births.

We hear with great pleasure that the manufactory of spurious university diplomas at Philadelphia has been broken up. Several barrelsful of fraudulent medical diplomas, just ready to be issued have been seized, and five " bogus " colleges, are in danger of suppression.

According to the " Medical Press and Circular " a belief is spreading in Hampshire that the luminous properties of Balmain's phosphorescent paint are due to an admixture of human fat !

It is not generally known that the superstitious practice of hoplochrism still prevails in Suffolk. If anyone injures himself with a tool or weapon he is at once exhorted to apply some healing ointment, not to the wound, but to the blade or point. The belief that stones are capable of growth is also still entertained in the Eastern counties.

The Birmingham Science College will open on October 1st, Prof. Huxley delivering an inaugural discourse. The course of Biology will be conducted by Prof. T. W. Bridge, F.Z.S.

Mr. J. S. Kingsley, writing in the " American Naturalist," pronounces the American Association for the Advancement of Science " a drag rather than an aid to progress." He adds that " aside from its grants of money to the Zoological and Geological Records, and to specialists, to enable them to carry out certain lines of investigation, its British prototype is no more worthy of its pretentious name."

Prof. Cope suggests that the mutilations and strains which plant-using animals have for long periods inflicted on the flowering organs may, as in some similar cases in the animal kingdom, have originated peculiarities of structure.

According to R. E. Kunze, in the " American Naturalist," the potato-beetle (*Doryphora decemlineata*) is poisonous to man. The usual result from handling the crushed beetles, as well as from inhaling the fumes arising from vessels in which they have been scalded, has been likened to serpent and scorpion poisoning.

Where death has followed, the blood has become disorganised the same as from septicæmia. (Can the poisonous principle be cantharidine? In America this question would admit of an easy answer.

It has been pointed out that the use of palls and mourners' cloaks at funerals furnishes a very efficient medium for the spread of infectious diseases, and their use has on that account been formally abolished among the Jewish community.

A writer in "Baily's Magazine" makes some very curious statements concerning the physiological action of different alcoholic liquids, if taken in excess. Good wines and malt liquors cause a man to fall on his side; whiskey produces a fall on the face; and cider and perry throws the subject on his back! To decide this question by experiment would, we suppose, be considered vivisection.

Abstinence from pork is no safeguard against trichinosis. Two French soldiers have lately died of this disease at Thionville, in consequence of having partaken of the flesh of geese. Trichinæ have also been detected by Dr. Clendenin in a pike caught near Ostend.

If the future career of the Manchester University bears any relation to the class of persons who figured at its formal inauguration, Science will occupy but a very subordinate position.

According to Dr. Vincenzo Peset y Cervera the crystals of hæmoglobulin obtained from the blood of different animals have forms so distinct and characteristic that the origin of a sample of blood may thus be determined. All that is required is to mix the blood with a little bile, when crystals not exceeding 0·003 metre in size are formed in the mass. The shapes of the crystals are said to be as follows:—Man, right rectangular prisms; horse, cubes; ox, rhombohedrons; sheep, rhombohedral tables; dog, rectangular prisms; rabbit, tetrahedrons; squirrel, hexagonal tables; mouse, octahedrons; &c.
[If these observations are confirmed they may serve for the solution of a most important question raised by Dr. Lionel Beale. If the theory of Evolution is true, the crystals obtained from animals which are nearly related should be either identical or such as are in form easily derived from each other. Should the hæmoglobulin crystals—*e.g.*, of the horse and the ass, of the dog and the fox, of the rabbit and the hare, or of the rat and the mouse—belong respectively to different systems, it will supply a serious argument in favour of independent creation.—ED. J. S.]

According to Prof. Claypole, of Ohio, many wild European plants have migrated into America, and have there become so common as to prevail over some of the indigenous plants of the country. Only two or three American wild plants have crossed the Atlantic and become naturalised in England. The difference of climates and the conditions of intercourse between the two countries do not furnish any satisfactory explanation of so marked a difference in the migratory power of the two floras.

According to Dr. Dujardin-Beaumetz the alkaloids of the pomegranate bark occasion paralysis of the motor nerves, whilst leaving the muscular contractility unimpaired. They do not affect sensibility, and may rank among the curarising poisons.

In a girl who died recently, at the Hôtel Dieu of Caen, the whole of the thoracic and abdominal organs were found completely inverted, the heart being at the right side, the liver at the left, whilst its usual position was occupied by the stomach and the spleen.

"Les Mondes" gives, from some source not specified, an account of a so-called electric girl, living at London, in Canada, whose hand cannot be touched save on penalty of an electric shock. She can give a violent shock to a chain of fifteen or twenty persons who join hands, and she has the power of magnetic attraction. Packets of needles, even if wrapped up in paper, hang suspended from her finger-ends. If she enters a room all the persons present undergo a perceptible influence ; some are made drowsy, and others feel indisposed and enervated until her departure. A sleeping child awakes at her approach, but a slight caress with her hand sends it to sleep again. [How, when her touch communicates an electric shock ?] Animals are equally influenced by her ; a favourite dog of the family remains for entire hours at her feet, as motionless as if dead. Very strange, if true.

Acccording to M. Bidard the plane tree (*Platanus*) exerts an injurious action on human health, sometimes to the extent of producing discharges of blood from the lungs. This action is said to be due to a white down which is formed on the lower surface of the leaves, and which is easily detached by the wind.

Mr. Tremlett, Consul at Saigou, gives particulars concerning *hwangnoa*, which is said to be an antidote for the bites of the most venomous serpents, and to have been used successfully in the treatment of cancer and leprosy, which is very common in Tonking.

A sum of £60,000 has been subscribed for the establishment of a college in Liverpool, on the principle of those of Leeds and Bristol.

A species of *Dasyurus* (*D. fuscus*) has been found in the Arfak Mountains of New Guinea—the first Australian carnivorous type found in that island.

MM. A. de Quatrefages and L. Hamy have communicated to the Academy of Sciences a memoir on the craniology of the various subdivisions of the Negro race.

The study of Anthropology is beset with difficulties. Even the skull of a savage killed in battle is begrudged to Science, and is made the subject of a " question " in Parliament.

A new poisonous alkaloid, of a pleasant odour, but of very deadly properties, has been discovered in the fumes of tobacco. It approaches closely to collidine.

An Acarus of the genus *Trombidium* has appeared in Vaucluse in the welcome character of a destroyer of the *Phylloxera.*

According to M. Dieulafait zinc can be detected in all rocks of primordial formation, and in the waters of the seas of all ages.

According. to M. H. Filhol a number of new mammalian fossils have been discovered in the phosphatic beds of Quercy. Among them may be noticed *Quercitherium tenebrosum,* a carnivorous species approximating to *Cynohyænodon.*

M. F. Peuch has proved, by a series of experiments described in the " Comptes Rendus," the fact of the transmission of tubercular disease by the milk of the animals affected.

A writer in the " Druggist's Circular " (American) complains that the English sparrows introduced into the United States have lately (?) acquired a taste for destroying the buds of the peach, pear, and apple trees, and even of the garden pink. The editor in reply gives the good, but scarcely practicable, advice to apply poison to the buds.

The " army worm," often mentioned in American agricultural literature as very destructive, is the larva of a Lepidopterous insect. It has a peculiar propensity for destroying wheat.

We regret to learn that in the State of Missouri the office of State Entomologist, which has been most ably filled by Prof. Riley, has been abolished.

The evil effects of opium smoking appear on closer examination to have been greatly exaggerated. Of the 17 million lbs. of opium yearly consumed in China, 5 millions are of native growth.

Dr. Sylvester Marsh, jun., has communicated to the Quekett Microscopical Club his experience of the processes used in bleaching and washing microscopical sections intended to be stained. The objection to alcohol as a bleaching agent is that it is very slow in action and uncertain in result. Chloride of lime and Labarraque's solution of chlorinated soda are liable to disintegrate and destroy, unless very cautiously used. Dr. Marsh prefers euchlorine. The apparatus consists of two of the bottles in which an ounce of quinine is usually packed : one is used as the gas generator, from which a bent glass tube passes nearly to the bottom of the second, containing water, in which the sections to be bleached are placed. In the first bottle a sufficient quantity of crystals of chlorate of potassa is placed to cover the bottom, and about a drachm of strong hydrochloric acid is poured upon them. The gas, saturating the water, safely and effectually bleaches the sections. A notch is cut in the cork of the second bottle, to allow the escape of the gas when the water becomes supersaturated. The process is conducted out of doors, and exposure during the night is sufficient to bleach the sections. The washing is conducted in a bottle of the same kind. A funnel passes through the cork, and is continued to the bottom of the bottle by means of a piece of india-rubber tubing. A small hole is drilled in the side of the bottle, about half an inch below the shoulder. The sections are placed in water in the bottle, a filter-paper is placed in the funnel, and the whole placed under a tap arranged to allow of a suitable flow of water. The washing will be found complete after a night's exposure. In the discussion which followed Mr. W. H. Gilburt remarked that the bleaching and staining processes were not altogether satisfactory, because, if the relation of the tissues to each other was to be observed, it was necessary to do something more than bleach them, and when cell-contents had to be examined it was necessary to do so in their natural condition, and the tissues should be as fresh as possible. Dr. John Matthews said there was a mode of differentiation of tissues which was too often overlooked, and that was the use of the polariscope. If the object under examination did not yield to one selenite, it usually would to another.

Mr. F. Barnard, of Kew, Victoria, writing to "Science Gossip," recommends the use of carbolic acid in place of turpentine for preparing objects for mounting in Canada balsam. It has the advantage of not rendering tissues brittle, as is the case with turpentine, and allows objects to be mounted in balsam without previous drying.

ERRATUM.

Page 406, line 13, *for* "coon" *read* "born."

THE

JOURNAL OF SCIENCE.

SEPTEMBER, 1880.

I. REPORT ON SCIENTIFIC SOCIETIES.
By Dr. C. K. AKIN.

(Concluded from p. 493.)

Pesth, July 5, 1870.

MY DEAR SIR,

IN my previous communications, at least as I take it, I have strictly kept to the terms under which the Royal Commission has been instituted, and in my interpretation of which I was mainly governed by the document printed among the reports of the British Association for 1869, and which, presented in the shape of a memorial to Government, led, if I am rightly informed, to the appointment of the Royal Commission. According to the document adduced, the objects of the proposed Commission were stated to be, to consider—

" 1. The character and value of existing institutions and facilities for scientific investigation ;
" 2. What modifications or augmentations of the means and facilities that are at present available for the maintenance and extension of science are requisite ; and—
" 3. In what manner these can be best supplied."

In the words of the report, also, in these points is " involved the whole question,"—to some discussion of which I may consequently, as I conclude, devote this letter, without being guilty of overstepping proper limits,—the " question," viz., concerning the mutual relations—

Of Science and the State.—I have lived long enough, and have had sufficient experience both of science and the world, not to over-estimate the value of knowledge to man or to mankind ; and although I should be sorry to back up the statement of Pascal that the search after truth were an occupation uncongenial to human nature, yet observation has convinced me that, in the actual state of society, it is not practicable, and therefore not desirable, that more than a very limited number of persons should follow, or be induced to follow, this calling. Now, if we wish to form a fair and unbiassed opinion or estimate of the importance of science and the recognition it receives, we ought to compare it exclusively with the remaining members of the sisterhood of, anciently so called, *liberal arts.* As the result of any such examination, we shall arrive at the conclusion that neither literature nor fine art is, upon the whole, so un-grateful as Science is to the man who wooes her, or rather whom she entices ; or, in plain words, with neither writer nor artist does society deal so unequitably as it does with the philosopher. For, in whichever aspect we may view Science, whether in regard to intellectual worth or practical utility, surely the discoverer of truth should obtain the same reward as the author of fiction ; as the former, to view only one side of the question, extends our knowledge of nature, and virtually expands man's intellect, while the latter pro-cures us but pleasant and fleeting emotions, or helps us to while weary hours away. Still fiction, if it does not obtain recognition in the shape of rank, reaps its recompense in wealth, favour, and fame, a single successful work being sufficient to procure lifelong independence and an enduring popular name : the most splendid discoveries, on the other hand, are rewarded, if at all, by medals of no intrinsic or outer worth, and the greatest philosophers have to spend habitually in penury lives of obscurity. True, the philoso-pher should work disinterestedly and seek for no worldly recompense, but so ought also the poet or artist. And if society spontaneously pay its debt to the latter through voluntary contributions, as it were, both of gold and of cele-brity, there is no reason why the public should not equally acknowledge its obligations to the former, if no other mode be found, by contributions levied by the State, and distinc-tions conferred by Government or the Crown.

Whatever may be said about the tendency of the age, the privilege of aristocracy still remains the greatest boon that can be bestowed upon any person who is subject of a State where such a distinction, by law or custom, is recognised.

For it gives that security to man in regard to the future welfare of his progeny which it is his highest ambition to achieve, and for which there is, perhaps, no other means equally efficacious ; inasmuch as it naturally opens out paths to distinction, an entrance to which others have to conquer, and rarely achieve with any ultimate success worthy the pains and struggles it has cost. Although illustration in any department is not wanting to the Upper House, yet eminence in science, literature, art, or even invention, has not been considered a sufficient claim for inscription upon its muster-roll, to which distinction in law, politics, or war, or even simple wealth, so frequently gives admission. In the Bill brought into Parliament, but subsequently dropped, by Earl Russell, concerning the creation of Life Peerages, the claim of scientific men to such a distinction was admitted ; but, in common with new life peers in general, they were to sign, previous to their admission to the House, a kind of self-given *testimonium paupertatis*. Now, in regard to life peerages—the grant of hereditary peerage being, as an undoubted prerogative of the Crown, always dependent on the mere will of the sovereign and his constitutional advisers—there appears to be a mode of naming, especially men of science to such a dignity, not liable to the objections which ultimately led to the abandonment of Earl Russell's Bill. Might not, indeed, new life peerages be based upon the principle admitted in the case of bishops, who now-a-days hold their peerages by Office, no more by land ? And might such persons, for instance, as the Master of the Mint (whose office has been taken from science in England, when, upon her example, it has become an appanage of science on the Continent), the Astronomer Royal, and even the Presidents of the Royal Society and British Association (who should then be elected for life, and by means of a *congé d'elire*), not be properly and regularly summoned to the Lords ? And instead of their being introduced, as it were, *in formâ pauperis* to the House, would it be squandering the public money if each of the above-named persons were endowed with an annual pension amounting not to the salary of a judge or bishop, but to that of a junior lord ?

If Science, however, is to obtain her fair share of recognition and due meed of reward from the State, men of science will have, as happens in the world, to exert themselves and prove themselves capable of holding their own. Otherwise, or if philosophers—out of too noble instincts which the world does not appreciate—spurn honours and disdain recompense, the spectacle, humiliating to Science and its representatives, will recur of a phenomenal intellect passing out of existence with scarcely any public notice being taken of it; and Chancellors of the Exchequer will continue to deny a national memorial to a philosoper who, in the world of mind and within a thousand years of his race's history, had no other peers but Bacon, Shakespeare, and Newton. On the other hand, the example of some living naturalists proves that by energy and perseverance much may be obtained of the State not simply in personal behoof, but for the interests of science; and the position held in public estimation by other workers proves that, if not always discriminating, society is not wholly indifferent to scientific merit placed—not under a bushel.

I trust that, neither in throwing out the preceding suggestions nor in making the above remarks, I shall be accused, upon *personal* grounds, of want of proper reserve; they occurred to me at a time when, residing in England, I naturally had changed somewhat my native horizon. In conclusion I will only further observe that, for none of the schemes explained or purposes adverted to in my previous letters, nor upon the whole, do I consider the appointment of a special Minister or Secretary of State for science or instruction at all requisite, desirable, or beneficial. On the contrary, I think it is very much to be wished that English science may continue in that state of independence and complete self-government which has become traditionary, and has produced such admirable—I might say without any exaggeration even unparalleled—results. All that is requisite, in my opinion, for the steady advancement of science, besides some *material* public aid, is such *co-operation* among scientific men themselves as in your speech at Exeter you justly laid so much stress on, as being one of the beneficial objects the British Association was designed to accomplish; and, further, such *a more systematic direction* as you similarly showed, from its Bye-laws, it was another aim the founders of the Association desired to realise in science, or in the activity of its cultivators. And in order that both these objects, as well as the results to which they should lead, be effected, it were but necessary that philosophers and scientific inquirers should adopt as their watchword *in practice* what is theoretically *supposed* to be their guiding principle in all their endeavours, and what, in the terms of a well-known formula, I will phrase by—

The Truth, the whole Truth, and nothing but the Truth.

Expressing to you my obligations for agreeing to act as the medium between the Royal Commission and myself, so far as these communications are concerned,

I remain, my dear sir,
Yours very sincerely,
C. K. AKIN.

G. G. STOKES, Esq., Sec. R.S., of the Royal Commission on Scientific Instruction and the Advancement of Science, London.

PS.—So far from being guilty of transgressing in my letters the scope of the Royal Commissioners' deliberations, I think that, on the contrary, I should myself notice the fact of having purposely confined myself to a consideration of one-half the questions only concerning which they are to make inquiry. Excepting a passing remark on the stunting effect of competitive examinations on invention, and another on the unphilosophic style of many scientific works designed for popular instruction, my observations had exclusive bearing on the advancement of science ; and I will here add but one other remark having reference, like the above quoted, to scientific instruction.

Owing, in part at least, to the really wonderful facts and instruments modern science has detected or invented, a tendency is spreading to make science altogether, if I may use the word, *sensational*. Experiment, which, in the words of Dr. Young,—that should be written up in conspicuous letters of gold in every lecture-room,—has its uses "in assisting the imagination to comprehend, and the memory to retain, what in a more abstracted form might fail to excite sufficient attention," instead of being employed as a means of instruction becomes *the* object of *attraction*. In superaddition to the drawback already entailed by our consitution—in consequence of which the senses perceive only projections of things, and the mind judges of these, to use the simile of Plato, as a man confined in a cave, into which passing objects threw their shadows, could judge of what is passing in the outer world—modern modes of scientific exposition have introduced for the representation of phenomena appearances not very superior to drawings in the aid they afford to the understanding, but of a *ghostly* character, and apt to make philosophy resemble in fact, as it anciently was likened in name, to magic. The desire to produce *excitement* leads also to worse deeds—such as the furbishing up of well-known facts, so as to give them the appearance of startling newly-discovered truths, or the throwing out of undigested hypotheses and baseless guesses as if they were accurate, well-tested theories. In this manner science becomes corrupted and debased, and the sense of truth, in lecturer and listener, blunted and confused; while precision, accuracy, and scrupulosity, being the chief ingredients of the philosophic mind, are looked upon as of no. importance or value. But here, also, it is not the interference of Government, but the sincerity and moral courage of true men of science, that alone would be capable to stem the current of the spreading evil.

C. K. A.

Pesth, July 21, 1870.

My Dear Sir,

In my preceding letters I have examined into the state of science in England, and of the institutions and means subservient to its advancement, and I have suggested several measures of reform or improvement. My object in writing to you this supplemental communication is to submit some remarks concerning the state of science on the Continent ; and if these do not lead me to make any additional proposals, they may yet convey information " how not to do it."

The scientific institutions most prominent in France and in Germany—and to these countries I intend to confine my remarks—are, relatively to the former, the *academy*, and relatively to the latter, the *universities;* and the sign characteristic of both is the principle of *authority* supported by a system of *centralisation*. It will be sufficient for my present purposes if my observations point to these several subject matters.

The French Academy is in some respects similar to the Royal Society, and the points in which it differs from the latter are not, in my opinion, to its advantage. In the first place, the members of the Academy are salaried by the Government, but their emoluments are not sufficient to live upon, or to keep them, so to speak, in working order ; nor do they perform any specific service to Science or the State for the money. The Academy, next, is divided into a certain number of sections, according to the several branches of science, and the number of members in each section is strictly limited. As that subdivision is invariable, while the relative importance of the sciences is fluctuating, the abuse has crept in of electing members into a wrong division. On the other hand, such a proceeding not being always practicable, highly distinguished men are excluded from the Academy for many years if their proper sections happen to be full; while if, from the dearth of cultivators or accidents of mortality, the number of vacancies happens to be great, the standard of admission is considerably lowered. The Academy publishes weekly its proceedings or "Comptes Rendus," which, from the celerity and regularity of their publication, are a valuable means of conveying rapid information; on the contrary, its transactions or "Mémoires" are issued in a very irregular and dilatory manner. The practice of examining and reporting upon communications submitted has fallen into almost complete disuse; and the prizes, which are in considerable number, are in great part awarded upon the antiquated principle of putting forth questions. I have thus rapidly drawn the most distinctive features of the French Academy, roughly yet faithfully; and I feel constrained to confess my inability to comprehend the enthusiasm which there appears to exist in certain quarters in England for this institution, and which shows itself in the desire to copy it. I have dwelt in a former letter upon the functions which any society should perform in order to be called useful, and I cannot bring myself to believe that those of the French Academy correspond in any way to the model.

I have spoken, in a former communication, in words of unavoidable eulogium of the German Universities and the position which they occupy among similar institutions in Europe. Still I do not find in their organisation anything that I should be prepared to recommend for imitation or adoption. I shall presently mention the mischievous effects which the Universities in Germany, like the Academy in France, exercise on scientific development, according to my belief, when I shall enter upon the discussion of the principles which underlie the organisation of both: here I wish merely to give an opinion upon the institution of so-called *privat-docenten*, which is generally considered as most characteristic of the German University system, and which has many admirers out of Germany. A *privat-docent* is simply a lecturer who, as a rule, receives no pay from Government or the University, but may take fees from the students: he is simply a private tutor, who, in consideration of having passed an examination or other ordeals before the proper authorities, is admitted to the use of the public lecture-rooms. In my opinion the fellowships in the English Universities—if only Fellows were elected upon a better principle—are much more advantageous; and if the now somewhat dormant institution of lecturers or prælectors in the colleges were more largely developed, the English Universities would have nothing to envy from, and much to boast over, those of Germany in this respect.

The principal aim of the German Universities, as well as of the French Academy, is to uphold the principle of *authority* in science, which has a great many effects that are detrimental to its progress. Authority in science means infallibility, and it means also stagnation. But the essence of science is development, which is identical with change, and variation from ancient theories or received doctrines. The French Academy has generally not been favourable to novelties started out of its own precincts, as is shown by its treatment of such men as Fresnel, Fourier, or Melloni. I know also of a case in which it was found impossible to get a correction or mention of mistakes, which one of its members had happened to make, inserted in the Proceedings of the Academy, notwithstanding repeated attempts. The desire to have that done was supposed to imply *naïveté*. In a similar way the German Universities enforce a certain uniformity in the preparation of scientific students, and they measure all ability by a fixed yet arbitrary standard. Investigation must be *schulgerecht*, as it is called,—for which the French have the word *classique*, but I doubt whether there be any real equivalent in English. A mind of independent character or original turn has thus a hard struggle for existence; for, in order to get recognised, it must be fashioned on the approved pattern. Men like Davy or Faraday are consequently unknown to the history of German or French science, as their irregular preparation would have debarred them from coming under notice, and still more so from making their way. On the other hand, great errors are propagated and kept up under the wing of authority; and if once a philosopher has obtained a certain sway, or formed a so-called "school," his teaching will be kept up long after its errors have been detected. Thus certain theories are still taught all over Germany in physics which are manifestly untenable, and to attack them is punished more severely than heresy is in religion nowadays. Theories propounded by new men are generally overlooked. On the other hand, I could tell an instance in the recent history of physical science where a discovery undoubtedly not novel and manifestly incomplete has been accepted on the

Continent as an unexpected revelation proof against all doubt, because it was appropriated by names possessing authority. What constitutes authority in science it were difficult to define; yet its worship, although it be opposed to the very spirit of science, is in Germany and France, so to speak, without bounds. It were easy to prove by example that the test of infallibility is not applicable, if such a thing could be imagined, with respect to a human mind. Not only are the instances numerous where the authorities of one age have been scouted by those of the succeeding; but even in the works of the greatest among them, whose reputations were acquired on the strength of real intellect and conspicuous services, schoolboys nowadays frequently may point out glaring mistakes committed or upheld by great masters only one generation behind.

I have mentioned in a former letter the well-known fact that a German philosopher who wished to bring out some novel theory in his country encountered so many difficulties that he absolutely went mad. Another who started similar ideas about the same time, having been repulsed in one quarter, took it for granted that the same had happened to him also in another, where it was not the case, so hopeless did he consider his endeavour to obtain a hearing. Actually these ideas took wing in England, but not before, communicated also to the French Academy, they had been allowed to rest unnoticed in its archives for years (like the memoirs of Abel), notwithstanding repeated instances to have them examined. I also have it out of the mouth of one, who is actually himself a chief authority on physical science in Germany, that an early work of his, now the principal foundation of his fame, had proved injurious to his university career, for being of too novel a character. It is a slight consolation to the individuals concerned, for the anxiety or pain they have suffered, to have had their names recently enrolled on the list of members of the French Academy, or to have received an honorary title from a German University; and the damage which is done to Science by such proceedings, in all cases serious, is in many irreparable. Authority, whether exercised by academies or universities, would have its uses if it facilitated the endeavours of students during the early and more trying periods of their career, in which encouragement and aid are most welcome and needed; but if, instead, it check or impede novices, and establish merely a kind of confraternity, the chief end of which is to keep new men out as long as feasible, and to uphold its own sway, I make bold to say that the liberty of thought reigning in England, notwithstanding its abuses, is a far more valuable safeguard for Science, the very life of which is progress. Now, if the Royal Society, transformed into or superseded by an academy, were to arrogate to itself that kind of domination which the Académie des Sciences exercises in France, or if the English Universities endeavoured to absorb all the intellectual life of the nation, or to fashion it in their own way, as is the case in Germany, the superiority of England which has made it the head-quarters of scientific progress, and the mother country of so many amateurs more distinguished in Science than most French academicians or German professors, would probably be gone.

These are the reasons which induced me to express in my last letter the hope that liberty and self-government may be preserved to science in England, henceforth as heretofore; and to deprecate, as intimately connected with the subject, the creation of a special department, or the appointment of a new minister for science or instruction, or, as it is sometimes significantly called, for education— tending towards *State centralisation*. It is altogether erroneous to suppose, as I could prove it, that Government supervision on the Continent has been in any way whatever to the advantage of Science. It would have been so if scientific institutions in Germany or France were better organised than in England; if there were a proper division of labour between them; and if each had its distinct special object. Without entering into any examination of facts, I will merely state—what must be evident to anyone who has studied the subject—that nothing, on the contrary, can exceed the confusion of the multiple scientific establishments at Paris, for instance, involving the frequent abuse called in French *double emploi*,—while the want of any real organisation in the higher branches of instruction, or the scientific institutions, is shown by the existence of the so-called *cumul*. The former defect implies the existence of several posts for the same thing or subject, and generally also the absence of a proper care for others; while the latter means the union of several offices, laboratories, &c., in the hands of one man, to the detriment of others who have none. The unsatisfactory state of science in France, the head-quarters of centralisation, has been frequently acknowledged of late by the very highest authorities of the country. It has led to the creation of a special establishment, as yet very imperfectly developed and not irreproachably planned, destined to procure some amelioration ; and perhaps it may not be the exclusive result of personal considerations that, beside the ministry of instruction proper, there has been recently created in France another public department, designated as the Ministry for Letters, Science, and Art. Hitherto Science had been neglected by the ministers who controlled it, but took principally care of Instruction. This would certainly happen also in England, if a so-called Minister of Education had to take care of both ; and I doubt whether Science would not be an absolute loser by getting the patronage of Government in any such form.

If the prevention of evil be an object of as great importance as the promotion of good, which you will probably allow, perhaps the warning I have ventured to convey in this letter may serve a purpose as useful as the positive suggestions my previous letters contained. Trusting to your kindness to communicate this letter also to the Royal Commission,

I remain, my dear sir,
Yours sincerely,
C. K. AKIN.

Prof. G. C. STOKES, Sec. R.S.,
London.

PS.—Perhaps I ought to have taken some notice in former letters of scientific establishments such as the Greenwich Observatory, the British Museum, the Kew Gardens, &c. I will now content myself with stating that they ought to be utilised in connexion with the proposed Royal College, in a manner self-evident, and which, therefore, I need not detail here.

C. K. A.

[Our readers are of course aware that the above letters were not published by the Royal Commission for whom they were originally written ; nor can it be said that their spirit has at all perceptibly influenced the " Recommendations " which the Commissioners finally issued. It will be seen that Dr. Akin is far from holding that English Science would be greatly benefitted by a wholesale adoption of the features of the French Academy of Sciences or of the German Universities. This is the more probable as in such borrowings and imitations we generally contrive to copy unimportant details, leaving out of sight what is of primary importance.

" Wie er räuspert und ure er spukt
Habt ihr ihm alle abgekuckt
Doch das Genie, ich mein den Geist
Man nicht auf der Wachtparade weist."

Who will not be reminded of an unfortunate attempt to ape Germany in matters military? Since the date of Dr. Akin's letters, however, matters have become in England much worse. Examinationism is more and more rampant, and is eating away the intellectual life of the nation. More and more we are producing crammed puppets instead of discoverers and inventors, and with a marvellous infatuation we seem well content with this state of things, and even, Heaven help us, entertain a sort of contempt for original research. Dr. Akin rightly holds that the mere admission of " Science " at University examinations, on the same footing as " Classics " and " Mathematics," will be of no avail. So long as honours and the more tangible prizes, such as fellowships, can be obtained by merely passing examinations in the researches of others, what the candidates are to be examined in is a matter of supreme indifference. But we do worse; we formally exclude original investigation. We do not allow it to pass current in the land, and we close our ears to all warning voices, whether, like that of Dr. Akin, they come from afar, or are uttered in our midst like that of Dr. H. C. Sorby.—Ed. J. S.]

II. MENTAL EVOLUTION.

By J. Foulerton, M.D., F.G.S.

IN a lecture recently delivered by Mr. Romanes, F.R.S., at the Royal Institution, on the subject of Mental Evolution, he said he had two objects in view— first, to show how Mental Evolution, supposing it to be true, could be traced through all the forms of animal life, from the earliest appearance of irritability, as manifested by the direct and immediate contraction of the simplest organic forms on the application of stimuli, to the highest manifestations of thought as exhibited by man ; and secondly, knowing that these last or highest phenomena of Mind were not admitted by a certain school, most ably represented by Mr. Mivart, to be the product of Evolution at all, but to constitute an entire break in a chain of Mental Evolution continuous up to that point, he would endeavour to prove that there was in reality no such break.

I will not stop to discuss the fact, admitted by all, that Mental Evolution is true up to a certain point, and that all the phenomena of sensation, reflex action, intelligence, &c., common to man and the lower animals, have been so brought about ; nor the way in which Mr. Romanes traced this through the various kinds of animals, from the lowest up to man ; but will confine my observations solely to those mental phenomena manifested by the human species, which immeasurably transcend anything of the kind to be found in the lower animals, and which the school of Mr. Mivart deny being the product of Evolution at all; for though I agree with Mr. Romanes in thinking that they are, I differ from him as to the proofs on which that opinion should rest.

It matters little for the present argument in what these phenomena consist, since the fact that there is a wide gap which separates the highest phenomena of the mind of man from those of the lower animals is admitted by all. Now the line of argument adopted by Mr. Romanes was to show *how* that gap might be *intelligibly* filled in by a series of hypothetical connecting-links,—highly-developed anthropoid apes, predecessors of man as we now find him pretty much the same all over the world, or at any rate without anything that can be called a break in his mental continuity,—and was of opinion that if he succeeded in doing so it was as good as showing that the gap had been filled in, taking into consideration that up to the point when the gap began Mental Evolution was an admitted fact and Organic Evolution was admitted in its entirety. It is not necessary here to say anything of the way in which he accomplished his task; we will admit that it was done as well as it was possible to do it, and that the process he described was one which might actually have happened, and that speaking apes really did exist. It may in passing be stated that, only a few days before Mr. Romanes told us of his hypothetical speaking apes, Prof. Dawkins told his audience, at the Geological Society, that certain flint implements found in strata anterior to the time when he believed it possible for man to have come into existence, must therefore have been fashioned by an anthropoid ape, and at all events held that this was more probable than that man was then in existence.

Now, supposing Mr. Romanes to have shown how the gap in Mental Evolution might have been filled in, I do not think this is the way to prove mental or any other kind of evolution. When Darwin, twenty-one years ago, endeavoured to prove the theory of Evolution as the origin of species, he did so not only with a full knowledge of the wide and apparently unbridgable gaps separating and isolating organisms, and of the then absence of connecting-links, but prophesied—and prophesied truly—that this would be the great difficulty in the way of the acceptance of his theory. But Evolution was true nevertheless, for it depended upon two facts known to everybody from all time, just as it was known that apples fell to the ground, but the significance of which had never been fully appreciated—first, that every organism is the offspring of a similar organism; and secondly, that the offspring is never exactly like the parent: these two factors, Descent with Variation, constitute the groundwork of Evolution. Connecting-links have during the past twenty-one years come in abundantly, and the apparently unbridgable gaps have, many of them, been bridged over, but they have in reality added nothing to strengthen the fact that Evolution must have taken place: all they have done—and it is no doubt very important, though of secondary consideration—is to show the method by which, and the order in which, Evolution took place. So much for Organic Evolution as the origin of species: now what I maintain is, that unless there is similar evidence for Mental Evolution it will be useless to endeavour to supply its place by any attempt to discover mental missing links. Mental Evolution must be as independent of Mental Palæontology as Organic Evolution of Organic Palæontology; and such is I believe the case, and that the evidence for both is precisely similar, viz., Descent with Variation. Everyone unhesitatingly feels and admits, in the daily affairs of life, that our thoughts and our actions—which are the outcome of our thoughts—are always preceded by motives, so that we have an unbroken descent for our thoughts during the whole of our lives, and as at birth the human species is as much wanting in those highest mental faculties of which we are speaking as any of the lower animals, it follows that these faculties are connected by an unbroken line of descent with those which man has in common with them, and which are admitted by all to be the product of Evolution. But motives do not always give rise to the same thoughts in different individuals, which is to be accounted for just as we account for variation in the lower animals—by that inherited difference which always accompanies descent, and without which there would be no Evolution.

There is also what may be called a material aspect of thought,—I mean that change in organic tissue which always accompanies thought; and it is just as certain that this change is the effect of another that preceded it, as it is that any motion whatever is the effect of a previous motion, but the change which preceded it and acted as the cause could only be that which accompanied the motion. Thus in whatever way thought be regarded, whether as a feeling or physical change, there is no escape from the conclusion that, like everything else, it is an effect of something that went before; but this is what the opponents of Mental Evolution are obliged to deny, for it is absolutely fatal to their hypothesis.

The analogy between Mental Evolution and Organic Evolution might be carried a step farther. I allude to the influence of Natural Selection. If there is one department of Mind more than another which is supposed to be in a peculiar manner not the product of Evolution, it is the moral faculties; yet it would not be difficult to show that they are just as much the result of Evolution with Natural Selection as anything else in the whole of biological science —and fortunately it is so, for it is the certain and sole guarantee of their ultimate supremacy, though the quality of the morality may be very different from what commonly goes by that name, and, instead of the angry and vociferous dogmatism of conflicting creeds, which is an unmixed curse, it will be a product of the still small voice of Natural Selection, and an unmixed blessing. The former had to precede the latter, just as the Stone Age had to precede the Iron, just as Savagery had to precede Civilisation : there is just as little doubt in the one case as in the other which of the two Natural Selection will seize hold of.

As the proofs of Mental Evolution are precisely similar to those of Organic, so likewise are the arguments employed against it. Indeed there is but one answer possible in either case, viz., Special Creation, which means that there are organisms without parents, thoughts without motives. The former hypothesis has, in a wonderfully brief space of time, been completely routed along the whole line; even the poor Bacteria—the last shred of a theory once so wide that it included every species past and present—have at last had to give up their claim to any special intervention on their behalf.

The latter hypothesis, however, still holds its ground, but is a product not of the common sense of mankind, which always acts on the belief engendered and verified by the constant experience of their own feelings, and what they see in others, that thoughts and actions are always the outcome of motives : indeed the whole business of life is based upon that supposition, so that the burden of proof rests entirely with those who maintain a contrary opinion, and clear and distinct evidence must be given of thoughts arising without motives,—an exploit which may reasonably be presumed to be impossible but which nevertheless is absolutely essential to any effective opposition to Mental Evolution, just as evidence of an organism not being the product of a previous organism would be necessary to disprove Organic Evolution.

It is not, however, on scientific grounds that the opponents of Mental Evolution really take their stand, but rather, as Mr. Romanes pointed out, from an idea that if it be true then the Soul cannot be Immortal. Now let us suppose any one were to maintain a contrary opinion, and to affirm that Mental Evolution is a proof that the Soul is Immortal, how could such a conclusion be shown to be any less true than its opposite? It would certainly be nothing new for a supposed opponent of religious dogmas to become a pillar of the church; in fact every scientific opinion has passed through these two stages. Astronomy, Geology, the Origin of Species, and a host of others have at first been regarded as implacable enemies of religious dogmas, and then as their staunchest allies, and there is no reason why Mental Evolution should not go through a similar process. The fact is that Science is neither the enemy nor the ally of religious dogmas, and hence the reason why it is so differently regarded by different persons and at different times. Science rests upon experience, and nothing is good in Science that cannot be proved by experience, which is the touchstone for trying every fact and theory; now religious dogmas have nothing whatever to do with experience.

The Soul may be said to be the central idea of all such dogmas, yet any attempt to define, in terms of experience, or, in other words, to materialise the soul, would be generally considered as most irreverent. In the June number of this Journal a review was given of a work in which such an attempt was made, and, though the reviewer did what he could to put it in the most favourable light, it only adds another instance to the many that have preceded it of the hopeless nature of what, were it a fact, would be too evident to admit of doubt. This is no argument whatever against the reality of the Soul, as there may be many things in heaven and earth which are not dreamt of in the philosophy of experience, but it is an absolute bar against any attempt to test the teachings of the latter by an appeal to the former, or *vice versa*; for what relation can there be between that which wholly ignores the tests of experience and that which knows no other test but experience?

III. A CHANGE OF FRONT.*

By J. W. SLATER.

THOSE authors and orators who consider it their duty to criticise the spirit and the results of modern Science,—especially of Biology,—from a religious point of view, have of late somewhat altered their position. Physical research is no longer declared necessarily, or even generally, atheistic in its tendencies. The past duration of the earth is no longer limited to the old " 4004 years before the vulgar Christian era ;" creation is now regarded as a continuous process, not as an engineering operation completed at a certain date according to contract. The cosmogony of Kant and Laplace—the so-called nebular hypothesis —is accepted as perfectly compatible with theism, and even with the tenets of revealed religion. The independent origin of every animal and every plant is not now insisted upon, and even the high antiquity of the human race is treated as an open question. This is much ; and if we look backwards for a few moments, and read the sermons, the speeches, and the review-articles which a quarter of a century ago were discharged at modern research, we must feel duly thankful for the progress that has been effected. But we should beware of supposing that all the fundamental doctrines of the New Natural History are fully and finally accepted by " parsons, poets, novelists, lawyers," and generally speaking by outsiders of culture and intelligence. There are certain reservations, unpleasantly like the " saving clauses " in an Act of Parliament, which greatly qualify, if they do not altogether annul, the meaning of the concessions above mentioned. Nay, it becomes sometimes doubtful whether these admissions are anything more than a change of front, — a " flank attack " substituted for the more direct opposition which Evolutionism has had, till recently, to encounter. In the works which, as types of their class, I have selected for the basis of my present considerations,

* The Spirit of Nature, being a Series of Interpretative Essays on the History of Matter from the Atom to the Flower. By HENRY BELLYSE BAILDON, B.A. Cantab., Member of the Pharmaceutical Society. London: J. and A. Churchill.

God and Nature. By the Right Reverend the LORD BISHOP OF CARLISLE. " Nineteenth Century."

these features are revealed, though in very different forms and degrees. Man is still regarded as differing from the rest of the animal kingdom not merely in degree, but in kind. Nature is still declared perfect, and defended against the charge of "cruelty;" pain and suffering, if they cannot be fairly denied, are explained away; beauty is pronounced substantially universal, and is supposed to have been called into being for the special delectation of man ; and last, but not least, the purposes of God are assumed as fully known and understood, according to the old use and wont of teleologists.

Between the two treatises before us there is a well-marked difference. The Bishop of Carlisle approximates much more closely to the position of modern man of science than does Mr. H. B. Baildon. He may, indeed, justly claim our warm thanks for having drawn a distinction which, though it has escaped the notice of mankind for ages, must, when once pointed out, be accepted as strikingly natural. We hear, from time to time, complaints of the "atheistic tendencies" of modern Science, and of the dangers to be apprehended in consequence. The Bishop replies that "We want a new word to express the fact that all physical science, properly so called, is compelled by its very nature to take no account of the being of God : as soon as it does this it trenches upon theology, and ceases to be physical science. If I might coin a word I should say that science was *atheous*, and therefore could not be *atheistic ;* that is to say, its investigations and reasonings are by agreement conversant simply with observed facts and conclusions drawn from them, and in this sense it is *atheous*, or without God. And, because it is so, it does not in any way trench upon *theism* or *theology*, and cannot be *atheistic*, or in the condition of denying the being of God."*

The distinction thus drawn is perfectly satisfactory ; but the practical application of the principle turns upon the recognition of the boundary line between the spheres of theology and of physical science, between the *natura naturans* and the *natura naturata*. The necessity of respecting such a "scientific frontier" the Bishop sees and enforces. He holds that it is "not a mere arbitrary line to be laid down by treaty, but is like one of the great watersheds of Nature which no human arrangements can alter." This we fully grant, but we fear that it will be less easily traced out and recognised. Where is the legitimate boundary of physical

---

* It must be admitted that Mr. Baildon draws a similar distinction, though he does not employ the happy term " atheous " as contrasted with atheistic.

science ? According to the author a naturalist has no right, "upon the strength of investigations purely physical, to deny the existence of moral order, or of beneficence as an attribute of the Creator, if a Creator there be." Such denial, he contends, is an overstepping of the boundary.

I must here draw a distinction : the denial of beneficence in the Creator, on the strength of scientific researches, is doubtless illegitimate, and may be viewed as atheistic. But so long as the naturalist merely denies that he can find indications of such beneficence in the mutual relations of the organic world, he is, as it seems to me, on his own side of the frontier. And now for the converse transgression to that which the Bishop signalises. An author, knowing or believing from revelation, or from *à priori* considerations, that there exists a moral order in the universe, and that the Creator is beneficent, proceeds to use this belief as a datum in discussing the origin of species. If Ernst Haeckel crosses the legitimate boundary of physical science from the one side, so assuredly does Mr. H. B. Baildon from the other.

In marking out the frontier line which we are considering the Bishop of Carlisle insists, further, that physical science does not include the study of man, save "as a creature having certain material attributes, and leaving certain material marks of his existence in past ages." The science of necessity, he continues, "leaves out of consideration all that is most interesting to man or which makes man most interesting." Hence we are warranted in inferring that he would exclude from the domain of physical science all inquiries into man's moral nature, and would count attempts to explain the origin of our ethical codes as illegitimate.

But now comes the great difficulty : if a full consideration of one animal species—*Homo sapiens*—is not within the competence of Science, how is it with others ? If one part of Nature is relegated to the sphere of the metaphysician or the theologian, what of the rest ? The Bishop's reply is substantially that man is not an animal, not a part of Nature. "Putting aside all questions of immortality, it is not difficult to conclude that mankind possess attributes which do not belong to other creatures, and which make it necessary, in examining the world, to put man in a class by himself."

Here, then, all agreement ends. Whilst fully admitting the principle of a boundary line[*] I must hold man as an

---

[*] *See* Journal of Science, *passim.*

integral part of Nature, and maintain that as such he may be legitimately studied by the scientist. The Bishop's reasons for placing man " in a class by himself " have already been weighed elsewhere, and may safely be pronounced unsatisfactory. Will, purpose, thought, may all be claimed for the lower animals, though of course in a very much smaller degree than they are met with in man. But it is utterly unwarrantable to found absolute distinctions upon mere differences in grade.

From the author's comments on the materialist epigram, " Without phosphorus no thought," I do not dissent. It might be interesting to determine whether the proportion of phosphorus in the brains of different animal species, in that of one and the same species at different periods of maturity and decay, or, again, in individuals of the same species which during life have differed in mental power, bears any proportion to their respective degree of intelligence.

The Bishop says, with reference to the moral feelings, " I can discuss them, I can guide my conduct by means of them, I can feel ashamed of this or that failure in upright or high conduct." True, but domestic animals also can feel ashamed of wrong doing, irrespective of the prospect of punishment.* We read that a murderer " has no doubt as to the fact that the person who did the deed of darkness years ago is the same person as he who feels the pangs of remorse to-day.† Every material particle in his body may have changed since then; but there is a continuity in his spiritual being out of which he cannot be argued." In so far as this consideration is an evidence against what is commonly known as Materialism,‡ I appreciate its value ; but if it be urged in proof of the continuous personality of man as distinct from his fellow-animals, I must demur. Cases are known of brutes remembering both kindnesses and injuries for terms of years long enough to allow of the exchange of every material particle in their bodies; yet such an animal never doubts that it is the same being which was injured long ago, and in taking the opportunity for revenge it asserts its continuous personal identity just as distinctly as does the criminal supposed by the author. I see, therefore, here also no justification—far less necessity —for the old practice of putting man in a class by himsell. On the contrary, it will doubtless appear that by this very

* Journal of Science, 1875, p. 423.
† Savages do not seem capable of remorse.
‡ Hylozoism ; Apneumatism.

separation the study of human nature has been greatly obscured. In psychology, in sociology, in ethics, just as is the
case in morphology, embryology, and physiology, we shall
gain much by proceeding from the simple to the more complex, and in viewing man not as an isolated independent
being, but as a part of a coherent whole.

Let us turn now to a writer who speaks perhaps rather in
the interests of æsthetics and of natural theology than of
revelation, and whose attitude, as far as modern biology is
concerned, is far more hostile than that of the Bishop of
Carlisle. Mr. Henry Bellyse Baildon is probably a young
man, which may account for the general tone of dogmatism
and rashness which pervades his work, and which is rather
curious on the part of one who thinks it necessary to
describe himself as "naturally apprehensive, sensitive, and
timid." His declared object, a crusade against materialism,
is one with which I heartily sympathise. But I fear that
he includes under the term " Materialism " doctrines with
which it has no necessary connection. What, for instance,
must we think of the following passage ?—" My general aim
is not so much to discredit Darwinism proper, as held by
the original author of the doctrine, as to attack, and if possible demolish, that materialistic and atheistic system for
whose bricks Darwin himself has but supplied the stubble."
For such a task his preparation has been to " read a great
part "—not the whole—of the " Origin of Species." Again :
" Nor does Darwin in person head the crusade of atheism."
Further : " Thus in his anxiety to avow his deism (? theism)
he (Darwin) banishes the action of his Deity to a remote
period of the past, leaving Him as it were at the very verge
of His own universe, in such a position too that He must
recede continually before the advance of Science. That
Mr. Darwin by no means intended to leave the deistic
(theistic) idea in this perilous position may well be believed ;
it was his ill-timed zeal in giving his bow of belief (!) at
the end of a volume, which he could not but be aware was
of an atheistic flavour, that did all the mischief. His followers have seen the weakness of his position, and have
many of them gone on to atheism." From these passages,
and indeed from the whole tone of the book, it would seem
that Mr. Baildon views the New Natural History as favourable to atheism, and as hence deserving of hostility. He
gives in, truly, his formal adhesion to the great principle of
Evolution. He declares that " Evolution is, indeed, but the
generalisation or fulfilment of the dictum *Natura non habet
saltum,* wherein lies the alpha and omega of physical

science. Whoever admits this maxim to be universally true
is not only entitled but committed to a belief in Evolution,
whatever cause or causes he may have to assign for the
phenomenon." Yet at the same time, because man has not
succeeded up to the present date in resolving the chemical
elements into some one ultimate substance, he holds that
" The chemical analogy threatens continued hostility to
the theory of Evolution as long as that theory maintains a
unity of source for all life." But to what agencies does he
ascribe the development of species ? With the explanations
given by Mr. Darwin he is far from satisfied. As the main
points in favour of the great reformer of natural history,
he enumerates "the extensive modifications producible on
plants and animals under domestication with artificial
selection; some remarkable cases of reversion in various
breeds to certain characteristics of a common ancestor; the
close anatomical similarity in particular points observable
in creatures of extreme diversity; the existence of rudimentary organs; and last, but not least, the absence of any
definite opposition theory on an adequate scientific basis."
He omits in this summary the evidence drawn from the
distribution and the migration of species, from the geological record, and from embryology. He considers that the
" weakness in Darwin's Darwinism " (*sic !*) lies in the insuf-
ficiency of the survival of the fittest and of sexual selection
to account for the varied forms of life. In this opinion, as
our readers well know, he is not singular. Many competent
judges, whose writings have been criticised in the " Journal
of Science," take the same view. But he does not, like
Prof. Cope, raise the point that " Selection " and " Survival "
fail to explain the origin of the fittest. Nor do his moral
instincts revolt at the consecration which Darwin's theory
seems to confer upon the most painful phenomenon in
Nature—the struggle for existence. On the contrary, he
accepts the gospel of competition in its full extent:—
" Everything in Nature, man not excepted, is to be put on
its mettle; the great meaning and moral of Nature is
activity and progress, and one of her greatest functions is
to stimulate man, even by what seem hardships and cruel-
ties, to yet intenser and more divine activity."

As regards sexual selection, he considers that Mr. Darwin
" has forgotten to account for the origin of the instinct for
the beautiful thus assumed as existing in animals."* Else-

---

* He might be referred to the work of Mr. Grant Allen. *See* Journal of
Science, 1879, p. 395.

where he pronounces it "difficult to perceive and highly improbable that any animal except man has the faculty of discerning between what is beautiful and what is the reverse." Yet he admits that, according to the experiments of Sir John Lubbock, certain insects possess a sense of colour, and he even speaks of "dim instincts for beauty" as present in the lower animals. But letting this pass, what is his right to assume the absence of an æsthetic sense in the animal world? Is the mineralogist justified in asserting, without analysis, that a certain ore is free from arsenic or sulphur?

What, then, are the factors which the author invokes in place of natural selection and sexual selection? He does not, with the theistic evolutionists in general, hold that development takes place upon certain divinely ordained lines; he does not say that God has implanted in every organism a tendency to variation, which under certain circumstances becomes active, and which then gives scope for natural selection to come into play. He does not suspend judgment, and urge that we must here await the results of further investigation. He simply calls in the old Paleyan "design and contrivance!"

It will be at once perceived that this is no scientific explanation of the phenomena at all. We might as well say that the unsupported stone falls to the earth because God so wills, which, though theologically correct, is scientifically inadmissible. If God has "contrived" the existence of any animal, or any particular arrangement in such animal, has he done so arbitrarily, or upon some fixed principle? The former supposition surely is unworthy of Absolute Reason. If, then, the latter hold good, the task of Science is to discover what is the principle involved, and not to remain contented with such "brave words" as "design" and "contrivance."

But there is a further consideration : the great difficulty in the way of theism has been couched in the old question "*Si Deus est unde malum ?*" It is strange that the Darwinian theory, though accused of atheism, has supplied the most satisfactory reply to this perplexing query, by showing that God is—if I may use the expression—not responsible for the most striking and gratuitous forms of evil. In so doing it has, I submit, rendered a service to theism and to Christianity which will be recognised in the future. It may be useful to consider how the question had been previously dealt with. The shallow optimism of George Combe and his followers carefully confines itself to a certain class of evils, such, namely, as are founded in the very nature of matter and of physical energy. If a man steps over the edge of a precipice he must be dashed to pieces by the action of a law necessary for the very stability of the solar system. And, by multiplying examples of this kind, these writers draw away the attention of their readers from evils which cannot be thus explained or justified, and which mainly accrue from the action of certain portions of the organic world.

Another school of teleologists—still accepting man as the measure of all things, and upholding the "perfection of Nature"—seek to show that creatures openly and palpably injurious may possibly render us some secret and unimagined service, which latter is the purpose of their existence. To this contention the reply is easy: we condemn, and seek to supersede, any human contrivance if along with much good it effects even occasional harm. How, then, can we pronounce anything a Divine contrivance for our good which inflicts upon us great harm, and whose benefits are merely supposititious? It may further be asked, if seeming evil is only kindness in disguise, may not apparent good prove to be closely masked evil?

There is another theory of evil held by the religious world, which, though of late years it has been kept somewhat out of sight, harmonises better with facts and with the results of modern science than do the optimist dreams of the "natural theologians." According to this theory venomous serpents, locusts, mosquitoes, evil beasts of all kinds, are in their nature punitive—a consequence of the fall of man.* I have met with a naturalist who contends that the entire order of Diptera, or two-winged flies, is a Satanic creation, —a profane parody on the Hymenoptera, called into being, under Divine permission, for the chastisement of man,— and who hopes to find scientific evidence of that crisis which in Scriptural language is described as man's original apostacy.

There is a further explanation of some at least of the existing evils and imperfections which meet our gaze wherever we look with eyes not dimmed by preconceived notions. The earth is a raw material which man is to "subdue." He is to drain the swamp, to extirpate the tiger and the

* It would be ludicrous, were the subject less grave, to find men who accept the Bible as a Divine revelation, and who consequently must believe that the earth is cursed for man's sake (Genesis iii., 17), and that it therefore brings forth weeds—and if weeds, why not animal plagues?—to find such men still endeavouring to explain away the existence of evil in Nature.

wolf, and the cobra; he is to develope the flowers and the fruits to higher beauty and luxuriance. He is to supersede the struggle for existence, which may be characterised as the " old covenant " of the organic world; and in doing all this he is to show himself the joint worker, the co-evolver with God.

On the design and contrivance hypothesis parasitic worms, &c., " can only be regarded as instruments of torture devised by the Creator, and whose existence no writer of Bridgewater Treatises has yet even attempted to reconcile with His infinite wisdom and benevolence."*

Evolutionism gets rid of this difficulty, but not by showing that the existence of species and their properties is a result of blind chance. Fully admitting that the animal and the vegetable kingdoms have been developed according to God-given laws of which the most advanced of us all have as yet but very shadowy conceptions, I must call attention to the following considerations :—Every theist, and assuredly Mr. Baildon, holds that *e.g.* each individual man, though procreated by his parents, is created by God. But what if a man turns out a hunchback, a cripple, an idiot, a dipso-maniac, a kleptomaniac ? Shall we dare to say that this is God's contrivance ; that He has designed and willed deformity, idiocy, vice, and crime ? Just as little has He designed or contrived what we may call loathsome or criminal species, the aphis, the scale-insect, the screw-worm of Texas, the *Pulex penetrans*, the *Lucilia hominivora*, and legions more.

But Mr. Baildon, with a boldness that is perfectly amazing, comes forward to defend Nature against those whom he calls her " slanderers " and " libellers,"—in other words, those who cannot shut their eyes against facts. He evades the main question at issue by stating that he wishes to " confine the enquiry to sub-human Nature." Yet he elsewhere loses sight of this limitation when he says—" The impartial thoughtfulness of Nature puzzles us. We are offended to find the existence of a noxious weed, insect, or parasite cared for as tenderly as man's." But what should we think of a human government, very limited in power and wisdom, which should not merely fail to extirpate, but should knowingly and of full purpose cherish Thugs, Inquisitors, brigands, poisoners, promoters of bubble companies, and other the like human parasites and beasts of prey ? Yet Nature, which

* Habit and Intelligence. By J. J. MURPHY. *See* Journal of Science, May, 1879, p. 350.

acts exactly so, is not to be denounced either as a "bungler" or accused of cruelty. We who, in old-fashioned language, "wear the shoe" are not to be allowed to say where it pinches! Mr. Baildon contends that there is no unnecessary or useless pain, and enlarges on the ennobling influences of suffering. Did he ever hear of a man who was ennobled, morally, intellectually, or physically, by a brood of chigoes in his feet, by having his palate cut to pieces by the larvæ of *Lucilia hominivora,* or by having his rest broken night after night by mosquitoes? A word about these same mosquitoes, using the word in an unscientific sense to include all the blood-sucking Culicidæ and Tipulidæ: is there any necessity for their existence at all? If such there be, it is certainly not necessary that they should suck human blood, since millions upon millions of them live and die without ever settling upon man or any other mammalian. Would not a slight change in the odour of our juices, or, if that were too difficult to effect, a slight alteration in their own tastes, solve the difficulty, and deliver mankind from much torment without injuring the insects? Surely, then, we have here a case of useless, needless suffering. It has been further ascertained, contrary to what was once dreamed, that, so far from being of sanitary utility, they serve as transmitters of disease, and that their larvæ intensify the putrefaction of organic matter in the waters. We may similarly examine the properties and characteristics of the parasites which prey upon man, creatures doubtfully even conscious of their own existence. Yet it is to such beings that Nature sacrifices the weal of man, "her first mate and conqueror and monarch!" I do not, indeed, venture to pronounce all this suffering unjust; but may certainly conclude that if a maximum of earthly enjoyment and a minimisation of earthly suffering had been the objects of the Creator, the world would assuredly have been constituted very differently from what it is.

In another respect Mr. Baildon is at variance with the teachings of the New Natural History. He asserts, *con strepitu,* that the beauty of certain organisms exists for the edification of man, and, in a passage which I cannot pronounce other than deplorable, he speaks as if Mr. Darwin, in maintaining that "all this beauty" has arisen "without design or purpose,"—*i.e.*, without reference to man,—denied the very existence of the beautiful. Now facts do not harmonise with this purposive notion of beauty. We find multitudes of creatures exquisitely coloured — Actiniæ, Crustacea, and Mollusca—inhabiting the depths of the sea.

Have their hues and their designs been elaborated in order
that, after the lapse of myriads of years, some naturalist on
board a *Challenger* or a *Novara* might dredge them up and
be charmed with their appearance? We find the fossil remains of insects, *Buprestidæ, Malachii, Monanthia,* and
others, described in Heer's "Primæval World of Switzerland."* We have good reason to conclude, from their similarity to living members of these groups, that these creatures
were beautiful, and we may even argue from their existence
to that of a stately and beautiful vegetation. But did this
beauty arise and pass away in order that its existence might
be inferentially ascertained by a few zoologists in the nineteenth century? There are other fossils displaying exquisite
designs concerning which it has been well remarked that
not even Paley's watch-maker Deity would "contrive"
beauty in the Cretaceous epoch in order that a few traces of
it might be detected in the Post-tertiary. Again, as a correspondent not long ago pointed out,† if we take the hoof
of an ox, cut a thin section of it in certain directions, and
examine it under the microscope with the aid of polarised
light, a wonderful play of colour meets our eye. "Will any
teleologist maintain that the texture of such hoofs was
'contrived' with reference to the fact that, after expiry of
many thousand years, a few microscopists should detect all
this beauty?"

Further, many of the most lovely organic forms, especially animal, are decidedly rare. The adventurous naturalist
finds them in situations where human foot has never before
trodden. Nay, some thinkers even assert, though in my
opinion wrongly, that rarity is an essential feature in what
we regard as beauty. Something of this notion appears to
underlie the very word "homely," which our forefathers
applied to objects not perhaps positively ugly, but occupying
a neutral position. "Homely" surely means that which
surrounds us in our daily life, or, in other words, that which
is common. It is further to be noted that the most numerous species, whether animal or vegetable, those which meet
us everywhere, are precisely the least beautiful, and often
the most pernicious. The least attractive and the most injurious of our native butterflies are the cabbage whites.
Among the commonest beetles rank the cockchafer and
those "clickers" whose larvæ are dreaded as wireworms.
The most numerous of all insects are the house-fly, the

---

* *See* Journal of Science, 1877, p. 257.
† *See* Journal of Science, 1880, p. 343.

mosquitoes, the crane-fly, &c.—devoid of grace as of goodness. With plants the case is surely similar, even though few of them possess the intense ugliness so common in the animal kingdom. But if we clear a plot of ground, and abandon it again to the tender mercies of Nature, with what will it be covered in a few years' time? Shall we find there a primrose or a violet, an orchis or a foxglove, a wild rose or a honeysuckle, a royal Osmund or an oak-fern? No; the soil will be covered with nettles and docks and groundsel, with plantains and chickweed! It is as if the beautiful was hard beset to retain a foothold, even with our utmost aid, whilst the ugly and the hurtful increases and multiplies despite our labours.

In virtue of these considerations, to which many more of a kindred nature might be added, the beauty of the organic world where it occurs can no more be pronounced especially intended for man's delight than can the ugly sights, the doleful sounds, and the loathsome odours be considered as designed for his disgust. Man is placed in the world, and must take the rough with the smooth, avoiding or overcoming the former and making the best of the latter. To say that there is a soul of goodness in all things evil is merely to give a more poetical version of the old adage, "It is an ill wind which blows nobody good." How completely this scrap of "homely" wisdom decomposes our teleology! As for the problem of evil, we must await its solution hereafter.

Mr. Baildon's work, though in one sense evolutionist, furnishes proof how much evidence will have to be accumulated before the New Natural History can be definitely established, and to collect such evidence may be less difficult than to win for it a candid and patient reception.

IV. ON SOUND AS A NUISANCE.

FOR a long time it has been well known to the medical profession that in various critical states of the human system absolute silence, or the nearest possible approach to it, is not the least important condition to be secured. Accordingly muffled knockers, streets covered with straw or spent tan, and attendants moving about with noiseless step, are universally recognised as the signs and the requirements of severe disease. But the truth that noise is a contributor to the wear and tear of modern city life has scarcely yet been realised by the faculty, not to speak of the outside public. Consequently, whilst a zealous war is being urged against other anti-sanitary agencies, no general attempts for the abolition of superfluous noise have yet been made. We cannot, perhaps, give anything approaching to a scientific explanation why sound in excess should have an injurious effect upon our nervous system. Prof. Berthelot has recently shown, by a careful series of experiments, that sound-waves do not, like thermic and luminous vibrations, set up chemical changes in bodies submitted to their influence; but our inability to give an account of the fact does ist affect its existence. We feel that noise is distressing, exhaustive. The strongest man after days spent amidst noise and clatter, longs for relief, though he may not know from what. It may even be suggested that the comparative silence of the sea-side, the country, or the mountains, is the main charm of our summer and autumn holidays, and contributes much more than does ozone to restore a healthy tone to the brains of our wearied men of business. Indeed, if we consider, we shall find that this is the most unnatural feature of modern life. In our cities and commercial towns the ear is never at rest, and is continually conveying to the brain impressions rarely pleasant, still more rarely useful or instructive, but always perturbing, always savouring of unrest. In addition to the indistinct but never-ceasing sea of sound made up of the rolling of vehicles, the hum of voices, and the clatter of feet, there are the more positively annoying and distracting elements, such as German bands, organ-grinders, church-bells, railway-whistles, and the like. In simpler and more primitive times, and to some extent even yet in the country, the normal condition of things is silence, and the auditory nerves are only occasionally excited. It is scarcely to be expected that such a change can be undergone without unpleasant consequences.

The question has been raised, why should some noises interfere with brain-work by day and disturb our rest at night so much more than others? A strange explanation has been proposed. We are told that sounds made incidentally and unintentionally—such as the rolling of wheels, the clatter of machinery (except very close at hand), the sound of footsteps, and, in short, all noises not made for the sake

of noise—distress us little. We may become as completely
habituated to them as to the sound of the wind, the rustling
of trees, or the murmur of a river. On the other hand, all
sounds into which human or animal will enters as a necessary element are in the highest degree distressing. Thus it
is, to any ordinary man, impossible to become habituated to
the screaming of a child, the barking and yelping of dogs,
the strains of a piano, a harmonium, or a fiddle on the other
side of a thin party-wall, or the clangour of bells. These
noises, the more frequently we hear them, seem to grow
more irritating and thought-dispelling.

But whilst admitting a very wide distinction between
these two classes of sounds, we must pause before ascribing
these differences to the intervention or non-intervention of
will. We shall find certain very obvious distinctions be
tween the two kinds of sound. The promiscuous din of
movement, voice, and traffic, even in the busiest city, has e
it nothing sharp or accentuated; it forms a continuous
whole, in which each individual variation is averaged and
toned down. The distressing sounds, on the other hand, are
often shrill, abrupt, distinctly accentuated and discrete
rather than continuous. Take, for instance, the ringing of
bells; it is monotonous in the extreme, but it recurs at
regular intervals. Hence its action upon the brain is intensified, just as in the march of troops over a suspension-
bridge each step increases the vibration. The pain to the
listener is the greater because he knows that the shock will
come, and awaits it. Very similar is the case with another
gratuitous noise, the barking of dogs. Each bark, be it
acute or grave, is in the highest degree abrupt, sharply
marked, or *staccato*, as we believe a musician would term it.
Though the intervals are less regularly marked than in the
case of church-bells, we have still a prolonged series of
distinct shocks communicated to the brain. Well might
Goethe say—

> —— " vor allem
> Ist das Hundegebell mir verhasst;
> Klaffend zerreisst es das Ohr."

All the other more distressing kinds of noise possess the
characters of shrillness, loudness, and of recurrent beats or
blasts.

As an instance of an undesigned, unintentional noise
being distressing to those within ear-shot, we may mention
the dripping of water. A single drop, whether penetrating
through a defective roof, falling from the arch of a cavern,

or issuing from a leaky pipe, and repeated at regular intervals, is as annoying as the tolling of a bell, the barking of a dog, or the short sharp screams of a fretful infant. The only difference is that the noise is not heard as far. We may hence dismiss the " will " theory, and refer the effects of noises of this class to regularity, accentuation, and sharpness.

It is particularly unfortunate that the multiplication of sound should accompany, almost hand in hand, that increase of nervous irritability and that tendency to cerebral disease which rank among the saddest features of modern life. A people worn out with over-work, worry, and competitive examinations might at least be spared all unnecessary noise. Many persons cannot or will not understand how necessary silence is to the thinker. A friend of the writer's, engaged in investigating certain very abstruse questions in physics, is often compelled to throw aside his work when an organ-grinder enters the street, and suffers from acute pain in the head if he attempts to go on with his researches.

We should therefore propose, as measures of sanitary reform, the absolute prohibition of street-music, which is more rampant in London than in any other capital in Europe. The present law, which throws upon the sufferer the burden of moving in the matter, is a mere mockery. Another necessary point is the abolition of church-bells. In these days of innumerable clocks and watches everyone can tell when it is the time for Divine Service without an entire neighbourhood being disturbed for some twenty minutes at a time. Nonconformist places of worship collect their congregations without this nuisance. Further, all dogs convicted of persistent barking should be disestablished. And lastly, harmoniums, American organs, and wind-instruments in general should be prohibited, except in detached houses.

V. ON THE EVILS OF SECTARIAN MORALITY,

CONTRASTED WITH THE

BENEFITS OF THE SCIENTIFIC OR NATURAL SYSTEM.

By S. TOLVER PRESTON.

IT has been an idea with some that because the religions of sects have been naturally evolved accordingto the general theory of Evolution, that therefore they must have been fitting to some extent. It would seem to be overlooked here that error—connected doubtless with an imperfection or infirmity of the brain—is a concomitant to evolution as well as truth; that we cannot expect truth to prevail at once (though there are some recent signs of a rapid advance); and that the real fact may be that society has survived, not through sectarian ethics, but *in spite of it.* 1 do not allude here to the simple belief in a Supreme Being, against which there would be not one word to be said, but to the endless doctrines and sectarian *propaganda* (unfortunately tacked on to this simple belief), and which are unworthily classed as " religion." It is notorious that the most eminent minds have distrusted to the greatest degree the common idea of the salutary influence of popular religions, more especially the grand old Greek philosophers. It would be arrogant to suppose that we are so near perfection that there is not much to reform yet in doctrines and views; for progress consists in the continued elimination of error. The first step towards the establishment of truth must evidently be the abolition of sects; for truth can, of its very essence, never be sectarian. The foundation of morality upon an unsectarian or universal basis therefore becomes a pressing need. It cannot be doubted that the introduction of the new Morality now being unfolded by Natural Science may be accelerated by showing the conclusion to be well founded that the present sectarian morality, so far from conducing to stability in society, in reality tends to upset it (or constitutes an encouragement to wrong-doing), and that these doctrines have been mainly sheltered from attack by the unfortunate idea of their salutariness.

It is not doubted for one moment that there are a large number of well-meaning people who teach these doctrines with the best intentions. But unfortunately "well-meaning" people may be unconsciously great harm-producers, since there is nothing more detrimental than misdirected energy. Indifference is preferable. We want more "*well-acting*" people. When we observe, in the recent past, a large number of people devising an elaborate system for strangling trade, and, without even a feeble consciousness of the implied irony, *naïvely* calling the system "protection," we may be inevitably led to suspect that the "protection" of morals by sects may be as much a system for strangling them as in the case of trade ; and that the emancipation of morality from sectarian control may be a necessity for the national welfare of the same (but of a far more pressing) kind as the freedom of trade. It is a fact, notorious to any searchers after truth, that often things which when superficially looked at have one aspect, when deeply looked at have just *the reverse* aspect. This I maintain to be the case with the supposed "protection" of morals by sectarian doctrine, or, in other words, a little more than a superficial analysis will show that the present system actually sets a premium on wrong-doing. It would be scarcely reasonable to suppose that the present sectarian leaders (even if competent as a body to do so) should see it to their interests to submit this question to such a searching examination as to jeopardise their own *raison d'être*. This task must be left to the un-sectarian or independent students of Nature who are more concerned with the foundation of universal truths than of sects, but who (regretably) seem not quite to have escaped the contagion of the popular idea that unfounded doctrine can be salutary, or conduce to stability—as if ignorance were not far preferable to error.

While, however, the followers of Science are somewhat reluctant to come forward, under the influence of this idea, the "well-meaning" people are having it all their own way, and, with that overweening confidence invariably characteristic of incapacity, are issuing by tract and volume the literature called "moral" by the cartload,—for home and export,—to an extent not even feebly imitated by any other country in Europe.* It may be a fair question for the independent pioneers of knowledge, how long they intend to incur the responsibility of leaving practically the entire

* Because of the notorious fact that our insular position tends to restrict the immense benefits of the free interchange and conflict of ideas enjoyed by Continental nations, which is the surest safeguard for truth, we ought surely to be all the more careful and suspicious lest any errors may crystallise in our midst. Of course the fact that an error is thought to be good is not a remarkable thing, because that may be the chief reason *why* it is an error (or why it is an evil).

control of morality in the hands of interested sects. Signs are not wanting now that it is high time for Science to speak out boldly, and not be afraid of holding the mirror up to fallacies, whatever their nature. It may be a reasonable inquiry whether the dread of Science is not a great arrogance—just as if people constituted themselves self-appointed judges of what *ought* to be truth.

Strange as it may seem, it is none the less unquestionably true that a popular idea widely prevails that, by giving the name " religion " to certain doctrines, they become thereby privileged, or exempt from reasonable or independent criticism,—that these doctrines are not fit for Science to deal with. It must be admitted (as perfectly self-evident) that, if this contention were valid, there would be nothing to prevent giving the name " religion " to what might prove unconsciously to be the most destructive and pernicious errors which ever afflicted mankind, and thereby enthroning them as for ever exempt from the penetrating gaze of reason and of truth. We may perhaps recognise here without difficulty one cause for the indiscriminate use (or rather abuse) of the term " religion," as now applied to all kinds of doctrines. They become thereby " protected " as sacred,—just as if truth needed sanctification, or were incapable of standing on its own merits. It is certain that if a system wants a foundation, neither sanctimonious language, music, nor ceremonial will supply it, though unfortunately these may act with greater force upon inexperienced minds than proof itself. Even the noble science of music may admit of being perverted to a wrong purpose. It must not be a matter for surprise, therefore, that when (after the long period of protection) the system is thrown under a rigid analysis, some startling incongruities may disclose themselves.

My object will be to demonstrate the two following propositions, which are selected as primary ones of such a definite character as to avoid any possibility of entanglement :—The first proposition is, that the *effect* (discarding motives so as to ensure certainty) of the morality propagated by sects is invariably to secure the *domination** of the leader of the sect (or his emissaries). The second proposition is, that the conditions for determining this end must unavoidably (from a necessary property inherent in the conditions themselves, and apart from motives) be of such a character as to set a premium on wrong-doing. In order to make this quite clear, it will be desirable to recapitulate briefly one or two points contained in a foot-note (page 450)

---

* This first proposition has already been ably considered in some points by others.

in the former paper. on " Natural Science and Morality " (" Journal of Science," July, 1880).

It will be observed, on analysing the subject, that in principle the plan invariably adopted in the formation of sects is to attempt to enchain the interests of mankind as powerfully as possible by asserting the existence of some great danger or overhanging calamity, to magnify which adequately no range or license is set.   Thus, for instance, the asserted danger may be *eternal* punishment (in which case it may be observed that the range is made *infinite*, or the danger is magnified without limit).   To act upon the conceptions various similes are then used, of as terrifying a character as may be, by selecting well-known agencies which produce the most intense physical pain.   [It is apparently forgotten that an infinite—or everlasting—punishment necessarily implies an *infinite* insult to a just Deity; the magnitude of the insult being, mathematically estimated, in direct proportion to the magnitude of the violation of justice incurred.]   After taking this primary step, the next procedure is to suggest a remedy or means of escape from the asserted danger, coupled with a certain amount of spurious mystery, as this (as a well-known fact) tells most with the ignorant class.   The unfortunate man whose terrors have been worked upon by the varied images used to pourtray the nature of the punishment then looks upon the leader of the sect (or his emissary), who instructs him as to the means of escape, in the light of a benefactor and preserver.   Certain conditions are then stated to be contingent on being saved from the asserted future punishment, the most important one being implicit and unquestioning belief in all the statements of the leader of the sect (the validity of which are to be " justified by faith "—which latter quality, so far from being reprobated, is extolled as the highest virtue); together with absolute subjection to whatever course of conduct the leader of the sect (or his emissary) may please to command.   Here we have the achievement of *domination.**

* It would surely be an insult to the general public to make them as a body responsible for the *propaganda* of the originators of sects, or to indulge the dream for one moment that such monstrosities of doctrine could ever have had their origin in unbiassed persons, whose sole aim was the attainment of truth, in harmony with right and justice, and apart from personal ideas of ambition and self-advancement.

Precisely because such *propaganda* have been protected by the absurdity of calling them " religion," and by the (if possible) greater absurdity of supposing them beneficial to society, so may they be expected, when this spurious shield is removed, to crumble to pieces at the touch of a scientific analysis.   The full amount of unhappiness caused in the world by the originators of such doctrines will only be adequately guaged by the science of the future.   For just as the

In regard to our second proposition, viz., the premium *necessarily* set on wrong-doing by the above system,—independently of any motive to do so, but inherent in the system itself,—we may observe. first, that the fundamental principle involved here is that the escape from a certain imminent danger (such as a punishment of unlimited magnitude) is made contingent on certain *conduct.* The man appealed to therefore infallibly reasons that if the asserted danger be a myth, then such prohibited conduct must be in itself extremely desirable and profitable ; for he argues that it would be transparently absurd to hold out an enormous (infinite) punishment as a deterrent, unless the course of conduct itself were exceptionally attractive or conducive to profit. If, therefore, his belief in the asserted punishment in a future world is very shadowy (as is actually the case with the majority of mankind), he may be naturally induced to follow crime and vice in the belief that (in escaping the enormous future punishment) they must be extremely conducive to present interests. So far therefore from being taught that crime is unprofitable *in itself* (or brings its own punishment with it), he is in effect led to infer the exact opposite. Thus the highest premium (in fact by implication infinite) is set on wrong-doing by this doctrine, and, conversely, a deterrent set on virtue, which is made to appear opposed to present interests. For it may be justly argued (in an analogous way) that if an infinite reward in a future world be necessary for right conduct, then such conduct must be connected with great hardship and privation in this

---

advance of knowledge is sure to exhibit in a clearer light the happiness caused by the friends of science, so it will show more plainly the misery caused by its opponents ; and it is inevitable that in the purer intellectual atmosphere of the future the condemnation of these sectarian *propaganda* will be far more emphatic than has ever been penned yet. For the growing appreciation of the benefits of truth must unavoidably run parallel to that of the evils of error. It is a significant fact that those who opposed these *propaganda* most bravely in the past are becoming more and more recognised as the true champions of the happiness of mankind, as time advances. The severest judgment the public could pronounce on themselves would lie in their failure to appreciate the services of men who took so signal a part in the battle against fanaticism and superstition, one of the chief sources of human misery. To speak of the " conflict between religion and science " is to speak of ignominious defeat, or to degrade " religion " to doctrines unworthy of the name. The defunct Hebrew cosmogony once classed as " religion " may picture the fate of the Hebrew ethics of threats and promises. It may be superfluous to lay stress on the perfectly well-known fact that promises and threats have (all the world over) been the most commonplace devices adopted to influence mankind ; the only peculiarity here being that both the threats and the promises are magnified absolutely without limit, as if to strain the credulity of mankind beyond bounds.

life.   In fact the sectarian morality, when looked at a little below the surface, shows itself to be the *exact antithesis* of the scientific morality, founded on the study of Nature, which teaches that the path of virtue (or right conduct) actually coincides with that of self-interest, or present welfare, and that the path of vice is directly opposed thereto,—thus affording the most natural and healthy motive to right conduct conceivable, and an equally forcible deterrent against wrong conduct.   If, therefore, there be grounds for concluding that the scientific morality is good or beneficial, it must follow with equal necessity that the sectarian morality (as its antithesis) must be detrimental to a corresponding degree.   In fact the one *leads* mankind by enlisting his present interest on the side of virtue, the other attempts to coerce him by a low and degrading system of terrorism, which (like all such attempts) must inevitably recoil upon itself, and have the exact opposite effect to that intended.

This may perhaps form a not un-noteworthy instance of the fallacy of the idea of the salutariness of error,—so often warned against by a few eminent minds before, who have incurred *odium* on that account, but who have nevertheless boldly asserted that error never can do good, and never can conduce to stability in society, but brings the positive harm inseparable from error.   If a man is left in crass ignorance he has some chance, for he can inquire for himself; but when his brain is poisoned* (especially when young and extremely sensitive to first impressions) by the fearful doctrines of sects, he may be hopelessly ruined for life,—as has unquestionably occurred in many cases† under the wholesale

---

* The poisoning is here literal ; for it is a well-known fact that every sentence uttered physically affects the brain, and when reiterated may leave its permanent impress there.

† *Post-scriptum.*—I look back with an indignation difficult to repress at what was taught me when young [*i.e.*, at the system], and which almost proved my ruin, leaving its detrimental traces to the present day.   I am led to mention this in the hope of benefitting others—as one of a class who perhaps rarely pass through their experiences with strength of mind left to analyse the system which produced their undeserved grievances.   If error be not an evil, or if fallacies cannot cause injury, then what can ?—and errors of absolutely unlimited magnitude.   Some people, however, appear to have so restricted powers of realisation that they can actually profess to believe such fallacies without the faintest conception of their import.   Their inconsistencies are absolutely terrible (even if unconscious).   They can protest the Deity is a God of Love one moment, and ascribe the boundless Hate involved in an *eternal* punishment the next.   It apparently could not be of much consequence what was taught in their case.   Life can only be extinguished with the greatest effect when it burns most brilliantly.   The most sensitive children—those whose inquiring minds are readiest to appreciate the truth—are those who are most effectually strangled by fallacies of enormous magnitude.   Let it not be supposed that there is exaggeration here.   An intellect of high sensibility, while it will be

system of instilling the terrifying propaganda of half-
civilised Palestine into the fresh intellects of children of the
nineteenth century (aided by solemn music and ceremonial*),

---

beneficially affected to an exceptional degree by the inculcation of truth, cannot fail (unfortunately) to be injuriously affected to a corresponding degree by
the infusion of error. The greatest cruelty is to teach a child an untruth ; the
disastrous effect is not lessened by its being unintentional. And when the fallacies are numerous, of unlimited scope, and imprinted on the brain with the
greatest solemnity and ceremonial, the effect is necessarily proportional thereto.
The child rendered melancholy (in proportion to his intellectual powers of
realisation) by the perpetual oppression of monstrous and supernatural beliefs,
which disintegrate the mind, and cut off from the opinions of the intellectual
few, naturally judges of the world by his surroundings. Finding no sympathy,
and believing himself isolated in his opposition to these doctrines, the suspicion
of madness may at last present itself, the more he reflects and tries to reconcile the impossible. And this may be the end finally attained (as is known to
have occurred in many cases) under the utterly reckless system that prevails.
The miserable excuse of benefiting (!) society by the system has been urged in
palliation, just as if it required an array of argument to prove that error must
inevitably do harm to the whole bulk of society which it affects.

* It was remarked, on the occasion of a recent change of political rulers,
that a certain proclivity existed towards the adoration of the alphabet, such
that " a square inch of title was sometimes balanced against a square acre of
brains " (in influencing choice). If this be true, can we wonder that a vest-
ment, combined with a sanctimonious tone, should often have more effect than
plain and unadorned logic, and should even be capable of going far to supply a
deficient foundation of fact. Such is the limitless absurdity of the habits of
thought capable of being drilled into the brain in childhood (and strengthened
by the principle of inheritance through generations), against which reason may
protest in vain.

A passing reflection may even possibly not be wasted on the fact that the
irresponsible representatives of the alphabet domineer (in a special House)
over the elected representatives of the nation—a state of things which it will
be admitted (as palpably self-evident) no one in his sober senses would ever
have dreamt of *originating* in the present day. It is simply a case of habit or
custom transmitted down to us, and kept up by modes of thought inculcated
in early infancy, the custom itself having been *originated* by persons of whom
it may at least be said that they stood somewhat nearer to their probable
Simian progenitors than we do. So difficult is it to uproot a custom or a traditional belief when once it is set ; and this fact is one of general application.
The fallacy resembles in some respects a bodily infirmity, transmissible (or
affected in some degree) by inheritance, being no doubt connected with a defect
—or possibly torpid state—of the brain, which, like any other part of the organisation, is known to be subject to the laws of descent. It is true that a
particular brain process termed " reasoning " (when persistently and actively
entered upon) can obliterate the defect representing (say) an unfounded sectarian belief of long standing, and the physical effect of a paragraph of plain
logic on a brain may be like that of a scalpel on any other part of the organisation—in regard to the temporary pain caused and the permanent benefit
secured. Thus, by a single beneficial surgical operation of this kind, an error
or unfounded belief transmitted for centuries may sometimes be eradicated.
While the accumulation of error is in general slow, its destruction may be fortunately rapid. It may occupy but a short time to remove a tumour which has
taken perhaps centuries to develop. A single great and original work [of course
fitly received by a scream of opposition at first, since antagonism to set habits
of thought is the principal value of originality] has been known to change the
character of a nation within a relatively short period ; as nations, it is said,
are moulded by their greatest minds (of which their character bears the

and the vast quantity of dangerous literature relating thereto, which is sometimes almost forced into households by a special machinery, against their wills, kept alive in great part by the posthumous endowments of "well-meaning" and wealthy old ladies,* who are fed on the same literature, and whose minds have been so drugged by the process that their complete incapacity to appreciate the pure truths of

---

indelible impress). Not the least benefit conferred by the enormous power of knowledge acquired through the unfolding of the principle of Evolution will be, no doubt, the capacity it gives us to trace and nail error. The problem of the Evolution of Error will probably form a most instructive one for the future. in connection with the general theory of Evolution. Perhaps the greatest service that could be rendered to the rising generation would be the purely passive one of abstaining from teaching them error, leaving them a little more to their own ideas and natural resources. It might be a curious imaginative *à priori* problem to consider how far a generation would advance if all the books except those relating to demonstrated truths were burnt. It would not probably take them long to create a sufficient amount of fiction and speculative theories to amuse themselves with, which it would certainly not occur to them to teach to the succeeding generation as *truths* (as unscrupulous barbarians in the past might well be conceived to have done). The probability of would-be originators of sects obtaining a hearing would be (to say the least) extremely remote under the high intellectual *status* existing. Universal truth alone would be permitted to reign. This purely imaginary case would simply represent a step taken towards *cutting the link of error* which binds mankind to the barbarous past, and which (when unsevered) transmits by inheritance the interminable series of fictions inevitably attendant on evolution from a lower state to a higher—fictions which we should no more have dreamed of inventing ourselves than we should think now of turning to primæval clothing. It seems astonishing how little we distrust legacies bequeathed to us from a remote antiquity. We appear to forget that semi-barbarous races might not have been so particular about lying as we should be. One valuable feature of the doctrine of Evolution is the wholesome distrust it inspires for anything remotely ancient, and the corresponding lively hope it affords for the future. The doctrine breathes the very spirit of reform and liberalism. It is a common propensity of superficial people to have an intense reverence for the "damnably good old times " (as the late Charles Dickens called them).

* The system is worked (as is known) in this country by societies, receiving large funds out of the superfluous wealth characteristic of the nation—an evidently unhealthy system with the best intentions. For while scientific truth is mainly diffused by brain, these "moral " truths (!) [Heaven save the mark] seem to be mainly diffused by money. We might enlarge upon the evils of a sectarian system of Hebrew ethics which gives such an extraordinary prominence to alms-giving, setting it up indeed as the highest ideal of virtue (the practical carrying out of the text which makes the test of *perfection* the selling of all one's goods and giving the proceeds to the poor). The origin of our abnormal Poor Law, which has no parallel in Europe for its evils, may be naturally traced here. To inquire into the causes of the thriftlessness and improvidence which are so notorous in this country would be a burlesque in view of these facts. The huge list of alms-giving societies is culminated (as if in irony) by one whose special function appears to be to "organise " the disorganisation. The many excellences of the country are too notorious for it to fear its faults being pointed out ; indeed these appear all the more glaring by contrast. The national character seems to be made up of some curious contrasts and anomalies (naturally due, no doubt, to evolution in an *insular* position), and which may form a most fertile and interesting field for study.

Nature is only paralleled by their insensibility to errors of inconceivable magnitude (involving the most cruel violations of justice). If it were not for the fortunate fact that such literature (all founded on the ethics of terrorism of the barbarous Hebrew) fails to take root in the minds of the majority, the evil would be colossal.* As it is, the intelligent minority (or the young of exceptionally inquiring minds) are those most liable to be caught here. The known violent antagonism of the writers of such doctrines to demonstrated scientific truth, itself proves the magnitude of the error. The development of this species of literature has proceeded in a groove (a known characteristic of evolution) until it has attained gigantic proportions, forming a special feature unapproached by any other civilised country. Its complete collapse—together with the barbarous system of Hebrew ethics which it represents—will illustrate the fortunate property of error to defeat itself by its own growth, until it becomes a forgotten feature in the ever-advancing tide of progress. Unfortunately this tendency to forget the evils of the past, and the fearful state of religious fanaticism which once existed, by which progress and healthy life were strangled, makes mankind oblivious of the benefits conferred by those who fought against error in spite of popular prejudice, and through which conflict alone the present blessings of national prosperity and freedom have been won. Will it be supposed, however, for one moment that perfection is so nearly reached that more victories of the same kind have not yet to be achieved ? The new morality can only advance in triumph over the ruins of the old.

* If it be supposed for a moment that people whose antipathy to demonstrated scientific truth is notorious, can be active to the extent they are in the propagation of their ideas without doing harm, then one may be inclined to ask, What on earth could constitute an evil ? The ideas of some are so antithetical to natural truths that blame from them constitutes eulogy. What, for instance, could have been a more consistent eulogy of the great work of the French philosopher, Helvetius, than its denunciation by the Jesuits as an " *œuvre satanique.*" It is not denied that there may be some earnest people even among the Jesuits. But incapacity may be as fatal to the welfare of mankind as open hostility to right and justice. Having given in this essay every credit to good intentions, there is less reason for blinking the plain facts, which must be realised in the interests of truth. Nothing is more certain than that this world is not yet perfect; that errors tend to gravitate into fixed habits, the temporary pain caused in the eradication of which must be proportional to the permanent benefit secured ; and that future progress will certainly increase rather than diminish the estimate of past error.

ANALYSES OF BOOKS.

Gesammelte Kleinere Schriften Naturwissenschaftlichen Inhalts.*
Von CHARLES MARTINS.   Autorisirte Uebersetzung von
STEPHAN BORN.   1 Band.   Basel: Schweighauserische
Verlag's buchhandlung (Hugo Richter).

THE name of Charles Martins is so widely and so honourably
known that we shall not need to enter into any explanations on
introducing the German edition of his collected works to the
notice of our readers.

Of the six essays included in the present volume, three may
be considered of more temporary interest, and as dealing with
subjects sufficiently well known to the thoughtful and cultivated
portion of the English public.

The article on the " British Association for the Promotion of
Science " was written in order to promote the formation of a
similar body in France, and in this respect the labours of the
author and of his friends have been crowned with success.   A
" French Association," essentially on the same principle, has
been constituted, and is apparently running a course of increasing
utility.   We must not, however, forget that the German " Association of Naturalists and Physicians," called into existence by
the illustrious Oken, was the earliest society of this nature.

An account of the origin and constitution of the British Association, and of the proceedings at one of its annual gatherings,
needs of course little notice in a country where both the light
and the shadow side of this itinerant Academy are thoroughly
well known.   But we feel bound to acknowledge the accuracy
and fairness of Prof. Martins's descriptions.   It is sometimes
said that no Frenchman ever succeeds in fully understanding anything English.   To this rule, if rule it be, the author is an
exception.   We find here no mistakes which may rank higher
than clerical errors.

Another memoir is devoted to the exploring voyage of the
Challenger—a subject in itself of unquestionable interest, but
too well known to require special analysis, though it is introduced
by a brief historical survey of earlier scientific expeditions,
among which due attention is paid to the ever-memorable voyage
of the *Beagle*.   The author laments that the medical officers of the
French navy are rarely qualified by their previous training to
become successful observers in any department of science.

---

* The Collected Minor Works of Charles Martins on Subjects connected
with the Natural Sciences.

A third essay discusses the possibility of reaching the North Pole, and the causes of the failure of former attempts.

The three treatises which make up the first portion of the volume are of a wider and a more permanent interest. They are devoted to the " Theory of Evolution—the Value and Correspondence of the Evidence upon which it is founded ;" " Lamarck, his Life and Works ;" and " Floras, their Origin, Composition, and Migrations."

In ieading these essays we feel ourselves in the society of a philosophical naturalist of high order. He declares that " the theory of Evolution made known by Lamarck in 1809, philosophically comprehended by Goethe, definitely constituted by Charles Darwin and developed by his pupils, links together all the parts of natural history just as the researches of Newton harmonised the movements of the heavenly bodies. It is, indeed, closely conne&ed with the transformation of the physical forces." Here we miss the recognition of Erasmus Darwin, who certainly was not an imitator of Lamarck, and who yet came very near anticipating his illustrious grandson. The obje& of Professor Martins, in the treatise before us, is to show that the Evolutionist theory possesses all the chara&eristics of the Newtonian laws, and that like them it rests upon a daily accumulating mass of evidence. To this end he first expounds the continuity of creation and the phenomena of atavism. If each zoological epoch terminated, according to the catastrophists, in some convulsion of nature which made a clean sweep of all animal and vegetable life, how is it that plants which existed in the Tertiary period,— such as the pomegranate, the Judas tree, the oleander, the gincko, &c.,—though extirpated in their old haunts by reason of climatic alterations, still survive in warmer regions ? Hence we see that the flora of to-day is a continuation of the flora of the past, the forms being either identical or somewhat modified. The gincko has the leaves of a fern, the stem of a Conifer, the catkin male blossoms of the Amentaceæ (poplar, birch, &c.), and the naked seeds of a Cycad. These fa&s indicate that the ferns are the common ancestors of the gincko and of the Cycads.

In like manner the living fauna is the uninterrupted continuation of the fossil fauna. The animals which move and multiply around us are descendants of the beings whose skeletons or cases have rested for unnumbered ages in the rocks. As instances the author mentions the modern elephant, rhinoceros, tapir, &c., so little modified from their ancestors, and which therefore seem almost out of place among forms which have changed so rapidly as the deer, the antelope, and especially the horse. He points out that numbers of extin& species hold, as it were, intermediate positions, and form " missing links " between living groups whose chara&ers they combine.

The phenomena of atavism are no less striking among animals than among plants. Numerous forms display the atrophied and

now useless rudiments of parts or organs which in their forefathers exercised some important function.   Thus the rudimental
breasts of the human male are a far-off reminiscence of the
hermaphroditism which we meet in many lower animals ; the
pyramidal muscles, a remnant of those which close the pouch of
the marsupials.   The cœcum is another obsolescent organ which
in man, so far from being of any use, affords merely the possibility of a painful death.   There are more than twenty authentic
cases on record where the entrance of a grain of sand, a grape-
stone, &c., into this appendage has occasioned fatal peritonitis.

Prof. Martins does not omit to rebuke the unqualified writers
who impute to Darwin the assertion that man is descended from
the gorilla or the chimpanzee.

A second section is devoted to the transitions between organic
beings, and to the non-existence of species as natural objective
realities.   As a prolific hybrid he mentions *Ægilops triticoides*,
a cross between wheat and *Ægilops ovata*, often found in the
South of France.

In the third section the author gives a survey of the embryological proofs.   He concludes with a passage which we feel bound
to transcribe :—" The principle of Evolution is not confined to
organic beings ; it is a universal principle, governing all things
which have a beginning, a progressive duration, an inevitable
decline, and a prospective end.   The application of this principle
is ordained to facilitate the progress of all science, and to throw
a new light upon the history of humanity : the solar system, the
earth, organic beings, the human race, civilisation, nations, languages, religions, social and political order,—everything obeys
the law of Evolution.   Nothing is created,[*] everything is transformed.   Solomon probably knew this when he exclaimed ' Nothing
new under the sun !'   The evolution of living nature is the type
·and the pattern for everything which advances in the physical
as well as in the spiritual and moral world."   This is admirably
expressed.   But does not the author, when he formally proclaims
immobility and a definite retrogression impossibilities, forget
that the organic world affords instances of both, as has been well
shown by Dr. Dohrn and Prof. Ray Lankester ?[†]   Degeneration,
degradation may be traced in animal and vegetable life, in the
history of language, and in social science.

The essay on the life and works of Lamarck scarcely brings
out with sufficient distinctness his connection with Buffon.   It
contains one most painful passage.   On his death in 1829, at
the age of 85, his two daughters were left without resources :—
" I myself saw, in 1832, Miss Cornélie de Lamarck, for a paltry
remuneration, engaged in fixing upon white paper the plants in
the herbarium of the establishment where their father had been

---

[*] That is, no new element is flung, as it were, catastrophically among what
already exists.

[†] *See* Journal of Science, July, 1880, p. 465.

Professor. If they had been the daughters of a minister or a general the two sisters would probably have received a pension ; but their father was merely a great naturalist, and an honour to his country for all time."

Prof. Martins gives an exposition of the " Philosophie Zoologique " of Lamarck, showing how he approximates to the evolutionist writers of our day, and where he falls short of them for want of a knowledge of facts which have only been discovered since his death. The well-known argument of Cuvier for the permanence of species, founded on the identity of mummified animals with their still living descendants, is shown to prove too much. If of any weight at all, it would equally demonstrate the permanence and independent origin of races, breeds, and varieties. The ram depicted on Egyptian monuments is identical with the ram of Modern Nubia ; the small horse of the Lithuanian peasantry is the same whose skeleton is found in ancient tumuli. What, then, must we think of such writers as Dr. Bateman, who, in the face of such facts, refurbish this Cuvierian fallacy ?* Mention is made of an anatomical peculiarity of man, the posterior opening of the sacrum which under certain cases imperils the lives of great numbers of persons.†
Lamarck was fully aware that species, in the sense applied to to the term by Linnæus and Cuvier, do not exist. He admitted so-called " spontaneous generation," *i.e.*, the production of the lowest organisms from inorganic matter. But we shall do well to note that he ascribes the development not to chance or necessity, but to the will of the Creator of all things.‡ This agrees closely with the view of Lavoisier,‖ that " God, when He created light diffused over the earth the principle of organisation and of sensibility." Hence the theory of Lamarck on the origin of life, so far from being atheistic, agrees essentially with that asserted by orthodox Christians !

Space unfortunately does not allow us to enter into the interesting chapter on Lamark's psychology. In reference to the charge of Materialism raised against the great French naturalist, our author remarks :—" Materialism, spiritualism are meaningless words, which it is time to banish from the strict language of Science. What is matter ? It is impossible to define. What is spirit ? Another unanswerable riddle ? These words, which serve as the starting-points of antagonistic doctrines, produce vain discussions which can lead to no result."

It may perhaps be argued, in some quarters, that such essays as these which Prof. Martins has produced are no longer necessary. We do not accept this contention. So long as the Evolutionist doctrine is persistently misunderstood and mis-

* Journal of Science, December, 1879, p. 804. See also March, 1880, p. 166.
† BROCA, Rev. Anthropologique, t. 1, p. 596.
‡ Philosophie Zoologique, i., p. 74, and ii., p. 57.
‖ Traité de Chimie, i., p. 202.

interpreted, not merely by the jealousy and pseudo-conservatism of the old, but also by the rashness and sciolism of the young expositions of its evidences, its true character and its tendencies will be needed.

The German translator has performed his task ably and thoroughly.

A Treatise on Comparative Embryology. By F. M. BALFOUR, F.R.S. In two volumes. Vol. I. London: Macmillan and Co.

WE have here a complete and systematic treatise on one of the most important branches of Biology. Since the evolutionist theory of the organic world has become the accepted view of the majority of minds capable of judging, Embryology has undergone a remarkable increase in its interest and in the importance of its revelations. The resemblance of the embryonic and larval stages of the higher animals to the lower forms of their respective groups had long excited the attention of philosophical zoologists, and is now recognised as supplying a most valuable body of evidence in favour of the so-called doctrine of Descent, and against that of mechanical or contract Creation. For want, however, of such knowledge as the work before us will furnish, this parallelism between the evolution of the individual and of the group, if not overstrained, has been somewhat distorted. The higher embryo does not exactly resemble the lower adult. Nor would such close resemblance, if it existed, be in harmony with the requirements of the modern theory, which regards all organic groups as formed by a series of branchings out from one common stem. The human embryo, at a certain stage of growth, approximates very closely to those of apes, and even of carnivorous animals. But as the species approach maturity they diverge more and more from each other, so that the child has never passed through a stage resembling that of the mias or the dog. What really does occur, as stated by the author, is this :— " Each organism reproduces the variations inherited from all its ancestors at successive stages in its individual ontogeny, which correspond with those stages at which such variations appeared in its ancestors;" and to the law, as thus stated, we do not see that any exception need be taken.

But whilst the enhanced importance of embryology and the abundance of researches lately undertaken render a work like that before us desirable, they place, as the author fully recognises, serious difficulties in the way of its execution. He refers to the recent appearance of " a large number of incomplete and contradictory observations and theories." He might have mentioned that some of these researches have been undertaken by men

whose motives are open to question.   If a biologist enters upon
investigations not with a view to the discovery of the simple
truth, but in the hope of meeting with facts irreconcilable with
organic evolution, we are justified in regarding his results with
suspicion.   Nor is it certain that a contrary bias, in favour of the
doctrine of Descent, may not in other cases have been present.
Hence, as Prof. Balfour admits, it may be considered doubtful
whether the time has yet come for all these observations to harmonise and arrange in a coherent whole.   Still we think our
readers will agree with us that he merits the thanks of biologists
for the manner in which he has executed this difficult task.
The literature of the subject, scattered through the scientific
journals and the Transactions of learned societies, is of prodigious extent, as an examination of the bibliographical appendix
will show.   It conveys, at the same time, the unpleasant reflection how small a portion of the researches in this field has been
executed by Englishmen.

The present work is devoted solely to the embryology of those
animals now characterised as " metazoa," and the subject is
treated from a morphological rather than from a physiological
point of view.   The objects of Embryology are shown to be
twofold—to form a basis for phylogeny, the history of the development of the group, and for organogeny, the doctrine of the
origin and evolution of organs.   Among the subjects specially
discussed in the former department are the questions how far
embryology brings to light ancestral forms common to all metazoa ? how far some special embryonic larval form is constantly
reproduced in the ontogeny of the members of one or of more
groups of the animal kingdom, and how far such larval forms
may be understood as the ancestral type of such groups ? how
far such forms agree with adult forms, living or fossil ? how far
organs appear in the embryo or larva which in the adult state
are either atrophied or lose their functions, but which persist
permanently in members of other groups or in lower members
of the same group ? and how far organs in the course of their
development pass through a condition permanent in some lower
form ?   The author justly remarks that the solution of these
problems would be greatly simplified if each organism contained
in itself the full record of its origin.   But the law which we
have quoted above admits of what to the unscientific observer
seem exceptions.   It is, in other words, merely a statement of
what would occur were it the only influence under which organic
development takes place.   In actual life there are a number of
interfering conditions.   " The embryological record is almost
always more or less abridged in accordance with the tendency of
Nature (explicable on the principle of the survival of the fittest)
to attain given ends by the easiest means."   Secondary structural
features are introduced to adapt the larva or embryo for special
conditions of its existence, and the problems involved are thus
rendered exceedingly complicated.

The organogenic questions to be elucidated by embryology relate to the origin and homologues of the germinal layers, the origin of the primary tissues, and the origin and gradual evolution of the organs and systems of organs.

In expressing our general approval of Prof. Balfour's work we do not wish to be misunderstood. It is by no means a popular treatise. It contains no novel or striking generalisations, and to readers unacquainted with morphological zoology it will prove scarcely intelligible. But to the student and the professed biologist it may be recommended as affording a fair and comprehensive survey of the results hitherto reached in embryology, such as no other work in the language is calculated to afford. The illustrations are numerous and carefully executed, and the references to original memoirs accompanying the chapters and sections are ample. A second volume will treat of vertebrate animals, and with the special histories of the different organs.

The chief defect of the work is one which it shares with no small proportion of modern English biological works,—that is, the tendency to multiply technical terms of Greek origin. We see no good reason why we might not, like the Germans, employ technical terms of native origin. The days are passing away when every man of science may be assumed to be a classical scholar.

Practical Chemistry. The Principles of Qualitative Analysis. By W. A. Tilden, D.Sc., F.C.S. London: Longmans and Co.

The author tells us in his Preface that his book is intended for beginners, and that its production was prompted by the requirements of his own teaching, and " by the consideration that it is not necessary to learn the properties of any large number of substances in order to be in a position to understand the principles of chemical analysis." All this is perfectly true : but among the elementary works on chemistry which have been produced in such astonishing plenty during the last ten years, are there not several—to use the mildest expression—in which the consideration of the rarer elementary bodies is omitted on precisely the same grounds ? It cannot be expected that a chemist of Dr. Tilden's standing would be guilty of any error in compiling a book of this simple character, but we can scarcely see any *locus standi* for his work. Pending the avatar of some capital discovery which may essentially modify the outlines of the science, we think that there is no more need for the appearance of any further elementary treatises on chemistry.

Papers and Proceedings and Report of the Royal Society of Tasmania for 1878.　Hobart Town : Mercury Office.

THE work done by the Royal Society of Tasmania, if not very great of quantity, is of a sound, satisfactory character. The papers here reproduced deal with questions which can only be investigated on the spot.

Mr. R. M. Johnston contributes a valuable paper on the fresh-water shells of Tasmania. He has discovered one species belonging to a rare genus at present confined to Cuba, but he prudently declines theorising on this fact till our knowledge of the Mollusca of Tasmania and of the Australian mainland is more complete. As regards the distribution of fresh-water shells, he shows how the smaller species may easily be transported to vast distances by the agency of water-beetles and aquatic birds.

Dr. Morton Allport communicates some remarks on the habits of the *Platypus.* He finds that it feeds on caddis grubs, but suspects that it may likewise devour the ova of fish.

The Rev. J. E. Tenison-Woods continues his memoirs on the marine shells of the Tasmanian waters.

Mr. R. M. Johnston has studied the Tertiary and Post-tertiary deposits of the islands in Bass's Straits. He considers that an extensive upheaval of the ocean floor in the districts of South Australia and Tasmania has taken place, and has continued to a very recent period, if it be not still going on. Local encroachments of the sea may be in perfect harmony with a slow vertical movement of the land upwards. He considers that the evidences of Australian geology are in perfect harmony with the theory of Evolution.

The Bishop of Tasmania read a paper on " Water-supply in relation to Disease," full of forcible warning to the community.

We beg to congratulate the Society on the work it is doing, and hope it will persevere in the same direction. The scope for its exertions is practically boundless.

A Practical Treatise on Sea-Sickness ; its Symptoms, Nature, and Treatment.　By G. M. BEARD, M.D.　New York : E. B. Treat.

DR. BEARD takes here a new departure by regarding sea-sickness not as a mere disease of the stomach and liver, to be subdued by " capsicum, calomel, and champagne," &c., but a functional disease of the nervous centres. He utterly scouts the prevailing notion that the affection is in the long run beneficial to the system. He shows that the evil effects do not invariably cease when the sufferer arrives on land, but may remain for years. Even death is not entirely out of the question. The matter being so much more serious than it is vulgarly supposed to be, it is fortunate that a trustworthy remedy exists. Sodium bromide, or, in default, the corresponding potassium, ammonium, or calcium salts, given three times a day in large doses,—say thirty, sixty, or even ninety grains,—rarely fails to prevent the attack. Dr. F. D. Lente first proposed the use of this salt some three or four days before starting on a voyage, keeping it up until there is reason to believe that all danger is over. Failure is generally the result of too small doses, or of waiting till sea-sickness actually appears before the remedy is given. If vomiting has actually set in, the bromides, and indeed all medicines administered by the mouth, are useless, for the very sufficient reason that they are not retained in the stomach long enough for absorption. In such cases the author recommends the hypodermic injection of atropia—an operation which must, of course, be entrusted only to professional hands.

We hope that Dr. Beard's work will be widely circulated among the medical profession, as it will doubtless prove the means of dispensing with much needless misery.

The First Types and the Serial Succession of Insects in the Palæozoic Epoch.

THE well-known American naturalist, Mr. S. H. Scudder, has published, in the "Archives des Sciences Physiques et Naturelles," a summary of the occurrence of insects in the earlier geological periods. He concludes that, with the exception of certain wings of Hexapods, found in the Devonian strata, all the three groups of insects—the Hexapods, the Arachnides, and the Myriapods—make their appearance simultaneously in the Carboniferous system. The Hexapods, or true insects, may be divided into two classes—the higher (Metabola) comprising the Hymenoptera, Lepidoptera, and Diptera ; and the lower (Heterometabola) including the Coleoptera, Hemiptera, Orthoptera, and Neuroptera.

All the insects of the Devonian and the Carboniferous formations are Hetero-metabola, the Metabola making their first appearance in the Jurassic period. In the Palæozoic ages there occur numerous synthetic or generalised types, which combine the characters either of all the Hetero-metabola, or of the Orthoptera and Neuroptera, or of the true Neuroptera and the Pseudo-Neuroptera. The Devonian insects belong to types related to the two lower orders only, or they are Pseudo-Neuroptera with a lower organisation, which unquestionably inhabited the water during the earlier stages of their existence. The lower orders of the Hetero-metabola, the Orthoptera and Neuroptera, were much more numerous during the Palæozoic period than the higher orders, the Coleoptera and Hemiptera. Almost all the Palæozoic Orthoptera belong to the non-saltatorial families; they are Blattidæ, of low organisation. The true Neuroptera were also in those times far less numerous than the Pseudo-Neuroptera, whose organisation is less perfect. The general type of the structure of the wings of insects has remained the same since the remotest times. Excepting two Coleopterous and one Orthopterous species, it may be maintained that the anterior wings of Palæozoic insects were similar to the posterior or membranaceous pair; a difference only appears in the Mesozoic period. The venation in types of insects otherwise distinct was more uniform than it is at present. The Devonian, and even the Silurian, will probably be found to contain the remains of insects of a type still more generalised than has yet occurred in the Palæozoic strata. Most of the primeval insects were large, and many of them gigantic, and the American and European forms are strikingly similar.

The Food of Birds. By Prof. S. A. FORBES. ("Transactions of the Illinois State Horticultural Society.")

THIS memoir, which is abstracted and commented upon in that most ably-conducted journal the "American Naturalist," bears so forcibly upon a subject to which we called the attention of our readers in our April number that we feel bound to give it our consideration.

The utility of the small birds to mankind, in America certainly, and probably in England also, is far greater than we imagined. In Illinois there are estimated to be, during the summer half-year, three birds to an acre. The author considers that at least two-thirds of the food of birds consists of insects, and that this insect-food will average at the lowest reasonable estimate twenty insects or insects' eggs daily for each individual of these two-thirds, giving a yearly total of 7200 specimens per acre, or, for the State of Illinois, 250,000,000,000, a number which, placed one to each square inch of surface, would cover 40,000 acres. A careful estimate of the average number of insects per square yard, in this State, gives at the outside 10,000 per acre. Here we may remark that anything like exactitude is difficult. In seasons and districts where aphides, midges, sand-flies, &c., are abundant, this estimate seems to us too low.

To return: if the operations of the birds were suspended, the increase of the insects would be accelerated by about 70 per cent, and their number, instead of remaining at the present yearly average, would be increased each year by more than two-thirds. Multiplying in geometrical progression at this rate, in twelve years we should have the entire State covered with insects at one to every square inch of surface.

Mr. Walsh computes the average damage done by insects, in the State of Illinois, at 20 million dollars yearly. This means only about 2s. 4d. per acre. If in this country we run over the injuries occasioned by some few of the better-known of our insect enemies,—such as cockchafers, wireworms, weevils, turnip-fleas, cabbage butterflies, aphides, sawflies, &c., without the aid of locusts and Colorado beetles,—we shall certainly think Mr. Walsh's estimate within the mark. Prof. Forbes, however, only takes the damage at one-half, or 10 million dollars annually. On this basis he shows that if the efficiency of the birds could be increased by 5 per cent, 8½ million dollars would be added to the permanent wealth of the State. But the probability is that the number of insects destroyed by birds is far greater. According to Dr. Brewer a pair of European jays have been found to consume half a million caterpillars in bringing up their brood, and to devour a million eggs during the winter. A young mocking-bird, reared from the nest by Robert Forbes, a nephew of the author, ate about 240 red-legged grasshoppers daily.

Surely, in face of such facts, the very instinct of self-preservation—if it still survives among us—might lead us to suppress with some promptitude the bird-shooters, the bird-nesters, and the bird-trappers. But we allow the mischief to go on, and neither Science nor professional humanitarianism raises a protest.

Rough Notes on the Snake Symbol in India in connection with the Worship of Siva.

Prehistoric Remains of Central India.

Archæological Notes on Ancient Sculpturings on Rocks in Kumaon, India, similar to those found on Monoliths and Rocks in Europe.

Description of Stone Carvings collected in a Tour through the Doab from Cawnpore to Mainpuri. By J. H. RIVETT-CARNAC, F.A.S., F.G.S., F.R.A.S., &c.

THESE four pamphlets, all reprints from the "Journal of the Asiatic Society of Bengal, are interesting contributions to the archæology of India.

The cobra is worshipped as Mahades or the Phallus, the emblem of Siva. The author refers to the similarity between the menhirs of Carnac in Brittany and the Siva emblems in India.

The tumuli described in the second pamphlet bear a remarkable resemblance to the barrows of Europe. The inner colonnade at Stonehenge is a repetition of the horse-shoe temples of Ajanta and Sanchi, and the Buddhist tree-worship displayed in the Sanchi *bas reliefs* is the counterpart of the Druidical reverence for the oak.

The agate weapons found in India are supposed to have been shaped by human beings (or anthropoids?) whose reasoning powers were but very slightly developed, and point to an earlier stage of the existence of man than is yet known. The author thinks a geological inquiry into the age of the Nerbudda formation necessary. It is still doubtful whether these formations correspond to the Pleiocene of Europe, and whether these agate implements are really derived from the Nerbudda gravels or from some later deposit.

The Journal of the Royal Historical and Archæological Association of Ireland. Vol. v., Fourth Series, No. 39. Dublin: Ponsonby and Murphy.

WE find here but little matter which can legitimately come within our cognizance.

Mr. W. Noble, in a paper on the age of a bronze pin near Enniskillen, remarks that in a bog like that of Cavancarragh a formation of 6 feet would take about a thousand years to grow.

Mr. W. Gray communicates an elaborate illustrated paper on the character and distribution of the rudely-worked flints of the North of Ireland, chiefly in Antrim and Down. He draws no conclusions as to the time when Ireland was first inhabited, or to the character of the earliest settlers.

Archivos do Museu Nacional do Rio de Janeiro. Vol. ii. 1, 2, 3, and 4 Quarterly issues. Rio de Janeiro.

THIS volume of Transactions, bearing the date 1877, has only just reached us. As a matter of course the value of its contents, which are otherwise interesting, is considerably lessened by its late appearance, since the observations here recorded are already known to the scientific world.

The principal memoirs here included are—" On the Action of the Poison of *Bothrops Jararaca*, the Lance-headed Snake of

Brazil," by Dr. Lacerda Filho. The author enters upon, but does not decide, the important question as to whether the poison of snakes is a chemical individual, the action of which may be weakened to extinction by sufficient dilution, and which does not reproduce itself in the blood of the bitten animal ; or whether it is a ferment, soluble or figured ? He detects certain fusiform corpuscules in the poison, but he has not satisfied himself that these are the agents in the decomposition of the blood.

The illustrious biologist, Dr. F. Müller, who holds the appointment of " Travelling Naturalist of the National Museum," contributes four memoirs :—" On the Correlation between many-coloured Flowers and Insects ;" "On the Sexual Spots of Male Individuals of *Danais Erippus* and *D. Gilippus ;*" "On the Odoriferous Organs of the Species *Epicalia Acontius ;*" and "On the Odoriferous Organs in the Thighs of certain Lepidoptera."

There is also a paper " On the Geology of the Region of the Lower Amazon," by Prof. O. A. Derby, and a Summary of a Course of Anthropology delivered at the Museum by Dr. Lacerda Filho.

We are very much gratified to find so much attention paid to Biology in a region so fruitful in interesting forms of animal and vegetable life as in Brazil, and where so very much remains to be investigated.

Records of the Geological Survey of India. Vol. xiii., Part 2. 1880.

THIS issue contains notes, by C. L. Griesbach, F.G.S., on the Palæontology of the Lower Trias of the Himalayas.

Mr. W. King, the Deputy Superintendent of the Survey, communicates a paper on the successful Artesian Wells at Pondicherry, which he has carefully examined in order to throw light on the proposal of similar borings for the supply of Madras. The quality of the water used in many parts of India for domestic purposes, both by natives and British, is such as would be instantly condemned in England. Yet, unless the conclusions of modern sanitarians are entirely erroneous, the effects of the ingestion of decomposing organic matter must be far more deadly in a tropical climate. In India, if anywhere, our cities ought to be absolutely free from pollution in their soil, their air, and their waters.

Prodromus of the Palæontology of Victoria. Decade VI. By F. McCoy, F.G.S., &c. Melbourne: Ferris; Robertson. London: Trübner and Co.; Robertson.

WE have here carefully executed figures and elaborate descriptions of organic remains found in Victoria.

The first plate refers to *Macropus Titan*, a gigantic extinct kangaroo, which may be considered the representative of the "Old Man" kangaroo of the present epoch. *Percoptodon Goliah* is another gigantic species, which approaches the *Diprotodon*, and which differs from the true kangaroos in its more complete dentition, and its thicker, shorter, and more equal legs: its remains are found in the Pliocene Tertiary Clays of Lake Timboon.

The fourth plate is devoted to the ear-bones—so-called "cetolites"—of three extinct species of whales.

Plate V. represents the teeth of a gigantic fossil Spermaceti whale found in the Older Pliocene beds of Mordialloc.

Plates VI. and VII. give fossil Mollusca; Plate VIII. a new and abundant species of Hinnites from the Miocene Tertiaries. The last two plates represent sea-urchins from the same formations.

Four more decades are announced as forthcoming.

Notes on the Alluvial and Drift Deposits of the Trent Valley, near Nottingham. A Popular Lecture delivered to the Nottingham Naturalists' Society. By JAMES SHIPMAN. Nottingham: Norris and Cokayne.

THE author adduces visible and tangible evidence that the Trent has scooped out the valley in which it flows, and has left on its escarpments terraces marking the stages by which this process was effected. It is perhaps humiliating that such teachings should be needed, but we meet with persons who still entertain the belief that the beds of rivers were made not by them, but for them.

Several important books stand over for review, on account of the want of space.

CORRESPONDENCE.

₀ The Editor does not hold himself responsible for statements of facts or opinions expressed in Correspondence, or in Articles bearing the signature of their respective authors.

THE DILEMMA OF HUMAN LIFE.

To the Editor of The Journal of Science.

Sir,—I take the liberty of calling your attention to the subjoined remarkable passage from the " Medical Press and Circular ":—

"*Life and Death.*—This is the substance of a review of a book just published on the ' Destruction of Life by Snake-bites, Hydrophobia, &c., in Western India ':—' To know that twenty thousand people annually lose their lives in India from snake-bites ought to be enough to arouse the indignant remonstrances of philanthropic folk, especially when we reflect that a twentieth part of that number slain in fair fight would be held sufficient cause for which to hurl an able and useful Government from power. But in all seriousness, if the death-rate in India is to be persistently reduced by the prevention of famines, the destruction of wild beasts, the killing of snakes, the prevention of internal wars and feuds, the punishment for murder, improved sanitation, vaccination, better medical treatment, public dispensaries, purified water, drainage and removal of malaria, and all the other resources of advanced civilisation, some frightful calamity will be brought upon the country by the awful increase in the population without any possible means of increasing the food supply. If these energetic interferences with Nature's laws are to be extended, they must be accompanied by equally determined efforts to feed the additional millions preserved thereby ; otherwise we shall be merely creating mischief with one hand, while combating it with the other. The prevention of the twenty thousand annual deaths from snake-bites would increase the population of India by about ten millions every thirty or forty years ; and it would be interesting to get the ex-Commissioner's opinion as to the means of feeding this vast increase. And we say this while deeply regretting the loss of life now going on, and recommending his book to the careful attention of all who profess an interest in their fellow-creatures.' Flourish, then, malaria, snakes, and wild beasts, but save us from ourselves—from an increase of Indian population ! "

Surely what is here said of India, if true at all, is equally true of England, and indeed of all old-peopled countries. Will not the labours of sanitary reformers, temperance advocates, and philanthropists generally, if successful, result mainly in an intensification of the struggle for existence. Hence man is confronted with a dilemma from which there appears no escape. Malthusianism, which all political economists favour in their hearts, is useless, since the race or nation which adopts it simply effaces itself, and is overbalanced by its rivals. Is, then, the safety or the very existence of a commuity only to be bought by the misery of a large proportion of its component units.—I am, &c.,

EOSPHOROS.

CRUELTY OF THE STRUGGLE FOR EXISTENCE.

To the Editor of the Journal of Science.

SIR,—When we point out the harshness of the law of Natural Selection we are sometimes told that if the triumph of the strong involves a certain amount of suffering, the triumph of the weak would involve the degradation of the race. But we may surely ask, Is there any necessity that the weak should be called into existence to be trampled down? Suppose it were a law of " Nature " that an organism temporarily or permanently diseased or debilitated should be *ipso facto* temporarily or permanently incapable of reproduction,—would not the larger part of the weak never come into existence? would not the place which they now occupy until trodden down remain open for the strong, and would not an immense amount of misery be prevented? But to a very large extent the opposite rule prevails: sickly trees often yield the most numerous fruits or seeds, and consumptive persons have very frequently the largest families.—I am, &c.,

NEMO.

ABSTENTION FROM FLUIDS.

To the Editor of the Journal of Science.

SIR,—The question of the power of the human frame to endure fasting from solids, as exemplified in the case of Dr. Tanner, U.S., he taking only fluids in small quantities, is now amusing the scientific world. Accepting, in good faith, the experiment so for as the trial and his recovery has proceeded, we have on record the fact that man has the power of fasting for forty days without vital injury. The case of the convict at Cusana (reported in the Paris News in the 'Times,' August 10th, 1880), who starved himself, is in evidence.* He abstained from fluids and solids, and did not succumb until thirty days.

A case (for a long time) within my knowledge, the details of which are below, of a labouring man who abstained from taking fluids of any kind for two periods,—one of five and one of seven months,—appears to me of practical importance, especially in a military point of view, in cases where water is unattainable.

C— P—, in the peasant condition of life, a great advocate of the Temperance movement, in a moment of excitement upon the subject, maintained that not only could a man abstain from beer and alcoholic drinks, but that the drinking of any fluid was totally needless, and that life and strength could both be preserved without the imbibition of any fluid, provided suitable food, of a commonplace character, could be procured ; and that he in his own person would prove it.

C. P. is a labouring man (almost uneducated), but of an enquiring mind. The basis on which he built his theory was that all substances contain more or less of water, and thus the question was one of the selection of the food ; and when the proper selection of a common and easily attainable product was used, fluids were quite unnecessary to support life and to engender the energy needed in a laborious occupation. At the time, and some years before he entered upon his experiments, I should say that he was not only a teetotaller, but a vegetarian also.

On his first trial (five months) he fed wholly upon potatoes and butter, taking three meals a day, consisting (each meal) of three and a half pounds of potatoes, boiled and mashed with two ounces of butter : during this period he took *no fluid* of any kind, and only ended the probation when it became impossible for him to obtain potatoes in a fitting state. He then went on

---

* A mason, aged 26, condemned to four years' imprisonment at Cusano, has starved himself to death. From the day of his sentence he refused to take food, and, no compulsion being resorted to, he died at the end of thirty days. His funeral was purely civil, the priests considering him a suicide.

with his usual vegetarian diet, with which he drank water, sometimes milk, tea, or coffee.

The following year C. P. made his second experiment, during which time (*seven months*) he took three meals a day, each con sisting of twelve ounces of apples and twelve ounces of brown bread (made from the whole product of the grain), and as before, did not drink a single drop of fluid.

The experiments were made in the years 1863-64, when C. P. was about twenty-six years old. He states that during the time he dieted himself he was as able for his work as at any other period of his life. He is now forty-two—a stout, active man. He continues his teetotallism, but for the last three or four years has given up a strict vegetarian diet, thinking his powers were giving way, but which were wholly restored on his resumption of an *occasional* meat regimen.

In conclusion, I would say I have known C. P. for many years, and believe him incapable of a falsity of statement. He is strong, robust, and I should say an untiring worker.

To verify my facts I sent for C. P., who told me what I have stated above, and which statement is in accordance with my memory of the facts gleaned soon after they occurred. There are probably many other particulars which a physiologist might require, and should any of your readers, *for the purposes of Science*, desire to test the above narrative, the address of C. P. is at their service, and may be obtained on application to the Editor. I append my name as a proof of *bona fides* on my part. —I am, &c.,

S. BILLING.

August 12, 1880.

THE CONSTITUTION OF THE EARTH.

To the Editor of the Journal of Science.

SIR,—Geologists, having abundance of evidence that portions of the surface of Earth once beneath the waters of the ocean had been elevated into dry land, upon the supposition that the matter of our planet is a fixed quantity, naturally sought for evidence of a corresponding depression. Mr. Darwin was supposed to have found this evidence in the formation of coral reefs in the Southern Hemisphere. I have shown, by reference to the highly interesting discoveries made in connexion with the exploring voyage of H.M.S. *Challenger*, that there is "no necessity to assume that the submarine coral mountains of the Pacific or Indian Ocean afford evidence of a subsidence of the crust of the

Earth exactly representing its elevation elsewhere. (See "Constitution of the Earth," pp. 286, 287.) I am pleased to learn,
from the abstract of a paper read at the Royal Society of Edinburgh by Mr. John Murray, and published in "Nature" of
August 12th inst., that this seems to be the conclusion arrived
at by scientific members of the *Challenger* Expedition. The
object of the paper is "to show, *first*, that while it must be
granted as generally true that reef-forming species of coral do
not live at a depth greater than 30 or 40 fathoms, yet that there
are other agencies at work in the tropical oceanic regions by
which submarine elevations may be built up from very great
depths, so as to form a foundation for coral reefs ; *second*, that
while it must be granted that the surface of the Earth has undergone many oscillations in recent geological times, yet that all
the chief features of coral reefs and islands can be accounted for
without calling in the aid of great and general subsidences."—
I am, &c.,

ROBT. WARD.

August 18, 1880.

NOTES.

THE Birmingham Philosophical Society has established a fund for the endowment of research. Mr. Darwin has contributed £25.

The French Government has assigned to M. Pasteur the sum of 50,000 francs, to assist him in conducting his important researches on the contagious diseases of domestic animals.

M. Pasteur has communicated to the Academy of Sciences a memoir on the Etiology of Cattle-plague. He concludes that the germs of the peculiar bacteria which cause the disease, and which are of course present in the carcases of the animals which have perished, are brought to the surface of the earth by the agency of earth-worms. Hence he recommends that animals which die of "zymotic" disease should be buried in poor, chalky, or sandy soils, where earth-worms are rare. (Might not similar precautions be useful in disposing of the bodies of human beings who have died of pestilential diseases ?)

M. Max Cornu has observed the alternation of generations in the parasitic Fungi known as Uredineæ.

M. Vaillant has made a series of observations on the reproduction of *Pleurodeles Waltlii.*

M. N. Poletaien has demonstrated the existence of salivary glands in the Neuroptera, the existence of which had been long denied.

M. C. Richet, in a paper communicated to the Academy cf Sciences, gives an account of the action of strychnine upon the Mammalia. Under the influence of artificial respiration very large doses of strychine can be supported without the production of immediate death, and the phenomena produced differ decidedly from those observed after the ingestion of small doses. In very large quantities the poison acts somewhat like curare and somewhat like chloral.

" *Vivisection.*—The attentive reader of ephemeral literature cannot fail to be struck by the rapidly increasing number of journalists who weekly inveigh against the horrors of vivisection. It may be that the anti-vivisection fanatics are blameable for the excitation of popular indignation against a practice of which those most ignorant of physiological investigations consider themselves well able to be judges. The utter nonsense which greets the eye almost daily, and even in papers well informed on most ordinary subjects, though it receives a merited contempt from the instructed reader, is yet calculated to tell disastrously against the progress of research by stimulating a rabid opposition to scientific needs on the part of the multitude. As the supporters of the anti-vaccination movement have gathered strength from the unwise silence maintained by the profession in regard to truths whose incontestable nature they clearly apprehend, so it is with the vivisection question. In all controversy where fact and science are pitted against ignorance and assumption, the issue has never been nor will be in doubt. But the insidious influence of oft-repeated and uncontradicted assertions will do irreparable injury unless an effort ere long be made to stem the advancing tide of prejudice and ignorance." In making the above extract from the "Medical Press and Circular" we wish to point out that in all such agitations the party of attack has a great advantage. A falsehood can always be stated in fewer words than are required for its exposure, and this often in proportion to its very impudence. In this case, too, whilst the attack is made in journals read by the general public, in advertisements, and posting-bills, the defence has been almost exclusively confined to professional and scientific organs which never reach the multitude. Unless active steps are taken experimental biology will soon be entirely crushed, so far as the United Kingdom is concerned.

We are glad to learn that the Legislature is not wholly unmindful of the necessity of securing more efficient protection for wild birds. The existing Acts are to be consolidated and somewhat extended. The Duke of Argyll, however, pointed out that the new Bill had been carelessly drawn up :—" Of the sixty-eight birds enumerated therein fully half were simple synonyms, and one bird was in several cases mentioned two or three times over under different names." The omission of the woodpecker and the kingfisher is to be regretted. But the provisions of the Bill are of little moment unless when it becomes law it is to be strictly enforced.

Attention has been called by the "Standard" to the approaching extirpation of not a few of the most beautiful species of wild flowers, ferns, &c. We can, of our own knowledge, confirm many of the writer's statements. We know localities where the hart's-tongue, the beech- and oak-fern grew freely a dozen years ago, but where they would now be quite as difficult to find as would a deposit of diamonds.

According to M. Daubrée the Paris Museum receives from time to time spurious meteorites, sent in good faith, and sometimes by persons of scientific standing, such as Le Verrier and Becquerel.

The removal of the Natural History Department of the British Museum to its new site in Cromwell Road, South Kensington, has actually commenced. Though it was universally admitted that there was great lack of room in the old building, the situation of the new Museum, save in strictly official circles, is not admired. The respective keepers of the mineralogical and geological departments, Prof. N. S. Maskelyne, F.R.S., and G. R. Waterhouse, have resigned, and have been succeeded by Mr. Lazarus Fletcher, F.G.S.. and Dr. H. Woodward, F.R.S.

Mr. Searles V. Wood, in a letter to the " Geological Magazine," contends, in opposition to Dr. Croll's "eccentricity theory," that the climate of the Glacial period must have been an uniform diminution of mean temperatures as they now exist, and that the period hence resulted from a cosmical cause, wholly unconnected with geographical conditions and ocean currents,—that is to say, to a decrease in the heat-emitting power of the sun.

Dr. Croll, in the " Geological Magazine," considers that the perpetual snow of high mountains is due to the want of aqueous vapour in the upper regions of the atmosphere, where the heat received from the sun is radiated off into space.

According to Dr. Bogomelow and Dr. Koehler, *Blatta orientalis*, the Orthopterous insect absurdly called the " black-beetle " in England, is likely to play an important part in the treatment of dropsy and albuminuria. The insects are stupefied with chloroform, dried and ground to powder, and taken in doses of 3 decigrammes.

Dr. Kebler, of Cincinnati, has studied the physiological action of platinum. In frogs the chief symptoms observed were increasing paralysis of the voluntary movements, convulsive spasms of the extremities, loss of muscular irritability, and death. In Mammalia the direct action upon the muscles is not perceptible, but death ensues in consequence of paralysis of the abdominal vessels. For dogs five- to six-thousandths of their own weight is a fatal dose.

According to MM. Van Tieghem and Bonnier (" Bulletin de la Soc. Botan. de France," xxvii., p. 83) the seeds of *Acer pseudoplatanus* form an exception to the rule that seeds are not injured by cold.

The editor of " Science," a new and ably-conducted American contemporary, states that during the fifteen minutes he remained in company with the well-known Dr. Tanner, two opportunities were presented when he might have taken food unobserved by his watchers, who were not always in the same room.

In the brain of an adult chimpanzee, described by Dr. Spitzka, the right side presented a well-developed first transition gyrus, evident and exposed to view as in the human subject, whilst on the left side it was concealed as in the ordinary chimpanzee type.

A Society of American Taxidermists has been formed, and is about to hold an exhibition, in which, in addition to "Taxidermy proper," there will figure "ornamental articles in which portions of an animal are used,"—in other words, Taxidermy improper.

Mr. A. S. Packard, jun., calls attention to a curious case of mimicry, He observed an Egerian moth *Trochilium polistiforme,* so closely resembling the wasp, *Po*   , was at first afraid to handle it.

M. Magnin has observed the microbia found in different kinds of vaccine matter, taken respectively from the horse, the cow, and from a human subject. All three are exactly similar in form, but the last-mentioned kind is only one-fifth the diameter of the other two.

Carriere has repeated and confirmed the statements of Spallanzani and Schäffer, on the reproduction of amputated parts in snails. He finds that the eyes, the tentacles, and the lips were completely restored, but that an injury to the supra-œsophageal ganglion proved invariably fatal. The regeneration of the eye was quite similar to its first formation in the embryo.

Karl Pettersen (" Tromsö Museum Aarhefter," ii., p. 65), after a very extensive series of researches, concludes that the rounding and smoothing of the surfaces of rocks, the grooving and scoring of their sides, are not always of unquestionably glacial origin. They are often post-glacial, since it can be shown that similar phenomena are in the course of origination in the present littoral zones. The conclusions based upon the glacial theory, as to the alternations in the level of the Scandinavian peninsula during the post-tertiary times, cannot, as a whole, be pronounced scientifically justified. There is no direct proof that the Scandinavian peninsula during the last portion of the Glacial epoch was more elevated than at present, nor that it has been subject to a continental depression during the latter portion of the same epoch. It is, however, a fact that the peninsula has been rising in the Post-tertiary period.

Mr. H. D. Minot, writing in the "American Naturalist," thinks that in England wild birds are better protected than in America, and are consequently more easily observed by the naturalist. He considers the birds of England, as a whole, inferior to those of New England, both in plumage and song. He met with mosquitoes on the Derbyshire moors.

In the same journal Prof. A. N. Prentiss and Mr. W. A. Henry record some experimental attempts to destroy Aphides by sprinkling with yeast. The results were entirely negative.

Dr. H. Behr communicates to the " Rural Press " (Californian) an interesting paper on "Changes in Plant Life on the San Francisco Peninsula." Commenting on the herbaceous immigration from Europe and Africa, he remarks :—"Our original vegetation has very little power of resistance. Its very variation is a proof of a certain want of vitality, for any more vigorous organisation, by superseding the weaker ones, would have produced originally the monotony developed at present by the immigration of foreign plants." It is lamentable that plants triumph in the struggle for existence in an inverse ratio to their beauty and utility.

According to a report by Dr. Delpech, on Apiculture in Paris, it appears that the bees, when they penetrate into sugar-refineries, will not touch treacle if pure syrups are to be found, and in places where both cane and beet-root syrups are in store they completely ignore the latter.

The following mode of preparing glycerin-gelatine for mounting microscopical objects is given by Herr Otto Brandt, Secretary of the Berlin Microscopical Society :—A quantity of the best white gelatine is to be cut up coarsely, and laid overnight in a vessel with distilled water, so that it may swell up during the night. In the morning it is taken out, squeezed in the hand, and placed to melt (without adding fresh water) in a glass cup in the water-bath ; as soon as the mass has become fluid add to it (stirring continually) about one and a half times as much glycerin as was taken of the gelatine. Filtering is a matter of vital importance ; filtering-paper does not allow the fluid to pass through sufficiently, and flannel makes it worse than before. It is best cleared by being strained through a glass funnel with some spun glass pressed into the lower part of the cone : this must be kept warm during the process of filtration in a suitable water-bath. Some drops of carbolic acid should be added to the fluid product of the filtering.

Mr. R. Bullen Newton, Assistant Naturalist, under Professor Huxley, in the Museum of Practical Geology, Jermyn Street, has received an appointment in the geological department of the British Museum.

ERRATA.

Page 368, line 10 from bottom, *for* " temperate " *read* " temperature."
Page 370, line 21 from bottom, *for* " compact " *read* " impact."
Page 373, line 8 from bottom, *for* " 5 per cent " *read* " 5°."
Page 374, line 15, *for* " simplicity " *read* " ellipticity."

THE

JOURNAL OF SCIENCE.

OCTOBER, 1880.

I. THE VEHICLE OF FORCE.

By CHARLES MORRIS.

MODERN thought has made so many valuable conquests in the outlying districts of the kingdom of Science that it is daily drawing nearer, to the citadel of the kingdom, that abstruse question of the ultimates of Nature on which so much deep reasoning has been expended, yet which still remains unsolved. The whole process of thought in this direction has been a process of simplification. A century ago a very complex conception of these ultimates was entertained. To ordinary matter was added an indefinite number of imponderables, while the forces of attraction and repulsion were neatly parcelled out among these various forms of substance, until the whole affair became unpleasantly intricate. As for motion, there had been some inquiry into its laws, but very little consideration had been given to its office as a primary constituent of Nature.

We have changed all that. The imponderables have ceased to be substances, and have become special modes of motion; and Nature is being reduced in theory to two ultimates, matter and motion. True, matter has not quite reached this degree of simplification. There still underlies it a vague conception of an unparticled Ether, the supposed vehicle of radiant force. This ethereal constituent of Nature has been found necessary by two classes of reasoners,—the students of the science of Optics, and the promulgators of force theories. It has seemed impossible to explain Attraction and Repulsion as attributes of a single form of matter, and theorists have found themselves obliged to deal with two diverse substances, through whose aid they neatly make the whole business clear—to themselves. Another class of reasoners, the advocates of what we may call the pressure and shelter theory of matter, have sought to get rid of these forces altogether, and explain them as results of momentum. But in doing so they have raised new difficulties, almost or quite as great as the old ones they have avoided.

The true origin of force still remains a mystery, despite the theoretical attempts above mentioned. Yet the intense thought which has been expended on the effort to connect attraction and repulsion with matter, without satisfactory result, goes far to indicate that theorists have been clinging to a false conception, and wasting their powers on a stubborn effort to prove the impossible. There is, however, another point of view from which the question may be considered, which, strangely enough, seems to have never been attacked, and which I propose to consider in the present article.

If we take it for granted that attraction and repulsion really exist, as forces distinct from the impact force of momentum, certain necessary deductions follow. Not being motion they cannot be converted into motion, although long arguments have been made on the tacit assumption of such conversion. The quantity of motion in the universe is supposed to be unchangeable. Attraction and repulsion, not being motion, cannot add to nor take from this total quantity. And yet bodies, which were apparently at rest, start into visible motion under their influence, and our works on physical science are full of argument which amounts simply to a declaration that motion is created by attraction. For the theory of potential and actual force can be given no other interpretation. Disguise it as we may, potential force means simply possibility of motion. Actual force means real motion. Thus, in the accepted theory, there lurks a distinct claim of the creation of motion, the possible being converted into the actual under the influence of attraction and repulsion.

I have already dealt with this question in my article in the June number of the journal, on the modes of interchange between heat motion and mass motion. I have there sought to show that attraction and repulsion have not a creative— but simply a directive—power, and that all the motion which springs into visible existence had a previous invisible existence, the vibrations of particles being readily convertible into the onward movements of masses, and the latter as readily transformable into heat vibration.

We are thus enabled to somewhat simplify the question as follows :—There exists an unvarying quantity of matter, possessing, perhaps, the quality of existence only; an unvarying sum of motion, which yields matter the energy of momentum; and a directive principle, known as attraction and repulsion. No matter how vigorous this latter force, it is not motion, and therefore cannot add to nor subtract from motion, nor is it convertible into motion. It is limited in its action to the changing of the direction of moving matter, thus exerting a force, since matter retains its direction of movement with a certain energy.

The question to be solved is, What is the origin and character of this directive force? where shall we seek for that vital principle of Nature from which arise the various manifestations of directive influence known as Electricity, Magnetism, Chemism, and Gravitation? As we have already stated, it has always been assumed that this force is an attribute of matter, and men have shut their eyes to certain striking indications of nature, in their efforts to establish untenable hypotheses, based upon this false conception. Yet it is simply impossible to comprehend that two similar particles can now attract and now repel each other, from any principle inherent in themselves, while the idea of a self-repulsive ether and an attractive matter makes of nature a very intricate affair.

But it may be argued that the particles of matter are not similar, that they attract under one set of conditions and repel under another. Yet this does not dispose of the difficulty of their possessing two opposite forces as inherent attributes. It leads us, however, to the possibility of another interpretation of the question. For this difference in condition means, in many cases, and perhaps in all, a difference in motive relations. The matter may continue the same, but its internal movements have varied, and its force influence upon other substance varies accordingly.

Does not this yield us a new conception? If the attractive and repulsive energies of matter interchange as its motive relations vary, it naturally seems as if these energies had more to do with motion than with matter; as if, in short, attraction and repulsion were inherent attributes, not of matter, but of motion. And, again, if attraction and repulsion be simply directive energies, their whole influence consists in producing changes in the direction of motions; and it is certainly as easy to conceive that motions exert this influence upon each other, as to suppose that it is exerted upon them by matter. All the results of directive force are variations in motion, and it is very natural to conclude that the force itself is an attribute of motion.

If we approach this subject more closely, we at once perceive that at least two of the four special manifestations of this force spring from motive conditions. These four manifestations are Electricity, Magnetism, Chemism, and Gravitation. Chemism, however, has probably nothing special in itself, but is more likely a result of the other three modes of energy; as are also the cohesive and adhesive displays of force. But of the three other manifestations of directive force there is abundant reason to believe that electricity and magnetism are modes of motion, while we are as yet unaware of the cause of gravitation. If the electric and magnetic energies, then, arise from special conditions of motion, and if their sole manifestation—beyond that of motive vigour—is attractive and repulsive force, we certainly have some warrant to conclude that this force is an attribute of the special motions with whose birth it is born, and with whose death it dies.

We may safely take it for granted that the directive energy is not an attribute of both matter and motion, arising from the innate constitution of matter, and also from the innate energy of motion. Such an attribute cannot possibly belong to two distinct ultimates of Nature—to concrete matter and to abstract motion. It must appertain to one only. But if we impute it to matter there arise certain mysteries which still remain unsolved, despite numberless attempts to explain them :—First, how it can exist in two directly opposite forms ; secondly, how it can increase and diminish in quantity ; thirdly, how it can be transferred from one substance to another ; and finally, how it can utterly disappear in one form without any apparent change into another form. If, on the contrary, it belong to motion, all these difficulties at once vanish. Motion readily assumes directly opposite conditions ; it readily varies in quantity in any substance or part of a substance ; it is readily transferred from one body to another ; and, finally, it may so change as to produce neutral relations, with complete disappearance of directive force. Thus the many objections which apply to matter as the vehicle of force all disappear if we ascribe this attribute to motion.

Such a transfer of directive force is one of the most common manifestations of electricity. Static electric force, for instance, spreads from a point over the whole surface of a conducting body, and leaps from one conductor to another, carrying with it, in all its movements, attractive and repulsive vigour. Now the transfer of electricity is simply the transfer of a mode of motion ; therefore the force transferred with it must be the movement of a force pertaining to motion. For the outgoing motion could not conveniently sweep away inherent force from one substance to plant it in another; nor could a force attribute of matter well go to sleep in one mass the instant motion left it, and awake in another the instant motion entered it. Similar shifting phenomena of force might be shown in other instances of the transfer, or the sudden action, of motive energy, all tending to strengthen the position we have taken. But without specially considering them here we will try to deduce from the force manifestations of Nature the law which governs this directive principle of motion.

Our work is, in fact, largely done to our hand, by the investigators of electrical phenomena. We know that if electrical currents are sent through two contiguous wires, these wires exert a marked influence upon each other. If the currents proceed in the same direction, the wires are drawn together. If in opposite directions, they are mutually repelled. In this case substances, which previously had no apparent influence upon each other, are by the addition of motive energies exerted in similar directions, caused to approach, and to separate if the energies move in opposite directions.

In the case of magnetism the same result appears, since magnetism is believed to be an effect arising from the circling of motive vigour around the particles of matter, magnetic attraction arising from parallelism in these motions, and repulsion from reversal of their directions. The inter-actions of electric and magnetic force yield the same result, masses of matter being moved in the effort to produce parallelism of motion. The full expression of the principle involved is the following :—Motive currents which move towards or from the same point in space, atttract; while those that move one towards and the other from a point, repel.

May we not have here a ready solution of the whole difficulty of these forces? Matter cannot well change its character, but motion can very readily change its direction; and the above principle may cover the whole question. We may, in fact, formulate a law of directive force as follows :— All matter in motion exerts an influence over the direction of all other matter in motion, this influence proceeding from motion disappearing with the cessation of motion, and augmenting with its increase. Any two motions parallel, or tending towards or from the same point, are forced to approach each other. Any two motions reversely parallel, or

tending the one to, the other from, the same point, are forced
to recede from each other. And the vigour of this influence
is probably governed only by the energy of the motions and
the distance separating them,—not at all by their degree of
approach to parallelism.

All motions, whatever their direction, must, under the
operation of the above law, exert an influence, either attractive or repulsive, upon all other motions. There can be
no neutral state in which both these influences will disappear.
For a close observance of the possible directions of two
lines of motion, which end in or tend towards the same
point, shows that they must hold toward each other one of
the above relations, and that no intermediate relation is
possible. They must either both approach or both recede
from this point, or one must approach and the other recede.
But in the case of motions which do not tend to or from
such a point, and are not parallel, conditions of both attraction and repulsion might exist between their different parts.
A great variety of relations might thus arise, in some of
which the opposite influences might become balanced, and
neutrality arise through opposition, not through loss of
force.

Motions also may readily become reversed, with a consequent reversal of their forces ; and if this reversal is rapid
and incessant the effect of neutralisation will be produced.
Such is the case in heat vibration. A vibrating particle
reverses its relations of motion to all contiguous particles
at every change of phase in its vibration. Thus its influence
constantly shifts from attractive to repulsive, and the contrary, and its force effect is virtually neutralised. An
analogous condition may arise in the case of magnetic substance. When its internal movements are largely in the
same direction the magnetic phenomena appear. When not
largely in the same direction their external influence becomes
neutralised, the attractive and repulsive forces of the various
particles balancing each other.

But although positive neutrality cannot exist between two
motions, it may easily exist between motion and rest ; for if
force is an attribute of motion, changing in vigour as motion
changes in energy, a particle at rest must cease to exist so
far as the exertion of force is concerned, since with the disappearance of its motion its force display must likewise
disappear. In like manner a particle approaching rest must
decrease in force through the diminution of its momentum ;
at the same time the force exerted upon it by other matter
must similarly decrease. The force exerted between any

two particles, at a fixed distance apart, depends upon their momentums. If we assume a unit of momentum, then the force exerted must depend upon the number of these units of moving energy. Consequently a small quantity of matter might exert as effective a directive force as a large quantity, if the former move with sufficient rapidity to equalise their motive energies. In all cases they must exert an equal attraction or repulsion upon each other. For if one mass be conceived to have ten units of momentum, and another mass one unit, the latter acts with its unit vigour on each of the ten units, giving it ten elements of force. But each of the ten units of the larger exerts but one element of force on the one unit of the smaller, giving it likewise ten elements of force. The fact is that the force vigour exerted upon each other by any two masses in motion is always equal, but is not equally effective in speed production unless they are of equal momentum. The larger momentum resists change in direction more vigorously; but the quantity of effect produced upon the inertia of motion is always equal, whatever the difference in momentum of the acting masses.

Thus the utmost vigour of directive force is exerted when two bodies move with equal rapidity, and at the highest possible speed; it diminishes in vigour as one or both of the bodies lose speed; it becomes null when one or both attain a condition of rest; and it changes in direction when either body reverses its direction of motion.

It might, however, be deduced from the above argument that a vibrating particle, on coming to rest at one extremity of its vibration, could not move again, it losing, while in that state, all force relations with the neighbouring particles. Such might be the case had it no other motions; but it is at rest relatively only, not absolutely. Besides their motions of organisation—which we will not here consider—all particles are in constant general motion, and a relative position of rest is only a diminution of some of these common cosmical motions. A particle at rest, in the ordinary acceptation, is one obeying its cosmical influences only. One possessed of special motion is obeying some local influence. These common cosmical motions—such as the mutual movement of particles in the rotation of the earth on its axis, and in its revolution around the sun—must act to render the great sum of motions parallel, and the great sum of forces attractive. Repulsion is not, in the present condition of matter, a general affection of nature, at least in so far as affairs on the spheres are concerned. It appears only in local disturbances of the great sum of parallel motions. In fact all the special force phenomena of Nature are consequences of such local disturbances, and produce but temporary variations in the general vigour of attractive energy.

But if, as we have argued, both impact force and directive force are results of momentum, it may, at first sight, seem as if matter had something to do with them, for at equal speeds the vigour of momentum depends upon the quantity of matter. And yet, in reality, matter has nothing whatever to do with momentum, which is a property of motion alone. In momentum only the vigour of motion is concerned; in no sense the quantity of matter. For the motive vigour which moves one mass of matter at a certain speed will move half that mass at a double speed. And if this motive vigour be successively transferred to gradually decreasing masses it will move them more and more rapidly, the momentum remaining the same though the quantity of matter be infinitely varied. If, finally, the mass of matter be diminished to its vanishing point, the speed of motion will rise to infinity, the momentum continuing unchanged. Momentum, then, is not a property of matter, but of motion; and if impact and directive force are both controlled by momentum, they are attributes of motion only, and matter has no office in Nature save to serve as the vehicle of motion, it being the concrete substratum through whose aid motion attains manifestation, and becomes capable of assuming the most varied and complex arrangements, yielding intricate relations of force.

But there is a secondary result of the law of force which is one of the most active influences in Nature, and which we must now consider. Directive force, as we have argued, tends to draw bodies together or to drive them apart. But ordinarily this tendency is resisted and becomes ineffective. So far, for instance, as the mutual cosmical motions of particles are concerned, every particle may be so placed that the attractive energy arising from these motions affects it equally from opposite directions, and thus becomes neutralised. Only when some local aggregation of momentum causes the attraction in one direction to overbalance that in the other, does movement in response to attraction result. There is thus a constant tendency in the motions of particles to become parallel, in response to their cosmical relations, while their local movements are only minor deviations from this persistent line of direction.

But they not only tend to become parallel in direction,

but also equal in speed. Although the opposed attractions permit the particle to retain its general direction of movement, they tend to drag it back when its speed becomes too great, and to drag it forward when its speed becomes decreased, until equality of speed is attained. The forces which oppose and neutralise each other directly aid each other transversely, so that, while the direction of motion remains unchanged, its speed is controlled, the constant tendency being to produce equality of speed in all similarly moving particles.

Such is the influence at work in radiant impulses. Every new motion given to a particle exerts this influence on all neighbouring particles, and is transferred from particle to particle with wonderful rapidity. Its primary tendency is to drag neighbouring particles towards itself; but if this tendency is resisted through a similar drag in the opposite direction, it can only exercise its secondary tendency to drag the adjoining particles into equality of speed, and to yield to a reverse drag from them, the motive energy lost by the one particle being gained by the other. But it is only to the origin of this effect, not to its details, that we wish to here refer.

There remains another force agency to be considered— the generally diffused and unvarying attractive force of gravitation. Investigation has hitherto been principally directed to the effects, not to the origin of this force. Yet it seems as if, under the above hypothesis, gravitation may receive a definite place among the modes of motion.

The gravitative force of a particle has nothing to do with its fugitive special motions, which influence its behaviour towards its neighbours, but which vary in a manner to neutralise their influence on the great attracting masses of the universe. Its special motions, indeed, are but temporary deviations from its general cosmical motions, and the tendency to resume the normal state causes any deviation in one direction to be balanced by an equal opposite deviation, so that all effects in one direction are neutralised by equal effects in the opposite direction. But these cosmical motions may have a fixed and persistent gravitative influence. Thus the movements of the particles of the earth in its orbital revolution constitute a vigorous momentum, which should yield some degree of attractive influence upon the similarly rotating sun. In like manner the rotation of the sun upon its axis should yield a like attraction upon the similarly moving planets. The axial rotation of the earth should also create an attraction between its parallel lines of momentum, tending to draw all these lines of motion towards their general centre of influence.

If our principle be correct, these motions certainly have a gravitative influence. The equatorial regions of the earth rotate at a speed of about 1000 miles per hour. But the earth moves around the sun at a speed of over 1000 miles per minute—in round numbers, about 18 miles per second. Here is a vast mass of momentum all of whose influence must be attractive, since all its motions are in one direction. Its effect must be double. It must yield a centripetal influence, adding to the great sum of terrestrial gravity. It must also yield a centrifugal influence, extending to all similarly moving matter in space, and acting attractively upon the sun and its attendant planets. If such be the case the gravitative force of a planet could not be measured by its density or mass of matter alone, but must be to some extent influenced by its orbital momentum.

Yet this could be only an aid, possibly only a slight aid, to the full force of gravitation, since the latter must be largely due to atom attraction, proceeding from the deep-lying fixed motions supposed to exist in atoms.

It may be advisable here to consider the character of this supposed atom motion. The theory that atoms are minute vortex rings of ethereal substance has many points in its favour, although it is not easy to conceive how such vortex rings could arise in the original ether, since its conditions were certainly not those of ordinary fluids. Yet if the ether possessed inter-movement, with its consequent attraction, we can readily conceive of a condition arising similar to another probable condition of fluids, viz., an aggregation around a common centre of attraction, and possibly the formation of permanent spherical or discoidal masses through the influence of rotation about an axis. And upon such masses, if forced to move through the ether, secondary vortex-forming influences might act, similar to those which act upon moving fluids. Such a process of formation would give a double motion to the atom, it possessing not only the peculiar motions requisite to the vortex ring, but also a rotation of the ring as a whole around its axis, or the common centre of its material. In such a case the atom would be at once a collection of individual particles and a single mass, these particles pursuing the vortex movements as individuals, and the whole ring its axial rotation as a mass.

In such a rotating ring opposite influences must exist. Its parallel movements must yield an attractive force, and constitute centripetal influences, tending both to the centre and to the cylindrical core of the ring. But in every circle of motion the opposite arcs move reversely, and must therefore repel, producing a centrifugal force. Of these two forces the centripetal must be the strongest, since its influence is exerted at the least distances, but the density of the mass and the diameter of the ring must be a resultant of the action of these opposed forces.

The energy of such an atomic mass would depend not upon its quantity of substance, but upon its momentum. If all force is dependent upon motion instead of upon matter, two atoms widely differing in material contents might be equal in vigour if their momentums were equal. Atoms may thus be almost infinitely minute, and the total quantity of matter in the universe be insignificant. So long as a fixed energy of motion exists it is unimportant what quantity of matter serves as its vehicle.

This atom motion is possibly of extreme rapidity, constituting a vigorous momentum, and forming the main factor in the general sum of gravitative force. On the supposition that atoms are minute masses rotating around a central axis, they must in reality constitute small magnets, the directive force arising from their motion fulfilling all the requisites of the electric currents of Ampere. If one pole of the axis of motion be presented, the whole axial momentum of the atom must exercise an attractive energy. If the other pole be presented the momentum must act repulsively. (If the atom be really a vortex ring, its vortex motion must produce special deviations in these results.) If the equatorial plane be presented attraction and repulsion would not necessarily become neutralised. One half the momentum would oppose the action of the other half, but one half would be nearer the substance acted upon than the other half, so that one or the other energy would be in excess. It is probably this excess which constitutes the innate attractive force of atoms, considered without reference to their magnetic augmentation of force. The latter probably acts vigorously in the chemical relations of atoms, and causes the atoms composing molecules and masses to arrange themselves under the control of local influences, and without special regard to cosmical influences. Yet this would not necessarily hinder their cosmical relations continuing fixed in degree and direction.

For the combinations of atoms into molecules are never irregular. The crystallising tendency which controls the minor aggregations of masses doubtless controls still more vigorously the formation of molecules. The minute elements of molecules are regularly combined under special conditions of attraction, and their motions are forced to harmonise with each other, every deviation from the general direction of motion, caused by the exigencies of the crystalline relations, being balanced by an equal opposite deviation.

But the directive influence of these crystalline factors of masses acts upon and is reacted upon constantly by the immense directive vigour of the earth, which may exercise a leverage upon all molecules, forcing them to assume a general conformity in direction. Whatever change in position may take place in masses, their molecules do not necessarily change in accordance therewith. For molecules are not rigidly fixed elements of masses; spaces separate them, in which they possess a certain degree of liberty to move and rotate. Thus they may readily yield to the leverage of terrestrial force, and shift their axial directions so as to retain one fixed relation with this force in every change of direction in the masses containing them.

Yet, supposing molecules to be a combination of more minute atoms, it does not follow that these atoms would have an equal facility of movement. They must be more rigidly constrained, through their closer proximity to each other and their more vigorous attractive relations. Possibly different substances vary in the character of their relations to exterior force. In some the molecules of masses may be as rigidly constrained as we suppose the atoms of molecules to be. In such a case, if these molecules be made to conform in direction through exterior influence, they would continue to do so with an energy affecting the whole mass, and the phenomena of permanent magnetism occur. This magnetic condition may ordinarily exist in molecules even when not displayed in masses, the molecules assuming positions in accordance with terrestrial magnetism, except when they are acted upon by local magnetism. And there may be cases in which the axes of rotation of molecules assume directions, under exterior influence, not in harmony with the direction of their atom forces, thus yielding the diamagnetic repulsion.

If the internal conditions of bodies be such as we have here surmised, it seems as if the ordinary tendency of molecules must be to direct their axes north and south at the Equator; for we naturally conclude that the rotation of the mass of the earth around its axis must yield a directive energy forcibly tending to bring the motions of all its minor constituents into harmony therewith, the axis of motion of every molecule at the Equator becoming parallel to the earth's axis, and the direction of motion similar to that of the earth. Where these molecules are rigidly constrained elements of masses, as in permanent magnets, the axis of magnetism must assume a like parallelism, the whole mass being moved by this form of terrestrial force. And molecules and magnets alike must behave with the earth as one magnet would behave with another; that is, the magnetic dip must occur, the magnet becoming vertical at the pole, since there its axis must become virtually an extension of the earth's axis, to preserve the conformity in their movements. Local displays of magnetic energy must act like all other divergencies of motion from its normal relations, to temporarily vary the vigour of gravitative force. The normal conditions of motion are those possessed by spheres as wholes, and the rotations of molecules in harmony with those of their spheres. All other motions are discordant and temporary, their influence being to temporarily increase or diminish the gravitative force of the abnormally-moving particles. It is in these abnormal motions, not in gravity, that we see the full vigour of atom force. It is partly shown in magnetic action, but more fully in chemical action, in which its great energy is strikingly displayed.

The conformity in motion which we have thus supposed to exist throughout the earth probably extends throughout the solar system, and possibly through the whole astral system of which it forms a part. Yet, as the earth does not rotate in strict parallelism with the rotation of the sun, the molecules of the earth may be affected to some degree by this solar rotation, and their axes assume a position intermediate between the direction required by these diverse forces. In like manner the moon and the planets undoubtedly exert a similar influence, so that the true direction assumed by the axes of terrestrial molecules is a resultant of numerous diverse but closely-related influencing forces. And these forces are not fixed in direction and degree, but are constantly varying, there being daily, yearly, and other much longer periods of variation.

From these variations there may arise the regular variations in terrestrial magnetism, which have like periods of change. As for the irregular variations, known as magnetic storms, there is no reason to doubt that they are usually caused by sudden changes in the condition of the sun's surface, by convulsive movements which may well produce a striking temporary variation in the relations of the molecules over an extensive portion of the solar sphere, and thus in the sympathetically connected molecules of all the planets.

It would appear then, if our premises are correct, that the directive influence exerted by the cosmical motions of the sun and the planets upon any one of these bodies must act to cause all the molecules of that sphere to assume an axial direction in accordance with these influences, so that the movements of all the molecules of the solar system may be in general harmony, the motion of each being a resultant of the many directive forces acting upon it. From the attractive influence exerted between all these similarly-moving molecules, gravitative attraction would arise. If this be the case the vigour of attraction arising from the movements of the spheres would be exerted upon their molecules, in producing conformity of motion, while the force of these latter would react upon each other, and thus upon the spheres as wholes, forming the attraction of gravitation.

In regard to the relations of diverse solar or astral systems, repulsion may take the place of attraction, if their general motions be reverse to each other, so that the attraction of gravitation may not be a universal force, but may be neutralised or reversed between different systems.

II. EXPLORATION IN THE FAR EAST.*

$\mathfrak{S}$INCE the days of Sir T. Stamford Raffles, of Sir James Brooke (much maligned of anti-nationalist orators), and still more since the publication of Mr. Wallace's fascinating work, it may safely be said that no part of the world has so much attracted the attention of naturalists as the Eastern Archipelago. The number, the variety, and

* Der Malayische Archipel. Land und Leute in Schilderungen gesammelt während eines dreissig-jährigen Aufenthaltes in den Kolonien. Von C. B. H. VON ROSENBERG, Königl. Niederländisch-Ostindischer Regierung's-Beamter, Ritter hoher Orden, mehrerer gelehrter Gesellschaften Mitglied. Mit zahlreichen Illustrationen zumeist nach den Originalen des Verfassers, und einem Vorwort von Professor P. J. VETH, in Leiden. Leipzig: Weigel. Amsterdam: De Bussy. London: Trübner and Co.

(The Malay Archipelago. Land and People in Representations collected during a Thirty Years' Residence in the Colonies. By C. B. H. VON ROSENBERG, with numerous illustrations from the author's original sketches, and a Preface by Professor P. J. VETH, of Leiden.)

the beauty of both animal and vegetable forms of life, and still more the peculiarities of their distribution, have made these islands the very Mecca of botanists and geologists. The appearance of the present work has therefore been eagerly expected. The thought suggested itself at once— If Wallace has observed and recorded so much in a residence of eight years, what must not Von Rosenberg have effected in thirty, and with the additional advantage of visiting not a few islands where Wallace never set foot, and especially of examining a much larger portion of New Guinea !

Further reflection will show that these demands are exorbitant. Mr. A. R. Wallace's whole time was devoted to Science, and his going and coming were regulated accordingly. Von Rosenberg was for twenty years in the military service of the Dutch Government, and his opportunities for observation depended on his duties as an officer. During the last ten years of his sojourn in these lands of wonder and beauty he held, indeed, the appointment of Surveyor and Naturalist ; but a very great part of his time was compulsorily devoted to a topographical and statistical examination of the islands. Latitudes and longitudes had to be determined, bays and rivers measured and sounded, villages noted and their population enumerated. Great attention was also paid to the manners and customs of the inhabitants, their religion, architecture, language, &c., as also to the meteorological and sanitary conditions of every locality visited. That in this manner a great amount of useful information was collected—some of it of high scientific value—is manifest ; but the time and opportunity for observations in natural history, strictly speaking, were sadly narrowed.

The author's investigations extend to Sumatra, the smaller islands along its western coasts, Celebes, the Moluccas (including Ceram), the Aru, Ke, Matabello—or, as he writes it, Watubella—Islands, Waigiou, Ternate and Tidore, and New Guinea. On the coasts of Sumatra, and subsequently in the Moluccas, he observed a curious and obscure phenomenon :—At night the sea over the reefs emits a peculiar sound ; at the depth of 20 feet it is like the crackling of salt when thrown upon burning coals; at 50 feet it sounds like the ticking of a watch, and is repeated more or less rapidly according as the bottom consists of coral alone or of coral intermingled with sand and mud ; over mud-banks it resembles the humming of bees. Of the marine fauna of the Sumatran coasts he speaks with enthusiasm.

Brakian he describes as an El Dorado for ornithologists and entomologists. Among the more characteristic butterflies he mentions *Papilio palinurus, P. pompeius, P. antiphus, P. theseus, P. priapus,* and *P. memnon.* The European hawk-moth, *Sphinx ocellata (Smerinthus ocellatus)* occurs here.

The orang-utan of Sumatra is confined to the flat, swampy coast-forests extending from Tapanoli to Singkel. It is named by the inhabitants "Mawas," and may be distinguished from the Bornean variety by its more foxy-red colour. The remaining Sumatran apes are *Hylobates syndactylus* (the amang, improperly termed siamang) and *H. agilis* (ungko), inhabitants of the mountain-forests; *Semnopithecus cristatus* and *S. melalophus, Cercopithecus cynomolgus,* and *Innuus nemestrinus.* The bats (*Pteropus*), during the author's residence at Lumut, flew every evening in a swarm from the north-west towards the south-east, and returned before sunrise in an opposite direction. The author fired once at a female which was flying unusually low, when a young one which was clinging to her teats fell downwards: before it could reach the ground the mother darted after it, seized it with her teeth, and soared away with her rescued offspring.

The fort was almost besieged by tigers, so that no one ventured to go out after six in the evening. The animal differs in some respects from the Bengal race. The ground colour is less of a rusty red, and the black stripes are not so slanting, and do not display upon the back those sharp angles, directed forwards, which form a black dorsal stripe in the tiger of the Indian continent. In the Sumatran race the tail is shorter and thinner. Elephants, identical with the Ceylonese species, are tolerably numerous. The rhinoceros (*R. sumatranus*) is met with in the mountains as high as 6000 feet above the sea-level. The author gives a very extensive list of the birds of Sumatra, in many cases with remarks on their habits and structure, which, however, space does not allow us to notice.

The islands on the western coast of Sumatra are rich in Coleoptera and Lepidoptera of remarkable beauty. One species of the latter, *Hebomoia Vossii,* is peculiar to the Isle of Nias. The author observed no mammal or bird which is not common to Sumatra, whilst all the larger members of the former class and many of the Sumatra are wanting. There are neither tigers, bears, elephants, rhinoceros, or tapirs. Only two monkeys occur, and these are absent in Engano, the most southern, and in the author's opinion the most recent, of these islets.

Celebes has been very thoroughly investigated by Herr von Rosenberg. At Ajer-Pannas, on the north-east coast, are mineral springs, rich in calcium chloride, and of the temperature of 171° F. For the entertainment of distinguished visitors it is customary to capture crocodiles and plunge them into the hot water, where they speedily perish. At Sumalatta the chief object of interest is the deposit of copper and iron ores, alternating with auriferous pyrites. Latterly the workings have been abandoned, the miners have left the district, and the garrison has been removed.

At Linu is a lake of boiling mud. Here Count Vidua de Conzana, a distinguished Italian naturalist, perished miserably, having fallen into the mud up to his hips.

The author agrees with Mr. A. R. Wallace in regarding Celebes as the terminal point of the Indo-Malayan and the beginning of the Austro-Malayan fauna. He mentions *Cercopithecus cynomolgus* as very common in the southern part of the island, and extending into the territory of the Australian fauna. *Cynocephalus nigrescens* he considers a local variety of *C. niger*. It is very numerous, and when in large bands will sometimes attack a solitary traveller. Their presence in Batschian is explained by the fact that a pair of these apes were set at liberty by the grandfather of the reigning sultan. *Tarsius spectrum* is by no means so ugly as it is figured by Horsfield in the " Zoological Researches." A valuable catalogue of the birds of Celebes concludes this section. The author's collections and those of Bernstein have added eight species to those given by Wallace. He remarks that though the avi-fauna of Celebes is poorer, both in species and in brilliant colouration, than those of the Moluccas and of New Guinea, it surpasses those of Java, Sumatra, and Borneo in the development of certain families, especially the parrots, kingfishers, and pigeons.

On pedestrian tours across the island of Ceram he secured a rich harvest of insects, such as *Ornithoptera Priamus, O. Remus*, and *O. Hellene, Papilio Codrus, P. Sarpedon, P. Ulysses, P. Amphytrion, P. Severus, P. Deiophobus*, &c. These species haunt the banks of the streams which are filled with water-worn stones. Here it appears that he found Mr. A. R. Wallace, whose acquaintance he had previously made in New Guinea.

The ethnologist will here find an account of the Kakean League, a secret society which sprang up in the seventeenth century, and whose object is to prevent the spread of Christianity and Islamism. The Ceramese are Pantheists.

The Aru Islands were found especially abundant in Lepidoptera, including the magnificent *Cocytia d'Urvillei*.

The author's account of the birds of Paradise of Aru and the neighbouring islands is very complete. He describes *Paradisea apoda* and *P. regia*. We cannot refrain from expressing our fears that this splendid family will soon be extirpated, except the use of the birds for the decoration of persons, rooms, &c., be prohibited.

At Ternate the author visited the grave of his distinguished predecessor, Dr. H. A. Bernstein, who died at Batanta in 1865.

Herr von Rosenberg's visits to New Guinea comprise an examination of the south-west and north-east coasts, a journey to Geel Vink Bay, and an expedition to Andai by way of Batjan. He considers that only an insignificant fraction of the fauna, and still more of the flora, of this marvellous island is as yet known to Science. His expeditions appear to have been made chiefly in the rainy seasons, when both collection and observation are much impeded. Nevertheless he studied with care and success the splendid genus *Epimachus*. He is the first European who met with *E. speciosus* " in the flesh," all specimens formerly obtained having been skinned and spoiled by the natives, who are the vilest of taxidermists.

Our author's account of his visit to Java is substantially a description of the admirable Botanical Gardens of Buitenzorg, maintained by the Dutch Colonial Government. These gardens contain 251 families, 2305 genera, and 8506 species of plants. For the benefit of plants that require a cooler climate there are branch establishments in the mountains, one of them on the summit of Pangerango, 9600 feet above the sea-level.

An appendix contains a description of species of birds which the author has discovered, or at least has had the opportunity of examining more closely than had been done before.

In this short survey we have necessarily been able to refer only to a very small part of the interesting matter contained in this work, which we strongly recommend to all who are interested in the geography, the ethnology, and the natural history of the Far East.

III. THE PHENOMENA OF FLUORESCENCE.

By Edward Rattenbury Hodges.

NOW and then it happens that Dame Nature " lets the cat out of the bag," or, to speak more philosophically, she reveals some hitherto unnoticed and unknown fact or phenomenon which takes us by surprise.

The exhibition of this to the young or old, the learned or unlettered, is invariably followed by the ejaculation—How beautiful ! how wonderful ! No matter how diversified our habits, gifts, or tastes, we feel a common interest in the thing, and mutually share the sentiment of pleasurable surprise. We think it may fairly be said that the movement of a comet through space, or the motions of a creosote-drop upon a water surface, are alike equally worthy of our attention and our thought ; for He who made the glowing gases of the one also formed the oily particles of the other, and fully intends us to augment our knowledge and exercise our reasoning faculty as much by a careful contemplation of these as of other phenomena. By this delightful mental exercise we insensibly learn to observe and reflect at one and the same time, in consequence of being brought into direct contact with what a great leader in Science has aptly termed " objective realities." From an abstract point of view there is a curious correlation to be found in the examples just given. The astronomer makes his calculations upon the vast elliptical orbit of the former, while the physicist makes mathematics explain the fitful gyrations of the latter. And we might pursue the analogy still further were it necessary.

While candidly admitting that much is being done to popularise the facts of physical science, it is surely justifiable to ask that the *rationale* of many of its doctrines should also be made sufficiently simple as to find ready access to minds of ordinary capacity. To take any material object and detail its various known properties in plain language is by no means difficult ; but to describe the various known laws which govern its character is, on the other hand, not an easy task, for at this point we travel out of the region of the material into that of the immaterial. In other words, we move away from the domain of the concrete, and enter the kingdom of the abstract.

The bodily eye looks at the material, and then the mental one looks at the immaterial or invisible, and sees how inseparably connected is matter with force.

These observations may well apply to the remarkably interesting phenomena of Fluorescence, or "internal dispersion," for such are the terms given to the modifying effect produced by certain chemical substances upon a particular kind of light-energy which may be made to fall upon them. The phenomena appear to have been first observed in 1833,. by Sir D. Brewster, in an alcoholic solution of the green colouring-matter of leaves.

About ten years later Sir John Herschel made further observations, and subsequently published a paper on the subject in the " Philosophical Transactions " for 1845. Since then Prof. Stokes, of Cambridge, and other observers in England and on the Continent have been making further researches in Fluorescence.

Without detailing the history of the investigations connected with the subject, we give our readers some of the more interesting and remarkable facts which have been gleaned of late years, as well as some account of the earlier discoveries.

If we put some common paraffin oil, or a solution of sulphate of quinine, into a glass tube or other suitable vessel, and then look through it, the liquid will appear quite colourless; but if we allow the light to fall upon it, and then view it at a little distance and at a certain angle, some parts of the liquid will present a delicate sky-blue tinge. The effect in the case of quinine is heightened if the source of light is burning magnesium wire.

The large number of substances belonging to this class are termed fluorescent bodies, because they exhibit phenomena similar to the examples above given. The term itself, however, was suggested to Prof. Stokes by a particular kind of fluor-spar which shows this property.

Again, if we cause a room to be darkened, and allow only blue light (*i.e.*, by covering a hole in a window-shutter with cobalt-blue glass) to fall upon a glass vessel filled with water which has been standing some minutes, on floating a strip of horse-chestnut bark upon its surface, in a few moments a stream of bluish grey fluid (æsculin) will be seen slowly descending from the bark, hanging, in fact, like a bunch of barnacles from an old ocean waif. Or if, under the same arrangement of light, or by using even more powerful absorbents of the ordinary rays (such as a solution of ammonio-sulphate of copper or one of chromate of potash), we look at a piece of what is commonly termed canary glass,—*i.e.*, glass coloured with an oxide of the metal uranium,—it will be seen to glow as it were with rich greenish yellow rays, just as though it were itself a source of light; or if we take a solution of a uranium salt (the normal acetate) the phenomena are very striking when examined under the same conditions, and still more so by the electric light. But the salts of aniline—a substance which is the parent, so to speak, of mauve, magenta, and other brilliant colours—are singularly rich in exhibiting these effects.

A very beautiful experiment may be performed with the aniline red ink now so commonly in use. It affords, at one and the same time, an admirable illustration of Prof. Tomlinson's submersion figures and of the phenomena under consideration. If we take a long cylindrical glass vessel, or one with parallel sides, fill it with water, which is allowed to settle, and then gently deliver a drop of the red fluid to the surface, the drop begins to contract, and slowly from its centre descends in the form of a tube; the denser parts of the colouring-matter presently form a thick circular rim at the end of the tube,—but this is only for a moment, for a wavy edge appears upon this rim, then expands into a triangular parachute with a thickened edge, and from the extremity of each corner two or three smaller tubes descend; these in like manner pass through the same phases as the parent stem or tube. When this beautiful figure is viewed by transmitted light it exhibits the ordinary bright red colour; but when placed against a very dark background, and looked at from a little distance and at an angle, it is seen to be of a brilliant green,* very like a curious sea-weed serenely floating in a summer sea. Viewed vertically it presents a most singular shape, and glows with a fine golden tinge. Chlorophyll, the green colouring-matter of leaves before mentioned (those of the nettle are preferred), by reflected light is a bright green; but when a beam of white light concentrated by a lens is sent through it, the path of the beam is of a blood-red hue.

Crude resin oil, which is almost opaque, presents a delicate sky-blue fluorescence at its surface and the side opposite the light. One investigator has shown that some colouring-matters obtained from different woods, which though soluble in alcohol, alkaline solutions, or in alum, and not in this way sensitive, are very brilliantly fluorescent when dissolved

* The brilliancy of colour, of course, depends on the amount of diffused day light. The effects in each case are best seen on a bright day.

in castor oil : the most marked among these is camwood ; in this medium it shows a strong apple-green fluorescence.

When paper is written upon with a tincture of stramonium seeds the characters are quite invisible by ordinary light, but appear self-luminous when exposed to highly refrangible rays, such as those given out by burning sulphur in oxygen.

Fluoraniline (obtained by acting upon aniline with chloride of mercury) is said to be one of the most powerful fluorescent bodies at present known.

The tinctures of several vegetable substances have a marked effect on light in this way: that of stramonium gives a pale but lively green fluorescence ; guaiacum, a beautiful violet colour ; turmeric, a greenish tint. Two artificial colouring matters, fluorescein and eosin, exhibit this property in a truly remarkable manner; but the effect is most brilliant in the green and blue portions of the spectrum (*i.e.*, the rainbow band of colour obtained by decomposing light by a prism).

A decoction of madder in a solution of alum shows a high degree of sensibility; it displays a copious yellow light. A weak solution of archil, which is red or purplish by transmitted light, gives a sombre green fluorescence.

But Stokes found that the electric light (from carbon-points) gave a spectrum, by the aid of a train of quartz prisms, which was more than six times the length of the ordinary spectrum. This light is in fact very rich in what are called the highly refrangible rays, and, moreover, its effect on fluorescent bodies was strikingly beautiful, and especially in the case of uranium glass.

As to the explanation of these phenomena, physicists are still somewhat at variance. But the experimental researches and elaborate reasoning of Prof. Stokes, who is at present the chief investigator in this department of inquiry, led him to the conclusion that beyond the visible spectrum there are certain rays which ordinarily are not sensible to the eye, but when these highly refrangible or much-drawn-apart rays are allowed to fall upon certain substances the molecules of these substances are made to vibrate in slower periods than the exciting rays, these molecular vibrations react on the ether, and consequently the rays on emerging (*i.e.*, on reflection) are hindered in their course and lowered in their refrangibility, that on being thus reflected they are brought within the range of vision : that is to say, if a ray of white light be made to traverse a prism, all those rays which go to make up the beam of white light, instead of passing through and being bent at the ordinary angle of simple reflection, are spread out like a fan ; the degree of divergence from the line of simple reflection represents the amount of refraction. The difference in degrees of refraction constitutes the difference between one colour and another, and one shade of colour from that which is akin to it.

That some rays are bent at a more acute angle than others is only another way of saying that they are more spread apart, dispersed, or more refrangible than others. For example, the yellow part of the spectrum consists of more refrangible rays than the red, and the violet than the green, and so on. But beyond each end of this rainbow band, or visible spectrum, there are the invisible rays. Those at the violet end are of very high refrangibility, and their motion as ether waves is also of very quick periods, while those beyond the red end are of low refrangibility and of comparatively slow undulatory periods. Now inasmuch as ordinary white light, or even diffused daylight, contains within itself those other highly refrangible rays, as well as those which are visible, it follows that when this light falls on a fluorescent body the substance absorbs or quenches those rays, and gives back rays which are of slower periods. This, then, is a kind of *selective absorption ;* consequently those highly refrangible rays which, as it were, survive, in the course of their egress are evidently retarded within the molecules of the sensitive substance, to which molecules they impart their motion ; these in their turn and by their movements excite or react upon the ether, and cause it to vibrate in waves of slower periods than the incident waves, which latter may be of any degree of refrangibility. It is very difficult to account for the phenomena otherwise. Moreover, it was found that if the incident light was polarised, the fluorescent rays from this source showed no traces of polarisation ; they are, in fact, *always* unpolarised. Fluorescent rays were also found to be totally incapable of producing the same effect in a like sensitive body. And here it must be remarked that Prof. Stokes's original paper on the subject, entitled " On the Change of Refrangibility of Light " (" Phil. Trans.," 1852), furnishes one of the finest examples of inductive reasoning to be found in the literature of Physical Science. However, it must be borne in mind that some investigators are not satisfied with Stokes's theory, and especially when he laid it down as a law that the light emerging from a fluorescent body under excitation is *always* of lower refrangibility than the incident rays.

Thus Prof. Angström, while agreeing with Stokes in some respects, considers that the molecular motions of the sensitive substance, on the contrary, by acting on the luminiferous ether, originate rays of higher refrangibility, the analogy to which might be, for example, as a higher octave in musical sounds. But this is considered by another authority as amounting to negative fluorescence.

M. W. Eisenlohr is of opinion that the phenomena are caused by the interference of the bluish violet and the ultra violet rays, which, it is argued, would give rise to rays of lesser refrangibility, in the same way as "combination tones" are of a lower pitch than the tones taken separately. But Prof. Bohn considers that according to this opinion fluorescence could never be seen in a spectrum that was pure, because "*there* rays of only *one* degree of refrangibility are to be found at the same place." .

Prof. Lommel combines the theories of Stokes and Eisenlohr, but is of opinion that he has discovered a series of substances (anthracen-blue among others) which do not obey Stokes's law,—*i.e.*, that a body in the act of fluorescence never emits rays which exceed the refrangibility of the incident or exciting rays.

The results of Lommel's researches, it was reported, were coroborated by Brunner (of Prague) and Lubasch (of Berlin). But Hagenbach, the author of some very accurate studies of the phenomena, could not arrive at Lommel's results. M. Lamansky, however, after repeating the experimental work of his contemporaries, announced last year that, notwithstanding the facts and arguments adduced by Eisenlohr and others, he had nevertheless fully confirmed the validity of Stokes's law. This he accomplished by devising a means of *measuring the difference* between the incident or exciting rays and those simultaneously emitted by the substance under excitation,—*i. e.*, the fluorescent rays. This he ingeniously effected in the following way :—Reflecting the solar rays by a heliostat, and concentrating them upon a narrow slit by the aid of an achromatic lens, he then placed two flint glass prisms behind the slit and another achromatic lens, the latter being put at twice its focal distance from the slit. By this arrangement he obtained a spectrum of such purity that the chief lines were rendered visible. The spectrum was then thrown on the side of a box provided with a movable slit capable of adjustment to the several parts of the spectrum, and also provided with a means of widening or narrowing as might be required. Through this slit certain rays of the spectrum, which had been previously made to pass through a flint glass prism, entered the box, within which was a vessel full of fluorescent liquid : then by the aid of a reflecting prism these perfectly homogeneous rays were directed upon the liquid. An achromatic lens was then placed between the surface of the fluid and the slit: a coloured image of the slit upon the liquid surface was thus produced. With another reflecting prism the fluorescent light was directed into the slit of the collimator of a Brunner's spectrometer. On looking through the instrument two coloured images appeared in the field of view; one was the fluorescent light, and the other that reflected directly at the surface of the fluid. Nothing more then remained to be done than simply to measure the minimum deviation of these two images. Each of the experimental measurements showed invariably "that the fluorescent substance has a lower refrangibility than the incident light."

Quite recently, however, Hagenbach has announced that after a prolonged investigation he has fully come to the conclusion that at present there exists no reason for rejecting the law of Stokes, and, moreover, that Lommel's classification of fluorescent bodies is not based upon any essential difference in their behaviour.

But the question as to the practical utility of these discoveries naturally arises. Science has a satisfactory reply. When each of these fluorescent bodies are examined with the spectroscope they exhibit well-marked and characteristic absorption-bands or dark spaces; consequently we have here a most subtle means of detecting their presence in a mixed solution or in a free state. Not only so, but Prof. Stokes has pointed out that if, on examining a solution in a pure spectrum, "we find the fluorescence taking a fresh start with a different colour, we may be almost certain that we have to do with a mixture of two different fluorescent substances, the presence of which is thus revealed without any chemical process."

Nearly two years ago, however, a most singular inquiry was brought to a satisfactory issue by the aid of fluorescence. It was long a question whether or not the head waters of the Danube found their way through subterranean passages into the Aach. The problem was thus solved: 10 kilos. about 22 lbs.) of fluorescein were poured into the first-named river, and three days afterwards a splendid green colouration with golden reflections (*i.e.*, its fluorescent effect) was quite distinct in the waters of the Aach. It was estimated that in this way 200 millions of litres of water were coloured by the presence of this substance. Doubtless Science will apply these beautiful phenomena to other and even greater ends than those which have yet been attained.

IV. GILBERT WHITE RECONSIDERED.

THE " Natural History of Selborne " is a quite exceptional phenomenon in English literature.   We know
no second instance of a scientific treatise which has
become so general a favourite with the reading public, and
of which new editions are still welcomed nearly a century
after its first appearance.   This lasting popularity is the
more to be noted since the work is singularly free from all
extraneous allurements.   It is not seasoned with any of
those sectarian and political condiments so dear to the
British mental palate, and often obtruded upon us where
utterly out of place.   It is not garnished with tales of travel
and adventure.   It ministers little to the national love for
sport,—*i.e.*, killing for the sake of killing.   It is simply
made up of careful observations of animal and vegetable
life, told in the quietest and plainest language, and without
any obtrusion of the author's personality.   He is evidently
anxious merely that facts should be ascertained and recorded, not that any *kudos* should accrue to him as their
discoverer.

Herein lies his merit : he is a careful, patient, thorough-
going observer.   He seeks to know what birds are to be
met with in the sphere of his observations, the parish of
Selborne.   He aims at identifying each species, at ascertaining its food, its song, its manner of flight, its breeding
season, and mode of nesting.   If not to be met with all the
year round, he notes the times of its coming and departure,
and asks whither it goes during the winter.   Indeed, though
not without doubts and misgivings whether our martins and
swallows do not remain during the cold season in a state of
torpidity, concealed in hollow trees, caves, and such like
shelters, he throws a most important light on the migrations
of the small birds.   By the co-operation of his brother John,
who resided for some time at Gibraltar, and who left at his
death a MS. Natural History of that district, still unpublished, he ascertained that not merely the swallow tribe, but
bee-birds, hoopoes, orioles, and many of the   soft-billed

---

* The Natural History and Antiquities of Selborne, in the County of South-
ampton.   By the Rev. GILBERT WHITE, M.A.   The Standard Edition, by E.
T. BENNETT.   Thoroughly revised, with additional notes by J. E. HARTING,
F.L.S., F.Z.S., &c.   London : Bickers and Son.

summer birds, are seen crossing over from Africa in spring and returning thither in autumn. He points out that the difficulties of migration are thus far less than had been assumed by mere book-naturalists, since " a bird may travel from England to the Equator without launching out and exposing itself to boundless seas, and that by crossing the water at Dover and again at Gibraltar." He learns from his brother that the swallows are careful to select the easiest point, " sweeping low just over the surface of the land and water direct their course to the opposite continent at the narrowest passage they can find. They usually slope across the bay to the south-west, and so pass over opposite to Tangier, which it seems is the narrowest space."

He notes the absence of the nightingale in Devon and Cornwall, which " cannot be attributed to the want of warmth," but " is rather a presumptive argument that these birds come over to us from the continent at the narrowest passage, and do not stroll so far westward."

He concerns himself much with the peculiarities in the propagation of the cuckoo, and, though he fails to solve the difficulties which are here encountered, he refutes the conjecture of M. Herissant that the bird was disqualified for incubation by the peculiar structure of the digestive organs.

He observes a fact which some writers of books, even in our own days, have scarcely mastered, viz., that birds are much influenced in their choice of food by colour, and consequently must be able to recognise colour. He writes— " Though white currants are a much sweeter fruit than red, yet they [birds] seldom touch the former till they have devoured every bunch of the latter,"—an assertion which everyone who has frequented a garden will be able to confirm.*

He notices, at least as far as the swifts and swallows are concerned, the non-increase of birds. His explanation of this circumstance—*i.e.*, that the parent birds oblige their young to seek new abodes—is obviously one that will not bear scrutiny.

He notices the first appearance in his district of the cockroach,—an insect which had, indeed, been prevalent in London for a long time, but had not yet established itself in the country. It may not be here out of place to suggest how this vermin may have acquired its vulgar name of " black beetle." It may at first doubtless have been

---

* We have often asked why, in view of the superior quality of the fruit and its comparative immunity from birds, is the white currant so little cultivated ?

confounded with *Tenebrio molitor* or with *Blaps mortisaga,* both of which are true beetles, both black, and which haunt bakehouses.

But whilst White's writings teem with facts which had escaped all previous naturalists, we find that, with trifling exceptions, all relate either to the locality or the habits of animals. With their structure, and even their outward appearance, he is much less concerned. He complains, with justice, of the vagueness of the descriptions of species given by the systematic writers of his time; but he does not attempt to amend their defect. He assumes each bird, &c., as a something known, and goes on to treat of its doings and its occurrence. Classification was as foreign to him as morphology,

We naturally ask what was his position with regard to the great questions of zoological philosophy, such as the origin, the persistence, and the distribution of species, their relations to each other, their dependence on climate, and diet ? This enquiry appears the more reasonable if we reflect that he lived in the same century which produced the elder Darwin, and that he was acquainted with the writings of Buffon and Linnæus. Yet there is no evidence that he gave these subjects any especial consideration. He writes, indeed, that " Linnæus ranges plants geographically; palms inhabit the tropics, grasses the temperate zones, and mosses and lichens the polar circles ; no doubt animals may be classed in the same manner with propriety. This is merely an assertion of the fact that different climates have their characteristic forms of animal and vegetable life. He speculates also on the origin of " those genera that are peculiar to America, how they came there, and whence ?" Without hazarding an opinion of his own, he refers to the arguments of those writers who " stock America from the western coast of Africa and the south of Europe, and then break down the isthmus that bridged over the Atlantic,"—an hypothesis which he characterises as a " violent piece of machinery." Had he had any inkling of the doctrine of Organic Evolution he could scarcely have failed to bring it forward in this connection.

He observes, however, the influence of food upon the colouration of birds, and puts the question whether the variation in the colours of our domestic animals may not in part be due to more abundant and more stimulating diet ?

He notices the helplessness of young swifts,—a fact which might have served for the better instruction of those writers who dwell so persistently on the precocity of gallinaceous and natatorial birds and of ruminant mammals, as if common to the whole animal kingdom, save man only. Nor do we find him quite so often and so prominently as some of his contemporaries, posing in an attitude of wonder over the results of "instinct." He even notices the mistakes which birds may commit under the influence of this supposed faculty. By recording these imperfections and errors he strikes, in fact, a deadly blow at the existence of instinct as commonly defined and explained by the "philosophers," a definition to which he also elsewhere commits himself. In so doing he reminds us of Priestley, who, after he had discovered oxygen, remained an upholder of the Phlogistian system. He admits that "there are instances in which instinct does vary, and conform to the circumstances of place and convenience," and thus renders its diagnosis, as a faculty distinct from reason, impossible. Yet elsewhere he remarks, in self-contradiction, "Thus is instinct, taken the least out of its way, an undistinguishing limited* faculty." He gives one instance of a pair of birds devising an unusual expedient to meet a special and exceptional case in the nurture of their brood. Yet, considering how long and how carefully he observed the habits of animals, it is strange that he does not recount more cases of animal sagacity, or its opposite. Incidents narrated by him would have been valuable data, as we might have felt assured of his accuracy. In his day, however, the custom of referring all animal action to instinct was so universal that even the observant naturalists rarely troubled themselves to enquire further.

White must be pronounced less prone to teleology than most of his contemporaries—a highly creditable peculiarity if we remember the temper of the age in which he lived and wrote. In his sixth letter to Robert Marsham, of Rippon Hall, near Norwich, first published in the edition before us, he declares, indeed, that "physico-theology is a noble study, worthy the attention of the wisest man." We find also occasional quotations from, or rather references to, Derham. But those rhetorical outbursts on the harmonies and contrivances of the organic world which have played so decidedly into the hands of materialists, and injured the cause they were intended to serve, are here mainly conspicuous by their absence.

White was essentially a writer of the modern or real type, in contradistinction to the mediæval or verbal. These terms

---

* Is not reason a "limited" faculty, in many cases also "undistinguishing"?

may require some little explanation. Suppose that an author in the fourteenth, fifteenth, or even seventeenth century, had undertaken to write a history of the sand-martin. His first step would have been to collect all the names, learned or "trivial," by which the bird is known. He would have examined their origin, fancying thereby to throw some light upon the nature of the sand-martin, and quite forgetting that the name of a thing can tell us no more about such thing than was known to the person who first assigned it the name, and often embodies merely his erroneous notions. After having taken up a goodly space with this etymological game he would have made a laborious search through all accessible books, those especially of classical antiquity, for mention of the sand-martin. Every such passage he would have carefully quoted and commented upon. But it would never have entered his mind to examine a sand-martin,—to see with his own eyes wherein it agreed with, and wherein it differed from other birds, and in particular other members of the swallow group,—to observe its habits, its manner of feeding and nesting, and to note the localities where it occurs. Hence his learned and ponderous book would be a history not of the thing " sand-martin," but merely of the word " sand-martin," and of all the fables, dreams, and superstitions which had gradually adhered to the word, much like the droppings of rooks on the twigs below their nests. This manner of writing natural history originated with Pliny, and survived down to the middle of the seventeenth century. Gilbert White's method was the very reverse. Caring little as to what names might have been given to a bird, or what might have been dreamt about it by authors of old, he looked to the facts, observed patiently, and described clearly what he had observed. He cannot, indeed, be pronounced the pioneer of this reformation. Before him Marcgrave and Bontius had written local faunæ; Lister, Swammerdam, Willughby, and many others had studied Nature, not in books, but in things. Yet White succeeded in winning for his science a degree of popularity which in England at least it had never before enjoyed, and it thus fell to his lot to become the founder of a school of naturalists which has done good service, and which even yet is by no means extinct. We refer here not merely to his more direct followers, such as Jesse, Knapp, Jenyns, and Rusticus of Godalming. These writers have avowedly chosen White as their model, and acted upon his suggestions. Taking up some local centre, they have made its fauna and flora their especial study. We are far from undervaluing these authors. Even Jesse, the least trustworthy of the group and the most inclined to put faith in mere hearsay evidence, scarcely deserves the severe sentence passed upon him by Waterton. But besides these imitators—if the reader prefers so to call them—White's indirect influence pervades English natural history to the present day. That tendency to study local faunæ which has so powerfully tended to constitute the science of Animal Geography, and has furnished such important data for the elaboration of the Doctrine of Descent, is thoroughly Whitean. We recognise it in our travelling naturalists, as well as in stationary observers; in T. V. Wollaston, in Belt, Bates, Wallace, Darwin, no less clearly than in White himself. It is somewhat remarkable that whilst England has been so fruitful in faunæ, in works descriptive of the habits of animals, and in contributions to the philosophy of zoology, she should be so poor in systematic works. We have no English Linnæus or Oken, Buffon or Cuvier. This is the more remarkable since in chemical science the English press has literally teemed with systematic treatises.

The question may here be raised in how far White can be held answerable for that morbid and narrow localism known as the " British " mania, which leads so many of our naturalists, commonly so called, to confine their researches within a boundary which is neither natural nor political, but an unhappy mixture of both. Thus they will take cognizance of a species found in Ireland, the Hebrides, or the Orkneys, but any form peculiar to the Channel Islands they ignore as "not British." We think he must be held guiltless. His object was to study the fauna of a certain district. The true " British maniac " aims more at collection than study, and, provided some one will guarantee that a specimen was captured within the four seas of Britain, he cares little for its exact locality.

It may not be uninstructive to contrast White with one who was his admirer,—to some extent his disciple,—and who cultivated with success the selfsame department of Natural History, becoming very eminent as an observer, whilst paying little heed to system or to the philosophy of zoology. Both these men rank among the happy few who are able to devote substantially their whole time to their favourite science; but their opportunities were otherwise exceedingly different. Gilbert White was no traveller; his days were spent and ended in his native village. An occasional journey to London and a few excursions in the southern counties seem to have formed the extent of his rambles: he

does not even appear to have visited the New Forest.
Charles Waterton, on the contrary, passed many years in
some of the most interesting parts of South America ; he
visited Canada, the United States, and the West Indies.
He was familiar with Spain and Italy, and indeed with the
entire of Western Europe. His own park, where 200 acres
of land fenced in by a lofty wall afforded a happy shelter to
all the birds of the North of England, might be considered
an excellent ornithological station. He lived, too, in days
when every branch of biology had made great advances as
compared with the eighteenth century. But the results
which he has bequeathed to the world do not surpass those
left behind him by Gilbert White in the degree that might
have been expected, either in quantity or in quality. It is
probable, indeed, that Waterton must have made multitudes
of observations, which from eccentricity he did not think fit
to publish. He evidently paid a far greater attention to the
structure of animals than did White, having, as he tells us,
dissected 5000 specimens. But where are the results ?
Though second to no man, living or dead, in observing accurately anything that had actually passed before his eyes, he
was inferior to White in critical acumen, and presented an
odd mixture of obstinate credulity and of scepticism equally
stubborn. Witness his persistent denial of the facts that
serpents sometimes act on the aggressive, and that wolves
hunt in packs; of cannibalism, of the rationality of the
lower animals, &c. ; and, on the other hand, his faith in
phlebotomy and in the liquefaction of the blood of St. Janu-
arius, and in "instinct." Where facts are wanting or
doubtful White suspends judgment, but Waterton jumps to
a conclusion. White, again, is gentle, unassuming, little
disposed to criticise others, and eschewing controversy.
Waterton, on the other hand, is irascible, pugnacious, never
happier than when exposing the errors of Audubon or of
the Quinarians. It must be admitted, however, that his
temptations to severe criticism were such as never fell in
the way of the kindly recluse of Selborne. What he might
have said or done had he been shown a picture of a serpent
with its poison-fangs pointing outwards, or had he been
called upon to read and understand the works of Swainson,
is another question.

Not the least pleasing feature in the works of Gilbert
White is the thoroughness with which he, in old-fashioned
phrase, "sticks to his text." He writes to give us his observations on Natural History, and his opinions on other
subjects he keeps to himself. Though living in the stormy

and exciting times of the American War of Independence and of the first French Revolution, only once does a political reflection escape him. In the ninth of his recently discovered letters to Robert Marsham, he writes—"You cannot abhor the dangerous doctrines of levellers and republicans more than I do. I was born and bred a gentleman, and hope I shall be allowed to die such." Perhaps he was aware that political agitation is more hostile to scientific research than is even war. Waterton, on the other hand, is for ever obtruding his Catholicism and his somewhat anti-national politics upon our notice. We cannot help regretting that if he must write on such subjects he did not embody his opinions in distinct treatises, addressed to different classes of readers. Upon the whole we must pronounce White the superior, inasmuch as he made the better use of more limited opportunities.

Both these naturalists, it must be noted, protested—each in his distinct fashion—against the reckless waste of animal and especially of bird life, which still goes on. White complains bitterly of the persecutions to which our rarer birds are subject. We feel certain that the golden oriole, the hoopoe, and even the bee-eater, once ranked among our regular summer visitants. Perhaps their disappearance is due as much to the butchery which they suffer in Southern Europe as to their inhospitable reception on reaching our shores.

The present edition of "Selborne" presents certain recommendations. The ten newly-discovered letters to R. Marsham form a valuable feature, and the editor—unlike some of his predecessors—does not use the author's text as a peg on which to hang irrelevant notes for his own glorification.

V. THE FOUR FORCES IN NATURE.
By George Whewell, F.I.C., F.C.S.

THE forces in Nature may be divided into four kinds, and we assume that all atoms are endowed with these four forces, either in an active or latent condition.

We have ventured to call these four forces "Viva." The word Viva is used as a limited expression for the term "life," and means, in the physical world, the motion which produces expansion and contraction, which causes molecules to assume the form of solid, liquid, gas, or ultra gaseous.

In the case of animals and plants it means the force which causes the ascension of the sap of plants, the movement of the substances of the body, such as the circulation of the blood, the secretions, &c., the contraction of the muscles, the production of sensation and thought.

The four forces are Atomic Viva, Organic Viva, Animal Viva, and Mensic Viva (mind).

Previous to the Great Creator calling the world into active existence three of these forces were in a state of combination, and atomic viva alone was active.

The environment of the atoms in the process of evolution determined the development of organic viva ; then animal viva ; and, lastly, mensic viva (mind).

Life cannot exist without heat ; deficiency of vital heat is weakness or disease ; absence of vital heat is death to organic, animal, and mensic viva. Heat is a mode of motion, and caused by the vibration of atoms ; all atoms vibrate more or less. The ultimate composition of the universe is therefore *Vibration and Heat.*

That the forces of the physical world—such as light, heat, electricity, chemical action, &c.—are modes of motion are quite agreed to by philosophers. This mode of motion we have given the name Atomic Viva.

Atomic viva is therefore a mode of motion. Taking the embryo as a type of our second force, we find the first indication of animation is a minute pulsating point. It is the development of our second force, organic viva ; the cause of existence—the distinction between atomic and organic viva. This motion causes the formation of the young heart, which in its turn causes the propulsion of the infant stream.

With the motion of the heart begins life ; with its stoppage begins death. It is by this motion that arteries, veins, capillary vessels are formed and receive the vital fluid. With the circulation begins secretion, then digestion, : hen nutrition. These functions are closely related and dependent on each other ; one cannot stop without the whole circle stopping. It causes the formation of brain nerves and muscles. It is anterior to animal viva, which begins at birth. Organic viva begins with this first motion, is complete at birth and continues without fatigue, and is never for a moment suspended until the stoppage of the motion, which we call death.

Heat is produced by the first and subsequent pulsations. Organic viva is therefore a mode of motion.

At birth is developed our third force, animal viva. This force does not come into existence until the new being becomes detached from its mother. It then for the first time comes in contact with external objects; its organs being complete, all the functions of its economy can be carried on, and it enjoys independent existence. The power of the brain to control the actions of the voluntary muscles is weak at first, and gets strength slowly; it comes to perfection at adult age.

We have seen that organic viva exists before animal viva, and it may exist after animal viva has ceased. The two forces are distinct, and come into action at different periods. When sensation and the power of voluntary motion are abolished, animal viva is extinguished, yet the whole circle of the organic functions—such as circulation, respiration, secretion, excretion—may continue to perform their rightful duties. As in the case of apoplexy the greatest intellect and the strongest muscles may be rendered powerless, never to come into action again, it may totally extinguish animal viva, yet the involuntary contractions of the muscles and the action of the heart may not be impaired, but rather strengthened, and organic viva may go on for days.

Catalepsy may abolish sensation and volition, and cause the muscles to contract and remain fixed, yet the heart continues to beat, the pulse to throb, and the lungs to respire, an all the organic functions of the body continue in full force.

From the above it will be seen that animal viva may perish, both the sensitive and motive portion of it, and yet organic viva continue in operation and full force.

Animals possess three forces,—atomic viva, organic viva, and animal viva,—the latter being controlled by the will. Will is common both to man and animals, and will and mind are not one and the same thing. Animal viva produces heat, and heat is a mode of motion. Animal viva is therefore a mode of motion.

Man possesses the four forces—atomic viva, organic viva, animal viva (which includes will), and mensic viva (mind). This latter motion, although blended with and essential to a well-developed man, is independent and distinct from organic and animal viva. Organic viva is born at the moment of the development of our second force; animal and mensic viva not until a later period. These latter forces are feeble at birth, and acquire strength by slow degrees, and attain their ultimate perfection (in the case of animal viva at adult age) later in life.

Taking the lowest form of plant-life we find the lowest developed form of organic viva, the power of assimilation and growth, and ascending the scale we come to those plants possessing in a remarkable degree sensitiveness and the power of digestion and assimilation. Plants and animals possessing organic viva only have not the power of voluntary motion, but either remain fixed or—if free to move—have no control of their movements.

In the lowest forms of animal life we find the lowest developed form of animal viva, viz., sensitiveness, given by the possession of nerves and muscles. As we ascend the scale we find this sensitiveness concentrated, and forming what is called the brain. It is this sensitiveness which educates the brain, and causes some of the higher animals —such as the horse, dog, elephant, parrot, &c.—to have the power of remembering, and in some cases almost to think and to reason.

In the case of those animals which have memory we have the beginning of the development of mensic viva; and in those animals with both memory and slight reasoning we have a still further development of mensic viva, until we come to the lowest form of man, when we have the lowest form of mensic viva (mind), which is the power of reasoning, and is found in the greatest perfection in Christ, and in our Shakspeares, Miltons, Newtons, Davys, and Faradays. This power gradually diminishes as we descend the scale, until we come to those semi-animals who are the lowest form of human beings.

Man can have organic and animal viva well developed, and mensic viva strong, as in the case of the philosopher; weak, as in the case of the imbecile; or suspended, as in the case of the lunatic. The mind may for ever remain blank, as in the case of one born whose brain is not developed, or only capable of slight cultivation although fully developed, as in the case of one born deaf and dumb, and blind.

If we go a step further, and conceive a human being born deaf and dumb, blind, and without sensitiveness, such an individual might be fully developed, but yet would remain a perfect blank so far as possessing any knowledge of this life. On the other hand, man can have mensic viva strongly developed, and animal viva partially or wholly suspended, as in some cases of paralysis, when he has lost entire control of his limbs or some important part of his body, and still retains all his reasoning faculties. *Cæteris paribus*, the higher the mental power and the lower the organic and animal viva, and also the higher the mental power and the lower the power possessed by man of procreating his species.

Great men never have a numerous progeny. This is accounted for by the fact that the greater the mental power, and the more it is used, the more blood is required to renew the tissues of the brain ; consequently the generative apparatus suffers. On the other hand, low brain power and physical exertion determines the blood to the lower parts of the body, and increases the sexual desires. This is the reason why ignorant and low forms of human beings have a greater progeny, and bring them into the world with less pain and trouble than the more cultivated.

When, in the highest form of animal, mensic viva was sufficiently well developed, and the animal became responsible for his actions to the Great Creator (it was thus that man discovered his nakedness), then he acquired the power of transmitting the fourth force (mensic viva) to his offsprings.

In the same way the animal or plant has the power of transmitting, to a full-developed offspring, organic and animal viva without mensic, or organic without animal and mensic viva.

The missing link between the highest animal and the lowest man need not be looked for in his physical appearance, such as the wearing away of the tail of the monkey, but in the transition from animal to mensic viva. Will is common both to man and the higher forms of animals, and is possessed by those animals whose sensitiveness is concentrated, and forms what is called the brain. Will serves as a connecting-link between man and animals.

When man's sexual passions are excited, heat is developed at the back of the head (cerebellum) ; when man's virtuous passions are excited, heat is developed at the crown of the head ; when man's reasoning and thinking faculties are excited, heat is developed at the forehead.

When we think heat is produced we use up a portion of the substance of the brain; and there can be no heat and no waste of the brain without work, and there can be no work without motion.

Mensic viva is therefore a mode of motion.

ERRATA.

Page 381, line 10, *for* " lightning " *read* " lighting."
Page 383, last line, *for* " would require " *read* " would *not* require."
Page 390, line 5, *for* " proportions " *read* " propulsion."
Page 556, line 5 from bottom, *for* " motion " *read* " motive."

ANALYSES OF BOOKS.

Nature's Hygiene. A Series of Essays on Popular Scientific Subjects, with Special Reference to the Chemistry and Hygiene of the Eucalyptus and the Pine. By C. T. KINGZETT, F.C.S. London: Baillière, Tindall, and Co.

WE have here an interesting treatise on a subject which is by no means exhausted. The author contends that these are the great natural agents for the destruction of those mysterious entities—be they organised or not—which are the cause of epidemics. The first fact to be taken into account is the remarkable immunity from fevers and other pestilential maladies enjoyed[1] by the great island-continent of Australia. Such diseases, if introduced into Melbourne or Sydney, die out without spreading. This circumstance is the more remarkable if we reflect that no special care has been bestowed upon sanitary arrangements in the colonies, and that, judging from the latitude and the temperature, Australia might seem a fair field for cholera, yellow fever, or plague. Its soil and atmosphere a certainly dry, but these features are no less to be recognised in rsia, Asiatic Turkey, certain regions of India, North Africa, Mexi o, and La Plata, where pestilence is far from uncommon.

Of late years it has become pr valent to ascribe the remarkable sanitary advantages of Austr ia to its woods of different species of Eucalyptus—a vegetable enus not naturally occurring in any other part of the world. or can we pronounce this opinion baseless. It seems perfect well established that certain spots in Italy, Algeria, &c, previousl notorious for their unhealthiness, have been rendered perfectly fe by the introduction of this tree. The generally received ex lanation of the sanitary activity of the various species of Euc lyptus is that they exhale large quantities of a volatile oil. It ha further been discovered by the anthor that this oil, and indee all the oils upon which the aroma of herbs and flowers depen , generate peroxide of hydrogen by their action upon air and wate vapour. Now as peroxide of hydrogen acts very energetically on decomposing animal and vegetable matter, we can see why a Eucalyptus forest should have a beneficial effect upon the ealth of the surrounding district.

There are still, however, a few questions to be answered be we can lay claim to a full knowledge of malaria and of disi fection. We learn that in Queensland, in regions where the odour of the Eucalyptus is very distinct, ague is not unknown, though it is never either as prevalent or as intense as the "fever and chills" of Louisiana or Georgia. We further find that in South Africa—and not merely in the old Cape Colony, but in Natal—the so-called zymotic diseases are rarer than in most countries of a similar climate, especially those in the northern hemisphere. Yet here the Eucalyptus is naturally absent; nor are trees of the pine kind, if at all indigenous, found to such an extent as to affect the atmosphere. This leads us to another consideration. Mr. Kingzett classes the pine along with the Eucalyptus as a sanitary agent. He even says—"It is probable that the hygienic influence of the pine is much greater than that of the Eucalyptus, since it has a much wider and more extensive distribution in Nature." We ask, therefore, whether there is any country which enjoys in virtue of its pine-forests an immunity from pestilence similar to that which Australia derives, or is supposed to derive, from the Eucalyptus? Russia, Norway, British North America, and the more northern States of the American Union, are all, or were all, rich in pine-woods. From their latitude and climate they are not likely to suffer from yellow fever, nor, over the greater part of their extent, from ague; but they have been visited by cholera, typhus, and small-pox. Russia and Norway, like the rest of Europe, did not escape the "black death." Germany, Austria, Switzerland, and Italy, have all their pine-forests; but they have from time to time been scourged with plague, and in more recent times with cholera, dysentery, diphtheria,* and, at least as far as Italy is concerned, with malarial fever of the deadliest type. We naturally wish to learn whether the spread and the virulence of this disease has been found to be at all influenced by the local presence or absence of pine-forests. Has the sanitation of any unhealthy district been attempted by planting pine-trees? If so, with what success? and if not, why not?

We think it will be found that a country partially wooded, irrespective of the nature of the trees, will be found more healthy than one completely cleared, but otherwise similar, and that denudation is generally attended with an increase of sickness. The back-woods of South America are healthier than the towns and plantations. The Pontine Marshes are treeless. Woods regulate the temperature of any locality, and above all render its changes less abrupt, and by so doing may modify the character of the fermentations and putrefactions which organic refuse undergoes.

It must not be supposed, from these remarks, that we wish to dissuade those interested from planting the Eucalyptus in insalubrious districts. We wish these experiments to be multiplied, and we shall be very glad if any species of this tree can be found

---

* We learn from the "Victorian Review" that the town of Hamilton, near Bourne, is now suffering from a severe outbreak of diphtheria.

adapted to an equatorial in contradistinction to a merely tropical climate, and consequently suitable for introduction into Sierra Leone, the Gold Coast, and similar regions. Are there any species of Eucalyptus indigenous in New Guinea? If so, they will probably meet the emergency.

As regards the origin of the "zymotic" diseases, Mr. Kingzett does not accept the "germ theory," against which he, following Dr. Drysdale, advances arguments which cannot be ignored.

Believers in ozone will derive little edification from this work. They will learn—what is utterly beyond dispute—that most of the observations made as to the presence or absence of this agent are inconclusive, and that the commonly received notion that zymotic disease prevails inversely as its proportion in the atmosphere is simply baseless.

Mr. Kingzett's book may be read with great advantage, and will do good service in enlightening the public mind on many points connected with health and its preservation. An index would have been useful.

The Truth about Vaccination. An Examination and Refutation of the Assertions of the Anti-Vaccinators. By ERNEST HART. London : Smith, Elder, and Co.

THE Anti-vaccination movement is being carried on with a zeal and a pertinacity which, even in these days of professional agitation, must be pronounced surprising. We cannot help wondering where the promoters of the outcry—none of whom are known to us even by name—find the two essential requisites, dupes and funds. That money is forthcoming is evident, from the quantity of documents, pamphlets, and advertisements which the anti-vaccinators issue in support of their views, and which, if feeble in argument, are remarkably strong in assertion.

On Sunday, July 25th, emissaries were posted near the doors of a number of places of worship in the north of London, and tracts denouncing vaccination were handed to the congregations on leaving. We are bound to say that, to our knowledge at least, this proceeding did not in any way receive the countenance of the clergy and ministers. Surely, when such means are resorted to, it is incumbent upon the intelligent part of the community not to rely upon the inherent absurdity of the movement, nor to hope that if left alone it will die a natural death. Disease-germs, moral and social as well as physical, show a wonderful vitality. So long as an ignorant and excitable community coexists with eloquent sophists who can gain fame, power, and perhaps more tangible rewards, by organising ignorance and appealing to public " conscience," so long watchfulness and unceasing efforts will be required " *ne quid detrimenti respublica capiat.*" Mr. Hart has therefore done well in issuing the pamphlet before us, which we hope may be as widely circulated as the errors which it exposes and refutes.

Amongst the items of the " anti-vaccination creed " which the author endeavours to extract from the bundles of tracts sent him, one of the most prominent is that compulsory vaccination is an infraction of the " liberty of the subject." This objection, if it can be maintained, strikes at the root of sanitary legislation altogether. If a man has the right of neglecting a known precaution to diffuse small-pox among his neighbours, he has the same right by neglecting other precautions to make his person, his family, and his house a focus of diphtheria, scarlatina, typhoid, or cholera : he is justified in travelling in a public conveyance when suffering from infectious disease ; in allowing sewage, &c., to run into wells and streams used for the supply of drinking-water ; and, indeed, in setting fire to his house, regardless of the danger to his neighbours. All this, and much more, follows by logical sequence if we once deny the right of the community to coerce a refractory anti-vaccinationist. We cannot recognise the vested rights of disease. If vaccination is really a safeguard against small-pox, then the individual who rejects this protection is much more to be censured than the suicide. If the latter fails in effecting his purpose we punish him far more severely than the anti-vaccinationist, though his offence cannot possibly injure the public. No more valid is the " conscience " plea. It is a most remarkable fact that the greatest wrong-doers in the world—witch-burners, inquisitors, Thugs, Torquemada, Cotton Mather, Robespierre, and the like—have been conscientious men. But no well-ordered community would on that account give scope to their propensities. It seems to us that the magistrates, in dealing with " Peculiar People " who have allowed their children to die rather than call in medical aid, have been somewhat mistaken in entering upon this " conscience " question at all. If the neglect is once proved, why should the court inquire as to its motive ?

An objection is sometimes raised against vaccination to which Mr. Hart does not refer. We have heard it argued that the small-pox cleared away the sickly and debilitated part of the population, who, if suffered to survive, would have died of consumption, scrofula, general debility, and perhaps have become previously the parents of unhealthy children. But is it certain that the small-pox showed so much discrimination in the selection of its victims ? We suspect not : except we are strangely misinformed, healthy children, the offspring of sound and vigorous parents, were not unfrequently swept away.

It is sometimes asked, how it comes that small-pox epidemics are still possible, and that the disease has not continued to decline as vaccination became more general? Why, in spite of compulsory vaccination, did deaths from small-pox rise in the year 1872 to a number exceeding anything recorded in the previous seventeen years? Mr. Hart contends that the proportion of deaths from small-pox to the total pupulation from 1838 to 1853, being prior to compulsory vaccination, was 420 per million; whilst during the succeeding twenty-six years of compulsory vaccination, 1854 to 1879, the death-rate in question had fallen to 208·5 per million. He argues that a great many persons, though nominally vaccinated, have not in reality been successfully operated upon, and are consequently unprotected. The proportion of deaths from small-pox in the earlier part of the century is not given, and indeed cannot be accurately ascertained, as there was in those days no accurate and general registration of deaths and their causes. Still it might be estimated as closely as the mortality from 1728 to 1757, and 1771 to 1780, which is given as 18,000 per million. This omission in Mr. Hart's pamphlet is the more to be regretted as an impression prevails that the disease nearly died out from 1810 to 1830, and has since 1838 undergone a recrudescence.

The contention that vaccination may be the means of introducing other diseases, and especially syphilis, cannot be set aside as out of the question. J. Simon, F.R.S., in a paper on "Contagion" ("British Medical Journal," December 13th, 1879, p. 923) declares that "even in states of chronic dyscrasy, and even at times when the dyscrasy may be giving no outward sign, the infected body may be variously infective; . . . . the vaccine lymph of the syphilitic infant may possibly contain the syphilitic contagium in full vigour, even at moments when the patient who thus shows himself infective has not on his own person any outward activity of syphilis." Of course no medical practitioner in his senses would knowingly use lymph from a syphilitic infant for vaccination. But does he always know the health-record of the parents? It is therefore a wise precaution to use lymph which has not circulated through the human system. Had this step been taken years ago the only rational argument of the anti-vaccination party would have been rendered non-existent. It must, however, be noted that no case of the transmission of syphilis along with vaccine lymph has been demonstrated. Had such cases existed the agitators would not have been silent.

The Brain as an Organ of Mind. By H. CHARLTON BASTIAN, M.D., F.R.S. London: C. Kegan Paul and Co.

DR. BASTIAN is chiefly known to the general public as an advocate of the somewhat heterodox theory of spontaneous generation, or, as it is now somewhat more pompously called, abiogenesis. In opposition to Pasteur, Tyndall, and others, he maintained that under favourable circumstances life may take its origin, not from any ova, germs, or spores, but from lifeless matter. In how far he still upholds this view, now it has been shown that certain microbia are capable of existing at far higher temperatures than was once supposed, we are not aware. The present volume, however, is of a less speculative character. The author lays before his readers facts on which there is now little difference of opinion among men of science.

As may readily be supposed from the title, the work is very comprehensive in its scope. Starting with the uses and origin of a nervous system, Dr. Bastian gives a general account of nerve fibre-cells and ganglia, and of the use and nature of the organs of sensation. He describes in succession the nervous systems of the Mollusca, Vermes, and Articulata. He next describes the brain of fishes, Amphibia, reptiles, and birds. At this point of the animal scale we are introduced to the questions of the scope of mind, reflex action and unconscious cognition, sensation, ideation and perception, of consciousness in the lower animal, of instinct, nascent reason, emotion, imagination, and will. Returning to things visible and tangible, the author describes the brain of quadrupeds and some other mammals, and that of the Quadrumana. Surveying the mental capacities and powers of the higher brutes, he very justly gives the first rank to the anthropoid apes. Too many writers, we may here remark, confound docility with intelligence, and in enlarging upon the reasoning powers of the dog forget that he exhibits the hereditary influence of an education which has extended over many generations, whilst the ape, caught wild, and rarely surviving above a couple of years in captivity, is observed at a disadvantage.

Next we are introduced to the most interesting and important part of the subject—the human brain. Here the reader will meet with many facts perfectly well established, but scarcely in harmony with popular notions. He will learn that the brain of man belongs to the same general class as that of the apes and monkeys, and that, though the difference in weight between the brain of the large anthropoids—such as the orang and gorilla— is great, the range met with among individual men is still greater. Concerning the lack of symmetry in the corresponding convolutions of the two cerebral hemispheres, it is remarked that the same want of symmetry is met with in the elephant and the whale. Careful examination of series of skulls, mediæval and modern, show that the cranial capacity—the size of the space filled by the brain—has distinctly increased during the last seven centuries. Another remarkable fact, pointed out by Vogt and Le Bon, is, the difference between the sexes as regards cranial capacity increases with the development of the race, the male

European surpassing the female to a greater extent than is observed among savages. The difference between the average capacity of the skulls of modern Parisians, male and female, is almost double that which prevails between the skulls of male and female inhabitants of ancient Egypt. Further, as civilisation progresses, the range of variation among male individuals increases with the position of the race in the scale of civilisation. Thus among negroes large and small male skulls may vary by 204 cubic centimetres; among ancient Egyptians by 353; among twelfth century Parisians by 472; and among modern Parisians by 593. Hence Le Bon holds that the real test of superiority of one race over another in regard to cranial capacity is not to be ascertained by averages, which are often deceptive, but by observing how many individuals per cent in different races possess skulls of given volumes.

An important conclusion here at once suggests itself, viz., that the progress of the race is towards greater diversity, and that the revolutionary doctrine of equality has no natural foundation.

Some interesting conclusions are also reached concerning the average weight of the brain in different European nations. The mean weight of the English brain is given by Boyd at 47·8 ozs., and by Peacock at 49; that of the French, 47·9; of the Germans at 48·3; and of the Scotch at 50. On the other hand, the weight of the negro brain is given by three authorities as 44·2, 44·3, and 44·5 ozs. respectively. The brain weight of the male negro is therefore the same as that of the European female, and falls short of the male European by 5 ozs., or about 11 per cent. The Chinese brain is remarkably heavy. The eleven adult males whose brains were weighed belonged to the lowest class, and were all chance victims of the cyclone of 1874. Yet they showed an average weight of 50·45 ozs., exceeding therefore that of the English, Scotch, French, or Germans! The great size of the brain in the Lapps and Esquimaux is mentioned, but no weights or measurements are given.

Le Bon carried out a series of measurements of the heads of Parisians belonging to different classes of society. He found the largest size among the scientific and literary class; next followed the commercial world; then nobles of ancient families; and lastly domestic servants. We believe that in England also the heads of footmen, grooms, &c., have been found smaller than those of any other class.

A small size and weight of brain have been long ago laid down as ranking among the characteristic features, and indeed among the causes, of idiocy. Broca places the limit compatible with ordinary human intelligence at 37 ozs. for the male and 32 ozs. for the female. But in a number of cases given by Dr. Thurnam and others the brain-weight ranges from 35·76 down to 8·5 ozs., the scale among the anthropoid apes being from 16 to 12 ozs. The converse statement, that a brain of unusual size necessarily

accompanies and indicates intellectual superiority, is quite unfounded. Brains far above the average may be observed among
lunatics. At the Wilts County Asylum, in 10 per cent of the
males and 7 per cent of the females who died there, the brain-
weight exceeded the upper limit of the medium size, being
52¼ ozs. and 47¼ ozs. respectively, while in from 3 to 4 per cent
the weights were respectively 55 and 50 ozs. In an asylum in
one of the northern counties Dr. Clapham found, among seven
hnndred male brains, forty-three the weight of which exceeded
55 ozs, and reached in four of them 60 to 61 ozs. The heaviest
brain weighed by Dr. Thurnam (62 ozs.) was that of a butcher
who was just able to read, and who died of epilepsy combined
with mania. The heaviest brain weighed by Dr. Bucknill—
64·5 ozs., equal to that of the celebrated Cuvier—was that of a
male epileptic. The brain of a female monomaniac weighed by
Dr. Skae was 61·5 ozs., a monstrous weight for a woman.

Brains of unusual weight may also be found among persons
not indeed lunatics, but quite ordinary, common-place characters,
who have never displayed any high intelligence. The heaviest
human brain on record (67 ozs.) was that of a Sussex bricklayer
who could neither read nor write, and who died of pyæmia in
University College Hospital, in 1849.

Nor must it be forgotten that though certain eminent men
have had very large brains, others, perhaps not less eminent,
have but little exceeded the mean standard, or even fallen below
it. Hence, as the author remarks, there is no necessary relation
between the mere brain-weight of individuals and their degree
of intelligence.

A most interesting chapter is that headed "Phrenology, Old
and New." But into this, as into many other portions of Dr.
Bastian's work, space will not allow us to enter. For all who
wish to acquire a general knowledge of the structure and functions of the brain and nervous system, human and animal, without dissecting, or consulting the original documents scattered
through the scientific journals and the transactions of learned
societies, the book before us will prove a most valuable guide.

We cannot help pointing out that in one passage (p. 211) the
"Journal of Science" is mis-quoted as the "Quarterly Review
of Science."

---

*Abstract of a Discourse on Ornament, delivered at the Royal
    Institution, June 4th, 1880. By H. H. STATHAM.*

AN interesting pamphlet, but dealing with a subject as regards
which we must admit ourselves outsiders. The concluding
remarks have a special claim upon our attention :—" One aspect

of the subject may be touched upon in conclusion which seems
to connect it with the great modern all-pervading idea of Evolution.   For though we cannot historically trace back all the forms
of ornament to their origin, we can see enough to leave no doubt
that if we had all the connecting-links before us we should find
that many of the most admirable, widely used, and characteristic
forms of ornament originated not so much in any sense of beauty
as in mere superstition and grossness ; and that ornaments are
habitually used in our churches and public buildings and habita-
tions the actual though remote origin of which, were it hinted at,
would very much astonish those who execute and those who
admire them ; and it may, perhaps, be accepted as one more
illustration of the upward tendency of human development that
even the very knowledge of this uncomely side of the subject has
fallen away from all except those who have had special reason to
study its history, and that from these clods of earthiness and
superstition there has sprung this bright and innocent flower of
ornament-"
    A lesson worth pondering over !

***

*The Life of Thomas Wills, F.C.S., Demonstrator of Chemistry,
    Royal Naval College, Greenwich.* By his Mother, MARY
    WILLS PHILLIPS, and her Friend, J. LUKE. London : James
    Nisbet and Co.

BIOGRAPHIES have latterly become exceedingly plentiful, and we
not unfrequently open a goodly volume recording the being and
the doings of some person of whose very existence we were in
utter ignorance, and should have been contented so to remain.
Such is not the case with the memoir before us.   Though called
from our midst at the early age of twenty-eight, he had already
earned for himself an honourable name in Science, and had he
survived he would doubtless have bequeathed to the world a
treasure of valuable results.   His first paper communicated to
the Chemical Society was on the " Solidification of Nitrous
Oxide," February 20th, 1873.   On June 13th, of the same year,
he laid before the Society an account of an " Ozone Generator "
which he had devised.   In September, the same year, he entered
upon his duties at the Royal Naval College.   His lectures delivered before various societies and associations were numerous,
and his depth of thought and clearness of exposition were fully
recognised by his hearers.   His last paper read before the
Chemical Society, December 19th, 1878, treated on the " Production of the Oxides of Nitrogen by the Electric Arc "—a
subject which he was still investigating up to the time of his
death.

Thomas Wills, however, was by no means exclusively a chemist or a physicist. He was passionately fond of music, and even took pleasure in listening to speeches in the House of Commons. He was as active in connection with Sunday-schools and Young Men's Christian Associations as with the Society of Arts, the Royal Institution, or the British Association.

We think that had he survived his influence would have done much to allay that feeling of uneasiness, if not of actual hostility, with which the advances of Science are viewed by the religious world. He felt deeply that men of Science are doing God's work in every investigation and every discovery. We honour this spirit, and regret that it is not more general and better recognised where it does exist.

For the student the brief career of Thomas Wills is rich in useful lessons. May they be appreciated as they deserve!

Journal of the Society of Telegraph Engineers. Edited by W. E. AYRTON. London: E. and F. N. Spon.

THIS issue contains a goodly assortment of interesting communications. Among these particular attention is due to a paper entitled "A Decade in the History of English Telegraphy," by Mr. E. Graves; and one on "The Behaviour and Decay of Insulating Compounds used for Dielectric Purposes, by the President of the Society, Mr. W. H. Preece. This memoir includes a history of gutta-percha, with a notice of the causes which lead to its destruction. A minute insect, *Templetonia crystallina,* gnaws into it, and actually seems capable of thriving upon a substance which to us seems the very type of indigestibility.

In some localities the mycelium of a fungus has been supposed to be an active agent in the decay of the gutta-percha coating of underground telegraph wires, but on insufficient evidence. It seems that oxidation is the main cause of decay, and that it takes place wherever gutta-percha is exposed to air, or alternately to air and moisture. The outer layer of the substance becomes dry and brittle, and changes into a resin, soluble in alcohol. This fact alone explains the reason why gutta-percha has fallen into disuse as a material for jugs, funnels, measures for liquids, &c.

Histoire des Coléopteres de France. Par le Dr. SÉRIZIAT. Précédée d'une Introduction a l'étude de l'entomologie, par M. CH. NAUDIN. Ouvrage adopté par le Conseil de l'Université. Paris : Firmin-Didot et Cie.

To understand why, even within the limits of one and the same great zoological region, certain species stop short at boundaries not apparently marked with any great distinctness by a change of climate and productions, we require, in the first place, local faunas drawn out with precision. We seek to be enabled to trace upon the map the exact range, if not of every species, still of the representative forms of each group. Every step taken in this direction reveals to us relations previously unsuspected. We all know that within an epoch which geologically speaking is but as yesterday, England and France formed one country, connected together by land, where now roll the waters of the Channel. We know that in climate the southern counties of England differ little from the northern and north-western departments of its continental neighbour. Yet when we compare the insects of the two countries—and we select them for comparison as being less modified by human intervention than the mammals, birds, or reptiles—we perceive at once a most remarkable difference. Species entirely wanting in England make their appearance across the Channel ; others, rare in this country, become at once common; and yet others, present in both countries, are found to select different stations. It is especially to be noted that certain forms, which with us are almost confined to the sea-coasts, are found on the Continent in inland situations. The question then arises whether the species wanting in Britain had found their utmost northern limit before its separation from France, or whether they at one time occupied England and have since become extirpated ?

The work before us will prove a very useful document for the student who enters upon such enquiries. Though it does not profess to give an exhaustive catalogue of the Coleopterous fauna of France, similar to the " Manual of British Beetles " by Stephens, it furnishes an account of the characteristic forms of each group. Entomologists collecting in France will find it an excellent guide what to seek and where to seek it. The descriptions of species are generally more complete than those of Stephens, and the illustrations, which are numerous, will prove of great assistance to the beginner in the recognition of species. To some of them objection may be taken. Thus *Polyphylla fullo* is correctly described in the text as possessing very large antennæ, the club of which in the male is seven-jointed ; yet it is figured with smaller and less distinctly clavate antennæ than *Melolontha vulgaris. Calosoma sycophanta*, very rare and littoral in England, is described as nowhere common in France, though more plentiful in the south than in the north, and as abundant in Algeria. This fine species has therefore a wide range, since in some parts of Eastern Europe, though local, it may be gathered up in hundreds; and we fear that its splendid colouration may some day attract the notice of those who purvey decorative insects to milliners, and thus ensure its extirpation. It would almost seem as if, in the more eastern parts of Europe, the "Mediterranean" and "European" subregions of the Palæarctic region were less distinctly severed than in the more western parts. Is this because extreme climates are more favourable to insect life, or because in Eastern Europe the Mediterranean basin is not fenced off from the central plain by such unbroken barriers as the Alps and the Pyrenees?

It is painful to learn that the beautiful species *Carabus rutilans*, once common in the Eastern Pyrenees, has now become very scarce, and that *C. hispanus* is undergoing the same fate in the Cevennes.

The Buprestidæ, which in England are exceedingly rare, not a few species being doubtful, are in France sufficiently numerous to rank among injurious insects. Their larvæ are described as particularly numerous to pear trees. We find no mention of *Chrysobothris chrysostigma* and of *Ancylochira octoguttata*, both of which are found in Central Europe, and we should think might naturally be expected to occur in France.

We may strongly recommend this little work to all entomologists who wish to obtain a general oversight of the French Coleopterous fauna. We do not know any other work which, under this head, supplies so much information in so little compass.

We regret, however, to point out that, in the introductory chapter on the study of Entomology, M. Naudin goes somewhat out of his way to pronounce man entitled to rank in a separate kingdom, distinct from the animal world, and known as the kingdom of reason! We had hoped that the day for such consecrations of popular prejudice had passed away.

Revue Internationale des Sciences Biologiques. July 15, 1880.

WE give from this journal an abstract of a paper by Dr. Hovelacque, on the "Characteristics of the Inferior Human Races.' On the length of the cranial vault the writer lays no emphasis, remarking that the anthropoid forms of Africa (gorilla, chimpanzee) are dolicocephalous, whilst those of the Eastern isles (mias, gibbon) are brachycephalous. The cranial capacity he considers a valuable characteristic, the cubic contents in the white race sometimes exceeding 1600 cubic centimetres, whilst in more than one of the dark races it does not exceed 1200. It may here be remarked that in the Esquimaux the cranial capacity does not fall short of the average European standard.

A peculiarity of the lower races is the less degree of complication of the cranial sutures, which approximates them to the anthropoid apes. In them, as in the apes in question, the obliteration of the sutures proceeds from the front to the back,—that is to say, commencing with the region corresponding to the nobler faculties of the brain. It is the same in the apes, but not in the white man, in whom, further, the ossification of the sutures is notably less precocious. Among the dark races the frontal part of the cranium occupies less room, both in width and length, than it does in the higher races, which is again a Simian feature.

Another characteristic of the inferior races is the considerable development of the face in proportion to the total head. The lower races approach the apes by their wider sphenoidal angle, which measures 135° in the European, 145° in the negro, 150° in the chimpanzee, and 170° in the orang. The inferior races have also a wider nasal aperture, and the nasal bones become united at an earlier epoch—also a Simian characteristic. In the white races the parietal and the sphenoid are articulated together, whilst the frontal and the temporal are joined by a more or less horizontal suture, measuring 1 to 2 centimetres. In the apes this arrangement is reversed, the frontal and the temporal being articulated, whilst the parietal and the sphenoid are connected by a vertical suture—a structure frequently found in the dark races. These latter races are also characterised by their prognathism. At the same time the lower jaw sometimes even retreats as in the pre-historical specimen of Naulette, which is almost completely Simian. In the aborigines of Australia and New Caledonia the three large molars are nearly equal. This is a transition between the decreasing form in the white race and the increasing in the apes. The canines are also stronger in the Australian than in the European—another Simian feature. The elongation of the pelvis is another characteristic which approximates the lower human forms to the apes.

The author then justly and truly declares that the anthropomorphoid apes have, like man, two hands and two feet; they are neither quadrupeds nor quadrumanes.

If a brief digression may be permitted we would express our surprise that any one, after seeing the skeleton of a gorilla, can persist in calling this huge ape " quadrumanous." Its foot has a decided heel-bone, like that in the foot of man, which is wanting in its hands. But the hands and feet of the apes are, as we might expect, less specialised than those of man; yet in the lower human races the use of the foot is less exclusive than in the white man, and it is more capable of grasping. In certain inferior races the tibia is transversely flattened, as in the apes ; their anterior limbs are also longer in proportion to their hind limbs than in the higher races.

M. Hovelacque gives a further distinction which we cannot accept :—" A not less striking characteristic of inferiority is that of credulity, of faith, of religion, whatever it be. The idea of fetiches, of gods, of Deity, is the peculiar property of the lower races ; and the more or less gradual abandonment of these conceptions, puerile but dangerous, is an evident character of the superiority of a race. Faith in the higher races is merely an affair of education and of intellectual sloth."

Traité d'Anæsthésie Chirurgicale, contenant la Description et les Applications de la Méthode Anæsthésique de M. Paul Bert. Par le Docteur J. B. ROTTENSTEIN. *Paris :* Germer Baillière *et Cie.*

THE means of annihilating the pain formerly an invariable and fearful attendant upon all surgical operations have not unnaturally attracted a large share of public attention, and it might be possible to name reputations which have been cheaply earned by the persistent recommendation of new anæsthetics.

The author pronounces artificial anæsthesia to be " the greatest scientific conquest of our age, and the greatest discovery which has been made in Medicine, with, perhaps, the exception of vaccination." The glory of the invention he ascribes to Horace Wells. Still he quotes the words which Sir H. Davy had used in his pamphlet on Nitrous Oxide, published as far back as 1799 :—" Pure protoxide of nitrogen appears to possess, among its other properties, that of destroying pain, and may probably be employed with advantage in surgical operations not attended with a great effusion of blood." Well may Dr. Rottenstein exclaim " l'idée était là." Surely we may submit that the merit of Davy in connection with anæsthesia was not inferior to that of the man who forty-five years later put this published proposal into execution.

The principal object of the work before us appears to be the recommendation of the anæsthetic method of M. Paul Bert. It is found that, though the inhalation of nitrous oxide produces complete anæsthesia, this state cannot be prolonged for the time necessary for the majority of surgical operations, since symptoms of asphyxia soon appear, and prove rapidly fatal if the inhalation of the gas is continued. M. Paul Bert has overcome this difficulty by administering a mixture of equal volumes of nitrous oxide and of air under a pressure of two atmospheres.

The work is illustrated with cuts of the apparatus required for administering the various anæsthetic agents.

Report upon Cotton Insects, prepared under the direction of the Commissioner of Agriculture in pursuance of an Act of Congress approved June 19th, 1878. By J. H. COMSTOCK, Entomologist to the Department of Agriculture. Washington: Government Printing-Office.

SOME little time ago we attempted, with but scanty apparent success, to draw attention to the need of more efficient protection for our small insectivorous birds. Casting about for arguments and pleas more effective than any we had used, we come upon the work before us, and find here plain facts more than sufficient for our purpose. The damage done to crops by certain insects is here set forth in such a matter-of-fact style that any man of common sense can scarcely avoid two conclusions: that flies and caterpillars, and other the like buzzing, creeping things, are by no means the contemptible trifles commonly supposed, but are well worth the attention of practical men; and, secondly, that whatever natural agents can protect our fields and gardens from their ravages deserve not merely protection, but formal encouragement. It is somewhat strange that, though we have had acute observers of insect and bird life for more than two centuries, yet the public mind is less alive to these truths in England than in the United States. Though we still keep up the farce of professing to be a practical people, we accept the ravages of vermin as a necessity, and neither enquire what is the total of the damage sustained, nor if there is not the possibility of a remedy.

They manage these things differently in America, as Mr. Comstock's work most clearly shows. It appears that a single Lepidopterous insect, a mere moth, inflicts upon the cotton-growing States losses which are estimated at from 15 to 20 million dollars annually for the entire period since the Civil War, whilst there is good reason for supposing that prior to 1861 this insect was equally destructive. The same enemy, *Aletia argillacea,* infests the cotton plantations of South America and of the West Indies, and in some of the latter has absolutely caused the cultivation of this staple to be abandoned. If we call to mind that the figures above given show merely the loss inflicted by one insect upon one crop, we may form some faint idea of the world's vermin-bill, and the extent to which the human race is impoverished by creatures which some persons affect to consider worthy the attention merely of children or of industriously-idle old gentlemen.

The author's account of the natural enemies of the cotton-worm is exceedingly interesting. Among them figure swine, dogs, cats, racoons, the common "leather-wing" bat, domestic poultry, and almost all the insectivorous birds of the Southern States, among which the king-bird (*Tyrannus carolinensis*) and the mocking-bird (*Mimus polyglottus*) are perhaps the most efficient.

Some planters have been tempted to call in the aid of the English sparrow, the introduction of which into America may be regarded as one of the most ill-advised pieces of "acclimatisation" ever perpetrated. Every naturalist knows that this marauder, though it feeds its young on insects, is yet during its adult life by preference a devourer of grain and fruits. Their hostility to the harmless and useful swallow-tribe is ground enough for their death-warrant. We are therefore by no means surprised that the sparrow is decidedly condemned by such eminent authorities as Dr. Elliott Coues. Our American friends will find, however, that the elimination of this pest, wherever it is once introduced, is a matter of extreme difficulty, since every known scheme for its destruction acts also upon those species of birds which it is desirable to preserve.

Next follows an enumeration of the invertebrate enemies of the cotton-worm, including spiders, dragonflies, bugs, beetles, ants, wasps, and a group of two-winged flies, of which the principal species are *Erax apicalis* and *E. Bastardii*. These creatures are sufficiently powerful to overcome even bees, wasps, and hornets, whose sting seems to have little effect upon them. They certainly destroy numbers of the cotton-moth; but these flies occasion such losses to the bee-keeper that their services in other respects are more than outweighed. A fly of this kind has been known to kill more than a hundred bees in a single day. The *Erax*, though so fearfully carnivorous in its adult state, commences life as a vegetarian.

The interesting experiments made by the author to induce the destruction of the cotton-worm through the action of Fungi gave only negative results. He considers it, however, by no means improbable that some practical and economical method of parasitising noxious insects may yet be discovered.

Experiment with poisons have also been made. It seems to us that the wholesale use of such agents as arsenious acid and potassium cyanide is fraught with very serious danger.

It appears that while *Aletia argillacea* is ruinous to the cotton plantations in its larva state, when mature it occasions no little damage to peaches, grapes, figs, melons, &c. These epicurean propensities seem to point to the possibility of destroying the enemy by means of poisoned sweets, but no very decided success has been achieved.

The "boll-worm," *Heliothis armigera*, is considered as scarcely less to be dreaded than the cotton-worm, and it extends its ravages to maize, the tomato, peas, beans, vegetable marrow, hemp, tobacco, and lucerne.

An interesting essay is appended on the nectar of plants and its uses. The author remarks—"Nectar is secreted apparently to attract insects to a plant, and some of the insects so attracted oviposit on the plant, on the foliage, flowers, and fruit of which their larvæ feed. How could this secretion have been acquired by natural selection? It looks as if such an acquisition must imply the survival of the unfittest. . . . It appears that the secretion of these glands first attracts the worst enemies of the plant, and then attracts their enemies, which afford it partial relief from the misfortune that it has brought upon itself."

This essay will well repay thoughtful study, and must be pronounced as interesting in a theoretical light, as are the former portions of the book from a practical point of view.

What constitutes Discovery in Science? By GEORGE M. BEARD, A.M., M.D. New York.

IN this thoughtful and eloquent pamphlet the author contends that the first honour in Science belongs to him who organises. " To organise a science, to vitalise it, so that it may live and grow, is to make oneself expert in it, and to point out the way for others also to become experts." Dr. Beard holds that all new ideas before their full and final reception pass through three stages of evolution. In the first they are simply ignored; in the second they are denied; and in the third strenuous efforts are made to show that the fact or theory is old as the hills— "that Newton and Faraday, Hunter and Harvey, Fulton and Morse, were but feeble and conscienceless imitators, and that Edison is one of the few Americans who never invented anything." Every discovery dates from the time when it is made a part of the organised knowledge of men. That somebody may have dreamt of the thing long ago, or suggested it as one among many guesses, detracts nothing from the honour due to the man who first brings it out into full daylight. In all this we are thoroughly at one with the author. But when he says that those who are active in contesting the priority of discoveries, and who "drink the past to its dregs for proofs that the world's work has not been done by its workers" become themselves original in their search, we entertain strong doubts.

Elements of Chemistry, Theoretical and Practical. By W. ALLEN
MILLER, M.D. Revised and in great part re-written by
H. E. ARMSTRONG, Ph.D., F.R.S., and C. E. GROVES, F.C.S.
Part III.—*Chemistry of Carbon Compounds, or Organic
Chemistry.* Section I.—*Hydrocarbons, Alcohols, Ethers,
Aldehyds, and Paraffinoid Acids.* Fifth Edition. London:
Longmans and Co. 1880.

THE " Elements of Chemistry" of the late Prof. W. A. Miller
has for years enjoyed a high reputation, both as a manual for
the more advanced student and as a work of reference for persons
engaged in research, whether of a more theoretical or practical
character. Many a time, when some rather out-of-the-way
question has arisen and authorities were being appealed to, we
have heard the exclamation "To save time, look in Miller!"
Even yet, among the chemical "handbooks," "manuals," and
"elements," in which the English press has been so alarmingly
prolific, this our old favourite has continued to hold its own.
But in chemistry, more perhaps than any other science, with the
doubtful exception of biology, discoveries have been made so
numerous and important that no systematic work can continue
to meet the wants of the student unless it too advances in a
corresponding manner. The task which Dr. H. E. Armstrong
and Mr. Groves have undertaken, and which they evidently wish
to carry out without sacrificing those features which have rendered the work in its original form so valuable is not easy. New
discoveries are rarely mere additions which can be simply interca-
lated among what was already known ; they shed fresh lights upon
the older parts of the science, and necessitate modifications.
Prof. Miller's system of classification, based as it was mainly on
the sources from which the various compounds were obtained,
has been of necessity abandoned in favour of an arrangement
based on analogy in constitution and properties.

The edition before us, so far at least as the present volume is
concerned, may almost rank as a new work in which have been
retained all such portions of the original as have been not essentially affected by recent discoveries.

Among the portions entirely novel we may call attention to
Section XI., in which the hypothesis of isomerism proposed,
independently of each other, by Van't Hoff and Le Bel, is expounded and discussed, but with the conclusion that it is probably
insufficient.

The preface also, which contains some useful suggestions, is
worthy of careful notice.

The table of contents is copious and the index excellent—a
feature for which we have a special affection. We consider, in
fine, that the authors have succeeded in their object of laying

down a solid foundation for the chemistry of the carbon compounds, and we warmly recommend their work to the scientific
world.

=====

Bulletin of the Philosophical Society of Washington. Vols. I.,
    II., and III.

WE have here the proceedings of a learned society founded in
1871, and extending from that date down to June 19th of the
present year. The papers read are naturally of a very mixed
nature. Some which, judging from their titles, must be exceedingly interesting are merely mentioned,—*e.g.*, one on the
"Origin of the Chemical Elements," by Mr. L. F. Ward.
Others, on the contrary, which, whatever their value, seem to us
more suitable for a Chamber of Commerce than for a Philosophical Society, are inserted in full, and occupy a very large
share of space.

We may mention a memoir, very briefly given, on the "Geographical Distribution of Mammals," by Mr. T. N. Gill. The
author concludes that at a remote epoch Australia, South America, and Africa had been colonised from one common source,
and might be grouped into the division *Eogaea*, contrasted with
the rest of the world, named *Pleiogaea*. Of the Eogaean regions
Africa has received the greatest number of intrusive elements.

The Glacial epoch and its causes appear to have been the
subject of frequent discussion.

The paper "Common Errors respecting the North American
Indians," by Mr. Mallery, contains much interesting matter for
the consideration of ethnologists. It retains and consecrates,
however, the greatest error—the application of the term
"Indians " to the aborigines of the Western continent.

The following extract from the opinions of the late Professor
Joseph Henry, the first President of the Society, deserves quotation :—" When, from the likeness between the Infinite Mind and
the finite minds made in His image, it was sought by *à priori*
logic or by any preconceived notions of man to infer the methods
of the Divine working, or the final causes of things, he suspected at once the intrusive presence of a false as well as
presumptuous philosophism, and declined to yield his mind an
easy prey to its blandishments. To his eyes much of the free-
and-easy teleology with which an under-wise and not over-
reverent sciolism is wont to interpret the Divine counsels and
judgments seemed little better than a Brocken phantom."

United States Entomological Commission. No. 4. The Hessian Fly. Washington : Government Printing-Office.

THE Hessian fly ranks among the more destructive of the noxious insects of the United States, and pays its unwelcome attentions principally to wheat. So serious have been its ravages that the cultivation of wheat in the New England States was abandoned about twenty years ago, and is only now being resumed. The more interest attaches to this insect as it is aboriginal not in the United States, but in Central Europe, and only became noxious in America about a century ago.

The Hessian fly is as unlovely in its appearance as in its propensities, being, except as regards its antennæ, a miniature crane-fly. It has fortunately a diligent enemy in *Semiotellus destructor*, a parasitic ichneumon of the family Chalcididæ, which in some seasons disposes of nine-tenths of the Hessian fly.

Mineral Statistics of Victoria for the Year 1879.

FOR the first time for many years the annual yield of gold shows a slight increase on that obtained during the previous season, and amounts to 758,947 ounces. Extensive and highly auriferous quartz-veins have been discovered near Ballarat.

The total yield of silver was 23,729 ounces, all parted from gold, no silver-ore having been raised.

Only 24 tons of tin-ore have been obtained. The yearly produce of copper-ore has been 3862 tons, and of antimony 495¾ tons.

No lead-ore and coal have been raised, and only 120 tons of iron-ore.

The Journal of the Royal and Archæological Association of Ireland. Vol. V., No. 40. (Fourth Series.)

IT appears that at present the most scandalous destruction of important antiquities, cairns, cromleacs, &c., has been going on at Devenish and Bundoran. Mr. E. H. Cooper gives a cata. logue of fifty-one monuments, most of which have been damaged, and some entirely destroyed. We trust that Sir John Lubbock will before long succeed in getting his Bill for the preservation of such remains safely through Parliament.

Mr. R. J. Usher has discovered a submarine crannoge under high-water mark at Ardmore Bay. It is suspected to have existed from times when the British islands were connected with the Continent.

Fashion in Deformity. A Discourse by Prof. W. H. FLOWER, LL.D., F.R.S., &c., delivered at the Royal Institution, May 7th, 1880.

A POPULAR but correct account of the various mutilations and disfigurements of the human person practised at the dictates of fashion, from the perforated noses and lips or distorted skulls of savage tribes to the cramped feet and waists of so-called civilised Europe and America.

Let the British public read this pamphlet, and then believe, if it can, the assertion that fashion is an aiming after ideal beauty!

Several important books stand over for review, on account of the want of space.

CORRESPONDENCE.

₊ The Editor does not hold himself responsible for statements of facts or opinions expressed in Correspondence, or in Articles bearing the signature of their respective authors.

AN ANTI-MATERIALIST ARGUMENT.

To the Editor of The Journal of Science.

Sir,—Presuming that you do not object to a brief and temperate criticism on views advanced in your Journal, I venture to offer a remark on a passage in your September number. The Bishop of Carlisle is quoted as saying that a murderer " has no doubt as to the fact that the person who did the deed of darkness years ago is the same person as he who feels the pangs of remorse to-day. Every material particle in his body may have changed since then, but there is a continuity in his spiritual being out of which he cannot be argued." Your contributor apparently agrees with the Bishop, for he adds :—" In so far as this consideration is an evidence against what is commonly known as Materialism, I appreciate its value."

This argument, it seems, assumes that any impression made on a system of ever-changing material particles cannot be permanent ; and that the continuity of consciousness and memory proves the existence of an underlying spiritual—at least non-material—being.

But we find that impressions made upon our bodies, in spite of the continuous change of matter, may yet become permanent, the new particles being deposited in the old mould. To take an instance :—I have on my left wrist a small scar, the result of a mishap in my schoolboy days. Upwards of thirty years have since gone by, and every atom in my wrist has doubtless been replaced more than once. Still the scar remains ; as old matter is withdrawn new matter takes its place and formation.

If the experience of our lives leaves impressions of any kind upon the brain,—though I by no means maintain that such impressions are scars,—why may not permanence be obtained in an analogous manner ?—I am, &c.,

A Lucretian.

THE DEFENCE OF VIVISECTION.

To the Editor of the Journal of Science.

SIR,—I fear that the greatest peril to vivisection arises from the want of unity among its defenders. Medical practitioners are too apt to view it as an exclusively professional question. Hence as a body they refuse to co-operate with non-medical biologists, naturalists, &c., and to ignore the action of non-medical journals like your own. If they would consider the number of fanatics who are being enlisted by the " suppression societies," and the unscientific journals who give in their adhesion to the party of ignorance, they would, I think, welcome all allies.—I am, &c.,

GRADUATE.

"DR." TANNER'S FAST.

To the Editor of the Journal of Science.

SIR,—I am perfectly amazed to find that certain papers, supposed to address a class of readers not wholly devoid of culture, seem to believe that Dr. Tanner's success in his sensational experi-ment really proves, as he supposes, that the majority of mankind take too much food. Now, whether the fast was *bona fide* or not, and whether the conclusion is in itself true or false, its truth does not in the least follow from the successful issue of the experiment.—I am, &c.,

SCRUTATOR.

NOTES.

THE Swansea meeting of the British Association, from whatever causes, has been one of the least numerously attended of these annual gatherings since many years. A distinguished American scientist informed us that he had had the pleasure of reading a paper before the Chemical Section to an audience of eight persons! The Inaugural Address of the President, Dr. A. C. Ramsay, F.R.S., was naturally devoted to geology. He upheld the multiplicity of glacial epochs, with the inferences necessarily following, and gave in his decided adhesion to the doctrine of Uniformity, as will appear from the conclusion of his Address :—
" If the nebular hypothesis of astronomers be true (and I know of no reason why it should be doubted), the earth was at one time in a purely gaseous state, and afterwards in a fluid condi tion, attended by intense heat. By-and-bye consolidation, due to partial cooling, took place on the surface, and as radiation of heat went on the outer shell thickened. Radiation still going on, the interior fluid matter decreased in bulk, and, by force of gravitation, the outer shell being drawn towards the interior, gave way, and, in parts, got crinkled up, and this, according to cosmogonists, was the origin of the earliest mountain-chains. I make no objection to the hypothesis, which, to say the least, seems to be the best that can be offered, and looks highly probable. But, assuming that it is true, these hypothetical events took place so long before authentic geological history began, as written in the rocks, that the earliest of the physical events to which I have drawn your attention in this address was, to all human apprehension of time, so enormously removed from these early assumed cosmical phenomena that they appear to me to have been of comparatively quite modern occurrence, and to indicate that from the Laurentian epoch down to the present day, all the physical events in the history of the earth have varied neither in kind nor in intensity from those of which we now have experience. Perhaps many of our British geologists hold similar opinions, but, if it be so, it may not be altogether useless to have considered the various subjects separately on which I depend to prove the point I had in view."

Prof. F. M. Balfour, as President of the Anatomical and Physiological Department of the Biological Section, took for the subject of his opening address the recent progress in Embryology.

Mr. F. Galton, during his discourse on " Mental Imagery," exhibited the photograph of a " generalised " Welsh Dissenting

Minister, formed by taking the common features underlying a number of actual portraits.

M. A. Netter has addressed to the Academy of Sciences a note entitled "An Experimental Fact demonstrating that Ants have neither an Antennal Language nor any Exchange of Ideas." The nature of the fact has not yet transpired.

Dr. Sternburg, after examining microscopically the blood of yellow-fever patients at New Orleans, has been unable to find any "germs" or figured ferments which might be the cause of the disease.

Prof. Alexander Agassiz, in an able address delivered before the American Association for the Advancement of Science, objected to the "genealogical trees" of animal groups.

According to a memoir presented by M. J. Kunckel to the Academy of Sciences, the chrysalides of Lepidoptera suspend themselves by the hooks of the membranous anal feet, which are modified and adapted to particular biological conditions.

M. Alph. Milne-Edwards reports on the dredging operations conducted in the Bay of Biscay by the French ship *Le Travailleur*. As regards the Crustacea there are two supposed faunæ which do not mix, none of the species dredged up having been found on the adjoining coasts. The Mollusca found at great depths are identical with those of the Norwegian and Arctic Seas.

In "Science," July 24th, we find a most interesting paper by H. F. Jayne, on Monstrosities observed in North American Coleoptera. The most remarkable specimen figured is one of *Prionus californicus*, in which each femur bears two tibiæ, duly furnished with tarsi and claws. A hasty observer would pronounce the insect twelve-legged.

A race of domestic cattle found in Senegambia, belonging to the Zebu group (*Bos indicus*), present, according to M. de Rochebrune, the exceptional characteristic of a true horn on the nasal region, identical in its nature and mode of development with the frontal horns. It is common to both sexes, and is sometimes conical, but more generally takes the form of a four-sided truncated pyramid, about 2½ to nearly 3 inches in height. The skeleton differs altogether from that of the zebus of India and Madagascar. The incisors have on their outside a deep groove.

M. H. Filhol has presented to the Paris Academy of Sciences a memoir on new fossil mammals discovered in the phosphatic deposits of Quercy. The most interesting of these forms is *Ailurogala acutata*, a connecting-link between the Felidæ and the Mustelidæ.

The " Comptes Rendus " contains an account of deforming pilosism in plants. The author distinguishes physiological pilosism, the increase of hairs upon plants produced when they are removed from a humid to a dry medium ; teratological pilosism, which extends so far as to create suspicions of a new species, and due to disarrangement of the process of nutrition ; and vulnerary pilosism, produced by the stings of insects, as in galls. It is quite localised, and does not mark the physiognomy of the species.

M. E. Spée, in a memoir communicated to the Belgian Academy of Sciences, proposes that the so-called " helium " line, in the solar spectrum, D_3, is a hydrogen line.

P. Kaiser (" Botanische Zeitung ") shows that the trunks of trees are subject to a daily-recurring, regular change of diameter. which decreases from the early morning, and reaches its minimum in the early hours of the afternoon. Then a gradual increase begins, till a first or smaller maximum is reached at the approach of darkness. After a short decrease the diameter rises again, and reaches its larger maximum about dawn. The author considers that temperature is not the only factor concerned.

The maximum temperature in the shade registered at Melbourne Observatory has been 111·2° F., and the lowest 27°.

Mr. W. G. Lapham (" American Journal of Microscopy ") points out as a serious evil that many biological terms are incapable of definition.

Prof. H. L. Smith, in an article inserted in " Science," admits that the wax-cell is a failure.

The experiments of Prof. Prentiss (" American Naturalist "), of Mr. W. Trelease, and Prof. J. H. Comstock, on the use of yeast for the destruction of noxious insects, have had a negative result. On the other hand, Dr. Hagen and Mr. J. H. Burns have been to a certain extent successful.

Mr. C. O. Whitman (" American Naturalist ") maintains that flying-fish really flap their pectoral wings when in the air, and are capable of altering their direction without touching the water.

Mr. T. Mellard Reade (" Geological Magazine ") combats the notion of the general permanence of the present main features of the continents and oceans, as asserted by Agassiz, Dr. W. B. Carpenter, and others.

According to H. H. Howorth (" Geological Magazine ") the name Mammoth is a corruption of Behemoth, which the Arabs, who confound M and B, pronounce Mehemet.

At the Astor Library, New York, out of a total of 56,891 volumes given out for the year 1879, there were only 142 on zoology and 214 on biology, whilst theology numbered 2231, ecclesiastical history 2402, music 1462, and jurisprudence 3284.

The two heaviest gold nuggets found in Victoria from October, 1877, to June 30th, 1879, weighed respectively 20 lbs. 10 ozs. and 23 lbs. 6 ozs. troy.

Psilomelane containing from 1 to 7 per cent of cobalt is obtained in many localities in Victoria, and is profitably worked by a process devised by Mr. Malcolm Hills ("Geological Survey, Victo:ia," vi.).

According to the Rev. G. K. Morris ("American Naturalist") *Pheidole pennsylvanica*, an ant found in New Jersey, is a true harvester.

Mr. E. J. Hill, writing in the same journal, describes a non-poisonous snake which very successfully mimics the rattlesnake.

A honey-collecting ant with an enormously distended abdomen has been discovered in Australia.

M. Adalère Liénard has communicated to the Academy of Sciences of Belgium a memoir on the nervous system of the Arthropoda and the constitution of the œsophagian ring, in which he recognises four distinct types :—that of the Crustaceans ; that of the *Dytisci*, common to a great number of carnivorous insects ; that of *Cossus ligniperda*, which occurs in the larvæ of many moths, locusts, and herbivorous or lignivorous beetles ; and finally that of the suctorial insects. (It is strange to find the brain of *Necrophorus vestigator* and *N. germanicus* belonging to the first of these types, whilst the kindred form, *Silpha nigrita*, is assigned to the third.—Ed. J. S.)

According to the "Geological Magazine" an entirely new fauna has been detected, in the Menevian States of St. David's and of Dolgelly, which were considered wholly unfossiliferous. The fossils were discovered by looking at the bedding-ends of a number of the slates placed together in their natural position.

THE

JOURNAL OF SCIENCE.

NOVEMBER, 1880.

I. THE DUKE OF ARGYLL "ON THE UNITY OF NATURE."

By J. FOULERTON, M.D.

IN an article by the Duke of Argyll, in the "Contemporary Review" for September, 1880, the author explains what is meant by the Unity of Nature by examples drawn from natural phenomena, and gives certain speculative opinions with regard to its belief in very early times, and its relation to Monotheism.

The examples drawn from natural phenomena, though containing nothing that is actually new, are very forcibly put, but I can only give a very brief statement of them here, just sufficient to show the line he adopts.

First of all he refers to gravitation as a mysterious force linking the whole universe together, and not only keeping the heavenly bodies in their places, but intimately connected with our very existence and the due performance of vital functions.

He then considers that highly elastic body pervading space, the medium of light and heat, electricity and chemical action, and how the various wave-motions which constitute these all move with immense velocity through the ether without jostling or confusion, like the harmonious vibrations of sound in the atmosphere.

He also refers to the nature of Life, which he cannot regard, as some do, as simply a physical condition and property of protoplasm, but rather of the nature of a force or energy working on protoplasm, and that there is a unity in

this force which makes one piece of protoplasm develop into
a tree, whilst another not distinguishable from it may develop into a man.

Thus we see that the author has been dealing with some
of the latest acquisitions of natural science to exemplify the
Unity of Nature, which he defines as "that intricate dependence of all things upon each other which makes them
appear to be parts of one system," and adds "that it may
well be that the sense of Unity in Nature, which man has
had from very early times reflected in such words as the
'Universe,' and in his belief in one God, is a higher and
fuller perception of the truth than is commonly attained by
those who are engrossed by the laborious investigation of
details," and considers that "this is one of the many cases
in which the intuitions of the mind have preceded inquiry
and gone in advance of Science, leaving nothing for systematic investigation to do, except to confirm by formal
proofs that which has been already long felt and known."

My object will be to endeavour to show, first, that the
Unity of Nature and Monotheism have never been generally
believed by mankind; and, secondly, that, though the evidence given by the author does not prove the Unity of
Nature, however probable it may make it, and though
everything he says about gravitation, ether, light, heat,
electricity, life, &c., may be wrong,—we may reasonably
presume some of it is,—the Unity of Nature would still be
true, and that there is evidence which cannot be negatived,
and which no errors in detail can in any way affect.

First, as regards the evidence given by the author of the
general belief by mankind in the Unity of Nature from the
use of the word "Universe." I do not see how this can
prove it. The words "Universe" and "Unity," though
very near each other in the Dictionary, have quite different
meanings. One might say of the Laws of the Universe
that they acted inharmoniously and in antagonism to one
another, which is just the opposite of what is meant by the
Unity of Nature.

This is not the place to discuss the "intuitions of the
mind," or what may be called, except in metaphysical
language, inherited ideas; but such ideas or feelings always relate to actions of vital importance to the species, and which
have so long been performed by them that they have at length
become organised—such, for instance, as those prompted by
hunger, or the care of the young. But to what actions of
vital importance to the human species could ideas of the
Unity of Nature lead; a million years hence, perhaps, they

may, if competitive examinations last so long, and if
so they will become organised and take their place among
the " intuitions of the mind."

From what the author has said, in the passage quoted
above, one would suppose that general beliefs on abstruse
natural phenomena were frequently correct, and that all
Science had to do was to interpret the inspiration and set
its seal upon it. Now I think it would not be too much to
say that there never yet was a natural phenomenon, not
being about as obvious as that the sun is a source of light
and heat, that vinegar is sour to the taste, or that a blow is
painful,—that is to say, which did not directly impress itself
upon the senses,—that was not misinterpreted by general
beliefs.   A step or two in the process of deduction and general beliefs are a chaos of errors.   Indeed if one wants to
know for certain what is wrong with regard to any question
which is not immediately patent to the senses, he has only
to find out what the popular belief is or was.   I will give
one or two illustrations :—There were two ways at least of
accounting for the apparent movement of the heavens round
the earth, one of them being that the heavens actually did
move and that the earth was still ; so this erroneous theory
was adopted, and the correct one when discovered was rejected,
on account of the supposed great importance, morally, of the
earth and man; so that not only was a wrong theory accepted,
but the right one was rejected for a wholly irrelevant and
personal motive—a very common line of argument, as we
shall presently see.—Except in the case of direct physical
injuries, all pain and disease were accounted for by witchcraft and similar agencies, by which also many other natural
phenomena were accounted ; and we all know the erroneous
beliefs of mankind with regard to the nature and effects of
comets, eclipses, meteors, &c.—Illustrations could also be
given from religion and politics, but as these are still in the
region of popular beliefs there would be too much difference
of opinion for them to have any force.   There are, however,
two facts which may be said to have reached scientific precision which will forcibly illustrate the wrongheadedness of
popular beliefs ; one is the advantages of free trade over
fettered trade, the other is the advantages of tolerance of
opinion over intolerance ; yet how many ages have passed
and how much misery had to be endured, through the action
of the wrong theory, before the right one made any impression on popular beliefs.   In speaking of popular beliefs
it must not be supposed that I am alluding to the populace,
or lower classes, as distinguished from their so-called betters.

The errors of popular beliefs affected Popes and Kaisers just as much as they did the lowliest of their subjects; indeed if they had not, they would not have been Popes or Kaisers at all. So persistently, indeed, are popular beliefs wrong, even in matters intimately affecting the welfare of mankind, whenever it is necessary to take a step or two in the process of deduction, that if they were right —as the author thinks—with regard to the Unity of Nature, it would be a very extraordinary exception to the rule.

But besides all this, which may be called probable evidence, we have abundant positive evidence that the general belief of mankind has been, and still is, not in a unity, but in a duality of nature and supernatural beings.

In the same way as the author has stated some of the most recent scientific views as evidences of Unity of Nature, let us take a very brief survey of some of the principal beliefs with regard to the Universe which have been, and many of which are still entertained even by civilised nations as evidence rather of a belief in its duality, viz., that this world, though only a flat surface a few miles in area, bounded by the horizon, was the largest portion of the universe; that the sun, moon, and stars were mere appendages quite close to the earth, for the benefit of man—the central figure of the universe, and on whose account it was all brought into existence in a week; that in this world, which was practically the universe, it was the earnest wish of its Maker that there should be no pain or death, no sin or misery, but perpetual health and life, morality and happiness; that from the very beginning another personage appeared upon the scene with views totally antagonistic to those of its Maker, viz., that there should be disease and death, sin and misery; that these two powers have been in a perpetual struggle ever since, and that so far the evil power has been much the more successful of the two; that, during the struggle, miracles—breaks in the uniformity of Nature—are continually being performed by both parties, and are looked upon very much as are the stratagems of ordinary warfare; that prayers offered up to the beneficent power—in some communities they are offered up to the evil power, in order to propitiate him, the beneficent power being supposed not to need them—may at any time alter what would otherwise be the course of Nature; that thus the dead have been restored to life, the sick made well, the earth more fruitful; that wind or calm, sunshine or shadow, rain or drought, health and prosperity of ourselves and our friends, or death and confusion of our enemies, have all been obtained through their means, and contrary to what would have happened but for our prayers. But every such event is in fact a miracle,—a breach in the uniformity of Nature's law. This is a bare statement of facts which will be admitted by all, whatever may be their views otherwise with regard to prayer—a subject which does not concern us here.

No people are supposed to have been more firm in their monotheistic belief than the Jews, yet their whole lives were spent in such an atmosphere of miracles as it would hardly be possible for us now to realise.

It could easily be shown that ancient Pagans and modern Savages, just like ourselves, have regarded the universe solely from a personal point of view, and in a dualistic sense as it affected their joys or sorrows, without any thought whatever of a great law of uniformity.

If, notwithstanding such general beliefs with regard to the system of the universe, wholly erroneous as we now know them to be, without any knowledge whatever of those generalisations we call laws of Nature, such as are given by the author as evidence of its unity, and believing, moreover, that the ordinary course of Nature was continually being set aside, at one time by its Author, at another by his opponent, mankind still believed in the Unity of Nature, it would be a most marvellous and unaccountable circumstance, and would require much stronger evidence in support of it than merely calling it an " intuition of the mind,"—a most convenient psychological Cave of Adullam for the reception of all real or supposed mental phenomena which will not fit in anywhere else.

Theology is in the habit of twitting Science by saying " You admit there was a beginning to the universe; that of itself is a miracle; but if you believe in one, why not in others." Science knows nothing of the matter, and postulates nothing. Theology dogmatically affirms that the universe will have no end; suppose Science were to affirm, equally dogmatically, that it had no beginning? The one assertion is just as incomprehensible and unprovable as the other; but the modesty of Science is shown in admitting its ignorance, and its honesty and love of truth in not predicating anything under the circumstances.

The great mistake made by the author, as by many others, is in supposing that Theology has anything to gain from Science, or Science from Theology. Abraham was a sound theologian, with no more knowledge of Science than any other Bedouin of the Desert; and there have been

many able men of Science whose Theology was, as the
author remarks, weak. Theology and Science would
thus seem to be in inverse proportion to one another,
and the reason of this is to be found, I believe, in
the erroneous relationship in which they are supposed
to stand to one another. Theology anathematises Science
because it is not theological, and Science returns the
compliment because Theology is not scientific; whilst
there are those who, like the author, bless them both, and
would make out that they are indispensable to one another.
Perhaps some day it will be discovered that, on the
contrary, they are entirely independent of one another.
There are no doubt many things included under Theology,
and generally considered the most important part of it,
which have no more right to be so than paint and feathers
to be called man, though they too are often considered the
most important part of him. An enlightened common sense
may be trusted to make the necessary distinction.

Attempts are now frequently made to prove that miracles
are not breaks in the uniformity of Nature's laws, just as
they are made to show that seven days are countless ages,
and it is a hopeful sign to see this growing tribute to
great truths; but for the present their success does not concern us, but only this—that no one thought of such a
question formerly, there being no idea of any such uniformity any more than there was of the countless ages.

The author makes the following remark, which seems
somewhat singular from so firm a believer in the Unity of
Nature :—" A sense and a perception of the Unity of Nature
—strong, imaginative, and almost mystic in its character—
is now prevalent among men over whom the idea of a personal agency of a living God has, to say the least, a much
weaker hold." Why does he call a sense of the Unity of
Nature imaginative and mystic ? It can hardly be because
he has any doubt about it. The fact is that, like many
others, he seems to shrink from what he believes to be its
logical consequences on religious faith; but nobody need
trouble himself in the least about that : man's faith has
never yet been comprised within the three corners of a syl-
logism, whether they contained the Unity of Nature or
anything else.

The second object I propose to attempt is to show
that,—though everything the author says about gravitation, light, heat, electricity, life, &c., may be wrong,
without in the least affecting the truth of the Unity of
Nature,—there is such evidence of its truth as cannot be

refuted, and which no errors in detail can in the least affect.

The position of any planetary body in space, at any given moment, is the resultant of all the forces of gravitation that have acted on it. So, likewise, of all the planetary bodies.

Any particular organism—plant or animal—is the offspring and resultant of all the varying conditions of all its ancestors. So, likewise, all the organisms living at any particular time are the resultant of all those that have preceded them, and from which they have descended.

The present geographical and geological configuration and structure of the earth are the outcome of all those that have gone before, by an uninterrupted succession of gradual changes.

Thus, whilst the present position of the planetary bodies, the present organisms that inhabit the earth, and its present geographical and geological features may be very different from what they were millions of years ago, the great change was brought about by a succession of small ones,—each, indeed, so small as to be wholly inappreciable, the one condition gradually shading into the other. There is thus a continuity, unity, or what may be called an individuality, pervading each, similar to that which unites the old man to the child, but without anything to indicate either youth or old age. But if this be so with each separate thing, or system of things, it will also happen with everything and all systems; in other words, the condition of the whole universe at any particular moment will be the outcome or resultant of all previous conditions, and, however much it may differ to-day from what it was at any former period, a thread of continuity or unity will run through the whole. We talk very glibly of the beginning of the world, and think we have then got to the beginning of the universe; but to say nothing of the enormous antiquity of this world itself, compared to which the whole geological record from the earliest stratified rocks to the present day is probably a mere vanishing point, we would practically be no nearer the beginning of the universe, even if we could arrive at the commencement of this infinitesimal portion of it, counting from the time when it may be said to have had a separate individuality; for who shall tell the previous history of its component parts, and into the structure of how many worlds they may not formerly have entered? Science knows nothing of beginnings or endings in an absolute sense, whilst relatively it knows of nothing else, each following the other in unceasing succession. But as the total amount of matter in the universe is always the same, and as its condition at any particular moment is the resultant, the outcome, the effect of the previous condition, it follows that the effect contains the whole of the cause, neither more nor less. Take an example on a smaller scale. Oxygen, hydrogen, and motion are not only the cause, so to speak, of water; they are water. Cause and effect are therefore merely different aspects of one and the same thing. The universe of to-day is therefore one and the same with that of any previous time, however different may be the arrangement of its constituent parts, and the reason of this is neither more nor less mysterious than is the reason of the infinite succession of causes and effects which are going on every moment within us and around us.

I will briefly sum up thus :—1st. There is no evidence of a general belief in the Unity of Nature and Monotheism; the probabilities are all against it; and there is abundant evidence of a belief in a dualism both of Nature and of supernatural beings. 2nd. As the sum of all effects is = the sum of all causes, the universe of to-day is = the universe of any previous time, and this it is which constitutes the Unity of Nature. In order to· disprove it, the sum of an effect must be shown to be either greater or less than the sum of its cause, which is impossible.

The above was written before the issue of the October number of the " Contemporary Review," in which there is a second article on the Unity of Nature, by the Duke of Argyll, which I think fully bears out my remarks regarding the faulty nature of the method he has adopted as evidence of it; for if what he now states be true, it would show not a unity but a duality of Nature.

The first article was mainly occupied with the phenomena of non-living matter — with the movements we call heat, light, electricity, chemical action, &c., and the intimate relations existing between them; the present one treats of the phenomena of life, and shows their close relationship in all living things from the amœba up to man. This is no doubt all quite true; but of what use is it as evidence of the Unity of Nature, if, as he now says, there is a great gulf of difference, even of antagonism, between the living and the non-living. A chain is no stronger than its weakest link; and if the Unity of Nature can withstand one such antagonism, why not any number of them? Thus heat, light, electricity, and chemical action might have no sort of affinity with each other, on the one hand; or organisms on the other, without in the least affecting the evidence.

In the same number of the "Contemporary" there is also an article by the Hon. Justice Fry advocating similar opinions, and, consistently enough from this point of view, deprecating "that simplicity or unity which partly entices men to adopt the purely materialistic theory of things." With that _bête noire_—the materialistic theory—we have nothing to do. All I wish to call attention to is the fact that the Hon. Justice Fry finds that a gulf of difference between the living and the non-living is opposed to the theory of unity.

But after all may not this gulf of difference be only one of ignorance? Until quite recently birds and reptiles were supposed to be isolated by impassable gulfs, and this was considered by many to be a strong argument against organic evolution, or what may be called the Unity of Organisms. These gulfs have now been bridged over.—At one time the earth was too hot for any living organism; but when it cooled down sufficiently to allow of these, is it not quite possible that they may have been formed out of non-living matter, and if so what would become of the antagonism between them? There are some who think that this process of conversion of non-living into living matter has been going on ever since and is going on still; but whether that be so or not does not much affect the argument, for the conditions of heat, moisture, atmospheric composition, &c., which may have existed when organisms first made their appearance, may not be possible now. If the Unity of Nature be true, as most assuredly it is, then all the phenomena of Nature must hang upon one another, and, having got hold of a fundamental, all-embracing fact, which is none the less true because the knowledge of it is of recent growth, we must not mar its symmetry and beauty either through our ignorance or for the sake of any barren speculations about Materialism.

II. ON HEAT AND LIGHT.

By ROBERT WARD.

ACCORDING to the modern scientific idea **Heat is a** " mode of motion." Prof. Tyndall, who has done so much towards popularising this view of heat, describes it as " an accident or condition of matter ; namely, *a motion of its ultimate particles.*" He discards the material theory which supposes heat to be "a kind of matter—a subtile fluid stored up in the inter-atomic spaces of bodies." Yet for a long time the latter had the greater number of supporters eminent in Science.

It is not my intention to suggest that either the one or the other theory is devoid of truth. I have elsewhere shown that all things exist by virtue of their circumstances, and that, as the circumstances of all things are continually changing, all things must be continually changing (*see* " Constitution of the Earth," p. 4). That such change is always going on is rendered obvious by the amazing fact that everything *grows older*. Even wine, however preserved from contact with surrounding objects by being hermetically sealed in a glass bottle, undergoes a change well known to the toper, though undiscoverable by the analytical processes of the chemist. That this change is a never-ceasing one is as evident as that a body falling from an elevated position must pass through all the intermediate points of space before it reaches the ground. There may be no evidence of any substantive addition to, or withdrawal from, the contents of the bottle ; nevertheless the wine *has grown older*, as discoverable by the sense of taste, and by such change, up to a certain point, it is humanly speaking "improved." It cannot be doubted that, from the time a human being comes into the world till he passes out of it, the entire phenomena of life consists of an unceasing change. Infancy, youth, manhood, and old age are only terms by which we recognise the more prominent phases of a never-ending, still-beginning evolution. To indicate their appreciation of such change, chemists formerly told us that in seven years every molecule of the original substance of the human body had been removed and substituted by other molecules. The more modern idea is that this takes place at much shorter intervals ; but is it not obvious that a human being is never for a single moment exactly what he was the moment which immediately preceded ?

Now what is change but motion ? And yet it is *more* than motion : however imperceptible it may be, it is part of the process of *transformation* by which the wine is improved, and by which the child becomes a man. The effect is not a mere mechanical rearrangement of parts,—not only a " vibration " or " clashing of atoms,"—but a constitutional or radical difference, however small, between the original and present states. Either " something " has been taken or something has been added ; and is it not probable that something has been both taken and added, by which the change has been produced ? We may call that something " matter," or we may call it " an accident or condition of matter :" the change could not have taken place without motion, but *neither could it have been produced by motion alone*.

In the volume to which I have referred I have shown that, on the assumption that all things exist by virtue of their circumstances, it follows that " a change in the circumstances of things necessarily involves a corresponding change in the things themselves." This I have described as " the Law of Motion." Hence it is that—as Nature is never at rest, as worlds and systems of worlds are always moving, and as all the other bodies of space, small or great, are perpetually taking up new positions in relation to each other—every bit of matter in the universe, every atom or molecule, is perpetually undergoing a process of change. May we not assume that heat is one of the forms of this change ? It is undoubtedly true that, according to the extent or degree of this change, we must have motion ; and according as this motion in extent or degree appeals to our sense of feeling, by adding to or subtracting from the natural warmth of our bodies, we employ the terms heat or cold to express our estimate of the character of the phenomenon.

The truth is that all things in Nature are united together by a constitutional relationship, of which the action of heat upon our organism is one of the manifestations. We become mentally acquainted with this relationship by means of the senses of seeing, feeling, hearing, tasting, &c. When I look round the room in which I am sitting, I see articles of various forms, sizes, and colours,—*e.g.*, chairs, tables, pictures, &c. Why do I see them ? I may be told that it is because rays of light are reflected from the objects to my organs of vision ; and it is true that without light I could not see them. But neither could I see them without eyes; and yet it is evident that, whether I see them or not,— whether my eyes are open or shut,—the relationship between the objeats and my organism, *so far as it is mechanical,* must always be the same as long as they and I remain unmoved. The objeats, in faat, exist independent of either the light or my eyes, but the aatual perception of them is due to certain qualities which are impressed upon my organism when the necessary relationship is established. Thus there are three faators necessary to sight—eyes, objeats, and light. The qualities by which objeats are revealed to us are colour, form, and size. Really, however, so far as the visual per- ception of things is concerned, colour is the basis or founda- tion of sight, form and size being due to a perception of differences in colour, and to the limitation or extension of such differences. Thus I can only see the chairs, tables, piatures, &c., by virtue of the light which strikes upon their surfaces and reveals to me various colours, or shades of colour, the perception of which (by contrast) is the founda- tion of my knowledge.

As in the case of heat, Light was believed by Newton to be due to the emission of a subtile fluid, but modern Science repudiates the teaching of the great philosopher as to both light and heat, and they are now severally described as simple " modes of motion." Heat, we are told, is a clashing together of the ultimate particles of matter; whilst Light is due to the vibrations of a luminous ether wholly unknown to the senses, but which fills all space and penetrates be- tween the molecules of all bodies. White light, as it reaches us from the sun, is a compound or mixture of many colours. The different colours of bodies are said to be due to the way they refleat the different kinds of light, by creating in the imaginary ether waves of different lengths. These waves, though as infinitely numerous as the shades of colour, are supposed to move simultaneously and perfeatly distinat, so that each will produce its own separate and charaateristic effeat upon the organs of vision.

Such is the modern scientific theory as to the creation of colours. I must confess that my understanding is not so much enlightened by it as by the emission theory of Newton. The solar speatrum exhibits seven principal colours, but these imperceptibly merge into each other through an infi- nity of intermediate tints. I find it hard to imagine seven waves of different lengths, all moving with equal speed and striking the retina of the eye at the same moment; but I am totally unable to realise the idea of an *infinity* of different lengths of waves—representing the infinity of intermediate tints—striking the retina in the manner described. If I must penetrate the mystery, I confess that it seems easier to comprehend the emission than the undulatory theory. I know that all things are growing older, and therefore undergoing an unceasing change, and I cannot understand how such change can take place without the emission of *something*. What is that something? It seems to me that it is the cause of all the changes perpetually taking place in Nature. It is by this emission (or by this absorption) that I become acquainted with the several qualities of things and their relationship to my own organism. Something is constantly passing between all bodies, by which they act and react upon each other; by which they are at once separated and united; by which all change—mechanical, chemical, and physiological—is produced, and which is at the same time the origin of life and the cause of decay. This something appeals to our senses in many phases—has heat or cold, as odours, as sounds, as colours differentiated into forms or sizes, as tastes, as the various impressions of touch or feeling, &c. This something, in its many forms and qualities, in its attractions and repulsions, and, in short, by the infinity of its phenomena, constitutes the *reality* of which gravity is the *abstract conception*. For, as Newton himself suggested, gravity is due not to " occult qualities supposed to result from the specific forms of things, but *as general laws of Nature by which the things themselves are formed; their truth appearing to us by their phenomena*, though their causes be not discovered."

Nor are the impressions of distant objects communicated through the senses of no physical account in building up the living organism. Such impressions fix themselves more or less permanently in the memory, and assist in creating the brain tissue by which memory itself is established. The brain may, in fact, be described as a storehouse of such impressions, in which are gathered not only the experience of the living individual, but of all his ancestors to the remotest conceivable beginning. And as is the brain, so is the body. The inconceivably attenuated matter (or something) which strikes upon our senses from surrounding objects may be said to have in it " the promise and potency of every form and quality of life."

Of all the impressions received from outward objects, those of heat and light are the most active in producing change. Heat is the moving agent in all manufacturing processes having for their object the speedy conversion or

transmutation of bodies, and heat and light are the most
important elements in connection with all the phenomena
of vital action.   By their means, out of the same soil and
air is evolved the tender bud, the green leaf, the woody fibre,
every form and description of vegetation, every variety and
tint of colour, every kind of odour, and numerous other
qualities by which our several senses are either obnoxiously
or agreeably influenced.

And yet it is said that light and heat can create nothing or
destroy nothing; that they are mere modes of motion—the
action of the "eternal and imperishable" parts of the huge
machinery of the universe, which change only in their
relation to each other and not in their constitution.   If this
meant that heat and light can, in the absolute sense, create
or destroy nothing, no objection could reasonably be made.
We know nothing of either matter or motion beyond the
impressions which they communicate to our consciousness
through the senses.   The forces of nature may indeed be
described as only the agents by which the Supreme Creator
works out His purposes in the universe; but are we not,
therefore, all the less justifiable, when we see these wonderful creations of light and heat, in denying that they are,
what they appear to be, transmutations, in some degree, of
the several materials or objects out of which they have been
constituted?   That is, a transmutation (not of abstract
matter, which is only a metaphysical conception) of the
several qualities by which the objects are presented to our
consciousness, and by which alone they are known to us?

Whilst some scientists treat heat as only a mode of
motion, an accident or condition of matter,—others constantly refer to it as an entity of absolute power.   Hence
we read of calulations to determine how much of the sun's
heat is utilised in warming the planetary bodies, and how
immensely more is wasted in space.   I do not dispute that
the sun, like everything else in Nature, is undergoing
changes, or "growing older"; but let this be remembered,
that heat, such as it is known to us on the earth, may be
something very different at its source; that, in fact, there is
no reason to believe that the sun is being burnt up like the
coals upon our fires.   On the contrary, there is much reason
to believe the reverse.   Only a few miles up in the air it is
freezing cold; the light rays illuminate the regions above
the clouds, but the heat rays *are absent!*   Certainly no
evidence this of immutable heat (or even of heat such as it
is known to us by combustion) radiating from the sun into
space.   If we place a sheet of glass before the fire, the heat

rays are stopped till the glass itself becomes hot ; or if we place a sheet of ice in the same position, the same result ensues till the ice is melted. On the assumption that the sun is throwing off heat like a fire, the thin air of the upper regions acts like a sheet of glass or ice with this remarkable difference, that the heat never raises the temperature of the air, and yet (apparently, at least) *it gets through it,* as known to our consciousness on the earth's surface ! The heat gets through, but not as the heat known to our senses, for Rumford showed that calorific rays can be passed through a vacuum without losing their temperature in the passage. The question of the temperature of the sun has been the subject of investigation by many scientists. Newton, one of the first investigators of the problem, tried to determine it, and after him all the scientists who have been occupied with calorimetry have followed his example. All have believed themselves successful, and have formulated their results with great confidence. The following, in the chronological order of the publication of the results, are the temperatures (in centigrade degrees) found by each of them : Newton, 1,669,300°; Pouillet, 1,461°; Zöllner, 102,200°; Secchi, 5,344,840°; Ericsson, 2,726,700°; Fizeau, 7,500°; Waterston, 9,000,000°; Spoerer, 27,000°; H. Sainte-Claire Deville, 9,500°; Soret, 5,801,846°; Vicaire, 1,398°; Violle, 1,500°; Rosetti, 20,000°. The difference is, as 1,400° against 9,000,000°, or no less than 8,998,600° ! There probably does not exist in science a more astonishing contradiction than that revealed in these figures.

Another of the fallacies respecting heat upon which some curious theories have been founded as to the earth's future destiny, is that of assuming that all but its crust is a mass of liquid fire. In support of this intense central heat, it was at one time asserted that the temperature of deep mines invariably increased in proportion to the depth ; but doubts have, in late years, been thrown upon the statement. Still it is confidently asserted that the interior of the earth is in a red-hot molten condition, and that it is radiating its heat into space, and so growing colder. One of the results of the *Challenger* and other explorations of deep ocean is to determine that the water towards its bottom is freezing cold. Considering that the ocean covers nearly three-fourths of the entire globe, this fact certainly does not support the theory of central heat accompanied by radiation. The coldest water, it is true, usually sinks by its greater weight towards the bottom, and that, it may be said, accounts for its coldness ; but, on the theory of radiation, the water of the.

ocean has been for long geological ages supported on the thin crust of the earth, through which the central heat has been constantly escaping; and yet it is still of freezing coldness! Experience would say that the heat cannot have escaped through the water without warming it, because the capacity of water for heat is greater than that of any other substance. We can no more imagine such a radiation, and consequent accumulation of heat in the ocean, without the natural result of a great rise in temperature, than we can believe in a kettle resting for hours on a hot fire without the usual result of boiling water. We have no reason, therefore, to believe, as has been suggested, that the earth is growing colder, or that we, in common with all living things, are destined to be frozen out of existence and the earth itself finally swallowed up by the sun.

III. THE BACONIAN PHILOSOPHY OF HEAT.

By Dr. Akin.

IF pessimist politicians are prone to consider ordinary history as the history of practical human folly, pessimist philosopers might not inaptly describe scientific history (if such a thing existed) as the history of theoretical human error. The relative proportion of right and wrong committed, of which political history forms the record, is not inadequately matched by the ratio of truth and falsehood propounded, of which philosophical history would have to give an account; and if speculative politicians are inclined to despond upon comparing the actual state of society with what they recognise as its ideal, philosophical inquirers into the secrets of nature have little cause to feel cheerful when they similarly compare the task before them with what has been hitherto accomplished. Among the objects of which the sensations of the first human beings must have rapidly rendered them conscious, were probably light and heat, owing to the great, sudden, and frequent variations to which these conditions of human existence are liable in all climates —that is to say, if it be true, as is generally assumed, and apparently with reason, that consciousness is awakened by change. Later, when philosophers began to contemplate and ponder on the phenomena of Nature, heat, amongst others, naturally formed a principal topic of their speculations; and, accordingly, from the days of Aristotle, or his precursors—whom, in the phrase of Bacon, Aristotle "had destroyed, as Ottoman emperors do their brethren, for to reign in greater security"—through almost all subsequent centuries, we have numerous expressions of opinion and statements of arguments intended to describe or demonstrate the nature of heat. Yet how unprofitable and even depressing a task would it be to follow out the history of these statements and discussions, even if we confined our attention to such as proceeded from persons of great and well-merited fame! Galileo and Gilbert, Bacon and Descartes, Boyle and Newton, Boerhaave and Crawford, Lavoisier and Laplace, as well as a host of others, have simultaneously or successively put forth views, sometimes nearly coincident, sometimes most widely discordant, as to this prolific subject of debate and research; yet the upshot of all this was, in our own century, a stolid acquiescence partly in avowedly untenable theories, partly in seemingly irremediable ignorance. The most bulky works on the subject of heat produced within this last-named period—whether they boasted the title of "theories," like those of Fourier and Poisson, or were of a practical character, like those of Peclet and De Pambour, almost confessedly slurred over the question as to the nature, essence, or, in Baconian language, the "form" of heat. Their writers took it simply for granted that heat is a something somehow capable of measurement, as proved empirically, and of exhibiting certain phenomena into whose precise mechanism or origin it was needless to inquire, it being sufficient to know by experience that they exist. Thus, after more than two thousand years of meditation, contention, and investigation, we were landed again in total ignorance, at least so far as the purpose of philosophy is to comprehend, not merely to analyse, Nature. Yet within the intricate coil, made of cobweb speculation rather than of solid scientific fibre, into which had been spun the dicta of successive generations of physicists on the philosophy of heat, there may be traced—to borrow a German simile taken from an English practice—a red thread of truth, which, originating apparently with Bacon, had been continued uninterruptedly, though without gaining in consistency, down to our own period. For a long time as good as buried in the superincumbent layers of the twisted fabric of error, this unwarped thread of truth has happily been seized upon and strengthened by some philosophers flourishing at the present time, owing to whose labours the views of Bacon have not only been drawn forth from their unmerited oblivion, but, it is hoped, have now regained for ever their ascendency.

As already indicated, the fundamental notion of the nature of heat which it is our intention to follow through its successive stages of growth, proceeded from Bacon—the Chancellor, not the Friar. In his *Novum Organum* something like a chapter is devoted by way of illustration to an " inquisition into the form of heat "; and, whatever may be its value considered as an illustration, there can be no doubt that in this "inquisition" is contained the first, comparatively speaking, precise, if not positively well-backed, statement of what is now coming to be universally recognised as the true definition of the nature of heat. This definition, expressed in the somewhat peculiar language belonging to Bacon, is that " heat is a motion, expansive restrained, and acting in its strife on the smaller particles of matter "; to which he adds, by way of caution on what is called " sensible " heat, that " heat, as far as regards the touch of man, is a thing various and relative, insomuch as tepid water, for instance, feels hot if the hand be cold, but cold if the hand be hot." Views very similar to these were described somewhat later by Descartes, in his *Principia*; whence, probably, they were likewise adopted by Boyle, who, however, re-stated them in language much more precise and philosophical than had been used by either of his predecessors. This happened in the seventeenth century. At the beginning of the eighteenth, Newton again propounded the same doctrine, that heat is essentially but a motion of the molecules; but, unfortunately, whilst his authority contributed so powerfully to spread and sustain false ideas on the nature of light, no such overruling influence was allowed to his much sounder notions on the subject of heat. To some opinions of great novelty and importance put forth by Newton in connection with this matter, we shall have to advert to a later stage, but the merit of having advanced the Baconian theory beyond the immediate position in which it had been left by Bacon himself, by Descartes, and by Boyle, belongs to Hermann, the author of *Phoronomia*. In this work heat was for the first time defined, not merely as motion in a vague general sense, but as energy or *vis viva*, having a precise mathematical meaning. This same point of view was later adopted also by D. Bernoulli, who in his *Hydrodynamica* showed, moreover, how the elascity of gases might be accurately accounted for, instead of by the aid of an assumed repulsive force or expansive imponderable substance, as the result of the impact of the continually moving molecules on their collision with the walls of enclosing vessels. Other philosophers of eminence living in the last century, and amongst them notably Cavendish, concurred in the same view, or at least in that which generally averred the essence of heat to be motion; and whilst some defended it, moreover, by arguments, two investigators of the highest fame in philosophy endeavoured to sustain it by experiments also. Sir Benjamin Thompson (Count Rumford) and Sir Humphry Davy, just as the century drew to its close, proved by decisive experiments that heat might not only be transferred, but actually evolved, or in a manner created, by friction; which could not be the case if heat, as was then currently assumed, was a substance, but was intelligible enough if heat was allowed to be motion. The force of this argument as well as the naturalness of the whole view which made out heat to be motion, was well put in his celebrated "Lectures on Natural Philosophy" (delivered at the Royal Institution) by Dr. Young; who, having by signal discoveries reinstated the true philosophy of light, thus contributed also to keep in some measure alive the true philosophy on the kindred subject of heat. And, though somewhat tardily, the prophecy which he uttered in reference to this matter has come right after all. With singular foresight and trust in the ultimate triumph of truth, Dr. Young in 1802, being the very time when the Baconian theory had become most discredited, expressed himself as follows:—"It was long an established doctrine that heat consists in the vibrations of the particles of bodies, and is capable of being transmitted by undulations through an apparent vacuum. This opinion has been of late very much abandoned. Count Rumford and Davy are almost the only modern authors who appeared to favour it: but it seems to have been neglected without any good reason, and will probably soon recover its popularity."

In point of fact, it took the better part of a century to make this prediction true. Wearied by the ceaseless, and for the most part baseless, metaphysical speculations which characterised the first decades of this century on the one hand, and on the other, dazzled by the prodigious development of the power of mathematics, which occurred at nearly the same time, natural philosophers had begun to shun all topics bearing ever so slight a resemblance to metaphysics, taking refuge instead in the processes of mathematics, which had been found capable of evolving the greatest multiplicity of practical results from the smallest possible stock of physical notions. Yet though by tacit consent purely theoretical questions may be generally eschewed, the necessities of practice itself seldom fail to start them anew in some shape or other; and to this general rule the problem touching the nature of heat has not been formed an exception. Whilst Davy and Rumford had attempted to settle the point by an investigation into the excitation of heat due to mechanical force, as in friction, it was now the study of the reverse process, or of the development of mechanical force by heat, as in the steam engine, by which the question was again brought under discussion. The first to start this practically, no less than theoretically, important inquiry into the production of motive power by the agency of heat, which until then had been neglected, was S. Carnot; who, in an essay published now just forty years ago, proposed a theory according to which the evolution of force by heat was owing to a degradation of temperature, not of bodies, which would involve losses of heat, but of quantities of heat, which thus remained undiminished—similarly as masses of water, by a change of level, can impart move-ment to other bodies without undergoing any *material* diminution. This view, though radically false, as must be apparent to any one who analyses the simile just employed, suggested by Carnot himself—involves nevertheless an element of truth of great practical importance. And, in an equally singular and tortuous manner, the little pamphlet in which this theory was published contributed in a greater measure to revive, and ultimately re-accredit, sound notions on the nature of heat—including, by construction, the correct solution of the mechanical problem primarily attacked by its author—than almost any subsequent publication. Ere, however, the influence of Carnot's essay had made itself felt upon this wider basis, the more limited question which he had attempted to solve had been fully determined. To this result three investigators contributed independent and diversified shares. M. Seguin, fifteen years after Carnot's contrary assertion, maintained that, in the production of mechanical force through the intervention of heat, heat was absolutely lost, and that between the amounts of heat so lost and the amount of force so gained there existed an invariable numerical ratio, which he attempted to calculate, but only with partial success. Starting from more com-prehensive premises than Seguin, but agreeing with him in the particular question before us, Dr. Mayer, a few years later, succeeded in deriving the numerical value of the magnitude sought to be determined by Seguin, with as much accuracy as the then state of knowledge of the sub-sidiary data of his calculations permitted. Lastly, but a few months after Mayer, Mr. Joule, from experiments especially undertaken for the purpose, published values of his own of the same quality, previously but indirectly computed by Mayer.

It is curious to note the different points of view which these three authors, as well as their predecessor Carnot—working, as far as is known, independently of one another, and with such diversified results—assumed relatively to the question of the intimate essence of heat. Carnot, who throughout his pamphlet had reasoned as if heat were a fluid capable of producing motion (but it was not proved owing to what cause) in the same way as flowing water does, had nevertheless explicitly asserted that heat was but molecular motion. Seguin and Mayer, on the contrary, had considered heat and motion to be interchangeable; but nevertheless the first of them pronounced the question as to the nature of heat to be for the present inapproachable, whilst the latter had distinctly denied that heat could be motion. Mr. Joule, finally, who was agreed on the mechanical question with his two last-mentioned predecessors, asserted at the outset, like Carnot, that heat essentially could only be motion. And this foregone conclusion of Mr. Joule, as soon as he had proved by decisive experiments that a given quantity of motion produced under all circumstances the same quantity of heat, became, if not an irresistible, at least a most plausible, inference from facts. Mr. Joule had found that the energy which, when involving finite motion, is expressed as 772 footpounds may be converted into exactly one conventional unit of heat; which, if it does not prove, finds at all events its most natural explanation in, the assumption that the unit of heat expresses similarly a quantity of energy, involving, however, no longer the finite motion of complexes of masses, but the infinitesimal excursions of the molecules of masses. Accordingly, in the conversion of motion into heat, or *vice versa*, there is no transmutation implied of two heterogeneous things into one another, but only a transformation of finite excursions into infinitesimal, due to the transfer of velocities from entire bodies to their component molecules. If the measure of heat be only rational, or at least fixed, and thus indicate correctly the amount of motion gained or spent in each case, it cannot fail to result that a certain relation subsists between that measure and the measure of energies ordinarily applied to finite excursions. Actually, this relation is such, as demonstrated by repeated practical comparisons, that if a quantity of *water* be allowed to fall through the space of 772 feet, and the complexual velocity of 222 feet which it has then acquired be coverted into molecular velocity by arresting the further fall of the whole mass, the temperature of the water will have increased just 1° Fahr. This is what is understood by—and only in this way can be really understood—what is called the *mechanical equivalent of heat.*

Thus the most natural inference to be derived from the investigation of the generation of motion by heat, or *vice versa*, strongly corroborates the truthfulness of the Baconian notion of heat. Soon after Mr. Joule's experiments, this view was adopted by notable physicists and mathematicians, being almost immediately applied by Mr. Joule himself to the explanation of the theory of gases—a task in which he unconsciously only reproduced the previous results of Bernoulli, which later again were re-discovered by Dr. Krönig. This subject has since been fully developed and brought to a considerable pitch of perfection by several authors, to whose labours and authority it is not a little due that the Baconian philosophy has since found a great number of adherents. But whilst the researches we have hitherto mentioned all confirm the adequacy of the Baconian view to explain known facts, there is another field of investigation whence proof may be derived of, perhaps, its necessity. Even more intimately connected with the question before us than the phenomena hitherto discussed are those designated by the name of radiant heat ; and these, more than any other, afford us the direct clue to the nature of heat. Nearly two centuries ago, Mariotte apparently succeeded in separating the effects of light from those of heat, from which it seemingly followed that these agents must be different also as to their origin. On the other hand, numerous experiments had indicated a certain similarity between light and heat under particular conditions, especially as to their mode of prorogation ; hence it was concluded that heat was capable of manifesting itself under certain circumstances in a form which, from its similarity to radiant light, was called radiant heat. What radiant heat really was, very few philosophers even of the first half of this century understood ; yet Newton, nearly two hundred years ago, had given an accurate account of it. According to Newton, what is called radiant heat are the vibrations—or, more correctly, the undulations—of an all-pervading ethereal medium, concerned also in the propagation of light, and which, arising from the vibrations of the molecules of bodies called hot or warm (but whose temperature may, in reality, be any whatever), produce the effects of heat on those bodies which they impinge. So much being granted, the further question arises, what is the relation between the undulations just described, and those other undulations which are now generally recognised (although Newton would not allow it) to constitute light in its state of propagation ? It had been found that the invisible radiations which produce but effects of heat were liable to exactly the same vicissitudes as the visible radiations which, besides general effects of heat, produce also in our eyes the particular sensation denoted as light; and, from a great discovery made by Sir W. Herschel, it was known also that among invisible rays there exist varieties of refrangibility entirely similar to those which Newton had discovered among the visible—limited, however, by this, that the most refrangible of the invisible rays just equal in refrangibility the least refrangible of the visible rays. The natural inference from all this—which Herschel also at first had recognised, but from which, at a later period, owing to a strange misunderstanding, he dissented—was stated at the beginning of this century by Dr. Young in the following words :—" Suppose the undulations principally constituting [invisible] radiant heat to be larger than those of light; while, at the same time, the smaller vibrations of light, and even the blackening [sometimes called photographic] rays, derived from still more minute vibrations, may, when sufficiently concentrated, concur in producing the effects of heat. These effects, beginning at the blackening rays, which are invisible, are a little more perceptible in the violet, which still possess but a faint power of illumination; the yellow-green possess the most light; the red give less light but much more heat; while the still larger and less frequent vibrations, which have no effect on the sense of sight, may be supposed to give rise to the least refrangible rays, and to constitute invisible [radiant] heat." According to this view, invisible rays differ from visible intrinsically but in the same respect as the blue, for instance, differ from the green rays— namely, in point of degree ; although physiologically the difference both between visible and invisible rays, and between blue and green rays, is immense. After considerable hesitation and contention, the opinion of Young has now become almost universally adopted—it being proved, on the one hand, that invisible and visible rays are alike in nature, and on the other, that the pretended separation of light from heat reposes in reality only on the separation of invisible and calorifically strong from visible and calorifically less powerful rays. Now, if it be allowed, first, that invisible radiant heat and visible radiant heat (which is the same as radiant light) are essentially alike ; and secondly, that radiant light consists in undulations due primarily to molecular vibrations —which latter point was even allowed by Newton and is confirmed by signal discoveries made within recent years—and which differ from radiant heat that is invisible only in the physiological accident of perceptibility by the eye ; it necessarily follows that what under all circumstances when propagated from one body to another, produces effects of heat, and in animals sometimes produces the sensation of light besides, consists in those very molecular vibrations which, when of a certain pitch, are capable of giving us the sensation of light, but otherwise only the sensation of heat. In other words, from the correct interpretation of the phenomena of radiant heat it results that heat, " its essence or quiddity," as Bacon said, is motion.

The fruitfulness of the Baconian definition of heat has, since the brief period of its revival, shown itself already in various ways, and it is impossible to overlook how great its influence is becoming in remodelling our traditional systems —if systems they may be called—of philosophy. Considered in its most enlarged aspect, it suggests an almost entirely new view of the whole of nature. In the words of the essay, of which the present forms an abstract :—" If heat is motion—and this is the distinctive point of the new or revived philosophy—it follows, from all what we observe, that the condition of material existence in the universe is motion, not rest. We may imagine in our minds quiescent matter, but nature affords us no such example. Whatever may be thought of its horror of a vacuum, nature certainly seems to abhor rest. Not only is the condition of life in superior beings, as identified with the circulation of sap and blood, motion, but inanimate systems also seem unavoidably pervaded by motion. The earth as a whole moves, so do the other planets ; even the sun and the fixed stars move. And as in the solar and other starry systems in heaven each great component mass perpetually rotates round its central axis, revolves round its central body, and with it partakes of a motion in progression, or perhaps alternately backwards and forwards ; so every atom also in the most impalpable cluster of molecules—being, like all things in nature, impregnated with heat—rotates, revolves, advances, and retrogresses for ever in the infinitesimal space allotted as the scene of its ever-changing existence."

IV. MYSTICISM AND ASCETICISM.*

T HERE is something almost startling in the clash of
opinion that makes itself heard at the present day.
We find men of culture and standing, graduates of
universities and members of learned professions upholding
in evident good faith doctrines which differ from each other
respectively even more widely than from generally received
beliefs.   On the one hand we hear the calmly dogmatic
denial of a spiritual element in man, of a future life and of a
personal Deity; on the other we perceive a revival of what
our grandfathers derided as worn-out superstitions, a belief
in "occult science," in demoniacal possession and in
magic.

The world at large shakes its head at the Materialist,
sneers at the Mystic and passes on its way, muttering perhaps the old question "What is truth?" and leaves. the
decision of the great questions raised to the mercy of
accident.

Not all minds, however, can reconcile themselves thus
easily to the chaos of theories now prevailing.   If we are
so completely in error as both the Materialist and the
Mystic agree in asserting, not merely our philosophy but our
daily practices must be profoundly erroneous, and require
revision.   Our science, it must be confessed, is not only
very limited, but it involves a fearful amount of unproved
assumptions.   Naturally then, we cannot—ought not to—
shut our eyes against any additional light provided we can
feel sure that it is light.   Here then lies the difficulty.   Can
the professors of Occult Science make good their claims?
If they possess the knowledge and the power ascribed to
them by Dr. W. there can be little difficulty.   Thus we are
told that certain Hindoo adepts are able to raise themselves
up in the air, influence the weather, or to call up storms
unlimited in extent and violence.   Be it so: let then one of
these sages float, not in a building or in a cavern, but under
the open sky.   Let him, for instance, soar from the Monument to the Victoria Tower at Westminster, fly across the
Thames, or even hover along one of our main streets above
the heads of the crowd.   Or if, as the author hints, our

* Theosophy and the Higher Life, and Spiritual Dynamics and the Divine
and Miraculous Man.   By G. W ....., M.D., Edin., President of the British
Theosophical Society.   London: Trübner and Co.

climate may not admit of the display of such powers, let
the experiment be made at Calcutta or Bombay. If an adept
does so then assuredly no "scientist will desire either to
place him in a madhouse or hale him before the magistrate." He who is able to perform a crucial experiment and
refuses so to do has little room to complain if his doctrines
meet with no acceptance, and if their rejection is an injury
to the world he incurs, as it seems to me, a very grave moral
responsibility. When a man of science has obtained a novel
and important result he makes every step of his procedure
fully known, so that all who feel interested may themselves
judge whether he is in the right or in error. Suppose that
when Davy discovered potassium he had neither exhibited
his product before those capable of judging, nor described its
source and the manner in which it might be obtained, but
had contented himself with saying that he had produced a
metal so light as to float upon water and capable of taking
fire if moistened : had he done this he would have met with
not credence, but rebuke.

Again, if any Hindu adept can influence the weather, the
death of all his countrymen who perished of famine lies
at his door.

Further we read :—Hence the adept can consciously see
the minds of others. He can act by his soul force on external
spirits. He can accelerate the growth of plants and quench
fire, and like Daniel subdue ferocious wild beasts. He can
send his soul to a distance and there not only read the
thoughts of others, but speak to and touch these distant
objects ; (what objects ?) and not only so, but he can exhibit
to his distant friends his spiritual body in the exact likeness
of that of the flesh. Moreover, as the adept acts by the
power of his spirit he can as a unitive force create out of the
surrounding multiplex atmosphere the likeness of physical
objects to come into his presence."

It need scarcely be said that any person thus endowed
could, if he thought proper, demonstrate his pretensions quite
as easily as the engineer can exhibit the uses of some newly
invented machine or the chemist display the attributes of a
recently-discovered compound. Nay, the task of the adept
should be the simpler and easier, requiring as he does
neither models, apparatus, nor materials. The whole spirit
of the Inductive Philosophy,—which' is after all merely
organised common sense, bids us suspend judgment where
facts are not forthcoming. We accept the testimony of a
Hofmann, a Würtz, or a Bunsen as to a new chemical fact,
simply because we know that their experiments will have

been repeated in a hundred laboratories, and that these eminent men have been conscious that such scrutiny would follow. Surely, therefore, all who have new truths of any kind to bring forward should be prepared to meet this reasonable demand, and should be the more careful both as to the quantity and quality of their experimental evidence the stranger the conclusions which have to be demonstrated.

On the other hand, it must be owned with regret that facts which do not square with established opinions are received not with critical scrutiny, not with a philosophical suspension of judgment, but with an obstinate and prejudiced scepticism; with the wish and the determination to find nothing but jugglery and imposture. Such a spirit is, if possible, more misleading than the grossest credulity.

Any thoughtful person on taking up Dr. W.'s work, which differs little in its teachings from Bulwer-Lytton's "Zanoni," will above all things wish to know how the knowledge and power of the adept are to be obtained? He will ask if the teachings of occult science supplement and extend those of open and acknowledged science, and if its spirit and its methods are similar? The replies to these questions are not satisfactory. We may read the chapter "How best to become a Theosophist," not as sceptics or scorners but as earnest searchers for truth, yet we shall find in the instructions nothing feasible or tangible. No intellectual discipline is enjoined; indeed occult science seems more an affair of the will and of the moral faculties than of the reason. The course of training prescribed consists mainly in austerities and asceticism,—of which more anon—and in contemplation rather than speculation.* The adept is subjective rather than objective in his endeavours. He does not seek to sharpen his senses, or to improve his observant faculties, but the reverse.

As I gather from another source " he learns to live underground, and for this purpose he digs a subterranean cavern (the *gublia*) in which he passes most of his time. The temperature must be warm and perfectly even, and the cavern is entered only by a hole which can be closed with a stone. Indeed, the essentials of the mode of life are the complete exclusion of free air, impenetrable darkness, and an unbroken silence. He lies upon a pallet of cotton or wool—something warm and soft—at the bottom of this subterranean cell, and repeats from day to day the mystic

* I use this term in its true and original sense and not as an apologetic synonym for gambling and thieving.

word " Om," the Hindoo name of the great abstraction of
universal life—a being more transcendental than that of
Hegel. The devotee takes occasional walks, but is very
slow in his movements, so as to lessen the rapidity of the
respiration. He repeats his " Om" sometimes 10,000 times
a day, and has other syllables among which are " Bam,"
" Ham," " Lam," " Ram," ·· Soham," " Yam," of which
he performs endless series of repetitions, arranging them in
every order of which they are susceptible, and rigidly following a prescribed order for a given number of repetitions.
He trains himself to sit squatted for hours together in a
.certain peculiar attitude (the *siddhasana*), which consists in
doubling the left leg under the body, so as to rest upon the
heel of the left foot, while the right leg is extended forward.
In this position, with the right arm advanced, he holds the
big toe of the right foot in his right hand, and with the left
arm flexed under the body, grasps the big toe of the left foot.
This brings the lower part of the face to rest on the breast
bone. In this awkward and difficult attitude the fakir sits
for hours together ; that is, when he is not standing upon
his head or training himself to take a deep inspiration and
expel it slowly—taking twelve seconds to breathe it and
twenty-four to breathe out the cubic feet of atmosphere that
the lungs can contain. Besides these exercises, his tongue
has to be cut twenty-four times, so as to sever all the liga-
tures one by one, and enable him to flex it backward and
close the throat with its tip."

Dr. W . . . . does not, indeed, advise a similar way of
life for those who are aspiring after the " Christian adept-
ship of the West." But his recommendations plainly tend
o an intensification of that phase of our being which
concerns itself with moral good and moral evil. In the
modern savant as met with in Europe and America this
phase is comparatively less prominent than in ordinary men
of the world. He must, indeed, possess in an eminent degree
the virtues of industry, patience, and perseverance, and his
truthfulness must be beyond suspicion. Nor can he indulge
in sensual excesses of any kind without traversing his own
career. But his mind is mainly set and most intently fixed
on matters which are neither good nor evil. This peculiarity
was perhaps most clearly manifested in Henry Cavendish, a
man incapable of love or hatred, of fear or hope, and simply
absorbed in the study of certain classes of physical phenomena. There is probably no other known case of a complete
atrophy of man's emotional nature. Still we may learn from
this extreme instance. that the savant and the theosophic

adept represent opposite poles of humanity. The more closely we approach the position of the one the more widely we are separated from that of the other. Hence I see no hope that occult science even if it should succeed in substantiating its claims, can supplement and extend our present knowledge.

But I may raise the question whether a course of training such as that undergone by the fakir is not eminently calculated to promote hallucinations and delusions of every kind? One of the main points of his discipline is a reduction as far as possible of the respiratory process. He must learn to breathe slowly, and the atmosphere of his underground cell will undoubtedly become very foul.

As a matter of course the aëration and purification of the blood will be very imperfectly performed,—a state of things that cannot but react injuriously upon the brain, especially in conjunction with the exercise of standing upon the head. It would be hard to conceive a mode of life more unfitted for a sober truth-seeker. That by these means the adept ultimately becomes capable of existing in a state of trance for weeks or even months, during which circulation and respiration are apparently suspended, seems fully proven. But this sham-death is after all merely an artificial reproduction of the winter-sleep of certain animals and can scarcely be accepted as a higher life than that of waking man. Such an adept seems to me rather an object for observation and experiment than as being himself an investigator and discoverer.

There is another feature in the teachings of the Theosophists which must not be overlooked—the glorification of asceticism. I have very little sympathy with or respect for the sensualist, but the ascetic I regard with loathing mingled with dread. The austerities and torments which he inflicts upon himself he is ever ready, if opportunity offers, to force upon others. Almost every persecutor, every ecclesiastical tyrant, of whatsoever creed, is recorded as given to fasting and penances.

An every-day observation proves the demoralising tendency of asceticism. Every man is harsher, more severe, less obliging when hungry than when satisfied,—a truth which extends to the very beasts.* What then must be the accumulation of " envy, hatred, malice, and all uncharitableness " in a man who is always hungry ? The ascetic, if sincere, is a Torquemada, a Tilly, or a Cotton Mather; if

* The Lent of the Greek Church and the Ramadan of Islam are saturnalia of all crimes of violence and cruelty.

insincere he is an Angelo. What a grim choice! I have somewhere read that every virtue is the golden mean between
two equal and opposite vices which are alike to be avoided.
Now the ascetic not merely practises a vice, but, unlike the
sensualist, he even glories in it and pronounces it a virtue.
He calls evil good, and may thus be regarded as a demon-
worshipper.

I may venture to call attention to the following excep-
tionable passages in Dr. W....'s work.

To the adept, we are told "alcohol is strictly forbidden
and the flesh of animals." This is a strange declaration
from one who recognises Jesus of Nazareth as the highest
of adepts, though described as partaking both of wine and
animal food and of whom it is written* "The Son of Man
came eating and drinking."

We are next told :—"It is a remarkable fact that among
the lower animals, the female, who generally becomes pregnant by a single act, so long as she is pregnant, rejects all
further approaches of the male with anger and indignation.
Does nature not in this wonderful fact teach man?" This
"wonderful fact" does not hold good among the *Simiadæ*,
the animals, be it remembered, most nearly approaching
man.†

Elsewhere we read :—"We know that the air-plant
flourishes without any soil and that gold-fish flourish in
pure water without visible organic food." Can Dr. W. prove
that gold-fish elaborate from water and air all the elements
needed for a healthy life? Air-plants, too, receive nourishment from the juices of the trees or the decaying wood to
which they cling, or even from the dust, organic and
inorganic, floating in the atmosphere. It will be quite time
enough to assert that they can flourish without any soil
when they are found to grow and obtain all their necessary
elements from distilled water and air fitered through cotton-
wool.

These errors relate merely to illustrations and to points
of secondary importance, but they show a want of accuracy
and thoroughness which leaves a disagreeable impresion
upon the thoughtful reader. If the author is not to be trusted
in such simple and familiar questions will he prove a safe
guide in matters more recondite?

I cheerfully admit, indeed, that there is much between

---

* Matthew xi., 19.

† It breaks down in other quarters also. But may we not remark that the
very essence of asceticism is to disregard what Nature teaches.

heaven and earth not dreamt of in our commonly received philosophy. But I fear lest such teachings as those of Dr. W. may not tend rather to obscure truth and hinder its recognition. Above all I hold that every friend of humanity should wage a war of ·extermination against asceticism as one of the foulest survivals of ignorance and savagery.

FRANK FERNSEED.

V. THE SANITARY MILLENNIUM.*

NO opinion is more common among the educated classes than that man's control over the forces of nature is becoming almost complete and that if any ruinous catastrophe occurs it is because the authorities or the public do not make use of the power and the knowledge they possess.† It is hinted that a sunless, dripping summer is due in some mysterious manner to the malfeasance of an administration; and it is openly suggested that if cholera, or typhoid, or small-pox, or any other epidemic henceforth breaks out in a town the municipality or other governing body should be subjected to severe penalties as having neglected their duties. It is, of course, exceedingly satisfactory to read an account of the Black Death in the fourteenth centvry, of the Plague of London in 1665, or of Marseille in 1720; of the visitations of epidemic dysentery, of sweating sickness, and the other forms of pestilence which attacked the known world every few years from about the 7th century to the middle of the 18th—and then to be told on official authority that we have " changed all this ;"—that horrors of such kind are no longer possible.

For have we not now fifty years ago inaugurated " sanitary reform ? " Have not certain learned men discovered for us that the only cause of the so-called zymotic diseases

---

* Epidemiology ; or the Remote Cause of Epidemic Diseases in the Animal and in the Vegetable Creation. With the Cause of Hurricanes and Abnormal Atmospherical Vicissitudes. By JOHN PARKIN, M.D. Part II. Second Edition. London : David Bogue.

† Thus the editor of the *Lancet* (December 23, 1871) commenting on the fact that in the same year 13,174 persons had died of small-pox in the seventeen chief towns of England, pronounces the fact " discreditable to the intelligence of the rulers and people of this country."

is dirt? That if we take care to keep our drinking waters free from pollution, abolish cess-pools, ventilate our houses, and avoid overcrowding, we may laugh at epidemics, and may, if we wish, reduce our death-rate to five or six in the thousand? Have we not spent some 300 millions of money in carrying out these teachings and in paying the fees of our teachers?

That we have parted with the money is a truth that cannot be gainsaid. But in view of the importance of the subject it will not be deemed any undue outbreak of scepticism if we ask what we have purchased therewith, and whether our modern security is as well-founded as we think.

For a better understanding of the case let us first refer to the plague in London in 1665,—the last be it observed of 22 visitations of the same disease in England. We are commonly told that the great fire of the succeeding year thoroughly purged the city, which was rebuilt in such an improved manner as regards cleanliness that the recurrence of the disease has been thereby prevented. A little reflection will show us that here lurks an error. Plague had not been confined to London ; it had from time to time ravaged the provincial cities and even the villages of England. Surely the fire of London could not improve their sanitary condition or secure them from further visitations. Nor could it in any way explain the gradual disappearance of the scourge from all parts of western Europe. Its last outbreak in France occurred at Marseille in 1720. Now it cannot be contended that either in England or in France any very marked change had taken place betwen the years 1650 and 1750. Social and domestic arrangements had not been so modified as to explain this cessation of pestilence. No additional precaution had been taken, and the intercourse between different nations —and in consequence the supposed facilities for the transmission of disease—was all the time increasing. Since then the plague has been confined to Egypt and Turkey, and even there its ravages have become less frequent and less severe, and till within the past few years it seemed likely to cease altogether. Yet in the East nothing bordering upon sanitary improvement has been undertaken. Surely then it must be conceded that the gradual recession of the plague from Europe and its almost extinction in Egypt, Syria, and the adjacent regions cannot be attributed to any human agency.

Let us now return to England : according to Dr. Percival and Mr. Griffith Davis in 1757 the death-rate in Manchester was 1 in 25, or 40 per thousand. In 1770 it had fallen to 1 in 28, or 35·71 per thousand, and in 1811 to 1 in 74, or 13·51 per thousand. Here, then, is a distinct and rapid advance towards the much talked-of "sanitary millennium,"—a decrease of mortality to the extent of two-fifths! Since that date most remarkable changes have been effected. Medical science has improved, the condition of the people has been ameliorated, the hours of work in factories have been limited, vaccination, which, in 1811 was merely optional, has been rendered compulsory, and "sanitary reform" has been fully inaugurated. In all these benefits Manchester has participated, and since about 1845-50 it has enjoyed the special advantage of an ample supply of water drawn from the unpeopled and uncultivated moorlands of north-west Derbyshire, where organic pollution seems out of the question. We might reasonably then expect that the advance above mentioned would be sustained, and that the death-rate of the great cotton-city would have fallen to—at most—1 in 100 or 10 per thousand. But alas! in the ten years from 1851 to 1860 inclusive the mortality has been 31 per thousand!* In other words, in spite of all our modern sanitary reforms, the death-rate has increased to more than double its amount in 1811.

Hence we are again led to the conclusion that the reduction of mortality which characterised the latter half of the 18th century and the beginning of the 19th is not due to modern science, to advancing civilisation, or to any human agency. From the evident impotence of "sanitary reform" to maintain the improvement which had taken place we are also warranted in questioning the soundness of the principles on which it is based.

Amongst the diseases which had become less frequent and less severe, but which have since resumed an epidemic and highly dangerous character, a prominent place is due to small-pox, especially as its alleged preventive, vaccination, has taken rank among the political questions of the day. We are told that if this disease no longer carries off its victims by tens of thousands, as in the dark ages, the change is due to vaccination. But there can be not a shadow of doubt that small-pox had begun to decline long before the discovery of Jenner was introduced into practice.

In 1722 Dr. Wagstaffe wrote that the mortality among children did not exceed 1 per cent. of the cases. From 1796

---

* According to the returns of the Registrar-General (September 25th, 1880) it has been for the past week 27 per thousand.

to 1825 there was not a single epidemic of small-pox in England. Yet, according to a report published by the College of Physicians in 1807, only about 1½ per cent. of the population were vaccinated. Now if we admit that the immunity gained by this operation is absolute and permanent, how is it possible that 3 vaccinated persons out of every 200 would protect the remaining 197? At the present time about 97 per cent. of the population are supposed to be vaccinated. Yet so far from being able to protect the residual 3 per cent it is considered that they are imperilled by the obstinacy or neglect of this small minority. We have the lamentable fact that, whilst vaccination has become all but universal, small-pox has reappeared among us not in isolated cases but in epidemics, succeeding each other at short intervals, and each more deadly than the foregoing. Thus in the epidemic of 1857-58-59 the deaths were 14,244; in that of 1863-64-65 20,059, and in that of 188c-71-72 44,840. Thus in the first interval the deaths from this cause had increased 50 per cent, whilst the population had grown only 7 per cent. In the second interval the deaths from small-pox have risen by 120 per cent, but the population only 10 per cent. Another ugly fact is that the number of persons who have been vaccinated but who are subsequently attacked with small-pox is steadily on the increase. At the Highgate small-pox hospital from 1835 to 1851 the previously-vaccinated formed 53 per cent of 'the total small-pox cases admitted. In 1851-2 it rose to 66·7 per cent; in 1854-5-6 to 71·2 per cent; in 1859-60 to 72; in 1866 to 81·1 and in 1868 to 84 per cent. How are such facts to be reconciled with the orthodox theory that vaccination is a safeguard against small-pox? What would be the conclusion formed by an unprejudiced statistician if these figures were laid before him? If a grows more common as b increases in number and general distribution no man in his senses will argue that b is a hindrance to a. The very opposite conclusion, that b is causally connected with a would seem more legitimate. How the credit of vaccination is to be saved is not apparent. We cannot cut the knot by supposing that modern medical practitioners are less careful and skilled in the performance of the operation or less scrupulous in the selection of vaccine lymph. There remains, then, merely the conclusion that small-pox, too, has had a period of cessation during the latter part of the past century and the first quarter of the present;—that the apparent success of vaccination was mainly due to its coincidence with this temporary lull, and that the disease is now rapidly regaining its old virulence and re-assuming the pestilential proportions which it displayed in the days of our forefathers.

We can now appreciate one portion of Dr. Parkin's theory. He holds that there occur certain "pestilential epochs" during which the world is at frequent intervals devastated by epidemics which travel in a determinate direction from central or eastern Asia to the west of Europe and even to America; that during such epochs all diseases, even those not considered as communicable from one person to another, increase in frequency and in violence; that these epochs are further marked by epizöotics and by "blights" or widespread diseases in the vegetable world, and are attended by a general intensification of earthquakes, storms, floods, droughts, fogs, seasons of abnormal heat or cold, and other convulsions of inorganic nature. Such an epoch is generally ushered in by the appearance of new diseases or of the reappearance of maladies that had become obsolete. Dr. Parkin holds that the last great pestilential term began about the 7th century. The advent of plague was accompanied by measles, small-pox, and malignant sore throat. From that time to the beginning of the 18th century Europe and western Asia were almost incessantly ravaged by epidemics, among which the plague, or Black Death, was most prominent and most formidable. During all these hundreds of years there was no such freedom from pestilence as western Europe has enjoyed between 1750 and the advent of the cholera in 1830.[*]

This non-pestilental period presents some most remarkable features. The year 1759 saw the close of a violent epizöotic, and from that time till quite a recent date England has been free from any generally prevailing disease among cattle. About 1692 the yearly deaths from dysentery in London had exceeded 2000. In 1799 they only amounted to 13! Between 1728-38 the burials of children under two years of age averaged in London 10,000 yearly. Between 1790-1800 they had fallen to 6000.

The author further declares that after the great earthquake of Lisbon in 1755 terrestrial concussions declined in Europe and Asia both as regards numbers and violence. The season of 1766-7 was the last of those visitations of

[*] Nevertheless we must not forget the outbreak of yellow fever at Malaga in 1803, which carried off 36,000 persons and was accompanied, according to Waterton, who was then a resident in that city, by seven shocks of earthquake.

intense cold which survive in tradition as "old-fashioned winters."*

Dr. Parkin also holds that whilst the pestilential epochs are marked not merely by physical but by mental diseases and by delusions, the non-pestilential epoch is characterised by an awakening from errors and superstitions. He writes: —" The mists which had so long obscured the human intellect, particularly during the Dark and Middle Ages, became suddenly removed. . . . It was then also that the glorious galaxy of intellectual stars made their appearance—Galileo, Kepler, Newton, Descartes, Bacon, Locke, Shakespeare, Milton, Dante, Molière, Corneille, Goethe, and a host of philosophical and scientific writers to whom we are indebted for those great discoveries that have enriched the present century."

We are by no means disposed to deny that a pestilential epoch may be fertile in delusions, a subject to which we may return below. But the great French revolution which, with its attendant wars, entirely belongs to the non-pestilential epoch (1750 to 1830) was rich in delusions. With the "intellectual stars" above enumerated we feel perplexed. We do not see how any of the names mentioned, with the single exception of Goethe, can be considered to belong to the " non-pestilential epoch."

In support of the view that we have entered upon a new period of epidemics the author adduces as evidence the appearance of cholera, of which we have had three successive visitations, the recrudescence of small-pox, the advent of diphtheria, the prevalence of typhoid, the increase of carbuncle, typhus, scarlet fever, and diarrhœa, the recent epizöotics, rinderpest, foot and mouth disease, the increase of rabies, and the maladies among vegetables, such as the vine, the potato, and we may add the coffee tree. Delusions are also growing apace—Dr. Parkin selects as an instance ritualism, a subject which does not lie within our limits. He might perhaps have here brought forward vege-

---

* Here, again, we must make a reservation. According to White, the frost of 1768 was the " most intense that we had then known for many years, and was attended by an epidemic among horses." The winter of 1771 was long and severe, the snow lying for eight weeks in the island of Skye, where snow seldom lies at all. In the spring following most of the cows were barren, the sheep perished for want of food and the grass did not grow. In 1776 the roads in the west of England were blocked up with snow; at Selborne the thermometer fell to 7° and 6° Fahr., and in Kent to—2°. The summer of 1783 is described as " amazing and portentous." A reddish fog covered all Europe for weeks, irrespective of changes of the wind. Calabria and Sicily suffered from earthquakes, and thunderstorms were unusually violent and numerous.

tarianism, the woman's rights movement, Mormonism, Nihilism, and the general tendency to political agitation.

In dealing with the causes of epidemics, the author is, in our opinion, perfectly successful in disposing of the two most popular theories, the dirt doctrine of the sanitary reformers, and the notion of contagion. The facts which he mentions, most of which are common matters of history, are decidedly opposed to both. Thus, when pestilence reaches a great commercial centre—*e.g.*, the Black Death at Constantinople in the 14th century—it does not, as the contagion theory would require, radiate out on all the lines of traffic, but selects some single track. Even the Black Death did not pass northwards from Constantinople to Russia, but turning westwards coasted the Mediterranean, passed through France into England, thence to Norway and Sweden, though the intercourse was then exceedingly slight, and lastly passed into Russia. The cholera has been known to travel steadily for hundreds of miles in the teeth of a strong monsoon. It often works up a river, showing that it is not occasioned by infectious matter draining into the current. Pestilences have generally been found more violent in open airy places than in such as are close and sheltered. The yellow fever haunts the breezy Antilles and the coasts of Brazil and of La Plata, but penetrates inland only in proportion as the forest is cleared away. In the epidemic of cholera in Trinidad in 1854 the mortality at the leeward end of the streets was less by 5 per cent than at the windward end. Houses on open roads were attacked, huts in the bush escaped. Alike in epidemics of plague, cholera, and yellow fever, it has been found that classes of people who from occupation or habit were most exposed to the air suffered most, whilst those who kept themselves shut up escaped. How ill this agrees with the teachings of the sanitary reformers!

We must now turn to Dr. Parkin's own theory. He refers pestilence to what he calls "volcanic action," understanding thereby not merely eruptions and earthquakes, but the more quiet and continuous agency of which earthquakes and emissions of lava are the occasional and palpable results. He affirms that both earthquakes and epidemics are most common and most destructive in alluvial districts, and on the contrary rarer and less violent among the mountains. As regards earthquakes we cannot accept this opinion without hesitation. The western mountainous region of South America is far more troubled with earthquakes than the comparatively level lands of Brazil and Guayana, including the alluvial valleys of the La Plata, the Amazon, and the Orinoco. In Europe earthquakes have occurred chiefly in the mountainous peninsulas bordering on the Mediterranean, and the adjacent islands, whilst the great plain extending from northern France through Belgium, Holland, North Germany, Poland, and Russia, has, in historical times, suffered little from this agency.

Iceland, Jamaica, New Zealand, all mountainous, have all been the scenes of great volcanic activity. On p. 117 the author remarks that " in the region of the Andes, the oldest range of volcanos next to extinct volcanos, epidemics would appear to be unknown, while endemics are extremely rare. Dr. Bryson states that with some few exceptions in which ague and malarious fevers exist, both sides of this volcanic chain are extremely salubrious." Yet he elsewhere speaks of yellow fever as occurring at the port of Islay on the western side of the chain.

Dr. Parkin, if we understand him aright, considers that springs are more abundant in alluvial tracts than in rocky districts. This is exactly opposite to what we have always found. Springs are, *e.g.*, far more numerous in the mountainous districts of Cumberland, Westmoreland, and North Wales, than in the valleys of the Thames, the Trent, and the Mersey, or in the plains of Buckinghamshire, Essex, Suffolk, Huntingdon, Cambridge, Bedford, &c. This is a point of some moment if, according to Dr. Parkin's fourth law, " the effects of volcanic action—and if his theory be correct epidemic diseases—are always much greater and more perceptible near the sea, lakes, rivers, *springs*, &c."

There are, therefore, several points on which we should desire further light before we can accept the causal connection between pestilence, volcanic action, and abnormal meteorological phenomena as proven. Nor can we forget that much of the historical evidence brought forward as to the earthquakes, storms, floods, famines, and the like visitations in the Middle Ages, is not absolutely free from suspicion. Even if we absolve the old chroniclers of any intentional exaggeration, they had very scanty means of ascertaining the truth. There were in those days no observers armed with thermometers, barometers, wind and rain guages. The descriptions which have been handed down to us are merely the guesses or the random fancies of ignorant peasants transmitted by hearsay from district to district, and from kingdom to kingdom. It is only from the beginning of the 17th century that the record can be accepted as trustworthy.

We consider Dr. Parkin's work, however, worthy of the most serious attention. It literally teems with important facts and suggestive reflections. Nor should it be over-looked by the mass of the public as a treatise addressed solely or mainly to the medical profession or to those versed in biological science. Apart from the fundamental hypothesis that "volcanic" action is the main factor in-fluencing public health, and the condition of the weather, we find here very much which is assuredly true, and which ought to be known by all educated men. If the author is not mistaken in his forecasts political economists have little need to trouble themselves about the population question. Perhaps, too, some of the political complications of the day may receive a solution little dreamt of by statesmen.

ANALYSIS OF BOOKS.

Insect Variety: its Propagation and Distribution. Treating of
the Odours, Dances, Colours, and Music in all Grasshoppers,
Cicadæ, and Moths; Beetles, Leaf-insects, Bees, and Butterflies; Bugs, Flies, and Ephemeræ; and exhibiting the bearing of the science of Entomology on Geology. By A. H.
Swinton, Member of the Entomological Society of London.
London: Cassell, Petter, Galpin, and Co.

By way of introduction to the present work Mr. Swinton gives
us a history of the causes which led to its being undertaken.
The account of his entomological rambles gradually extending
from Hampshire, Surrey, and Kent to France, Italy, and more
distant regions, and of his researches into insect life, beginning
with the capture and determination of British species, and thence
passing to observations on the modifications of forms due to
differences in food or in climate, has about it the true aroma of
fields, the woodlands, and the mountains. To elder naturalists
especially it will suggest pleasant memories of the past. He
repeatedly laments the disappearance of many of the rarer and
more beautiful butterflies from England—an extirpation due not
so much to the mere spread of cultivation and enclosure as to
the greed of dealers and their emissaries and to the mania for
British specimens. He anticipates—not without reason—that
in the future our insect hunters will be "doomed to a sport composed of an influx of Colorado beetles, white butterflies, Hessian
flies, and woody oak-galls." He notices the strange fact how,
with little difference of climate, the fauna of the northern coasts
of France is so much richer and more diversified than that of
Kent or Sussex. Have the species thus missing in Britain
arrived in or been developed in France subsequent to the separation of the two countries, or have they since become extinct on
the northern side of the Channel? Italy he considers has fewer
attractions for the entomologist than for the artist and the antiquary. All the more attractive butterflies seem to keep to the
Swiss pastures and the Sicilian high grounds,—the latter a
locality where the traveller may be himself "collected" by
brigands. "The bare-backed Apennines and fertile Maritime Alps
are equally ignored by Italian savants; a few varieties of our
northern favourites perchance grace Florentine gardens, and
around Pisa you may just capture enough Diurni to convince
yourself you are abroad." The following passage will give our
readers a fair idea of the author's manner:—"The end of the

Blood Cicada was dire and classic. About the commencement of July there appeared as if by magic certain greyish insectivorous birds with a harsh and guttural note, among the sunny vines and the woody knolls where the Cicadæ had established their coteries, and these, sitting on the low brambles, sometimes two together, knavishly whistled a tune until an unwary songster was inveigled to respond, and so betray his hiding. The intruders then flew at him and brought him to the ground in their beak and claws, screaking most piteously *Whee ! Whee !* "

The author, like ourselves, feels surprised that so many naturalists should be busied with the physiology of the Mollusca and lower forms of marine life " while we possess so little definite acquaintance with the life-history and structure of insects," and that entomology is still in the hands of the majority of its followers a mere matter of nomenclature. He very judiciously advises collectors before putting a specimen to death to note its specific odour, if any, and to observe whether it stridulates.

With the following passage we cannot entirely agree : speaking of bees, wasps, and ants he states " The very excellence of the acts performed by these unreasoning creatures is, as Dr. Carpenter has remarked, a proof of a non-intelligent nature." We have always suspected that the learned and voluble Doctor when penning these words must have been in a state of " unconscious cerebration." By the very same course of argument higher intelligences—and we are old-fashioned enough to believe that such exist—might demonstrate the non-intelligent nature of man. Nor can we admit Lamarck's definition of *Insecta* as mere sentient animals unable to combine the simple ideas which they derive from their sensations.

A little further we find the interesting generalisation that the Lepidoptera most conspicuous for beautiful colour are the most lethargic.

The section on the odoriferous, corrosive, and coloured principles secreted by insects is exceedingly valuable. Much further research is here evidently necessary, the chemical portion of which will be impeded by the difficulty of obtaining a sufficient quantity of the material for analysis. We read that the caterpillar of *P. Machaon* gives off a powerful odour of fennel. This fact is not surprising when, as is often the case, it feeds upon fennel, but is less readily understood when its food-plant is the carrot, or the caraway. The electric shocks, said on the authority of a writer in the *Magazine of Natural History*, to be given by the caterpillar of the puss-moth require confirmation.

In describing the scent-organs of butterflies Mr. Swinton gives the needful caution that these insects may occasionally absorb the odours of the flowers they have visited.

We must take exception to the author's remark that the Diptera are unarmed : witness the genera *Stomoxys* and above all *Erax*.

Mr. Swinton enters at considerable length into the question of the function of the antennæ in insects. He does not, however, refer to certain experiments which to us appear crucial. Insects when their antennæ were coated over with wax or varnish have been found utterly unaffected by fumes from which under all other circumstances they shrank back.

Many interesting cases are given of the recognition of colour by insects, and we find here a suggestive remark that the gradual extinction of species attendant upon the clearing of forests, drainage of swamps, and cultivation of heaths may be in some measure due to the removal of mimetic shelter. Mr. Swinton considers that browns, blacks, chlorophyll tints, and purple " are the pigments we notice on protected surfaces," whilst the hues specially designed to evoke the passions consists mainly in white and its prismatic elements. We have long considered bright and pure hues* as connected with the highest vitality. The author considers that the hind wings of a beetle are universally colourless. This is not absolutely correct. The posterior wings, *e.g.,* of the Colorado beetle are pink ; those of *Lomaptera Urvillei* a beautiful violet blue; of *Belionota fallaciosa* a deep brown, of *Cotinis mutabilis* a greenish bronze on the anterior margin and of a blue shade on the posterior, the intermediate portion being of a reddish bronze. We have indeed noticed so many changes of decided colour in this part that we think the wings of every new or imperfectly known species should be carefully examined on capture.

Mr. Swinton raises further the very interesting question how the pattern is produced upon the wings of insects, especially butterflies. Does the secreted colouring matter " flow from the thorax through the tissues of the flaccid wings, and are the stains impressed by their membranous coverings in the manner of a printing press ? In support of the latter hypothesis he mentions that a specimen of *Leucania conigera* captured near the Welsh Harp Tavern to the north of London in 1877 was found to have the markings upon the upper wings reproduced on the upper surface of the left lower ; and a specimen of the common Meadow Brown taken near Oxford in 1878 showed the same transformation effected in the under surface of the same wings, accompanied with the development of an additional vein and consequent enlargement that seemed to afford a reason how the hind wing came to be folded so as to receive not alone the impress of the inferior side but the very eye-spot of the upper.

Neither of these hypotheses, however, is quite free from difficulties. We find in some cases, *e.g., Apostrophia charitonia, Papilio Podalirius,* and many more which might be easily selected, the under surface merely a somewhat dulled and subdued reproduction of the upper. In such a case we might readily

* Journal of Science, 1878, p. 37.

conceive the design as produced by the transfusion or application of a colour either from the thorax or from the coverings of the wings. But in some cases we find the underside of the wings patternless, and in others marked with a design totally different from that of the upper. This especially holds good with the hind-wings. Thus in *V. Atalanta* the fore-wings exhibit the same general features on both surfaces, whilst the under surface of the hind-wing presents a design not reducible to that of the upper surface. Do not the colouring secretions enter the wing from within and become deposited on its upper and under surfaces according to laws still to be discovered?

On the local variation of insects—especially butterflies—the author has brought together a great mass of interesting and suggestive observations, a large portion of which is original. He describes the seasonable varieties, those due to climates, to latitude, altitude, the proximity of the sea, geological formation, the neighbourhood of towns, and the action of the unnatural atmospheres here often encountered. He quotes the remarkable fact that from a dark female *Trephrosia crepuscularia* the off-spring in the first generation were one-half dark and one-half pale. In the second batch, the produce of dark parentage, they were dark in the ratio of two to one, and the third generation were all alike dark.

It is with regret that we must thus prematurely break off our examination of this work, valuable not merely as a good over-sight of modern results in what may be called philosophic entomology, but as pointing out in what directions the student may usefully direct his labours.

The question may of course be raised whether a plainer and quieter style, with less of poetical quotations and classical allusions, would not have been more suitable to the subject. We find, too, numerous typographical errors, such as Vitis Graber, *passim* for Vitus Graber, anthropods for arthropods, *C. cantenulatus* for *C. catenulatus*, chloric for hydrochloric, *Calopterus* for *Caloptenus*, &c.

We cannot conclude without expressing our obligation to Mr. Swinton for the great pleasure we have derived from the perusal of his work.

Annual Report of the Smithsonian Institution, showing the Operations, Expenditure, and Condition of the Institution for the year 1878.

THE principal features of this volume include a biographical memoir of the illustrious physicist Joseph Henry, drawn up by

Prof. Asa Gray, with a catalogue of his published researches. Next follows a eulogy on Condorcet by Arago, and of course delivered a considerable time ago. A great portion of this discourse is far too political in its tendencies to be dealt with in our columns. We must mention, however, an interpolation of the works of Pascal, of which Condorcet has been accused. The passage runs: "Speaking according to the natural light of reason, if there is a God, He is infinitely incomprehensible, since having no beginning and no end He can have no connection with us. We are then capable of knowing neither what He is *nor if He is.*" The last four words are not to be found in the older editions of Pascal's works, and Condorcet was therefore accused of a serious literary forgery. In 1803 Renouard even declared that "an obstinate search through the manuscripts of Pascal in the Royal Library had failed to discover the contested words." Nine years afterwards, however, he confessed that the words were really to be found on the fourth page of the manuscript exactly as they had been published by Condorcet, and they have further been discovered in an earlier edition of Pascal by Father Desmolets.

The volume also include a biography of Louis Agassiz, by E. Favre, written in a somewhat Cuvierian spirit. The author does not consider that his opposition to the doctrine of evolution resulted from prejudice.

. There is an interesting paper by F. A. Ober on the ornithology of the Caribbee Islands. In this narrow region he obtained eighteen species and varieties of birds new to science. Mr. L. Kumlein contributes also a short note on explorations in Greenland.

Notice must also be taken of Mr. D. G. Elliot's "Classification and Synopsis of the Trochilidæ."

Flora of Plymouth. An Account of the Flowering Plants and Ferns found within twelve miles of the town; with Brief Sketches of the Topography, Geology, and Climate of the Area, and History of Local Botanical Investigation. By T. R. ARCHER BRIGGS, F.L.S. London: Van Voorst.

WE have here a laudable, useful, and pleasant book, yet not without its painful side. Painful, we say, because, like every local flora and fauna in the United Kingdom—we fear it might even be said in the civilised world—it is a record of the gradual disappearance and approaching extirpation of harmless and beautiful species. The "ferny dells of Devon" have long been famous. But now, in a district comprising about one-ninth of that beautiful county, some of the loviest species, such as *Polypodium phegopteris* and *P. dryopteris*, are now described as "rare and local," "very rare."

Concerning *Osmunda regalis*, the king of our British ferns, it is remarked :—"Considerably reduced in quantity through the plants being dug up and brought for sale to Plymouth and other large towns. All who value and wish to preserve the natural floral and vegetable beauties of the neighbourhood should do their utmost to prevent and discourage the selling in the streets of recently uprooted plants of the *Osmunda* and of our other ferns and wild plants." This caution is the more reasonable since many specimens are torn up at the wrong season of the year and so carelessly that they generally perish. It will become necessary for naturalists who find a good locality for any of our rarer and more conspicuous wild plants, and indeed insects likewise, not to make the matter public, since information intended for the botanist or the entomologist falls into the hands of dealers, who at once send their agents to make a clean sweep.

As regards flowers and ferns, the recent London practice of buying pot-plants in the spring, letting them perish in the winter, and getting a fresh stock the next year, tells disastrously upon every species that cannot be propagated as easily as the scarlet geranium.

The lily of the valley, one of the loveliest members of our native flora, is still met with in the Plymouth District, but is very rare. We know an instance of a country gentleman who had noticed with regret the gradual disappearance of this plant from his grounds, on coming up to London and taking a stroll through Covent Garden was offered for sale a quantity of roots fresh from his own park. His gardener and gamekeeper were in the pay of a dealer !

It is remarkable how easily beautiful species are extirpated whilst noxious and unsightly weeds gain ground in spite of all human diligence.

The author sometimes mentions interesting facts relating to animals. Thus we learn that an oil-beetle, *Meloe proscarabæus*, or in more modern nomenclature *Proscarabæus vulgaris*, feeds upon the acrid and poisonous *Arum maculatum*.

The account of the climate of the district is very interesting. The annual average temperature is higher by 2·56° than that of Greenwich. Spring frosts are, however, very frequent and destructive, and mild as are the usual winters ever and anon a severe season makes havoc among tender plants, the cultivation of which is never even attempted in the open air in most parts of England. The air is more uniformly moist and the sky more generally over-cast than in the more eastern parts of our island, —circumstances which cannot fail to leave their impress on the local vegetation.

Mr. Briggs's work bears the impress of thoroughness and

accuracy, and will prove the more valuable on account of that disappearance of species which we heartily join with him in regretting.

The book is illustrated with a map showing a division of the area into six sub-districts.

The Morals of Evolution. By M. J. SAVAGE. London : Trübner and Co.

A FEW thinkers are now venturing to trace the origin and the growth of that moral code which is, in theory at least, recognised by the civilised world. They are seeking to prove that our ethical system meets with a complete sanction in the very constitution of the world around us, with which, like every other product of Evolution, it must, by the very fact of its existence, be more or less completely in harmony. Such an undertaking meets with but little favour among the generality of so-called orthodox theologians. For this jealousy there is, we submit, no just cause. If certain moral doctrines have, as the majority of Europeans and Americans believe, been revealed to us by the same Being who created the universe, and if such Being is infinitely wise, we may fairly assume that they will be, not arbitrary and accidental, but in full accord with the nature of man and with the circumstances in which he is placed. Thoroughly knowing the latter we might infer the former. Hence if contributions to evolutionist ethics happen to be atheistic this feature is not essential. We do not see that either Theism or Christianity stands on a less firm footing if it be shown that an approximately complete system of moral laws might be developed by man from observation and experience, just as a very incomplete and fragmentary trace of such laws has been reached among certain of the lower animals. A theory is not considered to be compromised if we find that it can be arrived at by more than one independent procedure. Nor will religion be discredited if it can be shown that even in the supposed absence of a life to come good conduct is sure to be rewarded, and transgression to be punished. Whether any one has fully succeeded in furnishing such proof is of course an open question.

Mr. Savage undertakes to show that the moral significance of the doctrine of Evolution has been misapprehended ; that it is " no hard and cruel force," but " the power and process by which, through the ages, the best is selected, preserved, and transmitted." It is, he contends, " of the very essence of Evolution to take account of Christianity." From this point of view he attacks the burning question of the day " Is life worth living ?" He considers morality and religion in the past, the origin and nature of goodness, the sense of obligation, the principles of selfishness and sacrifice, the relativity of duty, the real and conventional virtues and vices, the mutual bearings of morals and knowledge, moral sanctions, and the morality and religion of the future.

In dealing with such themes it would have been well if Mr. Savage had remembered that the position of the philosopher, like that of the poet, " should be higher than on the battlements of party." If we observe the men whom he refers to, the opinions which he cites, the instances of moral delusions which he selects, and note how he confers the proud title " a leading thinker of England," we must pronounce him one whose speculations lie to some extent outside of our jurisdiction. But we fear the elimination of the party element from philosophic works is becoming unfashionable. Were we not amazed lately to find one who has been looked on as the Aristotle of the 19th century using language only fit for professional agitators?

Mr. Savage protests with some force against a " misconception as to the moral significance of the doctrine of evolution." We may here remark that this misconception—if such it be— refers not to the doctrine of Evolution *per se*, but to the hypothesis of natural selection as the principle to which such Evolution is alleged to be due. But this by side let us ask whether those forms of life which survive in the struggle for existence are the " best " in any worthy sense of the word ? This struggle assumes two aspects according as it rages between members of one and the same species and between different species. In the former case it will doubtless be found that as a rule the strongest, most agile, bravest, and most intensely animated, will prove victorious. But will it not frequently happen that when the conflict is over the conqueror will have suffered so much injury that he will be inferior even to what the vanquished was at the outset ? Hence the notion that *e.g.* the combat of two male animals for a female must necessarily lead to the perpetuation of the species by the more vigorous is not necessarily correct.

As regards the contests between different species we need only refer to the numerous facts which have appeared in the *Journal of Science* for years past proving that in the contests between different species, animal or vegetable, it is not the most beautiful, certainly not the most useful to man, or as far as we can judge, to the world at large, which wins the day. In organic creation, just as in the competition of daily life, or in international struggles, it is the *Nachbar-fresser*—the most persevering and unscrupulous destroyer of all neighbours—which triumphs.

To the author's query " would it have been any better if the weakest had won and the unfit had survived ?" one sufficient

answer may be found in the letter of Nemo.*   A further reply
is that in many cases the triumph of the weaker, if it had been
possible, would have been very much better.   Suppose the place
in creation now held by the rat were occupied by a species
feebler, less courageous, less cunning, less prolific, would this
not be an immense gain?   The author says " if death is to
exist at all it is difficult to see why it is any more cruel to have
it come in this way than any other."   But the charge against
the " struggle for existence " is that it needlessly multiplies
death.   Nature calls a hundred species into existence, turns
them into a spherewhere there is only room for ten, and bids
them fight it out.   Man puts in a given space only so many
animals or plants as can there find room, and thus obtains far
finer products.   Much more might be said on this part of the
subject did space allow showing that " Natural Selection " has
little claim to the beneficent character with which the author
invests it.

Mr. Savage tells us that as the ages pass " physical force is
dethroned, and thought, first as cunning, afterwards as intelligence, sits on the throne."   It will be some time before physical
force is dethroned, or before thought in its higher manifestations
wins in the struggle for life.   " The nations discover," says
our author, " in the stern school of experience, that the people
which loves most is most closely bound together, developes
most of tenderness, pity, charity, and mutual help; that this
people wins, is mightiest, and so the fittest to survive."   But
what would be the fate of a nation which, like some of our
modern English " advanced thinkers," had outgrown patriotism,
and extended to all mankind alike that mutual help which others
confine to their own countrymen?   Just what would happen to
any individual man who should open his table and his purse
indiscriminately to all his neighbours whilst they rendered him
no similar favours.   Mr. Sydney Billing warns us that " A
nation actuated alone by moral law, with conscience as a regu-
lator or administrator, could not exist beside other nations im-
pulsed by a lower ideal because it would be the prey of instinctive rapacity."†

As regards the discussion of the question " Is life worth
living?" our author is scarcely fair.   He writes :—" That the
vast majority of men desire life—stay in instead of going out—
turns the question as to whether it is worth living into an absurdity."   To this argument it may be replied that men cling to
life from hereditary instinct, from hope or curiosity as to what
may befall them, from religious convictions, from a suspicion
that it is safer to " bear the ills we have than fly to others that
we know not of;" and, lastly, from a reluctance to abandon

---

* Journal of Science, Sept., 1880, p. 598.
† " Scientific Materialism and Ultimate Conceptions." *See* Journal of
Science, 1879, p. 665.

children or others dependent upon them to the mercy of the world. Such, it may be urged, are the motives which restrain the millions from suicide. Granting that the majority of the evils under which we groan are remediable, the individual is powerless against them. We know that whether we struggle to remove them or not they will remain to torment our descendants. Nay, so far as can be judged from what has taken place within the last half century, there is room to fear that life is becoming harder, that the beautiful is growing rarer, and that man has less and less time and opportunity to think of truth and goodness. Evolution does not imply that the progressive changes taking place are necessarily towards perfection. We trace instances of degeneration in the structure of animals. There is no reason why a similar phenomenon may not occur in social or national life.

In the chapter, " Morality and Religion in the Past," the author argues that the two have not been necessarily connected, but that religions have often sanctioned and encouraged immorality, whilst " every grand onward movement of the religious life of the world has been started as a moral protest." He maintains that the origin of goodness is a much more important question than the origin of evil. Moral perfection, he thinks, is not a condition from which man has fallen, but one towards which he must gradually approximate. He believes that " the moral law stands by virtue of its own right, and would stand just the same without any regard to the question of immortality or the discussion between theism and atheism." He does not, however, raise the point brought forward by Mr. S. Tolver Preston that the " dogma of infinitely lasting punishment is practically tantamount to teaching that the pursuit of vice must be infinitely profitable in this life."* He accepts happiness as an end. He protests against the ascetic figment, that suffering is something meritorious in itself, and justly pronounces it " everywhere, and always evil, and only to be accepted for a larger and a higher good." He censures the churches for having taught that " if there was no God and no future life the way to be happy would be to break all the moral laws, and rush into every kind of evil." But whilst teaching that virtue entails its own reward, and is the way to happiness, he guards against the error of supposing that its reward will consist in the ordinary prizes after which men strive. " Do not find fault," he says, " with this universe because righteousness is not paid in greenbacks or government bonds."

Mr. Savage shows that there was a time when polygamy, slavery, despotism, were good in comparison with the conditions and practices which they had superseded. He recognises the fact that the philosophy of evolution has reconciled and explained

* Journal of Science, 1880, pp. 448 and 450.

the old conflict between those who uphold innate ideas, and those who derive all our knowledge from individual experience. And by the aid of heredity he accounts for the sense of obligation. He considers that man will not be perfected in virtue until he has reached a state where the feeling of duty, the consciousness of obligation, is left behind, and we do what is right with an instinctive spontaneity.

In treating on " morals and knowledge " the author states the " difference " of view between the educationist party and their opponents. Of the former he says :—" They point to the fact that the criminal class, in almost every civilised nation of the world, is recruited from the ranks of the ignorant." Now it is true that the criminal has no love of knowledge for its own sake. But if you force education upon him will his inherited nature be thereby modified? Again we read :—" They point to the fact that no one can dispute that those nations that are the most illiterate are the ones that present to us the longest catalogue of crimes." This alleged fact we consider by no means unimpeachable. It may be said on the authority of one who has had full opportunities of observing,* that among certain savage tribes each man scrupulously respects the rights of his fellow, and an infraction of these rights rarely or never takes place."

Mr. Savage insists strongly on the evils resulting from well-meaning ignorance. " The ignorant, stupid good nature of the world has been guilty of the larger part of its infamies." One term here requires emendation : for " good nature " let us read " conscientiousness."

Here space compels us to close our survey of this book. The bulk of its contents do not admit of discussion in our pages, and from many of its teachings we must emphatically dissent. But there can be no doubt that it is a work which thoughtful men will read with great interest and advantage.

The Natural History of the Agricultural Ant of Texas. A Monograph of the Habits, Architecture, and Structure of Pogonomyrmex Barbatus. By HENRY CHRISTOPHER McCOOK. Philadelphia and London: J. B. Lippincott and Co.

NEARLY twenty years have now passed since the first account of the Texan Agricultural Ant was given by the late Dr. Gideon Lincecum, of Long Point, and S. B. Buckley. The notes of these naturalists, however, were received with very general

* Mr. A. R. Wallace.

scepticism in America as well as in Europe. Dr. Auguste Forel, the well-known historian of the ants of Switzerland, remarks that Lincecum's observations inspired him with little confidence. The original papers of this eccentric entomologist fortunately fell into the hands of the Rev. Dr. McCook, and led to his visiting Texas in order to settle the doubtful points, and confirm, if possible, the observations of his predecessors. The result of this expedition is the interesting and valuable monograph now before us. The author took up his residence under canvas at Camp Kneass, about three miles from Austin, a region swarming with these ants, and devoted his entire time to the investigation of their habits and architecture. After his return to Philadelphia he continued his studies by means of artificial formicaries.

It is, in the first place, now thoroughly established, that the Agricultural Ant, and not a few other species in Australia, India, and the countries bordering upon the Mediterranean, do amass grain and other seeds during the summer, and employ them as provisions during the cold season. Thus by the labours of Col. Sykes, J. Traherne Moggeridge, Dr. Thomson, as well as of Dr. Lincecum, and our author, the credit of Solomon as an observer of animated nature* is re-established, and Latreille, Huber, Gould, Kirby, and other eminent observers, are proved to have been in error. But what was the mistake of these eminent entomologists? They were not inaccurate in their observations and experiments, but they were deceived by arguing from the ants of England and northern France, which are torpid during the winter, and require no food, to those of warmer climates. They forgot that what may be true for certain species, and under particular circumstances, does not necessarily hold good for other species and for the whole world. Hence, by the way, our author's remark in his preface, that from one species the reader may learn all, is to be taken with a grain of salt.

Dr. McCook has watched the agriculturals collecting seeds, carrying them for a considerable distance to their city and depositing them in underground granaries. Here they were stripped of their husks which were then carried out as refuse. The author submitted to the observation of qualified botanists parcels of seeds taken from the ants on their way home, others from the granaries, cleaned and uncleaned, and finally samples of the refuse. The seeds were found to be those of *Buchlæ dactyloides*, two species of *Panicum, Aristida stricta, A. oliganthia* (the grass supposed to be harvested in crops by the ant), besides seeds of a *Croton*, a *Paspalum*, a *Malvaceous* plant, and a *Polygonum*. The refuse contained not a single perfect seed, being made up of shells, glumes, and imperfect seeds.

But a further and yet more interesting question remains.

---

* Proverbs, vi., 6, 7, 8.

Lincecum maintained that *Aristida stricta* or *oligantha*—ant-rice as he named it—was actually planted, cultivated, and reaped by these insects. It is a biennial species which, he states, is sown in time for the autumnal rains to bring it up. "About the first of November, if the season is favourable, a green row of ant-rice, about four inches wide, is seen springing up on the pavement (*i.e.*, the clearing round the city) in a circle of fourteen or fifteen feet in circumference. In the vicinity of this row the ants do not permit a single spire of any other grass or weed to remain a day, but leave the Aristida untouched till it ripens, which occurs in June the next year. After the maturing and harvesting of the seed the dry stubble is cut away and removed from the pavement, which is thus left unencumbered until the ensuing autumn, when the same species of grass, and in the same circle, appears again and receives the same agricultural care as did the previous crop." Dr. Lincecum did not see the ants actually sowing the seed, but he asserts that there is no doubt " of the fact that this particular species of grass is intentionally planted."

Dr. McCook's observations on this point are to the following effect :—The *Arista*, ant-rice, or needle-grass as it is locally called in Texas, is the only plant ever found upon the clearing round an ant hill. No other vegetation is tolerated, and the position of a formicary is visible from a distance by the belt of needle-grass surrounding it. But it is not universally present. In many cases the clearing around the city was entirely bare. Nor could the author detect any reason for this difference. As far as the evidence goes he concludes that the ants find it to their advantage to permit the *Aristida* to grow upon their discs whilst they root out all other herbage. The crop may be re-sown in a natural way by droppings from last year's plants, or by seeds lost by the ants on their way home. Still he does not venture to pronounce that there is anything "unreasonable or beyond the probable capacity of the emmet intellect in the supposition that the crop is actually sown." The proof is wanting. The Texan ant has never been observed by the author gathering or plucking seeds from the plant, it merely picks up such as have fallen to the ground. Another species found in Florida and Georgia—*Pogonomyrmex crudelis*—has been seen by Mrs. Treat as well as by the author cutting seeds from plants.

As regards the nests of the Agricultural Ant Mr. McCook recognises and describes six distinct types. He can detect no difference in the structure of the ants taken from these different kinds of nests. As far as his inquiries have gone it seems that cities of these different kinds are not promiscuously intermixed, but that those of similar structure are found near each other.

Concerning the co-operation of ants at work the author makes some important observations. He says :—" Another ant comes to the help of the first, a third soon follows, and before the gallery is sunken very far a fourth and a fifth join in the digging. There has been no communication that I can observe to secure this assistance, nothing like a call to her comrades by the original worker, nor by any of her successors. Her movement, if I may so speak, is contagious. Her excitement has been communicated by some imperceptible bond of sympathy to others of the colony who have joined her in the work, as if urged by some uncontrollable instinct. The excitement, however, has not captured all the little community. I believe that the digging is conducted by relays. The excitement which carries individuals into the trenches seizes all in turn, is intermitted and repeated as long as there is any necessity or opportunity for exertion."

It is evident that the industry of the working ants cannot be traced to the two causes which, according to some authors, underlie all animal activity, *viz.*, hunger and the sexual appetite.

Mr. McCook notices that ants—differing in this respect from our domestic vertebrates—never quarrel over their supply of food. There is none of the selfishness and bullying which is always witnessed when poultry, swine, or cows, are being fed.

As regards the intelligence of these ants the following statements are not without value. A young farm labourer informed Mr. McCook that his employer had tried to poison the Agriculturals. Arsenic spread upon bread was put in their way. This killed those that partook of it, but they soon abandoned the bread. The arsenic was then mixed with meal and put into their nests, but they separated the poison and carried away the meal. It was then mixed with treacle, but here again, after a few had been poisoned, the bait was left untouched.

It is further established beyond doubt that ants have funeral regulations. "All species whose manners I have closely observed are quite alike in their mode of caring for their own dead, and for the dry carcasses of aliens. The former they appear to treat with some degree of reverence, at least to the extent of giving them a sort of sepulture without feeding upon them. The latter, after having exhausted the juices of the body, they usually deposit together in some spot removed from the nest.

According to Mrs. Treat, *Formica sanguinea*, a slave holding species, never buries its own dead along with those of the slaves, *Formica fusca*.

The Agricultural Ant is described as not wantonly aggressive, unlike, *e.g.*, the fire-ants of Aveyros, of which Mr. Bates has given such a graphic account. Their sting, however, when inflicted, is very severe. The first symptom is a sharp pain resembling the sting of a bee. "Then followed twice, at short intervals, a nervous chilling sensation which seemed to sweep upwards, and was felt quite sensibly around the roots of the hair. Then followed a steady, heavy pain, about the wound, which continued from three hours more or less severely, a slight numbness accompanying, but no inconvenience thereafter." Ammonia and dilute tincture of iodine are used as remedies.

We must here, however, bring our examination of this admirable monograph to a close. No naturalist, we are convinced, can read it without great pleasure, full as it is of the most interesting and carefully-conducted observations.

Life and Mind, or the Basis of Modern Medicine. By ROBERT LEWINS, M.D. Edited by "Thalassoplektos." London; Watts and Co.

WE have here a small, but most remarkable book, noteworthy not so much for the doctrines it contains as for the outspoken manner in which they are asserted. We cannot better expound these views than by quoting some of the "ten theses" in which they are summarised by the author :—

"3rd. Natural Science is bound in conscience to divulge *all* her results, however much they may conflict with contemporary prejudices in order to satisfy the human mind and leave it free for the further pursuit and enjoyment of truth. Mental reservation, prevarication, and *suppressio veri*,—even from such respectable motives as Mr. Darwin confesses, in his 'Descent of Man,' to have influenced him for many years,—as habitually practised by contemporary English thinkers and *savans*, is disloyalty to humanity and reason, ruinous alike to their country and to the cause of civilisation and rational liberty throughout the world.

"4th. Natural Philosophy for 190 years past has rendered trite the axiom that everything in the universe proceeds by unalterable law—an axiom thoroughly fatal to any form of religion.

"5th. The world is from eternity to eternity. Nothing is ever created, nothing lost. Beginning or ending there is alike none. Only the form and condition of things is perishable. *Everything that exists dates from eternity. Ex nihilo nihil fit.*

"6th. The universe is boundless in space and time. Matter *can* have no limits, eternity in time and immensity in space being correlative.

"7th. As the logical inference from the above, millions and millions of millennia are before us, in which worlds and systems of worlds shall flourish and decay; at their lapse the universe *can* be no nearer its dissolution than at the present or any former period.

"8th. The so-called 'Personal God' is merely an idol of the human brain—a pseudo-organism of pre-scientific man endowed with man's attributes and passions, a remnant of Fetichism or Animism. . . Absolute Atheism is, however, no postulate of Science which does not venture to impugn the evidence of

Cosmical Design, or the existence of an unknown, inconceivable, intelligent 'First Cause' (the phrase is a misnomer), of whose Eternal Mind the Eternal Universe *may* be an hypostasis or emanation."

To examine all these theses, or even those we have quoted, would manifestly require far more space than is at our disposal ; but we may briefly direct our attention to the fifth, in which the eternity of the universe is plainly asserted. But as far as man has been able to penetrate into the mysteries with which he is surrounded, all natural forces tend to an equilibrium, which when reached all life, all activity, all motion must come to an end. Our planets must gradually become dried up like the moon, and will doubtless fall into the sun. The sun, which in the opinion of the ablest authorities has " seen his best days," must—in spite of this additional fuel and of the heat engendered by the conversion of mechanical motion—burn himself out. Even the falling of suns into each other cannot prevent the end for ever. There may, indeed, be a recuperative process at work in the universe tending to counteract all this levelling process ; but we have no physical evidence of its existence, and cannot even conceive its nature. Dr. Lewins, be it noted, does not venture to deny the possible existence of an Eternal Mind. Why, then, should he venture to deny finite minds, without which it is extremely difficult to give an account of many familiar phenomena. It will be perceived that there is in this work not the slightest attempt to show how physical energy is transformed into life, or how atoms of carbon, oxygen, &c., come to feel an interest in the position in which they are arranged.

The Appendix, from the pen of "Thalassoplektos," is an exposition of the treatise of Dr. Lewins, and contains much of which we can approve and no little against which we must protest. He condemns, very justly, our competitive examinations, but he does not honour the man of research and the specialist. What, for instance, must we think of such passages as the following :—" With such men as these [*i.e.*, country squires, cavalry officers in regiments rarely selected for foreign service, &c.] *savans* and literati cannot for a moment compare. The highest life, as compared with the narrow *epicier*-like existence of the specialist, despite his assumption, wherever possible of ill-sustained superiority, is that of the country gentleman, the dignified clergyman, and that of the well-fed contented peasant !"

Again :—" *Savans* may pile discovery upon discovery, until, as was the case with Michael Faraday, the desire to discover becomes a real mania ; but the life that is wholly given over to such pursuits is a mis-spent one, and is, as an example, positively injurious to society. The whole burden of 'Faust' is to trace the gradual development of a superior mind, a perfect man, from contemplative to active pursuits, from the library, laboratory, museum, and observatory, and finally to the execution of colossal works of engineering. In truth the perfect man is not given to dreaming or to unduly exercising any one portion of his nature. *Savans*, such as Faraday, who pedant-like stick throughout their whole lives to contemplative pursuits, are, as we have often asserted, inferior natures, like the famulus Wagner of Goethe's great tragedy!"

Surely the writer forgets that his supposed "perfect man," in forsaking Science, is under the especial guidance of the Evil One, and that the "active pursuits" beginning with seduction and murder go on to finance operations, whilst the "colossal works of engineering" are executed at the cost of expropriation and bloodshed, as we see in real life.

Goethe makes his Fiend give the student this advice :—

> "Verachte nur Vernunft und Wissenschaft
> Des Menschen allerhöchste Kraft!"

Thalassoplektos accepts this as *bona fide*, and would have us take Mephistopheles at his word !

We do not see that Science has anything to gain or to learn from this work. Nay, we fear that its spirit must be regarded as decidedly anti-scientific, and as a verification of the forebodings expressed from time to time in this Journal,—to wit, that *savans* may in future experience ill-will, and even persecution, not from ecclesiastics, but from the champions of "free thought."

We may well ask, who is "Thalassoplektos" that he should dare to pronounce Faraday an "inferior nature," and consider his life less perfect than that of such nonentities as country gentlemen and idle cavalry officers ? Might he not have gone a little step further in the same direction, and pronounced the career of a gorilla a still closer approach to perfection ? There is, further, nothing "dreamy," "contemplative," or "pedantic" in the scientific nature.

Into the religious bearings of this work we do not feel it within our province to enter ; but we would ask whether Pessimism is likely to be abated if we learn that there is no "land of the leal" where the weary may hope for rest ?

Both Dr. Lewins and "Thalassoplektos" predict the approaching downfall of England. Their prophecies may possibly be verified, but if so the crisis will be due not to the survival of the religious element among us, but to the decay of the "tribal instinct."

Several important books stand over for review, on account of the want of space.

CORRESPONDENCE.

. The Editor does not hold himself responsible for statements of facts or opinions expressed in Correspondence, or in Articles bearing the signature of their respective authors.

"THE DISESTABLISHMENT OF THE SUN."

To the Editor of The Journal of Science.

SIR,—Referring to the work bearing the above title, noticed in your August number, it seems to me that neither the author nor your reviewer has given full weight to one fact. The former argues that if the poles are cold because the sun's rays fall upon them obliquely, a spot on the equator at midnight ought to be colder still, because it is out of the sun's rays altogether. To this erroneous conclusion your reviewer merely replies :—" He forgets that at the poles the sun's rays are absent for one-half the year entirely, and fall during the other half obliquely and scantily upon the soil; whereas at the equator the surface of the soil has not time to cool down during a single night." This is all very well; but both you and the author overlook the fact that in places in the torrid zone where vegetation is wanting, and the absence of watery vapour in the air gives full scope to the loss of heat by radiation,—*e.g.*, in the Sahara, and in many parts of South Africa,—the cold in the night is very severe. There can be no doubt that if the return of the sun could be delayed but for a few hours, frost would be experienced in all arid regions within the tropics. I do not see how these night-chills can be explained on the author's theory.

As regards his assertion that glass is impervious to heat, how can he explain the arrangement adopted at all meteorological observatories ? The maximum temperature of solar radiation is observed by means of a thermometer placed horizontally on a wooden frame 5 feet from the ground, whose bulb is made of black glass, externally covered with fine lampblack, and enclosed in an outer and exhausted tube of transparent glass. This arrangement has been selected not on theoretical principles, but in accordance with experiment. The author does not remember the different behaviour of glass with radiant heat according as it is emitted from incandescent or non-incandescent bodies. The heat of a pipe conveying warm water or steam, of a hot-bed, &c., is to a great extent retained by a glass covering. Of the radiant heat from an Argand lamp 92·3 per cent passes through a plate of rock-salt, and 67 per cent through a plate of flint-glass. The higher is the heat of the source of the rays, the larger proportion is transmitted.—I am, &c.,

PHYSICUS.

NOTES.

Thé Benefits of Vivisection.—The " Popular Science Monthly "
gives a neat summary of Dr. Charles Richet's arguments in
defence of vivisection. He demands that it shall be judged by
its practical results, and claims that if it can be shown that we
have gained by this method of experiment the means of curing
one or two diseases of man it must be considered legitimete.
He cites a number of discoveries made through vivisection.
Among them is the discovery of the circulation of the blood.
" Galen established the fact that the arteries contained blood by
observations in the artery of a living animal ; Harvey opened
the chests of living animals, cut into the pericardium, observed
the contraction of the heart, and what was going on in the veins
and arteries, and deduced from what he saw his theory of the
circulation. Transfusion of blood, an operation resorted to in
extreme cases with the best results in saving life, was introduced
after its possibility had been ascertained from experiments upon
animals first made in 1664 by Lower, and afterwards by Denis.
' Experiment alone,' Dr. Richet says, ' will teach us precisely
what quantity of blood is necessary and what is harmful ; and if
over-sensitiveness forbids animal suffering for this end, then the
experiments would have to be made on human beings.' The
mode of death from the inhalation of carbonic oxide, and cor-
relatively, the method of avoiding or preventing death from
inhalation, have been made known only through vivisection. So
also ' all that we know in hygiene of the quantity of air necessary to support life is the result of experiments on dogs and
rabbits. Sometimes a precise knowledge of the conditions of
respiration has served to prevent men from perishing.' Only
two methods exist by which we may learn the conditions of
gastric digestion and collect its secretion, viz., by observation of
gastric fistulæ produced by chance in man, and by artificial
fistulæ in animals. The first method has been possible only in
three or four instances, but the effect of food on the gastric
secretion in dogs and cats has been largely observed ; and the
knowledge of the remedies which have been applied to the relief
of dyspepsia has been derived from such studies. Our knowledge of nutrition has been largely added to by means of experiments in which dogs and cats have been submitted to varied
alimentation, and from which the quantity and quality of food
necessary to sustain life have been deduced. What we know of
the nerves has been gained from studies of animals, as have also
the means of relieving neuralgias and paralysis, in which, thanks

to the scientific analyses of the vivisectors Fritsche, Hitzig, and Ferrier, ' we can pass from the effect to the cause, and assign to paralysis a central lesion at a well-determined spot, so that trephining at this spot may cause the paralysis to disappear.' The experiments of Galvani and his followers on frogs have taught us to estimate the effect of the electric current on nerve and muscle, and shown us how to apply galvanisation to the prevention of the paralysis which ensues from the destruction of the motor nerves. The numerous patients relieved of nervous diseases ' by this admirable therapeutic agent have no call to speak ill of such vivisectors as Galvani, Aldini, Volta, Magendie, Marshall Hall, Remak, Du Bois Reymond, and many others, since it is to their discoveries that the relief of their ills is owing. Would Galvani have made his discoveries had he refrained from dissecting frogs? Would the electric current have been applied to atrophied limbs if it had not been found that the action of this current in dogs was salutary and not dangerous?' Certain diseases of the urinary organs have been studied in animals. The treatment of sympathetic ophthalmia by section of the ciliary nerves of the diseased side has been shown to be advantageous by experiment, and the results yielded by experiments on dogs and rabbits have been applied to patients. The correct treatment of cataract has been similarly learned. Encouraging progress is made by vivisection in the study of the formation of callus, of pseudarthrosis, of osseous grafts, of regeneration of bone by periosteum,—subjects of great importance in surgery. The vaso-motor theory, which plays a large part in the medicine and surgery of the present day, has been established by experiments on the great sympathetic and the rabbit's ear. Dr. Brown-Séquard has furnished useful ideas relating to epilepsy and tetanus from the results of painful experiments on dogs and guinea-pigs. Trial on animals is useful to determine the action of new medicines, for ' we do not wish to experiment on man at the risk of poisoning him, where animals can be employed,"— so with poisons. Finally, if we deprive *savants* of the right to submit living animals to experiment, we shall go back beyond the days of Galen. ' If all those who have been relieved—verily made to live again,'—says Dr. Richet, ' by modern medicine and surgery, could speak, they would confound those who load vivisection with calumny, and they would hold that their own life and sufferings weighed more in the balance than the sufferings of those animals which have been sacrificed in laboratories to the lasting benefit of man.' "

Anti-Vivisection.—"We observe that a memorial recently presented to Mr. Gladstone, urging him to do all in his power for the absolute abolition of vivisection, was signed by ' one hundred representative men ;' among them Cardinal Manning, Prince Lucien Bonaparte, Alfred Tennyson, Robert Browning, James Anthony Froude, John Ruskin ; the head-masters of Rugby,

Harrow, and seven other large schools; twenty-one physicians and surgeons; and thirty-seven peers, bishops, and members of parliament. The memorialists take the ground that vivisection, even with anæsthetics, should by law no longer be allowed, and they quote the opinions of Sir William Fergusson, Sir Charles Bell, and Mr. Syme, that 'it has been of no use at all, and has led to error as often as truth.' They add that the utility, if proved, would not, in this case, excuse the immorality of the practice. Of the persons herein named as being in favour of total abolition of vivisection, Sir William Fergusson and Mr. Syme may be accepted as competent to express an opinion with as much authority as belongs to eminent surgeons of the past age, which believed in nothing new and practised rule of thumb. Of the remaining 'one hundred representative men' probably not one knows anything whatever of the practices which he seeks to abolish, or of the results of those practices. Most of them are essentially 'emotionalists' who would sign any earthly manifesto on a humanitarian subject if only their bowels of compassion were sufficiently harrowed up by the falsehoods of which we know by experience that anti-vivisection agitators are capable. They are all either professed philanthropists and religionists whose signatures to such a document are a matter of course, or else they are well-intentioned people who are simply bored into signing without inquiry or serious thought. As for the 'twenty-one physicians and surgeons' we covet further information. We should like to analyse the list to ascertain how many of them are simply traders on gushing philanthropy as a means of advertisement or of fee-getting; or how many of them are of the fossil *laudator temporis acti* class, who are still sceptical as to germ-theories, septicæmia, and evolution, and who believe in bleeding for a cold in the head. We wish the anti-vivisectors would offer us the name of even one single modern physician or surgeon of repute for learning and research who will go for total abolition of vivisection. One such signature would be worth the whole hundred of peers, members of parliament, poets, and schoolmasters, or of any number of the class of doctrinarians who are in the habit of making up their minds without thinking and uttering manifestoes without learning."— *Medical Press and Circular.*

M. Lortet describes in the "Comptes Rendus" (Sept. 13, 1880) the dredgings which he has undertaken in the Lake of Tiberias. This body of water has once been salt, and is still slightly brackish. The fauna holds an intermediate rank between those of fresh and of salt waters. Algæ and Confervæ were entirely absent.

M. P. Hautefeuille communicates to the Academy of Sciences the observation that ozone at a high tension appears of a beautiful blue colour, which deepens as the pressure increases.

The " Journal of the American Chemical Society " contain biographical notice of the late M. Tessié du Motay. It appe that he was not merely eminent as a chemist and a technolo but had earned some reputation as a poet, a dramatist, a mus cian, and a student of Oriental literature and mythology. H leaves behind him an unpublished poem entitled the " Expiatio of Faust."

A morning paper favours its readers with the following choice tit-bit of un-natural history :—" The only good thing to be said for the shark is that it knows its own friends, and studiously abstains from eating the pilot-fish, which guides it to its prey. It spares this useful little attendant, as the crocodile spares the humming-bird or the trochilus." The writer might find it diffi-cult to show what opportunities the crocodile has for sparing the humming-bird, or how the latter can ever be of use to him.

Another writer, in the self-same paper, speaks of the hamster and the marmot as identical !

M. Th. Lécard has communicated to the Academy of Sciences an account of a vine with tuberous roots and a herbaceous stem, which he has discovered in Soudan. It yields abundant and delicious grapes.

Dr. B. W. Richardson, in a paper read before the Sanitary Institute (Exeter, Sept. 20th, 1880), seems to approach the posi-tion of Prof. Jäger. He writes :—" Go into the wards of a lunatic asylum, and notice among the most troubled there the odour of the gases and the vapours they emit by the skin and the breath. That odour is from their internal atmosphere, their nervous ethereal emanation. They are mad up to suicide or murder, or any criminal folly. Can it be otherwise ? They have secreted the madness ; they are filled with it ; it exhales from them. Catch it, condense it, imbibe it, and in like manner it would madden any one." Is not this the teaching of Jäger and Dunstmaier, spiced to suit the audience and the occasion ?

The " Medical Press and Circular " learns with delight—which we fully share—that " Dr." Tanner's lecture on his forty days' fast, given in Booth's Theatre, New York, " did not draw an audience capable of half filling the house.'

According to the " Zeitschrift für Physiologische Chemie " (iv., p. 382) G. Hüfner has obtained from putrescent blood long purple-red crystals of hæmoglobine, often above a milli-metre in length.

M. E. Yung (" Comptes Rendus," August 30, 1880) has studied the development of the eggs of *Loligo vulgaris* and *Sepia offici-nalis* exposed to light of different colours. The development is hastened by violet and blue light ; retarded by green and red. Yellow light behaves like white light. Larvæ of *Çiona intesti-nalis* also grew most rapidly in the violet light. Development under the red and green lights, though retarded, was effected in perfection.

We learn that the Rev. W. H. Dallinger, of Liverpool, the distinguished microscopist, has accepted the appointment of Governor and Professor of Biology at Wesley College, Sheffield. This institution may be congratulated on the acquisition it has made.

We learn with regret that the robin-redbreast is eaten in Loraine in such numbers that they are becoming scarce. What is still worse, we hear that a certain French provincial *prefet* has given orders for the destruction of swallows and martins as injurious !

One George Aldersbury, an imitator of Dr. Tanner, has succumbed to the system of fasting which he had attempted.

According to the Berlin " Klinisches Wochenblatt " hæmoglobinuria may be caused by inhaling hydrogen arsenide, even in small proportions,—*e.g.*, hydrogen gas prepared with sulphuric acid and zinc contaminated with arsenic, as is often the case.

The " Medical Press and Circular " states that Mr. Malcolm Morris, Honorary Secretary of the London Medical Society, is issuing circulars requesting Fellows of the Society who have seen cases of arsenical poisoning by means of wall-papers to forward particulars to him at 63, Montague Square.

A paper " On the Function of Aperture in Microscopic Vision," by Prof. Abbé, of Jena, has come before the Royal Microscopical Society. As the paper would occupy about 150 pages of the Journal, the Council have decided to print it as a separate volume.

Dr. H. Stolterforth, M.A., in a communication to the Quekett Club recommends boiling in soap and water as a means of cleaning Diatomaceous materials, in all cases where it is available, in preference to the ordinary mode of treatment with acids and strong alkaline solutions. He considers it far less liable to do mischief to the delicate organic structures under treatment than the ordinary methods.

In a paper read before the Royal Microscopical Society, Mr. J. W. Stephenson demonstrated that the visibility of minute objects depended upon the difference between the refractive index of the medium in which they were viewed and that of the objects themselves, and, further, that it was useless to employ objectives of the large apertures now attainable unless the objects were mounted in media capable of utilising the whole of large pencil; by using them on objects mounted in air their effectiveness is reduced to the common level of $180° = 1·0$ numerical aperture. The following table of indices is given :—

Air	equal	1·00
Water	„	1·33
Diatomaceous silex ... ...⎫	„	1·43
Sulphuric acid ... ...⎭		
Canada balsam	„	1·54
Bisulphide carbon	„	1·68
Solution of sulphur in bisulphide car-		
bon (approximately)...	„	1·75
Sulphur...	„	2·11
Solution of phosphorus in bisulphide		
carbon (approximately)	„	2·10

Taking the difference between the refractive indices of Diatomaceous silex and the several mounting media, the following results are obtained :

Water	10
Canada balsam	11
Bisulphide carbon	25
Solution of sulphur in bisulphide carbon ...	32
Solution of phosphorus in bisulphide carbon	67

The practical result of the investigation appears to be—

That it is essential, if the whole aperture of an objective is to be utilised, to mount minute structures in some medium other than air.

That although the full aperture and revolving power are secured by mounting in balsam, it gives nevertheless nearly the faintest image of all.

That a solution of phosphorus is, as far as visibility is concerned, by far the most effective, but the difficulties attending its use must render it unpopular.

The next best is a solution of sulphur in bisulphide of carbon (although pure bisulphide is very good), and with these there is no technical difficulty whatever. This medium can easily be secured by using the solution employed by Mr. Browning in making his bisulphide prisms. A ring being made on the slide, and a drop of the sulphur solution or pure bisulphide being placed in its centre, nothing is necessary but to place over it the thin cover with its adhering diatoms, press it down on the still moist ring, running round it a somewhat copious margin of the cement. When dry, to protect it from the water of the ordinary immersion lenses, it is desirable to give it a coat of gold-size or shellac varnish.

THE

JOURNAL OF SCIENCE.

DECEMBER, 1880.

I. SPECTRUM ANALYSIS.

By Dr. Akin.

IT has been truly said of Dr. Young's "Course of Lectures on Natural Philosophy," delivered at the Royal Institution (it is to be feared before a somewhat unappreciating audience) in the first years of this century, that it contains the germs of many more discoveries than can be found foreshadowed in any other single extant book. In p. 438 of the first volume of that invaluable work (ed. of 1807) the author observes :—" In light produced by the combustion of terrestrial substances the spectrum is still more interrupted ; thus the bluish light of the lower part of the flame of a candle is separated by refraction into five parcels of various colours ; the light of burning spirits, which appears perfectly blue, is chiefly composed of green and violet rays, and the light of a candle into which salt is thrown abounds in a pure yellow, inclining to green, but not separable by refraction. The electrical spark furnishes also a light which is differently divided in different circumstances." Of this passage it may be asserted that it contains the germ, although it has not actually formed the starting-point, of researches which, prosecuted at first in a desultory manner, have since, after the lapse of half a century, given rise to a series of discoveries which must be classed among the most striking, if not altogether among the most novel, of our age. The above statement by Dr. Young was neither absolutely the first nor the only forerunner of those discoveries. It is said that already, " in 1752, Melville observed the effect of soda on flame ;" and

Dr. Wollaston, in 1802, had first indicated those "five parcels" of rays in candle-light mentioned by Dr. Young, as well as the existence of a diversity in the spectra of electric sparks in different circumstances. The celebrated Fraunhofer, in 1815, speaking rather vaguely of "the light of electricity," observed that it gave a different spectrum from those of "the sun and of fire;" whilst Sir H. Davy— viewing the subject in a much more philosophical manner— made, in 1822, the novel and important assertion that "the light generated in electrical discharges depends *principally* on some properties or substances belonging to the ponderable matter through which it passes." In the same year Sir D. Brewster and Sir J. Herchel had devoted some attention to the investigation of the spectra of coloured flames ; and in a paper by Mr. Fox Talbot,—of Talbotype celebrity,—of the year 1826, the author states that " the muriate, sulphate, or carbonate of soda " all gave the same yellow ray in the spectrum, while "the nitrate, chlorate, sulphate, and car- bonate of potash agreed in giving a bluish white tinge to the flame. Hence," he continues " the yellow rays may indicate the presence of soda, but they, nevertheless, frequently appear where no soda can be supposed to be present." In fact, "ignited sulphur produced the very same " rays. In 1827 Sir J. Herschel reverted to this subject in his celebrated treatise on Light, where, after describing the colourations which different "saline bodies," "in general," impart to flames, he observes that "the colours thus communicated by the different bases to flame afford in many cases a ready and neat way of detecting extremely minute quantities of them ;" and he assigns also reasons to show that "these tints arise from the molecules of the colouring-matter re- duced to vapour, and held in a state of violent ignition.

Thus the matter rested for a while. In the year 1834 the same indefatigable Mr. Talbot made the remarkable disco- very that the two bases, lithia and strontia, while both impart a red tinge to flame, may yet be distinguished by the different lines visible in their prismatically analysed light ; in consequence of which, and of some later observations, he in 1836 insisted "that an extensive course of experiments should be made on the spectra of chemical flames," as "the definite rays emitted by certain substances—as, for example, the yellow rays of the salts of soda—possess a fixed and in- variable character, which is analogous in some measure to the fixed proportion in which all bodies combine, according to the atomic theory." In the same paper Mr. Talbot describes also the spectra of gold, silver, zinc, and copper

leaves " deflagrated by galvanism." But a much more important—and perhaps, as regards real novelty, the most
important—series of researches on the subject of the spectra
of metals acting as dischargers of the electric current was
published, in the year preceding, by Professor Wheatstone.
From these researches Mr. Wheatstone inferred " that
electric light results from the volatilisation and ignition (not
combustion) of the ponderable matter of the conductor
itself ;" and that, as the spectra were altogether different
according to the nature of the latter, "by this mode of
investigation [that is, by the analysis of either the electro-
magnetic or voltaic spark] the metals may be readily distinguished from each other." Mr. Wheatstone found also
that, with the ordinary electric spark, the lines of the
spectra were " different in number and position in every
metal employed," without stating, however, whether they
coincided with those produced by the corresponding substances employed as galvanic or electro-magnetic electrodes.
This it would seem important to have remarked on, as the
author had observed that " the light which accompanies the
ordinary combustion of the metals " really gave " appearances totally dissimilar to the above." After these investigations nothing of any great importance was brought to
light concerning the spectra of sparks. M. Foucault, in
1849, made some observations which yielded no remarkable
results, except one to which we shall have to revert, and
which was extremely striking, but is not connected with our
present subject. M. Masson, in 1851 and 1855, published
researches which, however painstaking and elaborate, tended
rather to confuse than to elucidate the matter in hand.
Much more valuable and highly suggestive was a paper by
M. Angström, published in 1855, who for the first time
noticed the very interesting fact that, whilst part of the lines
of the spectra of sparks belonged to the metals which formed
the electrodes, others were of aërial origin, and could consequently be varied according to the nature of the gaseous
medium in which the discharge was taken. This fact was
noticed at about the same time in America by Mr. Alter, and
was further investigated in 1858 by M. Van der Willigen ;
whilst the subject of the spectra of the light of exhausted
tubes filled with gases of extremely small density, and traversed by induction-currents,—in other words, of gases made
glowing by electricity,—was elaborately examined in 1858-9
by M. Plücker.

From these various researches, and especially from those
of Mr. Wheatstone on the metals and of M. Plücker on

gases, the constancy and distinctness of the spectra of the chemical elements, when made incandescent by electricity, had pretty clearly resulted, and the expediency of their adoption as a chemical test had thus become manifest. A powerful impetus in this direction—or, rather, to the use of the optical phenomena of substances when rendered incandescent or self-luminous by whatsoever means, in order to their chemical discrimination—was provided by some researches on flames by Mr. Swan, first published in 1856. By this observer " absolute identity was shown to exist between the spectra of dissimilar carbo-hydrogen compounds," thus proving (not, however, in a manner altogether irrefragable) the dependence of the constitution of the spectra solely on the chemical composition of the compounds ; and, adverting in particular to the colouration of flames by soda, he stated that it formed such a delicate test for sodium that 1-2,500,000th grain could be detected by it.   Under the circumstances which we have attempted to describe—and among which we should include also the publication of some valuable researches on coloured flames by Prof. Miller, of King's College, in 1845—Profs. Kirchhoff and Bunsen were scarcely justified in proposing, in 1860, the optical characters of glowing gases as the basis of " an entirely new method of qualitative chemical analysis."   Nor, on the other hand, did they place the trustworthiness of the proposed method, whether new or not, beyond a doubt.   MM. Kirchhoff and Bunsen, in point of fact, had investigated the spectra of a number of binary and other compounds of the six metallic elements—sodium, lithium, potassium, strontium, calcium, and barium, as volatilised in flames ; and they found that these spectra were distinct from one another, and always the same, whatever the temperature of the flame, at least between the limits of from 2000° to 8000° C.   They asserted, further, that the spectrum of any one of those metals was independent of the nature of the other elements with which it formed a binary compound or salt, except that " those compounds of a metal gave the brightest spectra which were most volatile ;" and that, in fact, the spectrum of any such compound was the same as that of the uncombined metal itself, volatilised by the electric spark.   All this, however, was yet very far from demonstrating, as alleged, that these spectra might be relied on " as absolute proofs of the presence of the particular metal " in each special case.   In the first place, it was strange, as asserted, that the spectrum of any of the metals combined with oxygen or chlorine, for instance, should be altogether uninfluenced by the nature of the latter substance, unless it were supposed that the compound was decomposed in the flame, the reverse of which opinion Prof. Kirchhoff at least seemed openly to favour. Yet that such is really the case seems to be now established beyond doubt. But then it is evident also that, in the case of a compound undecomposable by heat, the spectra furnished by MM. Kirchhoff and Bunsen, and which really relate but to the uncombined metals, would be no guide whatever to the detection of the metallic element which it contains. In the next place, an even more important remark than the above (which, indeed, has to be practically considered principally in reference to the matter which later on will be treated of, but is of less import in ordinary chemical analysis) is suggested by the circumstance that it had nowise been proved that, of the fifty-six other elements then known, the spectrum of none coincided, either totally or in part, with any one of the six spectra which alone had been mapped and examined; and, in fact, M. Kirchhoff—who had already found that the influence of temperature is far from being so totally immaterial as at first had been supposed—actually met at a later period with examples in which lines in the spectra of two different substances coincided. What some would call a practical answer to these objections, which would certainly have been urged from numerous quarters had the subject been mooted by some fortunate, or rather unfortunate, novice,—an answer which, however, is not the least to the point,—was furnished by the discovery, on the part of Prof. Bunsen, of two new metals whose existence had been first revealed to him by additional lines in the spectra of known metals, together with which the former occurred as otherwise unsuspected impurities. A few more such discoveries have since been made which, indeed, prove beyond cavil the value of the method in question, practically at least, although not in theory; for, besides the difficulties which we have already pointed out, some other observations have since been brought forward which, whilst they may possibly open a wide and fruitful field to inquiry for the future, for the present throw doubt on the reliableness of the optical test as an absolute sign of chemical nature. Not to mention certain observations by Dr. Robinson, of Armagh, which bear in the same direction, it has been alleged by M. Plücker that an elementary gas, such as nitrogen, may possess three spectra of altogether varying character, depending upon "three allotropic states; and, on the other hand, an English chemist who may be taken to express the opinions of MM. Bunsen

and Kirchhoff, allows that "certain metals—calcium,
barium, and strontium—yield spectra of two kinds ; one of
these, seen at the lower temperature, and consisting of
broad bands, being resolved at a higher temperature into
bright lines.   These bright lines . . . characterise the true
metallic spectrum, while the band spectrum is probably produced by the incandescent vapour of a metallic compound."
The same chemist observes further that " it has been stated
that all the various forms of carbon-compounds, when in the
state of incandescent gas, yield identical spectra.   This
proves not to be the case."

Notwithstanding the difficulties adduced—which, on being
more accurately examined, seem calculated rather to lead to
new and important discoveries than to discredit those already
made, or to permanenly invalidate the method to which they
are owing,—the method of spectrum analysis prepared and
pointed out as important by many a previous investigator,
and brought into prominent notice by MM. Kirchhoff and
Bunsen, is certainly a most valuable accession to the chemical philosopher.   In fairness it should also be stated that,
although the researches of the two distinguished professors
had been, in almost every respect, anticipated by others, yet
such was the indecision prevalent on the subject before their
own publications that a physicist of such great eminence as
Sir D. Brewster—to whom optical science owes almost innumerable discoveries, and who, in 1842, had himself for the
second time made observations on this very matter of the
colouration and spectra of flames—gave out in 1860, when
MM. Kirchhoff and Bunsen's memoirs were yet unknown in
England, that, besides in that of soda, the " ray [D] is
prominent also, according to W. A. Miller, in the flames of
lime, strontia, baryta, zinc, iron, and platinum, and, according to Angström, in the electric flame of every metal
examined by him."   Yet the great stir which, upon the
introduction of the new method, ensued in the scientific
world, was perhaps in no small measure owing to its accidental connection with another series of researches, some
account of which we now proceed to give.

Wollaston, in 1802, noticed for the first time that the
solar spectrum was interrupted by seven black lines, five of
which he took to indicate the limits of the four (assumed)
primary colours. Fraunhofer, in 1815, with superior apparatus, of his own make and invention, numbered about 600
such lines, some 350 of which he laid down in his celebrated
map of the solar spectrum.   The light of the moon, of
Venus, and of Mars, according to Fraunhofer's observations

in 1815 and 1823, gave lines in strict coincidence with those of sunlight, whilst the lines in the spectra of such of the fixed stars as he examined proved to differ in part both from those of sunlight and among one another. The "light of lamps" showed no dark lines, but, on the contrary, one bright yellow line which corresponded in refrangibility to the black line marked by Frauenhofer as D in the solar spectrum.

Sir D. Brewster, in 1832, observed that, in transmitting light through nitrous acid gas, a great number of new black lines became apparent in the spectrum—a discovery "lying so close at the root of atomical science," its author a few years later emphatically remarked, "that I am persuaded it will open up a field of research which will exhaust the labours of philosophers for centuries to come." In 1833 Prof. Miller, of Cambridge, conjointly with Prof. Daniell, noticed a similar behaviour of bromine and iodine vapour towards light; whilst Sir D. Brewster, "the first and principal object" of whose inquiries was "the discovery of a general principle of chemical analysis, in which simple and compound bodies might be characterised by their action on definite parts of the spectrum," in 1836 added the important discovery that the number of dark lines in the solar spectrum depended on the thickness of the layer of atmosphere which it must traverse to reach the observer, and that, consequently, some of those lines were owing to a similar absorptive action on the part of the colourless gases of the atmosphere as he had previously discovered in red-coloured nitrous acid gas. In 1842, moreover, Sir D. Brewster was "surprised to find" that, "in the spectrum of deflagrating nitre, the red ray, discovered by Mr. Fox Talbot, . . . occupied the exact place of the line A in Fraunhofer's spectrum," and he was "equally surprised to see a luminous line corresponding with the line B of Fraunhofer." The phenomena of absorption of light, and the consequent production of dark lines in luminous spectra, were elaborately investigated by Prof. Miller, of King's College, in 1845; and various hypotheses were hazarded, notably one by Baron Wrede in 1834, on the mechanical origin of those lines. In 1849 M. Foucault made an observation which was calculated to throw considerable light on the subject, but the bearings of which he himself, as well as his contemporaries, excepting one, utterly failed to perceive. "I caused an image of the sun," relates M. Foucault, "to fall on the [voltaic] arc, and I convinced myself in this way that the double bright line of the arc coincides exactly with the double dark line [D] of the solar spectrum. This process of investigation furnished me matter for some unexpected observations. . . . It showed to me that this arc . . . absorbs the rays D, so that the above-mentioned line D of the solar spectrum is considerably strengthened when the two spectra are exactly superposed. Thus the arc presents us with a medium which emits the rays D on its own account, and which at the same time absorbs them when they come from another quarter. To make the experiment in a manner still more decisive, I projected on the arc the reflected image of one of the charcoal points, which, like all solid bodies in ignition, gives no lines ; and under these circumstances the line D appeared to me [dark] as in the solar spectrum."

In 1855 M. Angström published a memoir, remarkable in many ways, and to which we have already had occasion to refer. In speculating upon the origin of the absorption of light by bodies, the author observes :—"As according to the principle of Euler [Quemadmodum (ergo) corda tensa a sono ei, quem ea edit, æquali vel consono concitatur, ita particulæ illæ minimæ in superficie corporis opaci sitæ, a radiis ejusdem vel similis indolis, contremiscere, pulsusque undique diffundendos producere valebunt—as correctly applied by M. Angström] a body absorbs all the series of oscillations which it can itself assume; it follows from this that the same body, when heated so as to become luminous, must emit the precise rays which, at its ordinary temperature, are absorbed." This proposition, it was added, is in so far true as the diversity in "the condition of the heated body, as regards elasticity, [which] is altogether different from the state in which the light is supposed to be emitted," may be neglected ; but then it would have been more correct at once to refer the emitted light to the power of absorption actually obtaining under the same conditions, instead of at common temperatures. Adverting further to the spectra of the electric spark as compared with the solar spectrum, M. Angström observes that, " regarded as a whole, the first produced the impression of being the reverse of the other. I am therefore of opinion that the explanation of the dark lines in the solar spectrum embraces that of the luminous in the electric spectrum." Thus, by the observations of M. Foucault, conjoined with the above explanations, the origin of the dark lines in the solar spectrum was accounted for, supposing the two had been brought mutually to bear. Like M. Angström, Prof. Kirchhoff was not aware of M. Foucault's experiments when, in 1859, he published the following :—" On the occasion of an examination of the spectra of coloured flames . . . I made some observations which disclose an unexpected

explanation of the origin of Fraunhofer's lines, and au-
thorise conclusions therefrom respecting the material constitution of the atmosphere of the sun. . . . I formed a solar
spectrum by projection, and allowed the solar rays concerned
. . . to pass through a powerful salt-flame. If the sunlight
were sufficiently reduced there appeared in the place of the
two dark lines D two bright lines ; if, on the other hand, its
intensity overpassed a certain limit, the two dark lines D
showed themselves in much greater distinctness than without the employment of the salt-flame. The spectrum of
the Drummond light," originally continuous when transmitted through a salt-flame, exhibited " two dark lines of
remarkable sharpness and fineness, agreeing with the lines
D of the solar spectrum." All this was similar to the previous observations of M. Foucault ; but, whilst the latter
drew no inferences from his facts, M. Kirchhoff immediately
concluded from his observations "that coloured flames, in
the spectra of which bright sharp lines present themselves,
so weaken rays of the colour of these lines, when such rays
pass through the flames, that in place of the bright lines
dark ones appear as soon as there is brought behind the
flames a source of light of sufficient intensity." And he
adds with emphasis :—" I conclude, further, that the dark
lines of the solar spectrum which are not evoked by the
atmosphere of the earth, exist in consequence of the presence, in the incandescent atmosphere of the sun, of those
substances which, in the spectrum of a flame, produce
bright lines at the same place. . . . The dark line D in the
solar spectrum allows us, therefore, to conclude that there
exists sodium in the sun's atmosphere. Brewster has found
bright lines in the spectrum of the flame of saltpetre at the
place of Fraunhofer's lines A, a, B ; these lines point to the
existence of potassium in the sun's atmosphere."

Already, from the researches of Leslie, of the year 1804,
and still better from some observations by Mr. Balfour
Stewart published in 1858-9, the proportionality between the
absorptive and emissive powers of substances for rays had
resulted,—at least so far as radiant heat or invisible rays
are concerned ; and, whilst a mechanical explanation, applicable in general, of this law, as we have seen, had been
deduced by M. Angström, from a hypothesis of Euler's, Mr.
Stewart, in 1860, had roughly proved the same to hold good
also in the case of visible rays or light. M. Kirchhoff has
attempted to furnish a strict mathematical demonstration of
the necessity of its validity ; and whatever opinion may be
held of the value of this proof, the law itself is sufficiently

corroborated by the facts discovered by himself. By comparing the dark lines of the solar spectrum, of which he has formed a new and highly elaborate map, with the bright lines in the spectra of metals volatilised by the electric spark, M. Kirchhoff has traced many other such coincidences as have been above referred to in the case of the sodium and potassium lines; and he believes he can consequently assert with certainty the existence in the sun's atmosphere of the elements iron, calcium, magnesium, sodium, chromium, nickel, cobalt, barium, copper, and zinc, whilst gold, silver, mercury, aluminium, cadmium, tin, lead, antimony, arsenic, strontium, lithium, and silicium appear to be wanting. According to his later investigations a considerable number of other elements, besides those above mentioned, are present in the solar atmosphere. The spectrum of iron is remarkable for the great number of lines which it exhibits, and of these sixty were found to coincide with dark lines in the solar spectrum. Allowing for a possible inaccuracy in his observations, M. Kirchhoff makes out the probability of his having confounded nearly coincident lines with one another in 60 concomitant cases to be equal to 1 divided by 1, followed by 18 noughts—that is itself as good as nought. This, it is expected, will bring home conviction to the minds of the most doubting. Yet the inference from these coincidences to the necessary existence of iron in the solar atmosphere (or conversely) must remain inconclusive, if not inconsequential, as long as (besides the relation of the spectra of compounds to those of their constituent elements) another question is not cleared up, of which M. Kirchhoff justly observes that it is one " of great interest "—namely, whether the observed coincidences of lines belonging to two different substances, of which he has noticed some eleven, are real or apparent ones.

Another important question, although without immediate bearing upon the point mooted, is whether a really elementary body is capable of emitting several rays differing in quality, except such whose wave-lengths bear to each other the relation of harmonics, as indicated in the theory of Angström and Euler. An incidental result of the researches of M. Kirchhoff has been, or at least probably will be, the relinquishment of the somewhat fanciful notions concerning the internal structure of the sun hitherto current, and the re-adoption of the much more rational views of Newton and Galileo. Newton asked, "Are not the sun and the fixed stars great earths vehemently hot ... whose parts are kept from fuming away, not only by their fixity, but also by the vast weight and density of the atmospheres incumbent upon them ?"—and Galileo had even gone further, and declared sun spots to be clouds in the sun's atmosphere. M. Kirchhoff, in order to explain the dark lines in the solar spectrum,—which he severally refers to the absorption of such gaseous substances as at the same time emit rays of the same quality as they absorb, but of much less intensity,—finds it necessary to revert to these older views, which, on their part, are in so far corroborated by his terrestrially observed facts as the theory which he builds upon them is well founded. That such is the case seems to be open to little doubt.

Reviewing the whole of the facts stated in this paper, the ample recognition which Profs. Kirchhoff and Bunsen have received for their researches will appear well deserved. Yet philosophers may congratulate themselves that discoveries of this nature should have been brought forward by persons of such eminent or distinguished names, for—owing to the openings that they afford to scepticism—if first published by some novice they would have been either unduly assailed, or, as is not altogether unlikely (for such things have happened ere now), perhaps even ignored.

II. THE WORLD OF THE POETS.

A WRITER in this Journal (1879, p. 271), when discussing the question of the perfection of Nature, took occasion to examine some of the views of the world and of its constitution which still prevail among men of culture. Looking a little farther in the same direction, it may be not uninteresting to scrutinise the pictures which imaginative writers draw of the world as they would wish it to be. It may perhaps strike us as somewhat singular that the very class of men who are so ready to pronounce Nature perfect, and who can scarcely remain within the bounds of ordinary courtesy if anyone venture to maintain that a plant or an animal is ugly,—to question the utility of any organic species,—to term certain scenery devoid of beauty, or indeed in any other way to withhold his sweeping approval from things as they are, should yet be, on their part, so ready to improve upon this supposed perfection, and thus to paint the lily and gild the refined gold. In all these imaginings, whether they refer to a golden age in the distant past, or to a " good time coming " in some far-off futurity,— whether they are idealisations of this earth of ours, or dreams of some other and better world,—there is a singular sameness. The poets of the past and of the present, while feeling that much around them does not answer to man's wishes, seem yet to have formed but very crude notions as to where the mischief lies. Their amendments on reality are impracticable and inconsistent, and could they be carried into effect would often prove true calamities.

Among the natural phenomena complained of by the tuneful and irritable race is the night :—

> " Fair wert thou with the light
> O'er thy blue hills and sleeping waters cast
> From purple skies ne'er deepening into night."

wrote an English poetess in an ode to Elysium, and the same idea in substance recurs in many other " lays and prose imaginings." We may well ask, at the outset, whether, in point of mere beauty, the loss of sunrise, of sunset, and of moonlight would be any great gain to the world? Surely many effects, priceless to the poet and the artist, and indeed to the æsthetic sense of all of us, would thus be rendered unknown, and even inconceivable !

But more, without a complete modification of constitution, the nature and extent of which we cannot imagine, neither man nor beast would be benefitted, or could fail to be injured by one unbroken exposure to the light of day.

Nor must we forget the part which the stars and the moon have played in the intellectual training of our race. Had we lived in constant daylight, knowing no heavenly body save the sun, not merely the science of astronomy would have been wanting, but the development of mechanics and physics would have been gravely affected. The theory of universal gravitation would scarcely have been established; the speed of light would not have been measured, or, rather, the fact that it has a rate of motion would probably have escaped us. Our conceptions of time would be very vague, deprived as we should have been of the very notions of the day and the month.

How perpetual daylight was to be obtained we are not informed. In a world revolving on its axis in presence of a sole source of light, day and night must necessarily alternate. In systems lighted up by twin-suns the nights might be very much shortened, but even there they could not entirely vanish.

To what extent a planetary body can be self-luminous, and yet sufficiently cooled down to be the abode of organic life of a type even remotely approximating to our own, we cannot venture to guess. There is, however, one conceivable arrangement by which perpetual daylight might be secured over a vast extent of surface. Suppose, instead of planets revolving round the sun at different distances, there were one broad belt of solid matter. The concave side of such belt would be constantly irradiated, and might be the theatre of life. Or, going farther still, suppose the sun fixed in the centre of one vast hollow sphere, with a radius of say 100 million miles. In this manner all his heat and light would be utilised. What a glorious flora and fauna might be developed in such a world! Is it not possible that the dark masses supposed by some to exist in different portions of the heavens may be the outer shells of such systems?

More desirable and more feasible than perpetual day is perpetual summer, or at least the absence of low temperatures. Such a state of things exists even now in many parts of the globe, and in the Eocene and Miocene epochs it was probably universal. If the human species had begun its career so early the traditions of a golden age, of a Saturnian reign, of a lost paradise, might have a solid physical basis. We may imagine with what regret our forefathers, struggling with the horrors of the Glacial epoch, would dwell on the memories of the good old times when Central Europe enjoyed a semi-tropical vegetation, and when even Spitzbergen was clad with luxuriant forests. But not content with painting the lost Miocene epoch as it really was, they would idealise it, and, in addition to its perennially mild climate, they would invest it with other and often impossible attractions.

Thus among the foremost charms ascribed by the poets to their better worlds are "never-fading flowers." The trees bud, the blossoms expand, but, instead of withering and falling off, they are to become permanent. Yet clusters of luscious fruits are all the same included in the picture. The absurdities here implied are of a most complicated nature. The flowers are, of course, supposed to exist not in order to fulfil any function of vegetable life, but merely for the delectation of man. How the fruits are to be formed and to grow without the displacement of the blossom is difficult to comprehend. Still more are we perplexed to conceive of tissues so frail and delicate as flowers, and whose beauty depends in no small degree upon that very delicacy becoming proof against the influences of air and light. We may suppose storms, winds, and frosts banished, but even in the absence of all these ruder agencies we find flowers fade,—in one sense from the continued action of the very forces which have called them into existence, and which thus in succession play the parts of Brahma and Siva, the creator and the destroyer; but more strictly this fading and decay must be regarded as the advance to a higher state of development, like the falling off of the gills and tail of a tadpole, or the loss of the milk-teeth of a child.

But a still greater difficulty remains. Unfading flowers mean an arrest of the transformations and the circulation of matter. If every tree and plant is to put forth constantly fresh flowers, and, we presume leaves, none of which are to wither and be re-converted into their crude elements, a continual drain upon the soil must take place which would sooner or later end in exhaustion, unless the supply of plant-food were absolutely infinite.

Another feature of the idealised world is the disappearance of all carnivorous species, or their conversion to vegetarianism. Rapine and bloodshed are to cease, and the reign of universal peace is to be inaugurated. Here it may be objected that the mere elimination of carnivorous species would be very far from securing unbroken peace. Herbivorous animals fight abundantly and most viciously among themselves, and are not rarely guilty of "unprovoked assaults" upon other creatures. Very few male phytophagous beasts are to be trusted if they possess sufficient strength to be formidable to man, and some—such as the Cape buffalo—are as dangerous as the lion or the tiger. Hence a mere change of diet must be pronounced a very small reformation.

But further: if it is painful to our feelings to see or to know that at any moment numbers of animals—some of them harmless, beautiful, or interesting—are struggling in the grasp of enemies, or being actually devoured, is it not equally unpleasant to see the most charming flowers mutilated by slugs or earwigs, or trodden in the mire by oxen? Must we not regret when stately trees are stripped of their foliage by locusts or cockchafers, or when their stems are riddled through by the larvæ of *Cossus lignıperda?* Is it pleasant to see shrubs barked by hares, browsed down by goats, or uprooted by swine? It may, indeed, be contended that vegetables are not conscious of the injuries they suffer.

Granting, for argument's sake, this undemonstrated proposition, we may still doubt whether a considerable portion of the animal world do not share this unconsciousness. Further, so long as beautiful objects are destroyed, and we are conscious of such destruction, we cannot help experiencing regret, whether the beings that thus perish are sensitive or apathetic.

This elimination of the Carnivora, or rather the continued existence of the Herbivora, is also out of all harmony with the notion of deathless flowers. To what good end is the rose or the lily to be exempt from spontaneous withering, if it is still liable to be destroyed by beasts?

Some dreamers have gone a step further, and have banished from their ideal worlds all animals save man only, who is to be the sole spectator and enjoyer of the flora of the golden age. Even, however, if we overlook the important part which many insects play in the fructification of flowers, it seems to us that woods and gardens without birds and bees and butterflies would strike us as dead and dreary. We should miss the chirping and the warbling up among the branches, and the mellow hum over the flower-beds, and the glance of many-tinted wings in the sunlight; and the moment we restore these welcome sights and sounds the old game of eating and being eaten re-commences, save indeed on one supposition, which has not, we believe, found its way into imaginative literature. A world is conceivable where both animals and plants should draw their support directly from inorganic matter without preying upon each other. Whether such a world exists, or ever will exist, are questions beyond our reach. It would be no inconsiderable step in that direction if synthetic chemistry, which even now gives us alizarine without the aid of the madder-plant, should some day succeed in producing starch, sugars, oils, albumen, casein direct from their elements without the intervention of animal or vegetable life at all. Strange to say this harmless suggestion, having been once brought forward in conversation by a friend of the writer's, was denounced by a reverend gentleman present as "revolting blasphemy."

Another feature of the "better world" is immortality. Sometimes men alone, but more commonly all creatures, are described as exempt from the pains and the bereavements of death. But the poets have forgot one important condition. In their deathless world reproduction is still assumed. The plants are still to flower and to fructify. The animals, from "the small gilded fly" up to man, are still regarded as sexual.

Were such the case neither this world nor indeed any world could, for any length of time, supply materials for the frames of its organic population, or even room for them to occupy. Reproduction involves death as its necessary condition. Hence we are told that in the life to come mankind shall "neither marry nor be given in marriage." That in the present state of things the longest-lived animals arrive most slowly at maturity, and produce the fewest young at a birth, is almost a trite remark.

Thus, with the exception of the milder climate once doubtless possessed by our globe, the other features of the golden age are such as never could have existed, and may even in most cases be pronounced mutually contradictory. The origin of such traditions of the past, or anticipations of the future, is easily explained, if, as above said, they are idealised memories of the Miocene days. But that man actually existed in that epoch, and existed in a condition so far advanced as to hand down and to idealise legends, is utterly unproven. If our species is of more recent development, these stories of a better world do but reflect the unsatisfied longings of all generations.

B. B.

III. EXCEPTIONAL SEASONS AND WEATHER PROPHECIES.*

SOME time ago we were led, by the sunless summer of 1879, to discuss the well-known theory which traces a connection between weather-cycles and the periodicity of the sun-spots—a term of rather more than ten years. Whilst unable to accept this hypothesis, neither did we see our way to its definite rejection. In speaking of the lack of heat which characterised that summer we fell, however, into an error which we are anxious to rectify. We ascribed the visitation which has brought ruin upon so many of our farmers, and has occasioned in the United Kingdom alone a loss for which a hundred millions of money would very poorly compensate to a predominance of the polar current. Such was not the case. The perplexing feature in

* "The Anomalous Season." Journal of Science, July, 1879, p. 457.

the season was persistent cold brought by southerly and south-westerly winds. Whether there may have been a polar upper current which chilled the atmosphere, and precipitated the moisture held in suspension by the equatorial current, we are unable to state.

Latterly a totally different theory has been brought forward by Mr. R. G. Jenkins, F.R.A.S., which leaves the sun-spots and their periodicity out of the question. It has been proved by the Astronomer Royal that the planet Venus exerts a disturbing effect upon the earth, so as materially to interfere with its orbit. This action appears to follow a cycle of eight years, and in it Mr. Jenkins seeks for an explanation of the waves of cold which he considers occur on an average every eight years. Another consequence of this hypothesis, if correct, is that for the next forty years the temperature will fall below the average, whilst during the past forty years it has been in excess.

Mr. Jenkins concludes, further, that every twelve years a wave of unusual heat sweeps over the earth nearly contemporary with the arrival of the planet Jupiter in perihelion, and that we are on the eve of such a great heat-wave.

We will endeavour to trace in how far these views are in harmony with past experience. To find an unquestionable starting-point for the series of warm years we can do no better than take a season still fresh in the memory of all the middle-aged portion of the public—the glorious year 1868, characterised by its July harvest. Taking twelve-year intervals for the last sixty years we should have the following succession :—1820, 1832, 1844, 1856, 1868, and 1880. How does this series agree with facts? As far as we have been able to trace out, the years marked by exceptional warmth were 1818, 1825,—both of which brought July harvests,—1854, 1857, 1858, 1868, 1869, and 1870. Whether there was any unusual heat in 1844 we cannot ascertain. As we pointed out in the article above referred to the summer of 1857 was beautifully dry and warm, and was followed by a winter free from snow and ice, save a short attack in February, 1858. The succeeding summer was at least equal to that of 1857, and the only frost in the ensuing winter was from November 12th to 17th. After this open weather continued. This long continuance of favourable weather is not separated from the next warm year by a twelve years' interval.

The season of 1868, again, was immediately followed by two other years somewhat inferior, but still very much

better than any seasons we have since experienced. As for
the year whose end is now approaching, no one can
pronounce it exceptionally warm, though it may seem
favourable in comparison with 1879. It has brought no
sultry nights, and the proportion of northerly and north-
easterly wind has been uncommonly high. Winter may be
said to have set in on October 20th, and all appearances
seem to point not to a heat-wave, but to a winter of exceptional severity.

We turn next to the cold years. Mr. Jenkins gives the
following series :—1829, 1837, 1845, 1855, 1863. 1871, and
1879. The remarkably bad years on record are 1816, 1845,
1855, 1860-61, 1871, and 1879.

We are bound to admit that here theory and experience
coincide far better than in the case of the warm years.
Concerning 1829 and 1837 we can learn nothing ; 1845 and
1855 were both wretched, and in 1871 the snow which had
fallen a few days before the end of 1870 lay unmelted into
February, whilst the spring and early summer were unfavourable. August and September were, however, dry,
warm, and sunny. But 1863 breaks the series. It cannot
for a moment be compared with 1860-61. Christmas-day,
1860, was ushered in by probably the most intense cold ever
registered in England since accurate thermometers became
available. Well-grown apple-trees, and even oaks of a yard
in circumference, were killed. The havoc among the ornamental trees and shrubs in Chatsworth Park was terrible.
The succeeding spring and summer—called so merely by
courtesy—were even more ungenial than those of 1879 :
perhaps less wet, but certainly colder. In November we
saw the corn over extensive tracts of country rotting where
it had grown.

The bad season of 1816, again, cannot be harmonised
with 1829 by any multiple of 8. It will also strike the
reader that the interval 1845 to 1855 is one of ten, not of
eight, years. If, then, we add to 1845 2 × 8, we are brought
to 1861, the really worst season. It might, therefore, be
asked whether any cause could have retarded the cold-wave
so that it reached the earth in 1855 instead of 1853, which
latter was, however, a most ungenial year. We doubt,
meantime, how far the expression a "wave of cold," as
applied to the regions of space, is justifiable. And this
doubt leads us to another very important consideration : do
these warm and cold years extend simultaneously over the
whole earth ? If not, it is surely hazardous to assign them
to cosmic causes. It is understood that the season of 1879,

so disastrous in Britain and the west of Europe generally, had a very different character in the United States, where the harvest was exceedingly abundant. The question whether the whole earth in 1879 enjoyed less solar heat than in 1868 has scarcely been asked, and assuredly not decided. Suppose any change in the regions of space interfered with the action of the sun's rays upon the earth, or caused the heat which the planet absorbs to radiate away more freely, it is difficult to say how such a change, acting continuously for months, could bear upon Europe and leave Eastern America unscathed.

The periods of forty years alternately exceeding and falling short of the average temperature form another feature in the hypothesis of Mr. Jenkins. Have the years 1840 to 1880 been in any appreciable way superior to those from 1800 to 1840 ? We doubt it. The first-named period has brought us only one July harvest, and, on the other hand, a succession of bad seasons, 1872 to 1879 inclusive, such as are not often recorded in authentic history.

If we take *à priori* grounds we must certainly pronounce the eight years' cycle of Mr. Jenkins less probable than the sun-spot hypothesis. A change in the conduction of the sun's surface is much more likely to be a *vera causa*, as far as the earth's temperature is concerned, than a slight orbital perturbation. It may not here be out of place to remark that if Venus is the cause of these waves of cold the astrologers were fearfully mistaken in ascribing to her a benign influence.

We cannot here help referring to the predictions of an anonymous French meteorologist, which so far have proved but too true. He foretold severe winters for the years 1878, 1879, 1880, and 1881, the last being the culminating point, after which a series of favourable years were to ensue. As far as we remember he promised, after 1881, to reveal the principles upon which his forecasts are based.

IV. MODERN CYNOLATRY.

Dedicated to " Ouida " and Dr. Lauder Lindsay.

By FRANK FERNSEED.

PLEA for Dogs ?—and wherefore ? Have they not privileges enough ? Has not every one of them the indefeasible right by judge-made law to at least one bite of human flesh ? Can the reader defend himself or his child against a canine aggressor—possibly rabid, without the risk, firstly, of assault and battery from its owner, and, secondly, of a prosecution by the "Royal Society for the Suppression of Cruelty to Animals," who, being rich and obstinate, may, if the accusation is dismissed by a sensible magistrate, "ask for a case" for a superior court ? Have not dogs a practically unlimited right of trespass and intrusion everywhere, and of committing damage to property animate and inanimate ? Is not the responsibility of the dog-fancier for the misdeeds of his pet simply null ? Do we not find, in defiance of all regulations to the contrary, dogs introduced into railway carriages, tram-cars, and public gardens ? What, then, more or further can these creatures demand, save, perhaps, the Parliamentary franchise.

It is fully admitted that in days bye-gone dogs have played a very useful part. So long as these islands were infested with dangerous beasts of prey, so long as cattle and sheep roamed in unenclosed forests and wastes, and so long as the criminal classes had not learnt the art of fascinating and silencing the fiercest and most vigilant mastiff, there was a very obvious reason for the "friend of man." But now all these conditions have passed away it surely becomes questionable how far dogs should be gerally tolerated in a populous and civilised country, except where it can be shown that they perform some real, definite service. Yet there exists a noisy party who, in addition to the privileges already mentioned, claim for these animals exemption from taxation, the abolition of muzzles and chains, and open, recognised admission into all places of public resort.

The arguments and appeals of these agitators are in their way interesting.

What, *e. g.*, must any sober-minded man of the world think of the following tissue of passionate fallacy ? I do not say sophism, for the writer, wild as are her statements, is evidently sincere:—

"The mischief, which, is ever done by our four-footed friends, is, after all, very small in its sum. All the dogs in great Britain and Ireland never slew in ten years as many human beings as one lead mine, one lucifer match manufactory, one railway company, or one refuse-choked river slays in a year. When human life is held so cheap that capitalists are allowed to pursue manufactures which are calmly shown by statistics to kill all employed in them before they can reach middle age, and whilst such manufactures are deemed quite honest and justifiable, though destroying human bodies with a frightful celerity and a mathematical certainty, it it is unutterably absurd to see the whole hue-and-cry of a nation out against dumb* animals, whose uttermost possible average of crime is that one in a million may bite one out of ten millions."

Who can fail to draw the obvious distinction that in a settled country like England dogs are, to say the least, useless and exist merely for the caprice or amusement of their owners? On the contrary, the industries to which "Ouida" takes exception are useful, if not absolutely necessary, and could not be dispened with except at a great loss. A country without dogs would neither be the poorer nor the less happy. A country without railways would suffer far greater evils than the loss of the few lives which they sacrifice, and which, in proportion to the number of travellers, is smaller than was the case in the old days of coaching.

A refuse-choked river is certainly neither pleasant nor profitable, but, as all municipal authorities know, it cannot be got rid of except at the cost of a heavy outlay.

But "Ouida" overlooks a further distinction which is by no means to the advantage of her clients. No one is compelled to work at any unhealthy employment; those who do so act of their own free choice, and usually receive higher wages in consideration of the risk. Those who reside near a lead-mine can, if they dislike the neighbourhood, remove elsewhere. A riparian proprietor, if injured in his health, his pocket, or his æsthetic sense by a refuse-choked river, can obtain an injunction against those who pollute the waters. But how, when, or where, can we avoid the dog nuisance? "Gentle shepherdess tell me where?" Nay, is not the very gist of "Ouida's" plea to demand the removal of all the present nominal restrictions on their movements and actions?

---

* What is the exact relevance of the term "dumb" in this connection?

Here are the very words :—"Let the Legislature *forbid* muzzles and chains; let public parks in all cities, at least for certain early hours of the day, be open to *all* dogs for exercise and liberty."

And now as to the "utmost possible average of crime" of which dogs may be guilty. I find it on record that in the first six months of the year 1865 no fewer than 354 persons suffering from bites by dogs were treated in the London hospitals. This would give for the whole of the year 708, and if we suppose,—which is surely no exaggerated estimate— that as many more were treated in private practice, we shall have in round numbers a yearly total of 1400 persons injured by dogs in London alone. Or, taking the entire population of Great Britain and Ireland at ten times that of the metropolis, we find the gross number of the sufferers rise to 14,000 yearly;—decidedly more, I submit, than "one out of ten millions," and certainly more than the worst-managed lead-mine, or lucifer-match manufactory, or railway company injures in the same length of time, leaving "ten years" out of the question.

Nor are these injuries always trifling. Before me lies the account of a deplorable case which occurred at Salford on the 9th of July last. A large black Newfoundland dog, belonging to a grocer in Howard-street, suddenly sprang over the wall in front of its owner's premises, ran along the street, seized a little girl named Caroline Cobb, and began worrying her. Two men, after a great deal of trouble, succeeded in driving the brute off. The child was severely wounded upon her neck, face, and body, and was disfigured for life.* It was stated that the dog had on previous occasions bitten people, and frequent complaints had been made to the owner about its savage character. The usual defence —in my opinion utterly irrelevant save as a mere legal quibble—was attempted, to wit, that the dog had never, to its owner's knowledge, bitten any person. The judge gave a verdict for the plaintiff for £30 and costs. I should like to ask the mothers of England whether a mere money penalty, which was all the court could inflict, should in such a case be regarded as at all adequate? In my opinion the destruction of the noxious brute and a few months' hard labour for the owner would be certainly not too severe a punishment.

---

* What worse could happen in a country over-run with wolves? Moreover, he friends of the maimed child might shoot a wolf, but the law does not allow them to shoot its equally dangerous cousin. If they poison it the newspapers call them "cowardly assassins."

A few such sentences would protect our children from being worried by ferocious beasts for a century to come.

I have termed the plea that the dog in question had never before, as far as its owner was aware, made an attack upon any person, "irrelevant." The fact is that, save to those whom it knows, every dog of any considerable degree of size and strength must be regarded as dangerous. No one knows when or against whom it may suddenly conduct itself according to its true nature, as a beast of prey. Surely, then, those who persist in allowing such creatures to frequent public places should be held criminally responsible for any outbreak that may occur.

But it will be said these 14,000 persons, or one in about every 2300 of the entire population of the United Kingdom, thus wantonly maimed year by year for the sake of the dog-fanciers' amusement are only wounded, not slain outright.

I will turn, therefore, to the official statistics of hydrophobia in England and Wales. The number of cases, or, in other words, of deaths, for the eleven years 1866 to 1876 inclusive is given at 387, or on an average 35 yearly! Now, as the population of South Britain does not greatly exceed 25 millions, we have here one person done to death by dogs not out of every "ten millions," but out of every 700,000,— a somewhat different result.

During the past year no fewer than 103 persons were bitten by mad dogs in Paris and its suburbs. Of these 30 are known to have died from hydrophobia. If we estimate the population of the French capital at two millions, this gives a death-rate from this one utterly gratuitous cause of 1 in 66,000! Five hundred mad dogs and a score of mad cats were destroyed by the police in the course of the year. In the months of July and August alone 4730 stray dogs were taken to the *fourrière*, and all but 200 were killed. In consequence of this most "judicious butchery," if Dr. Lindsay will permit me thus to modify his utterance, there has been a great reduction in the number of persons bitten and in the death-rate from hydrophobia.

I am aware that a most fantastically impudent plea has lately been advanced in opposition to the plain moral to be drawn from these facts and figures. We are told that if persons are bitten by a dog and die subsequently with the well-known symptoms of hydrophobia, it is merely the effect of their "excited imagination."

Had they strength of mind to repress their fears, no harm, we are told, would ensue! In reply to this lamentable sophism, I may point out that cattle, sheep, and horses after

having been bitten by dogs in the state commonly known as
" mad," display the very same symptoms. Are we to ascribe
this result also to the workings of a morbid imagination ?
Many human sufferers, so far from being tormented with
constant apprehensions, are recorded as having quite forgotten the original bite till the approach of the deadly
symptoms recalls it their memory. How, again, would
dog-worshippers explain away the outbreak of *rabies* in animals, *e.g.*, rabbits, experimentally inoculated with the virus
of the dog? Is here also imagination at work, or have we
" epileptic convulsions."

Another device of the dog's advocate is to assert that *rabies*
and, indeed, irritability and the tendency to attack passers-
by, are produced by confinement, muzzles, and chains, and
that were the curs allowed to roam about at their own
sweet will and to obtrude themselves everywhere, these evils
would vanish of themselves. This is, to say the least, a
daring assumption. The wolves of the Continent and the
jackals of India enjoy their full freedom. Yet rabies is far
from rare among them. The dogs which invest every village
on the frontiers of Poland know nothing of muzzles and
chains. Yet this freedom does not sweeten their tempers or
prevent them from attacking every stranger—especially if re-
spectably dressed—who appears in the neighbourhood. Very
similar is the behaviour of the large and powerful mastiffs
who, in the Carpathians, are kept by the Slawack herdsmen
as a defence against wolves. Probably not one of them
has ever been tied up in his life. But woe to the unarmed
stranger who happens to come within a quarter of a mile
of these ferocious brutes.

Even if we, for argument's sake, admit that some of the cases
of supposed hydrophobia in man may be partially set down to
the workings of a morbid imagination, and that some dogs
destroyed as rabid may have been merely suffering from
" distemper," I do not see that this in the least alters
the case. Had the persons affected never been bitten,
the fatal symptoms, however explained, would never have
arisen. Had there been no dogs straying about it is equally
certain that the victims would never have been bitten.
Surely the man who, owning a dog, allows it liberty when
in a state capable of communicating rabies, incurs morally—
though, unfortunately, not legally—a responsibility as great
as if he were to discharge a rifle along a public thoroughfare.
Indirectly, but not the less certainly, he trifles with the lives
of his neighbours, not out of any necessity, but for a mere
whim. And yet he can dare to talk about justice !

Dr. Lindsay pronounces the destruction of dogs for " *mere* biting " an " injudicious butchery," and laments the condemnation of the stray dogs of Glasgow after the occurrence of three cases of hydrophobia in that city.

I cannot agree with him. It is only in case of man that we can truly say, "better ten ruffians should escape than that one unoffending citizen should be imprisoned." With the lower animals, as with inanimate matter, we must invert this principle. Better a thousand bales of suspected, though harmless, merchandise should be burnt than that one bale should pass free and introduce the plague into our midst. Better, were it possible, that all the serpents in India, many of whom are perfectly harmless, should be extirpated than that one cobra should survive. No one questioned the wisdom or the perfect justice of stamping out the cattle-plague by destroying every herd among whom the scourge appeared, though many among them were not attacked, and might possibly have remained unaffected. Yet this was a case where property only, and not human life was at stake. Surely, then, we are *a fortiori* justified in destroying the dogs of a district if we have reason to suspect that *rabies* exists among them. The fewer curs remain, the less opportunity is there for the spread of the disease. A man attacked by any animal has, I submit, a natural and indefeasible right to destroy it, and if the law interferes with this right in the case of tame animals it should merely undertake the task of destruction on his behalf.

But the annoyance inflicted upon the public by dogs, or, more accurately speaking, by dog-owners, is not confined to biting. I object—and I submit with full right—to be barked at when going peaceably about my business. I object to be deafened by one of these " noble animals " which runs yelping after a public conveyance in which I am travelling. When taking my walks abroad I object to find a dog smelling at me and wiping his snout upon my clothing.

As we hear so much of the docility and intelligence of these animals, surely their owners might teach them not to interfere with passengers in public thoroughfares. I do not learn that any attempt is ever made in this direction, or that the dog-owner ever conceives that he is, by permitting and occasioning such conduct, infringing upon the rights of his neighbours.

Nor must the injury to property occasioned by dogs be lost sight of. Perhaps, indeed, I ought to have given it the precedence, since in England it is protected far more carefully than is the human person, and since an attempt was recently made to diminish even the slight defence which the latter at present enjoys. I refer to Mr. Hopwood's bill for giving the ruffian and the "corner-man" the option of paying by weekly instalments the forty shillings fine which English law deems a sufficient penalty for maiming and perhaps crippling for life some person against whom he has no cause of quarrel.

I will for the present pass over such minor mischief as chasing and killing poultry, ravaging flower gardens, and polluting merchandise placed at shop doors. But the destruction of sheep is too serious to be overlooked. In the State of Georgia as is learnt from official documents, no fewer than 28,000 sheep are worried by dogs in one year, the consequence being that in many districts otherwise suitable, sheep-farming is abandoned. In the whole of the United States the annual loss from this source is estimated at upwards of a million dollars ! This grievance has even attracted the notice of mechanicians, who have sought to devise protective collars, the neck being the part generally attacked with the object of drinking the blood from the large arteries.

In England the damage experienced is smaller because we have a substantial—though in my opinion an insufficient—dog-tax. "Ouida," it appears, would make the tax merely nominal, and would thus expose our flocks to increased perils. Even at present farmers have too often good reason to complain. It is very satisfactory to learn that in consequence of the havoc committed in their neighbourhood, Messrs. Newton, Chambers, and Co., of the Thorncliffe Ironworks and Collieries, have issued a notice that they will not employ any persons, or allow any tenants to occupy their houses, who keep dogs. It is to be hoped that this example will be followed by other iron and coal-masters.

Few persons ever take into consideration the national cost of dogs. The number of licenses granted last year in England was 1,238,867, and in Scotland 153,309, making in all 1,392,176. It is no exaggeration to suppose that the unlicensed curs must swell the total number to a million and a half. Each of these " friends of man " will, on a very low estimate cost in food 9d. weekly,—probably more, as dogs are decided gluttons—or 39s. yearly. Hence, without taking any account of the waste and loss which he occasions, the " British dog " costs the country in round numbers three millions sterling annually ! Is the game really worth the candle ? Does any benefit at all commensurate with such a sum accrue from dog-keeping ? Might not those who cry out for "retrench-ment and reform" find here an appropriate point upon which to fix their attention ?

It is small consolation to learn that there is, at any rate, one country which exceeds us in this phase of waste. America has been congratulated on the fact that she has not to main-tain a "vast standing army." But in place thereof she cherishes dogs to the estimated number of eleven millions, and, if our former suppositions are correct, pays for them the enormous sum of 22 millions sterling annually. Let those who doubt whether dogs are proper subjects for taxation remember that but for the dog-tax we should no doubt be similarly over-run.

When the advocates of the dog find themselves encoun-tered by the irresistible logic of facts, as far as the danger to mankind and the economical worthlessness of their clients are concerned, they have recoure to some most curious argu-ments. In the first place comes the "poor man" plea. "Ouida" saith :—"Let not the poor man be made to think that a dog is a luxury of the affections (!) only possible for the rich; let not the poor man be taught to see that to have a pet for his children is a greater crime and one more heavily punished than wife-beating or bestial drunken-ness."

It is remarkable how the poor man is "trotted out".when-ever a nuisance is to be upheld or a folly defended. When intra-mural interment was first attacked we were told that it would be cruel to rob the " poor man " of his " nook in the parish churchyard." The drift of Mr. Hopwood's bill, above referred to, was that the "poor man" as well as the rich might enjoy the occasional luxury of committing a wan-ton assault without having to go to prison.*

Again, "Ouida" asserts that "nothing is surer to humanise the heart hardened by toil and privation than the innocent affection of a dumb animal." In the like vein, when it was last year proposed to extend the dog-tax to Ireland, a certain "honourable" member—Mr. Biggar, if my memory does not fail me—was very eloquent on the humanising influence of the dog, and when reminded of Bill Sykes and his bull-dog he waxed exceedingly indignant. Yet the Bill Sykes of fiction is but a poor shadow when compared with the Peace of fact,—robber, assassin, and dog-fancier. Let those who

---

* The admitted injustice of the present law would have been remedied by the total abolition of fines and the infliction of hard labour in *every* case of wanton assault.

would ascribe to the "poor man's dog" any elevating or humanising influence visit those towns and villages where the iron and coal trades flourish, and make good use of their eyes. Let them take note of the sullen, low-browed ruffians who lounge at the tavern corners or slink along the lanes, each duly attended by his lurcher, his terrier, or his "snap." Or, better still, let them on some public holiday, or even on a fine Sunday morning, go out to some open plot of ground in the suburbs. To give a single example, far from the worst of its kind, let our investigators go to Leeds on a Whit Monday, and take a stroll along the Knostrop Road, past the sewage works, and note the crowd there collected for the elevating sport of dog-racing, or perhaps of rabbit-worrying. If they do this they will form a fairly accurate notion of dog-keeping in humble life and of its tendencies. Yet Leeds is correct and refined in the manners and conduct of its people in comparison with some towns which might be easily named. We do not find the "poor man's dog" in the "home of taste," in the clean, trim cottage with flowers in the window-sill and books on the table. No, his place is amidst rags and squalor and scenes of brutality and drunkenness. The "pet" for the poor man's children, as "Ouida" terms the dog, is pampered at their expense. The dog-owner *is* the wife-beater and the drunkard. As a specimen of the human-ising effects of dog-keeping we may takes the following gem, which every one who knows the manufacturing districts will admit is typical rather than exceptional :—

"To-day, at Willenhall Police Court, a locksmith, named John Henry Williams, was convicted of an aggravated assault on his wife. He had kept a racing dog, which slept in bed with him, and was fed off joints of beef, while his wife and family were scantily fed. For complaining of this treatment he beat and kicked her, breaking one or two of her ribs. He was sent to gaol for two months."

The parish-surgeon for the district of Brampton, near Chesterfield, lately reported at the meeting of guardians that many persons receiving out-door relief deprived their families of food that they might pamper dogs. It must not be sup-posed that these creatures are fed upon the scraps and leav-ings from the family table. The very reverse is often much nearer the truth. I have not merely heard of, but have personally observed cases where racing dogs are feasted upon prime cuts of meat, and where young children are robbed of their allowance of milk that it may be given to the "bull-pup." The "poor man's dog," which we are asked not to tax, is the cause and the symbol of suffering and privation to thousands of women and children. To demand for the poor cheap food, clothing, shelter, instruction is wise and just. But, alas, when demagogues clamour for cheap folly, cheap vice, or cheap crime for the million !

The fact is that dogs, unless they can be kept entirely within private boundaries, have of necessity upon their owners an effect which is anything but elevating or humanising. They are, according to their advocate, " the most sensitive, vivacious, highly sensitive and animated creatures in creation." In other words, they are the most restless, meddling, obstrusive, and intrusive of brutes. Nothing, living or lifeless, are they willing to let alone. Hence the dog-owner who leaves his cur at liberty must either be constantly paying damages and apologising to his neighbours, or,— the more common alternative—he becomes utterly hardened, loses all sense of justice, and sets the injured parties at defiance, which, thanks to the absurdities of the law, is no very difficult matter. As an instance how indifferent dog-owners become to the rights of others, we may turn to " Ouida " herself. After recounting a case where, at the Lewisham Police Court, a certain person had been fined £10 because his retriever had bitten a woman, this writer, without a thought of pity for the wounded woman, exclaims that the owner, " *if fined at all*," should have been punished for not giving the brute more liberty ! " Poor, imprisoned, tortured dog " is her exclamation ! This is quite in harmony w th Dr. Lindsay's " mere biting," and leads me to believe that the exclamation ascribed to a lady of fashion whose dog had severely bitten a poor little child may be something more than a mere apologue.

It is perfectly true that many great men have been fond of dogs. But what strange propensities and freaks might we not justify by a perfectly analogous plea ? Goethe, the type of sterling culture, as was Earl Lytton the type of electro culture, wisely said, man loves the dog because both are " erbaermliche Schufte." If the predilection for these animals has been handed down from barbarous times into an age of civilisation it is because they have readily made themselves the accomplices of man's vices. Cruelty—by which I understand the infliction of pain for mere amusement—would lack its chief agent were it not for the dog. I will merely hint that this is by no means the sole moral evil in which he is made instrumental.

Some day, if the progress of which our moralists dream is not all a delusion, dog-fancying will be regarded as one of the strangest and least creditable survivals from pristine ages. Our descendents, after reading an accurate account of the manners and habits of the dog, will hesitate between disgust and laughter when they are told that in the Victorian era ladies who fancied themselves refined used to kiss these unclean and mal-odorous brutes.

" Love me, love my dog," say you ? As the smaller of the two evils I will decline to love you !

V. FOG LORE.

TIME out of mind the fogs of England, and especially of that conglomerate which does duty as its capital, have been a favourite subject for Continental wits. We have even, in a thoroughly characteristic manner, joked over this unpleasant feature of our climate. But the subject is fast becoming too serious for jesting, and in place of recommending our friends to fit up circular saws in order to cut away the opaque and scarcely breathable atmosphere from before our doors we have taken alarm, and are beginning to arrive at the very luminous conclusion that "something must be done." For, during the last winter, we learnt from experience that a dense fog may last for longer than was formerly thought possible. We saw the traditional gloom of November last far into February, and we could not help concluding that, given the conditions of a low temperature and a windless season, London might possibly be wrapped for consecutive weeks or even months in a gloom far more perplexing than that of a starless night, and against which even the electric lamp is powerless. Hence those who— doubtless from the purest motives—are anxious to "improve" every occurrence, have not overlooked so convenient a topic. They have painted the destruction of this great city, or at least of the majority of its inhabitants, not by earthquake, pestilence, cyclone, conflagration, or war, but by a doom more weird-like and horrible than any of these scourges. A fog denser and more enduring than any yet on record is to settle down upon us. Traffic is to be rendered impossible. The existing stocks of fuel and provisions being exhausted the miserable people will be driven by cold and hunger to attempt their escape. Abandoning their houses and possessions they will sally out into the suffocating air and wander through the endless labyrinth of darkened streets in the vain attempt to reach the open country, and, wearied and stifled, will at last lay down and die. Such, stripped of its more sensational features, is the prospect held up to our view. Unfortunately, whilst we may justly pronounce such a crisis very improbable, it is still within the verge of the possible.

Thus in 1783 a fog—to which we had occasion to refer in our last number*—is said to have covered nearly the whole of Europe for about two months. What security have we against the recurrence of such a phenomenon ?

But without speculating on a visitation so horrible, even a winter such as that of 1879-80 is not a trifling evil. The hindrance to intercourse, whether by rail, river, or road, and the consequent general stagnation of business ; the waste caused by the unusual consumption of gas, oil, and candles represent, together, a very palpable tax on the community. The average Londoner found his expenditure increased and his profits diminished. Most serious of all was the increased death-rate. The mortality from affections of the organs of respiration rose to a fearful point, and many who survived the fog had still received then a fatal injury. What wonder, then, that in prospect of another severe winter we are anxiously inquiring if there is no remedy ?

To begin, then, what is the nature of fog ? We are gene-rally told that fog, mist, and cloud are one and the same phenomenon, differing merely in degree, or in position, and being in every case formed of a multitude of excessively minute vesicles of water, filled with air and suspended in the atmosphere. Such a collection of watery vapour when float-ing aloft in the air we call a cloud, and when it rests upon the surface of the earth a fog. But this view is not universally accepted, or rather, perhaps, it seems not applicable to all cases.

Dr. R. Angus Smith, F.R.S., describes† a fog which he observed at Reikjavik, in Iceland. The particles of this fog were exceptionally large, perfectly spherical and not hollow. It rolled along the streets of the town like a cloud of dust.

There are, further, two distinct kinds of fog, the wet and the dry. The former, like clouds and mountain mists, which are in reality the same thing, deposit a dew on the leaves of

* Journal of Science, November, 1880, p. 706.
† Memoirs of the Literary and Philosophical Society of Manchester, Series II., vol. 5.

plants and on the hair of animals, and are connected by imperceptible gradations with drizzling rain. On the other hand, the fogs of last winter were scantily charged with moisture, and many persons who found their breathing affected obtained relief by letting the steam of boiling water escape into the air of their rooms. A difficult question is the distribution of fog—the capricious manner in which it follows certain boundaries with little apparent reference to the direction of the wind or the character of the soil. Why should Tottenham-court Road and the surrounding district be often fairly clear, when to the east, south, west, and north the metropolis is plunged in darkness? Another peculiarity is the recent extension of fog to outlying localities which till lately were exempt from the nuisance. A friend who has lived at Sewardstone for about twenty years tells us that he never experienced anything worthy the name of a fog previous to the last two seasons.

But the fogs of all our large cities may be described, so far as their worst features are concerned, as man-made evils. Natural fogs, with the exception of some exceptional cases apparently of volcanic origin, are white, and even when densest transmit a considerable proportion of light. The yellow or black fogs of London consist largely of the products of the combustion of coal—in a word, they are in great part coal smoke, which in certain states of the atmosphere cannot escape. This fact is plainly proved by the peculiar smell which such fogs impart to the hair, the beard, and to woollen clothing, and by the taste which they leave upon the lips.

Hence we may see good reasons why the London fogs should be more severe than they formerly were and worse than those of any other city. In this " province covered with houses " there is a larger quantity of coal consumed than in any other district of equal size in the world. Nowhere else is there along with abundant coal consumption such a large tract where the free sweep of the wind is checked to such an extent by an endless array of streets.

Further, the very plan of the entire mass—London in the rational though not in the legal sense of the word—is as though purposely contrived to give fulll scope to fog and to prevent the free circulation of air. We have very few broad straight streets of any great length. The city is not cut up into quarters by boulevards and avenues.* Even when the wind is high it occasions sudden gusts and eddies instead of

---

* Suppose an earthquake—a first shock having given the alarm and further concussions momentarily expected. How few and far between are the localities where the public could take refuge!

the long continued sweep which would be desirable. The original plan of New York, when under Dutch rule, was, as Washington Irving tells us, laid out by the cows in returning from pasture. The plan of London has been committed to much worse animals,—building speculators,—who, on our modern principle of sacrificing the permanent welfare of the community to the momentary gain of the individual, have wrought a world of folly, hideousness, and abomination in general.

Having thus given a brief sketch of our present knowledge —or might it not better be said of our ignorance ?—concerning fogs in general and London fogs in particular, we come to the vital point, the remedy. In all such cases there is reason to fear lest the " something to be done " may fail in every respect, save in adding to the very heavy burdens of the ratepayers.

There are proposals to prevent the formation of fog, to remove it, and, again, to alter its nature.

To effect the first-mentioned object an appeal has been made to the " unknown god " of modern days—electricity. We generally find that if some difficulty has to be overcome, the very nature of which is still obscure, it is proposed to apply electricity. But fogs are sometimes electro-positive and sometimes electro-negative. For the dispersion of the former class Dr. T. L. Phipson* considers that " it would be necessary to supply them with an abundant source of electro-negative electricity more quickly than the earth usually supplies it." Some propose a multitude of lightning‑conductors in order to convey the positive electricity of the air down into the ground. Others suggest that trees should be planted wherever possible. It is certain that trees where numerous bring down wet mists in the form of rain, but we have not been able to meet with any proof of their efficacy in case of dry fogs. The question has also been raised in this connection whether asphaltic pavements, being poor conductors of electricity, may not promote fog by preventing a free interchange between the earth and the atmosphere. But the filthiest Macadamised streets of north and south‑western London seem to suffer no less than the clean asphalted districts of the centre.

Another plan for dealing with fog is its mechanical removal. The originator of the scheme, who offers his process freely to the public, is Mr. W. A Gibbs, of Gilwell Park, Sewardstone, a gentleman honourably known for his agri-

* Familiar Letters on the Mysteries of Nature.

cultural drying machinery, for which he has received the gold medals of the Society of Arts and of the Highland and Agricultural Society of Scotland. He proposes to erect at the outfalls of the London sewerage systems at Barking and Crossness huge fans, to be set in motion by tidal power, and by their agency to cause an inrush of air into the sewers all over the metropolis, whilst fresh air would stream in from the country on all sides.

It will be admitted that to ventilate the sewers in this manner by the entrance of the air of the streets rather than by the egress of sewage gas would, if otherwise practicable, be a decided advantage, not merely in times of fog, but all the year round. It is also, we believe, conceded that the rise and fall of the tide might be made to yield power equal to the work required. But there are certain difficulties of a most formidable nature. The multitude of apertures by which air can penetrate into the sewers are so many that it becomes doubtful whether any conceivable exhaustion applied at Barking Creek would cause an in-draught into the sewers at Kensington, Pimlico, or Chelsea. It may also be asked whether the fog, if drawn out at the sewer mouths at Barking or Crossness, would not be carried back to the metropolis by the slight easterly current which generally prevails in foggy weather. We should very much like to see the experiment tried in some town of smaller extent than London where tidal power was available. It would be important to learn to what distance from the exhaustion-fans an inrush of air into the sewers could be traced.

There is another project of a far more gigantic nature, which, along with a number of other important objects, would have a two-fold action upon the fog difficulty. It has long been known that by boring to a sufficient depth water of a very high temperature can be obtained. The suggestion, therefore, lay very near that wherever a water supply sufficient in quantity and of a sufficient heat could be obtained entire towns might be warmed, as are now our conservatories by means of hot water pipes. Not only could dwelling houses thus be kept at a comfortable temperature all the year round without the combustion of fuel and the production of smoke, but the streets might in winter be so far heated that snow could not lie and that mud would be quickly dried up. This splendid project has, we understand, been to some extent reduced to practice in America. We should submit that a town thus warmed would be almost entirely free from fog, and that on three accounts. Ordinary fogs are phenomena connected with a very low temperature, and dissolve as the heat increases. Secondly, the warmed streets would occasion a current of air; and, thirdly, the decrease in the consumption of coal would greatly reduce the production of smoke. That this project, if it could be carried out in London, would put an end to the existing coal monopoly might be even a greater benefit than the abolition of fog. Still, even if it could be demonstrated that an inexhaustible store of hot water existed at an accessible distance underground, we do not suppose that it would ever be utilised in the metropolis.

We have yet to consider the schemes for attacking fog by reducing the quantity and altering the nature of the fuel consumed. Here again we are tantalised by seeing things perfectly feasible from a physical point of view rendered impracticable by social peculiarities. All engineers who have studied the application of heat will tell us that at least three-fourths if not more of the warmth generated by the fuel consumed in ordinary English houses is simply wasted. Hence it would seem a very trifling matter to economise three-fourths of the coal employed, thus ridding the atmosphere of three-fourths of the smoke and soot-flakes, and rendering London a far more pleasant and sightly place of abode. But there is a barrier which cannot be broken Our dwellings are built for the open stove system, and cannot be altered without considerable outlay. Upon whom is this charge to fall? The tenant can scarcely be expected, in addition to overcoming the English prejudice in favour of an open fire, to sink a serious amount of money in modifying premises in which he has, in a vast majority of cases, only a brief and doubtful interest. The landlord, so far from bearing the expense, would probably not allow the alteration without an undertaking on the part of the occupier to restore the old open grate on the determination of the tenancy.

The very same considerations show the injustice and the cruelty of applying, as some have proposed to do, the Smoke Consumption Act to private houses. To the so-called working classes and to the poorer proportion of the middle class, who together form much more than half the population, such a decree would mean little less than ruin.

Another suggestion is to prohibit the use of bituminous coal, and to require, in its stead, the use of anthracite. This change could scarcely be effected without a serious rise in the preposterous prices now charged for fuel in London. At present the railway companies substantially refuse to carry coal save for a select circle of of individuals. But if only one coal-field, *i.e.*, that of South Wales, and one railway were rendered available for the supply of fuel to the metropolis, it is surely reasonable so suppose that the present monopoly would be succeeded by one still tighter.

The use of gas as fuel would doubtless tend to the decrease of smoke, besides presenting not a few collateral advantages. But until the supply is placed in very different hands and the price reduced it will continue to play but a subordinate part.

Our prospects of relief from the fog nuisance are, therefore, about as bright as the fog itself. The means most likely to be adopted may safely be pronounced remedies worse than the disease.

VI. WORKING *V.* FIGHTING.*

By J. W. SLATER.

" TAKEN with other recent scientific contributions, including especially the work of Professor Huxley 'On the Cray Fish,' the addresses of Prof. Agassiz on the echinoderms, and of Professor Le Conte before the Entomological Club, show a certain change of attitude, which naturalists are now assuming, on the subject of the development of vegetable and animal forms. There was a time, not long ago, when every voice was strident in advocacy of evolution. Now evolution is received as an established fact, and scientific effort is directed in explaining the many difficulties that lie in the way of the special working and applications of the hypothesis. This is the better and the truer spirit; for to increase our knowledge we need research not polemics."

Thus writes, most justly, the *New York Medical Record* (Sept. 18). But what has so long delayed biologists from entering upon this "more excellent way" and compelled them to spend twenty years in controversy often of doubtful

* Die Natürlichen Existenz-Bedingungen der Thiere: Von KARL SEMPER, Professor an der Universität zu Wurzburg. Leipzig: F. A. Brockhaus.
   Island Life; or, the Phenomena and Causes of Insular Faunas and Floras. By ALFRED RUSSEL WALLACE. London: Macmillan and Co.

merit ? Not, assuredly, their own good pleasure. The polemics have been due to the "parsons, poets, artists, lawyers," metaphysicians, and the like, who, *as such*, and without the biological discipline needful for understanding the evidences of the question, have come forward to dogmatise on the formation and the correlations of species and have even laid the flattering unction to their souls that they, forsooth, were better able to appreciate facts than the men whose special task is observation and discovery.

These Babel-voices, literary, æsthetic, teleological, are now fading into a wholesome quiet, and Evolution is about to undergo its veritable trial—the scrutiny whether it can be fairly made to harmonise with and to account for the many puzzling phenomena which we recognise on every hand—whether the hypotheses which have been set up can be verified by exact research. If unable to stand this test they will be laid aside by naturalists without any bidding from the lay public.

The two works just mentioned, proceeding from authorities whose competence is beyond question, are valuable contributions to this great task, and though approaching the subject independently and from different sides, they are substantially in accord with each other. Prof. Semper—after a thoughtful and suggestive introductory section, of which anon — considers in detail the influence upon animal life of the environment in which it is placed, *i.e.*, nutriment, light, temperature, standing water, still air, flowing water, and living organisms. All these agencies may act upon animal species by transformation, by selection, and by dispersion, and their effects have as yet been but very slightly and imperfectly studied. Lamarck, indeed, as is well known, ascribed the development of plants and animals as we now find them to such causes as abundance, scarcity, or peculiar quality of food, excess or deficiency of moisture, and the like. But his views, however important, remained mere speculations, not verified by observation and experiment,— an undertaking which in his day would certainly have been found impracticable. "The task of the zoologist, therefore," says Prof. Semper, " is to examine how vital conditions act upon individual animals and their organs, in order to infer back to the physiological causes of the origin of different animal forms." It will be noted that he does not overlook the transforming or modifying agency of outward circumstances. Most naturalists now fully admit that " natural selection," or, indeed, selection of any kind, though it may preserve and increase advantageous modifications when once they have taken their rise, cannot possibly explain their first appearance. Hence, before we can account for the origin of species we must know whether such conditions are in any case found to set up a tendency to variation, of which selection may take hold.

Mr. Wallace treats of the origin, and especially the dispersal, of species from a different point of view, basing his considerations on the faunas and floras of islands. He classes islands as oceanic—those formed by volcanic or coralline action in the midst of wide seas—and as continental, *i.e.*, such as have evidently at one or other point of time formed part and parcel of some adjacent continent. This latter group he sub-divides according as their separation from the mainland is of recent or of ancient date. The three classes thus formed present respectively most remarkable differences in their animal and vegetable population, and in pointing out these distinctions and explaining how such regions have first received their inmates, he is led to conclusions which may be usefully compared with those of Prof. Semper. Both authors examine the influence of temperature and recognise the changes of climate which have undoubtedly occurred upon our earth as important factors in the origin of species. Mr. Wallace shows, however, that "we find no indication that the almost perfect similarity of climate and general conditions has any tendency to produce similarity in the animal world." Prof. Semper, who has conducted very important experiments on certain species of mollusks exposed to abnormal temperatures, comes to the kindred conclusion that the fossil fauna of any region affords no trustworthy evidence as to its climate at any particular past epoch. He brings forward the remarkable fact that parrots, distinctly tropical birds, accustomed to an average temperature of 80° Fahr., have been found capable of living and increasing in the open air in England with a mean temperature not exceeding 50° Fahr.

Such considerations ought to inspire geological speculators with a wholesome caution.

Mr. Wallace enters at length into the action of oceanic currents and of storms in the distribution of species, and by their means he explains, as it seems to me very happily, the peculiarities of the faunas of the Azores, the Bermudas, the Gallapagos, and other remote island groups.

· Professor Semper, too, examines very fully the agency of currents both as promoting and as limiting the distribution of animals, and as sometimes checking the formation of new species. As regards the action of the wind in regulating the fauna of the West African islands above referred to, he agrees substantially with Wallace.

With reference to the zoo-geographical classification adopted by our illustrious countryman, he remarks, however, that it is scarcely possible to conceive a greater contrast than that existing between the fauna of Hong Kong, Amoy, or even Siam on the one hand, and that of Borneo, Java, and Sumatra on the other,* a distinction which is repeated in a very striking manner in the Phillippines, whose northern portion shows an unmistakable approach to the true Chinese type, whilst the southern islands point very decidedly towards Borneo on the one hand, and towards Celebes and Gilolo on the other. These features may be traced in the mammals, fish, reptiles, insects, and very clearly in the mollusks. What may be called the endemic fauna of the islands occurs in its nearest approach to purity in the centre of the group. These characteristics he explains by oceanic currents and by the changes of the monsoons. He does not, however, suppose that affinities between locally remote faunas can be universally thus explained.

We meet, further, with a criticism of Wagner's " Theory of Migration," which Prof. Semper considers as not really in opposition to Darwinism, as its author supposes, but as therein included. It may be said, then, when Mr. Wallace declares that species can arise only where there is room for them, he gives all that is really valuable in this supposed new theory.

Prof. Semper deals thoroughly with the vexed question of hybridism, refuting *seriatim* the four positions of the old school, viz., that hybrids do not exist at all, save in a few exceptional cases ; that, if produced, they are always, or nearly always, unfruitful ; that hybridisation never occurs without human intervention ; and that species which have been known to produce fertile hybrids are mere varieties, morphologically but not physiologically distinct. All these popular delusions are completely shattered to pieces by the facts here adduced, observed among mammals, birds, reptiles, fishes, insects, and even mollusks. One of the most interesting cases is that of *Tetrao medius*, now known to be a hybrid between the great *Tetrao urogallus* (cock of the woods, or capercailzie) and the black-game, *Tetrao tetrix*, and which in many districts is fast superseding the latter. But the most important lesson hence to be learnt is that in hybrids there

---

* It must here be remembered that Mr. Wallace places the former realms in the Indo-Chinese, and three latter in the Indo-Malayan sub-division of his Oriental region.

appears not merely an intermixture in varied proportions of the parental attributes, but the latter are rendered unstable or mobile, and the young become more easily modified by the conditions in which they are placed. A splendid field for experimental research is here opened.

. Mr. Wallace does not discuss hybridism and its consequences. But he holds that species are not all and at all times equally variable, and he shows reasons for assuming that the present condition of the earth is one of exceptional stability as to climate, and consequently of exceptional stability of species. It need scarcely be said that these considerations supply the solution of certain serious difficulties. They meet the objection of the almost infinite lapse of time required for the evolution of·species, if its rate is supposed to be constant: they answer—if any answer is needed—the cavil that none of us have witnessed the evolution of a new species, and they will surely satisfy those " eingefleischte anti-Darwinianer" who still cling to the "Egyptian" fallacy.* It is remarkable that the the climatic conditions of Egypt have probably undergone less modification during the past three thousand years than those of most other countries. If, then, its present animals are exactly similar to those found as mummies or depicted on its monuments, the fact is in accordance with, and not in opposition to, the doctrine of Evolution.

Prof. Semper gives a much-needed caution against the too common error of arguing from the position of an animal in the zoological series and from the structure of its organs, *e.g.*, its dentition, to its habits and its selection of food. The well-known pond-snail, *Lymnæus stagnalis,* belongs not merely to a herbivorous group of snails, but its teeth are formed on the true plant-eating type. Yet it attacks and devours the small water-newt, *Triton tæniatus,* even in an aquarium full of flourishing water-plants. The present writer has seen the same species feasting on frog spawn, though there was at hand vegetable food in abundance. Some of the rodents are purely vegetarian; others, with ·similar dentition, are semi-carnivorous, like the rat or the squirrel, which the author very correctly pronounces "the greatest enemy of our song birds, whose eggs and young it devours in great numbers."†

He holds that "by far the greater number of animals

---

* Journal of Science, 1880, p. 166.
† As this marauder has also a decided taste for wall-fruit he ought to be extirpated wherever practicable.

depend upon light only through the mediation of the eyes."
With all due deference I must beg to point out that according to the researches of Moleschott and Fubini (*Mitheilungen aus
dem Embryologischen Institut zu Wien*, iv., p. 265) "the action
of light in promoting the metamorphosis of matter is exerted
not merely through the eyes but through the skin, and can
be traced even in blind frogs, birds, and mammals. If the
eye alone or the skin alone is stimulated by light, the increased escape of carbonic acid is smaller than when the
entire body is exposed to light.

With reference to the brilliant colouration of the sea-
anemones and the true coral-polypes often compared to
submarine flowers, Prof. Semper shows that this beauty is
not to be explained on the principle of sexual selection.
Both males and females being rooted fast cannot seek each
other, but emit their sexual secretions into the water, leaving
it to the currents to effect the fertilisation of the ova. As
little can the decoration of these polypes be classed among
the "protective" or the "warning" colours. They
are not concealed but rendered more conspicuous, both
to their possible prey and to their enemies, and the latter,
such as the *Scaridæ* and the *Diodontidæ*, are not in the least
deterred by the sight. We must, therefore, refer these
colours to the "typical" class described by Mr. Wallace.*

Prof. Semper is, however, inclined to hope that both the
production of pigments and their distribution in different parts
of the animal system may be soon rendered intelligible. He
admits the existence of mimicry, as established by Bates and
Wallace, and detects instances of it even among snails. But
he adds the caution that any agency which may, indeed, select,
but is unable to transform, cannot be regarded as the efficient
cause of any phenomenon. This warning, which the author
reiterates, marks his distinction from that of not a few naturalists who have dealt with the question of Evolution. But
the same lesson is to be found, though perhaps less explicitly,
in the work of Mr. Wallace. He, too, urges that geographical
and geological changes, "the alternations of warm and cold,
or of uniform and excessive climates—of almost perpetual
spring in arctic as well as in temperate lands, with occasional
phases of cold culminating at remote intervals in glacial
epochs," have played not a mere selecting but a modifying
part, and have thus produced "some of the more remarkable
changes in the specific character of organisms." Hence it
is plain that the co-discoverer of "natural selection" guards

* Macmillan's Magazine, Sept., 1877, p. 392.

us against the popular error of ascribing to such selection the origin of the variations upon which it has afterwards to act. In one case Mr. Wallace advances statements which have the semblance at least of discrepancy. On p. 73 he remarks that "reptiles appear at first sight to be as much dependent on land for their dispersal as mammalia, but they possess two peculiarities which favour their occasional transmission across the sea—the one being their greater tenacity of life, and the other their oviparous mode of reproduction." Yet on p. 319 we read:—"To this cause we must impute our comparative poverty in mammalia and reptiles— more marked in the latter than the former, owing to their lower vitality and smaller powers of dispersal."

Mr. Wallace's sections on the permanence of the distribution of land and water and on the causes of glaciation must be passed over, as I find that they will be considered separately in some future number of the "Journal of Science."

Prof. Semper's introductory chapter is fraught with useful lessons which are sometimes left out of view even by the more advanced student. The author considers the distinctions between adaptive and hereditary peculiarities. He shows that parts, which from their adaptive character have little or no meaning in determining the affinities of the higher systematic groups, acquire a very great diagnostic value for smaller divisions, within which they must rank as hereditary. Hence it is impossible to draw an *à priori* distinction between adaptive and hereditary characters. He shows the necessity for the naturalist, of a "physiology of organisms" superadded to the physiology of organs which has been vitiated by being ordinarily treated and, studied exclusively with reference to medical practice. This is a deplorable fact. It may be inquired what would be the state of chemistry if cultivated merely from a pharmaceutical point of view, or of astronomy if confined to the service of navigation ? If it be asked what is such a "physiology of organisms," the author replies that it comprehends all those laws which are perceived on investigating the vital relations of different animal species to each other, and to the conditions by which they are, as such, preserved, destroyed, or transmuted. As an instance of the problems with which this branch of science has to deal he mentions the discovery of the causes which have led, in vertebrate animals, to the development of the two pairs of limbs and two only.

There is very much more in this work well deserving of special notice, such as the account of the eyes—of the true vertebrate type—found in the back of the molluskous genus

Oncidium, and the author's speculations on their origin. But space forbids any further remark or exposition, save of Prof. Semper's hypothesis of coral formations, which cannot be passed over in silence.

Few men of general reading are in these days ignorant of the theory now generally received on this subject. It was first propounded by Mr. Darwin in his "Naturalist's Voyage," and explained more in detail in his subsequent treatise on the "Structure and Distribution of Coral Reefs," and was independently discovered by Prof. Dana.* Its leading points may be shortly recapitulated as follows :—There are three great classes of coral reefs, atolls, barriers, and fringing reefs. The first kind are ring-like reefs enclosing a lagoon or tract of water; the second class "either extend in straight lines in front of the shores of a continent or of a large island, or they encircle smaller islands; in both cases being separated irom the land by a broad and rather deep channel of water, like the lagoon within an atoll." The fringing reefs where the land slopes abruptly under water are only a few yards in width, but where the slope is gentle there may be a channel—usually narrow and shallow—between the coral formation and the land.† Mr. Darwin's theory is that the atolls and the barrier reefs are founded on and intimate areas of subsidence, the coral polypes building in shallow water and working upwards as the foundation subsides. Fringing reefs, on the contrary, show that the shore must either have remained stationary or have been elevated.

Professor Semper, during a somewhat prolonged residence in the Pelew Islands, has been led to decidedly different conclusions. In these islands he has found shallows without any reefs closely bordering on atolls. The west coast of this group is steep, whilst the eastern shores slope very gradually down. According to the Darwinian theory there should hence be a fringing reef on the western side and a barrier on the eastern. The reverse is the actual case. On the eastern side there is a fringing reef, whilst on the western is found a barrier reef four or five miles from the land, and interrupted by three channels which do not lie opposite the mouths of rivers, or even of streamlets. He considers that the occurrence of large detached blocks of coral at the outer margin of all the reefs on the western side, the large, almost horizontal submarine plain to the north of Peleliu, the uninterrupted con-

---

* See Journal of Science, 1875, p. 534.
† Journal of Researches into the Natural History and Geology of the Countries visited during the Voyage of H.M.S. *Beagle*. By C. DARWIN. P. 472.

nection of the eastern reefs of Pelelin and Kreiangel with the dead elevated coral cliffs all argue against any recent subsidence.   The author's own theory is that the growth of corals is chiefly influenced by the strength and direction of the oceanic currents. · Wherever constant and deep currents strike a coast tangentially the reefs are compelled to grow vertically upwards, and on the other hand many corals have the tendency to spread out equally on all sides, as far as possible, when shallow currents sweep over them. He considers that the peculiarities of the Pelew reefs can all be explained by the action of such currents during a period of elevation.

He observes that certain corals, such as the Astreæ and Porites, form circular masses of a regularly convex surface. The polypes on the summit are as well developed as those on the sides, so long as they are never laid dry at time of ebb.   Larger blocks of the same kinds become flat on the summit, which is exposed even at ordinary ebbs, the polypes being dead and covered with sand and sea-weed.   In still larger masses the surface is concave with an elevated margin, presenting, in fact, the likeness of an atoll in miniature.

It is obvious that if Professor Semper's theory holds good, much of our present conclusions concerning supposed areas of subsidence will require to be carefully reconsidered. On Mr. Darwin's hypothesis an atoll in the present day is the monument over, or rather around, the grave of a submerged island.   At an earlier date it must have been a barrier reef surrounding a small, generally mountainous, plot of land.   Would it not be important to search whether at or near the centre of a lagoon within such atolls there could be found traces of rocks other than coral or its transformations ? It will also be essential to examine the relations of the more characteristic atolls, barrier reefs, and fringes, to prevailing currents.

Regretting the necessary shortness of this survey of the teachings of two most significant works, I must recommend them to the careful study of the reader.

ANALYSIS OF BOOKS.

Our Summer Migrants. An Account of the Migratory Birds which pass the Summer in the British Islands. By J. E. HARTING, F.L.S., F.Z.S. London: Bickers and Son.

WE have here an account of forty-nine species of birds which regularly or occasionally visit this country in the summer season, and, after laying their eggs and rearing their young brood, depart to spend the winter in some milder climate. We say generally because the author includes in his survey two beautiful and harmless species, the hoopoe and the golden oriole, which are only occasionally to be seen in our woods and gardens, and are, for the most part, wantonly murdered as soon as they are recognised. The brutish idiot who has done the deed then writes to the papers, and conceives that he has rendered a service to Natural History. One wretch, as here recorded, finding an oriole's nest, near Sandwich, took the young birds and gave them to his ferrets. Thus the true naturalist is deprived of the opportunity of studying the habits of such species. It is remarked that the hoopoe and the oriole were formerly more numerous in Britain than at present, and unless the law is sharpened and public opinion becomes more enlightened both these birds will soon rank in the extinct list, as far as these islands are concerned.

The stork, which was once as regular a summer visitant in Britain as on the opposite shores of Holland and Belgium, has already disappeared from our list, and is not noticed by Mr. Harting.

Even the nightingale is in no little peril. We learn that in 1867 "three London bird-catchers between April 13th and May 2nd took 225 nightingales," and that the previous year "these same bird-catchers had supplied the dealer by whom they were employed with 280 nightingales." It is manifest that no species—save, perhaps, the sparrow—could resist such an addition to the natural causes of destruction against which it has to contend. The possessing or offering for sale a captive nightingale ought to be visited with a heavy penalty. Let once the bird-catcher and the dealer be fined in sums heavier than they can possibly realise by the sale of the unfortunate captives, and they will very soon drop their unhallowed business. As far as we remember, the penalty in Prussia for capturing nightingales during the breeding season is about £4 10s.

Mr. Harting's accounts of the various species deal principally with their habits, localities, times of arrival and of departure. Structural descriptions are not given, and are the less needed as the illustrations are from Bewick's designs, and need consequently no encomium.

The protracted discussion whether the young cuckoo does actually eject its foster-brothers or sisters from the nest is here decided in the affirmative. That the parent birds fail to distinguish their true offspring from this interloper, or recognising him do not interfere, is one of those strange instances of animal stupidity which it is hitherto impossible to account for. Perhaps, however, higher beings—if modern "advanced thought" will allow that any such exist in the universe—may be equally puzzled with the occasional stupidity of man. It is suggested that the nerves and muscles of the young cuckoo must be much more developed than its outward appearance would warrant. Yet along with the direct evidence that the cuckoo lays her eggs in the nests of other species (which has never been disputed), and that the young eject the rightful tenants, it is mentioned that— "In Fife, Forfar, and Tayside Mr. P. Henderson, from personal observation, has frequently found cuckoos' breasts bare of feathers, as if from incubation, and has observed the old birds feeding their own young." It would be interesting to inquire whether in none of the fifty-seven species of birds intruded upon the young have developed a means of defence against the operations of the nestling cuckoo?

Mr. Harting's work ought to be in the hands of every student of British Ornithology, and will prove an incitement to observation.

Betrachtungen über die Entwickelungs-Geschichte der Schmetterlinge und deren Variation. By A. KEFERSTEIN. Erfurt: Carl Villaret.

WE find here, in the unpretending form of a pamphlet of 116 pages, a collection of facts connected with the development and variation of Lepidopterous insects. Much of the information here given is rare and curious, not to be met with in ordinary entomological treatises, and has been selected by the author from the transactions of learned societies with great care and diligence. He surveys these insects in their four successive phases of egg, larva, pupa, and imago. Under the first of these heads he mentions the fact that cases of viviparous reproduction, though decidedly exceptional, are not unknown among moths (*Tinea, Oeketicus*). He mentions that the eggs of different species of the same genus do not appear to be similar in form,— those, *e.g.*, of *Vanessa Polychloros* being smooth and bottle-shaped, whilst the eggs of *V. Io* and *V. Atalanta* are ribbed and cylindrical,—a phenomenon which deserves more extended investigation.

He considers that the difference of sex is to be traced even in the egg, and is not, as some naturalists assume, influenced by the larger or smaller supply of nourishment during the larval condition. In connection with this subject he enters upon the question of parthenogenesis,—a phenomenon far from rare among certain Micro-Lepidopterous genera; *e.g.*, *Fumea* and *Solenobia*. He finds it thoroughly established that the females of species belonging to these groups, without impregnation, deposit eggs from which larvæ are produced. Such caterpillars yield, however, female moths only, which continue to reproduce themselves in the same manner. Whether this kind of propagation is limited to a certain number of generations or may go on indefinitely does not appear.

Turning to the larvæ of moths and butterflies, Herr Keferstein devotes much attention to the irritant and poisonous secretions of many species. In certain species the hairs appear to be hollow, forming veritable stings, and are connected with true poison-glands at their base. Naked caterpillars are also poisonous. Prof. Zeller suffered severely from touching the larvæ of *Depressaria venificella*, which he found feeding upon *Thapsia gargarica*, in the neighbourhood of Syracuse.

On the diet of Lepidopterous larvæ we find a great number of observations, far more species being carnivorous than it is commonly supposed.

It is remarkable that whilst the caterpillars of some butterflies can be reared up to maturity in utter darkness, those of *Parnassius Apollo*, according to Prof. Von Siebold, refuse to feed in the dark.

It is well known that the adult death's-head moth (*Acherontia Atropos*) utters a squeaking sound if touched, but, according to Brauer, its larva stridulates also.

That the caterpillas of certain species should in their earlier stages appear respectively undistinguishable, and should afterwards assume characteristic differences, is what we should expect according to the principle of Evolution. But the larvæ of *Danais Phidippus* and *Bombyx Vishnu*, two species remote from each other in any possible classification, can only be distinguished by their respective size and by the plants they feed upon. On the other hand, there are cases (*Acronycta Psi* and *tridens*) where the perfect insects are scarcely to be distinguished, whilst their respective larvæ are plainly different—a most perplexing fact, of which several instances are known.

As regards the pupa-state there are, as might be expected, fewer interesting phenomena to put on record. Pupæ are, however, by no means so helpless, motionless, and unconscious as they are often represented, and the notion—common among the lay public—that the pupa condition is a kind of death which should entirely dissociate the past and the future career of the insect, is ill-founded.

We come now to the imago condition. Here the author, with somewhat questionable prudence, attempts the definition of the illusive idea "species." He says :—"Hereby the notion of species is established that we say of butterflies which copulate with each other, lay eggs, and so produce descendants of the same character, which again give birth to similar beings, that they make up a ' species,' or ' belong to one species.' " He quotes also Piochard de la Brulerie, who defines species as the " totality of beings who are capable of producing to infinity beings identical to themselves." As we have not observed an infinite number of generations, either of butterflies or of any other animal, we cannot accept these definitions which assert the permanence of species, and overlook the cases of hybridism which are known to occur in Nature, and are probably far from rare. Herr Keferstein examines the various morphological characters by which species are distinguished,—such as size, antennæ, palpi, feet, wings, scales, pattern, &c.,—and finds them all liable to fluctuate.

Entering upon the different classes of varieties and their causes, he enumerates sexual variation, the distinction (where it exists) between the male and female of the same species ; generation-varieties—cases where, *e.g.*, the summer and winter or the spring and autumn brood differ ; climatic and local varieties, of which the author enumerates many interesting instances ; ordinary varieties, those which cannot be explained,—*e.g.*, the twenty-four varieties of *T. parmatana* which Fischer von Röslerstamm obtained from a single brood of larvæ. Hybrids and monsters he classes as aberrations. He enumerates a number of variations produced by unusual diet, though his own experiments in this direction have given merely negative results. He considers that the beginning of variation must be sought for in the egg.

In conclusion we find a few remarks on the odours, the luminosity, and the stridulation of different Lepidoptera.

We can commend this little work to entomologists as a treasury of interesting facts, and we feel certain that all will regret to learn that the author's researches are suspended by the severest affliction which can happen to an observer of Nature—a serious affection of the eyes.

A Reply to Criticisms on " The Problems of Insanity," with Remarks on the Gosling Case. Delivered before the New York Medico-Legal Society, April 16th, 1880. By G. M. BEARD, A.M., M.D.

IT appears that on March 3rd Dr. Beard read a paper before the above-named Society, on " The Problem of Insanity." At the April meeting of the Society Dr. Hammond read a paper on the " General Paralysis of the Insane," and the final discussion of Dr. Beard's paper was postponed to a special meeting, April 16th. In his concluding speech Dr. Beard defends his definition of insanity as a " disease of the brain in which mental co-ordination is seriously impaired." He maintains that a definition of insanity is very necessary, and that his definition or short description—though it does not penetrate to the essence of the condition, is a practical working definition, which may be fairly used either in the scientific study of the question or in a court of justice.

In some appended remarks on a celebrated case which had recently engaged the New York courts, the author makes the following just observation:—" In studying fever we visit the patient a number of times before we make out a diagnosis, but in studying insanity the assumption is that a single visit is sufficient, even when we know nothing about the previous carriage of the patient, and therefore have no means of contrasting his present with his past and habitual behaviour, which is the most determining element in establishing the diagnosis of insanity."

Among the many " happy thoughts "—rising almost to the value of aphorisms—with which this reply abounds, space allows us to select one only :—" We may be justified in condemning a man scientifically when we have no right to condemn him legally."

Charles Waterton. By JAMES SIMSON. Edinburgh: Maclachlan and Stewart.

THIS treatise, which can scarcely be termed an *éloge*, has been called forth by the re-publication of Waterton's " Essays " and " Wanderings," edited respectively by Mr. Norman Moore and the Rev. J. G. Wood, and contains not a few extracts from a previous work of Mr. Simson's which we regret has hitherto escaped our notice.

With the author's estimate of Waterton as a naturalist we find a difficulty in dealing, since in a number of important points it agrees very closely with our own, as expressed in the articles

" The Balance of Nature " and " Gilbert White Reconsidered."*
In these essays we have referred to Waterton's deficiencies as a
systematist, to the paucity of the work he actually produced if
taken in comparison with his splendid opportunities, to his inconsistency in denouncing scientific terminology whilst indulging
in needless classical quotations, to his errors on the skunk, on
serpents, on wild dogs and wolves, on the Quadrumana, &c.
We have noted his disposition to deny altogether any fact which
had never come under his own observation, and we have pronounced him " an odd mixture of obstinate credulity and of
scepticism equally stubborn." It is therefore with no small interest that we find our opinion confirmed by Mr. Simson, who has
quite independently arrived at conclusions very similar. Nor
have we overlooked that eccentricity which pervaded Waterton's
whole character, and which is the only conceivable excuse for
his suppression of the bulk of his own observations. What, for
instance, has he put on record concerning the five thousand specimens which he dissected ? We have expressed our regret that
Waterton should have obtruded his peculiar religious and political opinions upon our notice in works treating ostensibly on
Natural History. Nor have we overlooked his irritability and
pugnacity. Still it must be remembered that he was placed
under great temptations. He had to deal with rampant charla-
tans and their worshippers, and with one who, though no quack,
was the champion of that incomprehensible delusion the Quinarian system.

The utterance of Mr. Moore that " there is nothing on tropical
natural history which deserves to be named with the ' Wanderings ' " seems to us strange indeed, remembering that it was
written in 1871,—a date subsequent to the appearance of Mr.
A. R. Wallace's " Malay Archipelago."

We can scarcely agree with Mr. Simson in supposing that
Waterton's aristocratic standing went any way in securing his
position as a naturalist. Whatever influence he might have
possessed as the head of an ancient county family would be
more than lost by his religious and political peculiarities. Sympathy with the enemies of England, at home or abroad, was not
in his day a passport to favour and power.

A very large proportion of the present work is devoted to the
consideration of Waterton not as a naturalist, but as a man,—a
matter in which we feel but little interest. Whether the Squire
of Walton Hall managed the family estates in Demerara well or
ill, in what manner he ought to have prepared himself for the
post, and whether his conduct under Custom-House persecution
was judicious or otherwise, are points of small moment. Waterton himself would have found difficulty in saying anything too
strong of a Government which displayed the unwise meanness

* Journal of Science, 1877, p. 145, and 1880, p. 632.

of imposing a tax upon zoological or geological specimens. Twenty per cent *ad valorem* was a most oppressive exaction upon articles which have no regular commercial value, and which might be estimated at any arbitrary figure by officials, unless duly bribed. The first step towards abolishing such laws is to exhibit them and the persons engaged in executing them in the most hateful and contemptible light. We have heard, on good authority, that prior to the abrogation of these imposts many collections, that would otherwise have been brought to England, found their way to other countries, whilst numbers of travellers were deterred from accumulating specimens on account of the penalty which would await them on their return home.

Much as we differ from Mr. Simson on this point we shall be very happy to meet with him again as an author. It strikes us that he is one whose thoughts are well worthy of attention.

Die Mythologie der Griechen und Roemer, unter steter Hinweisung auf die Kunstlerische Darstellung der Gottheiten: as Leitfaden fur den Schul und Selbst unterricht bearbeitet. By Dr. OTTO SEEMANN. Leipzig: E. A. Seemann.

THIS book, we are told, is an abstract of a larger work by the same author, which appeared five years ago, under the title "Gods and Heroes of the Greeks." Dr. Seemann's point of view, as will be seen from the title, is in a predominant degree æsthetic. He treats the now deposed divinities of antiquity as subjects for the artist.

Considerations on the origin and meaning of the myths are meantime not wanting. An introductory distinction is drawn between myths—in the stricter sense of the term—and sagas, the former being restricted to tales of the gods and their doings, whilst the latter is applied to the stories of the demi-gods or heroes. Both are distinguished from fables as being not mere consciously fictitious products of the human imagination.

The majority of the myths, Dr. Seeman considers, took their rise in the contemplation of natural phenomena which primitive man is apt to refer to the intervention of divine beings. Gradually these personified forces of Nature were more and more invested with a human form and a corporeal character, and were viewed as a family or clan, presided over by Zeus, "father of gods and men." The gods of old Italian tradition underwent this personifying process less completely than those of Greece,

* The Mythology of the Greeks and Romans, treated with constant reference to the Artistic Representation of the Deities: a Guide for School and Private Study.

3 D 2

and the subsequent identification of the two mythologies was chiefly confined to the poets and the learned, and never probably penetrated to the lower classes of Rome.

The author then describes the cosmogony and theogony of the Greeks, a section of mythology which was never thoroughly naturalised in Italy. With reference to the conflicts of the Gods and the Titans and Giants, said to have raged in the fields of Thessaly, Dr. Seemann remarks that this district plainly bears the marks of former natural convulsions. He thus suggests that these myths may have had a foundation in fact. But it is no part of his more immediate purpose to explore these border-lands of geology and mythology.

We cannot here follow the author through his full, able, and clear account of the various deities of Greece and Rome—" gods on half-pay," as they were somewhat wickedly called by Heine. The attributes of each and the chief sites of his worship are clearly described. The illustrations, taken from the most celebrated statues and reliefs, are both numerous and admirable, and the work is thus especially adapted for students of the fine arts.

The author, or the publisher, makes use of the new German orthography, which, like American English, is an uncomfortable compromise between the phonetic and the traditional ways of spelling.

A Synopsis of Elementary Results in Pure and Applied Mathematics, containing Propositions, Formulæ, and Methods of Analysis, with abridged Demonstrations. By G. S. CARR, B.A. Vol. I. London : C. F. Hodgson and Son.

THIS work opens with a series of mathematical tables. We have the Centimetre-Gramme-Second system of units, including the values of the dyne, the erg, ohm, volt, weber, the electrostatic and electro-magnetic units, &c. Next follow physical constants and formulæ, including a table of values for the principal lines of the spectrum in air at 160° C. and 76 c.m. barometer. After this section come Burckhardt's factor tables for all numbers from 1 to 99,000. The remainder of the work is devoted to algebra, the theory of equations, plane trigonometry, spherical trigonometry, elementary geometry, and geometrical conics.

The author's object is to supply a compact summary of the fundamental theorems, formulæ, and processes in the chief branches of mathematics. His compilation will prove very useful to advanced students.

Nature's Bye-paths. A Series of Recreative Papers in Natural
   History. By J. E. TAYLOR, Ph.D., F.L.S., &c. London:
   David Bogue.

WE have here a collection of pleasant, chatty essays, whose connection with natural history in the ordinary sense of the term is
not always of the closest. Thus the chapter entitled the " County
Palatine " is devoted to a description of Lancashire life and
manners rather too eulogistic, but otherwise so correct that we
could almost suspect Dr. Taylor of being a native of the county
he celebrates. There is a very fair account of the working class
entomologists and botanists of the Northern counties, who, however, if we are not mistaken, are neither as numerous nor as
eager in their researches as was the case forty years ago.

The "Story of a Recent Scare " is devoted to the redoubted Colorado beetle. The author does not apprehend that it will become
naturalised itself in England. Indeed, it is remarkable how few
American species, animal or vegetable, have succeeded in establishing themselves in Europe. The current of emigration seems
to set in the opposite direction. To what extent our insectivorous birds would rid us of an invader which appears to be
poisonous is of course open to doubt. We may be very sure
that the sparrows will not attack him so long as there is corn or
fruit in the neighbourhood.

" Old Wine in New Bottles " is a dissertation, not on plants or
animals, but on popular traditions.

" Scientific Pilgrims," as the reader will doubtless expect, is
devoted to the British Association, painted, perhaps, rather too
strongly in rose-colour. The author seems to forget that if this
Association has proved a model for other countries to copy, it
was by no means the first body of its kind. As part of the
excitement created by these great gatherings, mention might
have been made of the sermons which are invariably preached *at*
—or rather *against*—Science and her followers, sometimes by
local divines, and sometimes by dignitaries who come down for
the especial purpose. We are thankful to find here no reference
to Prof. Tyndall's " Belfast Address."

A "Naturalist on the Tramp " opens pleasantly with an
account of the geology and botany, and the glorious, though
scantily visited, scenery on the west coast of Ireland. He takes
us then to North Wales, and expatiates rather too much on the
slate quarries—concerns very lucrative to their owners, but which
are disfiguring the country in a way not pleasant to contemplate.
Lastly the said naturalist finds—or loses—his way into the Black
Country, where the flora has disappeared, the fauna has been
reduced to colliers and bull-dogs, and where, save for the geologist, there is nothing left to observe.

Much more agreeable reading is to be found in the sections on
the "Geological Dispersion of Animals and Plants," on the
"Time of Catkins," the "Flowers of the Prime," "Violets

Blue," "Summer Meadows," and "Hedge Garlands." Dr.
Taylor is by no means the "dry" man of science whom the outside world regard with such horror. He has a keen eye and a
loving heart for whatever is beautiful in Nature, and his book
will, therefore, prove a pleasant companion.

Elements of Astronomy. By R. S. BALL, LL.D., F.R.S., Royal
   Astronomer of Ireland. London: Longmans, Green, and
   Co.

THIS book, though mainly intended for beginners, is a scientific,
not a popular treatise. The author not merely avoids the opportunities for somewhat sensational contemplation which his subject affords, but his physical descriptions of the nearer and better
known heavenly bodies are kept well within the limits of what is
actually demonstrable. Thus, in treating of Jupiter, he leaves to
other works the consideration whether this huge planet is or is
not, in addition to the light received from the sun, slightly self-
luminous, nor does he discuss the question of the solid or molten
character of its globe. For the physical theory of comets he also
refers the reader to the well-known researches of Professor
Zöllner.

T e work opens with an account of the principal astronomical
instruments, the method of their use, and the phenomena to be
observed. The evidence for the approximately spherical form of
the earth and for its diurnal and annual motions are given with
great clearness, and the various terms used by astronomers, such
as right ascension and declination, polar distance, &c., are explained. Dr. Ball then considers the sun, its apparent motion,
solar observations, apparent diameter, path in space, velocity,
dimensions, sun-spots, &c. In successive chapters the motion
of the earth round the sun, the moon, the planets, comets, and
meteors, universal gravitation, stars and nebulæ, and the
structure of the sun, are considered in turn. The remainder of
the book is devoted to astronomical constants. A useful feature
consists in the references to more extensive or to special works
subjoined to many of the sections.

Dr. Ball avoids the error of many writers on astronomy who
pre-suppose on the part of their readers very profound mathematical attainments. A knowledge of the elements of Euclid
a nd Algebra, and a rudimentary acquaintance with the geometry
of planes and spheres, is sufficient for a thorough understanding
of this book. It has the high merits of clearness, simplicity, and
soundness.

CORRESPONDENCE.

CHEMICAL NATURE OF GROUND IVY.

To the Editor of The Journal of Science.

Sir,—Can any of your readers inform me whether the chemical principle of the ground ivy (*Gleichoma hederacea*) has been isolated, and, if so, whether it is an alkaloid, essential oil, or otherwise? The medicinal effects of the plant (a specimen of which I enclose) are astringent and diuretic; it is also slightly tonic, and has been used as a lotion and internally with marked success in certain scorbutic affections, especially in cases of *Impetigo*; —I am, &c.,

E. R. H.

Newport, Isle of Wight.

THE SPIRIT OF NATURE.

To the Editor of the Journal of Science.

Sir,—When my attention was first directed to the article (in the September number of this Journal) bearing the above title I felt somewhat flattered to find my " Spirit of Nature " dividing with a honoured dignitary of the church the honours of so extended a notice. But on reading it I was not a little disappointed to find that while Mr. Slater had apparently done my book the honour of reading a great part of it, he had not done himself or his readers the justice of understanding it. While he might, by being at a little pains to unfold and clearly state the position I take up, have exemplified that courage and zeal for truth to which his class lay almost too vociferous claim, he prefers the method of the callow curate who dresses up a scarecrow " Darwin " or " Huxley " in order to pluck it mercilessly to pieces. He accordingly commences operations by drawing the following caricature :—

" Nature," he writes, on the second page of his paper, by way of elucidating our position, " is still declared perfect, and defended against the charge of ' cruelty;' pain and suffering, if they cannot be fairly denied, are explained away ; beauty is pronounced substantially universal, and is supposed to have been called into being for the special delectation of man ; and last, but not least, the purposes of God are assumed as fully known and understood, according to the old use and wont of teleologists."

Now, as a matter of demonstrable fact, not one of these clauses describes my position with any tolerable degree of accuracy. Nowhere in my work is Nature "declared perfect" in the sense that Mr. Slater seems eager to attach to the assertion. I distinctly, and in so many words, disclaim the attempt to prove this the best of all possible worlds, which, if Nature were perfect in the full sense of the word, it would be. My book is rather an attempt to discover the tendencies of Natural Laws, to ascertain whether they point to justice or to injustice, mercy or cruelty, beauty or deformity. From a synthesis of such tendencies I seek to arrive at some conception of the Spirit of Nature. Nor do I "explain away pain and suffering" any further than by pointing out that we probably exaggerate the sensibility of the lower animals, and that there are certain indications of an effort on Nature's part toward alleviation of pain in cases where it can no longer act as an announcement of injury or stimulus to escape. Nowhere do I assert that "beauty is substantially universal," being fully aware of the difficulty of effectively upholding such an assertion as regards the animal kingdom, although I maintain, and I think establish, the prevalence of beauty in the plant-world ; nor do I hold the tenable thesis that it is for man alone that such beauty was prepared. Beauty for its own sake seems rather to be Nature's aim. But the most wildly inapplicable of all is the last clause, in which we are accused of assuming the purposes of God as fully known and understood. As applied to my volume this statement has all the delicate fidelity which distinguishes the pencil-drawings of a child of four. And in saying this I do not forget that Mr. Slater has the Bishop of Carlisle's work also in hand, for what he there says ought to be true of both. Now the remainder of the article is so completely based upon the conception assumed here that its pertinence stands or falls with it.

As to the accusation of dogmatism (not to mention the humiliating and highly original charge of youth) it comes with scant grace from one who, throughout the article, shows himself nothing if not a dogmatist of the New Natural History. That the indictment is untrue I am convinced, not only from my own consciousness, but from the gratuitous statements of other critics as impartial if not so "superior" as Mr. Slater.

That he should confound my position with Paley's is possibly excusable, considering that he is probably one of the thousands who speak glibly of Paley's without any thorough acquaintance with his argument. None the less do I stoutly deny the identity

of my "pervasive intelligence" with the "watch-making Deity."
Somewhat of the difference between us may be thus indicated—
An intelligent critic of my work writes :—

"The presence of intelligence as the counsel and regulation
principle of the order of nature is the focus principle to which
Mr. Baildon brings all his illustrations." Now this description
would be very insufficient and incorrect as a summary of Paley's
argument. Where I am content to show the action of intelligence, Paley's goes on into teleology, attempting to demonstrate
definite intention and particular purpose. And herein lies the
weakness of his argument. He tries too often to give a single
definite answer to the child-like query "what is this for?" To
such questions there is seldom or never a single simple answer,
because the relations of any being to its surroundings are so
complex and numerous.

I am anxious not to encroach on your space, and I think I
have said sufficient to make it patent to your readers that the
article in question gives anything but a true idea of my book.—
I am, &c.,

H. BELLYSE BAILDON.

THE SPIRIT OF NATURE.

To the Editor of the *Journal of Science.*

SIR,—Never having previously come in contact with any of the *irri-
tabile genus vatum*, I had little notion of their power of inventing
grievances and of looking at facts, at opinions, and at persons,
by "the light that never was on sea or shore." In the foregoing
letter two charges are brought against me :—Misrepresentation
or "caricature" of Mr. Baildon's opinions, and discourtesy.
Turning to the former accusation, based upon a few lines at the
top of 560 ("Journal of Science" for Sept.), it is, I maintain,
incorrect. The passage in question, fairly construed in connection with the whole of the article, points out certain characteristics
of a school of thinkers which are to be met with in certain degrees
in the "Spirit of Nature." Mr. Baildon reminds us that he has
disclaimed attempting to prove "the extreme theorem that this
is the best of all possible worlds, or even that the process of
struggle and suffering that we see is inevitable;" but he adds,
sarcastically, "I must confess to so little acquaintance with any
other world or process* that *I cannot contradict either theorem!*"
Not merely so, but the whole book is filled with laudations of
"nature" and censure of those who do not share his sweeping
admiration. Referring to the opinion of John Stuart Mill that

---

* As if such acquaintance were necessary !

certain natural processes are unsatisfactory, he reproduces the sneering hope that this philosopher may have due warning "if plans for a new universe are required." I would, therefore, ask if a writer who goes so far is greatly wronged if he is said to uphold the perfection of Nature?" If I accuse B C of having stolen £100, has he any right to feel indignant if, on minute search, it appears that the sum pilfered was only £99 15s.?

He further writes:—"Nowhere do I assert that beauty is substantially universal." Yet on p. 59 he tells us that "natural objects are for the most part beautiful." Here again his own words fully bear out my construction. He denies having pronounced beauty as existing for man alone, but for its own sake— whatever that may mean. Yet on p. 59 we find him exclaiming against some who would have us believe that these things were in no wise made for beauty nor for the delight of man. The question implicitly put in my article ("Journal of Science," p. 570), for whom and for what ugly or loathsome things were made, he overlooks. If God has created beauty for his own delectation—which is, perhaps, the only rational interpretation to be put on the assertion that beauty exists "for its own sake"—"why," as Mr. Baildon asks (p. 61) "should the external aspect of an animal be so much more seemly than its internal? Are not both equally visible to God?

The attempt to deny or to explain away pain is so palpable in Mr. Baildon's seventh chapter that any attempt to prove it is utterly needless. I am aware that he professes a wish to "confine the inquiry to sub-human nature," but, as I pointed out in "A Change of Front," he loses sight of this limitation in the words:—"The impartial thoughtfulness of Nature puzzles us. We are offended to find the existence of a parasite cared for as tenderly as man." He says further, "unless *our* bodies were of invulnerable materials pain is indispensable," and he pronounces pain—if rightly or wrongly is not here the question—"an important factor in the evolution of *human* virtue."

He is particularly dissatisfied at the words, "the purposes of God are assumed as fully known and understood." Now in the "Spirit of Nature," p. 215, we read, "Nature is at once the Protagonist and Antagonist of man, at once his *Alma Mater*, his kindly mother, and the strong angel against whom he must for ever wrestle." Is not here the assumption of a knowledge of the Divine purposes, especially if we remember that one of the senses in which the word "Nature" is used (p. 95) is as a more reverent synonym for God.

The distinction between Mr. Baildon's "design and contrivance" and that of Paley is mainly that the latter goes more into detail. On p. 73 we find, however, a section which is thoroughly Paleyan.

Thus I submit that the representation which I have given of the author's views is essentially correct. Perhaps the offence which I have committed really lies in the unpleasant facts I have brought forward.

Now for Mr. Baildon's charges of discourtesy and for the personalities with which his reply is so largely interwoven. Such charges surely "come with a scant grace" from one who decorates his book with such flowers of rhetoric as "intellectually and morally imbecile," "vulgar and stupid fallacy," "crude, thoughtless, criticism," "the Darwinian with his lack of metaphysical acumen," "atheistic flavour," &c., and who (p. 17) ascribes "ignorance and pride" to those who do not share his sentiments. Yet this writer feels highly aggrieved because I declared his work to be pervaded by a general tone of dogmatism and rashness. This indictment he considers "untrue," among other things from his own consciousness! Is he not aware that men are often serenely unconscious of their own peculiarities? Is it not, however, rash, even flippant, when a man who, whatever his powers and attainments, has not, I believe, distinguished himself by any important researches, magnanimously tells us that his "general aim is *not so much* to discredit Darwinism proper as to," &c.? Elsewhere we read of subjects "beyond the range of Tyndall's mental vision." Perhaps a happier term than dogmatism might have been selected to characterise the tone of such a book, but it was at any rate the mildest possible.

In Mr. Baildon's reply I am told that the article "A Change of Front" proves me to be nothing if not a dogmatist of the New Natural History." This is a singular assertion to make of a writer who implicitly declares himself not satisfied with the laws of selection and survival, and who (p. 567) admits that "the most advanced of us all have as yet but very shadowy conceptions" of the God-given laws of animal and vegetable development.

Again, mention is made of critics as impartial if not "so superior" as Mr. Slater. It would puzzle Mr. Baildon to show where I have laid claim to any superiority or obtruded my personality upon the reader at all. Again, it is declared that I do not understand the book, and it is next insinuated that I have wilfully misinterpreted it, being deficient in truthfulness—accusations hard to be reconciled. Mr. Baildon speaks of a zeal for truth to which my "class lay almost too vociferous claim." I have laid no vociferous claims to truth, nor does it matter what may be done by some "class" to which Mr. Baildon is pleased to refer me.

Various other little outbreaks of bad temper, such as the charge of not understanding Paley, I shall not notice. They certainly will not hurt me and they may perhaps relieve Mr. Baildon's excited feelings.

For one oversight I beg most sincerely to apologise. I considered and stated that amongst the evidences in favour of Darwinism Mr. Baildon had omitted all reference to embryology. I find that I was mistaken.—I am, &c.,

J. W. SLATER.

NOTES.

THE meeting of the British Medical Association for the year 1881 will be held at Ryde, in the Isle of Wight, to which locality the Association has received a cordial invitation from nearly the whole of the medical profession in the island. Mr. Benjamin Barrow, an old and much-respected practitioner in Ryde, has been appointed President-Elect. The Council of the town of Ryde have passed a unanimous resolution that the whole of the Corporation Buildings, which are numerous and spacious, shall be placed at the disposal of the Reception Committee. There is every reason to believe that the many beautiful and private grounds in the Isle of Wight will be thrown open to the Association. The President will give a garden party to the members and residents of the island. A Soirée will be held in the Town Hall and adjoining buildings. The Address in Medicine will be given by Dr. John Syer Bristowe, London, of St. Thomas's Hospital; the Address in Surgery will be given by Mr. William Dalla Husband, of Bournemouth, Consulting Surgeon to the York County Hospital; and an Address in Obstetric Medicine by Dr. John Sinclair Coghill, of Ventnor. Such has been the spirit with which the movement has been taken up by the members of the Association and profession in the Isle of Wight and Ryde that the meeting bids fair to rival any previous meeting both in science and pleasure. The following grants in aid of scientific investigation for the year were made, viz.—Dr. McKendrick and Committee, Glasgow, for a continued investigation on anæsthetics, £25; Dr. Gerald Yeo, London, on the efficacy of the antiseptic method in injuries of the brain, £50; Dr. Thin, London, for a continued investigation on parasitic skin diseases, £25; Mr. W. North, London, for a continued investigation on the relations which exist between nitrogenous Egesta and muscular work, £50: Dr. D. J. Hamilton, Edinburgh, an investigation on the pathology of the brain, £30; Mr. Watson Cheyne, London, an investigation on the relation of organisms of septic diseases, £25; Dr. Augustus Waller, London, an investigation on the time and relation of muscular contractions in the human body in health and disease, £20; Dr. Alexander Ogston, Aberdeen, a continued investigation on the relation between Bacteria and surgical disease, £10; Dr. Newman, Glasgow, a renewed grant in aid of an investigation on the functions of the kidney, £10; Drs. Braidwood and Vacher, Birkenhead, to illustrate the third and final report on the life-history of contagium, £20.

MM. Couty and De Lacerda (" Comptes Rendus," xci., p. 549) have made a series of experiments on the absorption of the poison of a Brazilian serpent (*Bothrops jaracara*), and of the local effects produced. If injected under the skin there ensue merely inflammations or abscesses, though the dose was five to six times greater than would have proved fatal if introduced into a vein.

M. J. Chatin has presented to the Academy of Sciences (Sept. 27, 1880) a paper on the ciliated embryo of the formidable parasite *Bilharzia hæmatobia*.

The charters of the Eclectic Medical College of Pennsylvania and of the American University of Philadelphia have been cancelled by the Supreme Court of Philadelphia, for selling spurious diplomas. Dr. Buchanan is now awaiting his trial.

The Council of the Royal Microscopical Society have under consideration the subject of increasing the utility of their extensive library, by lending books to the Fellows.

According to " Science," Prof. Richard Owen, formerly State Geologist of Indiana, proposes the following general law for the distribution of land on the globe :—" The land shows itself above the ocean level in definite multiple proportions, by measurement; the unit is the angular difference between the axis of revolution and the axis of progression."

In a paper on the physiological action of *Conium maculatum*, communicated by M. Rochefontaine to the Academy of Sciences, it is concluded that *Conium* may act like *Curare*, but it produces, besides, phenomena which have not been observed in animals submitted to the action of *Curare*.

According to a daily paper, which we need not name, an English officer on the Gold Coast encountered a spider which it required three shots with a revolver to despatch, its skin being too tough to be penetrated with a clasp-knife !

According to the " Veterinary Gazette " numerous specimens of an extinct species of horse have been discovered in France, approaching the recent horse more closely than any extinct species hitherto discovered, and forming a connecting-link between the genera *Equus* and *Hipparion*. The lateral metacarpals are distinct from the great metacarpal, whilst in the existing horse they are co-ossified for the greater part of their length.

At the late meeting of the American Association for the Advancement of Science, Prof. Morse called attention to the existence of a number of species of land shells in Yesso identical with forms met with in New England. Two species of slugs also are common to Japan and New England.

We learn, from the " Rocky Mountain Medical Review," that one A. Wilford Hall, in a work bearing the ambitious title " The

Problem of Human Life," aims at refuting the doctrine of Evolution, and, as a preliminary step, attempts to disprove the vibratory theory of sound !

A peculiar frog, found in the *tierras templadas* of Columbia, is said to yield a poison whose physiological action is similar to that of curare.

A most painful scene has occurred at a session of the Paris Academy of Medicine. During a hot discussion on the "cholera of fowls" M. Guérin addressed M. Pasteur in the following words:—"You are a liar, sir; I will send you my seconds." The "Medical Press and Circular" adds that M. Guérin is an octogenarian, and M. Pasteur hopelessly paralysed on the left side.

It is declared by Dr. Maclaren that the types of insanity have changed in modern times, and that acute delirious mania is becoming rarer. On the other hand, mental enfeeblement, often accompanied by paralysis, is more and more common, due to the overwork and worry of modern life.

The "Nineteenth Century" speaks of Dr. Carpenter as "the great English physicist."

We learn that the London Decorating Company, Limited, has arranged with Mr. W. Crookes, F.R.S., to supervise the chemical tests and processes employed to ensure the freedom of their paints, paper-hangings, &c., from poisonous and deleterious ingredients.

M. A. Gaudry has communicated to the Academy of Sciences an account of a new and highly-developed fossil reptile found in the Permian strata of Igornay, and to which he has given the name of *Stereorachis dominans*. It must have been a carnivorous animal of large size, as one of its jaws, though somewhat broken, measures 0·18 metre. It has affinities with the Ganocephal and the Labyrinthodons, as well as with the Theriodontes and the Pelycosaurians described by Cope. The discovery of so perfect a reptile in the Permian formation renders it highly probable that the remains of its forerunners may be found in the Devonian.

M. Savage has described, in the "Comptes Rendus," a fossil serpent from the *Ostrea Columba* beds of the Charentes. It has received the name of *Similiophis Rochebruni*.

According to "Les Mondes" a woman, named Judith Singer, died recently at Glowitz, in Silesia, at the age of 112 years.

We learn that J. G. Romanes, F.R.S., will deliver two lectures on "Animal Intelligence" before the Society of Arts.

Prof. Max Müller, at the opening meeting of the Mason Science College, at Birmingham, made the following excellent remark:—

"The professors of a College of Science should not be conservative only, satisfied to hand down the stock of knowledge as they have received it, 'laid up in a napkin.' Professors must try to add something, however little it may be, to the talent they have received."

The "specific poison doctrine" is considered altogether inapplicable to the history of enteric fever in India.—*Medical Press and Circular.*

In his Address at the inauguration of Sir Josiah Mason's College, at Birmingham, Prof. Huxley pointed out "how erroneous was the idea that a classical education alone could make a man of culture, and that the struggle which had for years been going on must end in placing the study of science and modern languages at least upon an equal footing with classics as a means to culture. He believed if a man could not get literary culture out of his Bible, his Shakespeare, and his Milton, he would not get it from the profoundest study of Homer, Sophocles, Virgil, and Horace.

The "Monthly Magazine of Pharmacy" considers "Spiritualism like the confidence trick, so palpably a sham that those who are so weak-minded as to be deceived by it deserve all they get or lose."

H. Buchner has succeeded in effecting the reciprocal conversion of the Bacteria of splenic fever into those generally found in infusions of hay. He has not yet succeeded in producing splenic fever with the unmodified hay Bacteria.

According to A. Renouard ("Biedermann's Central-Blatt für Agrikultur Chemie), water in which flax has been steeped not only kills fish, but is injurious to vegetation.

The following two modifications of Wickersheim's fluid for the preservation of animal substances have been proposed, *a* being for injection and *b* for immersion :—

	a.	*b.*
Arsenious acid ...	16 ounces	12 ounces
Common salt ...	80 ,,	60 ,,
Sulphate of potash	200 ,,	150
Nitrate of potash...	25 ,,	18 ,,
Carbonate of potash	20 ,,	15 ,,
Water	16 pints	16 pints
Glycerin...	6 pts. 8 fl. ozs.	6 pints 8 ozs.
Methylic alcohol ...	24 fl. ozs.	16 fl. ozs.

On October 28th the Literary and Philosophical Society of Halifax celebrated its fiftieth anniversary with—an operatic performance !

M. Delbœuf considers that there is no infallible and universal sign which enables us to affirm with absolute certainty that a dream was a dream and nothing more.—*Revue Philosophique.*

The researches of Gustav Hausen (" Zeitschrift für Wissen Zoologie," vol. xxxiv., p. 367) seem finally to have decided the controversy concerning the function of the antennæ of insects. He has examined the behaviour of insects towards strongly odorous substances both before and after the removal of these organs, and when they had been coated with paraffin, and finds that after such operations they become indifferent to smells. Flies thus treated took no further notice of tainted meat. He shows also that the development of the antennæ in different insect groups decidedly harmonises with these experimental results.

According to the "Augsburg Gazette" the crayfish are perishing in numbers from an epidemic disease, the most striking feature of which is the development of white fungoid growths.

M. Etard (" Comptes Rendus ") considers that boron ought to rank among the elements of the phosphorus series, not far from phosphorus itself, and at the head of the vanadium family.

M. Faye, in a paper read before the Academy of Sciences, contends that pellagra is produced by the use in food of grain which has not undergone the action of yeast or leaven.

M. Poirot (" Comptes Rendus ") recommends wormwood as a remedy for the *Phylloxera*. It is said also to banish flies, ants, scorpions, and tarantulas.

In the earthquakes which ravaged Asia Minor in July last the fissures which opened in the soil are said to have emitted greenish black water and " sulphurous " fumes. It is much to be wished that these liquid and gaseous products could be submitted to a thorough analysis. Violent shocks were felt nearly at the same time in Manilla, the neighbourhood of Naples, and the Azores, where an island has risen out of the sea.

Mr. C. A. Feilberg, writing in the "Victorian Review," points out that the native vegetation of the inland parts of Queensland has a faculty of remaining dormant for months during the long and frequent droughts, and suddenly resuming active life on the fall of rain.

Mr. O. C. Marsh (" American Journal of Science," Ser. 3, vol. xx., p. 235), summing up the results of his researches on the mesozoic mammals of the Rocky Mountains, finds that they cannot be introduced into any existing order. With few exceptions the sixty species discovered are low, generalised forms, without decisive marsupial characters, and rather approaching the Insectivora.

THE

JOURNAL OF SCIENCE,

AND ANNALS OF

ASTRONOMY, BIOLOGY, GEOLOGY, INDUSTRIAL ARTS, MANUFACTURES, AND TECHNOLOGY.

(MONTHLY, FORMERLY "THE QUARTERLY JOURNAL OF SCIENCE.")

Vol. II. (Third Series)

No. LXXXIV.

DECEMBER 1880.

CONTENTS.

ARTICLES.   PAGE

1. Spectrum Analysis. By Dr. AKIN 735
2. The World of the Poets.. .. 745
3. Exceptional Seasons and Weather Prophecies 750
4. Modern Cynolatry. By FRANK FERNSEED 754
5. Fog Lore 764
6. Working v. Fighting. By J. W. SLATER 770

ANALYSES OF BOOKS.

Our Summer Migrants 779
Betrachtungen uber die Entwickelungs-Geschichte der Schmetterlinge und deren Variation .. 780
Charles Waterton 783

ANALYSES OF BOOKS—continued.   PAGE

A Reply to Criticisms on "The Problems of Insanity " 78
Die Mythologie der Griechen und Roemer, unter steter Hinweisung auf die Kuntslerische Darstellung der Gottheiten .. 78
A Synopsis of Elementary Results in Pure and Applied Mathematics.. 78
Nature's Bye-paths 78
Elements of Astronomy 78

CORRESPONDENCE.

Chemical Nature of Ground Ivy 7
The Spirit of Nature 789, 79

NOTES 79

LONDON:

3, HORSE-SHOE COURT, LUDGATE HILL.

Where Communications for the Editor and Books for Review may be addressed.

Seventeenth Year of Publication.

THE

JOURNAL OF SCIENCE

AND ANNALS OF

Astronomy, Biology, Geology, Industrial Arts, Manufactures, and Technology.

MONTHLY, FORMERLY "THE QUARTERLY JOURNAL OF SCIENCE."

PRICE, ONE SHILLING AND SIXPENCE.

This Journal is an established medium of communication between the Scientific Investigator and the Reading Public, as well as between Students in the various branches of Science. It also constitutes a Review of the Progress of Science in all parts of the world. Its past and present Contributors include the following names :—

F. A. Abel, C.B., F.R.S.
Prof. Ansted, F.R.S.
W. E. A. Axon
Prof. W. E. Ayrton
Rev. S. Barber
Prof. W. F. Barrett
Dr. Lionel S. Beale, F.R.S.
Dr. G. M. Beard
Thomas Belt
A. W. Bennett
Professor Bentley
E. W. Binney, F.R.S.
Dr. H. Carrington Bolton
F. W. Brearey
Sir Charles Bright
H. W. Bristow, F.R.S.
Capt. Richard F. Burton
Dr. W. B. Carpenter, C.B., F.R.S.
Prof. A. H. Church
Professor Coughtrie
James Croll, F.R.S.
William Crookes, F.R.S.
F. C. Danvers
W. Boyd Dawkins, F.R.S.
Dr. Henry Draper
Colonel Drayson, R.A.
Sir William Fairbairn, Bart., F.R.S.
Frederick Field, F.R.S.
Rev. O. Fisher
Camille Flammarion
Sir Charles Fox
Prof. Frankland, F.R.S.
Dr. Gladstone, F.R.S.
James Glaishier, F.R.S.
George Gore, F.R.S.
A. Vernon Harcourt, F.R.S.
Edward T. Hardman

Prof. Harkness, F.R.S.
Prof. W. Noel Hartley
Rev. S. Haughton, F.R.S.
Prof. F. V. Hayden
Sir John Herschel, F.R.S.
Rev. H. H. Higgins
J. R. Hind, F.R.S.
Prof. Hofmann, F.R.S.
William Huggins, F.R.S.
Prof. Hull, F.R.S.
Robert Hunt, F.R.S.
H. M. Jenkins
Prof. W. Stanley Jevons, F.R.S.
G. H. Kinahan
Prof. W. King
Sir Francis Knowles, Bart., F.R.S.
Dr. Lankester, F.R.S.
Prof. A. R. Leeds
J. Norman Lockyer, F.R.S.
Robert Mallet, F.R.S.
Dr. R. J. Mann
Prof. N. S. Maskelyne, F.R.S.
Prof. Thompson B. Maury
Prof. Henry Morton
J. Chalmers Morton
Prof. M. M. Pattison Muir
J. Munro
Edmund Neison
Prof. Simon Newcomb, F.R.S.
Prof. Odling, F.R.S.
Capt. Oliver, R.A.
Dr. J. Bell Pettigrew, F.R.S.
Dr. Paul

William Pengelly, F.R.S.
Prof. Phillips, F.R.S.
J. A. Phillips
Raoul Pictet
Mungo Ponton
S. Tolver Preston
T. Symes Prideaux
H. Baden Pritchard
R. A. Proctor
C. W. Quin
Prof. A. C. Ramsay, F.R.S.
G. F. Rodwell
Dr. Rolleston, F.R.S.
G. J. Romanes, F.R.S.
James Samuelson
Dr. A. E. Sansom
Dr. P. L. Sclater, F.R.S.
R. H. Scott, F.R.S.
Prof. H. G. Seeley, F.R.S.
J. W. Slater
Dr. R. Angus Smith, F.R.S.
Piazzi Smyth
H. C. Sorby, F.R.S.
Prof. Balfour Stewart, F.R.S.
W. T. Suffolk
Prof. Silvanus Thompson
William Topley
Prof. Tyndall, F.R.S.
Dr. Voelcker, F.R.S.
Alfred Russel Wallace
Dr. Wallich
Dr. W. M. Watts
Rev. J. Clifton Ward
Dr. F. C. Webb
W. Mattieu Williams
Henry Woodward, F.R.S.
Horace B. Woodward
Prof. Zöllner

LONDON: 3, HORSE-SHOE COURT, LUDGATE HILL,
AND ALL BOOKSELLERS.